Lecture Notes in Networks and Systems

Volume 450

Series Editor

Janusz Kacprzyk, Systems Research Institute, Polish Academy of Sciences, Warsaw, Poland

Advisory Editors

Fernando Gomide, Department of Computer Engineering and Automation—DCA, School of Electrical and Computer Engineering—FEEC, University of Campinas—UNICAMP, São Paulo, Brazil

Okyay Kaynak, Department of Electrical and Electronic Engineering, Bogazici University, Istanbul, Turkey

Derong Liu, Department of Electrical and Computer Engineering, University of Illinois at Chicago, Chicago, USA
   Institute of Automation, Chinese Academy of Sciences, Beijing, China

Witold Pedrycz, Department of Electrical and Computer Engineering, University of Alberta, Alberta, Canada
   Systems Research Institute, Polish Academy of Sciences, Warsaw, Poland

Marios M. Polycarpou, Department of Electrical and Computer Engineering, KIOS Research Center for Intelligent Systems and Networks, University of Cyprus, Nicosia, Cyprus

Imre J. Rudas, Óbuda University, Budapest, Hungary

Jun Wang, Department of Computer Science, City University of Hong Kong, Kowloon, Hong Kong

The series "Lecture Notes in Networks and Systems" publishes the latest developments in Networks and Systems—quickly, informally and with high quality. Original research reported in proceedings and post-proceedings represents the core of LNNS.

Volumes published in LNNS embrace all aspects and subfields of, as well as new challenges in, Networks and Systems.

The series contains proceedings and edited volumes in systems and networks, spanning the areas of Cyber-Physical Systems, Autonomous Systems, Sensor Networks, Control Systems, Energy Systems, Automotive Systems, Biological Systems, Vehicular Networking and Connected Vehicles, Aerospace Systems, Automation, Manufacturing, Smart Grids, Nonlinear Systems, Power Systems, Robotics, Social Systems, Economic Systems and other. Of particular value to both the contributors and the readership are the short publication timeframe and the world-wide distribution and exposure which enable both a wide and rapid dissemination of research output.

The series covers the theory, applications, and perspectives on the state of the art and future developments relevant to systems and networks, decision making, control, complex processes and related areas, as embedded in the fields of interdisciplinary and applied sciences, engineering, computer science, physics, economics, social, and life sciences, as well as the paradigms and methodologies behind them.

Indexed by SCOPUS, INSPEC, WTI Frankfurt eG, zbMATH, SCImago.

All books published in the series are submitted for consideration in Web of Science.

For proposals from Asia please contact Aninda Bose (aninda.bose@springer.com).

More information about this series at https://link.springer.com/bookseries/15179

Leonard Barolli · Farookh Hussain ·
Tomoya Enokido
Editors

Advanced Information Networking and Applications

Proceedings of the 36th International
Conference on Advanced Information
Networking and Applications (AINA-2022),
Volume 2

 Springer

Editors
Leonard Barolli
Department of Information
and Communication Engineering
Fukuoka Institute of Technology
Fukuoka, Japan

Farookh Hussain
University of Technology Sydney
Sydney, NSW, Australia

Tomoya Enokido
Faculty of Bussiness Administration
Rissho University
Tokyo, Japan

ISSN 2367-3370   ISSN 2367-3389   (electronic)
Lecture Notes in Networks and Systems
ISBN 978-3-030-99586-7   ISBN 978-3-030-99587-4   (eBook)
https://doi.org/10.1007/978-3-030-99587-4

This Springer imprint is published by the registered company Springer Nature Switzerland AG
The registered company address is: Gewerbestrasse 11, 6330 Cham, Switzerland

Welcome Message from AINA-2022 Organizers

Welcome to the 36th International Conference on Advanced Information Networking and Applications (AINA-2022). On behalf of AINA-2022 Organizing Committee, we would like to express to all participants our cordial welcome and high respect.

AINA is an international forum, where scientists and researchers from academia and industry working in various scientific and technical areas of networking and distributed computing systems can demonstrate new ideas and solutions in distributed computing systems. AINA was born in Asia, but it is now an international conference with high quality thanks to the great help and cooperation of many international friendly volunteers. AINA is a very open society and is always welcoming international volunteers from any country and any area in the world.

AINA International Conference is a forum for sharing ideas and research work in the emerging areas of information networking and their applications. The area of advanced networking has grown very rapidly, and the applications have experienced an explosive growth especially in the area of pervasive and mobile applications, wireless sensor networks, wireless ad-hoc networks, vehicular networks, multimedia computing and social networking, semantic collaborative systems, as well as grid, P2P, IoT, big data, and cloud computing. This advanced networking revolution is transforming the way people live, work, and interact with each other and is impacting the way business, education, entertainment, and health care are operating. The papers included in the proceedings cover theory, design, and application of computer networks, distributed computing, and information systems.

Each year AINA receives a lot of paper submissions from all around the world. It has maintained high-quality accepted papers and is aspiring to be one of the main international conferences on the information networking in the world.

We are very proud and honored to have two distinguished keynote talks by Prof. Mario A. R. Dantas, University of Juiz de Fora, Minas Gerais, Brazil, and Prof. Isaac Woungang, Ryerson University, Toronto, Ontario, Canada, who will present their recent work and will give new insights and ideas to the conference participants.

An international conference of this size requires the support and help of many people. A lot of people have helped and worked hard to produce a successful AINA-2022 technical program and conference proceedings. First, we would like to thank all authors for submitting their papers, the session chairs, and distinguished keynote speakers. We are indebted to program track co-chairs, program committee members and reviewers, who carried out the most difficult work of carefully evaluating the submitted papers.

We would like to thank AINA-2022 General Co-chairs, PC Co-chairs, and Workshops Co-chairs for their great efforts to make AINA-2022 a very successful event. We have special thanks to Finance Chair and Web Administrator Co-chairs.

We do hope that you will enjoy the conference proceedings and readings.

Organization

AINA-2022 Organizing Committee

Honorary Chair

Makoto Takizawa   Hosei University, Japan

General Co-chairs

Farookh Hussain   University of Technology Sydney, Australia
Tomoya Enokido   Rissho University, Japan
Isaac Woungang   Ryerson University, Canada

Program Committee Co-chairs

Omar Hussain   University of New South Wales, Australia
Flora Amato   University of Naples "Federico II," Italy
Marek Ogiela   AGH University of Science and Technology, Poland

Workshops Co-chairs

Beniamino Di Martino   University of Campania "Luigi Vanvitelli," Italy
Omid Ameri Sianaki   Victoria University, Australia
Kin Fun Li   University of Victoria, Canada

International Journals Special Issues Co-chairs

Fatos Xhafa   Technical University of Catalonia, Spain
David Taniar   Monash University, Australia

Award Co-chairs

Arjan Durresi   Indiana University Purdue University in
                                           Indianapolis (IUPUI), USA
Fang-Yie Leu   Tunghai University, Taiwan

Publicity Co-chairs

Markus Aleksy   ABB AG, Germany
Lidia Ogiela   AGH University of Science and Technology,
                                           Poland
Hsing-Chung Chen   Asia University, Taiwan

International Liaison Co-chairs

Nadeem Javaid   COMSATS University Islamabad, Pakistan
Wenny Rahayu   La Trobe University, Australia

Local Arrangement Co-chairs

Rania Alhazmi   University of Technology Sydney, Australia
Huda Alsobhi   University of Technology Sydney, Australia
Ebtesam Almansour   University of Technology Sydney, Australia

Finance Chair

Makoto Ikeda   Fukuoka Institute of Technology, Japan

Web Co-chairs

Phudit Ampririt   Fukuoka Institute of Technology, Japan
Kevin Bylykbashi   Fukuoka Institute of Technology, Japan
Ermioni Qafzezi   Fukuoka Institute of Technology, Japan

Steering Committee Chair

Leonard Barolli   Fukuoka Institute of Technology, Japan

Tracks and Program Committee Members

1. Network Protocols and Applications

Track Co-chairs

Makoto Ikeda   Fukuoka Institute of Technology, Japan
Sanjay Kumar Dhurandher   Netaji Subhas University of Technology,
                                           New Delhi, India
Bhed Bahadur Bista   Iwate Prefectural University, Japan

TPC Members

Admir Barolli	Aleksander Moisiu University of Durres, Albania
Elis Kulla	Okayama University of Science, Japan
Keita Matsuo	Fukuoka Institute of Technology, Japan
Shinji Sakamoto	Kanazawa Institute of Technology, Japan
Akio Koyama	Yamagata University, Japan
Evjola Spaho	Polytechnic University of Tirana, Albania
Jiahong Wang	Iwate Prefectural University, Japan
Shigetomo Kimura	University of Tsukuba, Japan
Chotipat Pornavalai	King Mongkut's Institute of Technology Ladkrabang, Thailand
Danda B. Rawat	Howard University, USA
Amita Malik	Deenbandhu Chhotu Ram University of Science and Technology, India
R. K. Pateriya	Maulana Azad National Institute of Technology, India
Vinesh Kumar	University of Delhi, India
Petros Nicopolitidis	Aristotle University of Thessaloniki, Greece
Satya Jyoti Borah	North Eastern Regional Institute of Science and Technology, India

2. Next-Generation Wireless Networks

Track Co-chairs

Christos J. Bouras	University of Patras, Greece
Tales Heimfarth	Universidade Federal de Lavras, Brazil
Leonardo Mostarda	University of Camerino, Italy

TPC Members

Fadi Al-Turjman	Near East University, Nicosia, Cyprus
Alfredo Navarra	University of Perugia, Italy
Purav Shah	Middlesex University London, UK
Enver Ever	Middle East Technical University, Northern Cyprus Campus, Cyprus
Rosario Culmone	University of Camerino, Camerino, Italy
Antonio Alfredo F. Loureiro	Federal University of Minas Gerais, Brazil
Holger Karl	University of Paderborn, Germany
Daniel Ludovico Guidoni	Federal University of São João Del-Rei, Brazil
João Paulo Carvalho Lustosa da Costa	Hamm-Lippstadt University of Applied Sciences, Germany
Jorge Sá Silva	University of Coimbra, Portugal

Apostolos Gkamas   University Ecclesiastical Academy of Vella,
                                        Ioannina, Greece
Zoubir Mammeri   University Paul Sabatier, France
Eirini Eleni Tsiropoulou   University of New Mexico, USA
Raouf Hamzaoui   De Montfort University, UK
Miroslav Voznak   University of Ostrava, Czech Republic
Kevin Bylykbashi   Fukuoka Institute of Technology, Japan

3. Multimedia Systems and Applications

Track Co-chairs

Markus Aleksy   ABB Corporate Research Center, Germany
Francesco Orciuoli   University of Salerno, Italy
Tomoyuki Ishida   Fukuoka Institute of Technology, Japan

TPC Members

Tetsuro Ogi   Keio University, Japan
Yasuo Ebara   Osaka Electro-Communication University, Japan
Hideo Miyachi   Tokyo City University, Japan
Kaoru Sugita   Fukuoka Institute of Technology, Japan
Akio Doi   Iwate Prefectural University, Japan
Hadil Abukwaik   ABB Corporate Research Center, Germany
Monique Duengen   Robert Bosch GmbH, Germany
Thomas Preuss   Brandenburg University of Applied Sciences,
                                        Germany
Peter M. Rost   NOKIA Bell Labs, Germany
Lukasz Wisniewski   inIT, Germany
Angelo Gaeta   University of Salerno, Italy
Graziano Fuccio   University of Salerno, Italy
Giuseppe Fenza   University of Salerno, Italy
Maria Cristina   University of Salerno, Italy
Alberto Volpe   University of Salerno, Italy

4. Pervasive and Ubiquitous Computing

Track Co-chairs

Chih-Lin Hu	National Central University, Taiwan
Vamsi Paruchuri	University of Central Arkansas, USA
Winston Seah	Victoria University of Wellington, New Zealand

TPC Members

Hong Va Leong	Hong Kong Polytechnic University, Hong Kong
Ling-Jyh Chen	Academia Sinica, Taiwan
Jiun-Yu Tu	Southern Taiwan University of Science and Technology, Taiwan
Jiun-Long Huang	National Chiao Tung University, Taiwan
Thitinan Tantidham	Mahidol University, Thailand
Tanapat Anusas-amornkul	King Mongkut's University of Technology North Bangkok, Thailand
Xin-Mao Huang	Aletheia University, Taiwan
Hui Lin	Tamkang University, Taiwan
Eugen Dedu	Universite de Franche-Comte, France
Peng Huang	Sichuan Agricultural University, China
Wuyungerile Li	Inner Mongolia University, China
Adrian Pekar	Budapest University of Technology and Economics, Hungary
Jyoti Sahni	Victoria University of Technology, New Zealand
Normalia Samian	Universiti Putra Malaysia, Malaysia
Sriram Chellappan	University of South Florida, USA
Yu Sun	University of Central Arkansas, USA
Qiang Duan	Penn State University, USA
Han-Chieh Wei	Dallas Baptist University, USA

5. Web-Based and E-Learning Systems

Track Co-chairs

Santi Caballe	Open University of Catalonia, Spain
Kin Fun Li	University of Victoria, Canada
Nobuo Funabiki	Okayama University, Japan

TPC Members

Jordi Conesa	Open University of Catalonia, Spain
Joan Casas	Open University of Catalonia, Spain
David Gañán	Open University of Catalonia, Spain
Nicola Capuano	University of Basilicata, Italy
Antonio Sarasa	Complutense University of Madrid, Spain
Chih-Peng Fan	National Chung Hsing University, Taiwan
Nobuya Ishihara	Okayama University, Japan
Sho Yamamoto	Kindai University, Japan
Khin Khin Zaw	Yangon Technical University, Myanmar
Kaoru Fujioka	Fukuoka Women's University, Japan
Kosuke Takano	Kanagawa Institute of Technology, Japan
Shengrui Wang	University of Sherbrooke, Canada
Darshika Perera	University of Colorado at Colorado Spring, USA
Carson Leung	University of Manitoba, Canada

6. Distributed and Parallel Computing

Track Co-chairs

Naohiro Hayashibara	Kyoto Sangyo University, Japan
Minoru Uehara	Toyo University, Japan
Tomoya Enokido	Rissho University, Japan

TPC Members

Eric Pardede	La Trobe University, Australia
Lidia Ogiela	AGH University of Science and Technology, Poland
Evjola Spaho	Polytechnic University of Tirana, Albania
Akio Koyama	Yamagata University, Japan
Omar Hussain	University of New South Wales, Australia
Hideharu Amano	Keio University, Japan
Ryuji Shioya	Toyo University, Japan
Ji Zhang	The University of Southern Queensland
Lucian Prodan	Universitatea Politehnica Timisoara, Romania
Ragib Hasan	The University of Alabama at Birmingham, USA
Young-Hoon Park	Sookmyung Women's University, Korea
Dilawaer Duolikun	Cognizant Technology Solutions, Hungary
Shigenari Nakamura	Tokyo Metropolitan Industrial Technology Research Institute, Japan

7. Data Mining, Big Data Analytics and Social Networks

Track Co-chairs

Omid Ameri Sianaki	Victoria University, Australia
Alex Thomo	University of Victoria, Canada
Flora Amato	University of Naples "Frederico II," Italy

TPC Members

Eric Pardede	La Trobe University, Australia
Alireza Amrollahi	Macquarie University, Australia
Javad Rezazadeh	University Technology Sydney, Australia
Farshid Hajati	Victoria University, Australia
Mehregan Mahdavi	Sydney International School of Technology and Commerce, Australia
Ji Zhang	University of Southern Queensland, Australia
Salimur Choudhury	Lakehead University, Canada
Xiaofeng Ding	Huazhong University of Science and Technology, China
Ronaldo dos Santos Mello	Universidade Federal de Santa Catarina, Brazil
Irena Holubova	Charles University, Czech Republic
Lucian Prodan	Universitatea Politehnica Timisoara, Romania
Alex Tomy	La Trobe University, Australia
Dhomas Hatta Fudholi	Universitas Islam Indonesia, Indonesia
Saqib Ali	Sultan Qaboos University, Oman
Ahmad Alqarni	Al Baha University, Saudi Arabia
Alessandra Amato	University of Naples "Frederico II," Italy
Luigi Coppolino	Parthenope University, Italy
Giovanni Cozzolino	University of Naples "Frederico II," Italy
Giovanni Mazzeo	Parthenope University, Italy
Francesco Mercaldo	Italian National Research Council, Italy
Francesco Moscato	University of Salerno, Italy
Vincenzo Moscato	University of Naples "Frederico II," Italy
Francesco Piccialli	University of Naples "Frederico II," Italy

8. Internet of Things and Cyber-Physical Systems

Track Co-chairs

Euripides G. M. Petrakis	Technical University of Crete (TUC), Greece
Tomoki Yoshihisa	Osaka University, Japan
Mario Dantas	Federal University of Juiz de Fora (UFJF), Brazil

TPC Members

Akihiro Fujimoto	Wakayama University, Japan
Akimitsu Kanzaki	Shimane University, Japan
Kawakami Tomoya	University of Fukui, Japan
Lei Shu	University of Lincoln, UK
Naoyuki Morimoto	Mie University, Japan
Yusuke Gotoh	Okayama University, Japan
Vasilis Samolada	Technical University of Crete (TUC), Greece
Konstantinos Tsakos	Technical University of Crete (TUC), Greece
Aimilios Tzavaras	Technical University of Crete (TUC), Greece
Spanakis Manolis	Foundation for Research and Technology Hellas (FORTH), Greece
Katerina Doka	National Technical University of Athens (NTUA), Greece
Giorgos Vasiliadis	Foundation for Research and Technology Hellas (FORTH), Greece
Stefan Covaci	Technische Universität Berlin, Berlin (TUB), Germany
Stelios Sotiriadis	University of London, UK
Stefano Chessa	University of Pisa, Italy
Jean-Francois Méhaut	Université Grenoble Alpes, France
Michael Bauer	University of Western Ontario, Canada

9. Intelligent Computing and Machine Learning

Track Co-chairs

Takahiro Uchiya	Nagoya Institute of Technology, Japan
Omar Hussain	UNSW, Australia
Nadeem Javaid	COMSATS University Islamabad, Pakistan

TPC Members

Morteza Saberi	University of Technology Sydney, Australia
Abderrahmane Leshob	University of Quebec in Montreal, Canada
Adil Hammadi	Curtin University, Australia
Naeem Janjua	Edith Cowan University, Australia
Sazia Parvin	Melbourne Polytechnic, Australia
Kazuto Sasai	Ibaraki University, Japan
Shigeru Fujita	Chiba Institute of Technology, Japan
Yuki Kaeri	Mejiro University, Japan
Zahoor Ali Khan	HCT, UAE
Muhammad Imran	King Saud University, Saudi Arabia

Ashfaq Ahmad	The University of Newcastle, Australia
Syed Hassan Ahmad	JMA Wireless, USA
Safdar Hussain Bouk	Daegu Gyeongbuk Institute of Science and Technology, Korea
Jolanta Mizera-Pietraszko	Military University of Land Forces, Poland

10. Cloud and Services Computing

Track Co-chairs

Asm Kayes	La Trobe University, Australia
Salvatore Venticinque	University of Campania "Luigi Vanvitelli," Italy
Baojiang Cui	Beijing University of Posts and Telecommunications, China

TPC Members

Shahriar Badsha	University of Nevada, USA
Abdur Rahman Bin Shahid	Concord University, USA
Iqbal H. Sarker	Chittagong University of Engineering and Technology, Bangladesh
Jabed Morshed Chowdhury	La Trobe University, Australia
Alex Ng	La Trobe University, Australia
Indika Kumara	Jheronimus Academy of Data Science, Netherlands
Tarique Anwar	Macquarie University and CSIRO's Data61, Australia
Giancarlo Fortino	University of Calabria, Italy
Massimiliano Rak	University of Campania "Luigi Vanvitelli," Italy
Jason J. Jung	Chung-Ang University, Korea
Dimosthenis Kyriazis	University of Piraeus, Greece
Geir Horn	University of Oslo, Norway
Gang Wang	Nankai University, China
Shaozhang Niu	Beijing University of Posts and Telecommunications, China
Jianxin Wang	Beijing Forestry University, China
Jie Cheng	Shandong University, China
Shaoyin Cheng	University of Science And Technology of China, China

11. Security, Privacy and Trust Computing

Track Co-chairs

Hiroaki Kikuchi	Meiji University, Japan
Xu An Wang	Engineering University of PAP, China
Lidia Ogiela	AGH University of Science and Technology, Poland

TPC Members

Takamichi Saito	Meiji University, Japan
Kouichi Sakurai	Kyushu University, Japan
Kazumasa Omote	Univesity of Tsukuba, Japan
Shou-Hsuan Stephen Huang	University of Houston, USA
Masakatsu Nishigaki	Shizuoka University, Japan
Mingwu Zhang	Hubei University of Technology, China
Caiquan Xiong	Hubei University of Technology, China
Wei Ren	China University of Geosciences, China
Peng Li	Nanjing University of Posts and Telecommunications, China
Guangquan Xu	Tianjing University, China
Urszula Ogiela	AGH University of Science and Technology, Poland
Hoon Ko	Chosun University, Korea
Goreti Marreiros	Institute of Engineering of Polytechnic of Porto, Portugal
Chang Choi	Gachon University, Korea
Libor Měsíček	J.E. Purkyně University, Czech Republic

12. Software-Defined Networking and Network Virtualization

Track Co-chairs

Flavio de Oliveira Silva	Federal University of Uberlândia, Brazil
Ashutosh Bhatia	Birla Institute of Technology and Science, Pilani, India
Alaa Allakany	Kyushu University, Japan

TPC Members

Rui Luís Andrade Aguiar	Universidade de Aveiro (UA), Portugal
Ivan Vidal	Universidad Carlos III de Madrid, Spain
Eduardo Coelho Cerqueira	Federal University of Pará (UFPA), Brazil

Christos Tranoris   University of Patras (UoP), Greece
Juliano Araújo Wickboldt   Federal University of Rio Grande do Sul
                                        (UFRGS), Brazil
Yaokai Feng   Kyushu University, Japan
Chengming Li   Chinese Academy of Science (CAS), China
Othman Othman   An-Najah National University (ANNU), Palestine
Nor-masri Bin-sahri   University Technology of MARA, Malaysia
Sanouphab Phomkeona   National University of Laos, Laos
Haribabu K.   BITS Pilani, India
Shekhavat, Virendra   BITS Pilani, India
Makoto Ikeda   Fukuoka Institute of Technology, Japan
Farookh Hussain   University of Technology Sydney, Australia
Keita Matsuo   Fukuoka Institute of Technology, Japan

AINA-2022 Reviewers

Abderrahmane Leshob   Baojiang Cui
Abdullah Al-khatib   Beniamino Di Martino
Adil Hammadi   Bhed Bista
Admir Barolli   Caiquan Xiong
Adrian Pekar   Carson Leung
Ahmad Alqarni   Chang Choi
Aimilios Tzavaras   Christos Bouras
Akihiro Fujihara   Christos Tranoris
Akihiro Fujimoto   Danda Rawat
Akimitsu Kanzaki   David Taniar
Akio Doi   Dimitris Apostolou
Akira Sakuraba   Dimosthenis Kyriazis
Alaa Allakany   Eirini Eleni Tsiropoulou
Alex Ng   Elis Kulla
Alex Thomo   Enver Ever
Alfredo Cuzzocrea   Eric Pardede
Alfredo Navarra   Ernst Gran
Amita Malik   Eugen Dedu
Angelo Gaeta   Evjola Spaho
Anne Kayem   Farookh Hussain
Antonio Esposito   Fatos Xhafa
Antonio Loureiro   Feilong Tang
Apostolos Gkamas   Feroz Zahid
Arcangelo Castiglione   Flavio Silva
Arjan Durresi   Flora Amato
Ashutosh Bhatia   Francesco Orciuoli
Asm Kayes   Francesco Palmieri

Funabiki Nobuo
Gang Wang
Goreti Marreiros
Guangquan Xu
Hideharu Amano
Hiroaki Kikuchi
Hiroshi Maeda
Hsing-Chung Chen
Indika Kumara
Irena Holubova
Isaac Woungang
Jana Nowaková
Javad Rezazadeh
Ji Zhang
Jianxin Wang
Jolanta Mizera-Pietraszko
Jordi Conesa
Jorge Sá Silva
Kazunori Uchida
Kazuto Sasai
Keita Matsuo
Kevin Bylykbashi
Kin Fun Li
Kiyotaka Fujisaki
Koki Watanabe
Konstantinos Tsakos
Kosuke Takano
Kouichi Sakurai
Leonard Barolli
Leonardo Mostarda
Libor Mesicek
Lidia Ogiela
Lucian Prodan
Luigi Coppolino
Makoto Ikeda
Makoto Takizawa
Marek Ogiela
Mario Dantas
Markus Aleksy
Masakatsu Nishigaki
Masaki Kohana
Mingwu Zhang
Minoru Uehara
Miralda Cuka

Mirang Park
Miroslav Voznak
Nadeem Javaid
Naeem Janjua
Naohiro Hayashibara
Nobuo Funabiki
Norimasa Nakashima
Omar Hussain
Omid Ameri Sianaki
Othman Othman
Øyvind Ytrehus
Paresh Saxena
Pavel Kromer
Philip Moore
Pornavalai Chotipat
Purav Shah
Quentin Jacquemart
Ragib Hasan
Ricardo Rodríguez Jorge
Rosario Culmone
Rui Aguiar
Ryuji Shioya
Safdar Hussain Bouk
Salimur Choudhury
Salvatore Venticinque
Sanjay Dhurandher
Santi Caballé
Satya Borah
Sazia Parvin
Shahriar Badsha
Shigenari Nakamura
Shigeru Fujita
Shigetomo Kimura
Shinji Sakamoto
Somnath Mazumdar
Sriram Chellappan
Stefan Covaci
Stefano Chessa
Takahiro Uchiya
Takamichi Saito
Tarique Anwar
Tetsuro Ogi
Tetsuya Oda
Tetsuya Shigeyasu

Thomas Dreibholz
Tomoki Yoshihisa
Tomoya Enokido
Tomoya Kawakami
Tomoyuki Ishida
Urszula Ogiela
Vamsi Paruchuri
Vinesh Kumar
Wang Xu An

Wei Ren
Wenny Rahayu
Winston Seah Isaac Woungang
Xiaofeng Ding
Yaokai Feng
Yoshitaka Shibata
Yuki Kaeri
Yusuke Gotoh
Zahoor Khan

AINA-2022 Keynote Talks

Data Intensive Scalable Computing in Edge/Fog/Cloud Environments

Mario A. R. Dantas

University of Juiz de Fora, Minas Gerais, Brazil

Abstract. In this talk are presented and discussed some aspects related to the adoption of data intensive scalable computing (DISC) paradigm considering the new adoption trend of edge/fog/cloud environments. These contemporaneous scenarios are very relevant for all organizations in a world where billion of IoT and IIoT devices are being connected, and an unprecedent amount of digital data is generated. Therefore, they require special processing and storage.

Resource Management in 5G Cloudified Infrastructure: Design Issues and Challenges

Isaac Woungang

Ryerson University, Toronto, Canada

Abstract. 5G and Beyond (B5G) networks will be featured by a closer collaboration between mobile network operators (MNOs) and cloud service providers (CSPs) to meet the communication and computational requirements of modern mobile applications and services in a mobile cloud computing (MCC) environment. In this talk, we enlighten the marriage between the heterogeneous wireless networks (HetNets) and the multiple clouds (termed as InterCloud) for a better resource management in B5G networks. First, we start with an overview of the building blocks of HetNet and InterCloud, and then we describe the resource managers in both domains. Second, the key design criteria and challenges related to interoperation between the InterCloud and HetNet are described. Third, the state-of the-art security-aware resource allocation mechanisms for a multi-cloud orchestration over a B5G networks are enlighten.

Contents

A Fuzzy-Based System for Determining Driver Stress in VANETs
Considering Driving Experience and History   1
Kevin Bylykbashi, Ermioni Qafzezi, Phudit Ampririt, Makoto Ikeda,
Keita Matsuo, and Leonard Barolli

Performance Evaluation of WMNs by WMN-PSOHC Hybrid
Simulation System Considering Different Instances: A Comparison
Study for RDVM and LDIWM Replacement Methods   10
Shinji Sakamoto, Yi Liu, and Leonard Barolli

Millimeter-Wave Dual-Band Slotted Antenna for 5G Applications   19
Prince Mahmud Ridoy, Arajit Saha, Khadija Yeasmin Fariya, Pranta Saha,
Khan Md. Elme, and Farhadur Arifin

NARUN-PC: Caching Strategy for Noise Adaptive Routing in Utility
Networks   31
Fabio Pagnotta, Leonardo Mostarda, and Alfredo Navarra

HYPE: CNN Based HYbrid PrEcoding Framework for 5G
and Beyond   43
Deepti Sharma, Kuldeep M. Biradar, Santosh K. Vipparthi,
and Ramesh B. Battula

The Multi-access Edge Computing (MEC)-Based Bit Rate Adaptive
Multicast SVC Streaming Using the Adaptive FEC Mechanism   55
Chung-Ming Huang and Kai-Jiun Yang

PROA: Pipelined Receiver Oriented Anycast MAC for IoT   68
João Carlos Giacomin and Tales Heimfarth

A Watchdog Proposal to a Personal e-Health Approach   81
Gabriel Di iorio Silva, Wagno Leão Sergio, Victor Ströele,
and Mario A. R. Dantas

Computation Offloading by Two-Sided Matching in Fog Computing . . .   95
Meng Wang and Minoru Uehara

**Distributed Log Search Based on Time Series Access and Service
Relations** 105
Tomoyuki Koyama and Takayuki Kushida

**Detector: Hierarchical Distributed Fault Detection Algorithm
for Lattice Based Modular Robots** 118
Edy Hourany, Benoit Piranda, Abdallah Makhoul, Julien Bourgeois,
and Bachir Habib

**ManufactSim: Manufacturing Line Simulation Using Heterogeneous
Distributed Robots** 130
Benoit Piranda, Ishan Gautam, Jerome Meyer, Anass El Houd,
and Julien Bourgeois

Sports Data Management, Mining, and Visualization 141
Bamibo C. Isichei, Carson K. Leung, Lam Thu Nguyen, Luke B. Morrow,
Anh Tuan Ngo, Trang Doan Pham, and Alfredo Cuzzocrea

Mining for Fake News 154
Renz M. Cabusas, Brenna N. Epp, Justin M. Gouge, Tyson N. Kaufmann,
Carson K. Leung, and James R. A. Tully

**Software Functional and Non-function Requirement Classification
Using Word-Embedding** 167
Lov Kumar, Siddarth Baldwa, Shreya Manish Jambavalikar,
Lalita Bhanu Murthy, and Aneesh Krishna

Topic Guided Image Captioning with Scene and Spatial Features 180
Usman Zia, M. Mohsin Riaz, and Abdul Ghafoor

**A Socially-Aware, Privacy-Preserving, and Scalable Federated
Learning Protocol for Distributed Online Social Networks** 192
Mansour Khelghatdoust and Mehregan Mahdavi

**A Multi-layer Modeling for the Generation of New Architectures for
Big Data Warehousing** 204
Asma Dhaouadi, Khadija Bousselmi, Sébastien Monnet,
Mohamed Mohsen Gammoudi, and Slimane Hammoudi

**Efficient Retransmission Algorithm for Ensuring Packet Delivery to
Sleeping Destination Node** 219
Ali Medlej, Eugen Dedu, Dominique Dhoutaut, and Kamal Beydoun

**The Development of an Elderly Monitoring System with Multiple
Sensors** 231
Yasunao Takano, Hiroyuki Adachi, Hiroji Ochii, Mikio Okazaki,
and Sena Takeda

Predicting Cyber-Attacks on IoT Networks Using Deep-Learning and Different Variants of SMOTE 243
Bathini Sai Akash, Pavan Kumar Reddy Yannam,
Bokkasam Venkata Sai Ruthvik, Lov Kumar, Lalita Bhanu Murthy,
and Aneesh Krishna

A Decentralized Federated Learning Architecture for Intrusion Detection in IoT Systems 256
Francisco Assis Moreira do Nascimento and Fabiano Hessel

Regression Analysis Using Machine Learning Approaches for Predicting Container Shipping Rates 269
Ibraheem Abdulhafiz Khan and Farookh Khadeer Hussain

Robust Variational Autoencoders and Normalizing Flows for Unsupervised Network Anomaly Detection 281
Naji Najari, Samuel Berlemont, Grégoire Lefebvre, Stefan Duffner,
and Christophe Garcia

Multiplatform Comparative Analysis of Intelligent Robots for Communication Efficiency in Smart Dialogs 293
Anna Pogoda, Ewa Lyko, Michal Kedziora, Ireneusz Jozwiak,
and Jolanta Pietraszko

Using Simplified EEG-Based Brain Computer Interface and Decision Tree Classifier for Emotions Detection 306
Rafal Chalupnik, Katarzyna Bialas, Zofia Majewska, and Michal Kedziora

Anomaly Detection from Distributed Data Sources via Federated Learning 317
Florencia Cavallin and Rudolf Mayer

On Predicting COVID-19 Fatality Ratio Based on Regression Using Machine Learning Model 329
Md. Mafijul Islam Bhuiyan, Mondar Maruf Moin Ahmed, Anik Alvi,
Md. Safiqul Islam, Prasenjit Mondal, Md Akbar Hossain,
and S. N. M. Azizul Hoque

Distributed Training from Multi-sourced Data 339
Ibrahim Dahaoui, Mohamed Mosbah, and Akka Zemmari

Viterbi Algorithm and HMM Implementation to Multicriteria Data-Driven Decision Support Model for Optimization of Medical Service Quality Selection 348
Jolanta Mizera-Pietraszko and Jolanta Tancula

Performance Evaluation of a DQN-Based Autonomous Aerial Vehicle Mobility Control Method in Corner Environment 361
Nobuki Saito, Tetsuya Oda, Aoto Hirata, Chihiro Yukawa,
Kyohei Toyoshima, Tomoaki Matsui, and Leonard Barolli

OpenAPI QL: Searching in OpenAPI Service Catalogs 373
Ioanna-Maria Stergiou, Nikolaos Mainas, and Euripides G. M. Petrakis

Sensor Virtualization and Provision in Internet of Vehicles 386
Slim Abbes and Slim Rekhis

**A Secure Data Storage in Multi-cloud Architecture Using Blowfish
Encryption Algorithm** 398
Houaida Ghanmi, Nasreddine Hajlaoui, Haifa Touati, Mohamed Hadded,
and Paul Muhlethaler

**Micro-Service Placement Policies for Cost Optimization
in Kubernetes** 409
Alkiviadis Aznavouridis, Konstantinos Tsakos,
and Euripides G. M. Petrakis

**A Differentiated Approach Based on Edge-Fog-Cloud Environment to
Support e-Health on Rural Areas** 421
Fernando de Almeida Silva, Walkíria Garcia de Souza Silveira,
and Mario Dantas

**Trustworthy Fairness Metric Applied to AI-Based Decisions in Food-
Energy-Water** 433
Suleyman Uslu, Davinder Kaur, Samuel J. Rivera, Arjan Durresi,
Mimoza Durresi, and Meghna Babbar-Sebens

New Security Protocols for Offline Point-of-Sale Machines 446
Nour El Madhoun, Emmanuel Bertin, Mohamad Badra, and Guy Pujolle

Building a Blockchain-Based Social Network Identification System 468
Zhanwen Chen and Kazumasa Omote

**Malware Classification by Deep Learning Using Characteristics
of Hash Functions** 480
Takahiro Baba, Kensuke Baba, and Toshihiro Yamauchi

Toward a Blockchain Healthcare Information Exchange 492
Ryuji Ueno and Kazumasa Omote

A Design Thinking Approach on Information Security 503
Lukas König and Simon Tjoa

**Modeling Network Traffic via Identifying Encrypted Packets to Detect
Stepping-Stone Intrusion Under the Framework of Heterogonous
Packet Encryption** 516
Noah Neundorfer, Jianhua Yang, and Lixin Wang

**A Study on Enhancing Anomaly Detection Technology with Synthetic-
Log Generation** 528
Takumi Yamamoto, Aiko Iwasaki, Hajime Kobayashi, Kiyoto Kawauchi,
and Ayako Yoshimura

Application of Hybrid Intelligence for Security Purposes 539
Marek R. Ogiela and Lidia Ogiela

**Semantic-Based Techniques for Efficient and Secure Data
Management** 543
Urszula Ogiela, Makoto Takizawa, and Lidia Ogiela

**CoWrap: An Approach of Feature Selection for Network Anomaly
Detection** 547
Anonnya Ghosh, Hussain Mohammed Ibrahim, Wasif Mohammad,
Farhana Chowdhury Nova, Amit Hasan, and Raqeebir Rab

**Attack Modeling and Cyber Deception Resources Deployment Using
Multi-layer Graph** 560
Amal Sayari, Yacine Djemaiel, Slim Rekhis, Ali Mabrouk,
and Belhassen Jerbi

**Quantum-Secure Aggregate One-time Signatures with Detecting
Functionality** 573
Shingo Sato and Junji Shikata

**Improving Robustness and Visibility of Adversarial CAPTCHA Using
Low-Frequency Perturbation** 586
Takamichi Terada, Vo Ngoc Khoi Nguyen, Masakatsu Nishigaki,
and Tetsushi Ohki

**Comparative Study of Ensemble Learning Techniques for Fuzzy
Attack Detection in In-Vehicle Networks** 598
Dorsaf Swessi and Hanen Idoudi

**ZeroMT: Multi-transfer Protocol for Enabling Privacy in Off-Chain
Payments** 611
Flavio Corradini, Leonardo Mostarda, and Emanuele Scala

**Prevention of SQL Injection Attacks Using Cryptography
and Pattern Matching** 624
R. Madhusudhan and Mohammad Ahsan

Monitoring Jitter in Software Defined Networks 635
Jithin Kallukalam Sojan and K. Haribabu

**Designing and Prototyping of SDN Switch for Application-Driven
Approach** 646
Diego Nunes Molinos, Romerson Deiny Oliveira, Marcelo Silva Freitas,
Natal Vieira de Souza Neto, Marcelo Barros de Almeida,
Flávio de Oliveira Silva, and Pedro Frosi Rosa

**SD-WAN: Edge Cloud Network Acceleration at Australia Hybrid
Data Center** 659
Junjie Wang and Lihong Zheng

**Decision Tree Based IoT Attack Detection in Programmable Data
Plane Using P4 Language** 671
Rahul Poddar and Hari Babu

**Prevention of DrDoS Amplification Attacks by Penalizing the
Attackers in SDN Environment** 684
Shail Saharan and Vishal Gupta

Author Index 697

A Fuzzy-Based System for Determining Driver Stress in VANETs Considering Driving Experience and History

Kevin Bylykbashi[1]([✉]), Ermioni Qafzezi[2], Phudit Ampririt[2], Makoto Ikeda[1], Keita Matsuo[1], and Leonard Barolli[1]

[1] Department of Information and Communication Engineering, Fukuoka Institute of Technology (FIT), 3-30-1 Wajiro-Higashi, Higashi-Ku, Fukuoka 811-0295, Japan
kevin@bene.fit.ac.jp, makoto.ikd@acm.org, {kt-matsuo,barolli}@fit.ac.jp
[2] Graduate School of Engineering, Fukuoka Institute of Technology (FIT), 3-30-1 Wajiro-Higashi, Higashi-Ku, Fukuoka 811-0295, Japan
{bd20101,bd21201}@bene.fit.ac.jp

Abstract. We have previously implemented an intelligent system based on fuzzy logic for determining driver's stress in Vehicular Ad hoc Networks (VANETs), called Fuzzy-based System for Determining the Stress Feeling Level (FSDSFL), considering the driver's impatience, the behavior of other drivers, and the traffic condition as input parameters. In this work, we present a modified version of FSDSFL, which considers the driving experience and history as an additional input. We show through simulations the effect that driving experience and history and the other parameters have on the determination of the stress feeling level and demonstrate some actions that can be performed when the stress exceeds certain levels.

1 Introduction

The highly competitive and rapidly advancing autonomous vehicle race has been on for several years now, and it is a matter of time until we have these vehicles on the roads. However, even if the automotive companies do all it takes to create fully automated cars, there will still be one big obstacle, the infrastructure. In addition, this could take decades, even in the most developed countries. Moreover, 93% of the world's fatalities on the roads occur in low- and middle-income countries [11] and considering all these facts, Driver Assistance Systems (DASs) and Vehicular Ad hoc Networks (VANETs) should remain in focus for the foreseeable future.

DASs are intelligent systems that are implemented in vehicles to increase driving safety by assisting drivers and can be very helpful in a variety of situations as they do not depend on the infrastructure as much as driverless vehicles do. Furthermore, DASs can provide driving support with very little cost, thus help the low- and middle-income countries in the long battle against car accidents.

VANETs, on the other hand, aim not only at saving lives but also improving traffic mobility, increasing efficiency, and promoting travel convenience of drivers and passengers. In VANETs, network nodes (vehicles) are equipped with networking functions to exchange information with each other via vehicle-to-vehicle (V2V) and with roadside units (RSUs) through vehicle-to-infrastructure (V2I) communications.

By leveraging the data acquired by other vehicles and infrastructure, DASs can make better decisions and offer more services, which provides drivers with enhanced applications and experience. These data range from simple information such as traffic and road condition messages to a complete perception of the surrounding environment obtained through cameras, thus improving road safety significantly.

Nevertheless, there are other determinants that affect the driving operation, such as the drivers and their behaviors. In fact, according to some traffic safety facts provided by a survey of the U.S. Department of Transportation [10], the drivers are the immediate reason for more than 94% of the investigated car crashes. While most of the errors committed by drivers are considered involuntary, other errors often come as a result of their behavior, and this must be utterly preventable. These errors are, in most cases, associated with the stress that drivers feel when they are behind the wheel. Determining the factors that cause the stress is consequently a need that requires careful and immediate work.

In [4], we have proposed an intelligent system based on Fuzzy Logic (FL) that determines the driver's stress in real-time based on factors such as the driver's impatience, the behavior of other drivers and the traffic condition. In this work, we present a modified version of our system that additionally considers the driving experience and history as an input parameter. A visualization of the proposed system is given in Fig. 1. We evaluate the proposed system by computer simulations and see the effect that driving experience and history and the other parameters have on the determination of the stress feeling level.

The structure of the paper is as follows. Section 2 presents a brief overview of VANETs. Section 3 describes the proposed fuzzy-based simulation system and its implementation. Section 4 discusses the simulation results. Finally, conclusions and future work are given in Sect. 5.

2   Overview of VANETs

VANETs are a special case of Mobile Ad hoc Networks (MANETs) in which the mobile nodes are vehicles. In VANETs, nodes (vehicles) have high mobility and tend to follow organized routes instead of moving randomly. Moreover, vehicles offer attractive features such as higher computational capability and localization through GPS.

VANETs have huge potential to enable applications ranging from road safety, traffic optimization, infotainment, commercial to rural and disaster scenario connectivity. Among these, road safety and traffic optimization are considered the most important ones as they have the goal to reduce drastically the high number

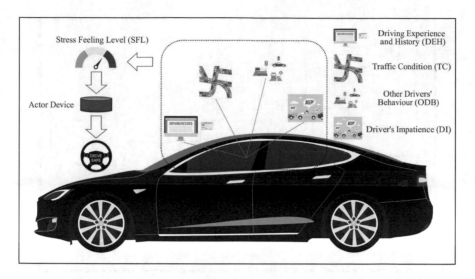

Fig. 1. Concept of proposed system.

of accidents, guarantee road safety, make traffic management, and create new forms of inter-vehicle communications in ITSs. The ITSs manage the vehicle traffic, support drivers with safety and other information, and enable applications such as automated toll collection and DASs [5].

Despite the attractive features, VANETs are characterized by very large and dynamic topologies, variable capacity wireless links, bandwidth and hard delay constraints, and by short contact durations which are caused by the high mobility, high speed, and low density of vehicles. In addition, limited transmission ranges, physical obstacles, and interferences make these networks characterized by intermittent connectivity.

Therefore, it is necessary to design proper systems, network architectures and applications that can overcome the problems that arise from vehicular environments.

3 Proposed Fuzzy-Based System

Our research work focuses on developing an intelligent non-complex driving support system that determines the driving risk level in real-time by considering different types of parameters. In previous works, we have considered different parameters, including in-car environment parameters such as the ambient temperature and noise, and driver's vital signs, i.e., heart and respiratory rate, for which we implemented a testbed and conducted experiments in a real scenario [2]. The considered parameters include environmental factors and driver's health condition, as these parameters affect the driver's capability and vehicle performance. In [1], we presented an integrated fuzzy-based system, which in addition to those parameters, considers the following inputs: vehicle speed, weather and

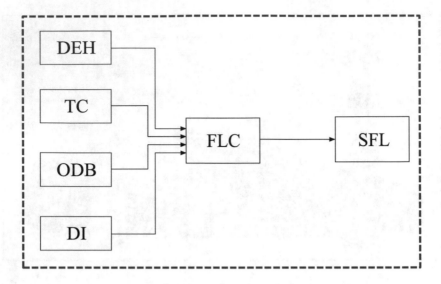

Fig. 2. A diagram of the proposed fuzzy-based system.

road condition, driver's body temperature, and vehicle interior relative humidity. The inputs were categorized based on the way they affect the driving operation. In a more recent work [3], we proposed a system that decides the driver's impatience since the impatient drivers are often an immediate cause of many road accidents. In this work, we propose a system that can determine the driver's stress feeling level based on the degree of impatience, driving experience and history, traffic condition, and other drivers' behavior.

We use FL to implement the proposed system as it can make a real-time decision based on the uncertainty and vagueness of the provided information [6–9,12,13]. The proposed system, called Fuzzy-based System for Determining Stress Feeling Level (FSDSFL), is shown in Fig. 2. FSDSFL has the following inputs: Driving Experience and History (DEH), Traffic Condition (TC), Other Drivers' Behavior (ODB), and Driver's Impatience (DI). The output of the system is Stress Feeling Level (SFL). The term set for each parameter is given in Table 1.

Based on the linguistic description of input and output parameters, the Fuzzy Rule Base (FRB) of the proposed system forms a fuzzy set of dimensions $|T(x_1)| \times |T(x_2)| \times \cdots \times |T(x_n)|$, where $|T(x_i)|$ is the number of terms on $T(x_i)$ and n is the number of input parameters. FSDSFL has four input parameters, one with two linguistic terms and three with three linguistic terms each, therefore there are 54 rules in the FRB. The FRB is shown in Table 2. The control rules of FRB have the form: IF "conditions" THEN "control action". The membership functions are shown in Fig. 3. We use triangular and trapezoidal membership functions because these types of functions are more suitable for real-time operation.

Table 1. Parameters and their term sets for FSDSFL.

Parameters	Term set
Driving Experience and History (DEH)	Bad (B), Good (G)
Traffic Condition (TC)	Light (Li), Moderate (Mo), Heavy (He)
Other Drivers' Behavior (ODB)	Very Bad (VB), Bad (Ba), Good (Go)
Driver's Impatience (DI)	Low (L), Moderate (M), High (H)
Stress Feeling Level (SFL)	Very Low (VL), Low (Lw), Moderate (Md), High (Hg), Very High (VH), Extremely High (EH)

Table 2. FRB of FSDSFL.

No.	DEH	TC	ODB	DI	SFL	No	DEH	TC	ODB	DI	SFL
1	B	Li	VB	L	Hg	28	G	Li	VB	L	VL
2	B	Li	VB	M	VH	29	G	Li	VB	M	Lw
3	B	Li	VB	H	EH	30	G	Li	VB	H	Hg
4	B	Li	Ba	L	Md	31	G	Li	Ba	L	VL
5	B	Li	Ba	M	Hg	32	G	Li	Ba	M	VL
6	B	Li	Ba	H	VH	33	G	Li	Ba	H	Md
7	B	Li	Go	L	Lw	34	G	Li	Go	L	VL
8	B	Li	Go	M	Hg	35	G	Li	Go	M	VL
9	B	Li	Go	H	Hg	36	G	Li	Go	H	Lw
10	B	Mo	VB	L	VH	37	G	Mo	VB	L	Lw
11	B	Mo	VB	M	EH	38	G	Mo	VB	M	Md
12	B	Mo	VB	H	EH	39	G	Mo	VB	H	VH
13	B	Mo	Ba	L	Hg	40	G	Mo	Ba	L	VL
14	B	Mo	Ba	M	VH	41	G	Mo	Ba	M	Lw
15	B	Mo	Ba	H	EH	42	G	Mo	Ba	H	Hg
16	B	Mo	Go	L	Md	43	G	Mo	Go	L	VL
17	B	Mo	Go	M	Hg	44	G	Mo	Go	M	Lw
18	B	Mo	Go	H	VH	45	G	Mo	Go	H	Md
19	B	He	VB	L	EH	46	G	He	VB	L	Md
20	B	He	VB	M	EH	47	G	He	VB	M	Hg
21	B	He	VB	H	EH	48	G	He	VB	H	EH
22	B	He	Ba	L	VH	49	G	He	Ba	L	Lw
23	B	He	Ba	M	EH	50	G	He	Ba	M	Md
24	B	He	Ba	H	EH	51	G	He	Ba	H	VH
25	B	He	Go	L	Hg	52	G	He	Go	L	VL
26	B	He	Go	M	EH	53	G	He	Go	M	Lw
27	B	He	Go	H	EH	54	G	He	Go	H	Hg

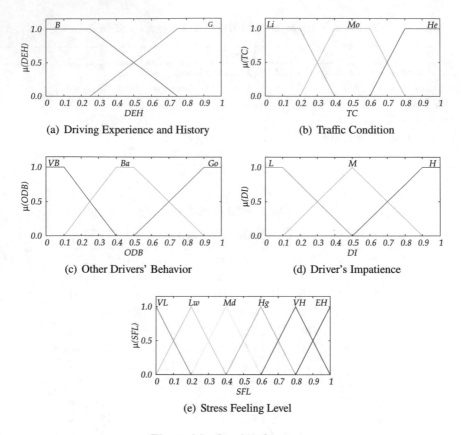

Fig. 3. Membership functions.

4 Simulation Results

In this section, we present the simulation results for our proposed system. The simulation results are presented in Figs. 4, 5 and 6. We consider DEH and TC as constant parameters. We show the relation between SFL and DI for different ODB values.

In Fig. 4, we show the results for DEH = 0.1 and change TC from 0.1 to 0.9. We can see that when there is no traffic congestion, the drivers experience severe stress only when they show high levels of impatience and when the behavior of other drivers is very bad. When the behavior of other drivers is good, and they seem to show patience, the stress is determined as even under the moderate level. However, when traffic is heavy, we can see that the drivers experience much more stress, with most scenarios involving very high stress levels. This can be attributed to the fact that they are still inexperienced and yet with a history of accidents.

The impact of better driving experience and history can be seen in Fig. 5 and Fig. 6. Figure 5 shows the results for experienced drivers who have been involved

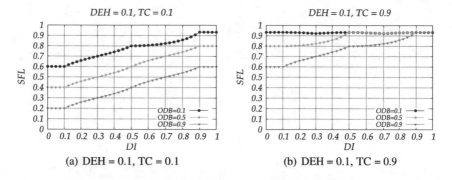

(a) DEH = 0.1, TC = 0.1   (b) DEH = 0.1, TC = 0.9

Fig. 4. Simulation results for DEH = 0.1.

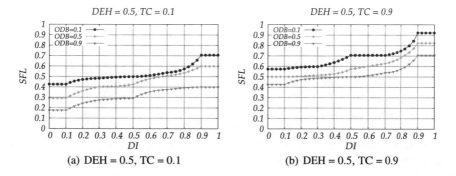

(a) DEH = 0.5, TC = 0.1   (b) DEH = 0.5, TC = 0.9

Fig. 5. Simulation results for DEH = 0.5.

in accidents in the past and inexperienced drivers, but with no bad records. They experience less stress but still in considerable values, with stress values above the moderate level accounting for most driving scenarios.

In the case of experienced drivers with no bad records (see Fig. 6), we can see that the drivers experience high stress only when they show high degrees of impatience while other drivers are violating traffic rules. All other scenarios indicate that the drivers are not experiencing stress that can cause a potential accident.

In the cases when the driver is considered to be under too much stress, the system can take several actions that can improve the driving situations and therefore reduce the risks of accidents. For example, the system can suggest the use of an alternative route that has less congestion, or it can adjust the interior environment of the car to help the driver manage the driving situations patiently.

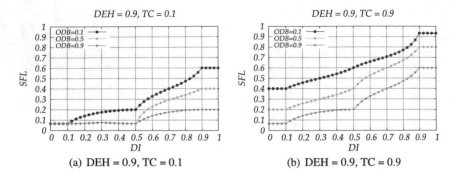

Fig. 6. Simulation results for DEH = 0.9.

5   Conclusions

In this work, we presented the implementation of an FL approach that determines the driver's stress feeling level by considering the driver's impatience, the driving experience and history, the behavior of other drivers and the traffic condition. We showed through simulations the effect of the considered parameters on the determination of the stress feeling level. The simulations show that when the traffic is heavier than usual, drivers tend to show an increased level of stress, especially if they are still inexperienced or have bad driving records. The experienced stress is even higher when other drivers violate the traffic rules. However, when the experienced drivers have no bad driving records and are driving with much patience, they can manage to drive smoothly in almost every driving situation.

In the future, we would like to make extensive simulations and experiments to evaluate the proposed system and compare the performance with other systems.

References

1. Bylykbashi, K., Qafzezi, E., Ampririt, P., Ikeda, M., Matsuo, K., Barolli, L.: Performance evaluation of an integrated fuzzy-based driving-support system for real-time risk management in VANETs. Sensors **20**(22), 6537 (2020). https://doi.org/10.3390/s20226537
2. Bylykbashi, K., Qafzezi, E., Ikeda, M., Matsuo, K., Barolli, L.: Fuzzy-based driver monitoring system (FDMS): implementation of two intelligent FDMSs and a testbed for safe driving in VANETs. Future Gener. Comput. Syst. **105**, 665–674 (2020). https://doi.org/10.1016/j.future.2019.12.030
3. Bylykbashi, K., Qafzezi, E., Ampririt, P., Ikeda, M., Matsuo, K., Barolli, L.: A fuzzy-based system for deciding driver impatience in VANETs. In: Barolli, L. (ed.) Advances on P2P, Parallel, Grid, Cloud and Internet Computing, vol. 343, pp 129–137. Springer, Cham (2022). https://doi.org/10.1007/978-3-030-89899-1_13

4. Bylykbashi, K., Qafzezi, E., Ampririt, P., Ikeda, M., Matsuo, K., Barolli, L.: A fuzzy-based system for safe driving in VANETs considering impact of driver impatience on stress feeling level. In: Barolli, L., Kulla, E., Ikeda, M .(eds.) Advances in Internet, Data & Web Technologies, vol. 118, pp 236–244. Springer, Cham (2022). https://doi.org/10.1007/978-3-030-95903-6_25

5. Hartenstein, H., Laberteaux, L.: A tutorial survey on vehicular ad hoc networks. IEEE Commun. Mag. **46**(6), 164–171 (2008). https://doi.org/10.1109/MCOM.2008.4539481

6. Kandel, A.: Fuzzy Expert System. CRC Press Inc., Boca Raton (1992)

7. Klir, G.J., Folger, T.A.: Fuzzy Sets, Uncertainty, and Information. Prentice Hall Inc., Upper Saddle River (1987)

8. McNeill, F.M., Thro, E.: Fuzzy Logic: A Practical Approach. Academic Press Professional Inc., San Diego (1994). https://doi.org/10.1016/C2013-0-11164-6

9. Munakata, T., Jani, Y.: Fuzzy systems: an overview. Commun. ACM **37**(3), 69–77 (1994). https://doi.org/10.1145/175247.175254

10. Singh, S.: Critical Reasons for Crashes Investigated in the National Motor Vehicle Crash Causation Survey. Traffic Safety Facts: Crash·Stats. Report No. DOT HS 812 506, National Highway Traffic Safety Administration (NHTSA), Washington, DC (2018)

11. World Health Organization: Global status report on road safety 2018: summary. World Health Organization, Geneva, Switzerland, (WHO/NMH/NVI/18.20). Licence: CC BY-NC-SA 3.0 IGO (2018)

12. Zadeh, L.A., Kacprzyk, J.: Fuzzy Logic for the Management of Uncertainty. Wiley, New York (1992)

13. Zimmermann, H.J.: Fuzzy Set Theory and Its Applications. Springer Science & Business Media, New York (1996). https://doi.org/10.1007/978-94-015-8702-0

Performance Evaluation of WMNs by WMN-PSOHC Hybrid Simulation System Considering Different Instances: A Comparison Study for RDVM and LDIWM Replacement Methods

Shinji Sakamoto[1(✉)], Yi Liu[2], and Leonard Barolli[3]

[1] Department of Information and Computer Science, Kanazawa Institute of Technology, 7-1 Ohgigaoka, Nonoichi, Ishikawa 921-8501, Japan
shinji.sakamoto@ieee.org
[2] Department of Computer Science, National Institute of Technology, Oita College, 1666, Maki, Oita 870-0152, Japan
y-liu@oita-ct.ac.jp
[3] Department of Information and Communication Engineering, Fukuoka Institute of Technology, 3-30-1 Wajiro-Higashi, Higashi-Ku, Fukuoka 811-0295, Japan
barolli@fit.ac.jp

Abstract. Wireless Mesh Networks (WMNs) have many good features and they are becoming an important networking infrastructure. However, WMNs have some problems such as node placement, security, transmission power and so on. To solve these problems, we have implemented a hybrid simulation system based on PSO and HC called WMN-PSOHC. In this paper, we evaluate the performance of WMNs by using WMN-PSOHC considering two instances: Instance 1 and Instance 2. Simulation results show that WMN-PSOHC performs better for Instance 1 compared with Instance 2. Also, RDVM performs better than LDIWM in this considered scenario.

1 Introduction

In this work, we deal with node placement problem in WMNs. We consider the version of the mesh router nodes placement problem in which we are given a grid area where to deploy a number of mesh router nodes and a number of mesh client nodes of fixed positions (of an arbitrary distribution) in the grid area. The objective is to find a location assignment for the mesh routers to the cells of the grid area that maximizes the network connectivity and client coverage. Network connectivity is measured by Size of Giant Component (SGC) of the resulting WMN graph, while the user coverage is simply the number of mesh client nodes that fall within the radio coverage of at least one mesh router node and is measured by Number of Covered Mesh Clients (NCMC). Node placement problems are known to be computationally hard to solve [14]. In some previous

© The Author(s), under exclusive license to Springer Nature Switzerland AG 2022
L. Barolli et al. (Eds.): AINA 2022, LNNS 450, pp. 10–18, 2022.
https://doi.org/10.1007/978-3-030-99587-4_2

works, intelligent algorithms have been recently investigated [3,9,10]. We already implemented a Particle Swarm Optimization (PSO) based simulation system, called WMN-PSO [6]. Also, we implemented a simulation system based on Hill Climbing (HC) for solving node placement problem in WMNs, called WMN-HC [5].

In our previous work [6,8], we presented a hybrid intelligent simulation system based on PSO and HC. We called this system WMN-PSOHC. In this paper, we evaluate the performance of WMNs by using WMN-PSOHC considering two instances and Normal distribution of mesh clients.

The rest of the paper is organized as follows. We present our designed and implemented hybrid simulation system in Sect. 2. In Sect. 3, we introduce WMN-PSOHC Web GUI tool. The simulation results are given in Sect. 4. Finally, we give conclusions and future work in Sect. 5.

2   Proposed and Implemented Simulation System

2.1   Particle Swarm Optimization

In Particle Swarm Optimization (PSO) algorithm, a number of simple entities (the particles) are placed in the search space of some problem or function and each evaluates the objective function at its current location. The objective function is often minimized and the exploration of the search space is not through evolution [4]. However, following a widespread practice of borrowing from the evolutionary computation field, in this work, we consider the bi-objective function and fitness function interchangeably. Each particle then determines its movement through the search space by combining some aspect of the history of its own current and best (best-fitness) locations with those of one or more members of the swarm, with some random perturbations. The next iteration takes place after all particles have been moved. Eventually the swarm as a whole, like a flock of birds collectively foraging for food, is likely to move close to an optimum of the fitness function.

Each individual in the particle swarm is composed of three $\mathcal{D}$-dimensional vectors, where $\mathcal{D}$ is the dimensionality of the search space. These are the current position $\vec{x}_i$, the previous best position $\vec{p}_i$ and the velocity $\vec{v}_i$.

The particle swarm is more than just a collection of particles. A particle by itself has almost no power to solve any problem; progress occurs only when the particles interact. Problem solving is a population-wide phenomenon, emerging from the individual behaviors of the particles through their interactions. In any case, populations are organized according to some sort of communication structure or topology, often thought of as a social network. The topology typically consists of bidirectional edges connecting pairs of particles, so that if j is in i's neighborhood, i is also in j's. Each particle communicates with some other particles and is affected by the best point found by any member of its topological neighborhood. This is just the vector $\vec{p}_i$ for that best neighbor, which we will denote with $\vec{p}_g$. The potential kinds of population "social networks" are hugely varied, but in practice certain types have been used more frequently.

In the PSO process, the velocity of each particle is iteratively adjusted so that the particle stochastically oscillates around $\vec{p}_i$ and $\vec{p}_g$ locations.

2.2   Hill Climbing

Hill Climbing (HC) algorithm is a heuristic algorithm. The idea of HC is simple. In HC, the solution s' is accepted as the new current solution if $\delta \leq 0$ holds, where $\delta = f(s') - f(s)$. Here, the function f is called the fitness function. The fitness function gives points to a solution so that the system can evaluate the next solution s' and the current solution s.

The most important factor in HC is to define effectively the neighbor solution. The definition of the neighbor solution affects HC performance directly. In our WMN-PSOHC system, we use the next step of particle-pattern positions as the neighbor solutions for the HC part.

2.3   WMN-PSOHC System Description

In following, we present the initialization, particle-pattern, fitness function and router replacement methods.

Initialization

Our proposed system starts by generating an initial solution randomly, by *ad hoc* methods [15]. We decide the velocity of particles by a random process considering the area size. For instance, when the area size is $W \times H$, the velocity is decided randomly from $-\sqrt{W^2 + H^2}$ to $\sqrt{W^2 + H^2}$. Our system can generate many client distributions. In this paper, we consider Normal distribution of mesh clients as shown in Fig. 1.

Particle-Pattern

A particle is a mesh router. A fitness value of a particle-pattern is computed by combination of mesh routers and mesh clients positions. In other words, each particle-pattern is a solution as shown is Fig. 2. Therefore, the number of particle-patterns is a number of solutions.

Fitness Function

One of most important thing is to decide the determination of an appropriate objective function and its encoding. In our case, each particle-pattern has an own fitness value and compares other particle-patterns fitness value in order to share information of global solution. The fitness function follows a hierarchical approach in which the main objective is to maximize the SGC in WMN. Thus, we use α and β weight-coefficients for the fitness function and the fitness function of this scenario is defined as:

$$\text{Fitness} = \alpha \times \text{SGC}(\boldsymbol{x}_{ij}, \boldsymbol{y}_{ij}) + \beta \times \text{NCMC}(\boldsymbol{x}_{ij}, \boldsymbol{y}_{ij}).$$

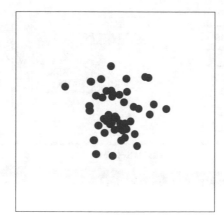

Fig. 1. Normal distribution of mesh clients.

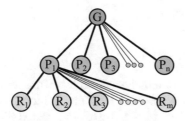

G: Global Solution
P: Particle-pattern
R: Mesh Router
n: Number of Particle-patterns
m: Number of Mesh Routers

Fig. 2. Relationship among global solution, particle-patterns and mesh routers.

Router Replacement Methods

A mesh router has x, y positions and velocity. Mesh routers are moved based on velocities. There are many router replacement methods in PSO field [2,11–13]. In this paper, we use Linearly Decreasing Inertia Weight Method (LDIWM) and Rational Decrement of Vmax Method (RDVM). In LDIWM, C_1 and C_2 are set to 2.0, constantly. On the other hand, the ω parameter is changed linearly from unstable region ($\omega = 0.9$) to stable region ($\omega = 0.4$) with increasing of iterations of computations [1,13]. In RDVM, PSO parameters are set to unstable region ($\omega = 0.9$, $C_1 = C_2 = 2.0$). A value of V_{max} which is maximum velocity of particles is considered. The V_{max} is kept decreasing with the increasing of iterations as:

$$V_{max}(x) = \sqrt{W^2 + H^2} \times \frac{T - x}{x},$$

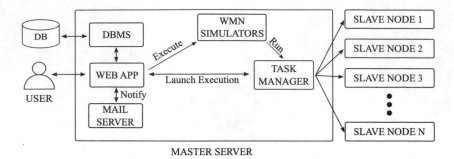

Fig. 3. System structure for web interface.

where, W and H are the width and the height of the considered area, respectively. Also, T and x are the total number of iterations and a current number of iteration, respectively [7].

3   WMN-PSOHC Web GUI Tool

The Web application follows a standard Client-Server architecture and is implemented using LAMP (Linux + Apache + MySQL + PHP) technology (see Fig. 3). We show the WMN-PSOHC Web GUI tool in Fig. 4. Remote users (clients) submit their requests by completing first the parameter setting. The parameter values to be provided by the user are classified into three groups, as follows.

- Parameters related to the problem instance: These include parameter values that determine a problem instance to be solved and consist of number of router nodes, number of mesh client nodes, client mesh distribution, radio coverage interval and size of the deployment area.
- Parameters of the resolution method: Each method has its own parameters.
- Execution parameters: These parameters are used for stopping condition of the resolution methods and include number of iterations and number of independent runs. The former is provided as a total number of iterations and depending on the method is also divided per phase (e.g., number of iterations in a exploration). The later is used to run the same configuration for the same problem instance and parameter configuration a certain number of times.

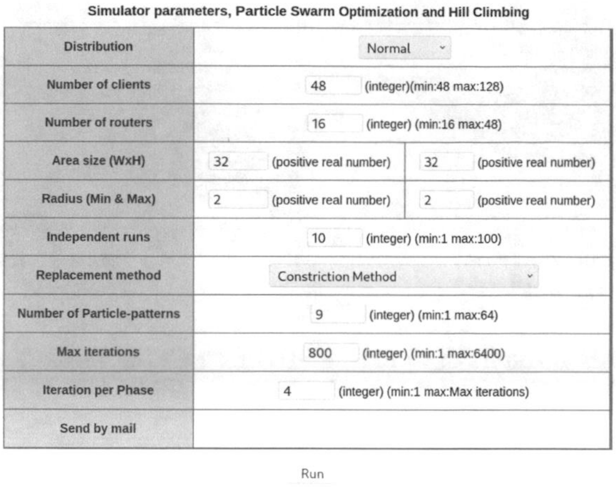

Simulator parameters, Particle Swarm Optimization and Hill Climbing

Distribution	Normal ⌄		
Number of clients	48	(integer)(min:48 max:128)	
Number of routers	16	(integer) (min:16 max:48)	
Area size (WxH)	32	(positive real number)	32	(positive real number)
Radius (Min & Max)	2	(positive real number)	2	(positive real number)
Independent runs	10	(integer) (min:1 max:100)	
Replacement method	Constriction Method ⌄		
Number of Particle-patterns	9	(integer) (min:1 max:64)	
Max iterations	800	(integer) (min:1 max:6400)	
Iteration per Phase	4	(integer) (min:1 max:Max iterations)	
Send by mail			

Run

Fig. 4. WMN-PSOHC Web GUI tool.

Table 1. Instances parameters.

Parameters	Instance 1	Instance 2
Area size	32×32	64×64
Number of mesh routers	16	32
Number of mesh clients	48	96

4   Simulation Results

In this section, we show simulation results using WMN-PSOHC system. In this work, we consider Normal distribution of mesh clients. We consider the number of particle-patterns 9. We conducted simulations 100 times in order to avoid the effect of randomness and create a general view of results. The total number of iterations is considered 800 and the iterations per phase is considered 4. We consider two instances: Instance 1 and Instance 2 as shown in Table 1. We show the parameter setting for WMN-PSOHC in Table 2.

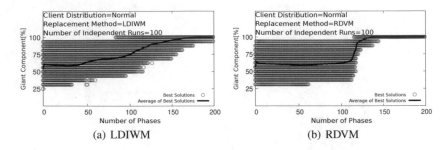

Fig. 5. Simulation results of WMN-PSOHC for SGC in the case of Instance 1.

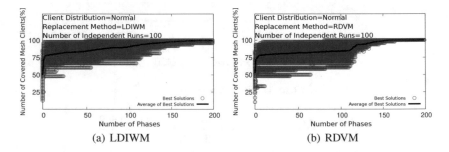

Fig. 6. Simulation results of WMN-PSOHC for NCMC in the case of Instance 1.

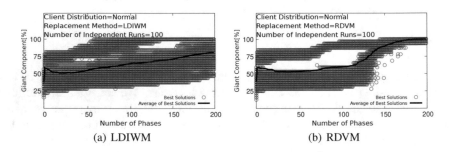

Fig. 7. Simulation results of WMN-PSOHC for SGC in the case of Instance 2.

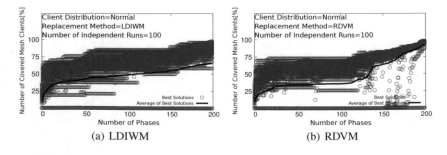

Fig. 8. Simulation results of WMN-PSOHC for NCMC in the case of Instance 2.

Table 2. Parameter settings.

Parameters	Values
Clients distribution	Normal distribution
Instances	Instance 1, Instance 2
Total iterations	800
Iteration per phase	4
Number of particle-patterns	9
Radius of a mesh router	From 2.0 to 3.0
Fitness function weight-coefficients (α, β)	0.7, 0.3
Replacement method	LDIWM, RDVM

We show the simulation results of Instance 1 in Fig. 5 and Fig. 6. The value of SGC and NCMC for 200 phases has reached 100% for both replacement methods. We show the simulation results in the case of Instance 2 in Fig. 7 and Fig. 8. Comparing with Instance 1, the performance of both replacement methods decrease because the search space of Instance 2 is larger than Instance 1. RDVM performs better compared with LDIWM for both parameters (SGC and NCMC) in the case of Instance 2.

5 Conclusions

In this work, we evaluated the performance of WMNs by using a hybrid simulation system based on PSO and HC (called WMN-PSOHC) considering two instances and Normal distribution of mesh clients.

Simulation results show that WMN-PSOHC performs better for Instance 1 compared with Instance 2. This is because the WMN-PSOHC needs more time to get the same performance with Instance 1. Also, we conclude that RDVM performs better than LDIWM in this considered scenario.

In our future work, we would like to evaluate the performance of the proposed system for different parameters and scenarios.

References

1. Barolli, A., Sakamoto, S., Ohara, S., Barolli, L., Takizawa, M.: Performance analysis of WMNs by WMN-PSOHC-DGA simulation system considering linearly decreasing inertia weight and linearly decreasing Vmax replacement methods. In: Barolli, L., Nishino, H., Miwa, H. (eds.) INCoS 2019. AISC, vol. 1035, pp. 14–23. Springer, Cham (2020). https://doi.org/10.1007/978-3-030-29035-1_2
2. Clerc, M., Kennedy, J.: The particle swarm-explosion, stability, and convergence in a multidimensional complex space. IEEE Trans. Evol. Comput. **6**(1), 58–73 (2002)
3. Ozera, K., Bylykbashi, K., Liu, Y., Barolli, L.: A fuzzy-based approach for cluster management in VANETs: performance evaluation for two fuzzy-based systems. Internet Things **3**, 120–133 (2018)

4. Poli, R., Kennedy, J., Blackwell, T.: Particle swarm optimization. Swarm Intell. **1**(1), 33–57 (2007)
5. Sakamoto, S., Lala, A., Oda, T., Kolici, V., Barolli, L., Xhafa, F.: Analysis of WMN-HC simulation system data using friedman test. In: The Ninth International Conference on Complex, Intelligent, and Software Intensive Systems (CISIS-2015), pp 254–259. IEEE (2015)
6. Sakamoto, S., Oda, T., Ikeda, M., Barolli, L., Xhafa, F.: Implementation and evaluation of a simulation system based on particle swarm optimisation for node placement problem in wireless mesh networks. Int. J. Commun. Netw. Distrib. Syst. **17**(1), 1–13 (2016)
7. Sakamoto, S., Oda, T., Ikeda, M., Barolli, L., Xhafa, F.: Implementation of a new replacement method in WMN-PSO simulation system and its performance evaluation. In: The 30th IEEE International Conference on Advanced Information Networking and Applications (AINA-2016), pp. 206–211 (2016). https://doi.org/10.1109/AINA.2016.42
8. Sakamoto, S., Ozera, K., Ikeda, M., Barolli, L.: Implementation of intelligent hybrid systems for node placement problem in WMNs considering particle swarm optimization, hill climbing and simulated annealing. Mob. Netw. Appl. **23**(1), 27–33 (2017). https://doi.org/10.1007/s11036-017-0897-7
9. Sakamoto, S., Barolli, A., Barolli, L., Okamoto, S.: Implementation of a web interface for hybrid intelligent systems. Int. J. Web Inf. Syst. **15**(4), 420–431 (2019)
10. Sakamoto, S., Barolli, L., Okamoto, S.: WMN-PSOSA: an intelligent hybrid simulation system for WMNs and its performance evaluations. Int. J. Web Grid Serv. **15**(4), 353–366 (2019)
11. Schutte, J.F., Groenwold, A.A.: A study of global optimization using particle swarms. J. Glob. Optim. **31**(1), 93–108 (2005)
12. Shi, Y.: Particle swarm optimization. IEEE Connect. **2**(1), 8–13 (2004)
13. Shi, Y., Eberhart, R.C.: Parameter selection in particle swarm optimization. In: Porto, V.W., Saravanan, N., Waagen, D., Eiben, A.E. (eds.) EP 1998. LNCS, vol. 1447, pp. 591–600. Springer, Heidelberg (1998). https://doi.org/10.1007/BFb0040810
14. Wang, J., Xie, B., Cai, K., Agrawal, D.P.: Efficient mesh router placement in wireless mesh networks. In: Proceedings of the IEEE International Conference on Mobile Adhoc and Sensor Systems (MASS-2007), pp. 1–9 (2007)
15. Xhafa, F., Sanchez, C., Barolli, L.: Ad hoc and neighborhood search methods for placement of mesh routers in wireless mesh networks. In: Proceedings of the of 29th IEEE International Conference on Distributed Computing Systems Workshops (ICDCS-2009), pp. 400–405 (2009)

Millimeter-Wave Dual-Band Slotted Antenna for 5G Applications

Prince Mahmud Ridoy[✉] ⓘ, Arajit Saha ⓘ, Khadija Yeasmin Fariya ⓘ,
Pranta Saha ⓘ, Khan Md. Elme ⓘ, and Farhadur Arifin ⓘ

American International University-Bangladesh, 408/1, Kuratoli, Khilkhet,
Dhaka 1229, Bangladesh
mahmudprince3358@gmail.com, arajitsahactg417@gmail.com,
fariya.khadijayeasmin@gmail.com, pranta.saha1907@gmail.com,
khanelme@gmail.com, arifin@aiub.edu

Abstract. This paper presents a millimeter-wave dual-band slotted antenna (26.28/37.72 GHz) for fifth-generation (5G) network. The dielectric layer of the antenna is made of Rogers RT 5880 material and has a thickness of 0.254 mm.

The resonant frequencies of 26.28 GHz and 37.72 GHz, among the most desirable frequency bands for 5G communication in the mobile network, are used in this study. The designed dual-band slotted antenna has an architectural structure of 8 mm × 6mm × 0.254 mm. It achieves a gain of 5.464 dB and 6.705 dB and directivity of 6.18 dBi and 7.538 dBi with the assistance of the Reflection Coefficient at resonant frequencies of –21.79 dB and –34.55 dB. With bandwidths of 0.867 GHz and 1.054 GHz, the proposed antenna's efficiency was successfully attained 84.81% and 82.55%, respectively, which is suitable and enough efficient for 5G communication. The Microwave CST Studio Suite tool is used to design and simulate this proposed antenna. According to the intended results, the suggested antenna works excellently, and due to its fundamental architecture, the proposed antenna will be suitable for mobile networks. The designed antenna can be appropriate for faster mobile networks and 5G communication technologies by estimating all the parameters.

Keywords: Millimeter-wave · Dual-band slotted antenna · 5G communication · VSWR · Gain enhancement · CST · Networking

1 Introduction

Increasing demand is a natural consequence of a growing population. Whether the speed, power consumption, access choices, or other factors, everyone is driven by technology to some extent. Customers expected mobile providers to deliver on their expectations. Since the fourth generation does not meet customers' expectations, scientists and researchers have conducted several experiments and produced a fifth-generation mobile technology (5G). In other places, it is still not delivering the speed customers demand for their applications. Fourth-generation cellular networks have several drawbacks, and as a result,

L. Barolli et al. (Eds.): AINA 2022, LNNS 450, pp. 19–30, 2022.
https://doi.org/10.1007/978-3-030-99587-4_3

they will be phased out soon. On the contrary, the demand for high data rates for high-definition video transmission and reception and massive traffic is constantly increasing.

As a result, fifth-generation 5G wireless communication is recommended to mitigate high data transmission and reception demand. The fifth generation's technology has several eye-catching features like increased speed, reduced latency, and increased bandwidth, including its application to IoT [1]. As long as the devices can communicate with the server, the user will control them from any location. There are many frequency bands in 5G communications but millimeter (mm) waves, with 26 GHz and 38 GHz core frequencies being the most promising options [2]. Millimeter-wave is primarily considered for 5G cellular communication and propagation models [3, 4]. Furthermore, given its importance in cellular communication services and multigigabit communication services, mmWave is a crucial part of fifth-generation (5G) communication [5, 6]. In some research, IoT was used to transfer real-time data to the authority for safety purposes [7, 8]. But choosing the right antenna for an Internet of Things (IoT) device is a key design challenge.

Slots are mostly utilized to produce new resonance. Slots in the bottom layers boost the antenna's gain. In many situations, slotting the patch reduces the gain at the expense of the increased resonance caused by the metallic mitigation. On the other hand, dual-band antennas can be configured to take up less than half the area required by two single-band antennas. Thus, by utilizing dual-band antennas, both cost and space may be minimized. In certain instances of antenna design, space control is critical in addition to cost reduction. Moreover, a slotted dual-band microstrip patch antenna is chosen for its smaller size, easy installation, lightweight, low cost, high efficiency, and easy fabrication [9]. In another research, a modified rectangular ring with a fork-shaped strip was used to generate higher band 5G frequencies while improving the antenna's return loss and bandwidth. This work achieved bandwidth of 1.37 GHz, 0.11 GHz and 2.42 GHz at the triple frequencies 28 GHz, 31.45 GHz and 34.6 GHz, which covers the higher band of 5G mobile application [19]. Besides, 26 GHz and 38 GHz are promising frequencies for 5G technology since they cover both of the millimeter (mm) Wave and K bands. From that point of view, a fork-shaped dual-band slotted microstrip patch antenna is proposed for 5G applications. The designed multi-band microstrip patch antenna meets the necessary criteria for addressing the problems of future 5G wireless communication systems. The suggested antenna model also has a higher realized gain and a more significant return loss than many other cited works, indicating that it is more efficient than the different designs under consideration.

Following this portion of the introduction, Sect. 2 highlights the antenna design and the intricacies of the geometrical structure. This section also discusses the appropriate material choices for antennas. Designed antenna's performance in terms of simulated outcomes and the antenna's relative characteristics is analyzed in Sect. 3. Section 4 summarizes the comparative analysis of the proposed work and in Sect. 5 future work for this proposed antenna is discussed. Finally, Sect. 6 contains the conclusion of this paper.

2 Antenna Design

A millimeter-wave dual-band slotted antenna for 5G technologies is proposed in this research. The radiation patterns of the designed antenna are approximately omnidirectional, and the polarization pattern is also linear.

This dual-band slotted antenna operates at 26.28 GHz and 37.72 GHz for simplified analysis and enhanced efficiency in the 5G Communications Systems.

2.1 Design Equations

The theoretical analysis is based on the transmission lines model. After the substrate material, Rogers RT5880 was chosen along with the operating frequency of 26.28 GHz and 37.72 GHz, which has been used to calculate the Width and Length of the Patch simultaneously using Eqs. 1 and 2 [10].

The equations for sequentially estimating the Width of the Patch (W) and the Length of the Patch (L) are given here.

$$W = \frac{c}{2f_0\sqrt{\frac{\varepsilon_r+1}{2}}} \tag{1}$$

Here, f_0, represents the resonant frequency of 26.28 GHz and 37.72 GHz.

Where c denotes the electromagnetic wave velocities in empty space, which is 3×10^8 m/s and ε_r denote the dielectric constant's relative permittivity, which is 2.2.

Now, the length of the patch was calculated approximately,

$$L = \frac{c}{2f_0\sqrt{\frac{\varepsilon_r+1}{2}}} - 0.824h\left\{\frac{(\varepsilon_{reff}+0.3)\left(\frac{W}{h}+0.264\right)}{(\varepsilon_{reff}-0.258)\left(\frac{W}{h}+0.8\right)}\right\} \tag{2}$$

Where,

- W = width of the patch antenna.
- L = length of the patch antenna.
- f_0 = resonance frequency.
- c = speed of light.
- ε_r = dielectric constant of the substrate.
- h = thickness of the substrate
- ε_{reff} = effective dielectric constant of the substrate.

Calculation of Effective dielectric constant ε_{reff}

$$\varepsilon_{reff} = \frac{\varepsilon_r+1}{2} + \frac{\varepsilon_r-1}{2} \times \left[\frac{1-12h}{W}\right]^{0.5}$$

Inset feed depth determination (y_0)

$$y_0 = \frac{\left\{\cos^{-2}\left(\frac{z_0}{R_{in}}\right)L\right\}}{\pi}$$

Where, z_0 represents the line impedance, R_{in} represents input impedance, and L is the length of the patch. The suggested millimeter-wave dual-band slotted antenna model was developed using the CST Studio Suite software by concerning all of the preceding equations.

2.2 Antenna Specification

The architecture of this simulated Antenna specification and its different viewpoints was determined using Computer Simulation Technology (CST) Microwave studio. Proper measurement has been established to accomplish the goal of this antenna. The substrate layer of the antenna (hs), which is 0.254 mm thick, is made of Rogers RT 5880 material. The ground layer (hg) of the antenna is 0.025 mm thick.

The antenna's geometrical model is shown in the Figs. 1 and 2.

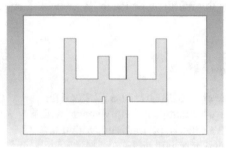

Fig. 1. A front view of the antenna.

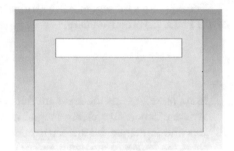

Fig. 2. Back view of the antenna.

Here, Fig. 1 illustrates the front view of the antenna, and Fig. 2 represents the back view of the proposed antenna.

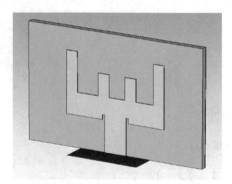

Fig. 3. Antenna perspective in X-Y-Z plane.

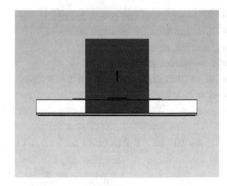

Fig. 4. Bottom outlook of the antenna.

Figure 3 shows the antenna perspective in the X-Y-Z plane in the free space of CST Microwave Studio software. It depicts the antenna from a 3D viewpoint (Fig. 4).

It is possible to derive that the dimension volume of the designed antenna is (8 (Width) × 6 (Length) × 0.254 (Height)) mm³. The resulting dimensions were achieved from the CST software. The parameter descriptions that were used to simulate the antenna, along with their symbols, are shown in detail in Table 1.

Table 1. Parameters of the proposed model.

Parameter indication	Parameter descriptions	Dimension (mm)
wg	Width of ground plane	8
lg	Length of ground plane	6
hg	Height of ground plane	0.025
hs	Height of substrate layer	0.254
wp	Width of patch layer	4.5
lp	Length of patch layer	3.7
t_{rns}	Transmission line width	1

2.3 Substrate Material Selection

The proposed antenna in this study utilizes the Rogers RT 5880 as a substrate material, which is more efficient at high frequencies than other substrates and is fed through a microstrip line. Rogers RT 5880 laminates are manufactured with PTFE glass fiber reinforced material with a low dielectric constant and dielectric loss, making them ideal for high-frequency applications. The dielectric performance of a printed circuit board (PCB) at fast speeds is necessary for usage in mobile and other industries. Rogers 5880 is a low-loss, high-performance dielectric material ideal for demanding PCB prototypes. Besides that, Rogers RT 5880 is an excellent choice for high humidity settings because of its low moisture absorption.

Also, Rogers RT 5880 is readily cut to the required form and is resistant to corrosion by all solutions and reagents used in the etching and plating of through-holes processes. As previously stated, Rogers RT 5880 is used as a Subtract material in this Antenna design for the reasons stated above and its better performance.

3 Result Analysis

By using Computer Simulation Technology (CST) software [18], total simulation work is done. Various parameters like VSWR, Return loss, Bandwidth, S-pattern, far-field patterns, and Efficiency are the main criterion to analyze a 5G application-based antenna. Similarly, all these parameters are analyzed after finishing the simulation.

3.1 Bandwidth

Figure 5 denotes the necessary information's to calculate the bandwidth of both bands. The bandwidth of the first band and second band are respectively 1.05 GHz, 0.87 GHz. With the mentioned equation, both bandwidths are calculated, which are acceptable for 5G communications.

$$Bandwidth = upper\ cutoff\ frequency - lower\ cutoff\ frequency$$

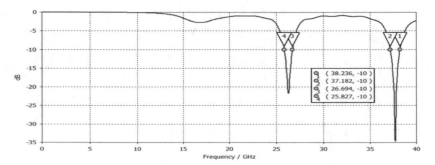

Fig. 5. S-parameter indicating bandwidth

3.2 2-D Radiation Pattern

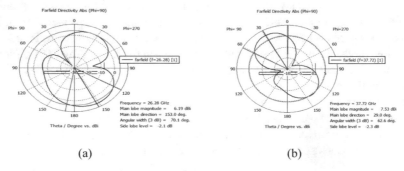

(a)   (b)

Fig. 6. Polar far-field pattern of the antenna at (a) 26 GHz (b) 38 GHz.

The Far Field Region follows the Near Radiative Near Field in the electromagnetic spectrum. Radiating fields dominate electromagnetic fields in this area. The far-field radiation in polar form was also analyzed for this dual-band antenna. The polar far-field pattern (for 90° polar form) at 26 and 38 GHz is illustrated in Figs. 6(a) and (b). Figure 6(a) depicts the far-field directivity at 26 GHz, that main lobe magnitude is 6.19 dBi in 153° direction, and Fig. 6(b) represents the far-field directivity at 38 GHz operating frequency, that main lobe magnitude is 7.23 dBi in 29° direction.

3.3   Directivity, Gain and Efficiency

The antenna's gain informs us how much power the antenna can provide in the stated direction. At the same time, directionality measures how concentrated the radiation is in the direction where it is strongest. On the contrary, efficiency is a key consideration when evaluating or assessing antenna performance. To calculate the efficiency, gain and directivity are most two important parameters. Because a well-known equation, efficiency $= \frac{Gain}{Directivity} * 100\%$ is used to evaluate the efficiency of the antenna.

At 26 GHz, resonant frequency directivity is 6.180dBi, and gain is 5.464 dB. Figure 7(a) and 7(b) represents the directivity and gain at 26.28 GHz operating frequency. Therefore the efficiency on that band is 88.41%.

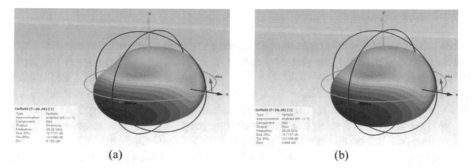

(a)   (b)

Fig. 7. 3-D radiation pattern showing (a) Directivity of the antenna at 26 GHz, (b) Gain of the antenna at 26 GHz

The directivity and gain at 38 GHz are extracted from Fig. 8(a) that indicates the 7.538 dBi directivity and Fig. 8(b) 6.705 dB gain sequentially. In accordance with the given formula, the efficiency at 38 GHz determined as 82.55%

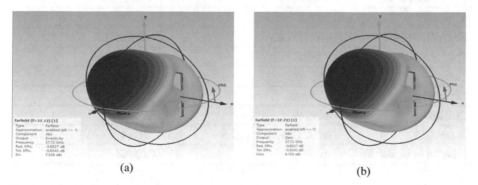

(a)   (b)

Fig. 8. 3-D radiation pattern showing (a) Directivity of the antenna at 38 GHz and (b) Gain of the antenna at 38 GHz.

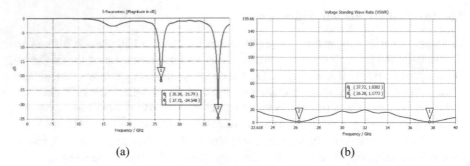

(a)   (b)

Fig. 9. (a) S-pattern at the operating frequency, and (b) VSWR at the operating frequency.

3.4 VSWR, Reflection Co-efficient or S-Parameter

VSWR is expressed as a voltage standing wave ratio that measures the mismatch between an antenna and the feed line that connects to it. Also called the Standing Wave Ratio, this ratio is used to measure (SWR). VSWR can have a value between 1 and infinite. Most antenna applications allow for a VSWR value of less than 2. VSWR nears 1 in both resonant frequencies for this proposed antenna, represented in Fig. 9(a).

S11 or Reflection Coefficient is another important parameter considered in antenna result analysis. It denotes how much power is reflected from the antenna. For the preferable antenna, the value is S11 ≤ –10 dB. The obtained S-pattern at the operational frequency for the specified antenna models are shown in Fig. 9(b). At 26.28 GHz and 37.72 GHz return loss is respectively –21.79dB and –34.548 dB.

Table 2. Summary results of the proposed antenna

Parameter of antenna	Achieved values at 26.28 GHz	Achieved values at 37.72
S-parameter (S11)	– 21.79 dB	– 34.55 dB
VSWR	1.1772	1.0382
Rad. efficiency	– 0.7157 dB	– 0.8327 dB
Total efficiency	– 0.7446 dB	– 0.8343 dB
Gain	5.464 dB	6.705 dB
Directivity	6.18 dBi	7.538 dBi
Efficiency	84.81%	82.55%
Bandwidth	0.867 GHz (867 MHz)	1.054 GHz
Fraction of bandwidth	0.0399	0.02794

4   Comparative Analysis

The obtained output efficiency of the designed antenna is 84.81% and 82.55%, at 26.28 GHz and 37.72 GHz Resonant Frequency. The comparison of obtained results with other research and comparison based on architecture is given in Table 3.

Table 3. Comparison of different antenna parameters with other reported antennas.

References	Resonant frequency (GHz)	Reflection co-efficient at resonant frequency (dB)	Bandwidth (GHz)	Gain (dB)	Efficiency (%)
[11]	28	<−10	0.847	7.93	78
	38	<−10	<1	6.9	86
[12]	28	− 23.6	1.49	5.41	90.33
	38	− 27.1	1.01	4.89	84.3
[13]	37	− 25.8	5.5	5	65
	54	− 27.8	8.67	6	75
[14]	28	<−10	1.44	2.25	93
	60	<−10	39.24	3.40	97.5
[15]	28	<−10	0.850	4	84
	38	<−10	0.750	4.5	91
[16]	28	− 28.1	4.5	4.54	94
	38	− 43.8	3.8	4.21	96.6
[17]	28	<−10	0.6	>6	>80
	38	<−10	1.1	>6	>80
Proposed antenna	26.28	− 21.79	0.867	5.464	84.81
	37.72	− 34.55	1.054	6.705	82.55

Table 2 shows the comparative analysis of the obtained result of the designed antenna with other published work. In this paper, different antenna properties such as return Loss, gain, bandwidth, efficiency, and resonance frequency, have been analyzed from previous research. After performing the simulation on CST at 26.28 GHz and 37.72 GHz resonant frequency, a better S1,1 parameter has been obtained compared to the other reference paper mentioned in Table 2. The designed antenna model has the S1,1 value of −21.79 dB and −34.55 dB, and the gain magnitude of 5.464 dB and 6.705 dB with the bandwidth of 0.867 GHz and 1.054 GHz in the free space. In this proposed design, the S1,1 has been obtained at a higher level than the other models, which is the most important for High-speed algorithms.

This designed antenna has obtained a 5G communication system with optimum gain, efficiency, and bandwidth. It can be said that the values of the proposed antenna design

Table 4. Comparison table based on architecture.

Antenna geometry (Length * Width * Substrate Height in mm)	Substrate material	Reference
30 × 24 × 4.5	FR406	[11]
13 × 11.25 × 0.787	Rogers RT 5880	[12]
5 × 7.2 × 0.787	RT-5880	[13]
2.9 × 1.6 × 1.3	FR-4	[14]
2.5 × 3.1 × 0.254	RT/Duroid 5880	[15]
12 × 12 × 0.237	Rogers RT 4003	[16]
8 × 8 × 0.26	RT Duroid 5880	[17]
6 × 8 × 0.254	Rogers RT 5880	Proposed antenna

have a better output compared to the other research works of different researchers that have been established recently.

Table 3 Represents the comparison table based on the architecture of the designed antenna with Different Existing research work (Table 4).

In this antenna architecture, the lg represents ground length, wg represents ground width, and hg represents the ground height. The antenna architecture and the substrate material of the proposed antenna model are shown in the table's last row. The dimension of the Proposed antenna is $8 \times 6 \times 0.254$ mm^3. The mentioned models are compared with other research works.

The analysis demonstrates that the proposed antenna provides better results than the other antennas mentioned in comparative Table 3. A compact architecture with a better output can be used for any antenna-related application in today's technology. In comparison to previous models, the suggested antenna would be a superior choice for 5G communication.

5 Future Work

The project prototype of the designed antenna is already in progress. Further modifications will be added to achieve better performance. Moreover, work is going on for the accomplishment of better efficiency, gain, directivity and incorporating new technologies. The proposed system is designed for 5G mobile communication, however, this antenna can also be adapted for bio medical applications. This antenna will be reconfigured by considering all those parameters and maintaining the cost and environmental impact.

6 Conclusion

For 5G connectivity, a novel fork-shaped dual-band patch antenna is proposed in this paper. The dimension of the proposed antenna is $6 * 8 * 0.254$ mm^3, and it operates in 26 GHz and 38 GHz. Antenna parameters such as S11, efficiency, gain, bandwidth, directivity, and VSWR are all metrics used to assess 5G mmWave multi-band antenna performance. A comparison table has been shown in table no 3 of performance and 4 of structure that shows the difference among the designed antenna and the existing work by performance and measurement respectively. The result shows that it has better bandwidth, gain, and efficiency according to its dimension compared with previous work. Considering all of these factors, our proposed antenna would be a good candidate for introducing 5G mobile communication systems.

References

1. Agiwal, M., Roy, A., Saxena, N.: Next-generation 5g wireless networks: a comprehensive survey. IEEE Commun. Surv. Tutor. **18**(3), 1617–1655 (2016)
2. Rappaport, T.S., et al.: Millimeter wave mobile communications for 5G cellular: it will work! IEEE Access **1**, 335–349 (2013)
3. Ge, X., Tu, S., Mao, G., Wang, C.-X., Han, T.: 5G ultra-dense cellular networks. IEEE Wirel. Commun. **23**(1), 72–79 (2016)
4. Zhang, J., Shafi, M., Molisch, A.F., Tufvesson, F., Wu, S., Kitao, K.: Channel models and measurements for 5G. IEEE Commun. Mag. **56**(12), 12–13 (2018)
5. Niu, Y., Li, Y., Jin, D., Su, L., Vasilakos, A.V.: A survey of millimeter wave communications (mmWave) for 5G: opportunities and challenges. Wireless Netw. **21**(8), 2657–2676 (2015)
6. Morgado, A., Huq, K.M., Mumtaz, S., Rodriguez, J.: A survey of 5G technologies: regulatory, standardization and industrial perspectives. Digit. Commun. Netw. **4**(2), 87–97 (2018)
7. Islam, M.H., Fariya, K.Y., Tanim, M.T.H., Talukder, T.I., Chisty, N.A.: IoT-based smart street light for improved road safety. In: Zhang, Y.-D., Senjyu, T., So-In, C., Joshi, A. (eds.) Smart Trends in Computing and Communications. LNNS, vol. 286, pp. 377–390. Springer, Singapore (2022). https://doi.org/10.1007/978-981-16-4016-2_36
8. Islam, M.H., Fariya, K.Y., Talukder, T.I., Khandoker, A.A., Chisty, N.A.: IoT based smart self power generating street light and road safety system design: a review. In: 2021 IEEE Region 10 Symposium (TENSYMP), pp. 1–5 (2021). https://doi.org/10.1109/TENSYMP52 854.2021.9550937
9. Imran, D., et al.: Millimeter wave microstrip patch antenna for 5G mobile communication. In: 2018 International Conference on Engineering and Emerging Technologies (ICEET), pp. 1–6 (2018). https://doi.org/10.1109/ICEET1.2018.8338623
10. Balanis, C.A.: Antenna Theory: Analysis and Design, 3rd edn. Wiley-Interscience, Hoboken (2012). ISBN-13: 978-0471667827
11. Chandra Sekhararao, K., Kavitha, A.: Circularly polarized dual band micro strip patch antenna design at 28 GHz/38 GHz for 5G cellular communication. JCR **7**(4) (2020). ISSN 2394-5125
12. Anab, M., Khattak, M., Owais, S., Ali Khattak, A., Sultan, A.: Design and analysis of millimeter wave dielectric resonator antenna for 5G wireless communication systems. Prog. Electromagn. Res. C **98**, 239–255 (2020). https://doi.org/10.2528/pierc19102404
13. Lodro, Z., Shah, N., Mahar, E., Tirmizi, S.B., Lodro, M.: mmWave novel multiband microstrip patch antenna design for 5G communication. In: 2019 2nd International Conference on Computing, Mathematics and Engineering Technologies (iCoMET), pp. 1–4 (2019). https://doi.org/10.1109/ICOMET.2019.8673447
14. Ibrahim, M.S.: Dual-band microstrip antenna for the fifth generation indoor/outdoor wireless applications. In: 2018 International Applied Computational Electromagnetics Society Symposium (ACES), pp. 1–2 (2018). https://doi.org/10.23919/ROPACES.2018.8364097

15. Aliakbari, H., Abdipour, A., Mirzavand, R., Costanzo, A., Mousavi, P.: A single feed dual-band circularly polarized millimeter-wave antenna for 5G communication. In: 2016 10th European Conference on Antennas and Propagation (EuCAP), pp. 1–5 (2016). https://doi.org/10.1109/EuCAP.2016.7481318
16. Sabek, A., Ibrahim, A., Ali, W.: Dual-band millimeter wave microstrip patch antenna with stubresonators for 28/38 GHz applications. J. Phys.: Conf. Ser. **2128**(1), 012006 (2021). https://doi.org/10.1088/1742-6596/2128/1/012006
17. Nosrati, M., Tavassolian, N.: A single feed dual-band, linearly/circularly polarized cross-slot millimeter-wave antenna for future 5G networks. In: 2017 IEEE International Symposium on Antennas and Propagation & USNC/URSI National Radio Science Meeting, pp. 2467–2468 (2017). https://doi.org/10.1109/APUSNCURSINRSM.2017.8073276
18. CST STUDIO SUITE Student Edition, 3DS Academy (2022). https://edu.3ds.com/en/software/cst-studio-suite-student-edition
19. Subramanian, S., Selvaperumal, S., Thangasamy, V., Nataraj, C.: Modified triple band microstrip patch antenna for higher 5G bands. In: 2018 Fourth International Conference on Advances in Electrical, Electronics, Information, Communication and Bio-informatics (AEEICB) (2018). https://doi.org/10.1109/aeeicb.2018.8480917

NARUN-PC: Caching Strategy for Noise Adaptive Routing in Utility Networks

Fabio Pagnotta[1]([⊠]), Leonardo Mostarda[1], and Alfredo Navarra[2]

[1] Computer Science Division, University of Camerino, Camerino, Italy
{fabio.pagnotta,leonardo.mostarda}@unicam.it
[2] Mathematics and Computer Science Department, University of Perugia,
Perugia, Italy
alfredo.navarra@unipg.it

Abstract. In a smart meter network, accounting information are gathered from meters and sent to a collector node. In smart cities, a trade-off between the location of meters and the best quality communication signal is often difficult to achieve for urban restrictions and other transmitted signals. Meters can also be constrained in terms of memory and CPU. This paper proposes NARUN with Path Cache (NARUN-PC), an extension of the Noise Adaptive Routing for Utility Networks (NARUN) for improved performance and routing in a dense mesh utility network. In NARUN, the collector calculates the path with the least noise to reach the destination meter. A weighted network graph that shows the connections among meters is used, where an edge weight defines the corresponding link failure index ranging from one (the link admits no noise) to infinite (the link is broken). No control messages are used to keep the weights updated. Meters report link failure index back to the collector by means of ordinary reading messages. NARUN-PC introduces a caching strategy in NARUN. NARUN-PC can use a previously cached routing path instead of always selecting the best path. NARUN-PC improves NARUN performance in terms of message overhead, failure rate and reading rate.

1 Introduction

Nowadays, automatic reading of gas and electricity meters is achieved through communication networks. Automatic reading can help consumers and resource providers in reducing energy consumption. Consumers can get aware of their consumption data to change their behaviour accordingly while resource providers can reduce their maintenance costs and enable real-time data analysis. Automatic reading of accounting data can be achieved through smart metering systems. Similar to a sensor network (e.g. [7,8,14]), a smart metering system is composed of a collector and a set of meters connected via a communication network. The usage of these smart metering systems is required by the Directive

Work supported by the Italian National Group for Scientific Computation GNCS-INdAM.

L. Barolli et al. (Eds.): AINA 2022, LNNS 450, pp. 31–42, 2022.
https://doi.org/10.1007/978-3-030-99587-4_4

2009/72/EC of the European Parliament to raise the awareness of the consumers in the electricity and gas supply markets [10]. In such systems, the collector gathers accounting information from the network. These data are measured and sent via distributed meter devices. A meter can be a battery-powered device with limited capabilities in terms of memory and computational power. Therefore, the highest cost is not the device itself but the communication network [17]. The European standard protocol for smart metering networks is Meter-bus (see [9]). The protocol defines the physical, data-link, and application layers. Routing protocols are not part of the Meter-bus stack. A wireless version of Meter-bus is called Wireless Meter Bus (WM-Bus). The stack of WM-Bus also includes the physical and data-link layer, node relaying, transport and security services. Network coverage can range from a few hundred meters up to some kilometers according to the used communication frequency.

1.1   Research Problem

In smart cities, a trade-off is required between reliable connectivity and the setup cost [2]. Meters need to be in the coverage area with the best quality of signal available. This can be expensive and sometimes not feasible due to urban constraints This is the case of dense networks such as urban cities where different devices are nearby to each other causing many communication signals to coexist at the same time. This generates interference and therefore poor quality links [1] which require higher communication attempts. As a consequence, the lifetime of battery-operated meters is reduced. This problem has been addressed in the context of direct collector-meter communication [11,12]. Related works often assume channels with Gaussian noise using different error recovery strategies for improving communication performance. NARUN [6,16] is a Noise Adaptive Routing for Utility Networks protocol where communication occurs in a mesh topology. NARUN assumes that static collectors and meters are connected through wireless links. The collector keeps a weighted network graph that shows meters connections. The edge weight defines the link failure index which is infinite when a link is broken, one when the link is perfect, while a number greater than one when the link is noisy. The collector calculates the routing path to a meter by minimising the failure index. NARUN does not use control messages to update the edge weights. Meters eavesdrop on surrounding communication and report back to the collector link failure information via ordinary reading messages. Although NARUN has been proved to be effective when compared to the state of art protocols such as the Dynamic Source Routing (DSR) [13], it 'suffers' for unstable links, i.e. links frequently broken. As NARUN always selects the best available path, unstable links might be frequently selected, hence generating failing paths. This increases the number of messages.

1.2   Paper Contribution

In this paper, we propose the addition of a caching strategy to NARUN. We call it, NARUN with path cache (NARUN-PC). The protocol can use a previously cached routing path instead of always selecting the best path. A new path is selected only when the cached path fails to reach the destination meter. NARUN-PC has been compared using simulation with the following routing protocols: (i) NARUN where the network graph has edges equal to one (the link is working) or infinity (the link is not working); (ii) ECC-NARUN where the network graph has weights between one and infinity and the Hamming code can be used to detect and recover errors; (iii) DSR [13] which is an efficient protocol suitable for wireless sensor networks [5,15]. NARUN-PC-ECC denotes the addition of caching into ECC-NARUN while NARUN-PC is the addition of caching into NARUN.

For simulations, we select a real topology from the San Paolo district, Camerino (MC, Italy). We have compared the protocols by using the collector failure rate, the collector reading rate and the traffic load. The failure rate measures the number of failing paths that are selected by the collector. This can choose a failing path when its network graph is not updated. The reading rate is the number of meters that are correctly read. The simulations are performed by varying the noise power for 30% of noisy links. We also use an ON/OFF setting where 30% of the links are unstable. At -70 dBm noise power when 30% of noisy links are considered NARUN-PC-ECC decreases the failure rate of ECC-NARUN by 2.29% and generates 31% traffic less. When unstable links are considered, NARUN-PC-ECC decreases the failure rate of ECC-NARUN by 2.1% and generates 13% of traffic less.

2   NARUN and NARUN-PC Protocols

In what follows, we denote by $N = \{n_0, \ldots, n_z\}$ the set z nodes (meters); by M a WM-Bus message. A message is composed by three fields: the header $M.h$, the payload $M.d$, and the footer $M.o$. A network graph is denoted with $G(N, E, w)$ where $E \subseteq N \times N$ defines the communication links. $w : E \longrightarrow R$ is the edge weight function. NARUN defines a *connection-based* weight function w^r where $w^r(n_i, n_j) = \infty$ when the link (n_i, n_j) is broken while $w^r(n_i, n_j) = 1$ when the link is working. NARUN defines also a *Hamming-based* weight function w^h where $w^h(n_i, n_j)$ is calculated by using the Hamming code. In particular, the header $M.h$ and the payload $M.d$ of a frame M are divided into l equal parts $M_1, \ldots, M_l$. The sender of M calculates the Hamming code $hmc(M_i)$ of each part M_i and adds it into the footer $M.o$. The format of the message can be summarised as follow: $M = M_1 || \ldots || M_l || hmc(M_1) || \ldots || hmc(M_l)$. The correction function $R(M_i)$ is defined as follows:

$$R(M_i) = \begin{cases} 0 & M_i \text{ is received with no error} \\ \infty & M_i \text{ has a non-recoverable error} \\ 1 & M_i \text{ has a recoverable error} \end{cases} \quad (1)$$

and the Hamming-based weight function $w^h(n_i, n_j)$ by the following equation:

$$w^h(n_i, n_j) = \begin{cases} \infty & \text{IF } \sum_{i=0}^{l} R(M_i) = \infty \\ \frac{\sum_{i=0}^{l} R(M_i)}{l} + 1 & otherwise \end{cases} \tag{2}$$

We can observe that $w^h(n_i, n_j) = \infty$ when M has a non-recoverable error, $w^h = 1$ when no error is recovered, $w^h = 2$ when each part M_i has a recoverable error. Effectively, $w^h(n_i, n_j)$ measures the link failure index lfi between n_i and n_j.

We denote with $P_{n_k}(G) = \{n_0, \ldots, n_k, n_{k+1}, \ldots, n_0\}$ (with $k > 0$) a path that starts from the collector node n_0, reaches the node to be read n_k and returns to the node n_0. The path cost of P_{n_k} is the following:

$$W(P_{n_k}(G)) = \sum_{i=0}^{i=k-1} w(n_i, n_{i+1}) \tag{3}$$

We assume symmetric links thus the path P_{n_k} is palindrome (i.e. the paths from n_0 to n_k and n_k to n_0 are the same). $G_{n_0}(N_0, E_0, w_0)$ denotes the collector node network graph. Each meter node n_s also defines a local meter network graph (referred to as projection graph) which is denoted by $G_{n_s}(N_s, E_s, w_s)$. N_s is the set of neighbours of n_s; E_s contains communication edges of the type (n_s, n_j) and (n_i, n_s), where the node n_s takes the role of sender and receiver, respectively. A projection graph is used by the meter n_s to store the failure indexes of its local communication links. These are sent back to the collector with ordinary reading messages and can be used to update the collector network graph. A collector keeps a timestamp t that is increased by one when a meter reading is attempted. The collector node n_0 reads all meter nodes $n_1, \ldots, n_z$ in turn (this is called **round**). A timestamp t_q denotes the q-th attempt made by the collector to read a meter. Two consecutive timestamps t_q and t_{q+1} may refer to the same meter node n_k when the routing path selected at time t_q fails to reach n_k. In this case, a different path can be tried at time t_{q+1}. In the rest of the paper, $G_{n_k, t_q}(N_{k, t_q}, E_{k, t_q}, w_{k, t_q})$ denotes a meter graph last updated at time t_q. Each $n_j \in N_{k, t_q}$ is a neighbour of n_k added at time t_u with $t_u \leq t_q$. The link $(n_k, n_j)_{t_u}$ would also be added to E_{k, t_q} at time t_u with the related weight $w_{k, t_u}(n_k, n_j)$.

Figure 1 sketches a collector round. The round starts by setting the id of the meter to be read to zero. This is stored inside the variable k. A clone of the collector network graph $G_{n_0, t}$ is put inside the variable $G1_{n_0, t}$. The collector will proceed by increasing the variable k by one so that the first meter ($k = 1$) will be read. A reading operation is described inside the pink square of Fig. 1 and is composed of various read meter attempts. A meter attempt is performed by using the procedure $READMETER$. The variable i stores the number of read meter attempts the collector performed on the same meter. We denote with a_i^k the i-th attempts on the meter k. The reading attempt a_i^k is equal to 1 when the meter n_k is successfully read, zero otherwise. Effectively, a collector reading operation can be formally defined as a finite sequence of reading attempts $r^k = a_1^k a_2^k \ldots a_i^k$ ($0 < i < max$) where max is the maximum number of consecutive attempts the collector can perform on the same meter. A reading operation r^k fails to read n_k when is composed of a sequence of max consecutive zeros. A reading

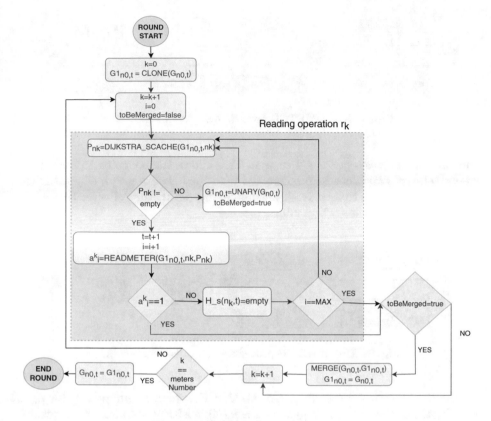

Fig. 1. NARUN collector behaviour

operation r^k successfully reads n_k when is composed of a sequence of zeros (also an empty one) that ends with 1 (the length cannot exceed max). Equation (4) formally defines the reading operation failure rate. $F(r^i)$ measures the number of failed attempts during the reading operation. Equation (5) formally defines the successful reading operation rate when reading meter i.

$$F(r^i) = \frac{\sum_{k=0}^{max}(1 - a_k^i)}{max} \quad (4) \qquad O(r^i) = \frac{\sum_{k=0}^{max} a_k^i}{max} \quad (5)$$

We highlight that a 100% reading rate does not mean a 0% failure rate. For instance, in the sequence of attempts 0000000001 we have 90% failure rate and 100% reading rate. During a reading attempt, the collector uses its network graph to find a path that leads to n_k (that is the meter to be read). When no path is found, the collector creates a unary clone of its network graph. This is a graph that has all edges equal to one. The use of the unary graph is twofold: on the one hand, it allows the collector to retry edges that have their weights set to ∞; on the other hand, ensures connectivity. When the unary graph is generated,

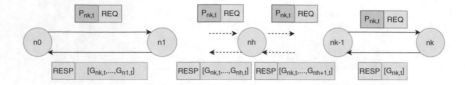

Fig. 2. READMETER procedure successful execution

the variable *toBeMerged* is set to true. This allows the collector to note down
that the unary graph needs to be merged with the network graph $G_{n_0,t}$. In fact,
the unary graph will contain fresh weights discovered by the collector that needs
to be stored in the network graph $G_{n_0,t}$.

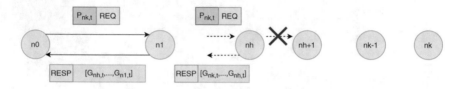

Fig. 3. READMETER procedure unsuccessful execution

Figure 2 shows a successful $READMETER$ execution where a meter n_k is
read by the collector n_0. This generates a REQ WM-Bus standard packet which
contains the path P_{n_k} to reach n_k. Each node in the path receives the REQ
message and forwards it to the next-hop by using the path P_{n_k}. The destination
node receives the request and builds a response message $RESP$. This contains
the reading and its local network graph $G_{n_k,t}$ with the quality of all links n_k
was able to eavesdrop from the neighbour communications. The response packet
is sent back to the collector node and at each hop, each meter adds its local
network graph. Effectively, the collector $READMETER$ procedure receives the
list of all network graphs and updates its local graph $G_{n_0,t}$ with all newer link
failure indexes. Figure 3 shows an unsuccessful $READMETER$ operation. In
this case, a timeout will be triggered at the node that does not receive a response
packet $RESP$. This node will send back to the collector node the reading failure
information and its local network graph.

NARUN-PC is a simple but effective improvement of NARUN. It substitutes
the $DIJKSTRA$ procedure of Fig. 1 with a DIJKSTRA_SCACHE($G1_{n_0,t}, n_k$)
procedure (see Fig. 4). This includes a simple hashmap function $H_s(n_k, t)$ that
stores a meter/path value. The function $H_s(n_k, t)$ can store the following values:
(i) a path $P_{n_k,q}$ that was successfully used to reach the node at time q with $q < t$;
(ii) a path $P_{n_k,q}$ that is return by the $DIJKSTRA$ procedure; (iii) empty when a
the $DIJKSTRA$ procedure finds no path. Figure 4 shows the modification to the
round reading operation of Fig. 1 in order to add cache management. The cache

Algorithm 1. NARUN-PC

procedure DIJKSTRA_SCACHE($G1_{n_0}, t, n_k$)
  Path $P_{n_k, q} \leftarrow H_s(n_k, t)$
  if $P_{n_k, q}! = empty$ **then**
    Return $P_{n_k, q}$
  end if
  $H_s(n_k, t) \leftarrow$ DIJKSTRA($G1_{n_0}, t, n_k$)
  Return $H_s(n_k, t)$
end procedure
procedure DIJKSTRA_ACACHE($G1_{n_0}, t, n_k$)
  Path $P_{n_k, q} \leftarrow H_s(n_k, t)$
  if $P_{n_k, q}! = empty$ **then**
    $res \leftarrow$ CHECK($G1_{n_0}, t, P_{n_k, q}$)
    if $res! = empty$ **then**
      Return $P_{n_k, q}$
    end if
  end if
  $H_s(n_k, t) \leftarrow$ DIJKSTRA($G1_{n_0}, t, n_k$)
  Return $H_s(n_k, t)$
end procedure

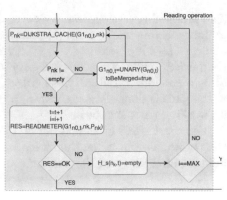

Fig. 4. Cache addition into the collector round reading operation

path $H_s(n_k, t)$ is updated with *empty* when the procedure *READMETER* has a result which is not ok.

This simple cache strategy has been proved to be not effective since can try paths without checking their current state. A path can currently contain some broken links, thus extra overhead messages can be generated. Our solution is a slightly more sophisticated caching strategy that checks the validity of the current stored path. Procedure *DIJKSTRA_SCACHE* shows the implementation of this strategy. Function *CHECK* is used to verify when the stored path contains a broken link (i.e., a link with weight set to infinity). In this case, a new path is stored inside the hashmap.

3 Simulation

NARUN has been simulated in the San Paolo district of Camerino City (MC, Italy). A free space propagation model is used.

The characteristics of this topology are summarised in Table 1. In our simulation, we consider the WM-Bus frequency of 868 MHz. For channel modelling, we used the Additive white Gaussian noise. The transmitter is assumed to use Frequency Shift Keying (FSK) modulation, transmission power of 10 dBm, and an antenna with 0 dB gain. Equation (6) shows the path loss computation formula, which depends on the frequency f, the distance between the source and the destination d, and the gain of antenna g.

$$Pathloss_{dB} = 20 \times \log_{10}(d) + 20 \times \log_{10}(f) - 27.55 - g \tag{6}$$

With the path loss value, we can compute the signal to noise ratio at the receiving end and the bit error rate as we in Eq. (7). This determines the success of communication.

$$BER = \frac{1}{2} \times erfc\left(\sqrt{\frac{SNR_{rec}}{2}}\right) \tag{7}$$

The probability of a success, a recoverable, or unrecoverable packet by assuming a binary symmetric channel [3] can be calculated as $(1-r)^n$, $n \times r \times (1-r)^{n-1}$ and $1-(1-r)^n - n \times r \times (1-r)^{n-1}$, respectively, where n represents the size of the packet, whereas r is the bit error rate. This means that a reading attempt a_k^i is a random variable. The DSR attempt may include the discovery of new paths (this is done via broadcasting). These additional messages are sent when no working path is known.

Our basic unit of simulation is a round $R = \{r^1, \ldots, r^z\}$ where each r^i is the reading operation on the meter i. Equations (8) and (9) formally define the failure round rate and the reading round rate, respectively.

$$F(R) = \frac{\sum_{i=0}^z F(r^i)}{z} \quad (8) \qquad O(R) = \frac{\sum_{i=0}^z O(r^i)}{z} \quad (9)$$

We can define a simulation run that is a sequence of h rounds ($h = 50$ in our simulation). This is denoted with $U = \{R_1, \ldots, R_h\}$. Equations (10) and (11) formally define the failure run rate and the reading run rate.

$$F(U) = \frac{\sum_{i=0}^h F(R_i)}{h} \quad (10) \qquad O(U) = \frac{\sum_{i=0}^h O(R_i)}{h} \quad (11)$$

An experiment is a sequence of q runs ($q = 50$ in our simulation). This is denoted with $E = \{U_1, \ldots, U_q\}$. Equations (12) and (13) formally define the failure experiment rate $F(E)$ and the reading experiment rate $O(E)$, respectively.

$$F(E) = \frac{\sum_{i=0}^q F(U_i)}{q} \quad (12) \qquad O(E) = \frac{\sum_{i=0}^q O(U_i)}{q} \quad (13)$$

We can finally define a simulation that is a sequence of p experiments. This is denoted with $S = \{E_1, \ldots, E_p\}$. Equations (14) and (15) formally define the failure rate $F(S)$ and the reading rate $O(S)$, respectively.

$$F(S) = \frac{\sum_{i=0}^p F(E_i)}{p} \quad (14) \qquad O(S) = \frac{\sum_{i=0}^q O(E_i)}{p} \quad (15)$$

The experiments $S = \{E_1, \ldots, E_p\}$ are a sequence of independent and identically distributed random variables. In order to stop a simulation, we use the criteria that has been presented in [4]. We stop when for a sequence of k consecutive experiments (i.e., E_t, $E_{t+1} \ldots E_{t+k}$) the following conditions hold: (i) $|F(E_i) - F(E_{i+1})| < \epsilon$; (ii) $|O(E_i) - O(E_{i+1})| < \epsilon$ (with $t < i < t+k$). This stop criteria makes sure that the sample average converges within a threshold ϵ.

Table 1. Network characteristics

Network characteristic	Value
Collector latitude	43.1465094238089
Collector longitude	13.0615996612235
Number of meter nodes	254 nodes
Node to node minimum distance	8.86 m
Node to node maximum distance	249.991 m
Node to node average distance	151.141 m
Collector to meter minimum path length[a]	1 hop
Collector to meter maximum path length[a]	4 hops
Collector to meter average path length[a]	2 hops
Network radius	778.10 m
Network density	134.283 nodes/km^2

[a] Path lengths computed by minimising the number of hops

We compared the following protocols: (i) NARUN which uses a connection-based weight function; (ii) ECC-NARUN which uses a Hamming-based function and the Hamming code can be used to detect and recover errors; (iii) DSR [13] which is an efficient protocol suitable for wireless sensor networks [5,15]. NARUN-PC-ECC denotes the addition of advanced caching into NARUN-ECC while NARUN-PC is the addition of caching into NARUN.

3.1   Experimental Results

We compare the protocols in the case where a certain percentage of the links is affected by an increasing amount of noise (i.e., from -70 dBm to -80 dBm). At each run U_i, we randomly pick a subset of the links. These have a noise power equal to Y (with $-69 < Y < -81$) while all the rest of the links can always deliver messages. We have performed simulations for 30% of noisy links.

Figure 5 shows the failure rate with 30% of noisy links. When the noise power is higher than -78 dBm, all protocols have a reading failure rate close to 0%. When the noise power is between -70 and -74 dBm, NARUN and DSR have always the highest failure rate. At -70 dBm the failure rate of DSR becomes stable around 35% while for NARUN increases to 42%. NARUN indirect observations cause NARUN to retry paths that have noisy links but they also produce very low overhead. Figure 6 shows the reading rate (the total number of meters is 254) when 30% of links have noise. When the noise power is higher than -72 dBm, all protocols can read all the 254 meters. When the noise power is -70 dBm the reading rate of the NARUN drops by 2.5%. All protocols have a good reading performance since they can discover paths with no noisy links. Contrary to NARUN, DSR floods the network to discover less noisy paths. After various message flooding phases, DSR eventually converges to the least noisy paths. This behaviour has been validated by checking the total number of messages received by the meters in a round (see Fig. 7). At -70 dBm DSR generates

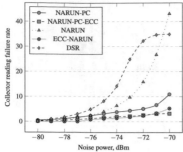
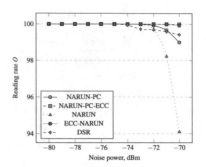

Fig. 5. Collector reading failure rate with 30% of noisy

Fig. 6. Reading rate O with 30% noisy links

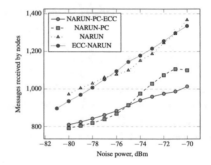

Fig. 7. Average messages received by the meters with 30% of noisy links using DSR protocol

Fig. 8. Average messages received by the meters with 30% of noisy links

14825 messages (that is an average of 58.3 messages per reading operation) while NARUN 1387 (that is an average of 5.4 messages per reading operation). The use of Hamming and caching seem to be effective in terms of failure rate and traffic overhead. At −70 dBm, ECC-NARUN has a failure rate of 5.29% with 1335 messages. NARUN-PC and NARUN-PC-ECC have a failure rate of 10.93% and 3% with 1100 and 1014 messages (see Fig. 8). We can conclude that NARUN-PC-ECC has the lowest failure rate and the lowest traffic. The caching strategy generates the least traffic while the use of Hamming seems to be very effective for decreasing the failure rate and traffic overhead. Similar results can be observed for the case with the 15% of noisy links (for the sake of presentation 15% results are not presented).

We also compare the protocols when links are *unstable*. At the beginning of an experiment, we randomly select a subset of the links and we alternate a run where the selected links are broken and the next run where all such links are working. Figures 9 and 10 show the reading rate and the failure, respectively, when 30% of the links are unstable. Figures 12 and 11 show the traffic load.

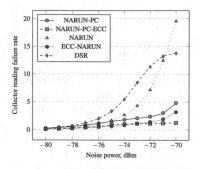

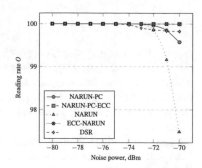

Fig. 9. Collector reading failure rate with 30% of noisy links in ON/OFF simulations

Fig. 10. Reading rate O with 30% noisy links in ON/OFF simulations

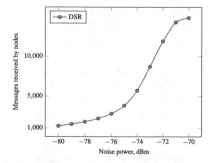

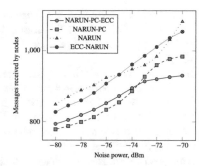

Fig. 11. Average messages received by the meters with 30% of noisy links using DSR in ON/OFF simulations protocol

Fig. 12. Average messages received by the meters with 30% of noisy links in ON/OFF simulations

4 Conclusions and Future Work

The WM-Bus is a widely used meter protocol in utility networks. However, in a noisy environment, faults become a serious issue resulting in excess volumes of transmissions. The consequently increased energy consumption is also problematic for battery-operated nodes. This work proposes NARUN-PC that improves NARUN with the addition of a caching strategy. While NARUN always selects the best available path, NARUN-PC adopts a caching strategy that converges to stable paths avoiding the selection of unstable links. We have validated NARUN-PC on a real topology. At −70 dBm noise power when 30% of noisy links are considered NARUN-PC-ECC decreases the failure rate of ECC-NARUN by 2.29% and generates 31% traffic less. When unstable links are considered NARUN-PC-ECC decreases the failure rate of ECC-NARUN by 2.1% and generates 13% of traffic less.

References

1. Abate, F., Carratù, M., Liguori, C., Paciello, V.: A low cost smart power meter for iot. Measurement **136**, 59–66 (2019)
2. Al-Turjman, F., Mostarda, L., Ever, E., Darwish, A., Khalil, N.S.: Network experience scheduling and routing approach for big data transmission in the internet of things. IEEE Access **7**, 14501–14512 (2019). https://doi.org/10.1109/access.2019.2893501
3. Bazlov, Y.: Coding Theory, Part2: Hamming Distance - 2010. Lecture from The University of Manchester, School of Mathematics, Manchester (2010)
4. Brentnall, A.: Discrete-event system simulation (international edition). J. Simulat. **1**(3), 223 (2007). https://doi.org/10.1057/palgrave.jos.4250022
5. Cacciagrano, D., Culmone, R., Micheletti, M., Mostarda, L.: Energy-efficient clustering for wireless sensor devices in Internet of Things. In: Al-Turjman, F. (ed.) Performability in Internet of Things. EICC, pp. 59–80. Springer, Cham (2019). https://doi.org/10.1007/978-3-319-93557-7_5
6. Culmone, R., Pagnotta, F.: Energy efficient light routing in utility network. In: Barolli, L., Takizawa, M., Xhafa, F., Enokido, T. (eds.) WAINA 2019. AISC, vol. 927, pp. 745–754. Springer, Cham (2019). https://doi.org/10.1007/978-3-030-15035-8_72
7. D'Angelo, G., Diodati, D., Navarra, A., Pinotti, C.M.: The minimum k-storage problem: complexity, approximation, and experimental analysis. IEEE Trans. Mob. Comput. **15**(7), 1797–1811 (2016). https://doi.org/10.1109/TMC.2015.2475765
8. Das, S.K., Ghidini, G., Navarra, A., Pinotti, M.C.: Localization and scheduling protocols for actor-centric sensor networks. Networks **59**(3), 299–319 (2012). https://doi.org/10.1002/net.21454
9. European Committee for Standardization. Communication systems for and remote reading of meters. Standard EN 13757 (2005)
10. Parliament, E.: Council of European Union: directive 2009/72/EC of the European parliament and of the council of 13 July 2009 concerning common rules for the internal market in electricity and repealing directive 2003/54/EC. Off. J. Eur. Union L **211**, 55–93 (2009)
11. Jacobsen, R.M., Popovski, P.: Data recovery using side information from the wireless M-bus protocol. In: IEEE Global Conference on Signal and Information Processing, pp. 511–514 (2013)
12. Jacobsen, R.M., Popovski, P.: Reliable reception of wireless metering data with protocol coding (2013)
13. Johnson, D.B., Maltz, D.A., Broch, J., et al.: DSR: the dynamic source routing protocol for multi-hop wireless ad hoc networks. Ad hoc Netw. **5**(1), 139–172 (2001)
14. Micheletti, M., Mostarda, L., Navarra, A.: CER-CH: combining election and routing amongst cluster heads in heterogeneous WSNs. IEEE Access **7**, 125, 481–125, 493 (2019). https://doi.org/10.1109/ACCESS.2019.2938619
15. Mostarda, L., Navarra, A.: Distributed intrusion detection systems for enhancing security in mobile wireless sensor networks. Int. J. Distrib. Sens. Netw. **4**, 83–109 (2008). https://doi.org/10.1080/15501320802001119
16. Pagnotta, F., Mostarda, L., Gemikonakli, O., Culmone, R., Cacciagrano, D., Corradini, F.: Narun: noise adaptive routing for utility networks (to appear). Int. J. Web Grid. Serv. (2022)
17. Saputro, N., Akkaya, K., Uludag, S.: A survey of routing protocols for smart grid communications. Comput. Netw. **56**(11), 2742–2771 (2012)

HYPE: CNN Based HYbrid PrEcoding Framework for 5G and Beyond

Deepti Sharma[1]([✉]), Kuldeep M. Biradar[1], Santosh K. Vipparthi[1,2],
and Ramesh B. Battula[1]

[1] MNIT, Jaipur, India
{2019rcp9155,2018rcp9503,rbbattula.cse}@mnit.ac.in
[2] Indian Institute of Technology Guwahati (IIT Guwahati), Guwahati, India
skvipparthi@iitg.ac.in

Abstract. 5G and beyond (B5G) technologies have emerged with the increasing demand for higher data-rate wireless communication with low latency. The B5G network utilizes hybrid beam-forming (HBF) as a promising solution to provide large bandwidths with directional communication. Conventional HBF techniques are computationally complex and unable to fully exploit the spatial & partial channel state information, which results in very low spectral efficiency. Hence, this paper proposes an optimized framework, HYPE, integrating the convolutional neural network resulting in the reduced complexity for the HBF technique. A two-phase analog shifter is used to maximize the spectral efficiency of the system by resolving the constant modulus constraint. Experimental results justify the enhanced performance of the proposed framework than conventional algorithms. Extensive ablation studies on the proposed work were carried out to analyze efficiency more in detail.

1 Introduction

The future is envisioned with billions of connected devices and intense bandwidth-hungry applications like the internet of vehicles (IoV), unmanned aerial vehicles (UAV), multimedia and interactive 3D videos, etc. The recent challenge of bandwidth shortage demands next-generation wireless networks that are fifth-generation (5G) and beyond, providing peak data rates by exploiting the higher band of the frequency spectrum. However, the communication at these bands is highly sensitive to environmental conditions. The small wavelength of the signal at a high-frequency band results in interaction with the environmental molecules limiting the communication distance. More concentrated and narrow beams are required to provide feasible communication at a distance. It leads to the emerging beamforming (BF) technology for 5G and beyond as it focuses a signal to a directional beam choosing the most-efficient route to a particular user.

BF brings technological advancements for 5G and beyond wireless communication, allowing the transmitter to adapt the antenna radiation pattern. For communication involving large antenna arrays at transmitter and receiver, BF helps with increasing spectral efficiency. Presently, hybrid beamforming (HBF)

L. Barolli et al. (Eds.): AINA 2022, LNNS 450, pp. 43–54, 2022.
https://doi.org/10.1007/978-3-030-99587-4_5

design, combining analog and digital precoding, has received immense popularity as it is providing the better BF gain in less hardware cost and better power consumption [9,10]. In the HBF design, digital precoding generally has a closed-form solution [2,7,11]. However, the analog precoder has a constant modulus constraint [2], which leads to the non-convex and complex optimization problem. Many previous works propose different algorithms to resolve the non-convexity issue but increase computation time and give poor system performance.

Nowadays, machine learning and deep learning (DL) solutions are emerging to optimize wireless communication and networking problems. In [5], support vector machines (SVM) based sub-optimum approach is given for analog beamforming vector selection. A DL based precoding framework is proposed in [3] and [4] with a network architecture that is based on multi-layer perceptrons. But, DL networks cannot effectively extract the hidden features inherent in the input data. Convolutional Neural Networks(CNN) is a more powerful DL model for feature extraction directly from the raw data and utilizing fewer parameters than neural networks. Unlike DL, the CNN model learns the important feature without human supervision. In [12] and [15], the author proposed a CNN based precoder but with an assumption of perfect channel state information (CSI). However, perfect CSI is an idealistic assumption that is practically infeasible, especially in the case of large-scale antenna arrays. Currently, the works integrating channel estimation and hybrid precoding utilizing a combination of DL and CNN are proposed to improve the precoding performance on partial CSI [13,14,16,17].

Though the DL based algorithms can help optimize the wireless algorithms, they cannot resolve the architecture-based limitations. All the earlier defined works failed in resolving the constant modulus constraint. Besides, majorly the literature considers an assumption of perfect CSI, making algorithm practical feasibility complicated at high frequencies communication.

This paper proposes HYPE: CNN based HYbrid PrEcoding framework for 5G and beyond. HYPE aims to remove the constant-modulus constraint of analog precoder to enhance system efficiency with large antenna arrays. It also integrates a convolutional neural network (CNN), which is more potent in extracting input features for optimization. The proposed framework integrates four modules named Channel Estimator (CE), Noise and Phase Corrector (NPC), Two-phase Shifter (TPS), and Spectral Efficiency Optimizer (SEO) to resolve the above limitations. To sum it up, the following contributions are made in this work:

1. The TPS module introduces a two-phase shifter for analog precoding to resolve the constant-modulus constraint for maximizing the spectral efficiency of the system.
2. A lightweight HP-CNN module is proposed to extract the maximum power transmission signal through partial CSI for realistic scenarios and capture the structural information of the hybrid precoding scheme.

3. To increase the convergence and enhance the learnability of the model, a new activation function rectified-rectified linear activation unit (RReLU) is used in HYPE.
4. Extensive ablations are done at the TPS module, activation function, and NPC module to study the proposed framework's efficacy.

For keeping the practical feasibility of the system, the NPC module takes partial CSI as input and removes CSI distortions by using CNN properties of feature extraction. Simulation results validate the proposed framework and show that it outperforms the traditional algorithms.

The rest of the paper is organized as follows; Sect. 2 provides the system model of the working scenario with channel model and optimization function. Section 3 describes the proposed HYPE framework elaborating its different modules. Section 4 discusses the simulation results and discussion with the ablation study to evaluate the effect of different parameters on the proposed network. Finally, Sect. 5 concludes our work with future scope.

2 System Model and Problem Formulation

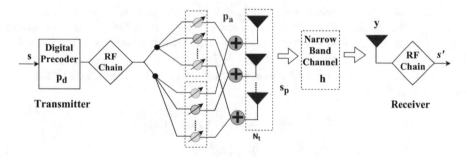

Fig. 1. Hybrid precoding system model

This paper considers a downlink multiple-input single-output (MISO) communication system as shown in Fig. 1. A transmitter with large-antenna arrays (N_t) and a receiver with single antenna communicate in a downlink MISO narrowband channel. A data signal s passes consecutively through digital precoder, radio frequency (RF) chain and analog precoder to finally get phase-shifted signals at different transmitting antennas. The final precoded data symbol $s_p = \mathbf{p_a} p_d s$ where, $\mathbf{p_a}$ is a vector representing $N_t \times 1$ analog precoder and p_d is a constant scalar digital precoder as single RF chain is present in the system [7]. The analog precoder is implemented using two-phase shifters. The final precoded signal passes through the narrowband channel, $\mathbf{h}$, with added η denoting the additive white Gaussian noise (AWGN) with zero mean, σ^2 variance and reaches a receiver. The final signal to reach the receiver is:

$$y = \mathbf{h}^{\mathbf{H}}(\mathbf{p_a})(p_d)s + \eta \tag{1}$$

The widely used Saleh-Valenzuela channel model [3] is used with L paths having $L = 1$ for line-of-sight (LOS) path and remaining $L - 1$ non-line of sight (NLOS) paths. It is given as:

$$\mathbf{h}^{\mathbf{H}} = \sqrt{\frac{N_t}{L}} \sum_{l=1}^{L} \alpha_l \mathbf{a_t^H}(\phi_t^l) \tag{2}$$

Where, α_l denotes the complex gain of lth path between transmitter and receiver, $\mathbf{a_t^H}$ is an array response vector at transmitter, with ϕ_t^l as the azimuth angle of departure associated with the lth path. $\mathbf{h}^{\mathbf{H}}$ defines the conjugate transpose of $\mathbf{h}$.

With the received signal at the user in Eq. (1), the achievable spectral efficiency (ω) of the downlink system is:

$$\omega = \log_2(1 + \frac{1}{\sigma^2}||\mathbf{h}^{\mathbf{H}}\mathbf{p_a}p_d||^2) \tag{3}$$

Our goal is to optimize the precoding vector to maximize the spectral efficiency. With single RF chain system, for getting the optimal value of scalar p_d, assuming $\mathbf{p_a}$ constant and considering maximum transmit power constraint $||\mathbf{p_a}p_d||^2 \leq P$, the value of p_d given as $\sqrt{P/N_t}$.

The final optimization problem constraint on $\mathbf{p_a}$ with two-phase shifter can be stated as follows:

$$\begin{aligned} \underset{\mathbf{p_a}}{\text{maximize}} \quad & \log_2(1 + \frac{\rho}{N_t}||\mathbf{h}^{\mathbf{H}}\mathbf{p_a}||^2) \\ \text{subject to} \quad & |[\mathbf{p_a}]_i| \leq 2 \\ & \text{for } i = 1, ..., N_t. \end{aligned} \tag{4}$$

Where $\rho = \frac{P}{\sigma_2}$ denotes the signal-to-noise ratio (SNR).

To resolve the problem defined in Eq. (2), conventional hybrid precoding algorithms utilize high complexity algorithms to find a conditionally optimal analog precoder. Moreover, those conventional hybrid precoding design schemes are based on perfect CSI. However, accurate channel acquisition will cause large training overhead and the hybrid precoding structure also brings challenges for channel estimation. In order to tackle this problem, in the next section, we propose a CNN based optimized hybrid precoding framework with imperfect CSI and two-phase analog shifter.

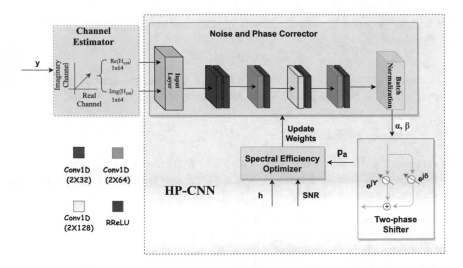

Fig. 2. The HYPE framework

3 HYPE: CNN Based HYbrid PrEcoding Framework for 5G and Beyond

This paper proposes a HYPE framework that aims to design the optimal analog precoder utilizing partial CSI via a CNN-based scheme. The precoder implementation utilizes a two-phase shifter to resolve constant modulus constraint, maximizing spectral efficiency. The overall structure of the proposed framework is depicted in Fig. 2. The framework consists of a channel estimator (CE) as an input module and a Hybrid precoding convolutional neural network (HP-CNN) network module for providing the precoding vector. CE is utilized to process the input signal (y) and result in the estimated channel ($\mathbf{H_{est}}$). HP-CNN uses the feature extraction power of CNN, resulting in the precoding vector ($\mathbf{p_a}$) with an unsupervised loss function using the input ($\mathbf{H_{est}}$). The detailed discussion of each module is described in the following subsections.

3.1 Acquiring Partial CSI

The channel estimator (CE) module is used to output the $\mathbf{H_{est}}$ for providing the partial CSI, which increases the practical feasibility of the proposed HYPE framework. The channel estimator is based on the pre-defined codebooks ($\mathbf{F}$) for estimating the channel. The estimator utilizes equal pilot signal y_p to estimate the steering matrix ($\mathbf{\hat{a}_t}$) and path gain ($\hat{\alpha}$) vector. Exploiting the sparse nature of the channel and vectorizing, we get the resultant pilot signal as [9]:

$$y_p = \sqrt{P}[\mathbf{F}]\mathbf{\hat{a}_t}\hat{\alpha} + \eta \tag{5}$$

Utilizing the compressed sensing tools on the vectorized pilot signal and approximating different parameters of the channel paths, the resultant estimated channel is evaluated. Utilizing $\hat{\mathbf{a}}_\mathbf{t}$ and $\hat{\alpha}$, the complex $\mathbf{H_{est}}$ values are calculated as:

$$\mathbf{H_{est}} = diag(\hat{\alpha})\hat{\mathbf{a}}_\mathbf{t}^\mathbf{H} \tag{6}$$

The resultant of $\mathbf{H_{est}}$ is further forwarded to NPC module by dividing it's complex and real part to get an optimized phase for communication.

3.2   Structure of HP-CNN

The structure of HP-CNN consists of an NPC module with convolutional layers and two specially designed lambda layers (TPS and SEO module). The details of each module are as follows:

Noise and Phase Corrector (NPC): The NPC block adapts the real-time scenario using partial CSI as input. This module removes the distortions and noise from the channel coefficient matrix to define the best phase for communication between transmitter and receiver.

Firstly, the responses of H_{est} are normalized to reduce the internal covariance shift and hence increase the optimization performance.

The normalized response passes through a lightweight shallow network to extract the maximum power transmission signal by reducing the noise. This network consists of a simplified four convolution 1D (Conv1D) layers to reduce communication's complexity and inference time. The nonlinear activation function, RReLU, follows these Conv1D layers. After extracting the essential features from the convolutional layers, the best estimated directional vectors are optimized by normalization before passing through the TPS module.

Let $\mathbf{H_{est}}$ be an input vector of channel coefficients having size 2×64. After applying batch normalization (Ψ) on $\mathbf{H_{est}}$, the resultant normalized vector $\mathbf{H_{nor}}$ is:

$$\mathbf{H_{nor}} = \Psi[\mathbf{H_{est}}] = \Psi[\Re(\mathbf{H_{est}})\Im(\mathbf{H_{est}})] \tag{7}$$

With $\Re$ and $\Im$ represents the real and imaginary part of input vector ($\mathbf{H_{est}}$) respectively.

The, output of the NPC module α and β are computed by recursively convolving normalized input with different filters as shown in Eq. (8).

$$\alpha, \beta = \Psi[\xi_{RR,64}(\xi^{RR,128}(\xi^{RR,64}(\xi^{RR,32}(\mathbf{H_{nor}}))))] \tag{8}$$

Where, $\xi^{RR,N}$ represents convolutional layer function, with, N is depth and RR represents non-linear activation function RReLU.

RReLU is rectified ReLU activation function, which is an upgraded nonlinear activation function used to improve the performance of the model. It is applied to increase the convergence and enhance the learnability of the model [1]. RReLU is a restricted network to vanishing gradient and dying ReLU problem.

After the NPC module, two specially designed Lambda layers are added, introducing a two-phase shifter and an unsupervised loss function as spectral efficiency optimizers.

Two-Phase Shifter (TPS): This module depicts the first Lambda layer, which ensures that the output precoding vector, $\mathbf{p_a}$ is complex-valued implemented using two-phase shifters. We define a two-phase shifter for each coefficient of the analog precoder, which alters the constraint to be of non-constant magnitude. Using two phase shifter, the constant modulus constraint of analog precoder ($\mathbf{p_a}$) has changed from single-phase to double phase that is from one to two as $|[\mathbf{p_a}]_i| \leq 2$ for $i = 1, ..., N_t$. Two different structures of the two phase shifter are proposed as shown in Fig. 3.

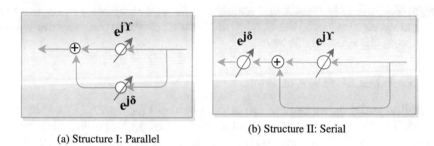

(a) Structure I: Parallel

(b) Structure II: Serial

Fig. 3. Design of 2-phase analog shifter

The output beamformer using Structure I and Structure II of two-phase analog shifter are given in Eq. (9) and (10) respectively:

$$\mathbf{p_a} = e^{j(\gamma+\delta)} + e^{j(\gamma-\delta)} \tag{9}$$
$$= e^{j\gamma} \times 2\cos\delta$$

$$\mathbf{p_a} = e^{j(\delta)}(1 + e^{j(\gamma)}) \tag{10}$$
$$= e^{j(\delta)}(1 + cos\gamma + jsin\gamma)$$

Here, γ and δ represents two phases of the analog shifter. γ handles the channel linearity and δ handles the signal amplitude. The $\mathbf{p_a}$ vector values has constrained with value two and not higher values as increasing the phases further will increase the hardware complexity.

The two phases (γ and δ) for the final precoded vector utilizing α and β from Eq. (8) are given as:

$$\gamma = \tan^{-1}\frac{\beta}{\alpha} \tag{11} \qquad\qquad \delta = \cos^{-1}\frac{\upsilon}{2} \tag{12}$$

Where, $\upsilon = \sqrt{\alpha^2 + \beta^2} \in [-2,2]$, $\gamma \in [0, 2\pi)$ and $\delta \in [0, \pi]$. The final output of the TPS module ($\mathbf{p_a}$) can be given using the evaluated phases (γ and δ) and putting in Eq. (9 and 10).

Spectral Efficiency Optimizer (SEO). The SEO module aims to maximize the spectral efficiency (SE) from the final precoding vector. To perform the maximization on SE, this module introduces an unsupervised loss function which will relate to the optimization problem Eq. (2) introduced in System Model. The loss function given as:

$$Loss = -\frac{1}{N} \sum_{n=1}^{N} \log_2(1 + \frac{\rho}{N_t} ||\mathbf{h^H p_a}||_n^2) \tag{13}$$

Where N is the batch size while training. The weight updates using the back propagation mechanism to the NPC module for minimizing the loss function and maximizing the spectral efficiency.

3.3   Training Strategy of HP-CNN

For off-line training, we use the imperfect channel $\mathbf{H_{est}}$ in Eq. (6) as the input of HP-CNN. Since most of the convolutional neural networks can only process with real values, $\Re(\mathbf{H_{est}})$ and $\Im(\mathbf{H_{est}})$ are adopted as input for the HP-CNN's NPC module. The output provides an optimized precoding vector $\mathbf{p_a}$. The entire framework is trained to find the optimal solution of a logistic regression task, which is a nonlinear mapping from estimated channel $\mathbf{H_{est}}$ to the best analog precoder, which maximizes spectral efficiency. A loss function measures our model's accuracy for the desired problem. It estimates the learning capability of an optimizing problem. This paper uses an unsupervised loss function stated in Eq. (9), which does not require labels. The model performs in two phases: the off-line training and the deploying stages. In the off-line training stage, the perfect channel matrix ($\mathbf{h}$) is available, which is used in the loss function to optimize the precoding vector $\mathbf{p_a}$. In the deployment phase, the model can directly estimate the analog precoding vector using partial CSI without the requirement of the perfect channel.

4   Simulation Results and Discussions

For this study, we have randomly generated the channel samples for $N_t = 64$ and $N_r = 1$ using the SalehValenzuela channel model in Eq. (2), where L is set to 3, i.e., the channel contains one LOS path and two NLoS paths. To evaluate $\mathbf{H_{est}}$, the conventional channel estimation algorithm in [9] is used based on the pilot signal. One more point to notice here is that the applied channel estimator uses pilot signals for estimation. So, defining PNR as pilot-to-noise ratio and taking it as channel estimation level, the data samples are generated for different PNR values that are -20 dB, 0 dB, and 20 dB [3]. For training and testing 10^5 and 10^4 data samples are used. Two state-of-the-art HBF algorithms in the particular case of one RF chain are considered for comparison, i.e., a beamforming neural network (BFNN) in [3] and the iterative HBF algorithm in [7].

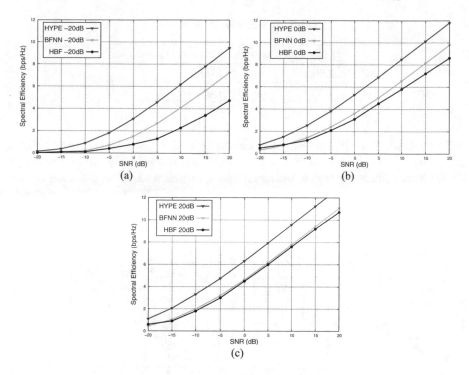

Fig. 4. HYPE comparison with BFNN and HBF at PNR $= -20$ dB, 0 dB, 20 dB

Figure 4 shows the SE versus SNR performance under three-channel esti-mation levels, PNR $= -20$ dB, 0 dB and 20 dB. The results are based on the assumption that the estimation of the number of channel paths and SNR value (ρ) are correct, i.e., $L_{est} = 3$ and $\rho = \rho_{est}$. The HYPE framework was trained and tested with the specific PNR value samples for getting the best results. The provided results depict that with partial CSI, the proposed HYPE frame-work significantly outperforms traditional BFNN [3] and HBF [7] algorithms. The major factors behind the improved performance of HYPE are: first, the use of a more powerful network, that is HP-CNN, which is more efficient in feature extraction from the raw wireless data. The use of CNN makes the frame-work learn to adjust from the partial CSI and hence reduce the complexity and provide robustness. Secondly, the use of two-phase analog shifters removes the non-convexity of the analog-precoder resulting in a simplified optimization prob-lem and enhances spectral efficiency. Table 1, defines the complexity analysis of proposed framework HYPE with traditional BFNN. It can be the seem from the results that utilizing a shallow network reduces the inference time hence improv-ing the communication efficiency. Also, the parameter complexity justifies the simplified network architecture resulting in a lightweight network.

Table 1. Complexity analysis

	Inference time (secs)	Parameter complexity
BFNN	3.50	70k
HYPE	2.65	40k

Influence of Two Phase Shifter: Changing the structures as given in Fig. 3, impacting the performance of the framework as shown in Fig. 5. The reason is that the two phases γ and δ are quantized and hence showing different performances for different structures.

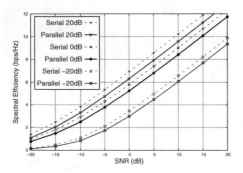
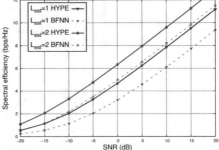

Fig. 5. Impact of design of phase shifter

Fig. 6. Impact of L_{est}

Influence of L_{est}: In the real-time scenario, the channel estimator may have errors in estimating the number of total paths L. Due to the sparsity of the mmWave channels and considering the estimation complexity, L_{est} is preset to a small value. Fig. 6 depicts the effect of different values of $Lest$ on spectral efficiency for a fixed PNR value $= 20\,$dB. The result depicts the improved performance of the HYPE framework even for inaccurate values of L_{est}.

Optimal Design of NPC Module: Depicted by Fig. 7(a), Fig. 7(b), there should be a optimum value of number of layers in the model. Adding more number of layers than optimum value increases the complexity of the model and removing the layers decreases the learnability of model. Hence, both actions decreases the performance of the model.

Influence of Different Activation Function: The activation function are affecting the overall performance of the framework. The detailed visualization are shown in Fig. 8. As depicted, the new activation function outperforms the results as compared to state-of-the-art functions.

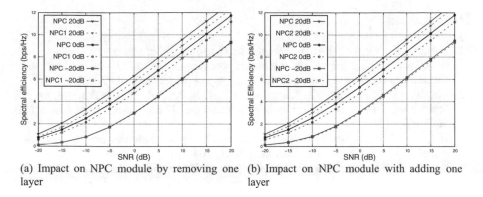

(a) Impact on NPC module by removing one layer   (b) Impact on NPC module with adding one layer

Fig. 7. Ablation study on NPC module

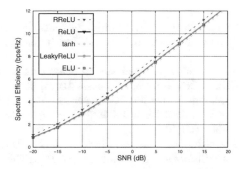

Fig. 8. Impact of RReLU

5   Conclusion and Future work

The proposed HYPE framework considered the problem of designing an optimal hybrid precoding vector to maximize the spectral efficiency of the MISO communication system. A CNN-based hybrid precoding framework, HYPE, is proposed, which reduces the system's computation complexity and improves communication efficiency. A two-phase shifter is used for implementing analog precoder for improving spectral efficiency. The system utilizes partial channel state information as input to improve the practical feasibility of the system. Simulation results justified the performance of the proposed framework compared to the conventional algorithms. In the future, an end-to-end precoding system can be designed for the 6G dataset by developing the 6G channel model. Also, the framework may be extended for more critical precoding issues providing low complexity and more efficiency.

References

1. Verma, M., Vipparthi, S.K., Singh, G.: Non-linearities improve OrigiNet based on active imaging for micro expression recognition. In: 2020 International Joint Conference on Neural Networks (IJCNN), pp. 1–8. IEEE (July 2020)
2. Lin, T., Cong, J., Zhu, Y., Zhang, J., Letaief, K.B.: Hybrid beamforming for millimeter wave systems using the MMSE criterion. IEEE Trans. Commun. **67**(5), 3693–3708 (2019)
3. Lin, T., Zhu, Y.: Beamforming design for large-scale antenna arrays using deep learning. IEEE Wirel. Commun. Lett. **9**(1), 103–107 (2019)
4. Huang, H., Song, Y., Yang, J., Gui, G., Adachi, F.: Deep-learning-based millimeter-wave massive MIMO for hybrid precoding. IEEE Trans. Veh. Technol. **68**(3), 3027–3032 (2019)
5. Long, Y., Chen, Z., Fang, J., Tellambura, C.: Data-driven-based analog beam selection for hybrid beamforming under mm-wave channels. IEEE J. Sel. Top. Sig. Process. **12**(2), 340–352 (2018)
6. Alkhateeb, A., Leus, G., Heath, R.W.: Limited feedback hybrid precoding for multiuser millimeter wave systems. IEEE Trans. Wirel. Commun. **14**(11), 6481–6494 (2015)
7. Sohrabi, F., Yu, W.: Hybrid digital and analog beamforming design for large-scale antenna arrays. IEEE J. Sel. Top. Sig. Process. **10**(3), 501–513 (2016)
8. Yu, X., Shen, J.C., Zhang, J., Letaief, K.B.: Alternating minimization algorithms for hybrid precoding in millimeter wave MIMO systems. IEEE J. Sel. Top. Sig. Process. **10**(3), 485–500 (2016)
9. Alkhateeb, A., El Ayach, O., Leus, G., Heath, R.W.: Channel estimation and hybrid precoding for millimeter wave cellular systems. IEEE J. Sel. Top. Sig. Process. **8**(5), 831–846 (2014)
10. El Ayach, O., Rajagopal, S., Abu-Surra, S., Pi, Z., Heath, R.W.: Spatially sparse precoding in millimeter wave MIMO systems. IEEE Trans. Wirel. Commun. **13**(3), 1499–1513 (2014)
11. Ghosh, A., et al.: Millimeter-wave enhanced local area systems: a high-data-rate approach for future wireless networks. IEEE J. Sel. Top. Sig. Process. **32**(6), 1152–1163 (2014)
12. Elbir, A.M.: CNN-based precoder and combiner design in mmWave MIMO systems. IEEE Commun. Lett. **23**(7), 1240–1243 (2019)
13. Ma, W., Qi, C., Zhang, Z., Cheng, J.: Sparse channel estimation and hybrid precoding using deep learning for millimeter wave massive MIMO. IEEE Trans. Commun. **68**(5), 2838–2849 (2020)
14. Lu, Q., Lin, T., Zhu, Y.: Channel estimation and hybrid precoding for millimeter wave communications: a deep learning-based approach. IEEE Access **9**, 120924–120939 (2021)
15. Faragallah, O.S., El-Sayed, H.S., Mohamed, G.: Performance enhancement of MmWave MIMO systems using deep learning framework. IEEE Access **9**, 92460–92472 (2021)
16. Unnisa, N., Tatineni, M.: Adaptive deep learning strategy with red deer algorithm for sparse channel estimation and hybrid precoding in millimeter wave massive MIMO-OFDM systems. Wirel. Pers. Commun. **122**, 3019–3051 (2022)
17. Li, Z., Gao, W., Zhang, M., Xiong, J.: Multi-task deep learning based hybrid precoding for mmWave massive MIMO system. China Commun. **18**(10), 96–106 (2021)

The Multi-access Edge Computing (MEC)-Based Bit Rate Adaptive Multicast SVC Streaming Using the Adaptive FEC Mechanism

Chung-Ming Huang$^{(\boxtimes)}$ and Kai-Jiun Yang

Department of CSIE, National Chen Kung University, Tainan, Taiwan
{hungcm,yangkj}@locust.csie.ncku.edu.tw

Abstract. This work proposed the SVC streaming method for n users based on the Multi-access Edge Computing (MEC) architecture. In the proposed method, a group member's handheld device is selected as the Header Handheld Device (H-HD) to be in charge of downloading video contents from the MEC Server and then multicasting downloaded video contents to the other group members' Receiver Handheld Devices (R-HDs). The video bit rate that can be adopted in the next downloading cycle is derived by combining the proposed adaptive Forward Error Correction (FEC) scheme with the estimated bandwidth. To have the smooth streaming, the Currently Downloading Segment Section (CDSS) mechanism, for which the segments inside CDSS can be downloaded at the same time in a downloading cycle, was proposed. The performance evaluation results shown that the proposed method can utilize the network bandwidth more effectively to improve video playback quality in the multicast environment.

1 Introduction

Let a group of n persons be traveling together and watching the same movie x in the train during their journey. If each person has the video streaming for x individually, i.e., each one downloads the video content individually, then it needs to download n copies of x from the remote video server. The individual downloading scenario not only results in the huge load of transmitting the duplicated video contents but also causes these persons to pay so much money to the cellular network operators to download the same movie x. The aforementioned phenomena can be solved using the cooperative streaming networks [1, 2], in which one person's handheld device in the group is in charge of (i) downloading the video content x through the 4G/5G cellular network and then (ii) multicasting the downloaded video content of x to the other group members' handheld devices using the Device-to-Device (D2D) communication technique.

In this work, a cooperative Scalable Video Coding (SVC) [3] - based adaptive streaming method using the Multi-Access Edge Computing (MEC) architecture [4] was proposed. Let (i) each Base Station (BS) y of the 4G/5G cellular network be associated with an MEC server w and (ii) the MEC server w can cache the video content, which is requested by those users' handheld devices that are connected with y, from the remote video servers. Additionally, let a group member's handheld device be selected as the

L. Barolli et al. (Eds.): AINA 2022, LNNS 450, pp. 55–67, 2022.
https://doi.org/10.1007/978-3-030-99587-4_6

Header Handheld Device (H-HD), which is in charge of (i) downloading the video content x from the MEC Server using the 4G/5G cellular network and then (ii) multicasting the downloaded video content of x to the other group members' Receiver Handheld Devices (R-HDs) using the multicast technique over Wi-Fi network. In the cooperative SVC video streaming scenario, the adaptive video streaming is more complicated because multiple R-HDs using multicast may experience different packet error rates and received rates. Thus, (i) the Forward Error Correction (FEC) technique, which is based on all users' experienced different packet error rates, is adopted to resolve the packet error situation because multicast is not suitable to use packet retransmission [5] and (ii) a compromised video transmission rate is derived based on all users' experienced different received rates [6, 7].

This work proposed the MEC-based Multicast for the Cooperative SVC Streaming (MM-CSVC) method to resolve the following two technical issues for having the cooperative SVC streaming. (1) How to derive the estimated bandwidth based on multiple R-HDs' networking situations for the cooperative SVC adaptive streaming? (2) How to have the cooperative SVC streaming's quality adaptation considering both estimated bandwidth and the buffering situations in the multiple R-HDs' client sites?

The remaining part of this paper is organized as follows. Section 2 presents related works. Section 3 introduces the proposed schemes of estimated bandwidth and adaptive FEC. Section 4 presents the proposed method in detail. Section 5 presents the performance evolution results. Finally, conclusion remarks are given in Sect. 6.

2 Relative Work

In [8], the authors proposed a method to determine which layers of the segments within the considered range are downloaded with the higher priority to achieve SVC quality adaption. The proposed method determines which layers of the currently downloading segment can be downloaded based on the quality difference between adjacent segments that have been downloaded in the buffer. Two parameters used are (1) the one for deciding the quality difference threshold between the currently downloading segment with its previous segment and (2) the one for deciding the quality difference threshold between the currently downloading segment with its next segment. Through these two parameters, it can ensure the segments that are closer to the playout time have the higher quality to be downloaded and it also affects the quality smoothness and variation among segments.

Wu et al. [7] proposed an adaptive video streaming scheme using the Named Data Networking (NDN) Multicast into the Adaptive Bit Rate (NM-ABR) algorithm, where some NDN multicast groups are dynamically formed based on the network situation. In the proposed NM-ABR method, a multicast data rate selection algorithm was devised to maximize the receiving rate of multicast members within a group according to the number of group members and the corresponding receivable data rates. Moreover, different transmission polices were adopted to deal with the packet loss problem based on the multi-layer structure characteristics of SVC. To ensure the reliability of the BLs, each video chunk of BLs is retransmitted until the corresponding video data have been received; on the contrary, ELs have a maximum retransmission limit to avoid a significantly increase of the stalling time. The experimented results have shown that the proposed adaptive video streaming scheme can improve video bitrate, startup delay and stalling time for SVC streaming over multicast.

In comparison with the aforementioned work, the considered scenario of this work is a group that consists of fixed members, e.g., a group of friends have a journey together in the bus, train, etc.; additionally, our proposed method derived the predicted bandwidth based on the past experienced bandwidth of multicast members and adopted the FEC mechanism for dealing with the packet loss problem.

3 Bandwidth Estimation and Adaptive FEC

In the proposed method, let the request SVC video be divided into s segments and each segment be encoded into ℓ layers, including a Base Layer (BL) and $(\ell - 1)$ Enhancement Layers (ELs), for which each segment contains several video frames. Let $Chunk(S_i, L_j)$ represent the jth layer of the ith segment. When H-HD finishes downloading $Chunk(S_i, L_j)$, it needs to decide to download $Chunk(S_i, L_{j+1})$ or $Chunk(S_{i+1}, L_{j'})$ according to the network situation and the configuration of the client's SVC video buffer. That is, it can select to download (i) more ELs of the same segment, i.e., $Chunk(S_i, L_{j+1})$, to improve the video quality or (ii) the SVC video layer(s), which can be either a BL or an EL of the next segment, i.e., $Chunk(S_{i+1}, L_{j'})$, to buffer more presentation time and/or to balance the video quality difference among segments after downloading $Chunk(S_i, L_j)$. The choice can be done according to the estimated bandwidth and the current configuration of the client's SVC video buffer.

Let time period t denote the period between time points $[t - 1, t]$. On time point t, the video downloading bit rate of each R-HD used in time period t is measured. Then, the measured bit rates in time periods $t, t - 1, ..., 1$ are adopted to estimate the available bandwidth used in time period $t + 1$.

After H-HD forwarding the received video packets using multicast over the Wi-Fi network, the received packet rate of each R-HD is measured and reported to H-HD, which delivers the collected context reports to the MEC server. Then, the MEC server derives the average estimated bandwidth and the corresponding adopted video quality for the video data to be downloaded by H-HD in the next time period.

Equation (1) depicts the average receiving bitrate of all R-HDs' receiving bitrates for downloading chunk j in time period t:

$$R_j^t = \frac{1}{n-1} \sum_{i=1}^{n-1} R_j^t(i) \tag{1}$$

where $R_j^t(i)$ is the receiving rate of chunk j measured by R-HD i in time period t.

Equation (2) depicts the standard deviation σ_t^j of these average receiving bitrates for downloading chunks j:

$$\sigma_t^j = \sqrt{\frac{1}{n-1} \sum_{i=1}^{n-1} (R_j^t - R_j^t(i))^2} \tag{2}$$

Equation (3) is $BW^t_{measured}(j)$, which denotes the measured bandwidth for down-loading chunk j in the time period $[t-1, t]$ based on the average bitrate (R^t_j) and the standard deviation (σ^j_t):

$$BW^t_{measured}(j) = R^t_j + x\sigma^j_t \tag{3}$$

Since the packet receiving rates has the oscillation characteristic, an estimated resiliency, i.e., $x\sigma^j_t$, is added to derive the measured bandwidth given in Eq. (3).

Equation (4) is $BW^t_{estimated}(C_t)$, which represents the estimated bandwidth $BW^t_{estimated}$ that is to be adopted for downloading SVC video data in the time period $[t, t+1]$ started on time point t.

$$BW^t_{estimated}(C_t) = (1-w) * BW^t_{estimated}(C_t - 1) + w * BW^t_{measured}(C_t) \tag{4}$$

$BW^t_{estimated}(C_t)$ depicted in Eq. (4) is the average of the past measured bandwidth values for downloading chunks $1, 2,...,C_t$ during the time period of $[t-1, t]$, using the Exponentially Weighted Moving Average (EWMA) calculation way, where w denotes the weight of the most recently measured bandwidth of downloading a chunk, i.e., w is used to adjust the influence of the most recently measured bandwidth for the future estimated bandwidth.

This work proposes an adaptive FEC for multicasting SVC video data over the Wi-Fi network. Since there are many receivers in the multicast scenario and thus different receivers may have different receiving situations, this work estimates how many FEC-based packets should be sent in the next time period by averaging the receiving situations of all multicast receivers. The FEC mechanism can derive the transmission rate $BitRate_{FEC}$ containing (i) the original data, whose rate is denoted as $BitRate_{original}$, and (ii) the redundant data. $BitRate_{FEC}$ is represented as

$$BitRate_{FEC} = \frac{BitRate_{original}}{1 - ER}$$

where ER is the measured error rate during the previous transmission.

In order to consider the impact of each layer on the quality of the SVC-encoded video, the received rate of each chunk is calculated separately in this work. Let each chuck be contained in serval packets. It can calculate the average packet error rate (ER_l) of $(n-1)$ R-HDs for an SVC chunk as follows:

$$ER_C = 1 - \frac{1}{n-1} \sum_{i=1}^{n-1} \frac{Correct\_Packet^C_i}{No\_T\_Packet^C}, C = 1 \ldots C_t \tag{5}$$

where $Correct\_Packet^C_i$ is the number of chunk C's correctly received packets in R-HD i and $No\_T\_Packet^C$ is the number of chunk C's transmitted packets from H-HD, C_t is the number of video chunks that have been transmitted in the time period t.

The average error rate $(error_t)$ in the time period t is calculated as follows:

$$error_{C_t} = (1-w) * error_{C_t-1} + w * ER_j \tag{6}$$

where ER_j denotes the error rate of the jth downloaded chunk, which can contain a BL or an EL in the time period t.

The transmission rate in the $(t+1)^{th}$ time period using FEC, which is adapted to the loss rate of all R-HDs in the past t time periods is represented as follows:

$$BitRate_{FEC}^{t+1} = \frac{BitRate_{original}^{t+1}}{1 - error_{C_t}} \tag{7}$$

where the $BitRate_{original}^{t+1}$ denotes the transmitting rate of the original video data without containing the redundant video data for error correction using FEC in time period $t+1$. That is, $BitRate_{original}^{t+1}$ is the actual bitrate that can be used to determine the video quality to be adopted in the time period $(t+1)$.

In summary, $BitRate_{FEC}^{t+1}$ should be smaller than or equal to $BW_{estimated}$ the video quality that can be adopted in the time period $(t+1)$ is derived as follows:

$$BitRate_{FEC}^{t+1} = \frac{BitRate_{original}^{t+1}}{1 - error_{C_t}} \le BW_{estimated}(t),$$

$$BitRate_{original}^{t+1} \le BW_{estimated} * (1 - error_{C_t}),$$

$$\max \ n \ \text{s.t.} \ \sum_{i=0}^{n} \overline{R}(L_i) \le BW_{estimated} * (1 - error_{C_t}),$$

$$n \le \ell - 1 \tag{8}$$

where $\overline{R}(L_i)$ is the average bitrate of layer i of the SVC-encoded video segments to be downloaded, i.e., $\overline{R}(L_0)$ is the average bitrate of the BL and $\overline{R}(L_i)$, $i = 1 \ldots n$, is the average bitrate of EL i.

4   The Proposed MM-CSVC Method

This Section presents the proposed MEC-based Multicast for the Cooperative SVC Streaming (MM-CSVC) method in detail.

Two key issues of the proposed method are as follows.

(1) Download the SVC video data delimited in the Currently Downloading Segment Section (CDSS) at a time: In the proposed method, the number of video segments that can be downloaded currently is limited by R_{CDSS}, in which R_{CDSS} segments of video data delimited in the range of $[Index_{header}^{CDSS}, Index_{header}^{CDSS} + R_{CDSS} - 1]$ can be downloaded currently and $Index_{header}^{CDSS}$ denotes the first segment of the CDSS. CDSS can be moved forwardly based on the following principle: when (i) all of the layers of the first segment of CDSS have been downloaded or (ii) the first segment of CDSS is started to be playing.

(2) Decide to (i) download the higher layers of those segments whose BLs and/or some ELs have been downloaded to obtain the higher quality or (ii) download the BLs or some ELs of the subsequent segments of the currently downloading segment to increase the buffered presentation time and/or balance the quality difference among segments in CDSS according to the network's condition.

Let $Index_{cur}$ denote the currently downloading segment, and (i) $Index_{pre}$ denote the segment that has been downloaded before segment $Index_{cur}$ and (ii) $Index_{next}$ denote the segment that is after segment $Index_{cur}$. $Index_{cur}$, $Index_{pre}$ and $Index_{next}$ are moved within the CDSS according to the downloading situation. Let $NoLS_i$ denote the number of layers that have been downloaded in the ith video segment of the SVC video. It determines the next chunk to be downloaded by comparing $NoLS_{Index_{cur}}$ with $NoLS_{Index_{pre}}$ and $NoLS_{Index_{next}}$. Let (a) δ_{high} be used to control the difference between $NoLS_{Index_{cur}}$ and $NoLS_{Index_{next}}$ and (b) δ_{low} be used to control the difference between $NoLS_{Index_{pre}}$ and $NoLS_{Index_{cur}}$, i.e., to control how many layers segment $Index_{cur}$ can be downloaded.

The main principle of the downloading control is as follows. Compare $NoLS_{Index_{cur}}$, i.e., the number of downloaded layers of the currently downloading segment $Index_{cur}$, with (i) $NoLS_{Index_{pre}}$, i.e., the number of downloaded layers of segment $Index_{pre}$, and (ii) $NoLS_{Index_{next}}$, i.e., the number of downloaded layers of segment $Index_{next}$, after the downloading of a chunk is finished. If $NoLS_{Index_{cur}} - NoLS_{Index_{next}} < \delta_{high}$ and $NoLS_{Index_{pre}} - NoLS_{Index_{cur}} \neq \delta_{low}$, then it continues to download chunks of the currently downloading segment $Index_{cur}$, i.e., to increase the quality of segment $Index_{cur}$; otherwise, switch to download chunks in segment $Index_{next}$, i.e., to increase the quality of the next segment $Index_{next}$.

In the case of switching to download the next segment, indexes $Index_{cur}$, $Index_{per}$ and $Index_{next}$ are moved to the next segments synchronously and then it downloads the corresponding chunk and updates the value of $NoLS_i$. Thus, it would keep downloading video data of the segment to which $Index_{cur}$ points in CDSS until either one of the following two conditions is met:

(1) If the difference between $NoLS_{Index_{cur}}$ and $NoLS_{Index_{next}}$ is equal to or greater than δ_{high}, i.e., $NoLS_{Index_{cur}} - NoLS_{Index_{next}} \geq \delta_{high}$, then switch to download the chunks of the segment to which $Index_{next}$ points and adjust the indexes of $Index_{cur}$, $Index_{pre}$ and $Index_{next}$ accordingly. This constraint tries to have the quality of the currently downloading segment not be δ_{high} more layers than that of the next segment before downloading the subsequent segments.
(2) If the difference between $NoLS_{Index_{pre}}$ and $NoLS_{Index_{cur}}$ equals δ_{low}, i.e., $NoLS_{Index_{pre}} - NoLS_{Index_{cur}} = \delta_{low}$, then switch to download the chunks of the segment to which $Index_{next}$ points, i.e., adjust the indexes of $Index_{cur}$, $Index_{pre}$ and $Index_{next}$ accordingly. This constraint can prevent the video quality of two adjacent segments from deteriorating drastically to have the smoother video playing.

If $Index_{cur}$ is in the beginning/end of the CDSS, $Index_{pre}/Index_{next}$ is outside the current CDSS, then they are set as follows: (i) Set $NoLS_{Index_{pre}}$ to '0' and $Index_{pre}$ to 'NULL'; (ii) Set $NoLS_{Index_{next}}$ to '$-\infty$' and $Index_{next}$ to 'NULL'. If $Index_{cur}$ is the last segment of CDSS, then $Index_{cur}$ points to the first segment of CDSS when it needs to be moved to the next one; if all ELs of the first segment of CDSS have been downloaded at this time, it moves CDSS forwardly and $Index_{cur}$ points to the first segment of the new CDSS.

In the proposed method, the video playout can be commenced when the BL and all of the ELs of the 1st segment have been downloaded. Note that when the video playout is started, BLs of some segments in the CDSS have already been downloaded during the commencement control period. After the commencement control, moving CDSS forwardly is triggered when either one of the following conditions happens: (1) $Index_{header}^{CDSS} = Max_{layer}$, which denotes the number of layers in the SVC video. (2) Let $Index_{palyout}$ be the index of the currently playing segment. When the playout of the segment pointed by $Index_{playout}$ is ended, the next $Index_{playout}$ is the one pointed by $Index_{header}^{CDSS}$, i.e., the 1st CDSS's segment. Condition (2) denotes the networks' condition becomes bad and it needs to move CDSS forwardly, even if not all of the layers of the segment pointed by $Index_{header}^{CDSS}$ have been download.

When $Index_{playout}$ is switched to the head of the CDSS and it is downloading the chunks of the 1st segment of the CDSS and none of the chunks of the 2nd and after segments of the CDSS has been downloaded, which denotes that the network's situation is really bad, the video playout is frozen until the currently downloading chunk is complete. Then, the estimated bandwidth is calculated and the CDSS is moved forwardly to download the chunks of the subsequent segments according to the result of the estimated bandwidth's calculation, and the video playout is resumed.

An illustrated execution scenario of the proposed method is depicted in Fig. 1, in which the number in each chunk represents the downloaded sequence. Let δ_{high}, δ_{low} and R_{CDFS} be assigned as 2, 1, and 3 respectively; the number of layers, which is denoted as Max_{layer}, of this SVC-encoded video be 4, i.e., $Max_{layer} = 4$ and thus there be BL, EL_1, EL_2 and EL_3 in the video; the initial $Index_{playout}$ be '0' and $Index_{header}^{CDSS}$ be '1'.

At the beginning, $Index_{cur}$ is the first segment of the CDSS ($Index_{header}^{CDSS}$), i.e., $Index_{cur} = Index_{header}^{CDSS} = 1$ and $NoLS_{Index_{cur}} = NoLS_{Index_{header}^{CDSS}} = 0$, i.e., there is no buffered video data in H-HD; $Index_{pre}$ is NULL and $NoLS_{Index_{pre}}$ is '0' because $Index_{pre}$ is outside the CDSS; $Index_{next}$ is the second segment of the CDSS and $NoLS_{Index_{next}}$ is '0'. The video playout cannot be started until the commencement control is done. Initially, H-HD starts to download video data that is constrained in CDSS because the condition of (i) $NoLS_{Index_{cur}} - NoLS_{Index_{next}} < \delta_{high}$ and (ii) $NoLS_{Index_{pre}} - NoLS_{Index_{cur}} \neq \delta_{low}$ is satisfied, $Chunk(S_1, L_0)$ is downloaded and $NoLS_1$ is updated to 1 after the downloading of $Chunk(S_1, L_0)$ is finished, for which the result is depicted in Fig. 1-(a).

Referring to Fig. 1-(a), since the condition of (i) $NoLS_{Index_{cur}} - NoLS_{Index_{next}} < \delta_{high}$ and (ii) $NoLS_{Index_{pre}} - NoLS_{Index_{cur}} \neq \delta_{low}$ is satisfied, $Chunk(S_1, L_1)$ is downloaded and $NoLS_1$ is updated to 2 after the downloading of $Chunk(S_1, L_1)$ is finished, for which the result is depicted in Fig. 1-(b).

Referring to Fig. 1-(b), since $NoLS_{Index_{cur}} - NoLS_{Index_{next}} = \delta_{high}$ and thus the condition of (i) $NoLS_{Index_{cur}} - NoLS_{Index_{next}} < \delta_{high}$ and (ii) $NoLS_{Index_{pre}} - NoLS_{Index_{cur}} \neq \delta_{low}$ is not satisfied, $Index_{cur}$, $Index_{pre}$ and $Index_{next}$ need to be mov ed to the next segment, i.e., $Index_{pre}$, $Index_{cur}$ and $Index_{next}$ are changed to 1, 2 and 3 respectively, and $NoLS_{Index_{pre}}$, $NoLS_{Index_{cur}}$ and $NoLS_{Index_{next}}$ are updated to 2, 0 and 0 respectively.

After updating, since (i) $NoLS_{Index_{cur}} - NoLS_{Index_{next}} < \delta_{high}$ and (ii) $NoLS_{Index_{pre}} - NoLS_{Index_{cur}} \neq \delta_{low}$ is satisfied, $Chunk(S_2, L_0)$ is downloaded and $NoLS_{Index_{cur}}$ is updated to 1 after the downloading of $Chunk(S_2, L_0)$ is finished, for which the result is depicted in Fig. 1-(c).

Referring to Fig. 1-(c), since $NoLS_{Index_{pre}} - NoLS_{Index_{cur}} = \delta_{low}$ and thus the condition of (i) $NoLS_{Index_{cur}} - NoLS_{Index_{next}} < \delta_{high}$ and (ii) $NoLS_{Index_{pre}} - NoLS_{Index_{cur}} \neq \delta_{low}$ is not satisfied, $Index_{cur}$, $Index_{pre}$ and $Index_{next}$ need to be moved to the next segment, i.e., $Index_{pre}$ and $Index_{cur}$ are changed to 2 and 3 respectively, and $NoLS_{Index_{pre}}$ and $NoLS_{Index_{cur}}$ are updated to 1 and 0 respectively. However, since $Index_{cur}$ is equal to $Index_{tail}^{CDSS}$, i.e., the last segment of CDSS, $Index_{next}$ is outside the CDSS. Thus, $Index_{next}$ is updated to NULL and $NoLS_{Index_{next}}$ is updated to '$-\infty$'.

After updating, since (i) $NoLS_{Index_{cur}} - NoLS_{Index_{next}} < \delta_{high}$ and (ii) $NoLS_{Index_{pre}} - NoLS_{Index_{cur}} \neq \delta_{low}$ is satisfied, $Chunk(S_3, L_0)$ is downloaded and $NoLS_{Index_{cur}}$ is updated to 1 after the downloading of $Chunk(S_3, L_0)$ is finished, for which the result is depicted in Fig. 1-(d).

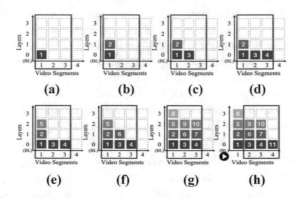

Fig. 1. An illustrated execution scenario of the proposed method.

Referring to Fig. 1-(d), since $NoLS_{Index_{cur}} - NoLS_{Index_{next}} > \delta_{high}$ and thus the condition of (i) $NoLS_{Index_{cur}} - NoLS_{Index_{next}} < \delta_{high}$ and (ii) $NoLS_{Index_{pre}} - NoLS_{Index_{cur}} \neq \delta_{low}$ is not satisfied, $Index_{cur}$, $Index_{pre}$ and $Index_{next}$ need to be moved to the next segment. Since $Index_{cur}$ is currently equal to $Index_{tail}^{CDSS}$, $Index_{cur}$ is changed to 1, $Index_{pre}$ is set to NULL because it is outside the CDSS and $Index_{next}$ equals 2. $NoLS_{Index_{pre}}$, $NoLS_{Index_{cur}}$ and $NoLS_{Index_{next}}$ are updated to 0, 2 and 1 respectively.

After updating, since (i) $NoLS_{Index_{cur}} - NoLS_{Index_{next}} < \delta_{high}$ and (ii) $NoLS_{Index_{pre}} - NoLS_{Index_{cur}} \neq \delta_{low}$ is satisfied, $Chunk(S_1, L_2)$ is downloaded and $NoLS_{Index_{cur}}$ is updated to 3 after the downloading of $Chunk(S_1, L_2)$ is finished, for which the result is depicted in Fig. 1-(e).

Referring to Fig. 1-(e), since $NoLS_{Index_{cur}} - NoLS_{Index_{next}} = \delta_{high}$ and thus the condition of (i) $NoLS_{Index_{cur}} - NoLS_{Index_{next}} < \delta_{high}$ and (ii) $NoLS_{Index_{pre}} - NoLS_{Index_{cur}} \neq \delta_{low}$ is not satisfied, $Index_{cur}$, $Index_{pre}$ and $Index_{next}$ need to be moved to the next segment, i.e., $Index_{pre}$, $Index_{cur}$ and $Index_{next}$ are changed to 1, 2 and 3 respectively, and $NoLS_{Index_{pre}}$, $NoLS_{Index_{cur}}$ and $NoLS_{Index_{next}}$ are updated to 3, 1 and 1 respectively.

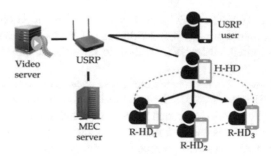

Fig. 2. The abstract configuration of the experimental environment.

After updating, since (i) $NoLS_{Index_{cur}} - NoLS_{Index_{next}} < \delta_{high}$ and (ii) $NoLS_{Index_{pre}} - NoLS_{Index_{cur}} \neq \delta_{low}$ is satisfied, $Chunk(S_2, L_1)$ is downloaded and $NoLS_{Index_{cur}}$ is updated to 2 after the downloading of $Chunk(S_2, L_1)$ is finished, for which the result is depicted in Fig. 1-(f).

After a number of follow-up execution depicted in Fig. 1-(d), (e) (f), Fig. 1-(g) shows the buffering condition when $Chunk(S_3, L_2)$ is downloaded. At this moment, $Index_{cur}$, $Index_{pre}$ and $Index_{next}$ are 3, 2 and 'NULL' respectively; $NoLS_{Index_{cur}}$, $NoLS_{Index_{pre}}$ and $NoLS_{Index_{next}}$ are 3, 3 and $-\infty$ respectively. Since $NoLS_{Index_{cur}} - NoLS_{Index_{next}} > \delta_{high}$ and thus the condition of (i) $NoLS_{Index_{cur}} - NoLS_{Index_{next}} < \delta_{high}$ and (ii) $NoLS_{Index_{pre}} - NoLS_{Index_{cur}} \neq \delta_{low}$ is not satisfied, $Index_{cur}$ needs to be moved to the first segment of CDSS. However, the first segment has the full quality, i.e., $NoLS_{Index_{cur}}$ is the 1st segment of CDSS and $NoLS_{Index_{cur}}$ is equal to Max_{layer}, it needs to move CDSS forwardly. After the movement, the commencement control is complete because all of the layers of the 1st segment have been downloaded and thus the video is ready to be played.

After CDSS being moved forwardly, $Index_{header}^{CDSS}$ and $Index_{tail}^{CDSS}$ are 2 and 4 respectively, and the BL of the last segment in the new CDSS is downloaded, i.e., increase $NoLS_{Index_{tail}^{CDSS}}$ by 1 after the downloading of $Chunk(S_4, L_1)$ is finished, for which the result is depicted in Fig. 1-(h). After that, it keeps requesting chunks with the value of $Index_{cur}$ and $Index_{next}$ equaling 2 and 3 respectively, $NoLS_{Index_{cur}}$ and $NoLS_{Index_{next}}$ ar e updated to 3. Since $Index_{pre}$ is outside the CDSS, $Index_{pre}$ and $NoLS_{Index_{pre}}$ are modified to NULL and 0 respectively.

5 Performance Evaluation

This Section presents the performance evaluation for the proposed method.

Referring to Fig. 2, the experimental environment was built using two computers, a Universal Software Radio Peripheral (USRP) and five Android handheld devices. Both computers use the Ubuntu 16.04.7 LTS Operation System. One computer plays the role of the video server and the other one plays the role of the MEC server. USRP is an open-source framework that can simulate the LTE network. In our experiment, USRP is connected and controlled by a computer, which plays the role of the MEC server. Two handheld devices are connected with USRP in which one is H-HD and the other one is the sink of the background traffic. The multicast group was made up of four handheld devices, including H-HD. The MEC server can use the reported context information to execute the proposed estimated bandwidth calculation scheme, the adaptive FEC scheme and the proposed MM-CSVC method to have the adaptive cooperative SVC video streaming.

In the adopted video, each video segment was encoded in four quality layers, i.e., BL, EL_1, EL_2 and EL_3, using JSVM 9.19. Table 1 shows the individual bitrate and the cumulative bitrate of each SVC-encoded video layer used in the experimental environment. The adopted video was encoded into 14315 video frames, whose frame rate is 24 frames per second (fps) and video presentation time is about 10 min. Except the last segment, each segment is a collection of exactly 48 frames, i.e., the time length of each video segment is 2 s.

The proposed MM-CSVC method were compared with the following two methods:

Table 1. The individual and cumulative bitrates of each SVC-encoded layer used in the experimental environment.

Video layers	Individual bitrate	Cumulative bitrate
Base layer	1.45 Mbps	1.45 Mbps
Enhancement layer 1	1 Mbps	2.45 Mbps
Enhancement layer 2	1.7 Mbps	4.15 Mbps
Enhancement layer 3	2.21 Mbps	6.36 Mbps

1) 1-to-k QA [8]: It is a method determining which layer to be downloaded based on the difference between adjacent segments. Comparing with our proposed MM-CSVC method, this method does not adopt the adaptive bitrate for multicast receivers. Its corresponding value for δ_{high}, δ_{low} and R_{CDSS} are set with the static values of 2, 1 and 3, respectively.

2) 1-to-k NM-ABR [7]: It is a method using different transmission polices for different SVC layers. When data loss happened in BL, this method retransmitted the corresponding video chunk until the video data have been received; however, video chunks of ELs were retransmitted for at most three times.

The proposed MM-CSVC adopted the adaptive FEC mechanism for the multicast group and its value of δ_{high}, δ_{low} and R_{CDSS} are set as 3, 1 and 6, respectively.

Four performance metrics used for comparison are as follows:

1) Average Playout Stalling Time ($\overline{T_{stalling}}$): It is equal to the cumulative time length of playout stalling ($T_{stalling}$) divided by the amount of video playout stalling events ($Count_{stalling}$). The formula of $\overline{T_{stalling}}$ is as follows:

$$\overline{T_{stalling}} = \frac{T_{stalling}}{Count_{stalling}}$$

2) Average Playout Layers (μ_q): It is equal to the sum of playout layers of all segments divided by the number of segments in the video. The formula of μ_q is as follows:

$$\mu_q = \frac{1}{S} \sum_{s=1}^{S} playout_s^q$$

where $playout_s^q$ denotes the playout layers of segment s and S denotes the number of segments in the video.

3) Quality Difference between Two Consecutive Segments ($Diff_q$): It is equal to the square root of the sum of the square of quality difference between two consecutive segments divided by the number of segments minus 1. The formula of $Diff_q$ is as follows:

$$Diff_q = \sqrt{\frac{1}{S-1} \sum_{s=2}^{S} (playout_s^q - playout_{s-1}^q)^2}$$

4) Frequency of Quality Switching without Considering the Video Playout-Freezed Event (f_{qs}): It is equal to the amount of quality switching events between two consecutive segments ($Count_{qs}$) divided by the total number of video segments (S) minus 1. The formula of f_{qs} is as follows:

$$f_{qs} = \frac{Count_{qs}}{S-1}$$

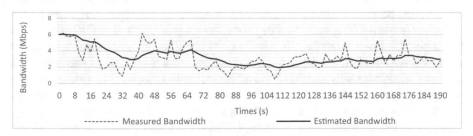

Fig. 3. The measured bandwidth and estimated bandwidth in the experiment.

Figure 3 shows the measured bandwidth and the estimated bandwidth that are derived using Eqs. (3) and (4). Table 2 depicts the results of the compared performance metrics using these 3 methods for the experiment.

Referring t o $\overline{T_{stalling}}$ of Table 2, it can be observed that 1-to-k QA and 1-to-k NM-ABR have some video stalling events, but MM-CSVC did not have. The reason is that MM-CSVC considered the buffer situation for downloading the follow-up segments and thus it can have enough buffered video to avoid the buffer-empty situation.

Referring to Table 2, it can be observed that MM-CSVC has the highest μ_q, i.e., the highest playout layers. The reason is that (i) 1-to-k QA does not have the adaptive bit

Table 2. The results of the performance metrics using these three methods.

	MM-CSVC	1-to-k QA	1-to-k NM-ABR
$T_{stalling}$	0	2	2.667
μ_q	3.209	2.527	1.187
$Diff_q$	0.367	0.656	0.289
f_{qs}	0.189	0.444	0.222

rate control and (ii) 1-to-k NM-ABR has the packet re-transmission when packets were lost. That is, MM-CSVC utilized the network bandwidth more efficiently.

Referring to Table 2, it can be observed that MM-CSVC and 1-to-k NM-ABR have the better performance of $Diff_{qs}$, i.e., they have the lower quality difference when the quality switching event happened. Nevertheless, MM-CSVC has the smaller f_{qs}, i.e., the lower frequency of quality switching. The reason is that MM-CSVC adopted the FEC mechanism to reduce the packet error rate, and the frequency of quality switching between two consecutive segments can be reduced.

6 Conclusion

This work has proposed the MEC-based Multicast for the Cooperative SVC Streaming (MM-CSVC) method to have the cooperative SVC adaptive streaming among n group users who are watching the same video on the same time in the moving vehicle environment. The proposed MM-CSVC method can derive the corresponding video bit rate for the next downloading cycle by integrating the proposed adaptive FEC scheme and the proposed estimated bandwidth calculation scheme in the multicast situation. The performance evaluation results have shown that the proposed method can reduce the frequency of video quality switching and utilize limited bandwidth to improve playback quality effectively for the cooperative SVC video streaming among n group users. The main future work is how to have the adaptive adjustment for the values of δ_{high}, δ_{low} and R_{CDSS} with the networking situation, CDSS downloading situation and buffering situation.

Acknowledgement. This work was supported by the Ministry Of Science and Technology (MOST), Taiwan (R.O.C.) under the grant number MOST 108-2221-E-006-006-MY3.

References

1. Elgabli, A., Felemban, M., Aggarwal, V.: GroupCast: preference-aware cooperative video streaming with scalable video coding. IEEE/ACM Trans. Netw. **27**(3), 1138–1150 (2019). https://doi.org/10.1109/tnet.2019.2911523

2. Zhang, T., Mao, S.: Cooperative caching for scalable video transmissions over heterogeneous networks. IEEE Netw. Lett. **1**(2), 63–67 (2019). https://doi.org/10.1109/lnet.2019.2911972
3. Schwarz, H., Marpe, D., Wiegand, T.: Overview of the scalable video coding extension of the H.264/AVC standard. IEEE Trans. Circ. Syst. Video Technol. **17**(9), 1103–1120 (2007). https://doi.org/10.1109/tcsvt.2007.905532
4. Abbas, N., Zhang, Y., Taherkordi, A., Skeie, T.: Mobile edge computing: a survey. IEEE Internet Things J. **5**(1), 450–465 (2018). https://doi.org/10.1109/jiot.2017.2750180
5. Huo, Y., Hellge, C., Wiegand, T., Hanzo, L.: A tutorial and review on inter-layer FEC coded layered video streaming. IEEE Commun. Surv. Tut. **17**(2), 1166–1207 (2015). https://doi.org/10.1109/comst.2015.2392378
6. Yang, S.H., Liu, T.W.: Quality control for hybrid unicast and multicast video transmission systems. In: 2020 IEEE International Conference on Consumer Electronics - Taiwan (ICCE-Taiwan) (September 2020). https://doi.org/10.1109/icce-taiwan49838.2020.9258044
7. Wu, F., Yang, W., Ren, J., Lyu, F., Ding, X., Zhang, Y.: Adaptive video streaming using dynamic NDN multicast in WLAN. In: IEEE Conference on Computer Communications Workshops (INFOCOM WKSHPS), IEEE INFOCOM 2020, July 2020 (2020). https://doi.org/10.1109/infocomwkshps50562.2020.9162662
8. Ullah, S., Kim, K., Manzoor, A., Khan, L.U., Kazmi, S.M.A., Hong, C.S.: Quality adaptation and resource allocation for scalable video in D2D communication networks. IEEE Access **8**, 48060–48073 (2020). https://doi.org/10.1109/access.2020.2978544

PROA: Pipelined Receiver Oriented Anycast MAC for IoT

João Carlos Giacomin[1] and Tales Heimfarth[2(✉)]

[1] Computer Science Department, Universidade Federal de Lavras, Lavras, Brazil
`giacomin@ufla.br`
[2] Applied Computing Department, Universidade Federal de Lavras, Lavras, Brazil
`tales@ufla.br`

Abstract. Environmental monitoring applications for IoT networks employ small electronic devices with constrained power resources. Aiming to extend network lifetime, duty cycle is adopted as the main strategy, at the cost of increasing end-to-end latency. This article introduces PROA: a low-latency asynchronous protocol for low-power IoT. It combines medium access control (MAC) and routing functions. Some strategies are employed to overcome the duty cycle drawbacks and reduce the latency: an anycast communication pattern, a medium reservation scheme to transmit segmented data in pipelined fashion. Our protocol was compared in simulation with others from literature, achieving a latency gain greater than 30%.

1 Introduction

At the beginning of the 21st century, the Internet-of-Things (IoT) emerged as a new technology that employs low-power electronic devices, with the ability to interact and collect information from the environment [2]. IoT systems can be broadly classified into macro-IoT and micro-IoT [4], the latter being the most suitable for applications in restricted areas such as smart buildings and precision agriculture. This class includes Low Rate Wireless Personal Area Networks (LR-WPAN), which employ a series of devices with limited power, processing and communication resources, such as those specified in the IEEE 802.15.4 reference model [1]. This kind of IoT has special characteristics, they have a uniform composition, relying on similar devices that work collaboratively and communicate by multi hops, forming a subnetwork. This network sends the collected data to an Internet access gateway.

Due to the low availability of energy to keep the sensor nodes in operation, the economy of this resource is always taken into account when developing communication control protocols for LR-WPAN's. The use of duty cycle, coordinated by the Media Access Control (MAC) layer, is the most efficient method for energy savings [5,11,15]. The sensor nodes adopt sleep/active cycles, being kept in a low power state, with radios turned off, most of the time. Periodically they are reactivated to perform measurements, processing and communication.

L. Barolli et al. (Eds.): AINA 2022, LNNS 450, pp. 68–80, 2022.
https://doi.org/10.1007/978-3-030-99587-4_7

Kumar [11], presents a review of MAC protocols for wireless sensor networks, a kind of IoT system with application in environmental monitoring. MAC protocols are grouped in contention-based and schedule-based. Contention-based methods do not require strict synchronization and are more scalable and more flexible. Among contention-based protocols, asynchronous MACs are less complex and have reduced power consumption for low-traffic networks [15]. Each data packet is preceded by a long preamble to establish communication between nodes. Duty cycles are used to save energy, but there is no synchronism in the active periods of nodes.

Despite the advantages of preamble sampling protocols, a sensor node that has a message to send must wait for the receiver to be awake, resulting in the sleep-delay [14,15]. An *anycast* communication pattern is employed to reduce this problem [7,13]. A set of nodes close to the transmitter is selected according to some metrics, composing a *Forwarding Candidate Set* (FCS). The FCS node that wakes up earlier becomes the forwarder. Another strategy is to transmit data packet from source to destination node in a *pipelined* fashion [6]. The routing nodes skew their wake-up times to occur in sequence [5]. This method is used by the anycast PAX-MAC [9], our protocol proposal prior to PROA.

The purpose of PROA is to go beyond PAX-MAC to further reduce its latency, dividing the data packet into smaller units, denominated segments, which are programmed to be transported through the channel reserved by preambles. Preambles are transmitted ahead in asynchronous phase, as in PAX-MAC, while, in the synchronous phase, a sequence of n segments of data are transmitted simultaneously by routing nodes. PROA was compared with other *anycast* MAC protocols. For the evaluated scenarios, PROA presented a gain in latency greater than 30%.

2 Related Work

GeRaF [16] is an asynchronous protocol whose contribution was an innovative use of the *anycast* communication pattern, with the objective of reducing latency in asynchronous LR-WPAN's based on preamble sampling. To take part in the FCS, a node within radio range of the transmitting node must be able to decrease the distance to the destination node. The CMAC [13] protocol is an improvement over GeRaF. It imposes a minimum advance (r_0) a neighbor node must provide in sink node direction in order to join the FCS.

AGA-MAC [7] follows the same principle as the CMAC, imposing a minimum advance towards the destination node at each hop. The threshold value for advance depends on the size of the data message. This protocol was an innovative approach in routing since it includes data length in decision process.

Another approach to achieve latency reduction is channel reservation, using two phases, like in R-MAC [6]. In the first phase, a control packet is sent in advance to schedule the communication time of the nodes that will participate in the transmission of the data packet. The routing nodes wake up in a staggered manner to transmit data synchronously in the second phase. Data packet is transmitted from the source node to the destination in a pipelined fashion, with the shortest delivery time.

The cross-layer PAX-MAC [9] protocol combines geographic routing and preamble sampling medium access control (MAC) functions. It uses the anycast communication pattern from GeRaF [16] and the two-phase transmission mode of R-MAC [6]. In the first phase, preambles are sent ahead in order to select the routing nodes, while synchronizing their wake-up times. In the second phase, the data is sent synchronously by the staggered routing nodes a few hops behind the preamble, simultaneously. This parallelism in the transmission reduces end-to-end latency.

The protocol proposed in this article achieves a lower communication latency by increasing the parallelism of communications in LR-WPAN's. The data message is divided into several segments, which are transmitted in a staggered manner. Preamble and segments travel in different areas of the network at the same time.

3   PROA Protocol

PROA is a *cross-layer* asynchronous protocol encompassing functionalities of the medium access control (MAC) sub-layer and network (NET) layer. In order to reduce the end-to-end latency, the protocol employs anycast communication pattern and a geographic routing, like GPSR (Greedy Perimeter Stateless Routing) [10].

Sensor nodes employ duty cycle to save energy, alternating long low energy (sleep) states with short active states, when they perform sampling and communication duties. There is no synchronization among duty cycles. In order to establish communication between nodes, the transmitter T sends a series of short preambles. After each small packet, the transmitter configures its radio in reception mode and waits for a confirmation packet. In the case of reception, the communication has been established, and the transmitter stops the preamble sequence. The overhead caused by the described procedure is known as *sleep-delay* phenomenon.

3.1   Anycast Communication

In order to reduce the latency caused by this problem, our protocol employs anycast communication pattern combined with geographical routing. In the anycast communication, a Forwarding Candidate Set (FCS) is selected as good routing nodes at each hop. The first node of the set to wake up is elected next forwarder.

To carry out anycast communication, the strategies employed to select FCS members can be divided into two groups depending on which side decisions are taken [9]: sender-side or receiver-side. In the former a transmitting node (T) selects the members according to some metrics and sends this information with the preamble packet. In the latter one, T node uses the preamble to inform its neighbors about the metrics so that each one decides for itself its inclusion in the FCS.

The receiver-side decision process was considered in Split-MAC [8], our previous protocol. The necessity of knowing the neighborhood brings the disadvantage of increased overhead (discovery packets). It is also not suitable for dynamic network topologies. In this paper, PROA utilizes the receiver-side decision process. The only metric considered is the proximity of the neighboring node to the destination (D) node, to which the message is addressed. In such geographical routing, nodes are aware of their geographical position using estimation algorithms or GPS.

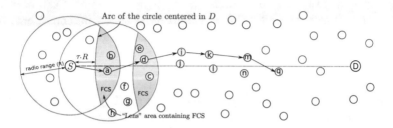

Fig. 1. PROA routing, presenting the first 6 hops. Gray circles represent FCS members.

Figure 1 exemplifies the new protocol operation. Data message transmission process from source node S to the destination D is done by multiple hops. In each hop, a transmitter T sends the packet for the next forwarder F, chosen from the FCS.

The figure presents the first six hops routing. In this example, the transmitter node informs the threshold value, τ, in the preamble, determining the minimum advance a neighbor node must provide to be considered an FCS member. This minimum advance towards D is computed as $\tau.R$, where R is the radio range of the nodes. Nodes c, d and e are members of the FCS of the node a. Nodes f and g were not chosen because they are further away from the destination and do not surpass the minimum advance, $\tau.R$. Node d became the next forwarder F after node a (acting as T here), since it woke up before c and e and sent back an eACK (early ACK [3]). After that, node a enters the inactive state until its time to receive data from S.

It is noteworthy the difference between this routing process and the one employing the sender-side decision. If a sender-side process were used, considering two members in the FCS, the node e would not be included in FCS of a. On the other hand, the use of the receiver-side decision method does not guarantee the desired number of members in the FCS. Eventually, a smaller number will be available to offer an advance greater than $\tau.R$. It is possible that none of the neighbors of T will be able to carry the message beyond $\tau.R$. This condition is denominated "empty lens". In this case, T will send all preambles within a cycle time and will not receive any response. In this case, another cycle is started with a null limit, $\tau = 0$.

3.2   Two Phase Transmission Process

In order to speed up message propagation, they are transmitted in a process similar to a pipeline [6]. Unlike other anycast protocols, PROA does not send data right after receiving the eACK signal. The communication process is split in two phases: one asynchronous and one synchronous. In the asynchronous phase, preambles are sent ahead to establish the route from S to D and to adjust the data packet transmission times. Preambles carry the information of the time when each node on the route must wake up to receive the data packet. In the synchronous phase, data is transmitted by the route nodes with staggered times. Each node on the route towards D wakes up after the previous one, with a difference δ between the times. This interval is set accordingly to accommodate a data packet transmission.

It is noteworthy that there is no synchronization between all nodes in the network, there is only a momentary synchronization in the passage of the preamble. Both phases run at the same time on the network, but in different locations. While the asynchronous phase is performed by a routing node, the synchronous phase occurs at another point in the network, a few hops back.

In order to avoid collisions between preamble and data packet, the source node S must separate transmissions of these packets for a suitable period of time. This is achieved by inserting an initial gap, Δ, placed between preamble submission and data submission. Δ is calculated as a function of the desired FCS cardinality, which is defined accordingly to give the preambles an average speed compatible with data speed. Since the interval between the transmission of preambles and data is small, the clock deviation due to drift, in the synchronous phase, is considered negligible.

PROA does not define a special eACK package like CMAC [13] and X-MAC [3]. When an FCS member wakes up and receives a preamble, it starts to submit its own series to make contact with one of its FCS members. The first pream-

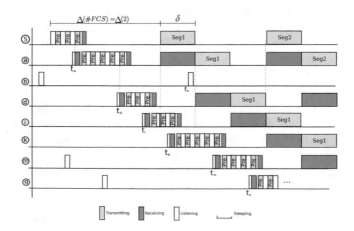

Fig. 2. Timeline of PROA communication, presenting the first 6 hops and simultaneous transmissions of preambles and two data segments. The expected FCS cardinality is #FCS = 2.

bles are understood by the previous sender as an eACK, causing it to interrupt its transmissions.

As example, the FCS of the source node S in Fig. 1 is composed of the nodes a and b. The timeline presented in Fig. 2 shows that at t_a, some times after S started sending preambles, node a turned its radio on. This happens before node b wakes up, on t_b. In the PROA anycast protocol, the first FCS node that receives the preamble assumes the role of the next router (transmitter T) towards the destination D. Immediately, it starts sending a new series of preambles to signal its intention to broadcast to a member of its own FCS. In this example, node a is trying to establish contact with c, d or e. The first preamble of a also acts as eACK for node S. At the same time, the sensor node a is scheduled to wake up just before the source node S starts sending data in the future. This process continues as seen in Fig. 2.

Node d becomes the next router, after node a, since it wakes up before nodes c and e. Next, node d starts sending preambles and schedules its time to receive data packet. As explained, the first preamble is recognized by node a as eACK. After three preambles, node i enters the active period and assumes the role of the next message forwarder, starting its sequence of preambles. While i transmits its preambles, node S starts transmitting the data packet to node a, which has already woken up to receive data. From that moment on, a simultaneous transmission of data and preambles takes place. It is noteworthy that there is no interference between these communication signals, as soon as they occur in distant places.

3.3   Segmented Transmission

In order to further reduce the delivery time of data to the destination, long messages are divided into smaller pieces, denominated segments, which are sent separately in the synchronous phase of the protocol. Several segments of the data message can travel along the route between S and D simultaneously, in different places in the network. Simultaneous transmissions of consecutive data segments by routing nodes can be carried out successfully only if a distance greater than two hops between the nodes is assured. In order to maintain this distance, the source node S separates the transmission of two consecutive segments by an interval greater than 2δ.

As an example, the timing diagram in Fig. 2 is presented. This figure shows the transmission of two segments of a data message by the source node S. Node k wakes up at instant t_k and receives a preamble from node i, immediately starting its own preamble transmission. Part of these transmissions occur simultaneously with the sending of the segment Seg_1 from a to d. Next, node d sends Seg_1 to node i, while m looks for contact with q. After that, the node S understands that the next transmission of Seg_1 does not interfere with its next communication, since a time interval greater than 2δ has already elapsed. Then S starts transmitting the second data segment to node a. At this moment, three transmissions occur simultaneously to establish communication between S and D: node q transmits preambles, node i transmits Seg_1 to k and S transmits Seg_2 to a.

A network with small packets can also have its performance improved by PROA, making a transmission where each small packet acts as a segment of a larger packet. Small messages are those whose transmission time is less than 33% of the cycle time, while large ones occupy the channel longer. This definition is related to the cardinality of the FCS.

3.4 Imminent Collision

Due to the asynchronous nature of PROA, the number of preambles spent on each hop to establish contact with the next router can vary, as shown in Fig. 2. The source node must carefully calculate the delay interval (Δ) to make the data transmission take place in a region far from preambles, with a distance greater than twice the radio range. The process of sending preambles ahead while data packets are transmitted few hops back continues until it reaches the destination node D. A collision could still happen if, for example, the node d preamble series took longer to establish contact with a member of its FCS (i or j). A message collision would occur at node a involving the data segment Seg_1 coming from S and the preamble from d. For these cases, PROA has a recovery method denominated *identification of imminent collision*. Since node d is aware of the time when a will receive the first **segment** of data coming from S, d interrupts its transmission of preambles and waits for the arrival of all segments. After that, node d restarts the message sending process as node S did. This is a situation that has a high latency cost.

It is important to note that collisions from other data transmissions are not considered here, we are just looking at the behavior of only a single message being transmitted on the network at a time. In the case of concurrent flows in a network, other recovery mechanisms must be used, such as the adoption of random intervals (*back off*) to retry transmission. We assume that every sending node finds at least one neighbor to send the packet in the destination direction. If this condition fails, other algorithms should be used, such as the *right hand rule* [10].

In order to avoid imminent collisions, the propagation speed of the preambles must be, on average, equal to or greater than that of the **segment**. The average time needed to find an awake node in the FCS can be regulated by adjusting its cardinality [9]. Thus, the number of FCS members is determined according to the size of the data **segment**, the smaller the size, the greater the cardinality of the FCS. The average distance between the node that sends the segment and the node that transmits the series of preambles is constant.

3.5 FCS Selection

Routing decisions in PROA are made locally, in a non-centralized manner. Each sensor node knows its own position and the position to which the data packet is addressed, called the destination position (D). The knowledge of neighbors' positions are not necessary since FCS membership is decided in the receiver's side. A node with a queued data message informs to its neighbors the position of D and the threshold value, τ. Each node in the neighborhood that receives a preamble package decides, based o those parameters, if it is an FCS member. When this node receives a preamble and starts sending its own preambles, we say the preamble has made progress. The average time required to get a preamble progress depends on the cardinality of the FCS. This time is associated with the number of preambles that must be sent before an FCS member sends a response. This number is calculated as [9]:

$$r(v) = \sum_{i=1}^{N_p} \left(\frac{i}{N_p} \right)^v \tag{1}$$

N_p is the number of preambles a sensor node can send in one cycle time. This equation can be used to obtain the cardinality (v) of the FCS that fits the average number of preambles ($r(v)$). PROA determines the value of τ appropriately to control the mean cardinality of the FCS, since the network density is known. As a consequence, the progress of a preamble takes, on average, a time equal to or less than what it takes to transmit a `segment` of the data message.

4   Experimental Results

This section presents the results and discussions of the evaluation of the protocol proposed in this work. The metrics evaluated were latency and energy consumption.

Nodes use the same duty cycle with independent activity time. Our proposal was compared with different asynchronous protocols: PAX-MAC [9], X-MAC [3], GeRaF [16] and AGA-MAC [7]. The simulations were performed with the GrubiX wireless sensor network simulator, which is an extension of the Shox simulator [12].

4.1   Setup

The IEEE 802.15.4 [1] reference model was adopted to characterize the radio transceivers, with fixed transmission power, bidirectional links and free space propagation model for isotropic diffusion in an ideal propagation medium. The radio range was set at 40 m for a unit radius propagation model. The energy consumption is considered the same ($P = 60$ mW) in active state, for transmission, reception, listening and idle modes.

The sensor nodes were randomly distributed in the simulation area. Nodes are static and aware of their own positions. PROA is a crosslayer protocol which combines anycast MAC with a geographic routing protocol based on the GPSR [10].

All parameters in different protocols have been set to the same values for fairness purposes. The GeRaF protocol was modified to avoid the use of `CONTINUE` packets, in order to improve its performance, becoming similar to the CMAC protocol, without using the limit (r_0).

All protocols are asynchronous and use preamble probing technique. Table 1 presents the parameters used for evaluation. The evaluated packet sizes were set to have transmission times proportional to 25%, 33% and 50% of the cycle time, which in the simulations was 100 ms. The distance between source and destination was 1000 m and one message was sent in each simulation. Each experiment was performed 200 times.

Table 1. Parameters employed in the simulations

Symbol	Parameter	Values
R	Transmission radio range (m)	40
P	Transmission power (mW)	60
$\overline{SD}$	Sender to destination distance (m)	1000
η	Node density ($\frac{nodes}{m^2}$)	8×10^{-3}
N_p	Maximum number of preambles	87
t_{cycle}	Cycle time (s)	0.1
t_{data}	Data length duration ($\%t_{cycle}$)	25, 33, 50
t_{pre}, t_{eACK}	Preamble and eACK duration (s)	0.512×10^{-3}
t_{cw}	Contention window interval (s)	1.024×10^{-3}
Δ	Initial waiting time (PROA, PAX-MAC) (s)	$8 \cdot t_{seg}$
ζ	Inter-segment interval (PROA) (s)	$5 \cdot t_{seg}$
n	Number of segments (PROA)	1, 2, 3, 4
B	Transmission rate	250 kbps

4.2   Results

The first experiment was carried out to verify the relationship between the latency obtained by the protocol and the number of segments used. Each data packet was left whole or split into 2, 3 or 4 segments before its transmission by PROA. The results of the experiment are shown in Fig. 3.

It can be observed that, for all packets, as greater is the number of segments into which the packet is split, the lower is latency in transmission. This can be explained by a greater parallelism in the network. It can also be seen that, doubling the number of segments, we do not have half the latency, which we would expect in an ideal case. This is due to the *overhead* of the protocol, with preambles traveling ahead and also imminent collisions. The gain does not scale indeterminately. This is because, with a greater number of segments, the cardinality of the FCS increases, leading to incomplete FCS, increasing the imminent collision probability. Short hops increases also the probability of interference between preambles and the various segments.

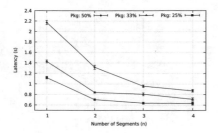
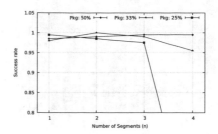

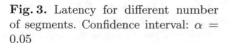

Fig. 3. Latency for different number of segments. Confidence interval: $\alpha = 0.05$

Fig. 4. Success rate of transmissions for different number of segments and packet sizes.

Figure 4 presents the success rate of the transmissions, where $1 = 100\%$. It is possible to notice that up to three segments, a very high success rate was obtained for all packet sizes. However, for the small packet (25% of cycle time), when 4 segments were employed, a very rapid decrease in success rate was observed. This can be explained by the necessity of more nodes than available in the FCS in order to avoid imminent collisions. Moreover, the small progression obtained increases the interference between different segments and preambles.

For the next result, PROA has been configured to use the number of segments that gives the best result for each package size with a delivery rate above 95%. Figure 5 shows the latency obtained for the transmission of a message using asynchronous protocols that use preambles from the literature. Results can be compared with PROA. It can be seen that protocols with advanced preamble (PROA and PAX-MAC) obtained the best results for all packet sizes. This can be explained by the chaining of nodes to receive the data packet, which, together with the advanced preambles, allow for parallelism in the network. AGA-MAC also achieved a good performance due to the anycast feature which considerably reduces the *sleep-delay*. As one might expect, latency varies with packet size. It can also be observed that PROA demonstrated a great reduction in latency when compared to other tested protocols. This reduction was at least 30% when compared to the PAX-MAC protocol.

For sensor networks with sporadic transmissions, the PROA protocol had a reduced latency compared to several other protocols tested. This makes it possible to use an asynchronous protocol, which saves energy by not requiring node synchronization, in scenarios where network latency plays an important role. Our protocol demonstrated even lower latency than a scenario without duty cycling (always on). In this setup, the message can be forwarded without any preamble or delay, since nodes are always energized and waiting to receive a packet. However, the parallelism achieved by our protocol among different segments and preambles was able to achieve lower latencies.

Figure 6 shows the energy consumption of the different protocols for packets of different sizes. Only the energy spent to transmit packets was computed (energy normally spent in the duty cycle without transmissions was ignored). We can observe that the GeRaF anycast protocol achieved, by a large margin,

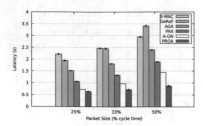

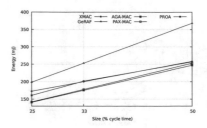

Fig. 5. Latency for different state-of-the-art protocols. Confidence interval: $\alpha = 0.05$

Fig. 6. Energy expenditure for different protocols and packet sizes.

the highest energy expenditure for all message sizes. This is due to the large number of nodes in FCS, which reduces the average size of each hop, increasing the total number of hops.

As shown in [7], a large number of nodes in FCS have a positive effect when very small packages are used, which can be seen in the slope of the GeRaF curve. PROA and X-MAC present a similar energy profile, followed by the best energy performance (AGA-MAC and PAX-MAC). Nevertheless, the difference in energy expenditure among PROA, X-MAC, AGA-MAC and PAX-MAC were low.

Given the results achieved, it is concluded that the PROA obtained, on average, a considerable gain in latency with an energy cost compatible with other protocols.

5   Conclusion

Asynchronous protocols use a series of preambles to establish communication between nodes. The increase in latency occurs because the transmitting node must wait until the receiver is awake to carry out data transmission. In order to alleviate this disadvantage, the *anycast* technique is used: instead of a single message forwarder, a set is used, which reduces, on average, the *sleep-delay*. One way to further reduce latency in *anycast* protocols is to send preambles and data separately. Preambles travel ahead in the network programming the time to receive the data packet, which travels at the same time some *hops* late.

In the present work, a deepening of this parallelism was proposed: the data packet is divided into several segments, which travel in parallel in different places on the path between origin and destination, using spatial multiplexing for this. In addition, the decision of the FCS membership is taken on the receiver's side, reducing the information collected by the transmitter and allowing dynamic topologies.

Simulations performed showed that, for different packet sizes, the presented approach leads to a reduction of at least 30% in observed latency when compared to other state-of-the-art protocols. This demonstrates that, for long communication and low traffic scenarios, PROA can be used when low latency is required.

Acknowledgments. The present work was realized with support of Fundação de Amparo à Pesquisa de Minas Gerais (FAPEMIG) grant number APQ-03095-16.

References

1. IEEE standard for low-rate wireless networks. IEEE Std 802.15.4-2020 (Revision of IEEE Std 802.15.4-2015), pp. 1–800 (2020). https://doi.org/10.1109/IEEESTD. 2020.9144691
2. Almusaylim, Z.A., Alhumam, A., Jhanjhi, N.: Proposing a secure RPL based internet of things routing protocol: a review. Ad Hoc Netw. **101**, 102096 (2020)
3. Buettner, M., Yee, G.V., Anderson, E., Han, R.: X-MAC: a short preamble mac protocol for duty-cycled wireless sensor networks. In: Proceedings of the 4th International Conference on Embedded Networked Sensor Systems, SenSys 2006, pp. 307–320. ACM, New York (2006)
4. Davoli, L., Belli, L., Cilfone, A., Ferrari, G.: From micro to macro IoT: challenges and solutions in the integration of IEEE 802.15.4/802.11 and sub-GHz technologies. IEEE Internet Things J. **5**(2), 784–793 (2018)
5. Doudou, M., Djenouri, D., Barcelo-Ordinas, J.M., Badache, N.: Delay-efficient MAC protocol with traffic differentiation and run-time parameter adaptation for energy-constrained wireless sensor networks. Wirel. Netw. **22**(2), 467–490 (2016)
6. Du, S., Saha, A., Johnson, D.: RMAC: a routing-enhanced duty-cycle MAC protocol for wireless sensor networks. In: 26th IEEE International Conference on Computer Communications, INFOCOM 2007, pp. 1478–1486. IEEE (2007)
7. Heimfarth, T., Giacomin, J., De Araujo, J.: AGA-MAC: adaptive geographic anycast MAC protocol for wireless sensor networks. In: 2015 IEEE 29th International Conference on Advanced Information Networking and Applications (AINA), pp. 373–381 (2015)
8. Heimfarth, T., Giacomin, J.C.: Split-MAC: um protocolo assíncrono de baixa energia e latência reduzida para RSSF. In: Anais do XXXVIII Simpósio Brasileiro de Redes de Computadores e Sistemas Distribuídos, pp. 421–434. SBC, Porto Alegre (2020)
9. Heimfarth, T., Giacomin, J.C., de Freitas, E.P., Araujo, G.F., de Araujo, J.P.: PAX-MAC: a low latency anycast protocol with advanced preamble. Sensors **20**(1), 250 (2020)
10. Karp, B., Kung, H.T.: GPSR: greedy perimeter stateless routing for wireless networks. In: Proceedings of the 6th Annual International Conference on Mobile Computing and Networking, MobiCom 2000, pp. 243–254. ACM, New York (2000)
11. Kumar, A., Zhao, M., Wong, K., Guan, Y.L., Chong, P.H.J.: A comprehensive study of IoT and WSN MAC protocols: research issues, challenges and opportunities. IEEE Access **6**, 76228–76262 (2018)
12. Lessmann, J., Heimfarth, T., Janacik, P.: ShoX: an easy to use simulation platform for wireless networks. In: Tenth International Conference on Computer Modeling and Simulation, UKSIM 2008, pp. 410–415 (2008)
13. Liu, S., Fan, K.W., Sinha, P.: CMAC: an energy efficient MAC layer protocol using convergent packet forwarding for wireless sensor networks. In: 4th Annual IEEE Communications Society Conference on Sensor, Mesh and Ad Hoc Communications and Networks, SECON 2007, pp. 11–20 (2007)
14. Qin, Z., Li, F.Y., Li, G.Y., McCann, J.A., Ni, Q.: Low-power wide-area networks for sustainable IoT. IEEE Wirel. Commun. **26**(3), 140–145 (2019)

15. Siddiqui, S., Ghani, S., Khan, A.A.: PD-MAC: design and implementation of polling distribution-MAC for improving energy efficiency of wireless sensor networks. Int. J. Wirel. Inf. Netw. **25**(2), 200–208 (2018). https://doi.org/10.1007/s10776-018-0393-4
16. Zorzi, M., Rao, R.: Geographic random forwarding (GeRaF) for ad hoc and sensor networks: multihop performance. IEEE Trans. Mobile Comput. **2**(4), 337–348 (2003)

A Watchdog Proposal to a Personal e-Health Approach

Gabriel Di iorio Silva, Wagno Leão Sergio, Victor Ströele$^{(\boxtimes)}$, and Mario A.R. Dantas

Institute of Exact Sciences, Federal University of Juiz de Fora (UFJF), Rua José Lourenço Kelmer, Juiz de Fora, MG, Brazil
`{iorio,wagno.leao.sergio,victor.stroele,mario.dantas}@ice.ufjf.br`

Abstract. During everyday activities, stress and anxiety are considered factors that influence users' behavior and, if recurrent, can bring different risks to the individual's health. This work proposes an architecture to assist the users in detecting and controlling emotions during their daily activities. The proposal was implemented using smartbands as body sensors to collect the data, machine learning algorithms to detect moments of stress, and a smartphone app for monitoring the user's environment. The proposal evaluation was carried out by collecting real data from a user for one year. Based on that, we detected frequency peaks and, using the location information, enriched the data to send recommendations. The results indicates the feasibility of the proposal since it was possible to identify stressful moments considering the environment that the user wanted to be monitored.

1 Introduction

Diverse data collection methods and tools are becoming even more popular between citizens and corporations [1]. Understanding this phenomenon can be associated with the importance of all information that can be extracted from data. This type of investigation detects patterns, enriches knowledge about a certain matter, and even predicts occasions with historical data. That is why the sentence "Data is the new oil" is becoming so well-liked and known [1].

However, analyzing information, detecting how it is distributed, and predicting cases can be pretty challenging. First, depending on how often new data is collected, it is required to store this information losing the least possible data. Secondly, this new information that has just arrived needs to be used in both parts: To actualize the predicting model and be used as a model variable to predict the result. In this way, it is easy to perceive that a robust architecture that embraces these prerequisites is needed.

The use of body sensors has also increased recently with the technology improvement and cost reduction [2]. This context presents a reality where new information is sent with very small granularity. Depending on what will be studied, that small data sending interval may be determinant to identify body changes. Heart rate, for example, is a type of data that is always altering and impairs or explains many different body reactions, such as if a person is under a stressful situation [3].

During anxious circumstances, heart rate tends to increase, provoking several body reactions. Therefore, analyzing heart rate is very helpful to understand a person's routine and, thus, recognize if it is a common condition or something more punctual. This investigation makes it possible to assist users in reducing stress and getting better results in those stressful situations since this can impact their quality of life.

Location information is useful to understand the user's context since stressful moments can happen in specific places, such as workplaces, classrooms, or meeting rooms. In addition, this information can help better understand which spots users tend to have more peaks of heart rate measures and even serve as an indicator of where assistance is needed.

Then, the parallel between the robustness mentioned previously, and the body data and location information analysis is easily noticed, being a challenge to processing this data volume. Cloud computing is a paradigm that is increasingly being used in systems and applications. It consists of a model where both the collected data and the processing of the generated results are the responsibility of a central computing unit. However, this type of approach offers some obstacles and challenges regarding performance and scalability since situations where a large volume of data needs to be handled by the system can generate latency instability and, consequently, a delay in response time as a whole [4].

In this context, a Fog computing-based architecture can be an effective alternative to address this problem. Fog Computing aims to offer a paradigm of decentralization of both the network and its processes. Fog Computing is a structure where data, computing, and communications are distributed modularly between data sources and the cloud. In this way, the data that is received by the Fog can be pre-processed before being sent to the cloud, thus reducing network latency and avoiding possible network overloads and delays. Thus, this paradigm is a promising alternative for developing real-time support and monitoring systems, as it allows us to handle data in a highly efficient, modular, and scalable way.

This paper proposes an architecture to assist and analyze users' body and location data on detecting heart rate peaks that indicate stressful situations that may impact their quality of life. The use of bodily data in a computing approach, respecting the ethical boundaries, presents itself as a viable approach with great potential when dealing with stress and anxiety. The study followed four main steps: (i) literature reviewing; (ii) define a proposal with a well-defined and concise architecture to ensure the fulfillment of the desired purpose of monitoring and assisting users; (iii) conduct studies with real data from body sensors and location to observe the feasibility of the proposal for the highlighted purpose; and (iv) research about the best tools to implement the architecture proposed.

This work carries several improvements and advances related to previous experiments made in [18]. In this article, we are using real user data; the architecture has been studied and carefully decided; its implementation began with development and choices regarding the best technologies used in each step. Hence, improved progress was made compared to the previous work to achieve the work goals in this article.

The article is organized as follows: In Sect. 2, the related works are discussed; in Sect. 3, the functioning and specificities of the proposed architecture are presented; in Sect. 4, the experimental results to validate the proposed architecture are mentioned, and in Sect. 5 we have the conclusions and future work.

2   Related Work

This section presents some studies that address the main elements that make up our research. It was investigated which concepts could be used as a reference for the execution of our research and which techniques could be explored even further. The main issue was establishing a baseline to understand the models used to perform real-time monitoring and their characteristics. To this end, searches were conducted concerning two main topics: real-time monitoring applications and monitoring system architectures.

Regarding to monitoring experiments, [5] presents a series of wearable sensor-based health monitoring systems. A systematic review was done, analyzing characteristics such as sensor types, processing techniques, and portability. Nevertheless, the authors identify that there are still some challenges to be overcome in the analyzed systems, such as for security, efficiency, and interoperability.

Considering the challenges pointed out previously, the authors in [6] research the different types of ambient sensors used for daily activities monitoring of older people in their homes. They point out the advantages of using ambient sensors, saying that it gives independence and comfort to the users, besides having more stability in the measurements taken. Further, we evaluated what types of available sensors could play this role, thus deciding to use location data from mobile devices.

In [7], the authors elaborate a study where they correlate psycho-social factors to cardiovascular problems through heart rate variation. Also, in [8], the authors propose an experiment to establish a relationship between the user's heart rate measurements and their 'biological' stress. The results of these studies showed that such data can be an efficient metric for stress monitoring. These researches were important in defining the type of metrics used to evaluate the user's mental state in developing the proposed system.

A system for real-time stress detection through the use of a SmartWatch was proposed in [9]. The results found in this work reinforce the decision of this type of sensor for experiments to detect the user's mental state. Such a conclusion supported the decision to use this type of sensor in our experiments.

Once we have established a general understanding of the user monitoring experiments in previous works, our next goal was to research system architectures that efficiently implements real-time systems. In [10], the authors survey the state-of-the-art architectures related to real-time systems. The investigation allowed us to understand the features of real-time systems and their possible components. Also, it was possible to identify the potential challenges encountered while implementing such a system. Thus, we were able to identify what would be feasible in implementing the system described in this article.

In [11], it is proposed a new architecture model for IoT applications, having as main feature a flexible and reconfigurable integration between Fog and Cloud computing. The presented paradigm demonstrates that it is possible to overcome some of the problems presented in cloud-based systems [4], highlighting, however, the utility of Cloud computing for high data volume processing. Furthermore, this work indicated that Fog and Cloud computing integration offer greater modularity and performance, leading us to implement an analogous system.

Lambda architecture is presented in [12], defining a parallel data processing channel, both for computing high data volume and for fast and efficient delivery of the generated data. The authors prove the usefulness of Lambda architecture to handle a range of input sensors simultaneously, making us decide to adopt this architecture into our research.

The authors in [13] propose a framework that makes use of Fog-Cloud computing concepts and Lambda architecture for the detection of unusual patterns in monitored homes. The use cases conducted reinforce the idea that the integration of the Fog-Cloud paradigm and Lambda architecture is an efficient solution for activities monitoring.

In order to extract a pattern of users' behavior, it is necessary to have some strategy to process this information. Well-known machine learning techniques have proven to be efficient for pattern recognition tasks. A study comparing several classification algorithms on the task of identifying stress or relaxation states of volunteers through data from electroencephalogram signals was performed in [14]. Although this type of sensor is not used in our work, this study highlights the usability of machine learning techniques for predicting the monitored data.

In our work, we propose a system for real-time user monitoring. Similar to [13], the architecture is defined using both the Fog-Cloud paradigm and the concepts of Lambda architecture. Despite this, the sensors used for data collection are both wearable and mobile. Furthermore, the heart rate data processing was implemented to generate predictive models to analyze the individual's state. Through the components studied in this research, we conclude that it is possible to develop a monitoring system that is useful for the end-user decision-making.

3   Stress and Anxiety Prediction Architecture

In this paper, the implementation of a system to monitor users in real-time at specific locations through sensors is presented. In order to develop such a tool capable of monitoring people's activities to improve their mental and physical health, we decided to collect their data in different settings and backgrounds. With these aspects in mind, we collected body and location data to create a model with enriched information about the user's profile and context.

Considering the large volume of data generated continuously by users and the complexity in the computation flow of the data, we perceived advantages in using the Fog-Cloud paradigm for data acquisition and processing. This paradigm reduces the possibility of network bottlenecks, bringing the processing closer to the data extraction nodes and enhancing the overall responsiveness of the system [15]. The data generated by the sensors are produced by an edge device and sent to a Fog processing node [10].

It is assumed that both the edge and fog nodes have the required computing power to handle the data that is constantly being sent and received. Furthermore, the system design was also based on the concepts of the Lambda architecture, where the received data is sent simultaneously to two parallel workflows, where both a batch data processing and a streaming routine are implemented in a hybrid way. Finally, the results generated by this process are sent to the cloud to be visualized and analyzed. Figure 1 provides an overview of the proposed architecture with its layers and workflow.

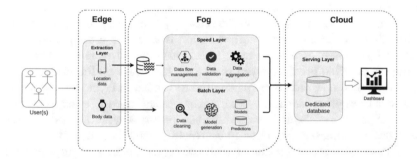

Fig. 1. Proposed architecture with the main elements and workflow.

The computational flow performed by the presented system starts with the extraction of the data generated by the sensors of one or more users simultaneously. For such a task, the **extraction layer** has this responsibility, being composed of sensors from each user that gather the information required for monitoring. It is important to note that, although the collection of this data is done continuously and in real-time, the sending and processing of this information is done asynchronously, being only necessary to store the measurement time sent by sensors.

After collection, the data is sent to the Fog node to be pre-processed. The proposed architecture relies on two different types of data from its users: their location as a set of categorical values and their cardiac rhythm. When this information arrives at the Fog node, it is processed in parallel, and the location data is previously stored in an SQL database that applies the same principles as a *Data Lake* [16]. We have adopted a Data Lake to allow the storage of a large amount of data at high speed to provide continuous availability of its contents.

When performing research on similar works, we decided to use the lambda architecture because it shows satisfactory outcomes in resolving problems concerning both performance and scalability. As mentioned in [12], the lambda architecture is centered on providing high scalability and having a low delay in both data queries and updates, also being able to deal with either human or hardware errors. The presentation of this architecture details a distributed structure of three main layers: the batch layer, the velocity layer, and the service layer.

Following the concepts outlined in the Lambda architecture, the **batch layer** performs all the high-cost computing in the system so that the output results produced are stored and are always up to date and available to the other elements. This is done to guarantee that the system will not be slowed down, especially by a delay in response due to possible processing overhead. The batch layer also enables us to have linear scalability of the data capacity that the system can support. After this process, we have in hand a securely stored collection of processed and enriched data that users can access at any time at a low compute cost. For the proposed system, the batch layer takes the body data from a specific user and updates its classification model through machine learning training and evaluation methods to provide increasingly accurate predictions about the individual's state. The generated information is stored in a dedicated database. The model generated using the new data is also stored to ensure the system's tolerance to possible prediction errors or even hardware malfunctions.

The time that new data is being treated in the batch layer, the same data set is also sent to another stream, a component named **speed layer**. The primary function of the speed layer is to reduce the response time to system requests since it deals exclusively with the most recently sent data from the extraction layer. The speed layer is developed with only low-latency queries and compute functionalities to ensure that monitoring users have real-time information. It can also have the ability to use the enriched data produced by the batch layer. The responsibility of the speed layer in the proposed system is to ensure consistent data flow management while also performing the necessary transformations and validations to generate the most meaningful information possible and support decision-making while monitoring the behavior of individuals in a given environment. In addition, this layer performs update procedures at periodic intervals, recovering location data stored in the Data Lake to verify that a user's most recent entries have been changed. If the system detects that the individual's status has not changed for a reasonably long period, the individual's location is assumed to be unknown.

The next step is to utilize the user state data already processed in the batch layer to combine the user location and the classification performed, generating an even more rich dataset. The two kinds of information are grouped to correlate them using the same period the measurements were taken. Once the necessary conversions have been made, the new data is saved in a specialized database in the cloud. It is important to emphasize the response time reduction gain achieved by implementing this paradigm. Since most of the information has already been computed by the batch layer, the speed layer can significantly reduce overall response time, ensuring scalability and low latency.

On completion of this processing flow, the produced outputs are sent to the cloud in the **service layer**. The main task of the service layer is to make the generated information available and present it to end-users and make sure the data is constantly updated. An interactive dashboard was implemented for users monitoring and analysis with the most relevant visualizations, being updated regularly and accessed remotely by any person with the allowed credentials. At the end of this entire processing flow, the system is intended to provide the appropriate tools to contribute to the decision-making process of monitoring the activities of individuals in real-time in a secure and cost-effective approach.

The Speed and Batch components have been configured on the Fog node. We decided on this configuration because the Batch requires more processing time and a higher volume of data access, resulting in latency problems and processing costs if configured in the Cloud. The Speed needs to process the new data quickly, and it uses the model trained and stored by the Batch. So, we chose to configure it on the Fog node too.

As the Serving component is responsible for presenting the data to end-users, we chose to configure it in the Cloud. We believe that in this way, we guarantee greater security of raw data and give access to only permitted information. Also, it facilitates the use of data analysis and query tools.

4   Experimental Results

Given the complexity and scope of data handling, we must consider different scenarios to carry out the proposed research. When dealing with data from corporal sensors, we must collect them to analyze the feasibility of methods to reach a viable solution to the problem at hand. Hence, it is necessary to carefully study each of the captured signals used as input variables for the machine learning model. As such, we must analyze each of them for possible correlations, missing data, patterns, and intrinsic insights about the data in question [17].

However, to enrich the body sensor data with information about the location of individuals within their home surroundings, we must consider a few aspects. Firstly, we must understand how to collect such information in the most non-intrusive way. In addition, it is necessary to analyze in detail possible disruptions in transmission, data format, and storage techniques, which are important issues for the development of the work. Thus, it is possible to see two significant study fronts: collecting and analyzing data from body sensors and their enrichment with data obtained from the user's location.

4.1   Body Sensors Data

Working with real data is a challenge, bringing expertise about variations, a range of possible values, and how data behave in a real environment. More than just working with real data, comprehending the raw data is necessary.

In this work, we use data from an anonymous and volunteer person to obtain historical information about the problem, build and train a pilot model, and identify possible peculiarities during exploratory analysis. Therefore, from March 2020 to March 2021, the data was collected with a *Samsung's smartband* and were composed of: heart rate, pedometer, stress level, and blood oxygenation rates.

The data concerning the user's heart rate is broken down into useful information in addition to his actual frequency. These are the Maximum, Minimum, and Measurement date/time heart frequency, collected and transferred to the application in one-hour intervals. We analyzed 6,304 measurement occurrences in the entire database with a minimum recorded value of 55 bpm, a maximum of 164 bpm, and an average of 78,5 bpm with a standard deviation of 13,39. It is also worth noting that the values for the quartiles are: 69 bpm for the first quartile, 77 bpm for the median, and 85bpm for the third quartile. Also, we enriched these primitive data by calculating the rate of increase and the acceleration of the rise in heart frequency to obtain more variables that seek to explain the variation in heart rate collected from the user.

4.1.1   Machine Learning Setup

With the data collected, processed, and properly aggregated with more useful information, the study's principle began to be developed. At first, it was defined as explanatory variables, some derivations of the primary data collected regarding heart rate. The correlation study between these variables was measured to identify strong correlations between them and the target variable: heart rate. In addition, it was verified strong correlations between explanatory variables that must be treated, enhancing the model's accuracy.

It is also important to highlight that missing heart rate values were not treated in any way since no matter what technique was implemented would infer information about the user's body behavior. After a carefully exploratory analysis to understand the heart rate distribution, growth and decay speed, and several other investigations, the first implementation of the model was fitted and tested with an accuracy of 94% using cross-validation.

However, this great accuracy led the model to overfit the training data. Since the proposal was to predict heart rate values and compare them with the actual heart rate measured to verify the distance between predicted and measured values, the model didn't perceive these stress and anxiety moments, interpreting it as normal. Therefore, we added a new attribute related to the next heart rate measure to solve this problem. The heart rate measured at a given moment was seen as an explanatory variable and not as a dependent variable anymore. The new dependent variable became the next heart rate measurements to be observed since the information that arrived regarding the user's body and local context aimed to predict the next measurement.

Therefore, the algorithms *Decision Tree, KNN Regressor, Random Forest Regressor, and Gradient Boosting Regressor* were used. Tools provided by the *Scikit-Learn* Python library such as *GridSearchCV* were also employed to search for the best parameters for each algorithm and obtain the best possible result from them. The parameters for each algorithm are passed with each value that must be tested by *GridSearchCV*. Therefore, an analysis with all possible combinations of parameters is studied to return the best combination found together with the cross-validation technique to avoid biasing the results. Also, a pre-processing data work was developed to scale the information, in addition to a data transformation work using the *boxcox* technique to work with data closer to a normal distribution.

The pre-processing approaches, verification of the best parameters for each algorithm, and data processing were performed. Therefore, the results concerning this standard development process were more promising. The results performed by the algorithms with this type of treatment made the model predict reasonable values and without the normalization of the anxiety peaks found in the first attempt. Therefore, in its vast majority of predictions, we observed that the model tried to predict values with a variation range always around 10 bpm (beats per minute). This variation is not highlighted as a sign of stress or anxiety by the system, which will alert the user only in higher variation cases. Thus, the results found for the *Decision Tree, KNN Regressor, Random Forest Regressor, and Gradient Boosting Regressor* algorithms were respectively: 46%, 45%, 45%, and 48%, using the R score technique.

Figure 2 shows the actual heart rate and the values predicted by the machine learning model. It is interesting to note that values near the bottom-left corner give us low predicted and real heart frequency values. On the other hand, the middle-right region of the graph shows points where a frequency peak is noticed, being necessary to analyze it with caution. For these, it is interesting to recommend, for example, some action to relieve stress and, consequently, lower their heart rate at that moment. All types of activities and possible actions for recommendation are based on a previous study of our research group [19].

Also, we can see that algorithms such as Random Forest and Gradient Boosting managed to organize the data in a better-distributed way when we look at the general format in which the points are distributed along the axes. Thus, it is interesting to note that these stand out as the best candidates for developing the batch layer.

Posteriorly, through validation of the results obtained, the model needed to be deployed so the other parts of the architecture could use its predictions. The *Heroku* platform was used to perform this task since it is a free tool along with the fact that it has a big acceptance among users that intend to provide their model online. We also built an automatic update through *Git Hub Actions*, so when the model is rebuilt and uploaded to *Git Hub*, an automatic deployment is made on *Heroku*. The *API* created and working can be seen in Fig. 3.

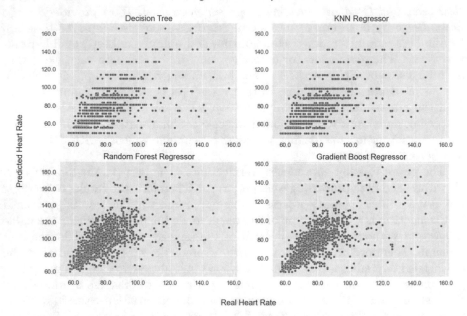

Fig. 2. Algorithms comparison between the predicted and real heart rate

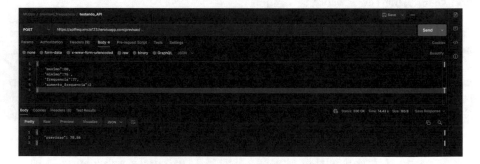

Fig. 3. API test made on postman platform

4.2 Monitoring Environment Solution

The architecture implementation was done by creating isolated but interconnected components to configure a system where feasibility and performance tests
can be conducted to monitor users continuously. Furthermore, we also defined
deploying the system in the users' homes to reduce latency in the data collection
process, taking advantage of the batch layer. The components that composes the
proposed project were installed through the use of Docker Containers [20] on a
remote server. In all, five interconnected containers were created to perform the
deployment of the system: (i) An API and a SQLite database for the location

data generated by the FIND3 application. (ii) The execution of the Node-RED application [21]. (iii) An API and a SQLite database for managing heart rate data. (iv) The execution of the Grafana application. (v) A Postgres relational database.

The processing flow developed in *Node-RED* first makes a request to the Data Lake for the most recent locations of all users and converts the response to *JSON*. The next step is to perform a probability check of how confident the sensor measurement was, if this probability is below 50%, the user's current location is unknown. This way we can ensure a more reliable database. After this, a routine is started to generate a custom query with the received information. This query is then sent for execution in the *Postgres* database by the serving layer.

The location sensor was deployed using a smartphone app FIND3 [22], while for the wearable sensor, a *SmartBand* was used. For the creation of the *Data Lake*, a *SQLite3* relational database was used, and the dedicated database was implemented using *Postgres* in the serving layer, where a dashboard shows the consolidated information to the users. Requests for the data stored in the *Data Lake* are made through a *Web API*. In addition, the dashboard was created using the analysis and monitoring tool *Grafana*.

A *Web API* has been developed in the form of a micro-service using the Flask framework to manage the heart rate data obtained. It can receive the previously collected data and store the database, so the final users can query it for analysis. Also, the *Web API* has the capabilities to receive and store a previously trained machine learning model files. The models will be loaded and used to automatically make predictions about the data received from a specific user, storing the estimation result soon afterward in the database. The *Web API* also has an administrative user interface to manage all the data collected. In this way, we can perform online data processing efficiently. Figure 4 shows the technologies used in the development and how they are correlated to each other. As can be observed, all the processed data that the system is dealing with are stored in the *Postgres* database to be consulted later by the monitoring tool *Grafana*.

It is worth noting that the location data is not collected in the form of *GPS* information but with the use of categorical values of previously classified locations. The measurement requests made by the location sensor are executed periodically, and can be adjusted according to the observed needs.

For the monitoring, it is important to emphasize that the location must be registered by the user. This is done through classification training, where data from *WiFi* and *Bluetooth* wave frequencies are captured by the sensor to be used as the training data of the location. From the results found by experiments, we observed that a user would need to train a specific location for at least one hour for a reliable measurement. In this article, the data of users in their houses were used, where it was defined that the location should be monitored in three places: the living room, bedroom, and kitchen. For this experiment, users themselves made the appropriate classifications with the sensor in the

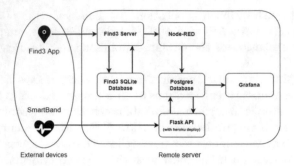

Fig. 4. Implementation diagram.

environment to perform the measurements. We also made a technical report to guide users with this configuration.

The resulted dashboard consists of three main panels: a representative image of the environment with the sensors of the most recent measurements taken in each room, a state transition panel showing a timeline of the user's movement through the environment, and a table with all the measurements taken so far. Figure 5 shows how the dashboard and its panels work. The data visualization is also periodically updated to present the most recent possible data.

In addition, it is necessary to emphasize that measurements taken at unspecified locations are not recorded or taken into consideration. As an example, suppose that a measurement has been taken on the balcony of the user's home (Fig. 5). Since it has not been previously classified, the system will consider that the user is in an unknown location or does not want to be monitored at that moment, which would be a different situation if the user is in the living room. The data is then stored in Postgres and consulted in Grafana to create the dashboard.

The variables used for monitoring the user's location are the data returned from the Data Lake, and these variables showed to be sufficient for user location

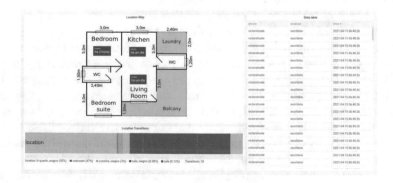

Fig. 5. Dashboard preview image showing the user location monitoring panels.

monitoring. In total, 112 entries were recorded in the cloud database, where 64 correspond to the user's bedroom, 23 to the kitchen, and 15 to the living room. The remaining 10 entries corresponded to unknown locations and were not considered in this study.

5 Conclusions and Future Works

This work presented a proposal capable of helping users during stress periods. An architecture was proposed to detect situations of stress and anxiety by using location monitoring and body sensors data. We used smartphone apps and smart bands to collect the data, and through machine learning algorithms, we identified stressful moments, enabling us to send recommendations for the user. An evaluation was carried out using real data collected from a user, and its results show that the technologies and algorithms used are viable and promising. A dashboard was implemented to enable users to visualize and monitor their data, verifying the environment where the monitoring is carried out and the user's stress level.

Some limitations were found during this work, such as the fact that the data exported by the *Samsung Health* app has a granularity of hourly heart rate measurements. It is also worth highlighting issues involving computational resources for testing more algorithms with more possibilities for their parameters. In future work, we will apply our solution with more users from our department and other institutions. Furthermore, new bracelets (*Wearfit 2.0*) have already been bought so that we can obtain less granularity between measurements, giving us more control over the extraction data step. Also, we will use machines with greater processing power, improving the Machine Learning algorithms' performance even more. Finally, we will enhance the combined use of location and body data to improve the proposal's performance.

References

1. Hirsch, D.D.: The glass house effect: Big Data, the new oil, and the power of analogy. Me. L. Rev. **66**, 373 (2013)
2. Mukhopadhyay, S.C.: Wearable sensors for human activity monitoring: a review. IEEE Sens. J. **15**(3), 1321–1330 (2014)
3. Kim, H.G., et al.: Stress and heart rate variability: a meta-analysis and review of the literature. Psychiatry Investig. **15**(3), 235 (2018)
4. Dillon, T., Chen, W., Chang, E.: Cloud computing: issues and challenges. In: 24th IEEE International Conference on Advanced Information Networking and Applications, vol. 2010, pp. 27–33. IEEE (2010). https://doi.org/10.1109/AINA.2010.187
5. Pantelopoulos, A., Bourbakis, N.G.: A survey on wearable sensor-based systems for health monitoring and prognosis. IEEE Trans. Syst. Man Cybern. Part C Appl. Rev. **40**(1), 1–12 (2009). https://doi.org/10.1109/TSMCC.2009.2032660
6. Uddin, M., Khaksar, W., Torresen, J.: Ambient sensors for elderly care and independent living: a survey. Sensors **18**(7), 2018 (2027). https://doi.org/10.3390/s18072027

7. Gorman, J.M., Sloan, R.P.: Heart rate variability in depressive and anxiety disorders. Am. Heart J. **140**(4), S77–S83 (2000). https://doi.org/10.1067/mhj.2000. 109981

8. Santhanagopalan, M., Chetty, M., Foale, C., Aryal, S., Klein, B.: Relevance of frequency of heart-rate peaks as indicator of 'biological' *stress* level. In: Cheng, L., Leung, A.C.S., Ozawa, S. (eds.) ICONIP 2018. LNCS, vol. 11307, pp. 598–609. Springer, Cham (2018). https://doi.org/10.1007/978-3-030-04239-4_54

9. Ciabattoni, L., et al.: Real-time mental stress detection based on smartwatch. In: 2017 IEEE International Conference on Consumer Electronics (ICCE), pp. 110–111. IEEE (2017). https://doi.org/10.1109/ICCE.2017.7889247

10. Gomes, E., et al.: A survey from real-time to near real-time applications in fog computing environments. In: Telecom, vol. 2, no. 4. Multidisciplinary Digital Publishing Institute (2021). https://doi.org/10.3390/telecom2040028

11. Munir, A., Kansakar, P., Khan, S.U.: IFCIoT: Integrated Fog Cloud IoT: a novel architectural paradigm for the future Internet of Things. IEEE Consum. Electron. Mag. **6**(3), 74–82 (2017). https://doi.org/10.1109/MCE.2017.2684981

12. Kiran, M., et al.: Lambda architecture for cost-effective batch and speed big data processing. In: 2015 IEEE International Conference on Big Data (Big Data). IEEE (2015). https://doi.org/10.1109/BigData.2015.7364082

13. Larcher, L., et al.: Event-driven framework for detecting unusual patterns in AAL environments. In: IEEE 33rd International Symposium on Computer-Based Medical Systems (CBMS), vol. 2020. IEEE (2020). https://doi.org/10.1109/ CBMS49503.2020.00065

14. Aditya, S., Tibarewala, D.N.: Comparing ANN, LDA, QDA, KNN and SVM algorithms in classifying relaxed and stressful mental state from two-channel prefrontal EEG data. Int. J. Artif. Intell. Soft Comput. **3**(2), 143–164 (2012). https://doi. org/10.1504/IJAISC.2012.049010

15. Deng, R., et al.: Optimal workload allocation in fog-cloud computing toward balanced delay and power consumption. IEEE Internet Things J. **3**(6), 1171–1181 (2016). https://doi.org/10.1109/JIOT.2016.2565516

16. Miloslavskaya, N., Tolstoy, A.: Big data, fast data and data lake concepts. Procedia Comput. Sci. **88**, 300–305 (2016). https://doi.org/10.1016/j.procs.2016.07.439

17. Klein, A., Lehner, W.: Representing data quality in sensor data streaming environments. J. Data Inf. Qual. (JDIQ) **1**(2), 1–28 (2009). https://doi.org/10.1145/ 1577840.1577845

18. Di iorio Silva, G., Sergio, W.L., Ströele, V., Dantas, M.A.R.: ASAP - Academic Support Aid Proposal for student recommendations. In: Barolli, L., Woungang, I., Enokido, T. (eds.) AINA 2021. LNNS, vol. 226, pp. 40–53. Springer, Cham (2021). https://doi.org/10.1007/978-3-030-75075-6_4

19. Silva, G., et al.: Hold up: Modelo de Detecção e Controle de emoçães em Ambientes Acadêmicos. In: Brazilian Symposium on Computers in Education (Simpósio Brasileiro de Informática na Educação-SBIE), vol. 30, no. 1 (2019). https://doi. org/10.5753/cbie.sbie.2019.139.

20. Anderson, C.: Docker [software engineering]. IEEE Softw. **32**(3), 102-c3 (2015). https://doi.org/10.1109/MS.2015.62

21. Node-RED. https://nodered.org/. Accessed July 2021

22. FIND3.   https://www.internalpositioning.com/doc/tracking_your_phone.md. Accessed July 2021

Computation Offloading by Two-Sided Matching in Fog Computing

Meng Wang[✉] and Minoru Uehara

Faculty of Information Sciences and Arts, Toyo University, Tokyo, Japan
{s3B102010016,uehara}@toyo.jp

Abstract. Fog computing extends cloud computing to the edge of the network and reduces latency. In fog computing, tasks generated by the user are allocated to fog devices to maintain load balancing. In this paper, a SPA two-sided matching algorithm, based on the student project allocation algorithm, is proposed for allocating user tasks to fog devices. The two-sided matching is intended to improve the stability of matching, and its feasibility is verified through experiments. The experimental results show that the proposed method can satisfy the requirements of matching and maximize its benefits.

1 Introduction

Cloud computing technology has already become a mature technology and has been extensively developed. With the rapid development of Internet of Things (IoT) [1] technology, the number of devices connected to the cloud data center continues to increase, causing the load on the cloud data center to continue to increase. Cloud computing data center has strong computing power and fast data processing speed, which can be applied to various application services. When multiple client request services from the data center at the same time, the transmission of large amounts of data increases the pressure on network bandwidth, resulting in a decrease in transmission rate and a potential decrease in the quality of service. The client's request must be processed by the remote cloud data center. Some delay-sensitive applications cannot be responded quickly, and the quality of service cannot be guaranteed. In this case, the concept of fog computing is proposed [2].

Fog computing is a computing paradigm, different from cloud computing, that has Received international attention. According to the report "Fog Computing Is the Scale and Impact of a Project", published in 2017, In the nearly five years, the computing hardware of fog computing has occupied a relatively high share in the global market, and it is predicted that it will play an important role in public utilities such as medical treatment and public transportation.

2 Related Research

2.1 Fog Computing

Fog computing is a new computing paradigm, as a highly virtualized platform, located in cloud data centers and with client user that provides computing, storage, and network

L. Barolli et al. (Eds.): AINA 2022, LNNS 450, pp. 95–104, 2022.
https://doi.org/10.1007/978-3-030-99587-4_9

services. Fog computing can be understood as the cloud located locally. Fog computing is deployed at the edge of the network, enabling fog devices to perform calculations and storage directly at the edge of the network without uploading all of them to the cloud, thus reducing latency and alleviating broadband pressure.

As a consequence of these characteristics, fog computing is widely used in the following scenarios: (1) Internet of vehicles; (2) wireless sensors and actuators; (3) smart grid; (4) medical monitoring; and (5) real-time video analysis.

Bonomi et al. [5] first introduced the concept of fog computing, explained its characteristics, and illustrated its role in application scenarios, with examples of car networking, wireless sensor networks, and smart grids. Liu et al. [6] described the computing architecture and network architecture of fog computing, provided detailed descriptions of the hardware platform, software platform, and wireless technology involved, and pointed out the challenges facing fog computing. Chiang et al. [7] presented the series of challenges faced by IoT and pointed out the advantages and challenges of fog computing in the IoT environment. Aazam et al. [8–10] conducted research on resource management and evaluation in fog computing. Other aspects of fog computing, such as resource allocation and scheduling [11, 12] and resource supply [13], are also being studied.

In this paper, the classification used is based on the distance from cloud (Fig. 1). In the following text, "cloud" stands for cloud computing, "mist" stands for fog computing, and "droplet" stands for client. In cloud computing, the cloud manages multiple mist nodes, and a mist node manages multiple droplets. A droplet is a device managed by a mist node. We also say that a mist node is a cloud of a droplet.

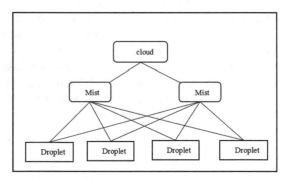

Fig. 1. The calculation hierarchy of mist

2.2 Computation Offloading

Computation offloading [3, 4] is a key technology of edge computing; which can provide computing resources for resource-constrained devices to run computationally intensive applications, speed up computation, and save energy. Computing offloading in edge computing is to offload the computing tasks of mobile terminals to the edge of the network to solve the deficiencies of equipment in terms of resource storage, computing performance, and energy efficiency.

From the operator's perspective, computation offloading allows applications to run in the edge network closer to the edge user's device, thereby reducing network congestion. From the user's perspective, resource-constrained edge users are liberated from computationally intensive and time-sensitive applications, thereby improving the quality of service for users.

Offloading can be seen as a strategy for improving mobile computing and storage capabilities. The computing task is forwarded to a computer with more abundant resources. There has already been substantial research related to computation offloading: researchers started to study computation offloading before 2000.At the beginning of the 21st century, algorithmic research began, that is, what kind of uninstall decision is made to be most beneficial to users. At the same time, the development of cloud computing allows resources to be offloaded to multiple servers and also promotes the development of computing offloading.

2.3 Matching Theory

The earliest application of matching theory was in economics, where it was used to describe the formation of mutually beneficial relationships over time. In 1962, Gale and Shapley first studied the problems of stable marriage and college admissions that is, through a trusted third party, collecting information, matching calculations, solving the problems of ladder, congestion and explosive quotation in the free markets. This algorithm is called the Gale–Shapley algorithm or the delayed acceptance algorithm.

According to the different preferences and corresponding relationships of the preference algorithms, the matching algorithms are classified, as follows:

(1) According to the correspondence relationship, algorithms can be classified as performing one-to-one matching, many-to-one matching, or many-to-many matching.
(2) According to the preference classification of matching objects, algorithms can be classified as performing bipartite matching of bilateral preferences, bipartite matching of unilateral preferences, or non-bipartite matching.

In the process of allocation, the aim is to allocate each droplet to a mist node with the best possible computing performance for processing. However, in our method, we expect each droplet to unilaterally select a mist node and expect each mist node to select the tasks generated by the droplet, according to certain rules. This is a two-way selection process.

3   Student Project Allocation

In the problem described in this paper, each droplet should be matched with mist nodes that have high computing power, and each mist node should also maximize its performance. A mist node may handle more than one task, and a task may also be offloaded to multiple mist nodes for processing, by computation offloading; therefore, there is a many-to-many match between droplets and mist nodes.

To enable a request sent by the client to be processed by the most suitable mist node, we use the student project allocation (SPA) algorithm [16]. SPA is an algorithm that performs many-to-many matching.

Each lecturer has multiple projects and each student is assigned to a different project. The constraints are that each project should be allocated to a good student and each student wants to be assigned to a good project. There are three roles in SPA: lecturer, student, and project. These correspond to the droplet, mist node, and task request issued by droplet, respectively, which are used in this paper. By establishing a good preference list and applying the SPA algorithm, a two-sided matching is established between the droplet and mist node.

In some schools, projects are assigned to students to complete as part of the curriculum. A project may be assigned to one or more students to complete. Each lecturer provides multiple projects: the number of projects is generally greater than the number of students, but it cannot expect all items to be occupied. Each student has their own preferences for the desired project, and the lecturer will also develop a list of preferences of students who are willing to guide and the number of instructors based on the students' situation.

Preference List. The degree of matching between the remaining resources of mist node m and the resources required by droplet n is called the preference value of droplet n to mist node m, $n \in N$, $m \in M.N$ represents the collection of all droplets, and M represents the collection of all mists.

Use $p_{n,m}$ to indicate the preference of the m-th mist to the number of CPU cores of the droplet n, Further definition, $\alpha_{(n-1),(m-1)}$ represents the number of CPU cores assigned to the droplet $(n-1)$ on the m-1th physical machine, $b_{(n-1),(m-1)}$ indicates whether the task generated by the droplet $(n-1)$ is assigned to the mist $(m-1)$, $\alpha_m + u_m$ Indicates the total number of CPU cores of the mist m, then:

$$p_{n,m} = \frac{\alpha_m - \sum_{n=1}^{N} b_{(n-1),(m-1)}\alpha_{(n-1),(m-1)} - \alpha_{n,m}}{\alpha_m + u_m} \tag{1}$$

Use $Q_{n,m}$ to express the preference of mist node m to the memory of droplet n, and $\beta_{(n-1),(m-1)}$ means that the n−1th droplet is allocated to the memory on the m−1th physical machine, and $b_{(n-1),(m-1)}$ indicates whether the task generated by the $(n-1)$th droplet assigned to the $(m-1)$th mist node, $\beta_m + v_m$ represents the total memory size of mist node m, then.

$$Q_{n,m} = \frac{\beta_m - \sum_{n=1}^{N} b_{(n-1),(m-1)}\beta_{(n-1),(m-1)} - \beta_{n,m}}{\beta_m + v_m} \tag{2}$$

From Eqs. (2) and (3), if $p_{n,m}$ and $Q_{n,m}$ is negative, it indicates that mist node **m** does not have enough material resources to handle the tasks generated by the nth droplet, this mist node is placed at the end of the preference list of the nth droplet. If no mist nodes can handle this task, then the task will wait for the next round of matching until an appropriate mist node appears that can process it.

The preference matrix between mist node **m** and droplet **n** can be expressed by A, which is defined as.

$$A = \begin{bmatrix} p_{11} & q_{11} \\ \vdots & \vdots \\ p_{1,m} & q_{1,m} \end{bmatrix} \tag{3}$$

We only consider two physical resources of mist nodes: the number of CPU cores and the memory size; therefore, **a** and **b** are used to represent the relative weights of the number of CPU cores and the memory size, such that $\mathbf{a} + \mathbf{b} = 1, 0 \leq \mathbf{a} \leq 1$, and $0 \leq \mathbf{b} \leq 1$. The preference vector of droplet **n** to mist node **m** is defined as follows:

$$\lambda = \begin{bmatrix} p_{11} & q_{11} \\ \vdots & \vdots \\ p_{1,m} & q_{1,m} \end{bmatrix} (\mathbf{a}, \mathbf{b})^{\mathrm{T}} = \left(\lambda_{11}, ..., \lambda_{1,\mathbf{m}} \right) \tag{4}$$

where λ represents the preference of the nth droplet to mist node **m**. It follows from Eq. (4) that a smaller value of λ indicates that droplet **n** is more suitable for matching with mist node **m**. In addition, stable matching is beneficial to both the droplet and the mist node. The stability of matching is defined as follows:

Stability. If there is no blocking pair in a matching set K, then K is stable. The following conditions need to be satisfied:

(1) Mist node **m** finds that droplet **n** is an acceptable matching object.
(2) In matching set **K**, droplet **n** is not matched, or mist node m has a higher preference ranking than the matched object **K(n)**.
(3) Or the following must be satisfied:

   (3.1) Mist node **m** has not been completely matched and satisfies the following three conditions:

   (A) **K(n)∈M**, and droplet **n** prefers mist node **m** to the existing matching object **K(n)**.
   (B) **K(n)∉M**, and droplet **n** is not matched.
   (C) **K(n)∉M**, and droplet **n** has been matched. Mist node **n** prefers droplet **m** to the worst matched object.

   (3.2) Mist node **m** has been matched. Mist node **m** prefers droplet **n** to the worst matched object and satisfies the following two conditions:

   (a) K(m)∉N.
   (b) **K(m)∈N** and the nth droplet prefers mist node **m** to the existing matching object **K(m)**.

The description of the calculation process of the droplet matching a preference list of the mist is over, and the mist matching the droplet will be much simpler. The remaining processing time of the task being processed is summed with the estimated processing time of each task to obtain the preference list $S = \left[s_{m,1}, s_{m,2}, ..., s_{m,n},\right]$ of the tasks generated by the m-th mist for the n-th droplets..

In the classic marriage matching problem, the male group and female group are perfectly matched, and there is no blocking pair; therefore, blocking pairs are not allowed in set **K**. In contrast, to find a stable match in fog computing, we must first establish a preference list of matching agents, such as droplets and mist nodes. The preference list of each agent is an ordered list, ordered by the preferences of another group of agents that considers acceptable.

In the first stage of matching, according to the preference list of the tasks generated by mist node **m** for N droplets, tasks with a relatively short processing time are found from the preference list for each execution. For droplet **n**, which has been assigned to mist node **m**, the most suitable mist node must be selected according to the comparison of the preference lists of the M mist nodes. If droplet **n** has been matched, the worst physical machine previously matched needs to be replaced according to the preference list. The time complexity of this stage is $O(n + m)$, and the second stage continues with a new round of cycles, so the overall time complexity of SPA is $O(n^2)$.

The following is a two-sided matching algorithm for fog computing, based on the SPA algorithm.

Input: N,M,λ,S
Output: Matching K
Initialization: Set K = $\emptyset$, and set all PMs to free;
while some λ m is free and m has a nonempty preference list do
for all m in M do
m proposes to the first entity n in λ ,and removes n from λ;
K $\leftarrow$ K $\cup$(n,m)
end for
for all n in N do
while n is oversubscribed do
find the worst pair (n_{worst}, m_{worst}) assigned to n in the list of PMs;
K $\leftarrow$K/(n_{worst}, m_{worst});
end while
end for
end while
terminate with matching K

4 Experiment and Results

The load balancing of tasks between droplet, mist, and cloud is an important part of resource allocation and management in cloud computing and fog computing. The problem is to select a suitable mist node for the client according to certain method and strategy. First, the resource requirements of the droplet need to be considered. Second, under this premise, the following conditions should be satisfied by load balancing:

(1) Reduce costs. Make full use of the resources of the mist node, to save energy and reduce power, cost, and network traffic.
(2) Service quality. As a result of balancing load, the user experience on the user side is improved, reliability is enhanced, and the average response time is reduced.

Load balancing is also a resource optimization problem, which plays a key role in improving user experience, enhancing reliability, and reducing cost. Through comparative experiments, we demonstrated that SPA has preferable stability in load balancing. We used random allocation and unilateral matching algorithms to verify that, in the droplet–mist structure, two-sided matching is more stable and complies more with the matching requirements. For example, a mist node with more computing resources will not be matched with a small task from droplet. Therefore, the computing resources of mist nodes can be allocated in a balanced manner, thereby reducing the waste of computing resources.

The unilateral matching performed in the experiment used the TTCM algorithm. For convenience, we assumed that each mist node had the same total number of CPU cores and the same total memory. The size of the task generated by each droplet, the memory required to process the task, and the number of CPU cores are different. When calculating the preference list of droplets matching a mist node, the CPU and memory were given the same weight.

Figures 2 and 3 show the number of CPU cores and the memory usage of each mist the abscissa represents the task of the droplet, and the ordinate represents the usage rate of CPU cores or memory.

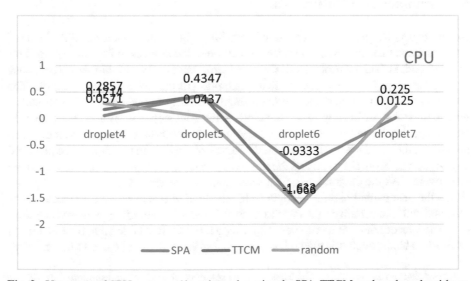

Fig. 2. Usage rate of CPU cores used by mist nodes using the SPA, TTCM, and random algorithms.

After the matching is over, the ratios of the remaining computing resources of each mist occupying the total resources are compared. The larger the ratio, the more sufficient

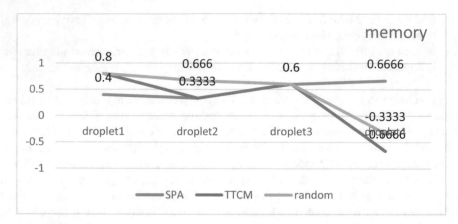

Fig. 3. Memory usage of mist nodes using the SPA, TTCM, and random algorithms.

the use of the resources of the mist is. At the same time, we also need to compare the stability of the matching through the average situation after each node is matched. Although the results of TTCM were better than those of SPA with respect to the matching results of some nodes, TTCM wastes computing resources and increases the probability of being unable to process a certain task. Therefore, the experiment proves that the SPA algorithm is more stability.

5   Summary and Future Work

In this paper, we proposed a two-sided matching algorithm, based on SPA, for load balancing and task distribution between the user and the devices in fog computing. This method both satisfies a droplet's desire to match a mist node with higher performance and realizes a mist node's desire to use its computing resources as much as possible. The proposed method is simple, easy to understand, and easy to operate, and implements a new method for performing load balancing. In the next step, I want to propose a credit function, using the credit value as currency to circulate in the droplet-mist-cloud. As a consumer, the droplet uses the basic credit value obtained from the cloud to request and pay for service from the mist, and the mist as a producer provides services to obtain the credit value. As the central bank, mist sends most of the credits obtained to the cloud. When the mist cannot handle a certain task, the cloud will also provide services for the mist and collect a certain credit value as a reward, and then according to the activity level of the droplet continue to distribute credits. keep looping. It is hoped that the system can be more stable through this method, and the goal of energy saving can also be achieved.

Acknowledgments. Thanks to all the teachers and students who helped and supported me during my studies. Especially my teacher, Uehara teacher. Whether it is learning or essay guidance, the teacher has devoted a lot of time and energy to take this opportunity to express gratitude to the teacher.

References

1. Atzori, L., Iera, A., Morabito, G.: The Internet of Things: a survey. Comput. Netw. **54**(15), 2787–2805 (2010)
2. Evans, D.: The Internet of Things: How the Next Evolution of the Internet is Changing Everything. Cisco Systems (2011)
3. Du, J., Zhao, L., Feng, J., et al.: Computation offloading and resource allocation in mixed fog/cloud computing systems with min-max fairness guarantee. IEEE Trans. Commun **66**(4), 1–1 (2017)
4. Chang, Z., Zhou, Z., Ristaniemi, T., et al.: Energy efficient optimization for computation offloading in fog computing system. In: IEEE Global Communications Conference, pp. 1–6 (2017)
5. Bonomi, F., Milito, R., Zhu, J., et al.: Fog computing and its role in the internet of things. In: ACM Edition of the MCC Workshop on Mobile Cloud Computing, pp. 13–16 (2012)
6. Liu, Y., Fieldsend, J.E., Min, G.: A framework of fog computing: architecture, challenges and optimization. IEEE Access **5**, 25445–25454 (2017)
7. Chiang, M., Zhang, T.: Fog and IoT: an overview of research opportunities. IEEE Internet Things J. **3**(6), 854–864 (2017)
8. Aazam, M., Huh, E.N.: Fog computing micro datacenter based dynamic resource estimation and pricing model for IoT. Strojarstvo Časopis Za Teoriju I Praksu U Strojarstvu **51**(5), 687–694 (2015)
9. Aazam, M., St-Hilaire, M., Lung, C.H., et al.: MeFoRE: QoE based resource estimation at fog to enhance QoS in IoT. In:IEEE International Conference on Telecommunications, pp. 1–5 (2016)
10. Aazam, M., St-Hilaire, M., Lung, C.H., et al.: PRE-Fog: IoT trace based probabilistic resource estimation at Fog. In: IEEE Consumer Communications & Networking Conference, pp. 1–17 (2016)
11. Do, C.T., Tran, N.H., Pham, C., et al.: A proximal algorithm for joint resource allocation and minimizing carbon footprint in geo-distributed fog computing. In: IEEE International Conference on Information Networking, pp. 324–329 (2015)
12. Name, H.A.M., Oladipo, F.O., Ariwa, E.: User mobility and resource scheduling and management in fog computing to support IoT devices. In: International Conference on Innovative Computing Technology, pp. 191–196 (2017)
13. Skarlat, O., Nardelli, M., Schulte, S., et al.: Resource provisioning for IoT services in the fog. SOCA **11**(4), 427–443 (2016)
14. Gu, Y.: Matching theory framework for 5G wireless communications. Dissertation, University of Houston (2016)
15. Bonomi, F., Milito, R., Zhu, J., Addepalli, S.: Fog computing and its role in the internet of things. In: Fog Computing and Its Role in the Internet of Things (2012)
16. Lin, L., Liao, X., Jin, H., Li, P.: Computation offloading toward edge computing. Proc. IEEE **107**, 1584–1607 (2019). https://doi.org/10.1109/JPROC.2019.2922285
17. Bermbach, D., et al.: Towards auction-based function placement in serverless fog platforms. In: 2020 IEEE International Conference on Fog Computing (ICFC), pp. 25–31. IEEE (2020)
18. Zhu, H., Huang, C., Zhou, J.: EdgeChain: blockchain-based multi-vendor mobile edge application placement. In: 2018 4th IEEE Conference on Network Softwarization and Workshops (NetSoft), Montreal, QC, pp. 222–226 (2018). https://doi.org/10.1109/NETSOFT.2018.846 0035
19. Pan, J., Wang, J., Hester, A., Alqerm, I., Liu, Y., Zhao, Y.: EdgeChain: an Edge-IoT framework and prototype based on blockchain and smart contracts. arXiv:1806.06185 (2018)

20. Sardellitti, S., Scutari, G., Barbarossa, S.: Joint optimization of radio and computational resources for multicell mobile-edge computing. IEEE Trans. Signal Inf. Process. Netw. **1**(2), 89–103 (2015)
21. Deng, R., Lu, R., Lai, C., Luan, T.H., Liang, H.: Optimal workload allocation in fog-cloud computing towards balanced delay and power consumption. IEEE Internet Things J. **3**(6), 1171–1181 (2016)
22. Mukherjee, M., Kumar, S., Shojafar, M., Zhang, Q., Mavromoustakis, C.X.: Joint task offloading and resource allocation for delay-sensitive fog networks. In: ICC 2019 - 2019 IEEE International Conference on Communications (ICC), Shanghai, China, pp. 1–7 (2019)
23. Labidi, W., Sarkiss, M., Kamoun, M.: Energy-optimal resource scheduling and computation offloading in small cell networks. In: 2015 22nd International Conference on Telecommunications (ICT), Sydney, NSW, pp. 313–318 (2015)
24. You, C., Huang, K.: Multiuser resource allocation for mobile-edge computation offloading. In: Global Communications Conference, pp. 1–6 (2017)

Distributed Log Search Based on Time Series Access and Service Relations

Tomoyuki Koyama[✉] and Takayuki Kushida

Graduate School of Computer Science, Tokyo University of Technology, Hachioji, Tokyo, Japan
`g21210247f@edu.teu.ac.jp, kushida@acm.org`

Abstract. Distributed tracing helps administrators to analyze root causes of microservices under system failure. It enables tracking procedures by log messages. Distributed trace log searches require short response times. Therefore, this study proposes a log search method with fast response time to search queries. Log messages are stored on several nodes as blocks grouped by date/time and service-name. The search method focuses on time-series access patterns and service relations. It decreases the number of accessed log messages per query on search. Experiment results show that the proposed method is maximally 0.91 s faster than the parallel method 'all parallel'.

1 Introduction

Microservice architecture consists of several services, which communicate themselves on a network. When one microservice responds error to another microservice, a new error occurs in another microservice [1]. A step of error propagation is called cascading failure. Since the architecture consists of a single part of the application, the root cause of errors is identified by intercepting or tracing (e.g., stack trace) in the monolithic architecture [2]. On the other hand, the stack trace can not apply in the microservice architecture since the architecture is built on several programming languages, frameworks and platforms [3]. The monolithic architecture is superior to the microservice architecture in traceability for root cause analysis.

Distributed tracing is widely used in a microservice architecture for root cause analysis [4,5]. Each service in a microservice generates log messages, and when distributed tracing, each message has an identifier. The identifier is generated on the service, directly receives the user's request, and can be passed to other services during the communication process. When system administrators track user requests in root cause analysis, they can find log messages that match request identifiers. As the number of microservices increases, the total number of requests for service-to-service communication increases. In addition to increasing the number of communications, the total number of log messages increases. Therefore the response time is required short with large-scale logs in search.

The scatter-gather pattern is an approach for large-scale data processing. This pattern enables large-scale data processing by sharing tasks into distributed

L. Barolli et al. (Eds.): AINA 2022, LNNS 450, pp. 105–117, 2022.
https://doi.org/10.1007/978-3-030-99587-4_10

workloads [6]. There are two types of nodes in this pattern, root nodes and leaf nodes. Root node receives tasks from users and scatters tasks to leaf nodes. Leaf nodes receive tasks from root node and respond result to it.

Prerequisite

This section describes prerequisites and a use-case scenario. The target log for this study is the web access log that enables distributed tracing on microservices. Code 1 shows the log format for distributed tracing generated from Envoy built-in Istio. It means that one microservice sends an HTTP request to another microservice.

Code 1. Example log message for distributed tracing.

```
[2021-10-22T00:27:09.383Z] "GET /paper/0416f705-df88-4d5f-82e8-095d4bd89e37/
   download HTTP/1.1" 200 - via_upstream - "-" 0 736954 134 133 "-" "
   Python/3.9 aiohttp/3.7.4.post0" "11c0553b-e1cd-9044-b4ce-49576dcbae6c"
   "paper-app.paper:4000" "10.42.2.65:8000" inbound|8000|| 127.0.0.6:37351
   10.42.2.65:8000 10.42.2.64:44452 outbound_.4000_._.paper-app.paper.svc
   .cluster.local default
```

A use-case of the log is root cause analysis on microservice architecture. System administrators utilize the distributed tracing log to find the root cause of system failure. When trouble is reported from a user on a web service, the system administrator executes the following search queries.

- $status\_code = 400$ & $service\_name =$ front
- $request_id = $ 1c0553b-e1cd-9044-b4ce-49576dcbae6

Figure 1 shows an example of microservices. A rectangle on the figure means a microservice (e.g., front-admin, front). A cylinder means Datastore (e.g., MongoDB, Minio). This paper defines a user who accesses web services as an End User and an administrator as Admin. A microservice connects to multiple microservices in order to respond to requests from users. For instance, when End User sends a request to 'front', 'front' sends several requests to 'author' service and 'paper' service. When 'paper' service receives a HTTP request from 'front' microservice, 'paper' service generates an ingress log message as shown in Code 1.

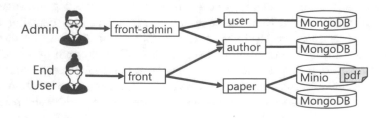

Fig. 1. Example of microservices for a paper publishing site.

Issue

The response time depends on the search query that the administrator sent in a log search on scatter-gather. This is because the search query differs from the pattern of accessing log messages on leaf nodes [7]. When the root node forwards the search query to leaf nodes by a search query, each leaf nodes finds log messages in a local volume. The placement of log messages on leaf nodes changes the search response time. Balanced log placement archives short search response time. When the log messages corresponding to a search request are unevenly distributed to several nodes, the search response time becomes slow. This is because the response time in scatter-gather takes the maximum value of the response time of the leaf node.

Figure 2 shows an issue of log search with scatter-gather. The figure has System Administrator, Root Node, and Leaf Nodes(A, B, C, D). The rectangle inside each Leaf Node is the log message. It has an identifier as an integer number such as 10. System Administrator sends Search Request to Root Node in order to find log messages which have specified request-id (e.g., ID=10). Root Node sends Search Requests to Leaf Nodes by mapping table that resolve log into Leaf Node. For instance, Root Node accesses Leaf Node A and B in the figure. Leaf Nodes find log messages that match the search request on local storage.

The time to find log messages depends on the number of log messages on Leaf Nodes. As the number of log messages increase, the disk I/O increase on a leaf node. The pattern of unbalanced disk I/O on Leaf Nodes produces high search latency since the total search response time is equal to the maximum search response time of Leaf Nodes. For example, The search targets are three in Leaf Node B, on the other hand, zero in Leaf Node C. This is called *straggler problem* in scatter-gather. The placement of log messages is essential for fast log search. On the other hand, an actual data access pattern corresponding to search requests does not balance Leaf Nodes.

An access pattern of log messages is different by conditions such as ID or Parameters(e.g., service-name) in search query on distributed tracing. Therefore, some existing methods that utilize simple conditions clustering produce high latency on log search. This paper proposes low latency log search, which specializes in two types of query: request-id based filtering, service-name based filtering.

This paper is organized as follows. Section 2 explains existing works on distributed tracing and Information Retrieval. Section 3 presents the method to speed up a log search engine to specialize in Distributed tracing. Section 4 displays the implementation of the proposed method and describes the environment for experiments. Section 5 evaluates the results of the experiments. Section 6 discusses points of improvement and parameters on the investigation. The last section shows the contribution and the conclusion of this study.

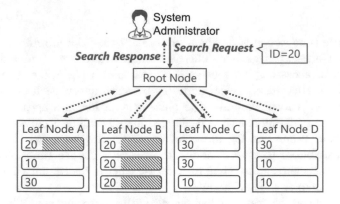

Fig. 2. Issue: access pattern and log placement.

2   Related Works

The scheduling approach to minimize job execution time finds a relation between replication factors and job execution time on Hadoop [8]. The factors in the study include the following parameters: available disk capacity, network throughput, data copying time. This approach decreases a job execution time by over 18–20

CDRM proposes cost awareness dynamic replication management of storage clusters on a cloud [9]. The study focuses on relations between availability and the number of replicas. The study calculates minimum replicas that satisfy availability requirements from node capacity and blocking probability. The proposed method improves data access latency. However, the technique could not figure out dependencies between each data. The study can improve access latency in distributed tracing since log messages have dependencies with other entries.

The heterogeneous database system produces high throughput database by SQL and NoSQL [10]. The proposal enables decreasing Disk I/O latency by Elasticsearch and MySQL. The characteristics of the proposed system (independence between transactions, single write) make it suitable for use with logs. Although, the study is not enough to be aware of the data access pattern. The method has an issue with access latency.

AptStore builds dynamic data management system for decreasing storage costs and improving Disk I/O throughput on Hadoop [11]. It enables the calculation of the probability of data accessing counts. The approach helps to match past access patterns and feature accessing design. However, the practice of data accessing is little in distributed tracing. The way has search throughput on distributed search.

The technique improves communication performance by data placement on InfiniBand [12]. The study decreases a memory registration overhead by reducing the page size of the file system. However, the works are not enough to be aware of data placement and data access patterns. Therefore, It has problems with search performance improvement.

3   Proposed Method

This study proposes the log search method for distributed tracing. The method fast responds to search queries as follows: request-id based filtering, service-name based filtering, date/time-based filtering. This study focuses on the relation between log access patterns and these parameters in search queries. The proposed approach determines the placement of log files in distributed nodes based on service-to-service relation graph and invocation order.

Figure 3 shows an overview of the proposed log search method. The figure consists of two phases as follows: *Store Phase, Search Phase.* Store Phase denotes procedures from the generated log to the log stored in Leaf Nodes and Search Phase denotes procedures to find log messages that match a filter condition in query. The figure has three types of nodes: Microservices, Root Node, and Leaf Nodes. Microservices provide several users with a web service over the internet. Root Node processes log messages, scans service relations, and forwards log messages to Leaf Nodes based on the placement rules. Leaf Nodes store log messages as log blocks and respond to search requests on log search. The flow of logs generated by the microservice and stored in Leaf Nodes is described below.

1. Microservices forward log messages to Root Node as log files.
2. Root Node converts log files into blocks.
3. Service Scanner discovers microservices.
4. Service Scanner generates Service Relations from discovered microservices.
5. Rule Generator loads blocks and service relations.
6. Rule Generator writes the placement rule.
7. Log Planner forwards blocks to Leaf Nodes based on the placement rule.

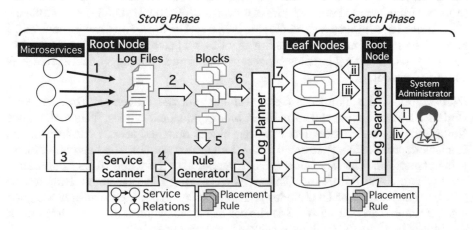

Fig. 3. The proposal overview.

The method works the followed flow when the System Administrator executes a search query with filtering conditions.

i. System Administrator sends a search query to Log Searcher on Root Node.
ii. Log Searcher scatters search requests to Leaf Nodes.
iii. Leaf Nodes finds log blocks and sends a search response to Log Searcher.
iv. Log Searcher gathers search responses and sends the search result to System Administrator.

3.1   Rule Generator

This section describes generating the log placement rule. The rule aims to decrease access blocks on log search for reducing search response time since the number of access blocks is equal to search response time. The Rule Generator gets two inputs such as **Service Relations**, **Blocks** and outputs the **Placement Rule**.

Service Relations: The Service Relations has a graph structure and represents relations in microservices. When one microservice α invokes another microservice β, microservice α has a service relation to microservice β. The relation displays as $\alpha \rightarrow \beta$. The proposed method utilizes a service-to-service relation within microservices for blocks placement. The way to get service relation is service mesh. A service mesh supports a service discovery that enables to find microservices and their relations [13]. Istio is used for service discovery in this study[1].

Blocks: The Rule Generator takes the Blocks as input. The Blocks are made from Log Files based on date/time and service-name to reduce the number of access blocks on search. For example, the service-name that the file name contains is "front," and the date/time are "2021-10-22T00:27:09.382Z" in Code 1.

A block length takes a fixed file size based on the date/time range, such as 12 hours. The steps of making blocks are as follows: 1) clustering log files by service-name; 2) concatenating log files per service-name and splitting concatenated files as a fixed size block.

Placement Rule: The Rule Generator outputs the Placement Rule for Leaf Nodes. The rule manages the pairs of blocks and Leaf Nodes. When the Root Node finds the leaf nodes that have log blocks matched search conditions, the Root Node utilizes the Placement Rule. The format for the rule has pairs of key-value structures. The key takes a block name, and the value takes node-name, service-name, and date/time. The steps of making the Placement Rule are as follows. The number of Leaf Nodes is defined as N. Each Leaf Nodes is assigned a serial number from 1 to N. B is the set of blocks as follows: $b \in B$. b is one of the blocks in B and is split by service-name and date/time.

---

[1] https://istio.io/.

1. Sort all blocks based on service-name and date/time in a log message.
2. Allocate blocks from the sorted block list to the leaf nodes while the number of iterators i increases. Select a block from the list. The leaf node for storing blocks is determined by modulo arithmetic. The serial number i that corresponds to a block is formulated as $mod(i, N)$.

Figure 4 shows an example of log placement where $N = 4$. The figure has three services as Service A, Service B, and Service C. A block in the figure has service-name and date/time range. For example, the top left block means that service-name is 'Service A', and date/time range is from 1 to 3.

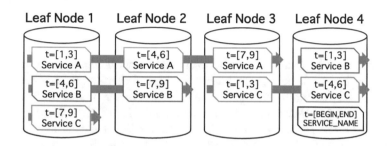

Fig. 4. Proposed log placement on 4 leaf nodes.

3.2 Log Searcher

This section describes the search method to decrease the number of accessed logs on log search. Figure 6 shows the method of proposed log search. The figure displays log blocks per service on microservices. The color of the block indicates whether to scan on the log search. The x-axis indicates elapsed time. The services such as Service A consist of a web service. The dotted lines indicate search target date/time range (e.g. from November 9, 2021, to December 7, 2021). When a user accesses Service A by Web Browser, Service A accesses Service B and accesses Service C to build a response. The following section describes the search procedure.

1. Find log messages in the block of Service A that locates the top of service relations when the system administrator finds log messages in log blocks parallelly.
2. Find the first message that matches the search condition in the search result of Service A.
3. Get the date/time from the first log message.
4. Find the log block that matches the first date/time in 'Service B' that is the next service from Service A.
5. Repeat steps until no next service.

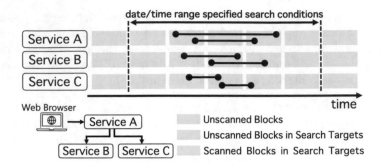

Fig. 5. Proposed log search.

Figure 6 compares the steps of block access in log search between the parallel method 'All Parallel Search' and the proposed method 'Proposed Search'. The proposed method finds only green blocks. The parallel method finds green blocks and red blocks. A number in blocks mean access order in search. The number of access blocks is related to the search response time. The proposed method reduces the search response time. The characteristic of this study reduces the number of access blocks on log search based on time-series access and service relations. Reducing access blocks enables to reduce the search response time since the time is related to disk I/O.

	All Parallel Search					Proposed Search				
Service A	1	1	1	1	1	1	1	1	1	1
Service B	1	1	1	1	1		2	2	2	
Service C	1	1	1	1	1		2	2	2	

time   time

Fig. 6. Flow of accessing blocks.

4   Implementation

Figure 7 shows a system architecture. The figure has two types of nodes: Root Node and Leaf Node. The Root Node generates Log Files and creates blocks from log files and forwards blocks to Leaf Nodes based on Placement Rule. The Placement Rule is implemented as JSON format as shown Code 2. This rule takes block name as key, '20.author_block_.log.004' and takes block metadata as value. Code 4 shows the block for 'author' service from 'begin_datetime' to 'end_datetime' is stored on 'koyama-log1'. The Leaf Nodes receive blocks from the Root Node and stores blocks in local storage. The effect of network delay is small since the traffic of node communication is less than network bandwidth.

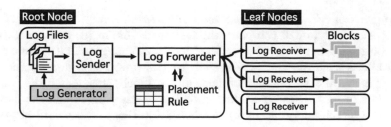

Fig. 7. The system architecture.

Code 2. Example of placement rule

```
{"20.author_block.log.004": {
    "begin_datetime": "2023-07-24T16:10:09",
    "end_datetime": "2023-12-28T18:42:21",
    "nodename": "koyama-log1",
    "servicename": "author"
}}
```

5   Experimental Results

Figure 8 shows an evaluation environment. The environment works on BareMetal Hardware with VMware ESXi 7.0 Hypervisor. Root Node runs on a single Virtual Machine. Leaf Nodes works on 13 Virtual Machines. Two types of nodes have homogeneous hardware resources (CPU: 1 [Core], RAM: 1 [GB], Storage: 30 [GB]). Log Generator expands log messages from 1,600 to 8,065,000. The log message reproduces the behavior of the microservices in Fig. 1 (i.e. 'front', 'paper' and 'author'). Figure 9 shows the evaluation method. The figure has two types of nodes: Root Node and Leaf Node. User sends a Search Query to the Root Node and receives the Search Result from the Root Node. The Root Node sends search requests to the Leaf Nodes (i.e. Leaf Node1, Leaf Node2) and receives search responses. The evaluation measures the search response time from the Search Requests sent till the Search Responses received.

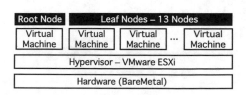

Fig. 8. Experiment environments.

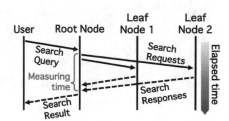

Fig. 9. Comparison of the block size.

Comparison of search response time and search target log bytes while block size increases: Figure 10 compares the response time in (a) and the maximum search target log bytes in (b) while the block size changes. The figure compares the parallel search method (*all-parallel*) to the proposed search method (*proposal*). The parallel search method finds all log blocks that match the date/time range. The proposed search method finds log blocks that match the date/time range. The search query has date/time-based filtering and keyword-based filtering as shown in Code 3. The length of a log message is around 400 [Bytes]. The response time is the median of 10 measurements, which is measured by the time module in Python.

Code 3. The search query for evaluation: filtering block size and date/time range.

```
s_bs=8, s_dt_begin="2021-12-13T10:21:50", s_dt_end="2023-02-13T10:21:50"
```

Figure 10**(a)** compares the search response time while block size increases. The x-axis is block size (unit: MB), and the y-axis is search response time (unit: seconds). The block size takes following values: {4, 8, 12, 16, 20, 24, 28, 32, 36, 40}. The response time of the proposed search method is 0.91 [s] faster than the parallel search method maximally, where the block size is 4 [MB]. The response time depends on the block size. As the block size increases from 4 [MB] to 24 [MB], the response time reduces. Response time gets shorter by reducing the total number of opened files. The existing study shows the same result [14]. As the block size increases from 24 [MB] to 40 [MB], the response time increases. The increase of response time results from increasing access target blocks per leaf node. As the block size decreases, the number of allocated blocks per leaf node decreases. On the other hand, increasing the block size increases the difference in the number of stored logs per leaf node compared to small block size. Thus, the difference of accessed blocks counts makes the response time increase.

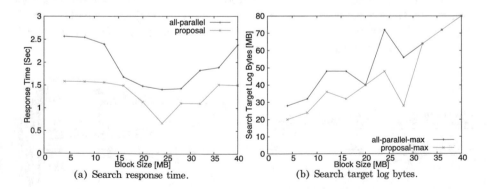

(a) Search response time.   (b) Search target log bytes.

Fig. 10. Comparison of the log placement method.

Figure 10(b) compares the maximum search target log bytes while block size increases. The x-axis is block size (unit: MB), and the y-axis is the maximum total search target log bytes in Leaf Nodes (unit: MB). The block size takes followed values: {4, 8, 12, 16, 20, 24, 28, 32, 36, 40}. As the block size increases, the maximum search target log bytes increases. The increasing response time is caused by the maximum search target log bytes increasing between 24 [MB] and 40 [MB] in Fig. 10(a).

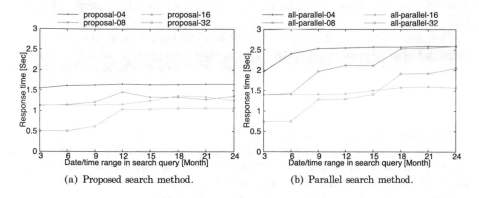

(a) Proposed search method.   (b) Parallel search method.

Fig. 11. Comparison of the log placement method.

Comparison of search response time while the number of target log bytes increases: Figure 11 compares the number of search target log bytes between (a) Proposed search method and (b) Parallel search method. The figure aims to recognize the effectiveness between the number of search target blocks and the search response time. The x-axis is the date/time range (unit: Month) which is specified in search query. It takes followed values: {3, 6, 9, 12, 15, 18, 21, 24}. For example, $x = 3$ means to find log messages within 3 months. The y-axis is response time. The legends indicate block size (unit: MB) and take followed values: {4, 8, 16, 32}. The response time is the median of 50 measurements, which is measured by the time module in Python. Figure 11(a) shows that the response time remains constant as the date/time range increases. The proposed method keeps the number of access blocks constant on search when the date/time range is extended in the search query. The search response time decreases as the block size increases from 4 [MB] to 32 [MB]. A large block size reduces Disk I/O overhead on a Linux file system. In order to decrease file system overhead, GFS which is created by Google adopts 64 [MB] as a block size [15]. Figure 11(b) indicates that the response time increases in the parallel search method with the date/time range increased. The response time depends on the number of accessing blocks on search. The number of accessing blocks increases on the parallel method as the search query's date/time range expands. Figure 11(a) and (b) indicate that the proposed search method faster than the parallel method in search response time. The reason of differ response time is that the proposed method is less than the parallel method in the number of accessing blocks on log search.

6   Discussion and Conclusions

The proposed method sets fixed block size. Thus, the number of log messages per block is homogeneous. While the data size of the block grows, the data size grows sequentially. The file size which can be read and written simultaneously depends on Disk I/O performance per node. Thus, block size has to be calculated from Disk I/O performance. One of the methods is *iostat* command which gets I/O performance.

Default length of block size is 16 [KB] on existing database systems such as MySQL (InnoDB) and PostgreSQL. This is because page cache size of file system uses multiple of 4 [KB]. For example, considering the case where a 16 [KB] block contains the format as shown in Code 1. When the length of a log message is 400 [KB] as shown Code 1, a block contains 40 messages $(16,000/400 = 40)$.

As block size increases, the response time is short. On the other hand, increasing the block size makes the placement problem. As the block size decreases, the number of log messages allocated on leaf nodes increases due to decreasing free space in the leaf node's local disk. Block allocation algorithm requires not only the block size but also free disk space on leaf nodes.

The issue is the slow response time for distributed tracing on the log search system. This study aims to reduce the search response time of two types of queries: filtering status-code, request-id. This study proposes a fast log search method for distributed tracing. The method makes the log blocks based on date/time and service-name in log messages. The log blocks are stored on several distributed nodes and retrieved on search. Experiment results show that the response time of the proposed method is maximally 0.91 s faster than the method of all parallel. This study contributes to reducing search response time on log search.

References

1. Montesi, F., Weber, J.: From the decorator pattern to circuit breakers in microservices. In: Proceedings of the 33rd Annual ACM Symposium on Applied Computing, pp. 1733–1735 (2018)
2. Mallanna, S., Devika, M.: Distributed request tracing using zipkin and spring boot sleuth. Int. J. Comput. Appl. **975**, 8887 (2020)
3. Fan, C.-Y., Ma, S.-P.: Migrating monolithic mobile application to microservice architecture: an experiment report. In: 2017 IEEE International Conference on AI & Mobile Services (AIMS), pp. 109–112. IEEE (2017)
4. Bento, A., Correia, J., Filipe, R., Araujo, F., Cardoso, J.: Automated analysis of distributed tracing: challenges and research directions. J. Grid Comput. **19**(1), 1–15 (2021)
5. Santana, M., Sampaio Jr, A., Andrade, M., Rosa, N.S.: Transparent tracing of microservice-based applications. In: Proceedings of the 34th ACM/SIGAPP Symposium on Applied Computing, pp. 1252–1259 (2019)
6. Alvarez, C., He, Z., Alonso, G., Singla, A.: Specializing the network for scatter-gather workloads. In: Proceedings of the 11th ACM Symposium on Cloud Computing, pp. 267–280 (2020)

7. Dan, A., Philip, S.Y., Chung, J.-Y.: Characterization of database access pattern for analytic prediction of buffer hit probability. VLDB J. **4**(1), 127–154 (1995)
8. Ciritoglu, H.E. , Batista de Almeida, L., Cunha de Almeida, E., Buda, T.S., Murphy, J., Thorpe, C.: Investigation of replication factor for performance enhancement in the Hadoop distributed file system. In: Companion of the 2018 ACM/SPEC International Conference on Performance Engineering, pp. 135–140 (2018)
9. Wei, Q., Veeravalli, B., Gong, B., Zeng, L., Feng, D.: CDRM: cost-effective dynamic replication management scheme for cloud storage cluster. In: 2010 IEEE International Conference on Cluster Computing, pp. 188–196. IEEE (2010)
10. Taware, U., Shaikh, N.: Heterogeneous database system for faster data querying using elasticsearch. In: 2018 Fourth International Conference on Computing Communication Control and Automation (ICCUBEA), pp. 1–4. IEEE (2018)
11. Krish, K., Khasymski, A., Butt, A.R., Tiwari, S., Bhandarkar, M.: Aptstore: dynamic storage management for hadoop. In: 2013 IEEE 5th International Conference on Cloud Computing Technology and Science, vol. 1, pp. 33–41. IEEE (2013)
12. Rex, R., Mietke, F., Rehm, W., Raisch, C., Nguyen, H.-N.: Improving communication performance on infiniband by using efficient data placement strategies. In: 2006 IEEE International Conference on Cluster Computing, pp. 1–7. IEEE (2006)
13. Li, W., Lemieux, Y., Gao, J., Zhao, Z., Han, Y.: Service mesh: challenges, state of the art, and future research opportunities. In: 2019 IEEE International Conference on Service-Oriented System Engineering (SOSE), pp. 122–1225. IEEE (2019)
14. Krishna, T.L.S.R. , Ragunathan, T., Battula, S.K.: Performance evaluation of read and write operations in hadoop distributed file system. In: 2014 Sixth International Symposium on Parallel Architectures, Algorithms and Programming, pp. 110–113. IEEE (2014)
15. Ghemawat, S., Gobioff, H., Leung, S.-T.: The google file system. In: Proceedings of the Nineteenth ACM Symposium on Operating Systems Principles, pp. 29–43 (2003)

Detector: Hierarchical Distributed Fault Detection Algorithm for Lattice Based Modular Robots

Edy Hourany[1], Benoit Piranda[1], Abdallah Makhoul[1(✉)], Julien Bourgeois[1], and Bachir Habib[2]

[1] FEMTO-ST Institute, University of Bourgogne Franche-Comté, CNRS, 1 Cours Leprince-Ringuet, 25200 Montbéliard, France
{edy.hourany,benoit.piranda,abdallah.makhoul,
julien.bourgeois}@femto-st.fr
[2] Holy Spirit University of Kaslik, 446, Jounieh, Lebanon
bachirhabib@usek.edu.lb

Abstract. Modular robots are robots built from multiple modules which can reconfigure the shape of the whole robot. This makes them a promising technology to realize highly adaptable and re-configurable robots. One of the main challenges in modular robotics is to detect and heal the system from a state of partial or complete disconnection. Partial disconnection can be achieved by removing the links between the modules. Complete disconnection can be achieved by removing all the connections between two group of modules. It is important for the modular robots to detect a failure between connections since the mobility and the functionality of the robot depend on it. This paper presents a novel approach to detect disconnections in lattice based modular robots. It detects the disconnections using a hierarchical algorithm. The leader of the system constructs a spanning tree, assigns a coordinate to each module and compares it to detect the disconnections. The simulations shows the ability to detect all partial disconnections. Furthermore, the complexity regarding the number of messages sent is linear against the number of modules in the system.

1 Introduction

The concept of programmable matter was first introduced by Christopher Langton in 1986 [9]. The term, "Programmable Matter" itself, was first introduced by Tommaso Toffoli and Norman Margolus [17] and has also been used in the context of artificial life, where it is referred to as digital organisms or artificial life.

Programmable matter is not a single robot or machine, but rather a network of all of the above [2]. It is a fabric that can change its physical properties in real time [11]. Programmable matter can respond to external or internal cues, or can be triggered by the completion of a task. It can grow and adapt, and self-assemble and self-repair [6, 18].

© The Author(s), under exclusive license to Springer Nature Switzerland AG 2022
L. Barolli et al. (Eds.): AINA 2022, LNNS 450, pp. 118–129, 2022.
https://doi.org/10.1007/978-3-030-99587-4_11

Programmable matter consists of a group of micro-robots, called granular computing elements. Each of these robots can move in space, sense the environment and communicate with each other. Each robot can communicate with its connected neighbor to exchange information about its own location and surroundings, and perform the computation in parallel. The robots are small, and can be manufactured in a highly parallel and distributed fashion. The difference between these robots and a traditional micro-robot is that these robots self-organize into a structure that is not readily predictable, and changes in response to a variety of environmental cues [18].

The benefit of programmable matter is that its capabilities are not limited to any particular design or configuration. The robots can be configured in a variety of shapes, sizes, and locations [6,11,18]. The robots can be added, removed or reorganized without having to re-design and re-manufacture the structure. The robots can even be built from other robots, or robots made from other materials. A programmable matter is more than a collection of objects with the ability to move and change shape.

A very frequent error in modular robots is disconnection. It can be caused by incorrect design, lack of care, or a defective module. Disconnections among robots can lead to collisions or a lack of mobility, which will make the robot unable to perform its function. Disconnections can also cause a loss of autonomy or even the robot's death. Therefore, it is crucial to detect disconnection among modular robots to prevent errors in the overall system. This paper proposes a novel hierarchical disconnection detection algorithm based on spanning tree that detects all partial disconnections in a lattice based modular robotic system.

The remainder of this paper is as follow: Sect. 2 presents the literature review, Sect. 3 presents the proposed algorithm, Sect. 4 shows the simulation results, and finally Sect. 5 concludes the paper.

2   Background

Modular robots are mobile robots built from multiple units, usually called modules or particles, capable of interactions with each other. These units are connected by links, which are able to transmit forces between the units. Several architecture arises from this technology such as hybrid architecture, chain architecture, and lattice architecture. Some examples of modular robots are shown in Fig. 1.

In modular robotics, the mobility of the modules is often not controlled by the computer itself, but the modules communicate among each other and decide among themselves how to move. Simple tasks such as collision detection, passing obstacles and shape formation become more complex with this approach.

Disconnections among modular robots can occur for a number of reasons: (1) a robot may move off-course, (2) a robot may collide with a fixed structure, (3) two robots may physically touch each other, (4) two robots may physically touch each other and then become disassociated from the system, and/or (5) a robot may disconnect from the system through no fault of its own. While

Fig. 1. Examples of modular robots: Fig. 1a represents a polybot introduced in [19]; Fig. 1b is a Tripod configuration of CONRO [3]; Fig. 1c is a Five YaMoR modules [10]; Fig. 1d is a lattice modular robot called ATRON [7]; Fig. 1e is a Rotating M-Block [13]; Fig. 1f is a 2D Crystalline [14] modular robots with extensible arms; Fig. 1g is a Superbot [15] modular robot in a humanoid configuration. Figure 1h M-TRAN [8] modular robot in 4-legged configuration; Fig. 1i is a SMORES [4] modular robot.

a modular robotic system is being operated, the system must ensure that all communications remains available. In order to achieve this, it is crucial to detect disconnections among the robots of the system. Such disconnections may, for example, result in the loss of valuable data that is being transferred among the robots, or in the loss of robot operation time.

A modular robotic system can be represented as an undirected graph, but traditional graph algorithms, such as Weakly Connected Components [5], can't be applied due to the system's distributed nature.

In [20], the authors presented a machine learning based algorithm to detect faults in distributed robotics system. Their work can be summarized as follows: Creation of data-sets of multi-robot system with faults and failures for the purpose of learning faults.

In [12], Fault detection over a wireless sensor network in a fully distributed manner. First, they proposed the Convex hull algorithm to calculate a set of extreme points with the neighbouring nodes and the duration of the message remains restricted as the number of nodes increases. Then they proposed a Naïve Bayes classifier and convolution neural network (CNN) to improve the convergence performance and find the node faults.

In [1] A distributed fault detection scheme for modular and reconfigurable robots (MRRs) with joint torque sensing was proposed. With the proposed scheme, the joint torque command is filtered and compared with a filtered torque estimate derived from the nonlinear dynamic model of MRR with joint torque sensing. Common joint actuator faults are considered with fault detection being performed independently for each joint module.

Modular robotics systems have limited memory and processing resources, a decentralized configuration, and a distributed architecture. Therefore, despite their accuracy and effectiveness, the previously described algorithms will not work.

This paper will be presenting a novel algorithm that detects disconnections in lattice based modular robots. The algorithm starts by constructing a spanning tree with the leader as root. Then, starting from the leaves, each module will send the list of none existing neighbors to their parents. Each parent will compare the received list with the list of the existing neighbors to detect any disconnections

3   Detector: Disconnection Detection

This section will introduce the proposed algorithm. Lets start by presenting some definitions and terminologies.

3.1   Definitions

Definition 1 (Leader). The leader is a module that was elected, using a leader election algorithm, before the execution of the proposed algorithm.

Definition 2 (Connected Modules). Connected modules are latched neighbors that are able to communicate throw their common interface.

Definition 3 (Partial Disconnection). A partial disconnection is a set of nodes that are not connected to each other but are able to communicate through other common nodes on the system. Figure 3a and 3b shows an example of partial disconnection.

Definition 4 (Complete Disconnection). A node is referred to be in a state of complete disconnection when it looses connections with all its neighbors. Making it impossible to be identified by the system and will be in a state of complete isolation. Figure 3c and 3d shows an example of complete disconnection.

Definition 5 (Lattice Architecture). The units of a lattice architecture link their docking interfaces at points into virtual cells of a regular grid. This network of docking sites is analogous to atoms in a crystal, and the grid to the crystal's lattice. As a result, the kinematical characteristics of lattice robots may be defined by their corresponding crystallographic displacement groups (chiral space groups). Typically, just a few units are required to complete a reconfiguration stage. Lattice designs provide a simpler mechanical design as well as a simpler computational representation and reconfiguration planning that can be scaled more readily to complex systems.

Definition 6 (Density). The density of a modular robotic system shows how scattered are the modules. The value of the density ranges between 0 and 1. A system is considered to be dense when its density value is close to 1. The coordinates assigned to the modules does not take into consideration the size of the module. Each module is considered to have one unit size. Equation 1 shows how to compute the density in a system. N represents the total number of modules. x_{max}, y_{max}, and z_{max} depicts the maximum coordinates in the lattice and x_{min}, y_{min}, and z_{min} depicts the minimum coordinates in the lattice.

$$Density = \frac{N}{(x_{max} - x_{min} + 1) * (y_{max} - y_{min} + 1) * (z_{max} - z_{min} + 1)} \quad (1)$$

Definition 7 (Potential Neighbors). A potential neighbor is defined as the not connected side of the robot. Two cases are presented, connection failure or absence of modules. Let's take module B in Fig. 2 as an example. Module B in this condition has two potential neighbors: The none existing one on the right side and the disconnected one-Module D-.

3.2 The Proposed Algorithm

This section will be describing the proposed solution in details. The algorithm is based on three phases.

- Phase 1: Spanning tree creation and coordinate distribution;
- Phase 2: Neighbors' Data Collection;
- Phase 3: Disconnection detection. Below is a detailed explanation of the phases.

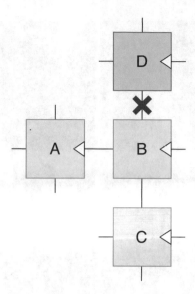

Fig. 2. Potential Neighbors

3.2.1 Spanning Tree Creation and Coordinate Distribution

The leader of the system initiates the algorithm. It will create a spanning tree containing all the modules in the system. The process starts by sending a recruitment message to all its neighbors. Once a module receives a recruit message, it will change the recruited flag to 1, to prevent future recruitment from other modules. Mark the interface that sends the recruit message as parent interface. Extract the coordinates from the received message and update the coordinates based on the received direction in order to generate its own coordinate. And finally, send a recruit message to all its neighbor except for its parent.

3.2.2 Neighbors' Data Collection

The proposed suggestion to find disconnections in a system is to gather information about the neighbours. The process starts with the leaves. Each leaf will generate the coordinates for its potential neighbors, add those coordinates into a vector - that will be called potentialNeighborsVector - and send this vector to its parent alongside their own ID and coordinates. Each parent will also generate the potential neighbors coordinates, check for possible disconnections - explained in Sect. 3.2 - then send the results to its parent. The same process will be repeated until the vector reaches the leader. Figure 4 shows an illustrative example of this phase.

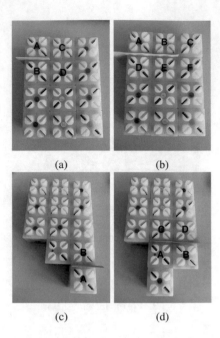

Fig. 3. Different examples of Partial and complete disconnections: Fig. 3a and 3b Represents a partial disconnection. In Fig. 3a. modules A and B are disconnected, but they still are part of the whole system since they can maintain their connection to the system through their neighbors C and D. Figure 3c and 3d are in a state of complete disconnection. In Fig. 3c, module A is disconnected from module B and module A is not connected to any other module. Therefore, module A is considered to be in complete disconnection from the system since there is no other way for the system to acknowledge its presence. Same case for Fig. 3d.

3.2.3 Disconnection Detection

Once a node receives from all its children the vector potentialNeighborsVector, it will compare with the list of coordinates of existing modules. If the node finds potential neighbors coordinates matching with existing modules, then there is a disconnection. Since each node have access to only its children, then the final decision of disconnection is actually at the leader's level.

4   Simulation Results

The proposed algorithm was implemented and evaluated using VisibleSim [16] a modular robots simulator. To study the performance of the proposed algorithm, the metrics that were taken into account in the simulations are the number of messages and the percentage of error detected (number of errors detected by our algorithm on the total number of errors in the system). Multiple simulations where done while varying different components at a time.

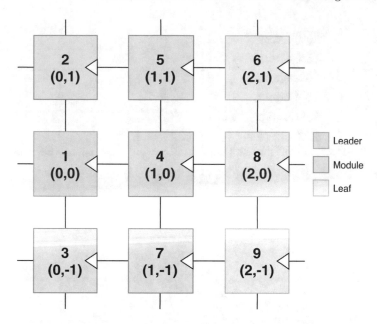

Fig. 4. Data collection in a dense system: Module 6 will add to the potentialNeighborsVector the following coordinates: $(2, 2)$ and $(3, 1)$, Module 8 will add $(3, 0)$ and Module 9 will add $(3, -1)$ and $(2, -2)$. They will then send this vector to their respective parents 5, 4, and 7. Module 5 will add $(1, 2)$, Module 7 will add $(1, -2)$, and Module 4 will not add anything since all its interfaces are connected. So on until this vector reaches the leader.

The first bash of simulations was for the same number of modules - In those simulations, the number of modules was 1000 -with different densities ranging from 0.18 to 1. All the shapes' configuration were randomly generated - For the exception of the cube since it is the only shape where the density is to its maximum-.

Figure 5 shows that when the density is increasing, the number of messages sent is also increasing. This is due to the fact that when the density is high, the modules are closer to each other, leading to an increased amount of neighbors per module wish will lead to a higher number of messages sent. Further more, Fig. 5 shows that whenever the percentage of errors in a system is higher, the number of messages sent is lower. The reason behind this is that when modules are disconnected they can not send messages which will decrease the total number of messages sent.

The second bash of simulations were done for the same density while increasing the number of modules. The chosen density for this simulation was 1, for the reason that it was the value where the system was sending messages the most. Figure 6a shows how the number of messages is increasing linearly with the number of modules. This shows that the complexity of the proposed algorithm, regarding the number of messages sent, is linear. It is also noticeable that

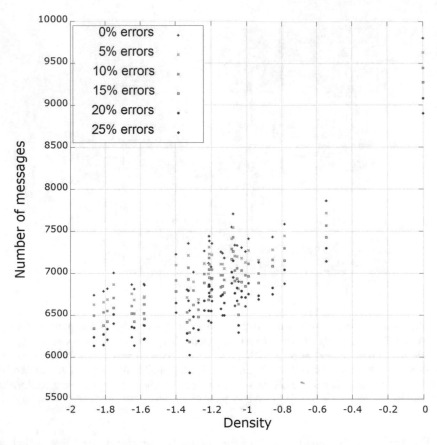

Fig. 5. This figure shows, on a logarithmic scale, how the number of messages sent in a 1000 3D system is changing with the density of the system. Each color represents the percentage of errors that existed in the system during the simulations. It is noticeable that the number of messages is increasing with the density. Furthermore, the number of messages decreases when the percentage of errors is increasing

when the number of errors is increasing in the system, the number of messages in decreasing. This is due to the fact that some messages are not able to be sent or delivered due to the errors encountered in the system.

During the previous simulations, the simulator collected the total number of errors detected by the proposed algorithm. Figure 6b shows the percentage of errors detected on the Y axes ranging from 90 to 100 against the density, on logarithmic scale, from -2 to -0.4. Five colors are shown on the graph. Each color represents the percentage of errors in the system. Figure 6b shows that for high densities, more errors are detected. A system with low density has less connected modules. Any occurring error might lead to a complete disconnection that is not detected by the proposed algorithm. Therefore, the higher the density the better the detection of errors is.

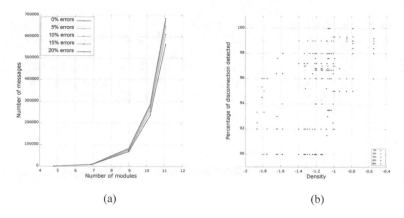

(a)   (b)

Fig. 6. These plots show: (a) the number of messages sent in the system according to the number of modules in a logarithmic scale, (b) the percentage of detected errors

5   Conclusion

This paper presented Detector, a novel distributed algorithm for a hierarchical fault detection in modular robots. The proposed algorithm is able to detect partial disconnections in a modular robotic system. To achieve such task, the algorithm creates a spanning tree having the leader as root. Each module generates its coordinates starting from the leader as origin. Then, each module - starting from the leaves - will send the potential coordinates of none connected interfaces to its parent. The parent will also generates the potential neighbors' coordinates and add them to the received list, compare the list of potential neighbors' coordinates with the list of the actual coordinates. Any matching coordinates means a disconnection is found. Then, the parent, sends the remaining results to its parent, so on until it reaches the leader who will have the final decision regarding the existence of disconnections if any.

The simulation results show an elevated rate of disconnection detection specially with a high density system. The accuracy of finding disconnections was ranging from 90% to 100%. Furthermore, the simulations also showed that the complexity of the algorithm regarding the number of exchanged messages remains linear.

Future work should consider the ability to detect if any disconnection happened during the construction of the spanning tree, keep the spanning tree updated during the movement of the robots and also the ability to detect the complete disconnections in a modular robotic system. For example, the addition of different method of communications such a wireless communication can help detect a complete disconnection that leads to an isolation of some modules from the whole system.

Acknowledgements. The authors would like to acknowledge the National Council for Scientific Research of Lebanon (CNRS-L) and the Holy Spirit University of Kaslik for granting a doctoral fellowship to Edy Hourany.

This work was partially supported by the ANR (ANR-16-CE33-0022-02), the French Investissements d'Avenir program, the ISITE-BFC project (ANR-15-IDEX-03) and the EIPHI Graduate School (contract "ANR-17-EURE-0002").

References

1. Ahmad, S., Zhang, H., Liu, G.: Distributed fault detection for modular and reconfigurable robots with joint torque sensing: a prediction error based approach. Mechatronics **23**(6), 607–616 (2013)
2. Bassil, J., Moussa, M., Makhoul, A., Piranda, B., Bourgeois, J.: Linear distributed clustering algorithm for modular robots based programmable matter. In: IEEE/RSJ International Conference on Intelligent Robots and Systems, IROS 2020, Las Vegas, 24 October 2020–24 January 2021, pp. 3320–3325. IEEE (2020)
3. Castano, A., Shen, W.-M., Will, P.: CONRO: towards deployable robots with inter-robots metamorphic capabilities. Autonom. Robot. **8**(3), 309–324 (2000)
4. Davey, J., Kwok, N., Yim, M.: Emulating self-reconfigurable robots - design of the SMORES system. In: 2012 IEEE/RSJ International Conference on Intelligent Robots and Systems, Vilamoura-Algarve, October 2012, pp. 4464–4469. IEEE (2012)
5. Dunbar, J.E., Grossman, J.W., Hattingh, J.H., Hedetniemi, S.T., McRae, A.A.: On weakly connected domination in graphs. Discrete Math. **167**, 261–269 (1997)
6. Hourany, E., Stephan, C., Makhoul, A., Piranda, B., Habib, B., Bourgeois, J.: Self-reconfiguration of modular robots using virtual forces. In: IEEE RSJ International Conference on Intelligent Robots and Systems (IROS 2021), Prague (2021)
7. Jorgensen, M.W., Ostergaard, E.H., Lund, H.H.: Modular ATRON: modules for a self-reconfigurable robot. In: 2004 IEEE/RSJ International Conference on Intelligent Robots and Systems (IROS), Sendai, September 2004, vol. 2, pp. 2068–2073. ISSN:null
8. Kamimura, A., Yoshida, E., Murata, S., Tomita, K., Kokaji, S.: A self-reconfigurable modular robot (MTRAN) - hardware and motion generation software. In: 5th International Symposium on Distributed Autonomous Robotic Systems, p. 10 (2002)
9. Langton, C.G.: Studying artificial life with cellular automata. Phys. D Nonlin. Phenom. **22**(1–3), 120–149 (1986)
10. Moeckel, R., Jaquier, C., Drapel, K., Dittrich, E., Upegui, A., Ijspeert, A.: YaMoR and bluemove - an autonomous modular robot with bluetooth interface for exploring adaptive locomotion. In: Tokhi, M.O., Virk, G.S., Hossain, M.A. (eds.) Climbing and Walking Robots, pp. 685–692. Springer, Heidelberg (2006). https://doi.org/10.1007/3-540-26415-9_82
11. Moussa, M., Piranda, B., Makhoul, A., Bourgeois, J.: Cluster-based distributed self-reconfiguration algorithm for modular robots. In: Barolli, L., Woungang, I., Enokido, T. (eds.) AINA 2021. LNNS, vol. 225, pp. 332–344. Springer, Cham (2021). https://doi.org/10.1007/978-3-030-75100-5_29
12. Regin, R., Suman Rajest, S., Singh, B.: Fault detection in wireless sensor network based on deep learning algorithms. In: EAI Endorsed Transactions on Scalable Information Systems (2021)

13. Romanishin, J.W., Gilpin, K., Rus, D.: M-blocks: momentum-driven, magnetic modular robots. In: 2013 IEEE/RSJ International Conference on Intelligent Robots and Systems, Tokyo, November 2013, pp. 4288–4295. IEEE (2013)

14. Rus, D., Vona, M.: A physical implementation of the self-reconfiguring crystalline robot. In: Proceedings 2000 ICRA. Millennium Conference. IEEE International Conference on Robotics and Automation. Symposia Proceedings (Cat. No.00CH37065), San Francisco, 2000, vol. 2, pp. 1726–1733. IEEE (2000)

15. Salemi, B., Moll, M., Shen, W.: SUPERBOT: a deployable, multi-functional, and modular self-reconfigurable robotic system. In: 2006 IEEE/RSJ International Conference on Intelligent Robots and Systems, Beijing, October 2006, pp. 3636–3641. IEEE (2006)

16. Thalamy, P., Piranda, B., Naz, A., Bourgeois, J.: Behavioral simulations of lattice modular robots with visiblesim. In: 15th International Symposium on Distributed Autonomous Robotic Systems (DARS 2021), Kyoto, Japan, June 2021

17. Toffoli, T., Margolus, N.: Programmable matter: concepts and realization. Phys. D Nonlin. Phenom. **47**(1), 263–272 (1991)

18. Xue, D., Qiang, L., Li, J.: Improving the morphology and control policy of self-reconfiguring modular robots in dynamic environment (student abstract). Proc. AAAI Conf. Artif. Intell. **35**, 15939–15940 (2021)

19. Yim, M., Duff, D.G., Roufas, K.D.: PolyBot: a modular reconfigurable robot. IEEE Int. Conf. Robot. Autom. **1**, 514–20 (2000)

20. Mahmoud Youssef, Y.: Inducing rules about distributed robotic systems for fault detection & diagnosis. In: Proceedings of the 20th International Conference on Autonomous Agents and MultiAgent Systems, pp. 1845–1847 (2021)

ManufactSim: Manufacturing Line Simulation Using Heterogeneous Distributed Robots

Benoit Piranda[1]([✉]), Ishan Gautam[1]([✉]), Jerome Meyer[1,2]([✉]),

Anass El Houd[1,2]([✉]), and Julien Bourgeois[1]([✉])

[1] University of Bourgogne Franche-Comte, FEMTO-ST Institute, CNRS, Montbliard, France
{benoit.piranda,ishan.gautam,jerome.meyer,anass.houd,
julien.bourgeois}@femto-st.fr
[2] Faurecia, Bavans, France
{jerome.meyer,anass.houd}@faurecia.com

Abstract. The creation of current assembly lines can benefit from the new advances made in the fields of Computer Science and the Internet of Things (IoT) to increase their flexibility and improve their reliability. There are assembly line simulators developed for this purpose. However, these simulators have been designed to model every detail of the line and take hours to be done. The aim of this paper is to introduce a faster and more accurate computer-based solution - *ManufactSim*- allowing the simulation of a real production system. This implementation derives from a behavioral modular robots simulator enhanced with a 3D display option. The results show that *ManufactSim*'s performances are above the standards with an execution time less than 11 s for 8 h shift running on a CAD computer. Our developed solution is able to face this challenge with an highly accurate and efficient simulator without compromise. The performed benchmarks show that we obtain a robust and agile tool needed for a global future solution based on Machine Learning. The benefits of this contribution will permit to automate the generation of industrial assembly lines while caring on multiple optimization criteria.

Keywords: Manufacturing · Industry 4.0 · Digital twin · Behavioral simulator · Distributed algorithms · Operational research · Optimization methods

1 Introduction

Industrial production is a constant adaptation process whose aim at increasing quality, speed or cost of the production, to cite only a few. Automation science has developed numerous theoretical tools which have been used to optimize the manufacturing process. However, designing a new manufacturing line still faces lots of challenges. A product is made of different parts assembled all together and the assembly order can vary, there are multiple choices of robots, some tasks can be made by humans and storage can placed between assembly units. The number of combination is therefore huge. Furthermore, the cost function of each solution has also many parameters like

L. Barolli et al. (Eds.): AINA 2022, LNNS 450, pp. 130–140, 2022.
https://doi.org/10.1007/978-3-030-99587-4_12

the employee cost, the machine cost or the production rate. Aside from these challenges, deep learning is becoming a prevalent solution in many fields of optimization. Its capacity to digest huge amount of data and to solve complex problems have broadened its scope of use. In manufacturing, deep learning has been successfully applied to vision, Automated Guided Vehicle (AGV), and in many other domains but rarely to the design of manufacturing lines. We propose to develop a complete environment for helping manufacturing line designers.

A machine learning algorithm will be used to generate the scenario and to optimize the different parameters (cf. Fig. 1). To do this, it will need to quickly evaluate the efficiency of the scenario. There exists plenty of manufacturing line simulation which could be used to do like [12] or [7], except that these simulators have been designed to model every details of the line and a simulation could take hours to finish. What we need, in this context, is a different type of simulator which does not exist to the best of our knowledge. That is why, we propose in this article to develop *ManufactSim*: a behavioral discrete event simulator able to simulate a real manufacturing line in a few seconds. The simulation and visualization engine is a fork of *VisibleSim* [10], a modular robots behavioral simulator and as a bonus, this visualization engine is able to display and animate the manufacturing line in 3D.

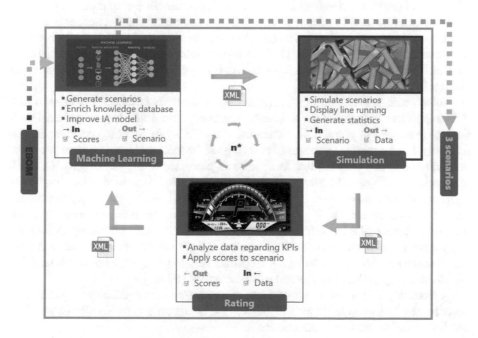

Fig. 1. Optimization solution architecture

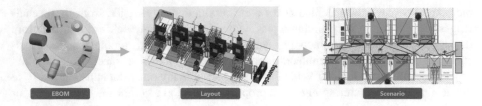

Fig. 2. Manufacturing environment

In Sect. 2, we will present some background information about the industrial context of this study, followed by a state of the art in manufacturing line simulation in Sect. 3 . Section 4 will present our approach to enable high-speed simulation of manufacturing lines and Sect. 5 will report our results showing that our environment is fast, precise and adaptable. Section 6 will conclude the article and will present some future work.

2   Industrial Context

Three components are identified in order to design a manufacturing process: the product's *Engineering Bill Of Material* (EBOM), the manufacturing environment and the scenario as presented in Fig. 2. The EBOM contains the list of parts to assembly with their constraints and parameters. The line layout sums up the physical manufacturing environment while the manufacturing scenario is mapping the entities interactions organized into process.

Today, each manufacturing scenario is designed manually by a *Process Architect*. This expert has to dispatch the assembly operations into a line layout and deal with an explosion field of possibilities. Then the scenario is presented to the *Plant Manufacturing Leader* who is in charge of carrying on the manufacturing process. Next the scenario is updated with the last feedback and presented to the job experts. Once all feedback are collected, the scenario get a last optimization for a final validation by the *Program Manufacturing Leader* who is in charge of respecting the engagement contracted with customers. The Fig. 3 illustrates three states of the combinatorial explosion for a seven parts EBOM assembled in four stations which results on five operations. So that for each state respectively loading number N_L and sub-assemblies number N_S we compute the pair $\{N_L, N_S\}$. We get the optimal solution $\{11, 5\}$, all possible loadings $\{64, 5\}$ and all possible loadings and sub-assemblies $\{64, 27\}$.

A assembly line is the result of a set of entities enumerated below, combined to describe how to assemble the pieces, manage the interactions between robots, workers, and finally define evaluation parameters.

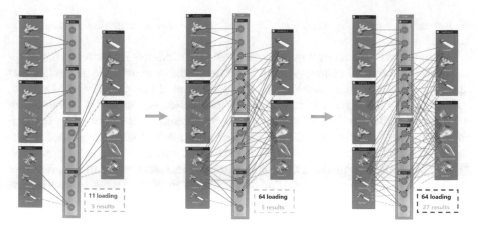

Fig. 3. Combinatory explosion

An EBOM is a list of sub components which are composing a manufactured product as illustrated in the Fig. 2. The assembly is following a defined process resulting from a multiple tasks succession. Many parameters and constraints are defining the EBOM such as a volume to produce per year, a target assembly cycle time, some customers datum.

A layout is a production line footprint describing the physical manufacturing environment aspects. Equipment and humans workers also called entities are listed in order to design the assemblies flows. The parameters and constraints of a layout can be defined by a list of equipment including cells, machines robots and tooling, a sum-up of storage and buffers, a way of working (clockwise, anti clockwise), an amount of workers.

A manufacturing scenario is a list of actions which defines the interactions over the line's layout entities. The scenario is a macro view of every steps needed to complete an assembly.

A flow is a continuity of actions performed by each resources. In our case we've got three types of interdependent flows such as human operator, robot and product. The KPIs for Key Indicators Performance are used in the industry for evaluating the efficiency rate. In an highly concurrency environment this criteria is required in order to keep an eye on the competitiveness. Our main KPI is focused on the actions duration performed by the line's layout resources. The load balancing is the way to evaluate the workload balancing between resources.

3   Related Work

Digital tools are currently used to manage the new processes creation such as "*3DEx-perience* platform" [2] from *Dassault Systems*. These tools are offering a wide range of services for industries by covering the product life cycle management. Considering the aim of this tools is to bring a complete overview of the process without generating optimized manufacturing scenarios, we assume that an extra automation tool might be helpful.

So that, several works were done by using Machine Learning methods and more precisely Reinforcement Learning. After a selection on the most accurate research, a first approach from 2018 was turned around the application of Google's Deep Q-Learning architecture by Waschneck et al. to optimize production scheduling in a multi-agent environment. Their method is applied to a dynamic and complex job shop environment consisting of work centers with different constraints, multiple machines of different types and multiple products [11]. Then a second research was done in 2019 on "Design, implementation and evaluation of reinforcement learning for an adaptive order dispatching in job shop manufacturing systems" [6]. In 2021, a work is covering the topic "Designing an adaptive production control system using reinforcement learning" [5]. In another contribution to scenario simulation and statistics generation, Kampa et al. [4] proposed an approach that allows a better representation of real production processes and assembly lines by using discrete event simulation methods.

A survey published in 2020 suggests that 75% of possible research areas in the application of machine learning to improve production planning and control are barely explored or not addressed at all [3], which is the case for the item "Smart Design of Products and Processes" in which we propose to develop this work.

4   Our Proposal

In order to simplify the complexity of the possible operations combinations (see Fig. 3), we have developed a manufacturing optimized scenario generation tool. We focus on the simulator which is a single part of a whole solution.

The goal is to output multiple manufacturing data for different cases as the volume of part produced or the shift duration. Three components are necessary to perform a simulation: a product, a line layout and a manufacturing scenario. Each pair of single manufactured product combined with different line layout will propose multiple manufacturing scenarios. Each of them will have to be simulated in order to evaluate them efficiency. In a addition to the simulation, a second API will analyze and score the generated data for each scenario. Then global solution will be able to perform a efficiency benchmark over each manufacturing scenario.

VisibleSim [10] is a behavioral simulator designed to simulate distributed programs running in modules composing distributed robots. This tool is useful to define large sets of modules placed in a lattice that are all running the same code. Each module is able to communicate with its neighbors, i.e. the modules that are placed in a direct neighbor cell.

This simulator really executes C++ codes that are ran by each module (we call them *Block Codes*) and simulates the real world, including point to point communications taking into account the delay of transmissions, sensors and actuators events due to the interaction with the simulated world.

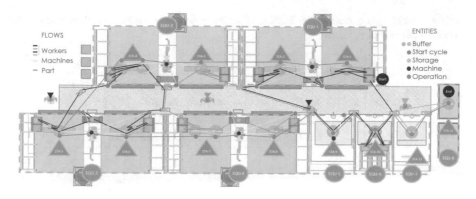

Fig. 4. Manufacturing line assembly flows example

VisibleSim has been evaluated with many kinds of robots, some of them exist in real and then allow to compare real and simulated value. For example, experiments in previous works [9] shows the very good capability of *VisibleSim* to simulate real distributed programs running in real *Blinky Blocks* robots.

We propose here to adapt *VisibleSim* in order to simulate precisely the production lines in order to extract many KPI values. The update concerns three items: the replacement of the lattice which allows to define which modules are connected, the way of writing the *Block Code* and the graphic representation of the system.

Concerning the definition of the graph, a connection must be defined between two elements that exchange parts of the conception during the assembly. These connections are bidirectional that allows to transmit a piece from a module to another one and receive an acknowledgment in the same canal. The list of connections that correspond to the edges of the flow presented Fig. 4 is enumerate in an XML files. As a consequence, the number of connections is much less important than the one defined by classic robots used in *VisibleSim* (they may have up to 12 neighbors by module). As the consequence, it reduces the number of resolution steps necessary to simulate the distributed system.

Programming robots with *VisibleSim* consists in writing a C++ code that describes what the module must do at starting and when receiving a message from a neighbor module or an event from a sensor. We need here define a more flexible method, adapted to the global workflow.

To do so in *ManufactSim*, we divide the treatment in two parts: first in the internal *Block Code* we describe the behavior of the agent using C++ for a list of actions, it mainly consists in sending messages or waiting for messages to communicate with the neighborhood. This part corresponds to the knowledge of the agent, for instance when a worker need to perform an action into a station, he has to first check if the door is open otherwise he has to wait. The second part is stored in an XML file, it consists in a succession of calls of the previous actions detailed for each agent.

Table 1. List of agents running on *Block Codes*, colors in the last column are the color of the corresponding object in Fig. 5

Agent	Layer	Description	Color
Station	0	Base component where parts are assembled	White
Buffer	1	Storage for sub assembly next to station	Green
Worker	2	Human workers interacting with elements	Gray
Robot	3	Welding machine which performs operations	Blue

Then in *ManufactSim*, the *Block Code* executes the list of actions which are stored in the XML file. This list is initially copied in the local memory of each module and then consumed respecting the order. This decomposition of the *Block Code* allows to evaluate different scenarios, described in the XML file without recompiling the simulator.

Concerning graphical representation of the configuration, *ManufactSim* shows the modules freely placed in several overlapping layers. For example a *robot* can be placed inside a *station* just by placing it in a higher layer than the one used by the *station*. Table 1 shows a part of the list of agents with their color and layer used in Fig. 5. This figure shows a screenshot of graphical interface of *VisibleSim*, where each agent is represented by a distinctive 3D model and a specific color. This scene is animated according to the message and events and can be freely observed during the simulation. Two simulation modes are available, one respecting the real time showing the actions played with the real duration and the other running the simulation as fast as possible to get the simulation results quickly.

The tasks list is the XML formalism of a manufacturing scenario attached to a line layout. Every action is mapped to a *Block Code* which can be an agent so that simulation can be performed. The XML code is organized with global parameters, such as the line details and the list of parts with them corresponding welding.

Fig. 5. *ManufactSim* graphic interface describing the arrangement of the manufacturing line.

The second part of the file describes the manufacturing scenario through a set of instructions according to the manufacturing equipment. Each equipment contains one or two stations which are assigned to a human operator. The file structure is respectively made of XML tags, i.e. 'vcell' containing 'station' and 'robot'. In addition, the tag 'action' defines and assigns the tasks to each resource by mapping it with the *Block Codes*.

The problem of finding optimal manufacturing scenarios is classified as NP problem: we can easily and quickly check if a candidate solution is indeed a possible scenario but we cannot prove if it is the unique best solution [1]. Our search for an optimal production scenario can be assimilated to a combinatorial optimization problem which consists in finding one of the best solutions in a set of feasible manufacturing scenarios [8].

A scenario is modeled by a set of parameters/variables presented by the following feature matrix M_s.

$$M_s = \begin{bmatrix} a_{11} & a_{12} & \cdots \\ \vdots & \ddots & \\ a_{K1} & & a_{KK} \end{bmatrix}$$

The feature vectors a_{ij} are chosen in such a way that they directly affects the KPIs, which means that a well-chosen combination of these parameters will result in optimal production performance.

5   Experiments

In this section, we are performing several experiments between our simulator and the *3DExperience* tool. This demonstrates the high speed and the accuracy as well as agility for future investigations. The experimental environment is powered by an Intel i7 CPU, 2.60 GHz 64-bit, Windows 10 operating system embedding 32 Gb of RAM memory.

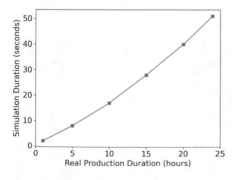

Fig. 6. Simulator execution time.

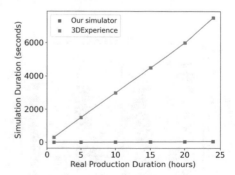

Fig. 7. Simulator versus *3DExperience*.

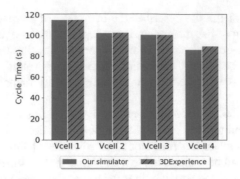

Fig. 8. Accuracy between simulator and *3DExperience*

This first experiment concerns the performance of our solution. Figure 6 highlights the fact that our simulator is fast regardless of the number of production hours simulated while Fig. 7 compares the simulation duration for both solutions. Regarding our experimental environment, we simulate 24 hours of production in 46 seconds while *3DExperience* takes upper than 2 hours. Therefore, our simulator proves to be at least 200 times faster than *3DExperience*.

In order to evaluate the accuracy of the proposed solution, Fig. 8 compares the results for both solutions. The Mean Square Error (MSE) is around 3,04 seconds resulting from updates applied on the tasks duration from the original *3DExperience*'s scenario.

Finally we can notice the agility of our method, Figs. 9 and 10 are showing the simulation results for 3 different line layout configurations by varying the amount of Vcells or operators. The first figure shows the impact on the workload rate while the second one is focused on the cycle time. Figures 11 and 12 are showing the simulation results for 3 different manufacturing scenario configurations by changing the job scheduling arrangement. The first figure presents the impact on the workload rate while the second one focuses on the cycle time.

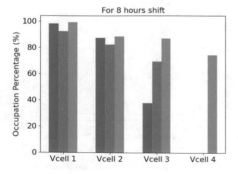

Fig. 9. Layout workload comparison.

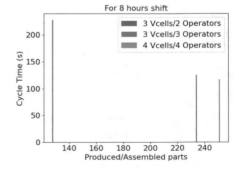

Fig. 10. Layout cycle time comparison.

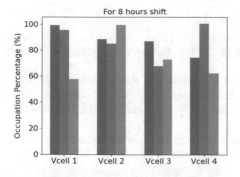

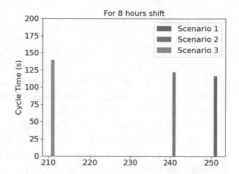

Fig. 11. Scenario workload comparison.   **Fig. 12.** Scenario cycle time comparison.

So that we show that our simulator is able to handle for the same EBOM, different layouts and scenarios to be generated by the artificial intelligence.

6   Conclusion

In summary, we can conclude that our manufacturing line simulation approach provides a stable, fast and accurate computer-based alternative ensuring a good yield for complex manufacturing flows.

The performed experiments show that our solution delivers accurate results for 8 hours shift during less than 11 seconds running on a standard CAD computer comparing to the current tools that can take more than 2 hours for equal results. In remains to be noted that the cycle time is only a study case that could be generalized to other KPIs without ensuring changes and adaptations. It should also to be noticed that an advantage of our solution is its configurability and its agility. Integrating other functionalities and methods will modify the simulator's parameters in order to adapt and optimize the whole system.

In our future work, we plan to implement additional technical features: First, embedding more KPIs for the assembly line scenarios evaluation. In this case, we might use a multi-objective evaluation that takes into account the economic constraints extending the cycle time: i.e. for the high cost labor countries, we will mainly seek to minimize the number of operators and vice versa. A other item concerns the automatic generation of scenarios to select the best ones using artificial intelligence techniques. This method should have the learning ability without relying constantly on experts. And as we will also implement a method for visualizing post-simulation results. We suggest to generate at the end of the simulation a complete report containing diagrams that allow a better analysis of the performance indicators of the scenario.

Acknowledgemnts. This work was done as a part of a CIFRE (2018-0927) project with Faurecia, funded by the Ministry of Higher Education and Research of France, managed by the Association Nationale de la Recherche et de la Technologie (ANRT).

References

1. Arora, S., Barak, B.: Computational Complexity: A Modern Approach. Cambridge University Press (2009)
2. Barth, A.: 3D experiences – dassault systèmes 3DS strategy to support new processes in product development and early customer involvement. In: Kovács, G.L., Kochan, D. (eds.) NEW PROLAMAT 2013. IAICT, vol. 411, pp. 24–30. Springer, Heidelberg (2013). https://doi.org/10.1007/978-3-642-41329-2_3
3. Cadavid, J.P.U., Lamouri, S., Grabot, B., Pellerin, R., Fortin, A.: Machine learning applied in production planning and control: a state-of-the-art in the era of industry 4.0. J. Intell. Manuf. 1–28 (2020)
4. Kampa, A., Gołda, G., Paprocka, I.: Discrete event simulation method as a tool for improvement of manufacturing systems. Computers 6(1), 10 (2017)
5. Kuhnle, A., Kaiser, J.-P., Theiß, F., Stricker, N., Lanza, G.: Designing an adaptive production control system using reinforcement learning. J. Intell. Manuf. 32(3), 855–876 (2020). https://doi.org/10.1007/s10845-020-01612-y
6. Kuhnle, A., Schäfer, L., Stricker, N., Lanza, G.: Design, implementation and evaluation of reinforcement learning for an adaptive order dispatching in job shop manufacturing systems. Proc. CIRP 81, 234–239 (2019)
7. Kursun, S., Kalaoglu, F.: Simulation of production line balancing in apparel manufacturing. Fibres Textil. Eastern Eur. 17(4), 75 (2009)
8. Papadimitriou, C.H., Steiglitz, K.: Combinatorial Optimization: Algorithms and Complexity. Courier Corporation (1998)
9. Piranda, B., et al.: Distributed prediction of unsafe reconfiguration scenarios of modular robotic programmable matter. IEEE Trans. Robot. 1–8 (2021)
10. Thalamy, P., Piranda, B., Naz, A., Bourgeois, J.: Visiblesim: a behavioral simulation framework for lattice modular robots. Robot. Autonom. Syst. 103913 (2021)
11. Waschneck, B., et al.: Optimization of global production scheduling with deep reinforcement learning. Proc. CIRP 72, 1264–1269 (2018)
12. Zahraee, S.M., Golroudbary, S.R., Hashemi, A., Afshar, J., Haghighi, M.: Simulation of manufacturing production line based on arena. In: Advanced Materials Research, vol. 933, pp. 744–748. Trans Tech Publ (2014)

Sports Data Management, Mining, and Visualization

Bamibo C. Isichei[1], Carson K. Leung[1(✉)] (iD), Lam Thu Nguyen[1],
Luke B. Morrow[1], Anh Tuan Ngo[1], Trang Doan Pham[1],
and Alfredo Cuzzocrea[2]

[1] University of Manitoba, Winnipeg, MB, Canada
kleung@cs.umanitoba.ca, Carson.Leung@UManitoba.ca
[2] University of Calabria, Rende, CS, Italy

Abstract. Data are everywhere. Examples include sports data. Embedded in these data is implicit, previously unknown and potentially useful information or knowledge to be discovered. In this paper, we present a solution for sports data management, mining and visualization. In particular, we focus on basketball data. Basketball is a culture and is respected by fans around the world. Ever since its birth, basketball has changed drastically. Under such effects, basketball discussion and analysis evolved as well. Our solution adapts three different approaches for predicting the win. Evaluation on real-life basketball data show the effectiveness of our solution.

Keywords: Data mining · Big data analytics · Social networks · Sports analytics · Basketball

1 Introduction

In the era of advanced technology, numerous amounts of data are generated and extracted from every aspect of life such as transportation [1–3], social networks [4,5], healthcare [6–8], as well as games [9] and sports. As an increase in data resources, people want to collect and analyze those to obtain some useful information—such as patterns, correlations—which can be used to predict outcomes. This calls for data science [10,11]—which makes good use of techniques like data mining [12–17], machine learning [18], and mathematical modelling [19]. Data have proven useful in providing unseen information that, when obtained, can worth more than a sum of money. People have used this to discover insights about many areas—namely web, social network and sports. Statistics and analytics have done a great job progressing and, with the help of data mining technology, we witness theories applied into real work and achieve meaningful results.

Sport data mining [20–22] has been continuously expanded throughout years along with prediction models. Numerous researches have been conducted to find out a precise system. These systems can calculate injury probability, coach's influence, especially the winning probability. We build our foundation on the

L. Barolli et al. (Eds.): AINA 2022, LNNS 450, pp. 141–153, 2022.
https://doi.org/10.1007/978-3-030-99587-4_13

work of Kubatko et al. [23] on the branch of quantitative analysis of sports, basketball being the center of focus. This technique makes use of mathematical and statistical perspective to understand the particular phenomenon. In other words, the situation and event are presented in terms of a numerical value. This representation is true to some degree. Thus, it is complicated to generate an accurate prediction. Naturally, we predict based on gut or experience, which sometimes can be groundless and unreliable. So, when using quantitative analysis, analysts can estimate better with the help of numerical and statistical data and hope to arrive at a reliable decision. We understand that researchers often evaluate basketball outcomes by investigating historical games. Moreover, to build reliability and accuracy, quantitative analysis makes use of other fields like psychology, economics, etc.

Kubatko et al. defined several terminologies and many other elementary concepts of basketball as well as some winning percentage formulas, which were invented a while back but are still valuable for any prediction models—namely Pythagorean method and bell curve method. As these are different methods, the effectiveness and efficiency also varies. When comparing them on different grounds, analysts are a step closer to the best truth. Hence, our focus is on winning percentage formulas and visualization of them to produce comparative factors between each formula. Data visualization represents data by statistics charts and graphics to investigate the sport events [24]. Different visualizations reveal blind spots and unthought of relationships while casual numbers on papers could not.

Given the applicability of data mining and the goal of Kubatko et al. to establish and ground several terminologies and definitions of elementary concepts, we are able to carry out our work on examining and surveying the effectiveness of original prediction models on a bigger sample size for years to come. Here, we investigate the Pythagorean winning percentage [23], which is highly efficient for its simplicity and widely adapted to other sports with various adjustment attempts from statisticians. Under further studies, it presents lower explainability and inspires researchers to close the gap between such a prediction with real life results as linear as possible with other models that perform less effectively but account for more factors and supply comprehensible patterns for investigation.

Our *key contributions* of this paper include our sports data mining solution. It adapts three different approaches for presenting, measuring and comparing prediction. We also evaluate our solution with real-life data, build visualization components to reflect their efficiency. Then, we analyze and report plot holes on where they fall short and derive at the correlations of factors that affect the prediction's accuracy to account for the foundations of past researches based from them.

The remainders of the paper are as follow. The next section gives some background information about basketball and introduces winning probability formulas, which are going to be compared. Section 3 describes our sports data mining solution. Section 4 presents our evaluation results. Conclusion and further study opportunities are presented in Sect. 5.

2   Background

In this section, we present three approaches for computing winning percentages. We adapt these approaches for our sports data mining, especially for predicting winning percentage for basketball data.

2.1   Bell Curve

Oliver [25] introduced a formula to calculate the winning percentage prediction based on statistics. He assumed that team's points scored and allowed are normally distributed. As normal distributions can be subtracted from each other to form another normal distribution, he took above normal distributions to create *net points* (which is also normally distributed). By using *Z-score*, the winning probability equals the probability of a random variable is distributed to a value less than Z-score, which can be computed as $Z\text{-}score = \frac{X-\mu}{\sigma}$, where (a) X is an element, (b) μ is mean value, and (c) σ is standard deviation. Z-score estimates the position of X from mean value. By setting X equals 0 for net points, Z-score calculates the position and find out the percentage of when team's points scored is larger than team's points allowed. The Z-score formula for net points is:

$$Z - score_{net\ points} = \frac{PPG_t - PPG_o}{\text{stdev}(PPG_t - PPG_o)} \tag{1}$$

where (a) PPG_t is points scored per game of team t, (b) PPG_o is points allowed per game (i.e., points scored per game by opponents), and (c) stdev($PPG_t - PPG_o$) is standard deviation of net points. So, the equation for predicting winning percentage is:

$$Win\% = NormsDist \left[\frac{PPG_t - PPG_o}{\text{stdev}(PPG_t - PPG_o)} \right] \tag{2}$$

where *NormsDist* function represents the area under the normal distribution to the left of the Z-score.

2.2   Linear Regression

Linear regression [26] is a statistical analysis, which depends on modelling the relationship between two variables. The main objective of using linear regression is to examine if (a) the independent variable is useful or successful in predicting the outcome variable and (b) which independent variables are significant in predicting the outcome. Examples include using physical appearance of soccer players (e.g., weight, height) and checking if there is a relationship with speed. From popular studies, shorter players tend to have more speed, and taller ones tend to be slower.

When a company identifies fraudulent activities online or business predicts what the demand for a product will be, data science is the oil that enables these predictions or findings. These types of data science innovations are based on regression. Regression analysis is the understanding the relationship between an input and output variables.

Regression helps to find a relationship between each individual statistics with the collective team's performance. Subsequently, an estimation to how the team should perform on that particular season can be computed. The accuracy of this method can be determined by comparing the estimation with the actual performance of the team.

2.3   Pythagorean Expectation

Pythagorean expectation[1] was originally devised—by American baseball writer Bill James—to estimate the percentage of games a baseball should have won using some of the team's statistics (e.g., runs they scored and allowed). The comparison of the actual percentage won by the team with the percentage gotten from the Pythagorean expectation, we can get the current form of the team—i.e., if the team is over-performing or under-performing. Hence, the win ratio for *baseball* can be computed as:

$$win\ ratio = \frac{runs\ scored^2}{runs\ scored^2 + runs\ allowed^2} \tag{3}$$

Observed that the New York Yankees (a professional baseball team) scored 897 runs and allowed 697 runs in 2002. Using Eq. (3), the Yankees should have a $win\ ratio = \frac{897^2}{897^2 + 697^2} = 62.35\%$ (i.e., should have won 62.35% of their games). In other words, based on 162 games in a season, the Yankees should have won 101.01 games but actually won 103 games and lost 58. So, to improve the accuracy, statisticians explored a better exponent for the equation and arrived at 1.83 to be used for baseball, i.e., $win\ ratio = \frac{runs\ scored^{1.83}}{runs\ scored^{1.83} + runs\ allowed^{1.83}}$.

Along this direction, an American sports executive Daryl Morey adapted Pythagorean expectation to basketball but with another exponent. The exponent 13.91 was deemed the best for basketball because it provided an acceptable model in the prediction of win-loss percentage and has been used internationally in that manner [27]. In other words, the modified equation for *basketball* became:

$$win\ ratio = \frac{points\ for^{13.91}}{points\ for^{13.91} + points\ against^{13.91}} \tag{4}$$

3   Our Sports Data Mining Solution

3.1   Data Collection

To mine sports data, our sports data mining solution collects data from a wide variety of rich data sources, which include statistic websites for sports. For instance, `basketball-reference.com` is a well-known data source for basketball. The request returned by basketball-reference has an easy to read and

---

[1] http://thegamedesigner.blogspot.com/2012/05/pythagoras-explained.html.

manipulate to extract table tags and get the necessary columns for calculation. For the Pythagorean expectation, we are especially interested in mining home team points scored and the points their opponent on that game scored. These values will go into the equation to compare with real results, which are game results on a whole season scale.

Challenges of managing data spanning over a long period of time (e.g., a 10-year span) include inconsistency in data (partially due to changes in franchise name throughout the course of history). For example, New Jersey Nets became Brooklyn Nets in 2012 when the team moved to back Brooklyn. As another example, the Charlotte Bobcats was renamed Charlotte Hornets in 2014. To handle these challenges, our solution keeps track of these name changes.

In terms of implementation, our solution first extracts columns of interest. It then puts these columns together in a DataFrame class in the Python Data Analysis Library (pandas) to generate tables for 30 teams—29 in the USA and 1 in Canada (namely, Toronto Raptors)—across several years.

3.2   Outcome Prediction

After collecting the data collection, our sports data mining solution adapts three prediction techniques. Specifically, it performs predictions with several years worth of games. For example, the National Basketball Association (NBA)—which is a professional basketball league in North America—consists of 30 teams. As each team playing 82 games per year, there are 2,460 games per year. Hence, for a 5-year period, the total game count is around 12,300 in the production of these predictions.

Bell Curve Prediction. First, our solution adapts Eq. (1) for bell curve prediction to compute the number for each team in a season from a period of several years. Each prediction is made by looking at 82 games for a given team in a given season. The prediction is measured against the actual win percentage that the team experienced that year:

$$Win\% = NormsDist \left[\frac{PPG_t - PPG_o}{\text{stdev}(PPG_t - PPG_o)} \right] \tag{5}$$

where (a) PPG_t is the sum of all the points the team scored in that season, (b) PPG_o is the sum of all the points that were scored on the team in question, and (c) stdev($PPG_t - PPG_o$) is the standard deviation of an array containing the difference in points the team scored subtracted by the points that were scored on them. The normal distribution function *NormsDist* assumes a location of 0, and a scale equals to the average number of points the team in question scored in that season. Using this equation, our solution calculates the predicted win rate for a team in a given season.

In terms of implementation, we first create a list containing the difference in points scored vs. points allowed for the team in question. Next, we use Python Data Analysis Library (pandas) built-in **sum** and **mean** functions on the points and opposing points columns, which give us the total points, total points allowed, and average points per game for a team over a given season. Using these variables, we calculate and format the expected win rate for said team in a single season using Eq. (5). Moreover, the standard deviation operation can be done by using Python's built in statistics module. The normal distribution function comes from the SciPy module, which is a Python library for scientific computing and technical computing. The normalization method required a location, which is defaulted to 0. We use a scale factor equal to the average number of points scored for the team in question for the season. The rationale of using this value for the scale factor is that (a) if a team is scoring more than average, then they would have better odds of winning, and (b) if they were scoring less than average, they would have a worse chance of winning.

Linear Regression Prediction. As an alternative, our solution adapts a linear regression model:

$$y = 0.50 + 0.000390561470 x_{PF} - 0.000390707588 x_{PA}. \tag{6}$$

This linear regression model was trained with two explanatory variables x_{PF} and x_{PA}, which are the *"points for"* (PF) and *"points against"* (PA), respectively. The response variable y is the win percentage.

Pythagorean Expectation Prediction. Furthermore, our solution adapts Pythagorean expectation as its third prediction technique. For make prediction for win ratio for basketball, we use Eq. (4), which with the exponent 13.91.

In terms of implementation, we reads the input files, extracts the elements needed, and feeds them into *Pythagorean _expectation(pts_scored, pts_allowed)* function to calculate measurements. Note that the real-life win rate can be calculated based on

$$win\ ratio = \frac{total\ win}{total\ win + total\ lost} \tag{7}$$

4 Evaluation

To evaluate our sports data mining solution, we conducted a case study by applying our solution to real-life basketball data.

4.1 Bell Curve Expectation

Our solution first adapted the bell curve to make prediction. Figure 1 shows side-by-side graphs to visualize the accuracy of prediction adapting the bell curve for a whole season versus real life percentage for all National Basketball Association (NBA) during 2015–2018. We observed an interesting trend that the bell curve expectation gave the lowest error when predicting win ratio 50%. Error was found to slowly build the further away from 50% the prediction gets. Moreover,

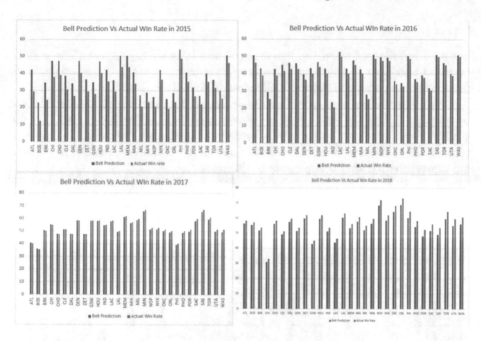

Fig. 1. Bell curve prediction and real win% of each team of NBA during 2015–2018.

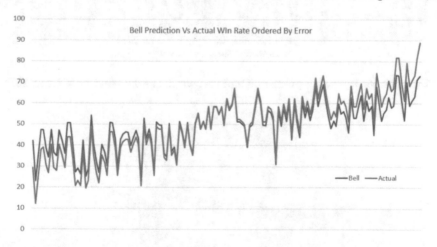

Fig. 2. Bell curve prediction vs. actual win% (ordered by error): from the most negative error (on the left) to the most positive error (on the right).

the bell curve expectation was observed to be a conservative predictor: The prediction is reliably high at low percentages, but the prediction is reliably low at high percentages. The results from the bell curve expectation function were close to the actual win ratio with an average error of 4.1%. See Fig. 2.

4.2   Linear Regression

Our solution then adapted the linear regression to make prediction. The results gave numbers close to the actual win rate for each NBA team when the data for all seasons are combined, which is illustrated visually in Fig. 3.

Linear regression was quite accurate for all the teams, as well as on an instance predicted the exact actual value in the case of the New York Knicks (NYK). The difference between the value predicted and the actual value was minimal. Factors for differences include changes in the management of the team, and departure of a major superstar in the team.

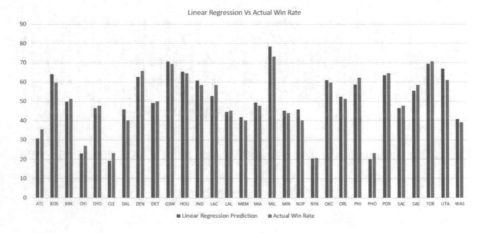

Fig. 3. Linear regression prediction and real win% of each team of NBA in 2018.

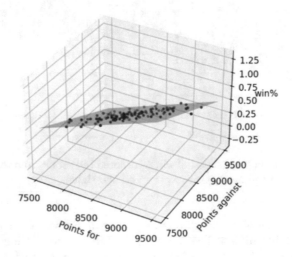

Fig. 4. Linear regression prediction: Win% vs. points-for (PF) vs. points-against (PA).

Figure 4 shows virtually how "points for" and "points against" affects a team's win percentage. It was observed that teams with high points-for (even if their points-against is a bit high) tend to do well.

4.3 Pythagorean Expectation

Moreover, our solution also adapted Morey's modified Pythagorean expectation, as shown in Eq. (4), to make prediction. Morey's exponential value for basketball of 13.91 was used in our evaluation. To assess the efficiency, we first derived at the sample mean and summed up individual deviation. We then established a table (e.g., Table 1) showing, for each year, this sum of deviations, along the sample size of teams (N = 30 teams), residual variance, and standard deviation (stdev) to account for how Pythagorean expectation performance behaves around the regression line.

The means of real win rate is the sum of all true win-rates over the total team numbers. Differences between each team predicted win-rate and the means of real win-rate is squared to attain variables for sum of deviations. This is because, along the regression line, predicted win rate sometimes over-predict or under-predict. Hence, their squared deviation from the mean over the sample size, the residual variance value, decides if the expected difference is minimized. From the basis of Bloome et al. [28], we investigated the size of residual variance value for large coefficients that determine a large variability within-group inequality. Caro et al. [29]—who frontiered in the same line of ideas—expressed in their work that there exists elements of decision that differs from professional games to collegiate ones. This motivated them to build upon other work. Polynomial differences between the residual variance and zero were closest to an exact prediction reflects such factors. Standard deviation is only the square root of residual variance. Low values of deviation suggests that Pythagorean expectation prediction is relatively close to the mean of the real winning set.

Similar to Fig. 1, we also show side-by-side graphs in Fig. 5 to visualize the accuracy of prediction adapting Morey's modified Pythagorean expectation for a whole season versus real life percentage for all NBA during 2015–2020.

Table 1. Pythagorean expectation prediction.

Year	Sum (deviation)	N	Residual variance	Stdev
2015	0.6433	30	0.021443	0.146435
2016	0.6827	30	0.022757	0.150853
2017	0.4607	30	0.015357	0.123922
2018	0.5001	30	0.016670	0.129112
2019	0.5533	30	0.018443	0.135806
2020	0.5480	30	0.018267	0.135154

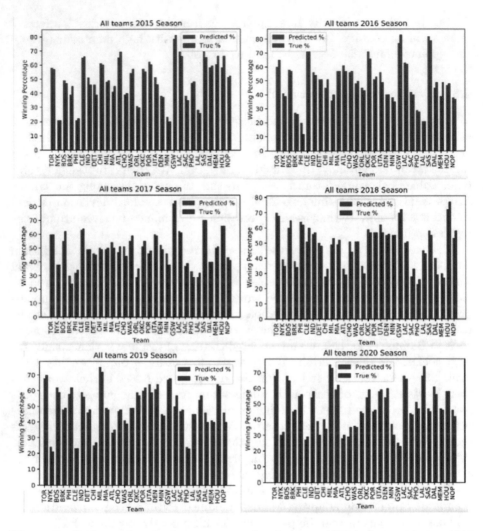

Fig. 5. Prediction and real win percentage of each team of NBA during 2015 to 2020 by adapting Pythagorean expectation

5 Conclusions

In this paper, we presented a sports data mining solution. It adapts three different approaches—namely, bell curve, linear regression, and modified Pythagorean expectation—to predict winning outcomes. For evaluation, we conducted a case study to apply our solution to real-life *basketball* data—more specifically, National Basketball Association (NBA) data from 2015 to 2020. The results show that these approaches lead to accurate predictions. Among them, Pythagorean expectation led to smaller average error than linear regression, which incurred smaller average error than bell curve expectation. For instance, the average prediction errors incurred by Pythagorean expectation, linear regression, and bell curve expectation is 2.3%, 2.6% and 4.6%, respectively, for the 2018 season.

As *ongoing and future work*, we would transfer knowledge learned from the current work to predict other sports and/or games. First, we would like to adapt Pythagorean expectation to other team sports such as American football, hockey, and soccer ball. Second, we would like to incorporate additional information in prediction. For instance, we would apply Pythagorean expectation to individual players to build up an expectation of a team based on its team's compiled performance. As another instance, we would incorporate other factors like funding, home vs. away games, and stadium status. Third, we would also like to modify Pythagorean expectation and/or other two prediction approaches for predicting the results for games (e.g., chess) in which scoring was determined differently. To elaborate, in chess, points may be awarded for pieces taken, but the number of pieces taken does not definitely determine a victory.

Acknowledgements. This project is partially supported by (a) Natural Sciences and Engineering Research Council of Canada (NSERC) and (b) University of Manitoba.

References

1. Argenzio, B., Amatucci, N., Botte, M., D'Acierno, L., Di Costanzo, L., Pariota, L.: The use of Automatic Vehicle Location (AVL) data for improving public transport service regularity. In: Barolli, L., Woungang, I., Enokido, T. (eds.) AINA 2021, vol. 3. LNNS, vol. 227, pp. 667–676. Springer, Cham (2021). https://doi.org/10.1007/978-3-030-75078-7_66
2. Leung, C.K., et al.: Data mining on open public transit data for transportation analytics during pre-COVID-19 era and COVID-19 era. In: Barolli, L., Li, K.F., Miwa, H. (eds.) INCoS 2020. AISC, vol. 1263, pp. 133–144. Springer, Cham (2021). https://doi.org/10.1007/978-3-030-57796-4_13
3. Xhafa, F., Aly, A., Juan, A.A.: Optimization of task allocations in cloud to fog environment with application to intelligent transportation systems. In: Barolli, L., Woungang, I., Enokido, T. (eds.) AINA 2021, vol. 1. LNNS, vol. 225, pp. 1–12. Springer, Cham (2021). https://doi.org/10.1007/978-3-030-75100-5_1
4. Leung, C.K.-S., Tanbeer, S.K., Cameron, J.J.: Interactive discovery of influential friends from social networks. Social Netw. Anal. Min. 4(1), 154:1–154:13 (2014). https://doi.org/10.1007/s13278-014-0154-z
5. Leung, C.K., et al.: Parallel social network mining for interesting 'following' patterns. Concurr. Comput. Pract. Exp. 28(15), 3994–4012 (2016)
6. Honda, M., Toshima, J., Suganuma, T., Takahashi, A.: Design of healthcare information sharing methods using range-based information disclosure incentives. In: Barolli, L., Woungang, I., Enokido, T. (eds.) AINA 2021, vol. 1. LNNS, vol. 225, pp. 758–769. Springer, Cham (2021). https://doi.org/10.1007/978-3-030-75100-5_64
7. Leung, C.K., Kaufmann, T.N., Wen, Y., Zhao, C., Zheng, H.: Revealing COVID-19 data by data mining and visualization. In: Barolli, L., Chen, H.-C., Miwa, H. (eds.) INCoS 2021. LNNS, vol. 312, pp. 70–83. Springer, Cham (2022). https://doi.org/10.1007/978-3-030-84910-8_8

8. Souza, J., Leung, C.K., Cuzzocrea, A.: An innovative big data predictive analytics framework over hybrid big data sources with an application for disease analytics. In: Barolli, L., Amato, F., Moscato, F., Enokido, T., Takizawa, M. (eds.) AINA 2020. AISC, vol. 1151, pp. 669–680. Springer, Cham (2020). https://doi.org/10.1007/978-3-030-44041-1_59

9. Braun, P., et al.: Game data mining: clustering and visualization of online game data in cyber-physical worlds. Procedia Comput. Sci. **112**, 2259–2268 (2017)

10. Anderson-Grégoire, I.M., et al.: A big data science solution for analytics on moving objects. In: Barolli, L., Woungang, I., Enokido, T. (eds.) AINA 2021, vol. 2. LNNS, vol. 226, pp. 133–145. Springer, Cham (2021). https://doi.org/10.1007/978-3-030-75075-6_11

11. Atif, F., Rodriguez, M., Araújo, L.J.P., Amartiwi, U., Akinsanya, B.J., Mazzara, M.: A survey on data science techniques for predicting software defects. In: Barolli, L., Woungang, I., Enokido, T. (eds.) AINA 2021, vol. 3. LNNS, vol. 227, pp. 298–309. Springer, Cham (2021). https://doi.org/10.1007/978-3-030-75078-7_31

12. Aggarwal, C.C.: Data Mining: The Textbook. Springer, Cham (2015). https://doi.org/10.1007/978-3-319-14142-8

13. Leung, C.K., et al.: Distributed uncertain data mining for frequent patterns satisfying anti-monotonic constraints. In: IEEE AINA Workshops 2014, pp. 1–6 (2014)

14. Leung, C.K., et al.: Fast algorithms for frequent itemset mining from uncertain data. In: IEEE ICDM 2014, pp. 893–898 (2014)

15. Liu, C., Li, X.: Mining method based on semantic trajectory frequent pattern. In: Barolli, L., Woungang, I., Enokido, T. (eds.) AINA 2021, vol. 2. LNNS, vol. 226, pp. 146–159. Springer, Cham (2021). https://doi.org/10.1007/978-3-030-75075-6_12

16. Ni, J., Yin, W., Jiang, Y., Zhao, J., Hu, Y.: Periodic mining of traffic information in industrial control networks. In: Barolli, L., Amato, F., Moscato, F., Enokido, T., Takizawa, M. (eds.) AINA 2020. AISC, vol. 1151, pp. 176–183. Springer, Cham (2020). https://doi.org/10.1007/978-3-030-44041-1_16

17. Ngaffo, A.N., El Ayeb, W., Choukair, Z.: An IP multimedia subsystem service discovery and exposure approach based on opinion mining by exploiting Twitter trending topics. In: Barolli, L., Takizawa, M., Xhafa, F., Enokido, T. (eds.) AINA 2019. AISC, vol. 926, pp. 431–445. Springer, Cham (2020). https://doi.org/10.1007/978-3-030-15032-7_37

18. Ahn, S., et al.: A fuzzy logic based machine learning tool for supporting big data business analytics in complex artificial intelligence environments. In: FUZZ-IEEE 2019, pp. 1259–1264 (2019)

19. Leung, C.K.: Mathematical model for propagation of influence in a social network. In: Alhajj, R., Rokne, J. (eds.) Encyclopedia of Social Network Analysis and Mining, 2nd edn., pp. 1261–1269. Springer, New York (2018). https://doi.org/10.1007/978-1-4939-7131-2_110201

20. Gao, X., Uehara, M.: Design of a sports mental cloud. In: AINA Workshops 2017, pp. 443–448 (2017)

21. Leung, C.K., Joseph, K.W.: Sports data mining: predicting results for the college football games. Procedia Comput. Sci. **35**, 710–719 (2014)

22. Takano, K., Li, K.F.: Classifying sports gesture using event-based matching in a multimedia e-learning system. In: AINA Workshops 2012, pp. 833–838 (2012)

23. Kubatko, J., et al.: A starting point for analyzing basketball statistics. J. Quant. Anal. Sports **3**(3), 1–24 (2007)

24. Perin, C., et al.: State of the art of sports data visualization. Comput. Graph. Forum **37**(3), 663–686 (2018)

25. Oliver, D.: Basketball on Paper: Rules and Tools for Performance Analysis. Potomac Books, Sterling (2004)
26. Upton, G., Cook, I.: A Dictionary of Statistics, 3rd edn. Oxford University Press, Oxford (2014)
27. Dewan, J.: STATS Basketball Scoreboard, 1993–94. Harpercollins Publishers, New York (1993)
28. Ritzer, G. (ed.): The Blackwell Encyclopedia of Sociology. Wiley, Hoboken (2007)
29. Caro, C.A., Machtmes, R.: Testing the utility of the Pythagorean expectation formula on division one college football: an examination and comparison to the Morey model. J. Bus. Econ. Res. (JBER) 11(12), 537:1–537:6 (2013)

Mining for Fake News

Renz M. Cabusas[1], Brenna N. Epp[1], Justin M. Gouge[1,2], Tyson N. Kaufmann[1],
Carson K. Leung[1(✉)] (iD), and James R. A. Tully[1]

[1] University of Manitoba, Winnipeg, MB, Canada
kleung@cs.umanitoba.ca, Carson.Leung@UManitoba.ca
[2] York University, Toronto, ON, Canada

Abstract. Fake news is an ever-growing concern in the modern age of the internet. Discerning fake information from the truthful is an important task given the simplicity of sharing information digitally. In this paper, we present a data mining solution to classify articles as real or fake by using bag-of-words (BoW) and sequential mining techniques, and compare reliability for detecting fake news on various datasets. Specifically, our solution first cleans the input news by normalizing words and removing "filler" words. It then uses the BoW or sequential mining techniques to vectorize cleaned data. Afterwards, it trains the classification models based on vectorized data and classifies unseen news as real or fake. Evaluation on real-life data shows the feasibility of our solution to mine and classify fake news.

Keywords: Data mining · Big data analytics · Social networks · Fake news

1 Introduction

Nowadays, advances in technology have led to generation and collection of numerous amounts of data in various real-life application areas such as transportation [1–3], social networks [4, 5], healthcare [6–8], and online games [9]. Hence, data (including news) are everywhere. Some of these news are real, while some are fake. To detect validity and veracity of the news (e.g., fake or not), data science [10, 11]—which makes good use of techniques like data mining [12–17], machine learning [18], and mathematical modelling [19]—is useful.

To elaborate, *fake news* are "news articles that are intentionally and verifiably false" [20]. They have become increasingly problematic in today's online world. With the rise of use in social media, sharing stories and news articles has never been easier. However, not all news articles are necessarily truthful. For the past century, many news companies have relied on advertisements. As their print sales lessen and as companies spend larger chunks of advertising budget on online advertisements, news companies aim to gain relevance online. With some popular companies (e.g., worldwide companies famous for social networking, search and services) dominate the online advertisement market, these reputable news corporations are incentivized to increase engagement while competing with websites that have no or little commitment to truthful journalism [21]. This unfortunately encourages publications to "bend the truth" or outright create news

L. Barolli et al. (Eds.): AINA 2022, LNNS 450, pp. 154–166, 2022.
https://doi.org/10.1007/978-3-030-99587-4_14

articles of events that never occurred in order to drive up website traffic [22]. Apart from the advertisement model, another reason companies may publish fake news is for indirect financial or political gain from the effect of the news on those who read it: news can drive sentiment affecting stock markets and political circles, motivating people to post news that benefit their interests, regardless of veracity [23].

The credulity of the public allows the spread of "fake news" and misinformation, which has the potential to harm individuals and society. Studies [24, 25] have shown that the most harmful effect of fake news is the devaluation and delegitimization of expert opinions, authoritative institutions and the concept of objective data—all of which undermine society's ability to engage in rational discussions rooted in shared facts. Hence, accurately detecting fake news content and classifying it as such is crucial to stop the spread of fake news that could negatively affect the opinions and trust of the public. Nonetheless, detecting fake news continues to be a complex task given how quickly and easily it can spread in social media.

Many methods [26] have been employed to detect fake news, ranging from simple approaches involving counting word frequencies in news articles to more complex neural networks and deep learning approaches that build off those initial word-counting methodologies. Many of these methods require information to be converted into vectors that can be used in data science tasks such as classification. Among them, *bag-of-words* (*BoW*) is a common technique used in the classification models for fake news [27]. It transforms the news articles into a frequency count of words present in the target dataset. As such, it does not take into account the context of the words. Nonetheless, the technique is used in natural language processing and information retrieval for the purpose of simplifying text and is commonly used for document classification and identification (hence, why it is useful for fake news detection).

However, the sequential mining of news articles is not a common practice. Hence, in this paper, we explore the potentials of sequence generation for vectorization. We examine if sequentially mining articles before applying machine learning algorithms could have improved efficacy on classification models for fake news detection over common approaches such as BoW. In particular, we implement a sequential mining approach to generate vectors that can be input into the same machine learning classifiers as a BoW approach. More specifically, we mine news articles for sequences of words using sequential mining, and we use these sequences as input to the same machine learning algorithm used for BoW. Hence, our *key contributions* of this paper include the feasibility study on incorporating sequential mining in fake news detection, as well as our design and development of a fake news mining solution that uses sequential mining to generate vectors of words and word counts for training fake news classifiers.

The remainder of this paper is organized as follows. In the next section, we discuss related work, including the brief discussion of shortcomings of previous work. Section 3 describes our fake news data mining solution, which includes (a) data cleaning techniques, (b) the vectorization process for the data such as the bag-of-words (BoW) technique and sequential mining algorithm, and (c) fake news classification. Section 4 shows our evaluation results on real-life data, which include a comparison of accuracy, F1-score, runtime and memory requirements. Finally, conclusions are drawn in Sect. 5.

2 Related Works

Although related works were very useful for the baseline of detecting fake news patterns, we focus the usefulness and efficiency of algorithms for fake news detection in the currrent paper. There are several approaches used for fake news detection and mining. For example, Pérez-Rosas et al. [28] focused on the automatic detection of fake news by using various linguistic features (including *n-gram*, which groups neighboring words in a contiguous sequence of up to length *n* items from a given sample of text) of articles on a linear support vector machine (LSVM) classifier in conjunction with five-fold cross-validation. They also presented various processes of collection and efficient validation of textual data in detail. Specifically, they examined key linguistic differences between fake and legitimate news sources by running experiments. The *n*-gram features gave results of up to 71% accuracy. Along that direction, they compared manual and automatic detection of fake news.

As another related work, Sriram [23] performed fake news detection and classification by using an LSVM classifier. She also used a cross-validation approach on a few datasets. Her evaluation results showed that the combination with the highest accuracy was the term frequency - inverse document frequency (TF-IDF) based LSVM classifier trained on some Kaggle competition datasets[1]. In general, *TF-IDF* is a text-feature representation technique that is used for converting studied text into vectors. The reason is that the text cannot be directly given as input to a machine learning classifier. She found that the traditional BoW approach generally worked better compared to word and sentence embeddings-based models.

Moreover, Shu et al. [27] mined disinformation and fake news. Social media can be considered as huge fake news outlets. Discerning between true and false content on social media can be a difficult task. They discussed various definitions and characterizations of information to sort the news as real or fake. Included with the survey of fake news presented, they also examined fake news characterizations on psychology, social theories, as well as existing algorithms from a data mining perspective, evaluation metrics and representative datasets. One of the main interesting pieces from her work is the differences between fake news on traditional media (e.g., psychology or social foundations, knowledge- or style-based news content) and fake news on social media (e.g., malicious accounts, echo chamber, stance- or propagation-based social content). They also suggested a few open research problems. These include different approaches for fake news detection:

- Data-oriented approaches, which explore temporal, and/or psychological characteristics of data;
- Feature-oriented approaches, which explore news content (e.g., linguistic- and/or visual-based feature extraction on source, headline, body text, image, and/or video) and/or social content (e.g., user-, post-, and/or network-based feature extraction);
- Model-oriented approaches, which explore supervised, semi-supervised, and/or unsupervised machine learning models; and

---

[1] https://www.kaggle.com/mrisdal/fake-news/kernels,   https://www.kaggle.com/c/fake-news/, https://components.one/datasets/all-the-news-2-news-articles-dataset/,   https://www.kaggle. com/mdepak/fakenewsnet.

- Application-oriented approaches, which explore fake news diffusion and/or intervention.

From the prospective of these approaches, our fake news mining solution in the current paper can be considered as a data-oriented approach with a supervised machine-learning model trained on the features extracted from the news content.

3 Our Fake News Mining Solution

To mine for fake news, our fake news mining solution follows a data-oriented approach with a supervised machine-learning model trained on the features extracted from the news content. It cleans the input data, vectorizes these data, trains the classifiers on the vectorized data, and applies the trained models to classify the input news as real or fake news.

3.1 Data Cleaning

Our fake news mining solution starts with cleaning article texts. By doing so, it reduces the search space by removing some words and clustering others together. Specifically, it:

- reduces the character set by (a) lowering all characters and (b) pruning all non-alphanumeric characters;
- removes all stop-words such as those connect-words that do not add context (e.g., "a", "be", "i.e.", "the")
- stems words by mapping words to their root (e.g., clusters {"writing", "writes", "written"} to become "write").

Table 1 shows an example of data cleaning. Note that, by cleaning the article texts, it reduces the search space. As a side-benefit, it also improve the runtime efficiency. A reason is that a smaller initial dataset means that the sequence generation algorithm—with an $O(n^3)$ runtime complexity (where n is the size of all words across all articles)—would initially have less data to work through. Moreover, cleaning the data also aids in the memory footprint of the algorithm. A reason is that mid-stages of the sequence generation would be looking over more similar data, and thus, there would be less sparsity in the returned sequences. Consequently, these slight differences in wordings between articles would not affect results because the cleaned versions may be picked up as the same sequences (rather than not).

3.2 Vectorization

After cleaning the article texts, our solution then converts the articles (in the datasets) into vectors through the bag-of-words (BoW) technique or sequential mining. Recall that *bag-of-words* is a technique for gathering features in text data, in which the features are the counts of individual words in a block of text with no regard to word order. Popularity of this technique is partially due to its simplicity, its speed, and its ability to gather features.

Table 1. An example of data cleaning: minimization of articles.

Original article	Cleaned article
Tesco will not pay. Tesco will not pay out any money to settle investigations by the Serious Fraud Office and Financial Conduct Authority into the 2014 accounting scandal that rocked Britain's biggest retailer It will pay 000 as part of a deferred prosecution agreement DPA with the SFO as this deal does not require court approval The DPA relates to Tesco subsidiary Tesco Stores Ltd The supermarket group has not agreed with the FCA to pay any compensation to the investors affected by a trading statement on 29 August 2014 that understated stated profits Tesco will not pay legal costs associated with the agreements and said the total exceptional charge was expected to be 000	**Tesco will not pay**. tesco pay money settle investigate fraud office finance conduct authorize 2014 account scandal rock britain biggest retail pay 000 defer prosecute agreement dpa sfo deal do require court approve dpa relate tesco subsidize tesco store supermarket group agree fca pay compensate investor affect trade statement 29 august 2014 understand state profit tesco pay legal cost associate agreement say total except charge expect 000

Example 1. Let us illustrate how BoW vectorizes article texts by transforming a sample text into a feature set. Table 2 shows BoW transforms the text "This is an example sentence for this example" into a feature vector by counting occurrences of individual words. For instance, words "this" and "example" occur twice, whereas the remaining words like "is", "an", "sentence" and "for" occur only once.

Table 2. An example of vectorization: transformation of the text "This is an example sentence for this example" into a feature vector by the BoW technique.

Word	Count
this	2
is	1
an	1
example	2
sentence	1
for	1

On the other hand, the *sequential mining* generates sequential pattern by taking a set of articles and mapping them 1-to-1 into a set of vectors, which represent the inclusion of some specific sequence within the model. Such a sequential mining-based vectorization can be preformed in a 4-step process:

1. It generates a *word map* to index every unique word in the initial dataset. Each newfound word is given an incremented indexing value. This is useful during the training stage of the process because it generates the word map to be used in the testing phase of the machine learning algorithm. Resulting vectors would be of the same format and size for both training and testing phases).
2. Once the word map is generated, our solution transcribes the words (in the data) into their indexed values.
3. Afterwards, it then feeds each index-transcribed article into a sequential mining algorithm, which would return the sequences found in the input article in $O(n^2)$ time (where n is the number of words in the article). To enable efficient processing of the entire article while reducing the search depth of sequential mining algorithm, our solution applies a sliding window feeder for the algorithm.
4. Finally, it builds a vector for each article by agglomerating uniquely occurring sequences from each article into a single map. The resulting vector indicates whether a mapped sequence occurs in the article or not.

Example 2. Let us illustrate how sequential mining vectorizes article texts by transforming a sample text into a feature set containing sequences of k words. Table 3 shows sequential mining transforms the text "This is an example sentence for this example" into a feature vector by counting occurrences of two-word sequences. For instance, a two-word sequence "this is" occurs once, whereas another two-word sequence "this example" occurs twice.

Table 3. An example of vectorization: transformation of the text "This is an example sentence for this example" into a feature vector by the sequential mining.

Two-word sequence	Count
this is	1
this an	1
this example	2
this sentence	1
:	:

3.3 News Classification

Finally, after the data area cleaned and articles are vectorized from the cleaned data, our solution builds and trains machine learning classifiers—such as naïve Bayes (NB) and linear support vector machines (LSVM). It does so by feeding the vectors as its input. Once the classifiers are trained, they can be used for predicting and classifying whether an unknown article is real or fake.

4 Evaluation

To evaluate our fake news mining solution, we applied it to some real-life datasets, which include a combination[2] of the FakeNewsAMT dataset and the Celebrity dataset [28]. Between these two datasets, the former was collected by crowdsourcing—via Amazon Mechanical Turk (AMT)—fake versions of legitimate articles from a variety of mainstream news websites and annotating both with the correct label. The dataset contains news about technology, education, business, sports, politics, and entertainment. The Celebrity dataset was a manually verified collection of articles about celebrities collected directly from the web. These articles were collected in pairs, with the legitimate articles from mainstream news websites and entertainment magazines websites, whereas the fake articles from gossip websites. We conducted empirical evaluation by using a machine with the following specifications: Intel Core i7-7700k CPU with 4 cores (i.e., 8 logical processors) and 24 GB of physical memory (RAM) with a solid-state drive (SSD) as a boot drive (i.e., swap space).

First, we evaluated the *data cleaning* of our solution. An observation from one of the datasets reveals that it drastically reduces the search space from 20,766 unique words down to 13,570 (i.e., a reduction of 1.5 ×).

Next, we evaluated the *vectorization* of our solution. Specifically, once data were cleaned, we randomly split the cleaned data into 80% for training set and 20% for testing. We also used 5-fold cross-validation to evaluate the classification performance. Quantifiable metrics include:

- Accuracy, which measures the percentage of correctly classified articles (i.e., the percentage sum of legitimate articles classified as REAL and illegitimate articles classified as FAKE out of all articles):

$$ACC = (TP + TN)/(TP + TN + FP + FN) \tag{1}$$

where TP is true positive, TN is true negative, FP is false positive, and FN is false negative;

- Precision (for the REAL news), which measures the percentage of articles classified as REAL are legitimate out of all articles classified as REAL:

$$Precision = TP/(TP + FP) \tag{2}$$

- Recall (for the REAL news), which measures the percentage of legitimate articles classified as REAL out of all legitimate articles:

$$Recall = TP/(TP + FN) \tag{3}$$

- F1-score for each label (FAKE or REAL), which is a weighted average of precision and recall:

$$F1 = 2TP/(2TP + FP + FN) \tag{4}$$

---

[2] https://lit.eecs.umich.edu/downloads.html,http://web.eecs.umich.edu/~mihalcea/downloads/fakeNewsDatasets.zip.

- Runtime, which measures time to run our solution; and
- Memory usage and requirements throughout the experiments.

For instance, Table 4 shows evaluation results for the quantifiable metrics when running the BoW for vectorization on both FakeNewsAMT Celebrity datasets.

Table 4. Evaluation results of the BoW for vectorization.

Dataset	Accuracy	F1-score (fake news)	F1-score (real news)	Runtime
FakeNewsAMT	69.8%	0.696	0.699	0.538 s
Celebrity	75.6%	0.755	0.756	0.551 s

For comparison, we also measured the quantifiable metrics when running the sequential mining with PrefixSpan [29] for vectorization. Table 5 shows a comparison between the BoW and sequential mining for vectorization on the Celebrity dataset. The results reveal that sequential mining (with PrefixSpan) led to higher accuracy, as well as F1-score for both fake news and real news, than the BoW.

Table 5. Comparison results on the BoW vs. sequential mining for vectorization on the Celebrity dataset.

Vectorization techniques	Accuracy	F1-score (fake news)	F1-score (real news)
BoW	75.6%	0.755	0.756
Sequential mining	78.6%	0.789	0.781

The price for the higher accuracy and F1-scores were higher runtime for sequential mining. For instance, sequential mining took 6.02 s for the FakeNewsAMT dataset.

Besides data cleaning and vectorization, we also evaluated *news classification* of our solution. Here, we used the BoW as the baseline, and varied different parameters for the sequential mining algorithm (for vectorization) to measure their impacts on news classification with the two classifiers—namely, naïve Bayes (NB) and linear support vector machines (LSVM):

- We varied the maximum sequence size from 1 to 2, 3 and 5 (i.e., 1-word, 2-word, 3-word and 5-word sequences)
- We also varied the maximum window size from 1 to 70.

We used the default parameters as per previous research [28] with scikit-learn [30] implementation of NB and LSVM. Again, we randomly split the data into 80% for training set and 20% for testing, used 5-fold cross-validation, and measured values for the aforementioned quantifiable metrics (e.g., accuracy, runtime, precision, recall, F1-score, memory usage and requirements).

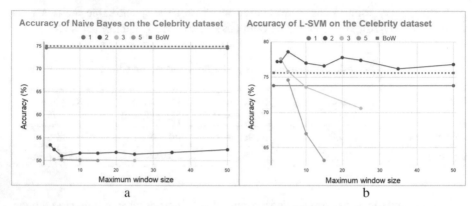

Fig. 1. Accuracy of (a) naïve Bayes and (b) LSVM on the Celebrity dataset.

In terms of accuracy, Fig. 1 shows line-curves for four different maximum sequences sizes (namely, 1, 2, 3 and 5) and for the BoW. The x-axis shows the maximum window size ranging from 1 to 50, and the y-axis shows the corresponding classification accuracy. Evaluation results in Fig. 1(a) on applying sequential mining for vectorization to the Celebrity dataset reveal that, the use of maximum sequence size of 1 (i.e., 1-word sequence) led to the higher accuracy by the naïve Bayes classifier than other sequence sizes. However, evaluation results in Fig. 1(b) reveal that the use of a maximum sequence size of 2 (i.e., 2-word sequences) led to the higher accuracy by the LSVM classifier than other sequence sizes or the BoW. Results in both figures also reveal that a small window size led to high classification accuracy. For instance, the LSVM classifier with a maximum sequence size of 2 (i.e., 2-word sequences) and a maximum window size of 5 led to a classification accuracy of 78.6% (cf. 71% accuracy when using n-gram with LSVM [28]). Note that a key difference between our solution and the use of n-gram is that we use PrefixSpan as a sequential mining algorithm that groups sequential words *regardless of neighbors*, whereas n-gram goups neighboring sequential words up to length N.

In terms of F1-score, the y-axis of Fig. 2 shows the F1-score for classifying news from the FakeNewsAMT dataset. In the figure, we used *solid* line-curves to represent F1-scores for classifying fake news and *dotted* line-curves to represent F1-scores for classifying real news. The line-curves represent four different maximum sequences sizes (namely, 1, 2, 3 and 5) and for the BoW. Figure 2(a) reveals that the NB classifier led to higher F1-scores for classifying fake news (as indicated by the solid line-curves) than real news (as indicated by the dotted line-curves). Similarly, Fig. 2(b) reveals that the LSVM classifier with a maximum sequence size of 3 (i.e., 3-word sequences) and a maximum window size of 10 led to a high F1-score for classifying fake news (as indicated by the yellow solid line-curve).

Figure 3 shows the F1-score for classifying news from the Celebrity dataset. Figure 3(a) reveals that the NB classifier with a maximum sequence size of 1 (and any maximum window size ranging from 1 to 50) led to a high F1-score for classifying

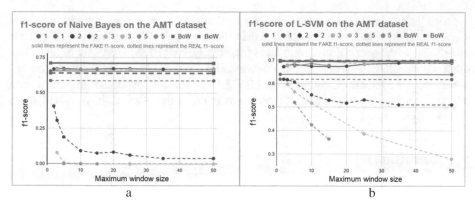

Fig. 2. F1-scores of (a) naïve Bayes and (b) LSVM on the FakeNewsAMT dataset.

fake news (as indicated by the blue solid line-curve). Such an F1-score was higher than those by the BoW. Similarly, Fig. 3(b) reveals that the LSVM classifier with a maximum sequence sizes of 2, 3 and 5 (i.e., 2-, 3- and 5-word sequences) and a maximum window size of 5 led to high F1-scores for classifying fake news (as indicated by the red, yellow and green solid line-curves). These F1-scores were higher than those by the BoW. Moreover, the figure also reveals that the LSVM classifier with a maximum sequence size of 2 (and any maximum window size ranging from 1 to 50) led to high F1-scores for classifying real news (as indicated by the red dotted line-curve) too. Again, these F1-scores were higher than those by the BoW.

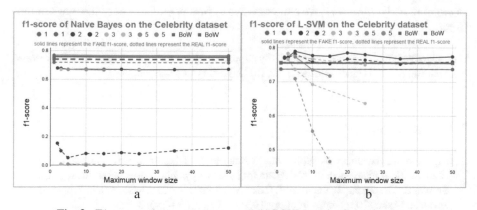

Fig. 3. F1-scores of (a) naïve Bayes and (b) LSVM on the Celebrity dataset.

As for the *runtime*, vectorization via sequential mining took longer than that via the BoW. The increase in runtime is mostly due to the generation of sequences (especially, k-word sequences for higher k). In other words, while the vectorization via sequential mining led to some gains in accuracy and F1-scores by some classifiers (e.g., LSVM classifier with maximum sequence sizes of 2 or 3 on the Celebrity dataset), these gains come at the price of the runtime.

Similarly, regarding the *memory consumption and requirements*, vectorization via sequential mining also took more space than that via the BoW. The increase in space is mostly due to the generation of sequences (especially, k-word sequences for higher k). In other words, while the vectorization via sequential mining led to some gains in accuracy and F1-scores by some classifiers (e.g., LSVM classifier with maximum sequence sizes of 2 or 3 on the Celebrity dataset), these gains come at the price of the time and space complexity.

5   Conclusions

In this paper, we presented a fake news data mining solution. It first cleans the input data news article by removing some words (e.g., non-alphanumeric characters, stopwords) and clustering others together (e.g., by mapping words to their root and groups the same roots). This reduces the search space. It then adapts bag-of-words (BoW) technique and sequential mining technique to vectorize the cleaned data. The resulting vectors capture bags (or sequences) of frequent words. By training the news classifiers such as naïve Bayes and linear support vector machine (LSVM) with these vectors, the trained classifiers in our solution could accurately classify unseen news as fake or real. Evaluation on real-life datasets shows the performance (e.g., accuracy) of our solution in mining and classifying fake news. It also demonstrates the feasibility of using sequential mining for vectorization. It also reveals that the LSVM classifier with certain parameters (e.g., sequence length, window size) led to more accurate classification than naïve Bayes for the evaluation datasets. The gain in accuracy comes at the price of higher time and space complexity. This illustrates a trade-off among accuracy, time and space. As *ongoing and future work*, we explore ways to further enhance our fake news mining solution.

Acknowledgments. This project is partially supported by (a) Natural Sciences and Engineering Research Council of Canada (NSERC) and (b) University of Manitoba.

References

1. Argenzio, B., Amatucci, N., Botte, M., D'Acierno, L., Di Costanzo, L., Pariota, L.: The use of automatic vehicle location (AVL) data for improving public transport service regularity. In: Barolli, L., Woungang, I., Enokido, T. (eds.) AINA 2021, vol. 3. LNNS, vol. 227, pp. 667–676. Springer, Cham (2021). https://doi.org/10.1007/978-3-030-75078-7_66
2. Leung, C.K., et al.: Data mining on open public transit data for transportation analytics during pre-COVID-19 era and COVID-19 era. In: Barolli, L., Li, K.F., Miwa, H. (eds.) INCoS 2020. AISC, vol. 1263, pp. 133–144. Springer, Cham (2021). https://doi.org/10.1007/978-3-030-57796-4_13
3. Xhafa, F., Aly, A., Juan, A.A.: Optimization of task allocations in cloud to fog environment with application to intelligent transportation systems. In: Barolli, L., Woungang, I., Enokido, T. (eds.) AINA 2021, vol. 1. LNNS, vol. 225, pp. 1–12. Springer, Cham (2021). https://doi.org/10.1007/978-3-030-75100-5_1

4. Leung, C.K.-S., Tanbeer, S.K., Cameron, J.J.: Interactive discovery of influential friends from social networks. Social Netw. Anal. Min. **4**(1), 154:1–154:13 (2014). https://doi.org/10.1007/s13278-014-0154-z
5. Leung, C.K., et al.: Parallel social network mining for interesting 'following' patterns. Concurr. Computat. Pract. Exp. **28**(15), 3994–4012 (2016)
6. Honda, M., Toshima, J., Suganuma, T., Takahashi, A.: Design of healthcare information sharing methods using range-based information disclosure incentives. In: Barolli, L., Woungang, I., Enokido, T. (eds.) AINA 2021, vol. 1. LNNS, vol. 225, pp. 758–769. Springer, Cham (2021). https://doi.org/10.1007/978-3-030-75100-5_64
7. Leung, C.K., Kaufmann, T.N., Wen, Y., Zhao, C., Zheng, H.: Revealing COVID-19 data by data mining and visualization. In: Barolli, L., Chen, H.-C., Miwa, H. (eds.) INCoS 2021. LNNS, vol. 312, pp. 70–83. Springer, Cham (2022). https://doi.org/10.1007/978-3-030-849 10-8_8
8. Souza, J., Leung, C.K., Cuzzocrea, A.: An Innovative big data predictive analytics framework over hybrid big data sources with an application for disease analytics. In: Barolli, L., Amato, F., Moscato, F., Enokido, T., Takizawa, M. (eds.) AINA 2020. AISC, vol. 1151, pp. 669–680. Springer, Cham (2020). https://doi.org/10.1007/978-3-030-44041-1_59
9. Braun, P., et al.: Game data mining: clustering and visualization of online game data in cyber-physical worlds. Proc. Comput. Sci. **112**, 2259–2268 (2017)
10. Anderson-Gregoire, I.M., et al.: A big data science solution for analytics on moving objects. In: Barolli, L., Woungang, I., Enokido, T. (eds.) AINA 2021, vol. 2. LNNS, vol. 226, pp. 133–145. Springer, Cham (2021). https://doi.org/10.1007/978-3-030-75075-6_11
11. Atif, F., Rodriguez, M., Araujo, L.J.P., Amartiwi, U., Akinsanya, B.J., Mazzara, M.: A survey on data science techniques for predicting software defects. In: Barolli, L., Woungang, I., Enokido, T. (eds.) AINA 2021, vol. 3. LNNS, vol. 227, pp. 298–309. Springer, Cham (2021). https://doi.org/10.1007/978-3-030-75078-7_31
12. Aggarwal, C.C.: Data Mining: The Textbook. Springer, Cham (2015). https://doi.org/10.1007/978-3-319-14142-8
13. Leung, C.K., et al.: Distributed uncertain data mining for frequent patterns satisfying anti-monotonic constraints. In: IEEE AINA Workshops 2014, pp. 1–6 (2014)
14. Leung, C.K., et al.: Fast algorithms for frequent itemset mining from uncertain data. In: IEEE ICDM 2014, pp. 893–898 (2014)
15. Liu, C., Li, X.: Mining method based on semantic trajectory frequent pattern. In: Barolli, L., Woungang, I., Enokido, T. (eds.) AINA 2021, vol. 2. LNNS, vol. 226, pp. 146–159. Springer, Cham (2021). https://doi.org/10.1007/978-3-030-75075-6_12
16. Ni, J., Yin, W., Jiang, Y., Zhao, J., Hu, Y.: Periodic mining of traffic information in industrial control networks. In: Barolli, L., Amato, F., Moscato, F., Enokido, T., Takizawa, M. (eds.) AINA 2020. AISC, vol. 1151, pp. 176–183. Springer, Cham (2020). https://doi.org/10.1007/978-3-030-44041-1_16
17. Ngaffo, A.N., El Ayeb, W., Choukair, Z.: An IP multimedia subsystem service discovery and exposure approach based on opinion mining by exploiting Twitter trending topics. In: Barolli, L., Takizawa, M., Xhafa, F., Enokido, T. (eds.) AINA 2019. AISC, vol. 926, pp. 431–445. Springer, Cham (2020). https://doi.org/10.1007/978-3-030-15032-7_37
18. Ahn, S., et al.: A fuzzy logic based machine learning tool for supporting big data business analytics in complex artificial intelligence environments. In: FUZZ-IEEE 2019, pp. 1259–1264 (2019)
19. Leung, C.K.: Mathematical model for propagation of influence in a social network. In: Alhajj, R., Rokne, J. (eds) Encyclopedia of Social Network Analysis and Mining, 2nd edn., pp. 1261–1269. Springer, New York (2018). https://doi.org/10.1007/978-1-4939-7131-2_110201
20. Shu, K., et al.: Fake news detection on social media: a data mining perspective. ACM SIGKDD Explorat. **19**(1), 22–36 (2017)

21. Whittaker, J.P.: Tech Giants, Artificial Intelligence and the Future of Journalism. Routledge, New York (2019)
22. Christin, A.: Metrics at Work: Journalism and the Contested Meaning of Algorithms. Princeton University Press (2020)
23. Sriram, S.: An Evaluation of Text Representation Techniques for Fake News Detection Using: TF-IDF, Word Embeddings, Sentence Embeddings with Linear Support Vector Machine. M.Sc. Dissertation, Technological University Dublin (2020). https://doi.org/10.21427/5519-h979
24. Hartley, K., Vu, M.K.: Fighting fake news in the COVID-19 era: policy insights from an equilibrium model. Policy Sci. **53**(4), 735–758 (2020). https://doi.org/10.1007/s11077-020-09405-z
25. Horne, B.D., Adah, S.: This just in: fake news packs a lot in title, uses simpler, repetitive content in text body, more similar to satire than real news. In: ICWSM 2017 Workshop W7 on NECO, pp. 759–766 (2017). https://aaai.org/ocs/index.php/ICWSM/ICWSM17/paper/view/15772/14898
26. Ibrishimova M.D., Li K.F.: A machine learning approach to fake news detection using knowledge verification and natural language processing. In: Barolli L., Nishino H., Miwa H. (eds) INCoS 2019. AISC, vol. 1035, pp. 223–234. Springer, Cham (2019). https://doi.org/10.1007/978-3-030-29035-1_22
27. Shu, K., et al.: Mining disinformation and fake news: concepts, methods, and recent advancements. In: Disinformation, Misinformation, and Fake News in Social Media, pp. 1–19 (2020)
28. Pérez-Rosas, V., et al.: Automatic detection of fake news. In: COLING 2018, pp. 3391–3401 (2018). https://aclanthology.org/C18-1287
29. Pei, J., et al.: Mining sequential patterns by pattern-growth: the PrefixSpan approach. IEEE TKDE **16**(11), 1424–1440 (2004)
30. Pedregosa, F., et al.: Scikit-learn: machine learning in Python. JMLR **12**, 2825–2830 (2011)

Software Functional and Non-function Requirement Classification Using Word-Embedding

Lov Kumar[1], Siddarth Baldwa[1(✉)], Shreya Manish Jambavalikar[1],
Lalita Bhanu Murthy[1], and Aneesh Krishna[2]

[1] BITS-Pilani Hyderabad, Hyderabad, India
{lovkumar,f20180914,f20181089,bhanu}@hyderabad.bits-pilani.ac.in
[2] Curtin University, Perth, Australia
A.Krishna@curtin.edu.au

Abstract. The classification of software requirements is an essential task in software engineering. Manual classification requires a large amount of efforts, time and cost. Hence, automated techniques are required to classify software requirements. This work aims to develop requirement classification models based on extraction of relevant features from requirement documents and thereafter classifying requirement into functional and non-function requirements. In this paper, different word-embedding techniques to extract numerical features, feature selection to remove irrelevant feature, SMOTE to balance data, and six different classifiers for models training. The experiments have been conducted on PROMISE software engineering dataset. The experimental finding indicate that Word2vec is best way to extracting numerical features from requirement documents, RANK-SUM test is best way to find important features, and SVM-R was found as the best classifier.

Keywords: Software requirement · Data imbalance · Feature selection · Word embedding

1 Introduction

Software requirements describe the features and functionalities of the software product, help understand user expectations and are a crucial part of the software development process. Software Requirements can be categorized into functional and non-functional requirements. Functional requirements define the components of the software system and are mandatory requirements specified by the end-user. Non-functional requirements define the quality attributes of a software system (such as security, reliability, portability, maintainability, etc.) and specify its operations. The automated classification of requirements written in natural language is necessary because its manual classification is tedious and time-consuming, especially for target softwares with a vast number of requirements. Classifying software requirements using machine learning algorithms is

L. Barolli et al. (Eds.): AINA 2022, LNNS 450, pp. 167–179, 2022.
https://doi.org/10.1007/978-3-030-99587-4_15

still a challenging task as requirements engineers and stakeholders have a different vocabulary for specifying the same requirement. The high inconsistency in requirements elicitation makes automated classification complex and error-prone, leaving us with the task of finding an optimal technique for automatic classification. The main contributions of this study are:

- Comparative analysis of five crucial text vectorization techniques, namely Term frequency–Inverse document frequency (TF-IDF), Continuous bag of words (CBOW), Word2Vec, Global vectors for word representation (GloVe) and Skip-Gram (SKG).
- Comparison between feature selection and dimensionality reduction techniques such as: Principal Component Analysis (PCA) and RANK-Sum Test Feature technique (RST), as opposed to models lacking feature selection.
- Comparison among six supervised classification algorithms, videlicet, K-Nearest Neighbor (KNN), Linear Support Vector Machine (SVM-L), Polynomial Support Vector Machine (SVM-P), Radial Basis Function SVM (SVM-R), Gaussian Naive Bayes (GNB) and Decision Tree (DT), and the best classifier is determined using F-Measure and AUC.
- To analyze the impact of Synthetic Minority Oversampling (SMOTE) on the performance of requirements classification.

The performance of classifiers is determined using F-Measure and Area under the ROC Curve (AUC). Friedman test is then performed in order to substantiate the most accurate classification structure based on the performance measures. The research presented in this study is conducted on the PROMISE dataset containing 625 labelled requirements written in natural language derived from a set of 15 projects. This work thoroughly compares different text vectorization techniques, feature selection techniques, and classification algorithms in more detail than any previous work. These comparisons will serve as a reference for industry professionals in choosing the most suitable methods for Requirements classification and act as motivation for further studies within and beyond the domain of text classification.

2   Related Work

This section details about the past studies available in the broad domain of software requirement classification.

The literature of software requirement classification can be broadly divided into three categories: a.) Functional vs Non-functional requirements, b.) functional requirement classification into various sub-categories (such as external communication, business constraints, user interactions, etc.), and, c.) non-functional requirements classifications into various sub-categories (such as security, reliability, performance, etc.). This work majorly deals with the classification of software requirements into two categories: functional and non-functional requirements. Canedo et al. [1] classified requirements into functional and non-functional requirements. They performed a comparative analysis among three vectorization techniques: Bag of words, TF-IDF and Chi Square. The experimental evaluation on PROMISE dataset shows that TF-IDF technique shows best classification performance. In this work, we compared the best performing vectorization technique (TF-IDF) with four other techniques.

Quba et al. [2] also classified software requirements using machine learning in functional and non-functional requirements. They used feature normalization, feature extraction and feature selection techniques along with machine learning classification. The BoW with SVM and KNN algorithms is used for classification. The experimental evaluation using PROMISE dataset shows that the use of BoW with SVM is better than the use of BoW with KNN algorithm. An average F-measure up to 0.74 has been obtained for classification.

Haque et al. [3] presented a comparative analysis on requirement classification using different feature extraction techniques such as BoW and TF-IDF with various classification algorithms named as Multinomial Naive Bayes (MNB), Gaussian Naive Bayes (GNB), Bernoulli Naive Bayes (BNB), K-Nearest Neighbors (KNN), Support Vector Machines (SVM), Stochastic Gradient Descent SVM (SGD-SVM), and Decision Tree. The experimental evaluation using PROMISE dataset shows that SGD-SVM classifier achieves best results with precision, recall, F1-score, and accuracy reported as 0.66, 0.61, 0.61, and 0.76 respectively.

From above literature work, it is observed that TF-IDF and BOW are commonly used by most authors. In recent years, word-embedding techniques have seen interest and their applicability across a wide range of problem domains. In this work, We have compared the predictive ability of different word-embedding techniques for software requirement classification. The above models validated with help of frequently used classifiers [1,3].

3   Study Design

This section presents the details regarding various design settings used for this research.

3.1   Experimental Dataset

The proposed solution for software requirement is validated with same datasets which was used by the Cleland-Huang and his team [4,5]. Cleland-Huang and his team extracted this data with the help of MS students at DePaul University and make available for public research at PROMISE repository[1]. The description of functional and non-functional (Quality) for this dataset is shown in Fig. 1. The first notable observation from Fig. 1 is that the number of function and non-function requirement are almost same. While, the number of non-function requirement or quality requirement are not same i.e., 382 out of 625 are quality requirement.

---

[1] http://promisedata.org/2019/index.html.

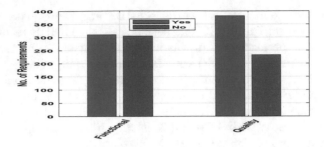

Fig. 1. Data-sets

3.2 Data Normalization

Data preprocessing is important prior to feeding data into the model as it helps in data standardization. The following steps have been taken in sequential order to Normalize the dataset:

- All the characters in the requirements documents are converted to lowercase letters.
- All alphanumeric characters and the symbols are retained while others are deleted
- Commonly used words such as 'a' and 'the' called the stop-words and the words with length less than 2 are removed owing to their insignificance.
- Finally, sentences in the corpus are tokenized into words.

3.3 Feature Extraction using Word Embedding

Word Embedding or Text Vectorization is the process of encoding text into numerical vectors to extract features from the source data in a format compatible with Machine Learning models. Vectorization techniques are beneficial in effective information retrieval and relevance ranking, which in turn help in improving the performance of Machine Learning Models. This study compares the performance of five word embedding techniques, namely Continuous Word2Vec, Skip-Gram (SKG), Bag of Words (CBOW), Fast text and Global Vectors for Word Representation (GloVe) in the classification phase. Each requirement, after being normalized, undergoes vectorization using the mentioned Vector Space Models, which converts them into a two-dimensional array with dimensions.

3.4 Feature Selection and Dimensionality Reduction

Feature selection is the process of reducing the number of input variables when developing a predictive model to prevent irrelevant or partially relevant features from negatively impacting the model's performance using some statistical approach like Rank-Sum Test in our case [6]. This step not only reduce overfitting and training time, but it also improves the accuracy of the model. Dimensionality reduction refers to techniques that reduce the number of input features in a dataset in accordance with feature selection. All the feature vectors obtained in the previous step undergo dimensionality reduction using PCA (Principal Component Analysis) [7] and RST (Rank-Sum test) to obtain final feature vectors ready for training the models. The original feature vectors which comprised of all the features were also maintained, to analyze the impact of Feature Selection.

3.5   Classification

Prior to fitting the Machine Learning model, Max-Min Scaling is performed to avoid the bias introduced by various features measured at different scales. The dataset is divided into training and testing subsets using K-Fold Cross-validation with the k value as 10 and classified using six machine learning algorithms: K-Nearest Neighbor (KNN), Linear Support Vector Machine (L-SVM), Polynomial Support Vector Machine (P-SVM), Radial Basis Function SVM (RBF- SVM), Gaussian Naive Bayes (GNB) and Decision Tree (DT) [8].

3.6   Minority Oversampling

Since the instances of class labels are not same i.e., the dataset is somewhat imbalanced. Imbalanced classifications are error-prone because of few examples of the minority class for a model to learn the decision boundary effectively. The skewed dataset can be balanced by oversampling the minority class. This work uses SMOTE (Synthetic Minority Oversampling Technique) [9], which resamples the minority dataset using the K-Nearest Neighbor algorithm to create synthetic data. We have also compared the models trained on SMOTE data with original data.

4   Research Methodology

The software requirement classification models depend on several factors such as extracting numerical features from requirement documents, selection of important features, data balancing, and machine learning techniques. To properly evaluate the effectiveness of our research work, we have designed framework illustrates in Fig. 2. From Fig. 2, we found that the proposed solution contains several major components. The first component is embedding components that helps to gathers numerical feature from document in terms of vectors. In this work, we have used five different embedding techniques to extract numerical features from requirement documents and also compared these approaches with frequently used techniques called TFIDF. After finding numerical representation of requirement documents, the second component called SMOTE has been used to balance data. The next component called feature selection techniques applied on both original data and balanced data to remove irrelevant features and select best combination of numerical features. In this experiment, we have used Rank-Sum test to find features having capability to differentiate between functional and non-functional requirements and Principal component analysis to improve variance of calculated numerical features. Finally, the requirement classifications are developed using six different classifiers and validated using 10-fold cross validation techniques. The performance of requirement classifications developed models are extracted in terms of F-Measure and AUC and compared using Box-plot and Friedman test [10].

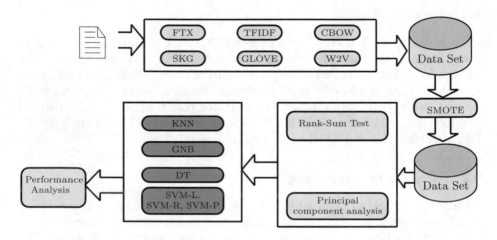

Fig. 2. Research framework

5   Empirical Results and Analysis

In this paper, we have trained software requirement classification models on six different classification techniques and evaluated them with using two afore-mentioned metrics such as: F-Measure and AUC value. The results of software requirement classification models are present in Table 1. Similarly, Fig. 3 shows the Area Under Curve (AUC) for the models trained on SMOTE data and TF-IDF features. The Area Under Curve (AUC) for other combinations are of similar type. The first notable result is that the models trained for software requirement classification have the AUC value greater than 0.7 for most of the cases represent that the trained models have ability to classify functional and non-functional requirements. The second notable result is that the software requirement classification models trained on SMOTE balanced data have the highest performance and Simulated F-Measure and AUC value of 0.91 and 0.98 respectively. Comparing results of different classifiers for software requirement classification, support vector machine with rbf kernel have higher F-Measure and AUC value. However, we have also observed opposite result where models trained using Gaussian naïve Bayes.

Table 1. AUC and F-measure values

		AUC											
		Original Data						SMOTE					
		All Features											
Embedding	Requirement	KNN	GNB	DT	SVM-L	SVM-P	SVM-R	KNN	GNB	DT	SVM-L	SVM-P	SVM-R
TFIDF	FUN	0.65	0.66	0.60	0.75	0.73	0.75	0.67	0.67	0.60	0.76	0.73	0.75
TFIDF	QUA	0.64	0.63	0.61	0.80	0.68	0.77	0.74	0.86	0.82	0.98	0.97	0.97
CBOW	FUN	0.67	0.55	0.57	0.72	0.70	0.74	0.69	0.55	0.55	0.73	0.71	0.75
CBOW	QUA	0.63	0.51	0.58	0.71	0.72	0.73	0.86	0.61	0.69	0.91	0.90	0.93
SKG	FUN	0.66	0.59	0.56	0.76	0.74	0.75	0.67	0.60	0.57	0.76	0.74	0.76
SKG	QUA	0.64	0.63	0.61	0.81	0.78	0.77	0.87	0.71	0.74	0.95	0.94	0.94
GLOVE	FUN	0.71	0.70	0.53	0.76	0.73	0.78	0.72	0.70	0.58	0.76	0.73	0.79
GLOVE	QUA	0.68	0.68	0.57	0.71	0.74	0.75	0.89	0.81	0.76	0.91	0.93	0.96
W2V	FUN	0.74	0.72	0.62	0.79	0.74	0.81	0.74	0.72	0.63	0.80	0.75	0.81
W2V	QUA	0.69	0.70	0.60	0.73	0.76	0.78	0.90	0.86	0.71	0.92	0.94	0.97
		RST											
TFIDF	FUN	0.70	0.78	0.70	0.86	0.81	0.84	0.70	0.79	0.68	0.85	0.80	0.83
TFIDF	QUA	0.73	0.73	0.74	0.87	0.81	0.89	0.90	0.88	0.84	0.96	0.94	0.94
CBOW	FUN	0.73	0.74	0.57	0.81	0.76	0.83	0.74	0.74	0.55	0.82	0.76	0.82
CBOW	QUA	0.69	0.72	0.56	0.83	0.81	0.82	0.88	0.74	0.71	0.92	0.91	0.94
SKG	FUN	0.68	0.69	0.56	0.79	0.69	0.79	0.69	0.69	0.60	0.78	0.71	0.79
SKG	QUA	0.65	0.73	0.61	0.78	0.79	0.78	0.88	0.80	0.74	0.92	0.93	0.94
GLOVE	FUN	0.74	0.78	0.59	0.79	0.73	0.81	0.74	0.78	0.60	0.79	0.72	0.81
GLOVE	QUA	0.69	0.77	0.58	0.76	0.75	0.78	0.90	0.84	0.75	0.91	0.91	0.95
W2V	FUN	0.77	0.81	0.64	0.81	0.81	0.85	0.78	0.81	0.64	0.81	0.81	0.85
W2V	QUA	0.74	0.82	0.62	0.82	0.81	0.84	0.90	0.89	0.74	0.93	0.93	0.96
		PCA											
TFIDF	FUN	0.66	0.67	0.62	0.63	0.64	0.71	0.66	0.67	0.61	0.64	0.65	0.73
TFIDF	QUA	0.66	0.67	0.65	0.69	0.70	0.75	0.70	0.76	0.77	0.93	0.93	0.94
CBOW	FUN	0.62	0.57	0.55	0.63	0.68	0.64	0.65	0.58	0.54	0.63	0.68	0.65
CBOW	QUA	0.62	0.54	0.62	0.63	0.71	0.62	0.74	0.64	0.71	0.85	0.9	0.88
SKG	FUN	0.68	0.64	0.61	0.75	0.72	0.76	0.69	0.64	0.59	0.75	0.73	0.76
SKG	QUA	0.70	0.64	0.57	0.76	0.79	0.80	0.89	0.75	0.76	0.92	0.93	0.96
GLOVE	FUN	0.67	0.62	0.55	0.70	0.70	0.74	0.69	0.63	0.55	0.70	0.70	0.74
GLOVE	QUA	0.67	0.62	0.57	0.70	0.73	0.73	0.86	0.76	0.70	0.90	0.91	0.95
W2V	FUN	0.74	0.67	0.56	0.76	0.74	0.77	0.76	0.68	0.57	0.77	0.74	0.78
W2V	QUA	0.73	0.68	0.60	0.72	0.76	0.74	0.87	0.80	0.73	0.91	0.92	0.96
		F-Measure											
		All Features											
TFIDF	FUN	0.61	0.66	0.54	0.65	0.44	0.57	0.64	0.68	0.56	0.67	0.47	0.61
TFIDF	QUA	0.66	0.74	0.75	0.79	0.77	0.78	0.19	0.84	0.83	0.91	0.89	0.91
CBOW	FUN	0.58	0.62	0.57	0.67	0.66	0.62	0.64	0.61	0.55	0.68	0.68	0.64
CBOW	QUA	0.76	0.5	0.68	0.76	0.78	0.77	0.68	0.51	0.68	0.85	0.84	0.86
SKG	FUN	0.62	0.42	0.55	0.68	0.68	0.62	0.63	0.43	0.57	0.7	0.68	0.63
SKG	QUA	0.74	0.75	0.72	0.8	0.79	0.77	0.75	0.72	0.73	0.87	0.85	0.88
GLOVE	FUN	0.63	0.64	0.5	0.66	0.63	0.66	0.67	0.66	0.57	0.67	0.65	0.68
GLOVE	QUA	0.75	0.72	0.68	0.74	0.76	0.76	0.77	0.76	0.75	0.86	0.86	0.88
W2V	FUN	0.66	0.61	0.62	0.69	0.65	0.71	0.69	0.62	0.64	0.72	0.68	0.71
W2V	QUA	0.76	0.76	0.71	0.77	0.78	0.77	0.78	0.79	0.7	0.87	0.88	0.89
		RST											
TFIDF	FUN	0.59	0.74	0.65	0.7	0.32	0.77	0.6	0.75	0.63	0.72	0.38	0.77
TFIDF	QUA	0.8	0.83	0.8	0.84	0.77	0.87	0.86	0.88	0.83	0.89	0.82	0.9
CBOW	FUN	0.64	0.69	0.57	0.74	0.73	0.73	0.68	0.69	0.57	0.75	0.72	0.74
CBOW	QUA	0.78	0.7	0.65	0.83	0.81	0.83	0.74	0.61	0.7	0.85	0.85	0.86
SKG	FUN	0.61	0.53	0.54	0.71	0.64	0.68	0.63	0.53	0.6	0.72	0.67	0.68
SKG	QUA	0.75	0.78	0.71	0.79	0.78	0.78	0.77	0.74	0.73	0.85	0.86	0.87
GLOVE	FUN	0.66	0.72	0.56	0.71	0.65	0.73	0.68	0.73	0.58	0.71	0.65	0.72
GLOVE	QUA	0.76	0.77	0.7	0.78	0.78	0.78	0.79	0.78	0.74	0.85	0.86	0.88
W2V	FUN	0.71	0.72	0.62	0.72	0.72	0.77	0.73	0.73	0.65	0.72	0.72	0.78
W2V	QUA	0.77	0.81	0.73	0.8	0.8	0.79	0.77	0.8	0.73	0.86	0.89	0.89
		PCA											
TFIDF	FUN	0.67	0.63	0.6	0.56	0.56	0.58	0.68	0.64	0.61	0.58	0.59	0.6
TFIDF	QUA	0.37	0.7	0.72	0.73	0.74	0.77	0.07	0.72	0.74	0.84	0.84	0.88
CBOW	FUN	0.65	0.62	0.55	0.58	0.65	0.55	0.69	0.62	0.53	0.59	0.66	0.56
CBOW	QUA	0.72	0.57	0.72	0.72	0.75	0.74	0.4	0.56	0.71	0.77	0.84	0.79
SKG	FUN	0.66	0.59	0.6	0.65	0.68	0.64	0.7	0.6	0.59	0.68	0.69	0.65
SKG	QUA	0.75	0.72	0.69	0.78	0.79	0.78	0.76	0.65	0.75	0.85	0.87	0.9
GLOVE	FUN	0.59	0.58	0.51	0.59	0.64	0.61	0.65	0.58	0.54	0.6	0.66	0.62
GLOVE	QUA	0.75	0.7	0.7	0.75	0.77	0.76	0.66	0.68	0.69	0.85	0.85	0.88
W2V	FUN	0.66	0.62	0.55	0.67	0.65	0.62	0.69	0.63	0.57	0.68	0.68	0.65
W2V	QUA	0.77	0.76	0.69	0.76	0.77	0.77	0.68	0.72	0.72	0.86	0.87	0.9

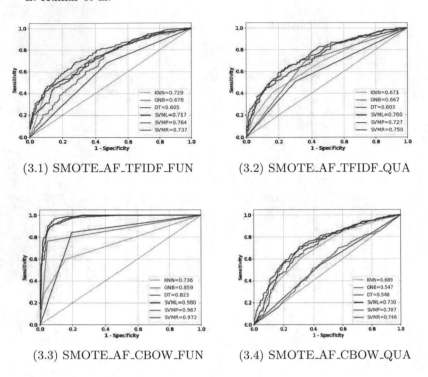

(3.1) SMOTE_AF_TFIDF_FUN   (3.2) SMOTE_AF_TFIDF_QUA

(3.3) SMOTE_AF_CBOW_FUN   (3.4) SMOTE_AF_CBOW_QUA

Fig. 3. ROC curve

5.1   Comparative Analysis

In this section, we further investigated the performance score of the software requirement classification models developed using different feature selection, SMOTE, word-embedding, and classification techniques. We have investigated the reason of selecting best techniques using box-plot diagram and Friedman test.

5.2   Word-Embedding

For our experiment, we have selected five different types of embedding techniques to serve as our experimental baseline for extracting features from software requirement documents. These extracted features are used as an input value of the software requirement classification models. In this work, we have investigated the predictive success of these techniques using Box-plot diagram as mentioned in Fig. 4a. Form this Fig. 4a, it has been interesting to observe that the models trained using word2vec word embedding performed better than the models trained using TF-IDF. While Comparing average AUC performance, we found that the models trained on Word2Vec have more than 0.75 average performance however, models trained on TF-IDF have less than 0.75 average performance. Similarly, we also found that the models trained on SKG and GLOVE have better performance as compared to TF-IDF models. This analysis would seem to indicate that the word-embedding have better ability to extract information from software requirement documents.

We have also presented the results of Friedman test in Table 2a on word-embedding techniques for deeper comparative analysis. The Friedman test is used to navigate our considered null hypothesis "The software requirement classification models do not depend on the features extracted form requirement documents by using different techniques". The above hypothesis is accepted if calculated Friedman p-value greater than 0.05. The first notable result is that the considered null hypothesis rejected i.e., the models developed for software requirement classification depends on the features extracted form requirement documents by using different techniques. The second notable result is that the word2vec has lower rank represent that the word2vect extract import information from requirement documents that helps to improve significantly performance of the software requirement classification models.

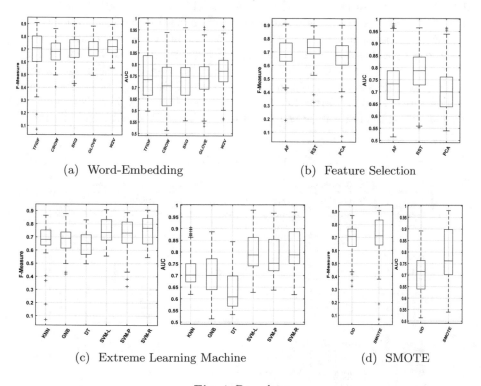

(a) Word-Embedding

(b) Feature Selection

(c) Extreme Learning Machine

(d) SMOTE

Fig. 4. Box-plot

5.3 Classification Techniques

In this section, we will broadly discuss the best classifier for software requirement classification. In this work, six different types of machine learning techniques such as Support vector machine with three different kernels, Gaussian
naïve Bayes classifier, k-nearest neighbor, and decision tree have been employed
to train requirement classification models by taking extracted information from
word-embedding as an input. The above trained models are validated with help
of 10-fold cross validation concept and the capability these techniques to classify requirements are examine using Box-plot diagram as mentioned in Fig. 4c.
The first notable finding from Fig. 4c is that the models trained using support
vector machine with linear and RBF kernel performed better than the models
trained using other techniques. The trained models using SVM-R have approximately 0.8 average AUC performance value, while models trained using decision
tree have only 0.6 average AUC performance value. Further, the performance
of trained models using different techniques have been investigated using Friedman test as results mentioned in Table 2b. This test is used to navigate the
null hypothesis "The software requirement classification models do not depend
on the training techniques". The first important finding from Table 2b is that
the above hypothesis is rejected because of P-value $<= 0.05$. Hence, we found
the that the software requirement classification models depend on the training
techniques. The second important finding from Table 2 is that models trained
using SVM with RBF kernel has lower rank represent that the performance of
the models significantly improvement after applying SVM with RBF kernel.

Table 2. Friedman test

(a) Word-Embedding

Technique	F-Measure Mean rank	AUC Mean rank
TFIDF	2.90	2.64
CBOW	3.65	4.25
SKG	2.92	3.13
GLOVE	3.51	3.32
W2V	2.01	1.67
P_value	$\leq$0.05	$\leq$0.05

(b) Classification Techniques

Technique	F-Measure Mean rank	AUC Mean rank
KNN	3.87	4.18
GNB	4.25	4.65
DT	5.30	5.77
SVM-L	2.50	2.28
SVM-P	2.85	2.68
SVM-R	2.23	1.43
P_value	$\leq$0.05	$\leq$0.05

(c) Feature Selection

Technique	F-Measure Mean rank	AUC Mean rank
AF	2.18	2.16
RST	1.41	1.27
PCA	2.41	2.58
P_value	$\leq$0.05	$\leq$0.05

(d) SMOTE

Technique	F-Measure Mean rank	AUC Mean rank
OD	1.84	1.86
SMOTE	1.16	1.14
P_value	$\leq$0.05	$\leq$0.05

5.4   Feature Selection Techniques

For our experiment, we have used extracted features from software requirement documents using different embedding techniques as an input. So, the performance of the requirement classification also depends on the feature sets. In this work, we further investigate the performance of the requirement classification models developed by taking all feature as input with two different feature selection techniques such as: PCA [7] and RANK-SUM test [6]. The above investigation is done using Box-Plot diagram and Friedman test as mentioned in Fig. 4b and Table 2c respectively. As evidenced by Fig. 4b, the requirement classification model trained by taking relevant sets of features computed using Rank-Sum test (RST) performed better. While Comparing the results with AF, we found that the models trained on RST gain 0.07 in average AUC value. Based on this results, we also found that there exist some irrelevant features in AF that reduce the performance of the requirement classification models. Further, we have also imported Friedman test to get better conclusion for requirement classification. The motive behind using this test to validate our considered null hypothesis: "The software requirement classification models do not depend on the input sets of features". Results are given in Table 2c. The lower p-value of Table 2c confirm that the above hypothesis is rejected i.e., the software requirement classification models depend on the input sets of features. The other important finding from Table 2c is that models trained by taking relevant sets of features computed with help of Rank-Sum test have significant improvement in their performance.

5.5   SMOTE

The final comparison of our experiment it to compare the capability of requirement classification models trained on original data with balanced data using SMOTE. The above comparison is done using Box-Plot diagram and Friedman test as mentioned in Fig. 4d and Table 2d respectively. As evidenced by Fig. 4d, the requirement classification model trained on sampled data using SMOTE performed better. Further, we have also imported Friedman test to validate our considered null hypothesis: "The software requirement classification models do not depend on the nature of the data i.e., imbalanced data". The first important observation from Friedman test results as mentioned in Table 2d that the requirement classification models depend on the nature of the data because of P-value $<= 0.05$. The other important finding from Table 2d is that models trained on balanced data using SMOTE have lower rank as compared to original data. The low rank of SMOTE confirm that the models trained on balanced data have significantly better capability to classify requirement.

6   Conclusion

Classification of functional and non-functional requirement based on requirement description is very important step for software development process that helps to improve quality like reliability, maintainability, portability etc.. The functional and non-functional requirement description in the requirement document should be analyzed carefully and mapped into their corresponding types. In this work, different types of embedding techniques with different classifiers have been incorporated to develop requirement classification models that help to automate the above process. The important finding which have been obtained from our experiment are summarized as follows:

- the models trained using word2vec word embedding performed better than the other embeddings.
- the models trained using SVM with RBF kernel performed better for requirement classification as compared to other techniques.
- the models trained by taking relevant sets of features computed with help of Rank-Sum test have significant improvement in the performance of requirement classification models.
- the models trained on balanced data have significantly better capability to classify requirement.

Based on this experiment, we motivated to explore other machine learning like ensemble learning and deep learning for requirement classification which can take advantage of text mining. It would also be of merit to investigate the performance of the models by changing sampling technique and feature selection techniques.

References

1. Canedo, E.D., Mendes, B.C.: Software requirements classification using machine learning algorithms. Entropy **22**(9), 1057 (2020)
2. Quba, G.Y., Al Qaisi, H., Althunibat, A., AlZu'bi, S.: Software requirements classification using machine learning algorithm's. In: 2021 International Conference on Information Technology (ICIT), pp. 685–690 (2021)
3. Haque, M.A., Rahman, M.A., Siddik, M.S.: Non-functional requirements classification with feature extraction and machine learning: an empirical study. In: 2019 1st International Conference on Advances in Science, Engineering and Robotics Technology (ICASERT), pp. 1–5. IEEE (2019)
4. Cleland-Huang, J., Settimi, R., Zou, X., Solc, P.: Automated classification of non-functional requirements. Requir. Eng. **12**(2), 103–120 (2007)
5. Lu, M., Liang, P.: Automatic classification of non-functional requirements from augmented app user reviews. In: Proceedings of the 21st International Conference on Evaluation and Assessment in Software Engineering, pp. 344–353 (2017)
6. O'brien, P.C.: Comparing two samples: extensions of the t, rank-sum, and log-rank tests. J. Am. Statist. Assoc. **83**(401), 52–61 (1988)
7. Ringnér, M.: What is principal component analysis? Nat. Biotechnol. **26**(3), 303–304 (2008)
8. Joachims, T.: Svmlight: support vector machine. In: SVM-Light Support Vector Machine, vol. 19, no. 4. University of Dortmund (1999). http://svmlight.joachims.org/

9. Fernández, A., Garcia, S., Herrera, F., Chawla, N.V.: Smote for learning from imbalanced data: progress and challenges, marking the 15-year anniversary. J Artif. Intell. Res. **61**, 863–905 (2018)
10. Chatfield, M., Mander, A.: The Skillings-Mack test (Friedman test when there are missing data). Stata J. **9**(2), 299–305 (2009)

Topic Guided Image Captioning with Scene and Spatial Features

Usman Zia[1(✉)], M. Mohsin Riaz[2], and Abdul Ghafoor[1]

[1] National University of Sciences and Technology (NUST), Islamabad, Pakistan
usman.phd@students.mcs.edu.pk, abdulghafoor-mcs@nust.edu.pk
[2] Comsats University, Islamabad, Pakistan
mohsin.riaz@comsats.edu.pk

Abstract. Automatic generation of captions for visual contents has recently emerged as a challenging research field due to it's enormous impact in areas like computer vision, information retrieval, autonomous vehicles and natural language processing. Traditional models mainly focus on single aspect of the visual features to generate descriptions. The proposed model incorporates spatial information of salient objects capturing detailed characteristics coupled with scene category to incorporate general image setting. These extracted features are processed by topic-aware attention-based language model to generate human like captions. Performance of the proposed model is compared with state-of-the-art research through evaluation on benchmark image captioning datasets. The experimental results depict the promising performance of the proposed model compared with the captioning models proposed in recent literature.

1 Introduction

Automatic generation of sequence of words known as caption for an image has emerged as a challenging research field due to it's enormous impact in areas like computer vision, information retrieval, autonomous vehicles and natural language processing (NLP) [1]. The image captioning models are essentially multi-classification problems with the objective to assign textual descriptions to an image based on the correlation between the textual and visual vocabulary [2]. Models proposed in recent literature exploit the correlation between the image features extracted through deep convolutional networks (such as VGG [3]) and the word vocabulary represented as embedding vectors to generate captions. These models employ an encoder-decoder framework. The input image is encoded into an intermediate representation of the information contained within the image. Encoder uses visual convolutional neural network (CNN) to output either a single feature vector or computes feature vectors of regions within the image. Subsequently, these features are decoded into a descriptive text sequence through sequence generating RNN [4–6].

Scene understanding by human visual cortex is strongly augmented by considering different levels of specificity such as a member of a semantic category

L. Barolli et al. (Eds.): AINA 2022, LNNS 450, pp. 180–191, 2022.
https://doi.org/10.1007/978-3-030-99587-4_16

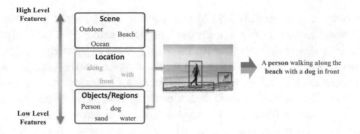

Fig. 1. Proposed model incorporates both scene understanding and salient objects

(e.g., kitchen, mall, hockey field); size, shape and location of objects (e.g., desert is identifiable given the presence of fine-grained yellow sand) [7]. [8] suggested that high level aspects of images are inextricably linked to salient objects present in the scene. [9] argued that the human visual system analyzes various aspects of scenes, ranging from salient objects to broader semantic and spatial properties (e.g. scene category). Furthermore, location of different objects also contributes considerably in describing an image, since it reflects the spatial relationships of objects in an image [10]. Furthermore, word embedding models like *Word2Vec* [11] or *GloVe* [12] utilized in traditional models are unable to detect contiguity within semantic space in the generated embedding and thus cannot differentiate polysemous words. Topic models [13,14] can associate each word with a unique topic based on probability distribution over context words to capturing the polysemy in vocabulary. The proposed model incorporates both scene and salient object features coupled with spatial information along with the topic discerning embedding for generation of captions (Fig. 1).

The model utilizes dual channel deep feature extractors for encoding scene category, object and spatial features. These features are consumed by RNN based decoder trained to generate word embedding using topic modelling. The attention mechanism used in decoder generates a spatial map from both scene and object feature vectors highlighting image regions relevant to each predicted word to improve caption generation.

2   Related Work

Sequence generation models [15,16], using encoder-decoder framework have been extensively used in caption generation problems. Model proposed in [15] estimated the probability of next word based on the image features from CNN and the previous word. Similarly, model proposed by [16] uses a combination of CNN and RNN to maximize a likelihood function to generate image description.

Visual attention [17–19] has recently been employed, to steer the caption generation by focusing on the image features when computing word sequence. [20] proposed adaptive attention model. In [4], the authors incorporated object level details to boost captioning. R-LSTM proposed in [21] generates captions by first detecting essential parts of image for caption generation. However, these models tend to maintain certain patterns in describing images by neglecting the overall scene information. [22] proposed model which focuses on the coarse features and translate them to a word sequence. [23] consists of stacked coarse decoder and a series of fine decoders. The model randomly samples the intermediate outputs of the first decoder which are not well-defined and thus prone to accumulate errors and hard to train. [10] augmented visual features with CNN output of images formulated by keeping the region of top-n salient objects' bounding box unchanged and setting remaining regions to mean value of the training set. However, replacing a major portion of an image with mean of training dataset can result into considerable loss of information. Furthermore, model augmented the decoder with only localization information and does not consider scene category.

[24] proposed Transformer-like region encoder to which fed LSTM decoder for description generation. Models proposed by [25–27] used Transformer for both encoder and decoder however transformer based architectures lack recursive computation [28]. [29,30] utilized an external tagger with the Transformer to refine generated captions. However, leveraging both scene and spatial feature information has not been explicitly explored in these models.

The proposed approach incorporates both scene category and salient object features coupled with location information along with topic modelling to capture semantic word relationships to feed sequence-to-sequence word generation task. Inspired by [31–33], the proposed model utilizes dual layered bi-directional LSTM layer for generating human like and natural image descriptions based on overall image setting and object level granular details.

3   Proposed Model

The proposed framework follows three stage encoder-decoder architecture as depicted in Fig. 2. In stage *I*, target image is processed through two parallel deep convolutional neural network tunnels for generating spatial object level and scene features. In stage *II*, the semantic relation in the word sequence is achieved through topic modeling of the vocabulary. By processing the training caption set, topic sensitive vocabulary is generated. This topic sensitive vocabulary along with spatial and scene features are in turn fed to a language generation model for computing topic sensitive captions in stage *III*. Each stage of the proposed architecture is explained in succeeding paragraphs.

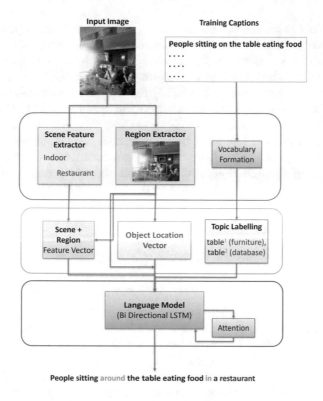

Fig. 2. Proposed model architecture

Suppose that caption C_i of target image I is to be generated from vocabulary of words represented as $\mathbb{V}$. Let each word of the generated caption C be represented as w_v such that $w_v \in \mathbb{V}$ where $\mathbb{V}$ comprises of topic aware word embedding. Let ν and ς represent scene embedding and spatial object features of the target image I.

3.1 Visual Feature Extraction

The proposed model simultaneously encodes scene embedding and salient object features of the input image I. Scene embedding establishes broad image settings while the salient object features expose the detailed object level information for generation of meaningful captions. Let scene embedding of image I be represented by N_ν-dimensional vector ν_I and N_ς-dimensional vector ς depict embedding of salient objects detected in image I. Each CNN tunnel in the proposed model detects local conjunctions of features using convolutional layers for mapping image to a feature map. Fully connected layers of each tunnel in turn compute stochastic likelihood representation using activation maps generated by the concatenation of convolutional, non-linearity, rectification and pooling layers [34]. These two parallel visual feature extraction tunnels process the target image I to generate scene and spatial feature embedding.

Proposed model follows Faster R-CNN [35] to detect top n box proposals represented as ς_i (where i = 1 to n). Region of interest (RoI) pooling is used to extract a small feature map (e.g. 14×14) for each box proposal. These feature maps are then batched together as input to the final layers of CNN.

Next, the spatial feature vector l_i of each region proposal ς_i is computed using the coordinates of bounding box along with the height and width of the region. Let $[x_i^1, y_i^1]$ and $[x_i^2, y_i^2]$ represent the coordinates of left top and right bottom corner of the bounding box, w_i depict the width and h_i represent the height of the region $\{\varsigma_i$. Spatial feature χ_i of each bounding box is computed as:

$$\chi_i = (\log(\frac{(x_i^1 - x_i^2)}{h_i}), \log(\frac{(y_i^1 - y_i^2)}{w_i})) \tag{1}$$

The spatial feature vector l_i is resultantly computed as

$$l_i = ReLU(Emb(\chi_i)) \tag{2}$$

where $Emb(\cdot)$ calculates a high-dimensional embedding following the functions PE_{pos} described in [36]. Finally, both image feature vector ς_i and spatial vector l_i of each bounding box are processed through fully connected layer to obtain N_ς-dimensional vector representing the salient object feature details.

Spatial feature vector represents intrinsic relationships among the objects in image. Proposed caption generation tunnel is augmented by incorporating scene information extracted through CNN trained on Places365 dataset [37]. Target image is resized to dimension 224×224 for the CNNs and fed to the pre-trained Places365 CNN model after data augmentation. N_ν-dimensional vector ν_i is generated for each target image I after processing by scene feature detector.

3.2   Textual Feature Extraction

Performance of traditional embedding techniques such as *Word2vec* [38], *GloVe* [39] and *Elmo* [40] is not encouraging for polysemous words like *solution* (which can represent both *a mixture of liquids* and *a way to resolve a problem*). Topic models enable representation of topics as multinomial distribution over terms resultantly capturing polysemous nature of words. The proposed model utilizes topic modelling technique proposed in [41] for generating captions.

Topic models assign a specific topic κ to each occurrence of word based on context. As a result, contextually similar occurrences of words are clustered together. Let $\mathcal{C}_I^j$ represent jth caption associated with Ith image in the training dataset where $\mathcal{C}_I$ represent the group of captions associated with the Ith image. The proposed model follows [41] to identify contextual topics of the words in vocabulary $\mathbb{V}$. Topic labelling extends the original vocabulary to $\acute{\mathbb{V}}$ which now contains word-topic pairs as:

$$w \leftarrow\!-- \{v_l, \kappa_p\} \tag{3}$$

This eventually increases the overall size of vocabulary as word v may now be included in multiple word-topic pairs due to association with multiple topics κ based on the context.

Following [42], vocabulary $\check{V}$ comprising of word-topic pairs (v_l, κ_t) is utilized to compute word embedding ε_j^i of (j)th word w_j belonging to caption C of Ith image. Model is trained to maximize the log-likelihood of the context words $v_l - n,...,v_l - 1, v_l + 1,...,v_l + n$ given word-topic pair (v, κ) represented as follows:

$$\Gamma_{HTLE} = \frac{1}{N_{\check{V}}} \sum_{l=1}^{V} \sum_{\substack{-c \leq d \leq n \\ j \neq 0}} \log p(v_{l \pm i} \mid v_l^{\kappa}) \tag{4}$$

where $N_{\check{V}}$ is the size of topic-labelled vocabulary, n_c depicts the number of context words considered for computing embedding and κ is the topic computed for each word occurrence v_l.

The goal of language model is to predict the next word in a sentence given the words and context seen previously. The proposed model implements decoder network optimized to maximize the probability of caption generation for ground truth image I given spatial (ν) and scene (ς) image features based on the context vector generated through the attention mechanism. Let C_i be the caption generated by the language model for Ith image, w_v^j be the jth word of caption C_i, n_i be the number of words in C_i such that $w_v^1 ... w_v^j ... w_v^n$ indicate the sequence of words in caption C_i. The proposed model follows chain rule to compute topic sensitive word at each time step. The computation by the language model can be represented as probability function over words in sequence represented as:

$$\log Pr(C|\mathbf{I}) = \prod_{t=1}^{N} Pr(w_t | I_{\nu,\varsigma}, w_0, ..., w_{t-1}) \tag{5}$$

where $I_{\nu,\varsigma}$ represent the feature vector formed by concatenating the scene (ν) and spatial (ς) feature vectors.

[43]. LSTMs [43] learn long-term temporal contexts in sequence generation problems thus avoiding exploding and vanishing gradient problems. Purposed model utilizes bidirectional LSTM to incorporate both the past and the future context information of the generated word sequence by feeding words of caption C to LSTM in forward and backward order. The forward LSTM starts at time $t=1$ to compute forward hidden sequences $\overrightarrow{h}$ while the backward LSTM computes backward hidden sequences $\overleftarrow{h}$ and starts at time $t=T$.

The language model consumes encoded scene (ν_i) and spatial (ς_i) features as input as well as topic sensitive word embedding vector ε_j^i one word at a time. The model is trained to predict the next word embedding ε_v^j, given a sequence of generated words. The log probability of generated sequence of words is given as:

$$\log Pr(C|\mathbf{I}) = \prod_{j=1}^{n_i} Pr(w_j | \mathbf{I}, w_0, ..., w_{j-1}) \tag{6}$$

where n_i is the number of words in caption C of image I.

The decoder network compromised of bi-directional LSTM generates forward and backward embedding for output word as follows:-

$$\overrightarrow{\psi} = \overrightarrow{\mathbb{L}}(\varepsilon_v^j, \varsigma_i, \nu_i; \Psi_L^{up}) \quad \overleftarrow{\mathbb{L}}(\varepsilon_v^j, \varsigma_i, \nu_i; \Psi_L^{dn}) \tag{7}$$

where $\overrightarrow{\mathbb{L}}$ and $\overleftarrow{\mathbb{L}}$ represents two layers of LSTM and Ψ_L^{up} and Ψ_L^{dn} are the trainable weights.

The model predicts a word w_t through scene and spatial image feature vectors of target image I while considering previous word context $w_{1:t-1}$ by computing $P(w_t|w_{1:t-1}, I)$ in forward direction and computing $P(w_t|w_{t+1:T}, I)$ in backward direction. Caption for given image is computed through summation of word probabilities in word sequences generated in both forward and backward direction as follows:

$$p(w_{1:T}|\nu_I, \varsigma_I) = \max(\sum_{t=1}^{T}(\overrightarrow{p}(w_t|\nu_I, \varsigma_I)), \sum_{t=1}^{T}(\overleftarrow{p}(w_t|\nu_I, \varsigma_I))) \tag{8}$$

$$\overrightarrow{p}(w_t|\nu_I, \varsigma_I) = \prod_{t=1}^{T} p(w_t|w_1, w_2, ..., w_{t-1}, \nu_I, \varsigma_I) \tag{9}$$

$$\overleftarrow{p}(w_t|\nu_I, \varsigma_I) = \prod_{t=1}^{T} p(w_t|w_{t+1}, w_{t+2}, ..., w_T, \nu_I, \varsigma_I) \tag{10}$$

where ν_I and ς_I are fine and coarse feature vectors of the target image I.

4   Experiments

Proposed model is evaluated on MSCOCO [16], Flickr8K [44] and Flickr30K [45] which are benchmark datasets in image captioning problems. Each image is randomly mirrored in horizontal or vertical direction and rotated 45°. The augmentation process increased the training dataset size four folds for MSCOCO, Flickr30K and Flickr8K dataset each. The train-val-test split is kept as [80-10-10]% following [42] and results are compared with existing state of the art models. Performance of the model is evaluated across different evaluation metrics including $BLUE - N(N = 1, 2, 3, 4)$ [46], $METEOR$ [47] and $CIDEr$ [48]. Model is trained for 50 epochs with batch size of 32. Training in MSCOCO dataset took around 48 hours to complete. Early-stopping technique is used to break the training iteration by monitoring validation loss convergence with patience of 3 epochs. Beam search is adopted to select best k generated words at a given time step for generation of next word where k is set to 3 for proposed model.

The experiments performed using the proposed model have been aimed to evaluate the performance of the model in generating human like captions with understanding of scene and spatial object level features present in the image.

Fig. 3. Examples of captions generated by the proposed model. Model learns to generate novel captions by incorporating scene and spatial information.

The analysis of generated caption shows some interesting patterns. (1) **Incorporate Word Topics**, the ability of the model to generate captions with word suitable placed as per their semantic relationship thereby incorporating topic modeling learnt by the LSTM; (2) **Generate Novel Captions**, considerable number of captions generated by the proposed model (around 75% of randomly selected 1000 images from MSCOCO Dataset) are novel and do not appear in training corpus although similar to the ground truth captions associated with the images as shown in Fig. 3. This illustrates the capability of the proposed model to generate novel captions by learning scene and spatial object level features of the input image.

Evaluation metrics including $BLEU - N (N = 1, 2, 3, 4)$, $METEOR$ and $CIDEr$ are computed for the proposed model using coco-caption code [?] and evaluated against state of the art caption generation models.

Proposed model achieves competitive performance compared to existing state-of-the-art research. The results of evaluation metric for test images from MSCOCO, Flickr30K and Flickr8K datasets are depicted in Tables 1 and 2.

Table 1. Evaluation of proposed model on 1,000 test images from MSCOCO dataset (Higher value represents better performance)

Model	MSCOCO					
	B1	B2	B3	B4	METEOR	CIDEr
Zia et al. [42]	72.1	53.4	40.6	24.1	20.1	67.3
Wang et al. [49]	68.5	51.1	36.9	26.7	24.7	–
Yao et al. [50]	78.7	62.7	47.6	35.6	27	116.7
Cao et al. [51]	72.3	54.8	40.2	29.3	24.4	97.5
Cheng et al. [52]	79.4	63.6	49.0	37.2	27.9	57.7
Gao et al. [53]	79.4	63.5	48.7	36.8	28.2	120.5
Zhou et al. [54]	**80.2**	–	–	38.0	**28.5**	109.0
Deng et al. [55]	76.7	60.8	46.9	36.0	27.8	114.9
Proposed Model	79.8	**63.9**	**49.3**	36.9	**28.5**	**121.2**

Table 2. Evaluation of proposed model on 1,000 test images from Flickr30K and Flickr8K dataset (Higher value represents better performance)

Model	Flickr30K				Flickr8K			
	B1	B2	B3	B4	B1	B2	B3	B4
Zia et al. [42]	71.3	53.6	21.6	16.6	71.7	54.6	26.6	17.6
Wang et al. [49]	60.7	42.5	29.2	19.9	–	–	–	–
Yao et al. [50]	60.7	42.5	29.2	19.9	–	–	–	–
Cao et al. [51]	–	–	–	–	59.7	40.5	**28.1**	17.4
Gao et al. [53]	73.8	55.1	**40.3**	**29.4**	–	–	–	–
Proposed Model	**73.9**	**55.6**	39.9	29.1	**73.7**	**58.6**	**28.1**	**17.9**

5   Conclusion

The proposed model utilizes dual convolutional tunnels to extract scene category and finer region based spatial visual features and label vocabulary with topics to feed language model. The language model is based on bidirectional LSTM which uses attention mechanism to generate novel human like image captions. The model shows promising performance on three benchmark captioning dataset comparable to existing models by incorporating scene and spatial visual features. Future research directions include integrating transformer architecture into image feature extraction.

References

1. Ling, H., Fidler, S.: Teaching machines to describe images via natural language feedback. In: NIPS (2017)
2. Ramisa, A., Yan, F., Moreno-Noguer, F., Mikolajczyk, K.: BreakingNews: article annotation by image and text processing. IEEE Trans. Pattern Anal. Mach. Intell. **40**(5), 1072–1085 (2018)
3. Simonyan, K., Zisserman, A.: Very deep convolutional networks for large-scale image recognition. In: ICLR (2015)
4. Fang, H., et al.: From captions to visual concepts and back. In: IEEE Conference on Computer Vision and Pattern Recognition (2015)
5. Tan, Y.H., Chan, C.S.: phi-LSTM: a phrase-based hierarchical LSTM model for image captioning. In: ACCV (2016)
6. Karpathy, A., Fei-Fei, L.: Deep visual-semantic alignments for generating image descriptions. IEEE Trans. Pattern Anal. Mach. Intell. **39**(4), 664–676 (2017)
7. Epstein, R.A., Baker, C.I.: Scene perception in the human brain. Annu. Rev. Vis. Sci. (2019)
8. Groen, I.I.A., Silson, E.H., Baker, C.I.: Contributions of low-and high-level properties to neural processing of visual scenes in the human brain. Philos. Trans. Roy. Soc. B Biol. Sci. (2017)
9. Epstein, R.A., Baker, C.I.: Scene perception in the human brain. Annu. Rev. Vis. Sci. (2019)

10. Yang, Z., Zhang, Y.J., Rehman, S., Huang, Y.: Image captioning with object detection and localization. Int. Conf. Image Graph. 109–118 (2017)
11. Mikolov, T., Chen, K., Corrado, G., Dean, J.: Efficient estimation of word representations in vector space. In: ICLR Workshop (2013)
12. Pennington, J., Socher, R., Manning, C.: Glove: global vectors for word representation. In: Proceedings of the 2014 Conference on Empirical Methods in Natural Language Processing (EMNLP), pp. 1532–1543 (2014)
13. Blei, D., Lafferty, J.: A correlated topic model of science. Ann. Appl. Statist. 1(1), 17–35 (2007)
14. Chen, B.: Latent topic modelling of word co-occurence information for spoken document retrieval. In: 2009 IEEE International Conference on Acoustics, Speech and Signal Processing, Taipei, pp. 3961–3964 (2009)
15. Mao, J., Xu, W., Yang, Y., Wang, J., Yuille, A.L.: Explain images with multimodal recurrent neural networks arXiv:1410.1090 (2014)
16. Vinyals, O., Toshev, A., Bengio, S., Erhan, D.: Show and tell: lessons learned from the 2015 MSCOCO image captioning challenge. IEEE Trans. Pattern Anal. Mach. Intell. 39(4), 652–663 (2017)
17. Xu, K., et al.: Show, attend and tell: neural image caption generation with visual attention. Int. Conf. Mach. Learn. (2015)
18. Lu, J., Xiong, C., Parikh, D., Socher, R.: Knowing when to look: adaptive attention via a visual sentinel for image captioning. In: Processing of IEEE Conference on Computer Vision and Pattern Recognition (CVPR), July 2017, pp. 375–383 (2017)
19. Chen, L., et al.: SCA-CNN: spatial and channel-wise attention in convolutional networks for image captioning. arXiv:1611.05594 (2016)
20. Gao, L., Li, X., Song, J., Shen, H.T.: Hierarchical LSTMs with adaptive attention for visual captioning. IEEE Trans. Pattern Anal. Mach. Intell. 42(5), 1112–1131 (2020)
21. Chen, M., Ding, G., Zhao, S., Chen, H., Liu, Q., Han, J.: Reference based LSTM for image captioning. In: Proceeding of 31st AAAI Conference, pp. 3981–3987 (2017)
22. Wu, C., Yuan, S., Cao, H., Wei, Y., Wang, L.: Hierarchical attention-based fusion for image caption with multi-grained rewards. IEEE Access 8, 57943–57951 (2020)
23. Gu, J., Cai, J., Wang, G., Chen, T.: Stack-captioning: coarse-to-fine learning for image captioning. In: AAAI (2018)
24. Huang, L., Wang, W., Chen, J., Wei, X.Y.: Attention on attention for image captioning. ICCV (2019)
25. Guo, L., Liu, J., Zhu, X., Yao, P., Lu, S., Lu, H.: Normalized and geometry-aware self-attention network for image captioning. CVPR (2020)
26. Li, J., Yao, P., Guo, L., Zhang, W.: Boosted transformer for image captioning. Appl. Sci. (2019)
27. Cornia, M., Stefanini, M., Baraldi, L., Cucchiara, R.: Meshed-memory transformer for image captioning. In: CVPR (2020)
28. Fan, A., Lavril, T., Grave, E., Joulin, A., Sukhbaatar, S.: Addressing some limitations of transformers with feedback memory. arXiv:2002.09402v3 (2021)
29. Li, G., Zhu, L., Liu, P., Yang, Y.: Entangled transformer for image captioning. In: ICCV (2019)
30. Liu, F., Ren, X., Liu, Y., Lei, K., Sun, X.: Exploring and distilling cross-modal information for image captioning. arXiv (2020)
31. Cheng, Y., Huang, F., Zhou, L., Jin, C., Zhang, Y., Zhang, T.: A hierarchical multimodal attention-based neural network for image captioning. In: Proceedings of 40th International ACM SIGIR Conference, pp. 889–892 (2019)

32. Hou, J.C., Wang, S.S., Lai, Y.H., Tsao, Y., Chang, H.W., Wang, H.-M.: Audio-visual speech enhancement using multimodal deep convolutional neural networks. IEEE Trans. Emerg. Topics Comput. Intell. **2**(2), 117–128 (2018)
33. Chen, H., Cohn, A.G.: Buried utility pipeline mapping based on multiple spatial data sources: a Bayesian data fusion approach. In: Proceedings of the 22nd International Joint Conference on Artificial Intelligence, pp. 1–9 (2011)
34. Deng, J., Dong, W., Socher, R., Li, L.J., Kai, L., Li, F.-F.: ImageNet: a large-scale hierarchical image database. In: 2009 IEEE Conference on Computer Vision and Pattern Recognition, pp. 248–255 (2009)
35. Ren, S., He, K., Girshick, R., Sun, J.: Faster R-CNN: towards real-time object detection with region proposal networks. Adv. Neural Inf. Process. Syst. 91–99 (2015)
36. Vaswani, A., et al.: Attention is all you need. NeurIPS (2017)
37. Zhou, B., Lapedriza, A., Khosla, A., Oliva, A., Torralba, A.: Places: a 10 million image database for scene recognition. IEEE Trans. Pattern Anal. Mach. Intell. (2017)
38. Mikolov, T., Corrado, G.S., Chen, K., Dean, J.: Efficient estimation of word representations in vector space. In: ICLR, pp. 1–12 (2013)
39. Pennington, J., Socher, R., Manning, C.: Glove: global vectors for word representation. In: Proceedings of the 2014 Conference on Empirical Methods in Natural Language Processing (EMNLP), pp. 1532–1543 (2014)
40. Peters, M., et al.: Deep contextualized word representations. In: Proceedings of the 2018 Conference of the North American Chapter of the Association for Computational Linguistics: Human Language Technologies, pp. 2227–2237 (2018)
41. Fadaee, M., Bisazza, A., Monz, C.: Learning topic-sensitive word representations. In: Proceedings of the 55th Annual Meeting of the Association for Computational Linguistics, pp. 441–447. Association for Computational Linguistics (2017)
42. Zia, U., Riaz, M.M., Ghafoor, A., Ali, S.S.: Topic sensitive image descriptions. Neural Comput. Appl. pp. 1–9 (2019)
43. Mikolov, T., Karafiát, M., Burget, L., Cernocký, J., Khudanpur, S.: Recurrent neural network based language model. In: Proceedings of the 11th Annual Conference of the International Speech Communication Association, pp. 1045–1048 (2010)
44. Hodosh, M., Young, P., Hockenmaier, J.: Framing image description as a ranking task: data, models and evaluation metrics. J. Artif. Intell. (2013)
45. Young, P., Lai, A., Hodosh, M., Hockenmaier, J.: From image descriptions to visual denotations: new similarity metrics for semantic inference over event descriptions. Trans. Assoc. Computat. Linguist. (2014)
46. Papineni, K., Roukos, S., Ward, T., Zhu, W.J.: BLEU: a method for automatic evaluation of machine translation. In: Proceedings of the 40th Annual Meeting on Association for Computational Linguistics (2002)
47. Banerjee, S., Lavie, A.: METEOR: an automatic metric for MT evaluation with improved correlation with human judgments. In: Proceedings of the ACL Workshop on Intrinsic and Extrinsic Evaluation Measures for Machine Translation and/or Summarization, pp. 65–72 (2005)
48. Vedantam, R., Zitnick, C.L., Parikh, D.: CIDEr: consensus-based image description evaluation. In: 2015 IEEE Conference on Computer Vision and Pattern Recognition (CVPR), pp. 4566–4575 (2015)
49. Wang, Q., Chan, A.B.: CNN+ CNN: convolutional decoders for image captioning. arXiv preprint. arXiv:1805.09019 (2018)
50. Yao, T., Pan, Y., Li, Y., Qiu, Z., Mei, T.: Boosting image captioning with attributes. In: ICCV, pp. 4904–4912 (2017)

51. Cao, P., Yang, Z., Sun, L., Liang, Y., Yang, M.Q., Guan, R.: Image captioning with bidirectional semantic attention-based guiding of long short-term memory. Neural Process. Lett. **50**(1), 103–119 (2019). https://doi.org/10.1007/s11063-018-09973-5
52. Cheng, L., Wei, W., Mao, X., Liu, Y., Miao, C.: Stack-VS: stacked visual-semantic attention for image caption generation. IEEE Access **8**, 154953–154965 (2020)
53. Gao, L., Fan, K., Song, J., Liu, X., Xu, X., Shen, H.T.: Deliberate attention networks for image captioning. In: Proceedings of the AAAI Conference on Artificial Intelligence, pp. 8320–8327 (2019)
54. Zhou, Y., Wang, M., Liu, D., Hu, Z., Zhang, H.: More grounded image captioning by distilling image-text matching model. In: CVPR (2020)
55. Deng, Z., Zhou, B., He, P., Huang, J., Alfarraj, O., Tolba, A.: A position-aware transformer for image captioning. Comput. Mater. Continua (2021)

A Socially-Aware, Privacy-Preserving, and Scalable Federated Learning Protocol for Distributed Online Social Networks

Mansour Khelghatdoust[1(✉)] and Mehregan Mahdavi[2]

[1] Umea University, Umeå, Sweden
mansour.khelghatdoust@cs.umu.se
[2] Sydney International School of Technology and Commerce, Sydney, Australia
mehregan.m@sistc.nsw.edu.au

Abstract. Online Social Networks (OSNs) have been gaining tremendous growth by attracting billions of users from all over the world. Such massive growth leads to scalability and data privacy concerns. Decentralized solutions still are not able to solve privacy and scalability problems efficiently. Hence, recently fully Distributed Online Social Networks (DOSNs) have been proposed. However, despite solving scalability and privacy issues, fully DOSNs impose difficulties in executing data mining and machine learning services which are vital for social networks. In the fully DOSN, each user has only one feature vector and these vectors cannot move to any central storage or other users in a raw form due to privacy issues. In addition, users can directly communicate only with their immediate neighbours/friends in a social network/graph. To cope with these problems, we propose a novel Federated learning algorithm for DOSNs based on the Gossip protocol. We propose a two layer protocol in which the underlying layer is a socially-aware gossip sampling protocol and the upper layer is a push-based merging gossip protocol. The former is responsible for creating a socially-aware random overlay network while the latter, utilizing the sampling protocol, does the training model. We implement our algorithm and through extensive experiments show that the algorithm trains the model up to 88% accuracy compared to the centralized approach.

Keywords: Social network · Federated learning · Big data · Gossip protocols

1 Introduction

Today, millions of people are using Online social networks, from friendship networks like Facebook, to professional networks like LinkedIn. Currently, most of the social networks operate in a centralized fashion with a central service responsible for providing the social network services. The service provider has access to large amounts of data, which can be used for many business-related purposes [1]. However, such unlimited access to private data for social network

© The Author(s), under exclusive license to Springer Nature Switzerland AG 2022
L. Barolli et al. (Eds.): AINA 2022, LNNS 450, pp. 192–203, 2022.
https://doi.org/10.1007/978-3-030-99587-4_17

providers raised privacy and access right concerns among users. Therefore, we are witnessing that researchers and the open source community have been proposing various decentralized solutions (e.g., [2,3]) that eliminate dependency on a centralized provider.

The backbone of an Online Social Network platform comprises of a variety of different services such as search, information dissemination, profile management, storage, application integration, etc. [4]. Data mining and machine learning services are, among others, one of the most important pillars in Online Social Network platforms. However, running such services in fully DOSNs is challenging for different reasons. *Firstly*, each social network user runs the application on its own machine/node and private data are stored in local nodes. Hence, training a machine learning or deep learning algorithms for DOSNs requires techniques to apply on multiple local datasets contained in local nodes without **explicitly exchanging data samples**. *Secondly*, to preserve scalability and privacy, DOSN users can be arranged in a friend-to-friend network with each user maintaining the information related to their direct connections. Such limitation and having **specific network topology** makes training a model even more difficult which may affect the performances of the iterative learning process.

Federated learning is a machine learning technique that trains an algorithm across multiple decentralized servers holding local data samples, without exchanging them. This approach differs from more classical decentralized approaches which often assume that local data samples are identically distributed. According to this definition, training machine learning algorithms in DOSNs can be formulated as a Federated Learning problem. Hence, In this paper, we propose a Federated Learning algorithm respecting DOSNs restrictions and is massively scalable. We introduce a gossip based learning algorithm. Due to DOSNs topology restriction, general gossip learning techniques [5] cannot be directly used. In DOSNs, users are able to communicate only with their immediate neighbors directly while gossip learning algorithms require users freely communicate with each other. To address the problem, we propose a new variant of gossip learning algorithm applicable for such restricted networks including DOSNs. At the heart of a gossip protocol sits a gossip sampling service responsible for providing a continuously changing subset of participating nodes for each peer so that gossip learning protocols utilizing this information, execute distributed learning algorithms.

First, we introduce a new gossip sampling service. Unlike general sampling services like [8], our protocol considers the limitations of communication between users in DOSNs and still can provide to each user, a subset of uniformly distributed IP of users. More precisely, The service provides every participating user with a set of uniform random users from the network, as well as efficient routing paths for reaching the nodes of those users via the restricted network. *Second*, we introduce a novel gossip learning service, which sits on top of the gossip sampling service and does the training task. Unlike general gossip learning, we propose a push-based merging gossip learning service. In our protocol, each node has three roles of **initiator, transmitter, and trainer**. Each network node holds private data and an initial model. At each round, nodes push the local model through the routing path to a random node given by a sampling service. The recipient node merges its local model with the received model and trains with local data. Transmitters are simply responsible for passing the message to the next network node in the routing path. Social networks have various levels of interconnection/interaction, ranging from the level of friends to that of nations and these interconnections can be of importance with regards to the flow of information. Social networks can be different types of networks e.g., scale-free networks [9]. Since gossip learning relies on our sampling service, it makes sure the convergence of the model for nodes even with very low degree distribution as all nodes treated equally for gossiping regardless of social network properties. However, as an amelioration technique, to speed up and ensure convergence for all nodes, the protocol is annotated with a pull-based mechanism for network nodes doesn't receive any model for some rounds.

We evaluate the algorithm through simulation experiments. We use the datasets of three different social networks and train a binary classification algorithm as a machine learning case study. The outline of the paper is as follows. Section 2 summarizes related work. Section 3 elaborates the whole algorithm. In more details, Sect. 3.1 describes socially-aware gossip sampling protocol, followed by an explanation of Push-based Merging Gossip Federated Learning in Sect. 3.2. Section 4 explains evaluation methodology. We discuss experimental results in Sect. 5. Finally, we conclude the paper in Sect. 6.

2   Related Work

In [5], the authors presented a gossip learning algorithm. Similarly to our algorithm they proposed for linear models over fully distributed data. However, it works for unrestricted networks and does not work when there is a restriction of social connectivity. In [17], the authors also introduce the possibility for gossip learning to store multiple data points at each node and to only communicate over a restricted network topology. However, the restriction assumptions is not applicable for social networks. In [12], the authors investigated the gossip learning algorithm over different assumptions such as distribution of the data, the communication speeds of the devices and the connectivity among them. Their results show that lifting these requirements can, in certain scenarios, lead to slow convergence of the protocol or even unfair bias in the produced models. However, we addressed some of these issues in this paper applicable for DOSNs.

Regarding the problem of achieving good information dissemination over a restricted communication topology, Kyasanur et al. [14] propose a technique to build an efficient random overlay over a restricted network, by routing communication through multiple hops. The resulting overlay, being a random graph, allows efficient gossip learning. The approach is able to identify intermediate nodes that are critical in achieving good dissemination. Similar techniques could be used to tune the performance of critical nodes.

3   Solution

First, we explain how we construct a random overlay taking into account communication limitations of DOSNs i.e., only friend-to-friend communication is possible. This serves gossip federated learning algorithm by continuously providing IP address of random network nodes as well as routing path of how to reach them. Next, we describe how gossip federated learning works in a fully distributed manner.

3.1   Construction of Socially-Aware Random Overlay

The sampling probability for social network users is defined by peer sampling algorithms that are used to implement the random walk. We apply gossip-based peer sampling with the advantage that samples are available locally and without delay. We model the underlying social overlay as an undirected graph with unweighted edges. We assume that every network node of a user has knowledge (IP address) of its immediate friends and friends-of-friends which is a common feature in today's OSNs. It facilitates users to expand their networks and does not undermine the scalability of the system. We assume that friends are trusted, non malicious and are willing to act as relays for forwarding messages.

Each node is assigned a unique identifier (such as an IP address) and maintains two groups of caches.

- **Social Cache**: The paths towards immediate neighbors (One hop), called friends and towards two hop neighbors, called friends-of-friends extracted from social overlay.
- **Random Cache**: A small, fixed-sized and continuously changing cache of C entries (typically with the size of 10, 20, 50) of paths towards peers that excerpted from random overlay.

Every time a node joins to the network, it fills in the random cache with entries of social cache starting with lowest degree friends to prevent initial bias toward higher degree friends. Nodes periodically exchange with each other the subsets of their caches, called *Swapping Cache*. In particular, in each round every node selects a copy of the longest waiting entry e from the cache, contacts it through routing path and sends to e a copy of swapping cache. Also, it receives a random subset of e's cache entries. Each node then updates its cache with the exchanged set. More precisely, each node updates its cache by replacing entries that are selected as swapping cache entries in its cache with the received entries.

However, since cache entries to the sampled nodes indicate paths which always start at the source node, after swapping, cache paths do not indicate a path from recipient node to the sampled node. A naive way to solve this issue is to merge reversed path from source node to destination node with the path towards the sampled node in the swapped cache. The problem of this approach is that such paths grow prohibitively large and do not scale. We propose an algorithm to prune swapping cache paths upon forwarding gossip messages at each

---

Algorithm 1. PATH CONSTRUCTION

1: **function** RECONSTRUCT PATH(*path*)
2:     resultPath = emptyList
3:     **if** self.Id **is** sourceNode **then**   ▷ Current node is gossip round initiator
4:         resultPath.addFirst(self.Id);
5:         **return** resultPath;
6:     **end if**
7:     **for all** id ∈ path.reverse() **do**
8:         **if** self.isNeighbor(id) **then**   ▷ id is immediate neighbor of current node
9:             resultPath.addFirst(id);
10:            **break**;
11:        **else if** self.isTwoHopNeighbor(id) **then**   ▷ id is two-hop neighbor
12:            resultPath.addFirst(id);
13:            resultPath.addFirst(self.getNeighbor(id));
14:            **break**;
15:        **else**
16:            **resultPath.addFirst(id);**
17:        **end if**
18:    **end for**
19:    **if** self.Id **is** relayNode **then**   ▷ Current node is within path acting as relay
20:        resultPath.addFirst(self.Id);
21:    **end if**
22:    **return** resultPath;
23: **end function**

---

node by exploiting the local knowledge of nodes and discovering shortcuts locally. In this algorithm, source node, destination node and relay nodes (nodes within path) that are involved in a gossiping round execute this algorithm once they have received swapping cache. Every node parses the current path, and prunes it if it can construct a shorter path by using knowledge of its own routing tables or the tables of its neighbors. The message is continued to be relayed through the updated path. The details of the algorithm is given in (Algorithm 1).

Since the underlying restricted overlay can be arbitrary, the resulting routing paths will inevitably exhibit all range of lengths, and can be as short as one hop. Such variation in routing path lengths imply that the communication times between nodes will vary greatly during the exchange process. This in turn will create a bias in selection of random nodes. To this end, we introduce a delay mechanism and define two system parameters called α (maximum threshold of path length) and β (maximum delay). Recipient nodes reject paths with larger length than α to ensure having short length paths. Furthermore, length of gossiping rounds are equalized using β that is preferably equivalent to α. In other words, a gossiping round is postponed to a time that is obtained from the difference between β and the length of the selected path. From practical perspective, each entry other than path consists of two variables. *WaitingTime*, represents time that entry is waiting to be selected for gossiping. *Swapped*, denotes the number of times that the entry was selected as one of the swapping cache entries in current node. The protocol is performed by letting the initiating peer P execute the following steps:

1. Increase by one *waiting Time* of all entries.
2. Select copy of entry Q with the highest *waiting Time* from the cache, and copy of S − 1 other random entries as swapping cache.
3. Increase by one the *swapped* field of all selected S − 1 entries within cache.
4. Set *waiting Time* entry Q within cache to zero.
5. Execute path construction algorithm for all entries of the swapping cache.
6. Wait w.r.t delay, send updated cache to next node of the path towards Q.
7. Receive from one of its social neighbors of reverse path a subset of no more than S of Q's entries and execute path construction algorithm for them.
8. Discard the entries with a path longer than α (Maximum path length).
9. Update P's cache, by *firstly* replacing the same entries already existed with longer path, (if any), and *secondly* replacing entries with the highest *swapped*.

On reception of a swapping cache request, node Q randomly selects a copy of subset of its own entries, of size S, execute step 3 for S entries, executes step 5 and sends it to one of its social neighbors in the constructed reverse path, execute step 7, insert sender entry to the received swapping cache and executes steps 8, 9. Any relay node involved in gossiping, execute path construction algorithm and cooperate in building reverse path.

Algorithm 2. Skeleton of the original push-pull based gossip learning protocol

---

1: **procedure** MAIN
2:     *currentModel* ← INITMODEL()
3:     lastModel ← currentModel
4:     **loop**
5:     WAIT(Δ)
6:     p ← RANDOMPEER()
7:     SEND(p, currentModel)
8:     **end loop**
9: **end procedure**
10: **procedure** ONMODELRECEIVED(m)
11:     *currentModel* ← Update(Merge(m, lastModel))
12: **end procedure**

3.2 Push-Based Merging Gossip Federated Learning

Gossip learning designed to train a global model over fully decentralized data using a gossip communication and in an asynchronous manner [5]. Conceptually, it is a push-pull gossip protocol in which network nodes start from a common initialization, and then multiple models perform random walks over the network, continuously learning from the data stored in each visited node. This is accomplished by having the nodes update the received models on their local data and then gossip them out to a randomly-chosen peer. Also, the models are also merged with each other along their walks in order to accelerate the learning process.

Algorithm 2 shows the generic skeleton of the protocol, The main loop performed by each device. More precisely, a random peer is chosen among the other participants in the network, and gossip the current model to it. On the other side, When a node receives a new model, it merges it with the last model previously received and then updates the resulting model by performing local training. The resulting model is stored locally for prediction and for gossiping with peers, until a new model is received and the process is repeated.

Two methods UPDATE and MERGE can vary depending on the kind of model used for training. In this paper, we use supervised learning models that is trained using Stochastic Gradient Descent (SGD). Also, we use a decaying learning rate and regularization as hyper parameters. Each model has weights and a timestamp representing as age of the model. Actually, Since at each round one data point is used to train the model, it can refer to the number of data points has been trained on. The MERGE method is simply the average of the model weights. Algorithm 3 represents the details.

Algorithm 3. Skeleton of the model update using SGD

1: λ ▷ The regularization parameter
2: x, y   ▷ The features and label of the local data point
3: **procedure** UPDATE(m)
4:     w, t ← m
5:     $\eta \leftarrow 1 / (\lambda \cdot t)$ ▷ The decaying learning rate
6:     $w \leftarrow (1 - \eta \cdot \lambda) \cdot w + \eta \cdot$GRADIENT(w, x, y)
7:     **return** (w, t + 1)
8: **end procedure**

The original gossip learning does not work for networks with restricted connectivity like DOSNs. It is crucial for gossip learning that every node at each round to get a global uniformly random peer to run learning algorithm. It is achieved by peer sampling service. However, existing sampling requires fully free connectivity between peers which is against the topology of DOSNs. In the previous section, we explained our algorithm for building a random overlay on top of restricted DOSN topology on the fly. We also adapt gossip learning algorithm utilizing virtual random overlay perform distributed training. Unlike gossip learning, our algorithm is a push based gossiping. Each node gets a random peer as well as routing path from sampling service and forward current local model to the next network node in the path. Once a network node has received the model, if it is the last node in the path, implying it is the target node, it executes update and merge functions and update its current model in the cache. Otherwise, it simply relay message to the next network node in the path. The main difference of the algorithm is that unlike original gossip learning ,which is

---

Algorithm 4. Skeleton of the push-based gossip learning protocol

---
1: **procedure** MAIN
2:     $currentModel \leftarrow$ INITMODEL()
3:     lastModel $\leftarrow$ currentModel
4:     **loop**
5:     WAIT(Δ)
6:     p $\leftarrow$ RANDOMPEER()
7:     SEND(p, currentModel, path)
8:     **end loop**
9: **end procedure**
10: **procedure** ONMODELRECEIVED(m, path)
11:     **if** path is empty **then**   ▷ Current node is target
12:         $currentModel \leftarrow$ Update(Merge(m, lastModel))
13:     **else**
14:         nextNode $\leftarrow$ path.removeFirst()
15:         SEND(nextNode, currentModel, path)
16:     **end if**
17: **end procedure**

---

a push-pull based, it is a push based algorithm and the recipient node does not exchange the model with the sender. As an amelioration technique and to speed up the convergence, each node, if does not receive any model from network nodes for a number of rounds, sends a pull request to one of its immediate neighbors and update its model.

4   Evaluation

We implemented the algorithm on PEERSIM [18], a discrete event simulator for building P2P protocols. Since social network graphs are the main target of this algorithm, the protocol is investigated on two data set of real world social networks including Facebook [21], and Wiki-Vote [19]. The experiments are also executed on collaboration network of Arxiv Astro Physics [20]. The largest connected component of the graphs are extracted for running the experiments.The detail information and properties of the graphs are given in Table 1. We also implemented a binary classification supervised learning that is trained using Stochastic Gradient Descent (SGD) optimisation algorithm. The data is fully distributed and each node holds one feature vector.

5   Experiment Results

5.1   Successful Construction of a Random Overlay on the Fly by the Sampling Service

To show that our sampling service construct a random overlay on the fly, we calculate the clustering coefficient (CC) and ensure that it is equivalent to the

Table 1. Data set

Data set	|V|	|E|	Type	Diameter
Wiki-Vote	7066	103663	Social	7
AstroPh	17903	197031	Collaboration	14
Facebook	63391	817090	Social	16

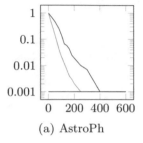 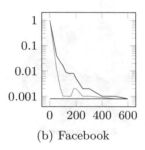 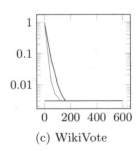

(a) AstroPh   (b) Facebook   (c) WikiVote

Fig. 1. Clustering Coefficient (CC).

CC of random overlay. It is formulated by division of C over N − 1 that C is cache size and N is network size ($\frac{C}{N-1}$). Neighbors in random overlay are target nodes of paths within caches. Two scenarios are executed ($\alpha = $ d, $\beta = $ d; $\alpha = $ 2d, $\beta = $ 2d) where d is diameter of the network. (C = 20, S = 5). Figure 1 shows clustering coefficient (CC). X and Y axis-es show cycle and average CC respectively. In blue and gray diagrams both α and β are set 2d and d respectively. Black line represents CC of random graph (Color figure online) As Fig. 1 exhibits, CC of the graphs converge to random graph ensuring global randomness. Larger value for α increases the speed of convergence but gives better local randomness (Fig. 1).

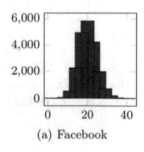 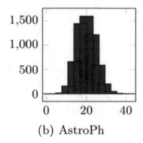 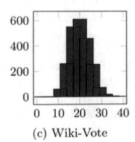

(a) Facebook   (b) AstroPh   (c) Wiki-Vote

Fig. 2. In-degree distribution. X-axis shows degree. Y-axis shows nodes count.

5.2   Removal of the Sampling Bias in Random Overlay

We evaluate In-degree distribution of nodes to see whether the sampling bias is removed in the random graph. We calculate degree distribution of target nodes of paths over random graph. The ideal case is to have a degree distribution with low standard deviation. It ensures an unbiased sampling independent of node degrees in social graph. In Fig. 3, the results show a normal distribution in which 70% of nodes have in degree 20 ± 20%, a value between 16, 24 (C = 20). Figure 3 shows model similarity. X-axis and Y-axis show cycles (x100) and Cosine similarity respectively. Left to right are Facebook, AstroPh, and Wiki-Vote. Circle Plots represent push-based technique and star plots represent augmentation with pull-based technique.

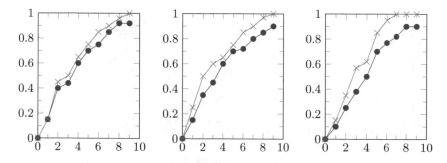

Fig. 3. Model similarity.

5.3   Convergence of all the Network Nodes to a Similar Model

In this experiment, we evaluate that all nodes converge to the same model. It indicates that our gossip learning algorithm works correctly. To do this, we calculate the cosine similarity of models at each round. We execute two separate experiments, once only push-based merging algorithm and then both push-based merging and its pull-based improvement to observe the impact of the amelioration technique. We configure each node to run pull-based learning if for 10 consecutive rounds it does not receive a model from the network. In Fig. 3, the results show a convergence of 90% to 95% for all three networks for applying only push-based merging algorithm. The main reason for such a high rate of convergence is because the random sampling of our sampling service. After applying the amelioration mechanism, we observe that all experiments converge 100% correctly. The speed of convergence is 600 to 900 rounds for different networks and it is obviously should be slower than general gossip learning as running each round of the algorithm requires traversing several network nodes through a routing path. However, because our sampling algorithm reduces the length of the path, it still provides a good convergence speed.

5.4   Comparison of Prediction Error with Centralized Training

It is important to see how precise the resulting model predicts. We train the model with the same algorithm and data in a centralized fashion and consider it as a baseline. For centralized mode, the number of iterations is considered as the number of rounds. Figure 4 shows prediction error. X-axis and Y-axis show cycles (x100) and average of 0–1 error respectively. Left to right are Facebook, AstroPh, and Wiki-Vote. Circle Plots represent social-aware gossip learning and star plots represent centralized training. The results show that our socially-aware gossip learning protocol predicts with an accuracy of 82% to 88% compare to the baseline.

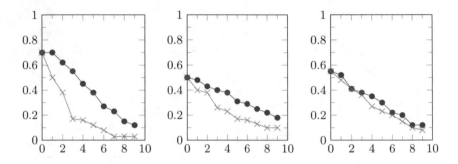

Fig. 4. Prediction error.

6   Conclusions and Future Work

In this paper, we proposed a federated learning algorithm based on the Gossip protocol for DOSNs. It enables training machine learning algorithms for DOSNs while users do not have to share or move private training data from personal nodes to preserve privacy. The algorithm respects topology restriction of DOSNs. We have shown the performance and correctness of the algorithm by running experiments on different social networks. As future work, we are going to test and adopt our algorithm considering different assumptions on the distribution of the data and the communication speeds of the devices as given in [12].

References

1. Debatin, B., et al.: Facebook and online privacy: atitudes, behaviors, and unintended consequences. J. Comput. Mediat. Commun. **15**(1), 83–108 (2009)
2. Koll, D., Li, J., Fu, X.: SOUP: an online social network by the people, for the people. In: Proceedings of the 15th International Middleware Conference (2014)
3. Nilizadeh, S., et al.: Cachet: a decentralized architecture for privacy preserving social networking with caching. In: Proceedings of the 8th International Conference on Emerging Networking Experiments and Technologies (2012)

4. Nasir, M.A.U., Girdzijauskas, S., Kourtellis, N.: Socially-aware distributed hash tables for decentralized online social networks. In: 2015 IEEE International Conference on Peer-to-Peer Computing (P2P), pp. 1–10). IEEE, September 2015

5. Ormándi, R., Hegedűs, I., Jelasity, M.: Gossip learning with linear models on fully distributed data. Concurr. Comput. Pract. Experience **25**(4), 556–571 (2013)

6. Jelasity, M., et al.: Gossip-based peer sampling. ACM Trans. Comput. Syst. (TOCS) **25**(3), 8-es (2007)

7. Khelghatdoust, M., Girdzijauskas, S.: Short: gossip-based sampling in social overlays. In: Noubir, G., Raynal, M. (eds.) NETYS 2014. LNCS, vol. 8593, pp 335–340. Springer, Cham (2014). https://doi.org/10.1007/978-3-319-09581-3_26

8. Jelasity, M., Voulgaris, S., Guerraoui, R., Kermarrec, A.M., Van Steen, M.: Gossip-based peer sampling. ACM Trans. Comput. Syst. (TOCS) **25**(3), 8-es (2007)

9. Baagyere, E.Y., Qin, Z., Xiong, H., Zhiguang, Q.: The structural properties of online social networks and their application areas. IAENG Int. J. Comput. Sci. **43**(2) (2016)

10. King, V., Saia, J.: Choosing a random peer. In: Proceedings of the 23rd Annual ACM Symposium on Principles of Distributed Computing (PODC 2004), pp. 125–130. ACM Press (2004)

11. Voulgaris, S., Gavidia, D., Van Steen, M.: CYCLON: inexpensive membership management for unstructured P2P overlays. J. Netw. Syst. Manag. **13**(2), 197–217 (2005). https://doi.org/10.1007/s10922-005-4441-x

12. Giaretta, L., Girdzijauskas, Š: Gossip learning: off the beaten path. In: 2019 IEEE International Conference on Big Data (Big Data). IEEE (2019)

13. McMahan, B., et al.: Communication-efficient learning of deep networks from decentralized data. In: Artificial intelligence and statistics. PMLR (2017)

14. Kyasanur, P., Choudhury, R.R., Gupta, I.: Smart gossip: an adaptive gossip-based broadcasting service for sensor networks. In: 2006 IEEE International Conference on Mobile Ad Hoc and Sensor Systems. IEEE (2006)

15. Montresor, A.: Gossip and epidemic protocols. Wiley Encyclopedia of Electrical and Electronics Engineering, vol. 1 (2017)

16. Alkathiri, A.A., et al.: Decentralized Word2Vec using gossip learning. In: 23rd Nordic Conference on Computational Linguistics (NoDaLiDa 2021) (2021)

17. Hegedűs, I., Danner, G., Jelasity, M.: Gossip learning as a decentralized alternative to federated learning. In: Pereira, J., Ricci, L. (eds.). DAIS 2019. LNCS, vol. 11534, pp. 74–90. Springer, Cham (2019). https://doi.org/10.1007/978-3-030-22496-7_5

18. Montresor, A., Jelasity, M.: PeerSim: a scalable P2P simulator. In: Proceedings of the 9th International Conference on Peer-to-Peer (P2P 2009), pp. 99–100, Seattle, WA, September 2009

19. Leskovec, J., Huttenlocher, D., Kleinberg, J.: Signed networks in social media. In: Proceedings of the SIGCHI Conference on Human Factors in Computing Systems, pp. 1361–1370. ACM (2010)

20. Leskovec, J., Kleinberg, J., Faloutsos, C.: Graph evolution: densification and shrinking diameters. ACM Trans. Knowl. Disc. Data (TKDD) **1**(1), 2 (2007)

21. McAuley, J., Leskovec, J.: Learning to discover social circles in ego networks. In: Advances in Neural Information Processing Systems, vol. 25, pp. 548–556 (2012)

A Multi-layer Modeling for the Generation of New Architectures for Big Data Warehousing

Asma Dhaouadi[1,2(✉)], Khadija Bousselmi[1], Sébastien Monnet[1],
Mohamed Mohsen Gammoudi[3], and Slimane Hammoudi[4]

[1] LISTIC Lab, Université Savoie Mont Blanc, France Annecy – Chambéry, 74940 Chambéry,
France
{asma.dhaouadi,khadija.arfaoui,sebastien.monnet}@univ-smb.fr
[2] RIADI Lab, Université Tunis El Manar, 1068 Tunis, Tunisia
[3] ECRI research Team, RIADI Lab, ISAMM, University of Manouba, 2010 Mannouba, Tunisia
gammoudimomo@gmail.com
[4] ERIS, ESEO-TECH, 49100 Angers, France
Slimane.hammoudi@eseo.fr

Abstract. With the explosion of new data processing and storage technologies nowadays, businesses are looking to harness the hidden value of data, each in their own way. Many contributions were proposed defining pipelines dedicated to Big Data processing and storage, but they target usually particular types of data and specific technologies to meet precise needs without considering the evolution of requirements or the data characteristics' change. Thus, no approach has defined a generic architecture for Big Data warehousing process. In this paper, we propose a multi-layer model that integrates all the necessary elements and concepts in the different phases of a data warehousing process. It also contributes to generate an architecture that considers the specificity of data and applications and the suitable technologies. To illustrate our contribution, we have implemented the proposed model through a Business model and a Big Data architecture for the analysis of multi-source and social networks data.

1 Introduction

The technological revolution is based on the techniques of knowledge extraction from the data shared in information systems, social networks, new communication systems and other content generators. One of the main challenges for companies is to create value from this knowledge and to exploit it in the best manner in order to achieve a specific business objective. It becomes a significant factor on which the competitiveness between companies is based nowadays. For this reason, in the last few years, many researchers have tackled data warehousing on a large scale due to the emergence of new internet services, mobile applications, social networks, such as Facebook, Twitter, Instagram and others.

To cope with the new challenges related to data sources (its heterogeneity, its variability, the velocity of its acquisition and obviously its volume), new technologies were proposed to meet the specific needs of companies and to adapt the specifics characteristics of data to be processed. These technologies, despite their advantageous offers,

L. Barolli et al. (Eds.): AINA 2022, LNNS 450, pp. 204–218, 2022.
https://doi.org/10.1007/978-3-030-99587-4_18

were designed to meet specific requirements and suffer from the lack of interoperability between them and especially the adaptability to evolutions like requirements or data types change [1–4]. Although a part of the research community has conducted comparative studies between several available technologies in this scope [5–8], they did not consider the importance of the conceptual modeling step when defining their business process and focused only on system level aspects. For instance, in [7] the authors presented a comparative study between Talend and Informatica. The authors of [5] presented, also, a comparison by focusing on the cost, the real-time analysis, the language binding, the performance optimization, the data lineage between the following ETL tools: Informatica, Datastage, AbInitio, Oracle Data Integration, and SQL Server Integration Services. In [6], the authors have proposed an approach to help an enterprise to choose a data warehousing tool by referring to some functionalities, such as Platform Architecture, Number of Clients Tools, Installation Effort, Additional Software requirements, Query ability on metadata, and others.

In summary, most of works proposing approaches for the acquisition, processing, storage, and analysis of data are focusing on the technologies, tools dedicated to ensuring these functions and to our knowledge, today there is no generic or rich enough model to present all the concepts, steps and notions of a process dedicated to big data warehousing.

The objective of this paper is to propose a generic model to support different processes of data storage and analysis dedicated to big data. To achieve our goal and to cover all the key concepts necessary for big data warehousing, we have browsed all the architectures already proposed in the literature for this issue. We thus propose a multi-layer model that covers all the necessary elements in the different modeling levels of an ETL process. For the validation of the proposed multi-layer model, we detail the different steps of a generic architecture implementation for a multi-source data analysis supported by this model. Indeed, we present a real use case for the analysis the Covid-19 pandemic impact on social networks, in particular on "Twitter". The corresponding business process was designed using our proposed meta-model to define the appropriate steps to realize for data warehousing and a generic architecture that implements it. We also present the results of the analysis carried out and a discussion regarding the different technologies used in our specific use case.

This paper is structured as follows. Section 2 review the related works. In Sect. 3 we present our proposed multi-layer model for big data warehousing. In Sects. 4 we present the proposed Business Model and from the Multi-layer model we detail the instantiation of all the elements necessary for the implementation of an architecture for multi-source data analysis. In Sect. 5, we detail the different objectives of the data analysis conducted and we provide analytical interpretations. In Sect. 6, we discuss the main features and limits of used technologies in our project. Finally, in Sect. 7 we summarize our contributions and give some perspectives.

2   Related Works

In this section, we present an overview of some proposed contributions to data warehousing in the context of Big Data. Among the first works proposing architectures devoted to Big Data, we mention [9] where the authors proposed a parallel and distributed ETL architecture, which is an 8-tier architecture with a 5-step process. This architecture is based on the MapReduce (MR) paradigm. They have proposed graphical notations to present the different process steps at the conceptual level, and in particular the transformations such as (cleansing/standardization) in the Map phase (M) and the merging and aggregation of data in the Reduce phase (R). Although the proposal of [9] showed the evolution of the classical DW architecture using Map Reduce, their results are validated on a sample of structured data.

In [1] the authors addressed ETL process design in the context of NoSQL storage. To do this, they proposed a set of rules for transforming a multidimensional conceptual schema into document-oriented system (MongoDB). As for [3], they presented a so-called BigDimETL architecture based on the MapReduce paradigm for the parallelization of data processing, and they used Hbase as a distributed storage mechanism. To validate their architecture, they tested it using Twitter data.

In the model-driven engineering domain, the Model Driven Architecture (MDA) approach proposed and fostered by OMG allow to represent systems at any level of abstraction or from different viewpoints, ranging from enterprise architectures to technology implementations [10]. In fact, the MDA multi-layered architectural framework is divided into 4 layers: 1-A Computation Independent Model (CIM) on the top of the architecture to describe the system requirements. 2-A Platform Independent Model (PIM) is the model of a subsystem that contains no information specific to the platform or the technology used to realize it [11]. 3-A Platform Specific Model (PSM) where the technology of the platform implementation is specified. 4-Code: this is the phase of generating code to be executed by the appropriate tool. Based on the MDA, several proposed approaches in literature have been presented in [12, 13]. However, each architecture allows to address a particular business need defined in the CIM layer. Moreover, they generally cover structured data. In this category of approaches, the genericity is only represented in the PIM where from a single PIM we can generate one or more PSMs using transformation techniques.

Otherwise, among the most popular standard architectures for the acquisition, storage, and retrieval of Big Data, we note Kappa[1], Lambda and Sigma architectures. 1-The Kappa architecture was first described in 2014 by Jay Kreps. It is a software architecture used for processing streaming data. 2-The Lambda architecture is simply the Kappa architecture plus a batch layer. In fact, this architecture proposed by James Warren and Nathan Marz in [14] is composed of three layers; a Batch Layer where the storage is immutable of all the data, a Serving Layer indexes the batch view, and a Speed Layer that contains recent data. For more details in [15] the authors have presented experiments involving the implementation of these architectures. Finally, 3-Sigma architecture is based on Lambda architecture by correlating the historical data and the data generated in real time to define simulated data [16].

Based on Lambda architecture and Hydre Architecture [2], the authors in [4] presented an extended version: Lambda + Architecture pattern to handle both exploratory and real-time analyzes, and to fit more various use cases than the Lambda Architecture [4].

---

[1] https://milinda.pathirage.org/kappa-architecture.com/.

Despite the efforts conducted by the community to propose dedicated approaches to Big Data warehousing, among the most redundant gaps we note that the heterogeneity of data sources is often not addressed. In fact, most proposed works target a particular type of data collected from a single source storage. In addition, the validation of such approaches is often performed on a simple sample of data and there is no visibility on the maintainability of the proposed process when the size or types of data evolve. Finally, the major shortcoming is that these solutions are generally architectural specific and therefore strongly tied to specific technologies to meet a particular need. As a result, to our knowledge, no research work has defined a sufficiently generic model to support multiple data warehousing processes suitable to specific application in the context of Big Data.

3 Contribution: A New Multi-layered Conceptual Model for Big Data Warehousing

The main contribution in this paper is the proposition of a multi-layer model (Fig. 1) that allows to integrate all the concepts, components and notions involved in the data warehousing process, from the requirements definition phase to the results representation and analysis phase. This model is the key to instantiate one or more architectures for data warehousing specific to different business requirements. For this reason, we have browsed many contributions: architectures and models, proposed in literature to gather all the key concepts necessary for data warehousing. Then, we have chosen the multi-layer Meta-Model representation to start from the most generic concepts to the specific concepts (instances) that will be subsequently linked to the technologies and meeting the business requirements of the company.

The proposed multi-layer model is composed of three layers. First, in the upper layer "Meta-Model Level", we define all the entities related to data warehousing and that can exist in such process, and in particular in a Big Data pipeline. These elements concern the requirements, the Implementation strategy, the logical modeling, the technologies, the processing, and others. Second, in the middle layer the "Meta-Model Specification Level" layer, we have specified possible instances of the classes of the upper layer (Meta-Model). These sub-classes are linked with "Is A" relations (the inheritance) with the generic classes of the upper layer. For example, "Processing" can be a "Batch" or "Stream" whereas a "Logical Modeling" can be an "OLTP" or "NoSQL". Third, the bottom layer concerns the "Conceptual Model Level" layer. In this layer, all entities are subclasses or instances of the classes presented in the "Meta-Model Specification Level" layer. These entities are presented by the relation "Instance of". For instance, "Document", "Graph", "Key-Value", "Column" are instances of the "NoSQL" class of the middle layer.

In addition, as the different elements modeled must have interactions between them to generate the data processing process, we modeled the upper layer of the multi-layer model by showing the interactions between the different entities. To do this, we have chosen the UML class diagram in figure Fig. 2 as it allows to define relations specification between modeled entities. Moreover, the different classes presented in this diagram and involved in the "Meta-Model Level" of this multi-layer model are quite generic to model

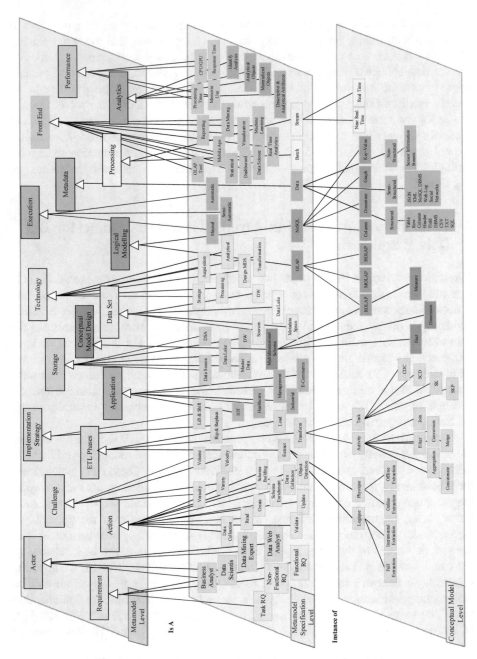

Fig. 1. A multi-layer model for big data storage and analysis.

any data warehousing process. Consequently, this Multi-Layer model implicitly allows us to create a process that meets specific needs by instantiation of this class diagram using the multi-layer model.

For example, an Actor defines one or more Implementation Strategies. An "Actor" can also choose 1 or more "Processing" types. Moreover, a "Technology" takes over 1 or more "Processing". Also, an "Actor" can specify only 1 "Logical Modeling".

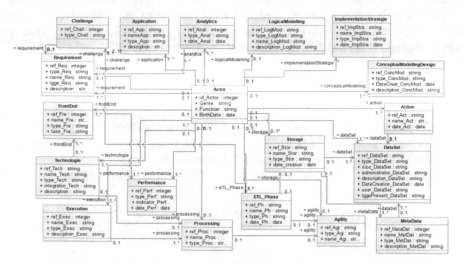

Fig. 2. An UML modeling of the generic architecture's meta-model

4 Validation of the Proposed Model: An Implementation for Multi-source Data Warehousing and Analysis

To illustrate our proposed multi-layer model, we present a real use-case for the analysis of data from different data sources and collected from social networks. The objective of the analysis is to study the impact of the evolution of the Covid-19 pandemic on social networks, and in particular on "Twitter". For this purpose, we referred to the steps to be performed when defining a data warehousing project in [17], namely: defining the business model involving the ETL process and implementing it using the appropriate technologies.

4.1 Proposed Business Model Process

The initial step in a data warehousing project is to determine the business requirements – the user requests [18, 19]. From these needs the designer will have to define the different elements to be involved in the business process, namely: the actor, the business challenges, data sources, conceptual model, and the involved technologies. All these elements are already modeled in our proposed meta model and especially in the meta-model layer. Then, we instantiated the needed elements according to the type of data sources and our initial Business Requirement. This operation consists in choosing the corresponding instances from the specification layer of our meta-model to each element. In fact, in our project, we started from the statistics presented on the web concerning the COVID-19 pandemic and the vaccination campaign results. Our objective is to analyze its impact on the community through the analysis of the data collected from Twitter. Then, we identify the Actor, which in our case could be a Data Scientist, a Business Analyst, or a Data Mining Expert. After that, we set the Challenges and the Application. The next step is to determine the candidate Data Sources to meet our objective which are COVID-19 statistics from the Web and community reactions from Twitter. Next, we design the conceptual and then logical model of each data storage, in particular the multidimensional schema of Data Storage Area (DSA) and Data Warehouse (DW) and the preparation of the OLAP data cube. To do this, we resorted to our proposed meta-model to determine the appropriate conceptual model specification described in the third layer. The final step consists on determining the technologies to implement the entire process from the data acquisition stage to the analysis, reporting and dashboarding. And we set the Analytics which are in our case an OLAP Tool, Statistical, Reporting, Dashboard, and Visualization. For this purpose, we tested different candidate technologies as explained in the following section to have a generic view of the business architecture.

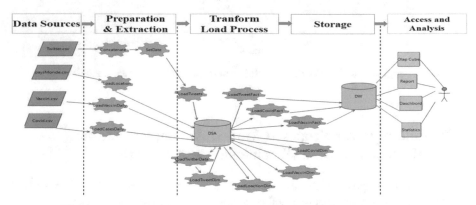

Fig. 3. A business model for multi-source data warehousing and analysis.

Figure 3 shows the Business Model of our project as generated from our UML model proposed in Fig. 2 and using the Talend Open Studio for Data Integration ETL tool as its our chosen ETL tool. In our context the Business Model is a business process applied to data warehousing allowing the processing of data from acquisition to the representation of results according to the ETL processing steps. For this purpose, we used the identified instances of our business process above and we defined the relationships between them using the proposed model using UML.

As an input to our business model, we identify the different source files. For the Twitter data, we referred to the Michigan State University[2] which traces the TweetIDs of COVID-19 shared since January 23, 2020. For data about the Covid-19 pandemic

---

[2] https://libguides.lib.msu.edu/covid-datasets-social-media.

and the evolution of the vaccination status in all countries of the world, we referred to the OurWorldInData[3] website. This site traces the daily status of the Covid since its appearance as well as for the vaccination stats. All input data was collected in the form of csv files to facilitate their preparation and integration.

Then, for the preparation and cleaning of data as indicated in the business model, we have performed the following steps: (i) we prepared the intermediate storage that will support all the data in the preparation phase. (ii) Then, we concatenated all the twitter files with the same date, and (iii) we inserted the date in each file in a new column. After that, (iv) we created jobs to load the Twitter, Country, Vaccination, and confirmed Covid-19 cases. These jobs are named respectively: SetDate, LoadTweets, LoadLocation, LoadVaccinationDaily, and LoadCaseDaily.

For the storage of data and respecting the chosen multidimensional "Snowflake Schema" model, we created four jobs to load the dimensions in the DSA by adding to each one a new column containing the Surrogate Key of each dimension table. Afterwards, the loading of the three fact tables in the final storage of the DW (LoadTweetFact, LoadCovidFact, and LoadVaccinFact) is necessary to interconnect to the different dimension tables and to guarantee a suitable modeling for the creation of the OLAP Cube in the following step.

Finally, to allow the access and analysis of stored data on the right side of the DW, we prepared an OLAP Cube allowing to perform different queries to explore the data model. We have also prepared reports detailing the evolution of the pandemic, the state of vaccination in the world and how much the community is interested in this actuality through the sharing of Tweets on social networks. We also created a dynamic dashboard connected to the DW providing meaningful visualizations and giving specific view on the evolution of the actualities. The results of this process can be exploited by the different types of actors presented in the Metamodel Specification level of our Multi-Layer Model previously presented.

4.2 An Architecture for COVID-19 Data Warehousing and Analysis

From the defined Business Model, we propose a specific architecture deduced from the generic architecture (Fig. 4) that can support the proposed model and in which we specified the different tools and technologies used to carry out our process for data acquisition, storage, query, and analysis.

The proposed architecture was designed for the analysis of shared data on Covid-19, and specifically to evaluate the impact of the pandemic and vaccination and the community's reaction on Twitter but could be applied to other types of data from other Web sources or social media like Facebook, Instagram or TikTok.

As shown in Fig. 4, this architecture is composed of five phases. First, in the Data Sources phase we selected the data sources that allow to fulfil the final purpose of the analysis. Second, it is the data preparation and extraction phase, we developed a python code for the concatenation of the Twitter files and used a varied set of functions available in Talend for data preparation and cleaning. The intermediate storage of this data was done in an Oracle database, and we used Oracle SQL Developer as our Database

---

[3] https://ourworldindata.org/coronavirus.

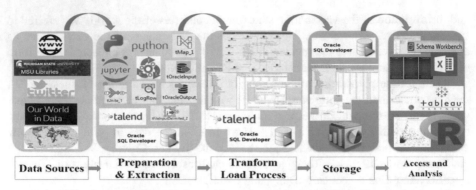

Fig. 4. An architecture for multi-source data warehousing and analysis.

Management System. Thirdly, in the phase of Transformation and Loading, we have relied on other components for the creation of Talend jobs allowing the final loading of the multidimensional model (Facts and Dimensions) in the fourth phase, which is Storage. Finally, in the fifth phase, we used the necessary technologies for OLAP navigation (Schema Workbench), reporting (Excel) and statistical computing (R language) and dashboarding (Tableau Software). Here, we note that the ETL process conducted could be supported by Pentaho Data Integration, but due to lack of space we showed only one ETL tool.

The choice of the technologies used is not always straightforward due to the heavy competition. But in our choices, we have relied on some characteristics, such as: the interoperability between the used tools, the simplicity of use and maintenance, the documentation, and the ease of solving execution errors, the satisfaction of the set needs and requirements, and the quality of the end results.

5   Results Analysis and Interpretation

Our use-case business objective is to show how social media track current events, more specifically how the rebound of COVID-19 pandemic and vaccination campaign impacted the community interactions in social media.

As social networks, we chose the Twitter because data related on COVID-19 are available since the start of the epidemic. To meet our business objective, and according to the collected data, we defined three axes of analysis: 1-The impact of the Covid-19 spread in the world over the community through the analysis of the number of tweets shared in the different peaks of the pandemic. 2-The evolution of the Covid-19 pandemic in the world since the appearance of this virus on January 23, 2020, versus the progress of vaccination campaigns. 3-The rebound of Covid-19 in relation with the progress of vaccination in different countries with comparing countries that started early the vaccination campaign (the United Kingdom and the France as samples), and countries that started much later (Tunisia as sample).

As results of for our first axis, and as shown by Fig. 5 on the left side, we state that the community talked very early about the pandemic which reflects the fear that was propagated in the world of this new unforeseen virus. It is clear that during the period from January 23, 2020, to April 18, 2020, the number of tweets shared in this context is much higher than the number of confirmed Covid-19 cases. Then people pursue their interests until they reach a good progress in the vaccination campaigns, where we notice a significant decrease in the number of Tweets from June 2021. Then with the appearance of the new Omicron variant of Covid-19 and several questions that arose around the effectiveness of the vaccines on new variants, we notice that the number of tweets bounces again. On the right side of the Fig. 5, we show the four peaks of the pandemic related to the four waves and it is clear that the world is being steered towards a fifth wave. In addition, we note that the vaccination campaign reaches its peak on June 21, 2021, and the effect of these campaigns can justify the drop in the number of cases, which reached its minimum on October 03, 2021. However, even though vaccination is continuing worldwide, the Covid-19 pandemic is recording a significant increase since the appearance of the Omicron variant.

For the third analysis question, from Fig. 6, we can see that, in the beginning, the Tunisia survived better to the Covid-19 spread thanks to the measurements token by its government too early, but it was the most impacted by the Delta variant as its vaccination campaign started later than the other countries. We can also see from the Fig. 7 those countries that first launched the vaccination campaigns were not privatized neither from the fourth wave nor from the fifth one. The new Omicron variant is forcing countries to launch a strong alert, revise protection measures and oblige citizens to take the third dose of vaccine.

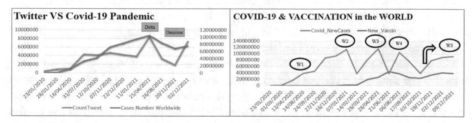

Fig. 5. The impact of pandemic Covid-19 on Twitter and vaccination progress.

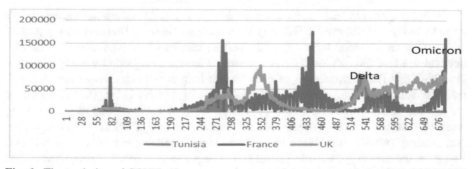

Fig. 6. The evolution of COVID-19 new cases in proportion to the population from 24/01/2020 to 12/12/2021

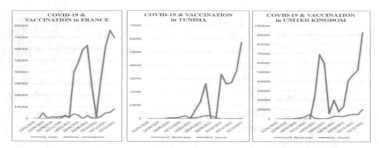

Fig. 7. The rebound of Covid-19 VS the progress of vaccination in different types of countries.

6 Discussion and Findings

As summary, through the work carried out in this project, we confirm the great importance that should be allocated to the modeling phase of the ETL process for the acquisition, processing, storage, and analysis of data as pointed out in [20]. In fact, the success of a data warehousing project is tightly based on a good modeling, and which is usually very costly in terms of time and effort. To alleviate the modeling task, the multi-layered model that we propose in this paper can help data experts to define all the concepts necessary for the design and implementation of an analysis and data warehousing architecture and consequently, it simplifies the steps necessary to model and then to implement the Business Model. Moreover, the proposed multi-layer model will allow the designer to design the system architecture according to his specific requirements. In addition, the meta-model provides the ability to guarantee the coherence between the different components of the implemented architecture.

Otherwise, from an architectural and technological point of view, we support [6]'s idea that the choice of ETL tool becomes more and more complicated when the options and features become multiple. For this reason, in literature it exists some works comparing the ETL tools and technological solutions. For example, [8] compared the performances of Microsoft's Azure Synapse (AS) and Azure Data Factory (ADF) ETL tools. The authors in [21] presented a comparative study of technologies regarding the Big Data characteristics: Volume, Variety, Velocity, Veracity, Value.

In our project, our main objective is to design a generic conceptual model and its corresponding architecture for ETL Big Data warehousing. For this reason, we studied and used a set of varied tools and technologies that allow to manipulate the data of our real use case through the different phases of the ETL project. Our study objective is to highlight the main features of each tool to help experts choosing those that suit their needs best. We summarized the studied tools in the table below (see Table 1). As we can see from this table, the best tools to be selected are those that make it easy to connect to others, especially to several data sources, the case of Talend Data Integration and Tableau Software. In addition, the documentation available around the tool makes it easier for the user to get started: Talend Data Integration, Tableau Software and Excel. However, the major problem of the tools is the large volume of data, and we notice from the presented table that TIBCO Jaspersoft Studio and QlikView are not adapted to acquire a large volume of data.

Table 1. The different tools investigated for the implementation of the Big Data Storage and Analysis architecture.

Phase	Tool/Technology	Advantages	Drawbacks
Extract transform load process	Talend data integration 7.3.1	Synchronize metadata from different data sources A graphical interface easy to use Ready-to-use. transformation and aggregation functions in the form of drag-and-drop components A rich documentation available in several languages	Unable to load large volume of data (139171 row) in Excel The component tCount does not return a number Does not allow a connection to HDFS or a Hadoop distribution Require knowledge of Java
	Pentaho data integration 9.2	Import data from different data sources Detailed documentation	Complex database connection
	CloverDX	The creation and maintenance of transformation tasks (graphics) are simple	No data synchronization No graphics and charts Does not support all NoSQL databases
Storage	Oracle SQL developer	Accessible by several tools for ETL and analysis Allow create and modify logical models Easy to install and to use Continuous improvement	Unable to run big queries (1134 Insert queries) Consumes too much memory Few documentations on its features Disconnection after a certain time of non-use
	PostgreSQL	It is well adapted to cloud applications	Queries on a large volume of data influence the processing speed
Access & Analysis	Mondrian schema workbench	Easy connection to the database The creation of a data cube is simple from a multidimensional schema	Only supports star schema

(*continued*)

Table 1. (*continued*)

Phase	Tool/Technology	Advantages	Drawbacks
	Microsoft excel 2016	Simple and easy generation of graphics It is accessible by the majority of technologies	It is sensible to the "." and "," in the numbers
	R Studio	Fast response time It offers several applications for machine learning and deep analysis	The GUI is not attractive and aesthetic
	TIBCO jaspersoft studio Professional 7.9.0	It is suitable for the generation of the reports	It is not well documented Problem solving is complicated It did not support our data volume
	QlikView	Easy to install Simplified presentation	It lacks the drag and drop facility Not suitable for Big Data
	Tableau software 2021.3	It is suitable for data visualization Provides guided dashboard creation Easy connection to multiple data sources	It does not detect the difference between date formats
	Pentaho report designer	The analysis reports are simple to generate	Not ergonomic

7   Conclusion

In this article, we have presented a multi-layer model supporting a generic architecture for the acquisition, processing, storage, and analysis of Big Data. The proposed model is composed of three layers. In the top layer we have included all the necessary concepts and notions involved in the data warehousing process, from the requirements definition phase to the results representation and analysis phase. Then, we applied the multi-layer model instantiation to answer a very important analysis topic for the community worldwide. We then chose to study the impact of the evolution of the Covid-19 pandemic on social networks, in particular Twitter to validate our model and to show its effectiveness. The nature of the analysis topic challenged us to address data from multiple data sources. We had to think early on about all the data pre-processing, cleaning and conversionning tasks before moving on to the loading stage in a Big Data Warehouse. For this reason, it was very useful to rely on the proposed Meta model that facilitated the task of instantiating the elements and concepts we needed to define the Business Model for the entire Big Data Warehousing process, from acquisition to data exploitation, visualization, and dashboard creation. Finally, we highlighted the limitations of different tools and technologies used during the implementation step of the architecture.

Our future works include instantiating the proposed model to define other Big Data warehousing architectures dealing with other types of data and responding to a different business requirement. Deploy other technologies to study their scalability and interoperability between them.

References

1. Yangui, R., Nabli, A., Gargouri, F.: ETL based framework for NoSQL warehousing. In: Themistocleous, M., Morabito, V. (eds.) EMCIS 2017. LNBIP, vol. 299, pp. 40–53. Springer, Cham (2017). https://doi.org/10.1007/978-3-319-65930-5_4
2. Leclercq, É.: Approches Multi-Paradigmes Et Contextuelles Pour La Gestion Des Masses De Données. Université de Bourgogne, Habilitation à Diriger des Recherches (2019)
3. Mallek, H., Ghozzi, F., Gargouri, F.: Towards extract-transform-load operations in a big data context. Int. J. Sociotechnol. Knowl. Devel. (IJSKD) 12(2), 7–95 (2020)
4. Gillet, A., Leclercq, É., Cullot, N.: Lambda+, the renewal of the lambda architecture: category theory to the rescue. In: La Rosa, M., Sadiq, S., Teniente, E. (eds.) CAiSE 2021. LNCS, vol. 12751, pp. 381–396. Springer, Cham (2021). https://doi.org/10.1007/978-3-030-79382-1_23
5. Mukherjee, R., Kar, P.: A comparative review of data warehousing ETL tools with new trends and industry insight. In: 2017 IEEE 7th International Advance Computing Conference (IACC), pp. 943–948. IEEE (2017)
6. Sureddy, M.R., Yallamula, P.: Approach to help choose right data warehousing tool for an enterprise. Int. J. Adv. Res. Ideas Innov. Technol. 6(4), 579–583 (2020)
7. Sreemathy, J., Brindha, R., Nagalakshmi, M.S., Suvekha, N., Ragul, N.K., Praveennandha, M.: Overview of ETL tools and talend-data integration. In: 2021 7th International Conference on Advanced Computing and Communication Systems (ICACCS), Vol. 1. IEEE. pp. 1650–1654 (2021)
8. Mayuk, V., Falchuk, I., Muryjas, P.: The comparative analysis of modern ETL tools. J. Comput. Sci. Instit. 19, 126–131 (2021)
9. Bala, M., Boussaid, O., Alimazighi, Z.: Extracting-transforming-loading modeling approach for big data analytics. Int. J. Dec. Supp. Syst. Technol. (IJDSST) 8(4), 50–69 (2016). https://doi.org/10.4018/IJDSST.2016100104
10. Model Driven Architecture (MDA), Object Management Group, and MDA Guide rev. 2.0, OMG Document ormsc/2014-06-01. https://www.omg.org/mda/specs.htm. Accessed 22 Dec 2021
11. Belaunde, M., et al.: MDA Guide Version 1.0. 1. OMG, Document Number: omg/2003-06-01, (2003). Accessed 22 Dec 2021
12. Maté, A., Trujillo, J.: A trace metamodel proposal based on the model driven architecture framework for the traceability of user requirements in data warehouses. Inf. Syst. 37(8), 753–766 (2012). https://doi.org/10.1016/j.is.2012.05.003
13. Lavalle, A., Maté, A., Trujillo, J.: Requirements-driven visualizations for big data analytics: a model-driven approach. In: Laender, A.H.F., Pernici, B., Lim, E.-P., de José Palazzo, M., Oliveira, (eds.) ER 2019. LNCS, vol. 11788, pp. 78–92. Springer, Cham (2019). https://doi.org/10.1007/978-3-030-33223-5_8
14. Warren, J., Marz, N.: Big Data: Principles and Best Practices of Scalable Realtime Data Systems. Simon and Schuster (2015)

15. Sanla, A., Numnonda, T.: A Comparative performance of real-time big data analytic architectures. In 2019 IEEE 9th International Conference on Electronics Information and Emergency Communication (ICEIEC), IEEE. pp. 1–5 (2019)
16. Antoniu, G., Costan, A., Pérez, M., Stojanovic, N.: The Sigma Data Processing Architecture: Leveraging Future Data for Extreme-Scale Data Analytics to Enable High-Precision Decisions (2018)
17. Sarbanoglu, H., Ottmann, B.: Business-Model-Driven Data Warehousing: Keeping Data Warehouses Connected to Your Business. White Paper (2008)
18. Fan, Z., Zhou, H., Chen, Z., Hong, D., Wang, Y., Dong, Q.: Design and implementation of scientific research big data service platform for experimental data managing. Proc. Comput. Sci. **192**, 3875–3884 (2021)
19. Yeoh, W., Popovič, A.: Extending the understanding of critical success factors for implementing business intelligence systems. J. Am. Soc. Inf. Sci. **67**(1), 134–147 (2016)
20. Trujillo, J., Davis, K.C., Du, X., Damiani, E., Storey, V.C.: Conceptual modeling in the era of big data and artificial intelligence: research topics and introduction to the special issue. Data Knowl. Eng. **135**, 101911 (2021)
21. Tardío, R., Maté, A., Trujillo, J.: An iterative methodology for defining big data analytics architectures. IEEE Access **8**, 210597–210616 (2020)

Efficient Retransmission Algorithm for Ensuring Packet Delivery to Sleeping Destination Node

Ali Medlej[1]([✉]), Eugen Dedu[1], Dominique Dhoutaut[1], and Kamal Beydoun[2]

[1] FEMTO-ST Institute/CNRS, Univ. Bourgogne Franche-Comté, 25200 Montbéliard, France
{ali_ghasswan.medlej,eugen.dedu,dominique.dhoutaut}@univ-fcomte.fr
[2] L'ARICoD Laboratory, Faculty of Sciences I, Lebanese University, Beirut, Lebanon
kamal.beydoun@ul.edu.lb

Abstract. The routing protocol plays a key role in allowing packets to reach their intended destination. We are interested in wireless nanonetworks (WNNs), which totally differ from traditional wireless networks in terms of node density and size, routing protocol used, and hardware limitations. This paper presents an enhanced retransmission algorithm used by the nodes in the destination zone, in combination with our previously proposed nanosleeping mechanism. This algorithm increases the chance of a destination node to capture the intended packet, while decreasing the number of participating nodes in the retransmission process. We evaluate the enhanced retransmission algorithm and show its effectiveness in reducing node resource usage while maintaining a high packet delivery to the destination node.

1 Introduction

Recent trends in telecommunication tend to promote work in wireless networks. This is due to several reasons, the most important of which is the easier installation and higher scalability compared to wired networks. Wireless networks include many types (Wi-Fi, Bluetooth, mobile communication 2G, 3G, 4G etc.) In the current paper we focus on *nanonetwork* communication paradigm.

Nanotechnologies promise new solutions for several applications in biomedical, industrial, and military fields [1]. Nanonetworks are built from tiny nodes, equipped with computing, sensing, and actuating devices. They usually have a small CPU, small memory, and low battery. The interconnection of nanonodes would expand the hardware capability of a single nanonode, and allow them to cooperate and share information. Those networks use electromagnetic waves in the THz band (0.1–10 THz) for their communications [3]. Due to the small communication range and power constraints, they need to use multi-hop communications to cover large areas.

Traditional communication technologies are not suitable for nanonetworks mainly because of the density, size and power consumption of transceivers and

L. Barolli et al. (Eds.): AINA 2022, LNNS 450, pp. 219–230, 2022.
https://doi.org/10.1007/978-3-030-99587-4_19

other components [6]. We have some knowledge about the main hardware components that constitute a nanodevice, and about network architecture [7].

To ensure that packets reach their destination, a routing protocol is required. Nanonetworks impose more constraints to it, and the routing protocol must take into consideration the nanoscale communication's characteristics. Traditional routing protocols are not adequate for wireless nanonetworks. Differences are in terms of bandwidth, energy, and node processing capability. Designing a routing protocol becomes a challenge in WNNs due to resource constraints on data processing, memory, and energy. While designing new routing protocols, the following points must be considered:

- *Energy efficiency*: Nanonodes are battery-powered. In low dense environments, and where there is a high rate of data exchanging, energy shortage is a major issue. Therefore, the routing protocol should be energy efficient [2].
- *Scalability*: Nanonetworks could be of different densities (low, medium, high, ultra-high, where nodes have numerous neighbours). Therefore, the routing protocol must support various network densities.
- *Complexity*: Due to limited hardware capability and resources, the complexity of a routing protocol may affect the performance of the entire WNN. The lower the complexity, the highest its effectiveness.
- *Delay*: In some applications, the delay, defined as the time taken to transmit the data from the source node to the destination node, is a key factor in message receiving or response. Therefore, the routing protocol should provide a reasonable delay.

SLR (Stateless Linear-path Routing) [8] is the protocol we use in our evaluation. It implements a coordinate-based routing, in which data packets are routed in a linear routing path. Nodes are assumed to be placed in a cubic space, distributed in zones. In the initial SLR phase, during network deployment, a few anchor nodes broadcast a packet (beacon) to the whole network. The hop counter in those beacons is used to define the coordinates of all nodes as a distance to the anchors. In the second phase, during data packet routing, nodes choose to forward a packet if and only if they are on the path between the source and the destination, based on the coordinates defined in the initial phase.

SLR protocol uses the TS-OOK (Time Spread On-Off Keying) modulation [9] to share the radio terahertz channel to nanodevices. Unlike traditional carrier-based network technologies, TS-OOK is pulse-based and consumes less energy. It is based on femtosecond-long pulses where packets are transmitted as a sequence of pulses interleaved by a given duration, cf. Fig. 1. "1" bits are encoded with a power pulse of duration T_p, and "0" bits are encoded as silence. Because sending consecutive pulses needs unavailable hardware and power at such small sizes, consecutive bits are spaced with a duration T_s which is usually much longer than the pulses themselves.

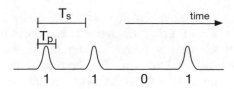

Fig. 1. TS-OOK pulse-based modulation.

In order to consume less resources (energy, memory etc.), a traditional mechanism is to make nodes sleep. However, in this case it might happen that the destination node be asleep when a packet arrives to it and its (destination) zone. To allow the destination node still receive the packet, one method is to make nodes at the destination zone retransmit the packet.

This paper presents an enhanced retransmission algorithm used by the nodes in the destination zone. This algorithm increases the chance of a destination node to capture the intended packet, while decreasing the number of participating nodes in the retransmission process.

The article is organized as follows. Section 2 presents the Related work, and Sect. 3 presents the Background. The probabilistic retransmission algorithm we propose is detailed in Sect. 4, and Sect. 5 evaluates it through simulations. Finally, Sect. 6 draws the conclusions.

2   Related Work

Several routing protocols have been proposed for sensor, ad hoc, and similar types of networks. Nanonetworks differ from those by:

- The limited processing power and memory available.
- The massive number of neighbors a node can have (thousands or even millions).
- The unavailability of node positioning mechanisms.
- The ability to multiplex many frames over the same period of time.
- The energy harvesting from the environment. This will lead to preserving nodes' resources and increasing network lifetime.

2.1   Pure Flooding

In ad-hoc wireless networks, multi-hop data broadcasting is an essential service. It is required by several applications, and used to broadcast information in the network (e.g. routing table updates, path updates, etc.)

Pure flooding is one of the traditional routing methods that has proven its performance in terms of delivery ratio, and delay in many always-awake network settings [11]. It is motivated due to its simplicity, which conforms to the constraint capabilities of the nanonodes.

Flooding is important especially in mobile ad-hoc networks (MANETs), which rely on it to perform routing discovery. It is an unreliable operation with no acknowledgment mechanism in place. In pure flooding, a node forwards each message (*without routing data*) received for the first time. However, this technique has drawbacks, the most notable is the generation of a significant amount of messages in the whole network. In dense networks, exponential propagation growth leads to a broadcast storm. Moreover, countermeasures have to be taken to prevent skyrocketing contention for channel access or collisions.

We argue, however, that these solutions suffer severe performance degradation (in both energy and time efficiency) if directly applied to low duty-cycle networks. It is very costly when energy consumption is considered.

2.2 Probabilistic Flooding

Many attempts have been made to optimize the pure flooding technique by selecting a subset of forwarding nodes. They are challenged by nanonode hardware limitation, either by the inability to build a complete map of even the direct neighbors, or because of too high memory requirements.

A common solution is to give each node a probability to forward a new packet (already seen packets by a node are discarded anyway.) The probability chosen could be fixed, or depend on several factors, such as density, distance, speed, and others. The most considered metric in calculating the probability is the number of neighboring nodes.

Probabilistic flooding greatly reduces redundant retransmissions and receptions compared to the pure flooding scheme. Several probabilistic flooding schemes have been proposed for wireless ad-hoc networks that require lightweight computing resources. Hence, they can be used for data dissemination in nanonetworks [12]. One of the most important defects they have is the die-out problem [13].

3   Background

3.1   Sleeping Mechanism

In networks where nodes have limited energy, the common technique to preserve energy is the duty cycling (sleeping). Nodes wake up from time to time to receive packets sent to them. Sleeping techniques used in the traditional networks are not adequate for nanonetworks due to communication peculiarities (pulse-based).

In our proposed fine-grained sleeping mechanism [10], all the nodes have the same awake-sleep cycle, equal to T_s. Inside the cycle, all the nodes have the same awake *duration* (or percentage of T_s), but the *beginning* of the awake interval is different for each node and is randomly determined. For this to work, all the flows must have the same spreading ratio $\beta = T_s/T_p$.

The normal purpose of a communication process is to deliver information to a destination. The definition of destination (zone or node) may change depending on the application. If the destination is defined as an SLR address (*zone*), this means that the packet should reach this SLR zone and at least one node must receive it (it does not matter which one). In that case, the mechanism we proposed in [10] is efficient.

However, if the packet needs to reach a specific node in the destination zone, then more aspects have to be taken into consideration. When a packet arrives at the destination zone, the destination node may indeed be asleep and would miss the packet.

For additional information about the nanosleeping mechanism, refer to [10].

3.2 Full Retransmission Algorithm

Nanonetwork applications can vary from biomedical (e.g. drug delivery) to agricultural (e.g. water and pesticide monitoring) and environmental (e.g. air pollution control) services. Nanonodes can be implanted into the environment, food, or the human body. Therefore, and in some particular applications, it is extremely important for the destination node to receive all the data.

To the best of our knowledge there has been no research in the literature on customizing a method to ensure that the packet reaches a sleeping destination node. For that reason, a retransmission algorithm was already proposed [5] in combination with the sleeping mechanism. The aim of this algorithm is to increase the destination node's chances of receiving the packet if it is asleep when the packet reaches the destination zone. It is worth mentioning that the algorithm is used only by the nodes at the destination zone.

In the absence of the retransmission algorithm, the destination node does not receive the packet if it was in sleep mode when that packet arrived at the destination zone.

Deep analysis for the retransmission behavior at the destination zone shows that after a segment of time all the nodes will participate in retransmitting the packet. For example, in a destination zone of 41 nodes, there are 41 retransmission attempts. This will lead to an increase in packet exchanging, therefore a waste of nodes' resources, and the occurrence of congestion phenomena. An enhancement to this algorithm is needed, without affecting the node's reception reliability.

The objective of this paper is to present such an enhancement.

4   Probabilistic Retransmission Algorithm

To the best of our knowledge, there is no similar proposed algorithm that takes into consideration achieving a reliable packet reception at the destination (zone/node).

In some applications, ensuring a reliable packet reception by a specific node is a key factor. For example, periodic car maintenance can be explained as a service/maintenance model, where a car undergoes a service/maintenance either after a certain specified time period or on the basis of a part getting faulty [14]. Car parts (brake, motor, etc.) are equipped with a sensor for monitoring and data collecting purposes. For some reason, if the brake sensor does not receive or collect information, this might put the driver at risk in case the brakes are faulty. Therefore, having a packet reception algorithm of high reliability becames a key factor in IoT applications.

Allowing all nodes at the destination zone to retransmit the packet leads to nodes' resources being exhausted. To avoid this problem, we propose a probabilistic retransmission algorithm, where not all the nodes participate in the retransmission mechanism. The number of participating nodes is determined based on a probability, calculated as follows:

$$probability = 1 - \frac{aD}{T_s} \tag{1}$$

In this formula, the retransmission probability is inversely proportional to the awaken duration percentage aD. Table 1 shows the expected number of participating nodes among various awaken durations.

No matter the network density, this algorithm never saturates the radio channel and does not require much memory, or computations. The only memory needed is the buffer to store the received packet to retransmit it at the end of the awaken duration.

In this algorithm, we took into consideration the case where the awaken duration spans over two time cycles. The variables used in the algorithms are the following:

- *waitingTime*: node waiting time before packet retransmission at the end of its awaken duration.
- *wT1ts*: node waiting time if the awaken duration range is 1 T_s.
- *wT2ts*: node waiting time when its awaken duration spans on 2 T_s.
- *aD*: node awaken duration.
- *aS*: node awaken starting time.
- *pcktrecp*: the time when the node receives the packet.
- *probaRNG*: a probability random number generator function (0,1).
- *proba*: the calculated probability based on node awakenDuration.

Table 1. The expected number of participated nodes in full and probabilistic retransmissions.

Awaken duration (%)	Full retransmission	Probabilistic retransmission
6	35	35–38
10	41	34–37
20	41	32–35
30	41	28–32
40	41	25–28
50	41	21–25
60	41	15–21
70	41	11–15
80	41	7–11
90	41	3–7
100	41	0

The node retransmits the packet if and only if the *probaRNG* random variable is less than the calculated probability. The enhanced packet retransmission algorithm is presented in Algorithm 1.

Algorithm 1. Probabilistic retransmission algorithm executed by nodes at the destination zone only.

---
alreadyseen = false
waiting $Time$
w T1ts = (aD - (pcktrecp - aS)) % T_s
w T2ts = - (aD - (pcktrecp - aS)) % T_s
if packet type is data **then**
    if packet !alreadyseen AND the received node is not the destination node **then**
        alreadyseen = true
        if pcktrecp % T_s ≥ aS **then**
            waiting $Time$ = w T1ts
        else // pcktrecp % T_s < aS + aD - T_s
            waiting $Time$ = w T2ts
        end if
        probaRNG = rand (0, 1)
        proba = 1 − (aD / T_s)
        if probaRNG < proba **then**
            the node will retransmit the packet at the end of its aD (now + waiting $Time$)
        end if
    end if
end if

---

5 Evaluation

This section evaluates the retransmission algorithm in improving packet reception reliability at the destination zone. As a detailed analytic study is not possible and nanomachines have not yet been manufactured, we evaluate the protocol through simulations. Technical details and information about the full reproducibility of our results are provided on a separate website[1].

We use BitSimulator[2] [4] to evaluate our proposed ideas. BitSimulator allows to simulate ultra-dense nanonetworks using TS-OOK modulation. It simulates applications and routing protocols while keeping a relatively detailed model for the MAC and physical layers. As such, it enables exploration and understanding of the effects of low level coding and channel access contention. It comes with a visualization program, VisualTracer, which displays graphically the simulation events, such as in Fig. 2.

---

[1] http://eugen.dedu.free.fr/bitsimulator/aina22.

[2] Free software, available at http://eugen.dedu.free.fr/bitsimulator.

In our simulations, the network topology consists of a homogeneous network as a 2D area of size 6 mm * 6 mm, cf. Table 2. One packet traverses the network; the source node is at the bottom left of the network, while the destination node is at the top right, cf. Fig. 2.

Table 2. Simulation parameters.

Parameter	Value
Size of simulated network	6 mm * 6 mm
Number of nodes	25 000
Communication radius	500 μm
Hops to reach the furthest node	17
AwakenDuration	6000 fs
T_p	100 fs
β (spreading ratio)	1000
Packet size	1000 bit

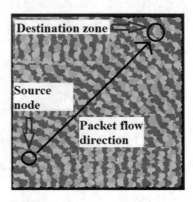

Fig. 2. The evaluated network.

In the following analysis, all the nodes use the sleeping mechanism. The simulation is repeated several times by changing the node awaken duration percentage, all the other parameters being kept identical.

The metrics used to analyze the algorithm efficiency are the number of nodes that retransmit the packet at the destination zone, and the reliability of receiving at least 1 copy of the packet by the destination node.

Determining a static awaken percentage for every node in the network (e.g. 20% is equivalent to 20 000 fs) means that all nodes will be awake for this percentage in a time duration equal to T_s. We recall that inside this cycle, all the nodes have the same awaken duration (or a percentage of T_s), but the beginning of the awake interval is different for each node and is randomly determined.

The probabilistic retransmission algorithm aims to decrease the number of participating nodes in packet retransmission at the destination zone. Figure 3 shows the efficiency of this algorithm compared to the full retransmission. We notice that for an awaken duration of 6%, both transmissions mechanisms have the same number of participating nodes (35). This is expected since for a low awaken duration the probability of retransmission is high.

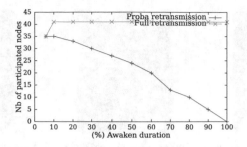

Fig. 3. Participating nodes in packet retransmission at the destination zone.

The relation between the awaken duration and the probability is inversely proportional. While the awaken duration increases, the probability decreases, therefore it reflects a decrease in the number of participating nodes while using the probabilistic retransmission algorithm. For example, for an awaken duration of 50%, the number of participating nodes is 25 with the proba algorithm, while with the full retransmission this number remains steady, 41 nodes, for all the simulated awaken duration (recall that the total number of nodes at the destination zone is 41).

Figure 4 is a sketch extracted from VisualTracer. The figure shows the benefit of applying the probabilistic retransmission against the full retransmission. (a) Shows that all nodes (41) participate in retransmission at the destination zone. Applying the proba retransmission in (b) while using the same awaken nodes percentage, shows a decrease of 68% of participating nodes. An increase in the awaken duration percentage (c) shows a higher decrease in the number of participating nodes (88%) compared to the full retransmission.

It is important to ensure that the algorithm does not affect packet routing in the previous zones. Figure 5a shows that the number of participating nodes in the packet routing does not change while applying the retransmission algorithm. Therefore, the previous zone is just playing the role of routing the packet and sending it to the next hop (*zone*).

Figure 5b shows the reliability of packet receiving by the destination node. In all simulations, and using several awaken duration percentages, the destination node is still able to receive at least 1 copy of the intended packet.

Using several flows in a network might affect the algorithm being applied. For that reason, it is necessary to evaluate our algorithm when using several flows too. Figure 5c depicts the results of a simulation with 15 flows from 15 source nodes sending packets to one destination node. For full retransmission, the number of retransmitted packet copies ($\approx$620) stays stable along all the awaken duration percentages used. Even if this number decreases, it achieves around 50% fewer retransmissions ($\approx$310) at 50% of node awaken duration, and 11% fewer retransmissions at 90% of awaken duration.

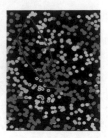
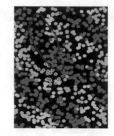
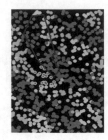

(a) Full retransmission, all nodes paticipating at the destination zone.

(b) Proba retransmission applied for 70% of node awaken duration.

(c) Proba retransmission applied for 90% of node awaken duration.

Fig. 4. VisualTracer sketch for the destination zone, for the number of participated nodes with full and proba packet retransmission.

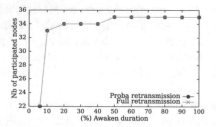
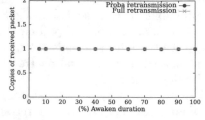

(a) Retransmission mechanism does not affect the previous zone.

(b) The reliability of receiving at least 1 copy of the packet by the destination node.

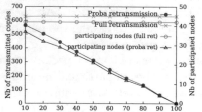
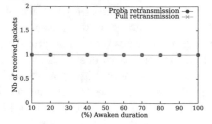

(c) Number of retransmitted packet copies and nodes handling.

(d) The destination node success to receive at least 1 packet copy of each flow.

Fig. 5. Probabilistic retransmission simulations results.

This is also reflected in the number of participating nodes, where all the nodes (42 nodes) are participating in the full retransmission. This number decreases according to the awaken duration percentage used in the proba retransmission (e.g. 20 nodes at 50% of awaken duration, and 5 nodes at 90% of awaken duration).

Figure 5d shows that the use of the probabilistic retransmission algorithm contributes to enhancing packet reception reliability: the destination node receives at least 1 copy of the retransmitted packet from each flow. Once again, the probabilistic algorithm proves its effectiveness even in case of several flows.

6   Conclusion

In this paper, we proposed and discussed a probabilistic retransmission algorithm at the destination zone where not all nodes participate in retransmission process. The goal is to ensure that the destination node receives the packet, while reducing the number of packets exchanged. In the proposed algorithm, the number of participating nodes is correlated to the percentage of nodes awaken duration.

The evaluations show that probabilistically retransmiting the packet at the destination zone ensures high reliability in packet reception. Therefore, the benefit from this algorithm can significantly vary from one application to another. The simulation results show that the destination node is still able to receive the intended packet while decreasing the number of retransmissions at the destination zone.

Besides the probabilistic retransmission, using the sleeping mechanism improves network behavior by limiting the amount of traffic an individual node can see. Traffic is statistically dispatched over all nodes, thus sharing the load. As individual nodes see less activity, they also use fewer resources (energy, CPU, memory), therefore the network lifetime will increase.

References

1. Akyildiz, I.F., Brunetti, F., Blázquez, C.: Nanonetworks: a new communication paradigm. Comput. Netw. **52**, 2260–2279 (2008)
2. Yin, G., Yang, G., Yang, W., Zhang, B., Jin, W.: An energy-efficient routing algorithm for wireless sensor networks. In: 2008 International Conference on Internet Computing in Science and Engineering (ICICSE), pp. 181–186. IEEE (2008)
3. Yao, X., Huang, W.: Routing techniques in wireless nanonetworks. In: Nano Communication Networks, pp. 100–113. IEEE (2019)
4. Dhoutaut, D., Arrabal, T., Dedu, E.: BitSimulator, an electromagnetic nanonetworks simulator. In: 5th ACM International Conference on Nanoscale Computing and Communication (NANOCOM), pp. 1–6. IEEE (2018)
5. Medlej, A., Dedu, E., Beydoun, K., Dhoutaut, D.: Self-configuring asynchronous sleeping in heterogeneous networks. ITU J. Future Evol. Technol. (ITU J-FET **2**, 51–62 (2021)
6. Jornet, J.M., Akyildiz, I.F.: Graphene-based nano-antennas for electromagnetic nanocommunications in the terahertz band. In: Fourth European Conference on Antennas and Propagation (EuCAP), pp. 1–5. IEEE (2010)
7. Piro, G., Boggia, G., Grieco, L.A.: On the design of an energy-harvesting protocol stack for body area nanonetworks. In: Nano Communication Networks, pp. 181–186. Elsevier (2015)
8. Ageliki, T., Christos, L., Dedu, E., Ioannidis, S.: Packet routing in 3D nanonetworks: a lightweight, linear-path scheme. Nano Communication Networks **12**, 63–71 (2017)
9. Jornet, J.M., Akyildiz, I.F.: Femtosecond-long pulse-based modulation for terahertz band communication in nanonetworks. IEEE Trans. Commun. **62**, 1742–1753 (2014)

10. Medlej, A., Dedu, E., Beydoun, K., Dhoutaut, D.: Scaling up routing in nanonetworks with asynchronous node sleeping. In: 2020 International Conference on Software, Telecommunications and Computer Networks (SoftCOM), pp. 1–6. IEEE (2020)
11. Miller, M.J., Sengul, C., Gupta, I.: Exploring the energy-latency trade-off for broadcasts in energy-saving sensor networks. In: 25th IEEE International Conference on Distributed Computing Systems (ICDCS), pp. 17–26. IEEE (2005)
12. Reina, D.G., Toral, S., Johnson, P., Barrero, F.: A survey on probabilistic broadcast schemes for wireless ad hoc networks. Ad Hoc Netw. **25**, 263–292 (2015)
13. Arrabal, T., Dhoutaut, D., Dedu, E.: Efficient multi-hop broadcasting in dense nanonetworks. In: 2018 IEEE 17th International Symposium on Network Computing and Applications (NCA), pp. 1–9. IEEE (2018)
14. Dhall, R., Kumar, V.: An IoT based predictive connected car maintenance approach. Int. J. Interact. Multimedia Artif. Intell. (IJIMAI) **4**, 16–22 (2017)

The Development of an Elderly Monitoring System with Multiple Sensors

Yasunao Takano[1]([✉]), Hiroyuki Adachi[2], Hiroji Ochii[3], Mikio Okazaki[3], and Sena Takeda[1]

[1] Kitasato University, 1-15-1 Kitasato, Minami-ku, Sagamihara-shi, Kanagawa 252-0373, Japan
tyasunao@kitasato-u.ac.jp
[2] Yasuragi, Daigo 888, Daigomachi, Kuji, Ibaraki 319-3526, Japan
[3] Elt Co., Ltd., Asahi Seimei Otemachi Building 3F, Otemachi 2-6-1, Chiyoda-ku, Tokyo 100-0004, Japan

Abstract. It is important to improve services in geriatric healthcare facilities in step with the aging of the population. However, it is difficult to secure human resources, and it is unclear whether the supply of human resources will be sufficient to meet the demand. Therefore, it is necessary to maintain the safety of services by looking after the care receivers with limited human resources. In this paper, we propose an elderly monitoring system using Pifaa which is designed by Murata Manufacturing Co., Ltd. Because the configuration of Pifaa can be changed by inserting sensors into the USB ports, we use it as a system that can be changed according to the caregiver's situation. This system is designed to monitor the state of beds of the care receiver, and to check the state of leaving and arriving at the bed using a PC or tablet application. This paper describes the current system configuration using multiple sensors, and also describes the future tasks, since we are still in the early stage of developing a system to store the results of multiple sensors on the cloud.

Keywords: Elderly care · Bed watching · Internet of Things · Multiple sensors

1 Introduction

In order to avoid social isolation of the elderly, it is becoming more and more important to improve services provided at long-term care facilities for the elderly. Although there is an urgent need to secure the nursing care workers for the improvement of services, it is unclear whether it is possible to supply the necessary workers to meet the demand. Therefore, it is necessary to look after the care receiver with limited human resources and maintain the safety of services.

© The Author(s), under exclusive license to Springer Nature Switzerland AG 2022
L. Barolli et al. (Eds.): AINA 2022, LNNS 450, pp. 231–242, 2022.
https://doi.org/10.1007/978-3-030-99587-4_20

On the other hand, there is a possibility that these problems can be improved by using IT technology. Some elderly care facilities are beginning to have network infrastructure, and it will become the standard in many facilities in the future. Therefore, it will be possible to develop softwares that take advantage of the network infrastructure.

In particular, with the development of artificial intelligence (AI) technology and Internet of Things (IoT) technology, it has become possible to automatically determine whether a person is present by analyzing camera images and sensor data. However, there are still some issues to be overcome in order to implement such a system with practicality. For example, there are problems such as false positives/negative, implementation costs, and scalability, although these problems will be dealt with appropriately by combining current technologies.

In light of this situation, this paper presents a case study of a system that uses multiple sensors to monitor elderly care receivers. Our proposed system is characterized by the fact that it can be used while changing the configuration of the removable sensors according to the care receiver's situation. This system is designed to monitor the condition of the bed of the care receivers and check the condition using an application on a PC or tablet. The system is based on **Pifaa** which is designed by Murata Manufacturing Co., Ltd.

This paper describes the current system configuration that uses multiple sensors, and also describes future issues, since we are still in the early stage of developing the system that accumulates the results of multiple sensors on the cloud.

2   Background

The Internet of Things (**IoT**) is becoming more and more common. For example, air conditioners that can be operated over networks, LED light bulbs that can be used to remotely control the lights in rooms, and cameras for security purposes have become commonplace in households. In this paper, we focus on the use of IoT in which various sensors are placed in rooms and their values are collected to the cloud.

When combining sensors in IoT, the following problems must be solved.

1. It is difficult to know how to interpret the raw data obtained.
2. Deciding on the right sensor is difficult.
3. It costs a lot.

Regarding the first item, even if data is obtained, it depends on the interpretation of the data as to what situation it indicates. For example, it is not easy to determine whether an object has fallen or a person has fallen just because a sound is heard, and some kind of devices is needed to increase the accuracy of estimation. Therefore, it is important to select appropriate sensors, but as mentioned in the second item, it is difficult to decide what to combine. There may be sensors that are necessary at the beginning, but are no longer necessary or become necessary over time or according to the situation. Such trial and error for sensor selection is costly, as mentioned in the third item.

We focus on the second and third of these problems, and deal with the system that can flexibly combine multiple sensors.

2.1 Pifaa

In order to solve the problems listed in Sect. 2, we use a system called Pifaa[1] for our elderly monitoring system. Pifaa consists of Pifaa Edge (hereafter referred to as **Edge**), a device that combines sensors and sends the data to the cloud, and Pifaa xSensors (hereafter referred to as **xSensors**), sensors that are connected to Edges.

Edge and xSensors are shown in Fig. 1. The Edge is connected to Wifi and uploads the data of xSensors to the cloud. By connecting xSensors to the Edge's USB ports, it is possible to freely select the necessary sensors according to the application. All we can do is connecting the required sensors to the Edge, which offers the following advantages

- cost can be reduced.
- trial and error is possible.
- data can be acquired in a uniform format from the cloud even if the sensor configuration is changed.
- multiple configurations can be mixed in a single system.

In the system described in this paper, the sensor configuration can be changed for each care receiver. However, even if Pifaa is used, the interpretation of the obtained data must be done by the system described in Sect. 4. As mentioned in Sect. 2, the problem of data interpretation in IoT sensors is not a problem that is addressed in this paper.

The available sensors are developed for each matter to be measured, as shown in each row of Table 1. The values of each xSensor can be uploaded at different frequencies through Edge, such as every second, every minute, etc., and the data can be retrieved with a unified interface from cloud.

---

[1] https://solution.murata.com/ja-jp/collaboration/theme/01_pifaa/.

Pifaa Edge

Pifaa xSensors

Connecting xSensors to the Edge

Fig. 1. Pifaa

Table 1. The example of xSensors

Sensor	Usage and function
Temperature Humidity Barometric pressure	Obtain the natural environment in which it is placed
Acceleration	Obtain the velocity per unit time in XYZ direction
Light	The illuminance of the place where the sensor is placed is acquired by the light receiving element
Distance	It can detect the proximity of an object in the direction of a sensor such as infrared or ultrasonic
Sound	The microphone is used to obtain the volume and height of the sound
CO_2 density	The wavelength of light absorbed by infrared rays is measured

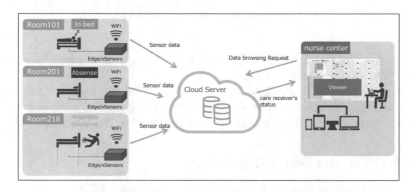

Fig. 2. Overall image of our system

3   Proposed system

3.1   Overview

Our system uses the Edges equipped with multiple xSensors to monitor the condition of the care receiver on the beds. The target of the system is to monitor multiple beds in an aged-care facility, and the system aims to be able to manage the condition of the beds in each room for the following purposes.

- Determine implantation/removal
- Pay attention to care receivers who have not been moving for a long time.
- Detect sudden movements of care receivers on the bed.

The entire image is shown in Fig. 2. We attach an Edge with a xSensor accelerometer and a xSensor sound sensor to each bed. The accelerometer can acquire the vibration of the bed. In addition to Pifaa, a camera is used to estimate the posture. The camera is connected to a Raspberry Pi, and the Raspberry Pi

Fig. 3. Attaching an edge to a bed

sends data to the cloud via Wifi. The camera information can also be uploaded to the Pifaa's cloud through an API.

This configuration is the initial configuration, and the sensor configuration will be changed as needed. For example, depending on the situation of care receivers, various bed configurations can be prepared in the system.

3.2   Selecting and Discarding Sensors

First, we attached an Edge to a bed, and then extended a USB cable to measure the vibration of the bed. Figure 3 shows how the Edge is installed.

The Edge needs a power source, so connect the cable from the bed to the power plug on the wall.

The vibration sensor should be placed between the mat and the bed frame as shown in the Fig. 4.

In the beginning of our project, each bed was equipped with five vibration sensors. The five vibration sensors were arranged as shown in Fig. 5. The preliminary experiment showed that the vibrations at each position were not significantly different, and it was determined that one value could be substituted for the other in terms of cost.

It is also known that the CO_2 sensor can be used to monitor human flow, but in nursing home for the elderly, multiple beds are contained in a single room (e.g., four beds in a room), so we decided that it was not necessary for this application. However, if it is needed, it can be connected a CO_2 sensor via USB using Pifaa, so it can be installed in a single bed room for flexible configuration. Another advantage of using Pifaa is that it allows for multiple types of Edge configurations.

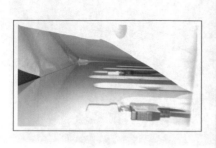

Fig. 4. Attaching the vibration xSensors

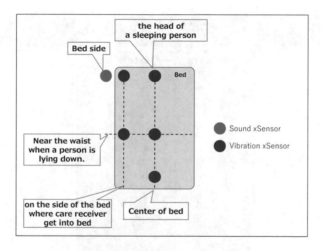

Fig. 5. Selecting a candidate for sensor placement

The advantage of using Pifaa is that the sensor can be attached and detached as described above, and trial and error can be performed.

3.3   Using Camera Images

By attaching a camera to the partition next to the bed, the camera can take pictures of the bed. The camera is placed at the bedside as shown in Fig. 6. Install a Raspberry Pi on top of the camera and send data to the Pifaa's cloud via Wifi. In consideration of privacy, the images are not sent to the cloud, but only the result of the Raspberry Pi's judgment as to whether the person is on the bed or not. We run Tensorflow's PoseNet posture control program on the Raspberry

Fig. 6. A camera to capture the scene on the bed.

Pi, and used it to control the care receiver's posture. For this application, about one frame per second is sufficient for use as a watchdog, so the Raspberry Pi 4 can be used. In addition, this camera supports infrared rays, so it can take pictures even at night.

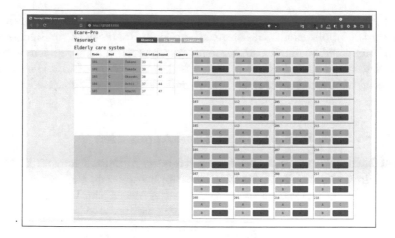

Fig. 7. Application to display the results

The posture of the patient is estimated using images from the camera. Initially, we used commercially available face recognition software to determine sleep and wakefulness based on whether the eyes were open or not, aiming to manage the sleep state of the care receivers. In many cases, it is not possible to set up a camera in a position where it can take a good frontal image as assumed by the face recognition software. Therefore, there were cases in which shadows such as wrinkles of pillows and blankets on the bed were recognized as a human face, and the system could not be put to practical use due to the problem of false recognition.

Therefore, we came to the decision to design the system not to make estimation based on camera images alone. As shown in the Sect. 3.1, our system estimate posture instead of facial expression by combining multiple sensors. Using the information from the sensors, the system can assist in determining whether there is a person on the bed or not. However, in order to measure sleep time, etc., it is necessary to use information such as whether the eyes are closed, as in the preliminary experimental system, so we decided to exclude this system.

4   Application to Display the Results

4.1   Overview

We need an application that can retrieve the data from the cloud, analyze it, and display the results. For example, the screenshot is as shown in Fig. 7. The right side shows the status of beds in the entire facility (in this case, there are four beds (A-D) in each room from 101 to 208, for a total of 28 rooms). In the upper left corner, the values of the individual beds are displayed. When a sudden movement is detected by sound sensor and vibration sensor, it is sorted to alert the user. The lower left corner displays logs.

The mapping between beds and the data from Edges will be maintained in JSON on the application. Therefore, at present, JSON data for bed positions must be managed manually, and consistency must be achieved when the bed is moved to other rooms. Since it seems to be difficult to obtain the positional information of the bed at present, a new and different mechanism will be necessary.

4.2   Native and Web Versions of the Application

We will have both an application that runs on Windows and an application that runs in a web browser. The Windows application and the web version have their own advantages and disadvantages. In the case of the Windows version, it is possible to create the application so that it can be accessed from within the facility. However, as an application, it is limited to devices such as Windows. The web version can use on portable devices such as tablets. Since it can be implemented using web technology, the cost of implementation can be also reduced even when multiple devices are targeted. On the other hand, access control must be designed appropriately.

4.3   Data Interpretation

Since the data sent by Edge is raw data, we use this application to estimate whether a person is on the bed or not

Currently, a threshold value is set to determine the presence of a person on the bed based on the vibration and sound values. The threshold is set so that when a sudden vibration is detected, it can be predicted that the person has fallen out of bed. However, it is difficult to adjust the current threshold value sufficiently to determine whether the patient is in or out of bed based on vibration and sound alone, and it is necessary to improve the accuracy by combining it with a camera.

In addition, it is necessary to design the application considering the selection of sensors in the future.

5   Related Research

A number of geriatric systems have been developed and are in the process of being commercialized [1–4]. Among them, [1] is a system that incorporates sensors in the bed to detect whether a person is on the bed or not, as in this paper. The system has very high functionality, including the ability to measure body weight, but it is very expensive because it is built into the bed. And also, it differs from our system in that it is not possible to add sensors later or change the configuration flexibly. Another problem is that a dedicated application must be used.

If we do not limit ourselves to detecting beds, there have been attempts to attach sensors to various objects to detect behavior [5–7]. Our study aims to solve the problem of elderly care by focusing on beds.

[8] is a study that uses multiple vibration sensors to estimate the state of a bed. In addition to this study, other factors can be investigated in a complex manner, but [8] can be used as a reference when estimating vibration.

6   Summary

In this paper, we presented a case study of a system that uses multiple sensors to monitor care receivers in elderly care facilities. Our proposed system is characterized by the fact that it can be used while changing the configuration of the removable sensors according to the care receiver's situation. By using Pifaa, we were able to build a foundation for configuring sensors according to the care recipient's situation. In addition, since the data is stored in the cloud, it can be checked from anywhere and with any device.

At the time of submission of this paper, the initial stage of the system has been installed in "Yasuragi", an elderly care facility in Japan. Therefore, we are planning to improve this system as future work. Specifically, the following points can be improved.

- Handling of bed movements
- Use of other sensors
- Data interpretation methods using AI

As for the first item, when we installed the system in geriatric care facility, we found that beds were frequently moved. Therefore, there are many cases where the power cord of the Edge is unplugged. In this case, we had to solve the physical problem that the power cord was forgotten to be plugged in and the data was not sent. It is possible to detect that the data has not been sent to the cloud and check it, but even in that case, the caregiver has to plug it again, which adds to the burden of the original care work. Also, since the mapping on the application becomes misaligned when the bed is moved, a mechanism to repair the misalignment will be necessary in the future.

For the second item, sleep can be measured using a heart rate sensor. In relation to the first item, we are also considering a method to estimate the bed movement from environmental information.

For the third item, since we have been able to accumulate data from xSensors as big data, we believe that it can be interpreted by AI. In that case, both supervised and unsupervised learning will be possible.

References

1. FRANCE BED CO., LTD.: Mimamoricare-system M-2. https://medical.francebed.co.jp/iryofukushi/mimamori_m2/
2. Ricoh Co., Ltd.: Mimamori bed sensor system. https://www.ricoh.co.jp/bedsensor/
3. NDSoftware Co Ltd.: Bed sensor. https://www.ndsoft.jp/product/sensor/
4. Future Inc.: Vital Beats. http://www.futureink.co.jp/vitalbeats/

5. Suryadevaraa, N.K., Gaddama, A., Rayudub, R.K., Mukhopadhyaya, S.C.: Wireless sensors network based safe home to care elderly people behaviour detection. Sens. Actuators A Phys. **186**, 277–283 (2012)
6. Awais, M., et al.: An Internet of Things based bed-egress alerting paradigm using wearable sensors in elderly care environment. Sensors **19**(11), 2498 (2019)
7. Steele, R., Lo, A., Secombe, C., et al.: Elderly persons' perception and acceptance of using wireless sensor networks to assist healthcare. Int. J. Med. Inform. **78**(12), 788–801 (2009)
8. Shao, S., Kubota, N., Hotta, K., Sawayama, T.: Behavior estimation based on multiple vibration sensors for elderly monitoring systems. J. Adv. Comput. Intell. Intell. Inform. **25**(4), 489–497 (2021)

Predicting Cyber-Attacks on IoT Networks Using Deep-Learning and Different Variants of SMOTE

Bathini Sai Akash[1($\boxtimes$)], Pavan Kumar Reddy Yannam[1],
Bokkasam Venkata Sai Ruthvik[1], Lov Kumar[1], Lalita Bhanu Murthy[1],
and Aneesh Krishna[2]

[1] BITS-Pilani Hyderabad, Hyderabad, India
{f20190065,f20190038,f20190017,lovkumar,
bhanu}@hyderabad.bits-pilani.ac.in
[2] Curtin University, Perth, Australia
A.Krishna@curtin.edu.au

Abstract. Predicting Cyber-attacks on IoT Networks using Machine Learning has a definite advantage over traditional methods because it helps secure against future attacks by identifying hidden patterns from past data, thereby improving the capability of a network. Thus automated systems are being developed which can be used to identify Cyber-attacks on IoT Networks using various machine learning techniques. In this work, three different types of features selection techniques were applied to the UNSW-NB15 data to find the best combination of relevant features. These selected sets of relevant features were considered to train five different deep learning architectures used to predict cyber attacks by varying the number of hidden layers. To handle the dataset's class imbalance problem, we have considered three different sampling techniques: SMOTE, Borderline SMOTE (BSMOTE), and Adaptive Synthetic Sampling (ADASYN). The experimental results on the UNSW-NB15 data highlight that the usage of considered feature selection techniques and class balance techniques does not significantly improve the predictive ability to detect cyber attacks. The results also suggest that variation in Deep learning Architecture impacts the prediction of cyberattacks.

Keywords: Cyber-attacks · Data imbalance methods · Feature selection · Classification techniques · Deep learning

1 Introduction

The use of IoT infrastructure has reached a new high. During the COVID-19 pandemic, there was a significant increase in the use of Smart Devices for domestic, medical, and industrial purposes [1]. The inevitable growth in the complexity of IoT infrastructures has raised unwanted vulnerabilities in various systems, which has led to a rise in Cyber-attacks on IoT Networks. This threat needs to be addressed since it could compromise the security of these networks and the infrastructure dependent on them [2,3]. For this purpose, Network intrusion detection systems (NIDS) are tasked with constantly surveying and detecting vulnerabilities simply by monitoring the network traffic. It was observed from the literature that Intrusion Detection Systems (IDS) use three different detection methods: signature-based detection, anomaly-based detection, and a machine learning-based approach to identify a potential intrusion [4]. Machine Learning-based IDS helps to detect novel mutated attacks by learning from heterogeneous data, and earlier forms of attack [5]. Consequently, researchers working in the area of IDS observed that Machine Learning based IDS is more resilient and does not need as many patches as in traditional techniques like signature-based detection [6].

Hence, using Machine Learning has a definite advantage because it helps stop future attacks on IoT Networks and thereby improves the capability of a network to handle future attacks by identifying the pattern of attacks. In order to detect cyber-attacks on IoT networks, several machine learning methods were considered in the literature, but low importance was given to finding the optimal combination of feature selection techniques and class balancing techniques to be used along with Deep Learning methods. Deep Learning techniques can identify hidden patterns in large amounts of data, i.e., big data. This study made use of the UNSW-NB15 [7,8] dataset for network intrusion detection to validate the proposed framework, with over 170,000 training records. The dataset contains modern network traffic, including low footprint attack types. However, it was observed that the considered dataset suffers from the class imbalance problem. A dataset is defined as a balanced dataset when roughly an equal number of input samples represents each target class. So, this study used three different data sampling techniques to balance the data with an objective to improve the predictive ability of the developed models. Finally, we also used different feature selection techniques to remove unimportant features. This helps reduce the computational complexity of the models and potentially improve the performance of Machine learning algorithms by extracting only the necessary features [9]. The key contributions of this research are as follows:

- Five different types of deep learning architectures were considered to develop a model for predicting Cyber-attacks on IoT Networks.
- Three different data sampling techniques were used to balance the data before applying the training algorithm.
- Three different feature selection methods were also applied to select the right combination of features with an objective to better the performance of deep learning models.

2   Related Work

K.A Tait et al. [10] provided a comparison of various machine learning techniques applied on the UNSW-NB15 data set. They reported that the random forest algorithm performed better than other methods in the case of binary classification, achieving an accuracy of 99.77%, while KNN gave the best results in the case of multi-class classification. C. Wheelus et al. [11] proposed the prepossessing step to tackle the class imbalance problem in cyber security datasets such as UNSW NB15 [7,8] and KDD-CUP99 [12]. The study re-processed the UNSW NB15 dataset using the attributes from the SANTA data set. They reported that SMOTE provided the most significant lift on average in the ROC-AUC values, compared to the baseline values. Nutan Farah Haq et al. [13] have analyzed 49 research papers related to using different intrusion detection classifiers between 2009–2014 and observed the need to remove redundant and irrelevant features for the training phase while stating that it is a crucial factor for system performance. Fatemeh Amiri et al. [14] exclusively considered the effect of feature selection techniques such as the forward feature selection algorithm, linear correlation-based feature selection, and a modified mutual information feature selection algorithm in preprocessing the KDD-CUP99 dataset. Finally, they have concluded that the feature selection algorithms can significantly improve classification accuracy.

To the best of our knowledge, all the related works in this field have notably used deep learning for cyber-attacks on IoT Networks. Our work analyses deep learning methods coupled with different feature selection and sampling techniques to predict cyber-attacks on IoT Networks.

3 Study Design

This section presents the details regarding the various design settings used for this research.

3.1 Experimental Dataset

This study used the UNSW-NB15 dataset, published in 2015 [7,8]. It has 49 features comprising nine modern attacks and class labels, synthetically produced in a test environment. Attack types such as Worms, Shellcode, Generic, Reconnaissance, DoS exploits, Fuzzers, and Backdoors are provided for multi-class classification. This study chose the UNSW-NB15 dataset over other recognized datasets such as the KDD CUP99 [12] or the NSL-KDD because it contains modern network traffic of normal and abnormal types, including low footprint attack types. In contrast, the other datasets mentioned above are relatively older and may not represent modern network traffic.

3.2 Feature Selection Techniques

Feature selection methods facilitate a way of reducing computation time and avoid the curse of dimensionality. This study analyzes whether these techniques improve prediction performance or mask critical features in the network intrusion detection domain. This work made use of three different feature selection techniques: Significance test, Cross-Correlation test, and PCA [15] with an objective to remove irrelevant features and select the most appropriate set of features for predicting cyber-attacks on IoT Networks.

3.3   Deep Learning Architecture

Deep Learning (DL) is widely known for finding complex patterns in data [1]. Deep Learning makes use of non-linear processing units, called neurons, stacked to form multiple layers, which collectively combine the input signals in complex ways to get an appropriate output. [16] This paper used five different types of deep learning architectures by changing the number of hidden layers with an objective to develop models for detecting cyber-attacks on IoT Networks. The architecture of the considered DL models (DL1, DL2, DL3, DL4, and DL5) is shown in Fig. 1. The considered cyber-attack prediction models were trained using a batch size of 128 and an early stopper which allowed a maximum of 1000 epochs. The dropout hyperparameter value used was 0.2.

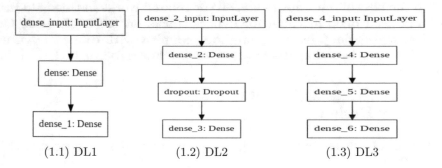

(1.1) DL1   (1.2) DL2   (1.3) DL3

Fig. 1. Deep learning architectures

3.4   SMOTE

The above-proposed models were validated with the help of the UNSW-NB15 dataset as mentioned in Sect. 3. However, the UNSW-NB15 dataset does not have an equal representation of attack and benign samples; in this case, only 1 out of 3 data points have a positive label and hence are classified under the attack category. So, this study used three sampling techniques: SMOTE, Borderline SMOTE (BSMOTE/BLSMOTE), and Adaptive Synthetic Sampling (ADASYN) [17] to overcome the problem of data imbalance [11]. In this work, we also validated the predictive ability of the models trained on the balanced data and original data(OD).

4   Results and Analysis

This section presents the performance of the trained models generated using different feature selection techniques, data sampling techniques, and different architectures of deep-learning models in terms of Accuracy and AUC scores. Figure 2 shows the steps followed to develop the prediction models for cyber-attacks on IoT Networks. These steps are as follow:

- Countvectorizer was applied to the dataset to convert the text features of the dataset to numerical features.
- The preprocessed data after the above step was passed as an input for different feature selection techniques to find the best set of features for predicting cyber-attacks on IoT Networks.
- Different data sampling techniques have been applied on the selected features with an objective to get an equal distribution of samples belonging to both classes.
- The resulting final preprocessed data was used as an input to five different deep-learning models with an objective to find patterns that help identify future cyber-attacks on IoT Networks.
- The models developed using the above techniques were validated with the help of the UNSW-NB15 data set.

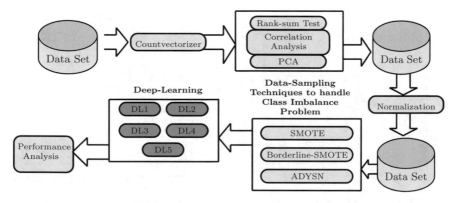

Fig. 2. Proposed framework

Three different combinations of features, four different sets of data, and five different variants of deep learning have been used to develop models for predicting cyber-attacks on IoT Networks. Therefore, a total of 120 ((3 sets of features) * (1PCA+1 Without PCA) * (1 Original Data+ 3 class balancing techniques) * 5 different classification techniques) distinct prediction models were generated in the study. The performance of these models is evaluated in terms of Accuracy, F-Measure, and AUC scores. Figure 3 and Table 1 show the performance of models trained on preprocessed data that involved PCA. The results for other cases are of similar type. The information present in Table 1 and Fig. 3 suggested the value of AUC ≥ 0.7 for all models confirms that the developed models have the ability to predict future cyber-attacks on IoT Networks. The information also suggested that the models developed using DL3 have a better ability to predict cyber-attacks on IoT Networks than others.

5   Comparative Analysis

This section presents the performances of various models obtained for intrusion detection. As explained in Sect. 3, this paper applies various techniques such as class balancing and feature selection on the UNSW-NB15 dataset to validate the proposed models. We have also investigated whether neural networks with different hidden layers and dropout regularization have any significant difference in their performance for the classification problem of intrusion detection. The results obtained in this study are summarized in the following subsections:

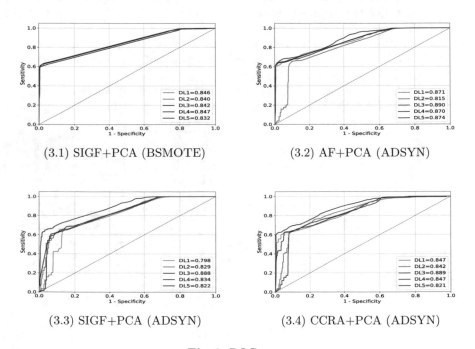

(3.1) SIGF+PCA (BSMOTE)   (3.2) AF+PCA (ADSYN)

(3.3) SIGF+PCA (ADSYN)   (3.4) CCRA+PCA (ADSYN)

Fig. 3. ROC curve

Table 1. Accuracy, F-Measure, and AUC Values

		Accuracy					F-Measure					AUC				
		DL1	DL2	DL3	DL4	DL5	DL1	DL2	DL3	DL4	DL5	DL1	DL2	DL3	DL4	DL5
OD	AF+PCA	70.00	71.16	73.54	72.63	72.78	0.78	0.79	0.79	0.79	0.78	0.82	0.82	0.87	0.87	0.83
OD	SIFG+PCA	70.29	69.20	70.73	71.95	69.50	0.78	0.78	0.78	0.79	0.78	0.83	0.84	0.84	0.83	0.82
OD	CCRA+PCA	71.23	67.35	71.92	71.71	71.02	0.74	0.75	0.79	0.79	0.78	0.80	0.80	0.84	0.84	0.83
SMOTE	AF+PCA	74.27	75.11	75.55	71.38	76.41	0.77	0.75	0.77	0.78	0.79	0.87	0.83	0.87	0.86	0.88
SMOTE	SIFG+PCA	73.08	74.77	73.35	71.96	71.61	0.75	0.75	0.76	0.78	0.77	0.83	0.83	0.85	0.84	0.83
SMOTE	CCRA+PCA	72.04	73.51	72.46	74.02	77.52	0.74	0.75	0.78	0.74	0.75	0.83	0.85	0.85	0.80	0.86
BSMOTE	AF+PCA	78.48	75.58	79.96	79.26	78.71	0.77	0.75	0.78	0.77	0.76	0.88	0.83	0.89	0.87	0.86
BSMOTE	SIFG+PCA	74.43	74.19	76.47	75.18	73.79	0.74	0.74	0.75	0.74	0.73	0.83	0.83	0.86	0.84	0.83
BSMOTE	CCRA+PCA	74.99	74.80	77.53	73.16	78.02	0.75	0.74	0.76	0.74	0.75	0.82	0.84	0.88	0.83	0.76
ADSYN	AF+PCA	78.24	74.79	79.82	78.93	78.81	0.77	0.74	0.78	0.77	0.76	0.87	0.82	0.89	0.87	0.87
ADSYN	SIFG+PCA	74.09	73.86	78.87	73.86	73.90	0.74	0.74	0.78	0.75	0.73	0.80	0.83	0.89	0.83	0.82
ADSYN	CCRA+PCA	75.07	75.57	77.90	75.55	75.46	0.74	0.75	0.76	0.75	0.74	0.85	0.84	0.89	0.85	0.82

5.1   Sampling Techniques

We have used three different types of sampling techniques: SMOTE, Borderline SMOTE, and ADASYN aiming to make an equal distribution of samples in both the categories. In this section, we aim to find the solution for *RQ1: "Is there a significant difference between models generated using various class balancing techniques?"* based on box plots and significant tests.

Box-Plot: Sampling Techniques: The performances of the models developed using different types of sampling techniques in terms of accuracy, F-measure, and AUC are represented in Fig. 4. The information present in Fig. 4 suggested that the models developed using original data have a better ability to predict cyber-attacks on IoT networks as compared to sampling techniques. As mentioned in Fig. 4, the models developed using original data achieved an average AUC value of 0.91, while the models developed using balanced data achieved an average AUC value of 0.82.

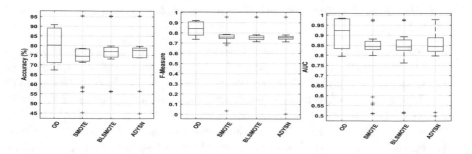

Fig. 4. Box-plot diagram: performance of different sampling techniques

Statistical Hypothesis Testing: In this work, we used statistical hypothesis testing to validate our considered null hypothesis, i.e., "There is no significant difference in the performance of the models across all the different class balancing methods". The above null hypothesis is accepted if the calculated p-value is greater than 0.05. The results of statistical hypothesis testing on the performance of different models trained using different data balancing techniques are mentioned in Table 2. From Table 2, it can be observed that all comparisons between SMOTE, BSMOTE, and ADASYN have P-values greater than 0.05, which means that the null hypothesis cannot be rejected in any pairing. This indicates that there is no significant difference in the performance of the models across all the different class balancing methods applied.

Table 2. Hypothesis testing: different sampling techniques

	OD	SMOTE	BSMOTE	ADASYN
OD	1.000	0.022	0.018	0.021
SMOTE	0.022	1.000	0.994	0.923
BSMOTE	0.018	0.994	1.000	0.947
ADASYN	0.021	0.923	0.947	1.000

5.2   Feature Selection Techniques

In this work, we have used three different feature selection techniques: Significance test-based feature selection (SIGF), features selected after cross-correlation analysis, and PCA with an objective to remove irrelevant features and select the right combination of features. In this section, we are going to find the solution for **RQ2: "What is the impact of using feature selection techniques for developing effective models?"** based on box-plot and significant test.

Box-Plot: Sampling Techniques: The performance of the models developed using different types of feature selection techniques in terms of Accuracy, F-measure, and AUC are visualized as a box plot in Fig. 5. The information present in Fig. 5 suggests that the models developed using Actual Features (AF - Baseline) have better ability to predict cyber-attacks on IoT networks as compared to feature selection techniques. The models developed using original data achieved an average AUC value of 0.98, while the models developed using feature selection techniques achieved an average AUC value of 0.76.

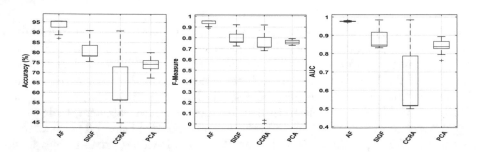

Fig. 5. Box-plot diagram: performance of different features selection techniques

Statistical Hypothesis Testing: Different Features Selection Techniques: In this work, we used statistical hypothesis testing to validate our considered null hypothesis, i.e., "The AUC performance value of the models developed for IDS prediction using neural networks used on datasets generated using different feature selection techniques is not significantly different". The above null hypothesis is accepted if the calculated p-value is greater than 0.05. The results of statistical hypothesis testing on the performance of different models trained using different feature selection techniques are mentioned in Table 3. From Table 3, it can be observed that the p-values are less than 0.05 for the pairings where one of SIGF, CCRA, and PCA are involved. This indicates that model trained using different features are significantly different.

Table 3. Hypothesis testing: different features selection techniques

	AF	SIGF	CCRA	PCA
AF	1.00	0.00	0.01	0.00
SIGF	0.00	1.00	0.00	0.02
CCRA	0.01	0.00	1.00	0.00
PCA	0.00	0.02	0.00	1.00

5.3 Deep Learning Technique

We have used five different types of deep learning architectures by changing the number of hidden layers with an objective to develop models for detecting cyber-attacks on IoT Networks. In this section, we are going to find the solution for *RQ2: "What is the impact of deep learning architecture on the performance of the models?"* based on box plots and significant tests.

Box-Plot: Deep Learning Technique: The performance of five different types of deep learning models in terms of Accuracy, F-measure, and AUC are visualized as box plots in Fig. 6. The information present in Fig. 6 suggest that all the deep learning model Architectures have a similar ability to predict cyber-attacks on IoT networks. The average AUC value of DL3 - (Two Hidden Layers) is slightly higher than other proposed architectures. Therefore, based on descriptive statistics DL3 Model performs better than other proposed models for cyber attack prediction.

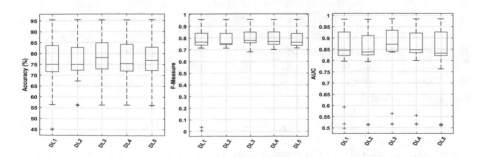

Fig. 6. Box-plot diagram: performance of different deep-learning models

Statistical Hypothesis Testing: Different Deep-Learning Models: In this work, we have used statistical hypothesis testing to validate our considered null hypothesis, i.e., "The AUC performance value of the models developed for IDS prediction using neural networks used on datasets generated using DL2 and DL3 are significantly different". The above null hypothesis is accepted if the calculated p-value is greater than 0.05. The results of statistical hypothesis testing on the performance of different models trained using different feature selection techniques are mentioned in Table 4. From Table 4, it can be observed that the p-values are less than 0.05 for the pairing with DL2 and DL3. This, coupled with the descriptive statistics values, which indicate that the AUC value of DL3 is the highest, implies that DL3 has the best performance. DL3 is a model of two hidden layers, while DL2 has only one hidden layer.

Table 4. Hypothesis testing: different deep-learning models

	DL1	DL2	DL3	DL4	DL5
DL1	1.00	0.73	0.15	0.53	0.84
DL2	0.73	1.00	0.04	0.22	0.93
DL3	0.15	0.04	1.00	0.23	0.10
DL4	0.53	0.22	0.23	1.00	0.30
DL5	0.84	0.93	0.10	0.30	1.00

5.4   Cost Benefit Analysis

This section presents the experimental setup used to find the effectiveness of the developed models using a cost-benefit analysis. The information presented in Table 1 suggested that the models trained using all features as an input have a better ability to predict attack as compared to the features selected using feature selection techniques. Also, computation time increases with an increase in the number of features processed as time complexity increases. Hence, the cost-benefit analysis has been used to find the best combination of features in this work. The formula used in this paper is given below. Further, the amount of space the data occupies is also directly proportional to the number of features present. Since time and space are crucial resources while training a model, they constitute the cost of training a model. The benefit of a model is how effectively it can categorize attacks and differentiate categories into their respective classes. This directly translates to a more efficient intrusion detection system (IDS). The formula used to calculate the cost and benefit of the developed models is mentioned below:

$$Cost_{M1} = \lambda_f * \frac{AF - SF}{AF} + \lambda_a * AUC_{M1} \qquad (1)$$

Where λ_f and λ_a are the weightage given to the number of features and performance of the models, respectively. Similarly, AF and SF are used to represent the number of all features and selected sets of features. In this work, we calculated the cost of models by varying the value of λ_f and λ_a. The result of cost-benefit analysis for different combinations of λ_f and λ_a are summarized in the Table 5. For example, when we assign 50% importance to the reduction in cost ($\lambda_f = 0.5$) and 50% ($\lambda_a = 0.5$) importance to the AUC value(the benefit), the models trained on feature exacted after applying PCA on AF have high benefit as compared to others.

Table 5. Cost-benefit analysis

λ_f	Rank 1	Rank 2	Rank 3
0.05	CCRA	SIGF	AF
0.1	CCRA	SIGF	AF
0.15	CCRA	SIGF	AF+PCA
0.2	CCRA	SIGF	AF+PCA
0.25	CCRA	SIGF	AF+PCA
0.3	CCRA	SIGF	AF+PCA
0.35	CCRA	AF+PCA	AF+PCA
0.4	CCRA	AF+PCA	AF+PCA
0.45	CCRA	AF+PCA	AF+PCA
0.5	AF+PCA	AF+PCA	CCRA
0.55	AF+PCA	AF+PCA	CCRA+PCA
0.6	AF+PCA	CCRA+PCA	AF+PCA
0.65	CCRA+PCA	AF+PCA	AF+PCA
0.7	CCRA+PCA	AF+PCA	AF+PCA
0.75	CCRA+PCA	CCRA+PCA	SGF+PCA
0.8	CCRA+PCA	CCRA+PCA	SGF+PCA
0.85	CCRA+PCA	CCRA+PCA	SGF+PCA
0.9	CCRA+PCA	CCRA+PCA	SGF+PCA
0.95	CCRA+PCA	CCRA+PCA	SGF+PCA

6   Conclusion

Modern IDS are capacitated in providing a wide range of services. However, even though significant work has been done in the field of IDS, they are still immature in many aspects, and certain problems are yet to be resolved. As a result, this research was on a Machine Learning-based IDS that can detect novel mutation attacks by learning from diverse data and previous forms of attack to make accurate predictions. Hence, this research study focused on evaluating whether specific methods in the preprocessing framework and Deep Neural Networks' architecture have a statistically significant impact on the performance of IDS compared to others. An empirical analysis was done using Descriptive Statistics and hypothesis testing to compare the performances of models generated across three data sampling techniques, three feature selection techniques, and different deep learning architectures. Our observations are the following:

- Our experiment results suggest that an increase in the number of hidden layers in Deep Neural Networks improves model performance.
- Our experiment results suggest that class balancing techniques such as SMOTE, Borderline SMOTE, and ADASYN do not significantly improve model performance.
- Our experiment results also suggest that significance test-based feature selection, cross correlation-based feature selection techniques, or PCA negatively affect model performance compared to the baseline of no feature selection (AF).
- The results of the cost-benefit analysis suggested that the models developed using PCA are computationally more efficient while giving reasonable performance compared to the case when all features are used.

Acknowledgment. This research is funded by TestAIng Solutions Pvt. Ltd.

References

1. Kumar, K., Kumar, N., Shah, R.: Role of IoT to avoid spreading of COVID-19. Int. J. Intell. Netw. **1**, 32–35 (2020)
2. Saharkhizan, M., Azmoodeh, A., Dehghantanha, A., Choo, K.-K.R., Parizi, R.M.: An ensemble of deep recurrent neural networks for detecting IoT cyber attacks using network traffic. IEEE Internet Things J. **7**(9), 8852–8859 (2020)
3. Stellios, I., Kotzanikolaou, P., Psarakis, M., Alcaraz, C., Lopez, J.: A survey of IoT-enabled cyberattacks: assessing attack paths to critical infrastructures and services. IEEE Commun. Surv. Tutor. **20**(4), 3453–3495 (2018)
4. Jyothsna, V., Rama Prasad, V.V., Munivara Prasad, K.: A review of anomaly based intrusion detection systems. Int. J. Comput. Appl. **28**(7), 26–35 (2011)
5. Al-Garadi, M.A., Mohamed, A., Al-Ali, A.K., Du, X., Ali, I., Guizani, M.: A survey of machine and deep learning methods for Internet of Things (IoT) security. IEEE Commun. Surv. Tutor. **22**(3), 1646–1685 (2020)
6. Sahu, A.K., Sharma, S., Tanveer, M., Raja, R.: Internet of things attack detection using hybrid deep learning model. Comput. Commun. **176**, 146–154 (2021)
7. Moustafa, N., Slay, J.: UNSW-NB15: a comprehensive data set for network intrusion detection systems (UNSW-NB15 network data set). In: 2015 Military Communications and Information Systems Conference (MilCIS), pp. 1–6. IEEE (2015)
8. Moustafa, N., Slay, J.: The evaluation of network anomaly detection systems: statistical analysis of the UNSW-NB15 data set and the comparison with the KDD99 data set, pp. 1–14, January 2016
9. Chandrashekar, G., Sahin, F.: A survey on feature selection methods. Comput. Electr. Eng. **40**(1) (2014)
10. Tait, K.-A., et al.: Intrusion detection using machine learning techniques: an experimental comparison. arXiv preprint arXiv:2105.13435 (2021)
11. Wheelus, C., Bou-Harb, E., Zhu, X.: Tackling class imbalance in cyber security datasets. In: 2018 IEEE International Conference on Information Reuse and Integration (IRI), pp. 229–232. IEEE (2018)

12. Tavallaee, M., Bagheri, E., Lu, W., Ghorbani, A.A.: A detailed analysis of the KDD CUP 99 data set. In: 2009 IEEE Symposium on Computational Intelligence for Security and Defense Applications, pp. 1–6. IEEE (2009)
13. Haq, N.F., Onik, A.R., Hridoy, M.A.K., Rafni, M., Shah, F.M., Farid, D.M.: Application of machine learning approaches in intrusion detection system: a survey. IJARAI Int. J. Adv. Res. Artif. Intell. 4(3), 9–18 (2015)
14. Amiri, F., Yousefi, M.M.R., Lucas, C., Shakery, A., Yazdani, N.: Mutual information-based feature selection for intrusion detection systems. J. Netw. Comput. Appl. 34(4), 1184–1199 (2011)
15. XIAOHUI XIE. Principal component analysis (2019)
16. Yin, C., Zhu, Y., Fei, J., He, X.: A deep learning approach for intrusion detection using recurrent neural networks. IEEE Access 5, 21954–21961 (2017)
17. Chawla, N.V., Bowyer, K.W., Hall, L.O., Philip Kegelmeyer, W.: SMOTE: synthetic minority over-sampling technique. J. Artif. Intell. Res. 16, 321–357 (2002)

A Decentralized Federated Learning Architecture for Intrusion Detection in IoT Systems

Francisco Assis Moreira do Nascimento[(✉)] and Fabiano Hessel

Pontifical Catholic University of Rio Grande do Sul - PUCRS, Porto Alegre, Brazil
f.nascimento@edu.pucrs.br, fabiano.hessel@pucrs.br

Abstract. Internet of Things (IoT) systems are vulnerable to several attacks, mainly due to the weakness of IoT devices, which have little computational and memory power, necessary for more sophisticated security features. In addition, IoT systems are distributed systems and thus inherit all problems related to the need to guarantee confidentiality, integrity, and availability. One of the traditional strategies to deal with these problems involves intrusion detection and prevention techniques. It is usual to implement them in a centralized way. In addition to not being scalable for IoT systems with an increasing number of devices, it implies an unacceptable single point of failure. Besides, sending all collected data to a centralized server in the cloud poses a great risk to the privacy of information. This paper presents a decentralized architecture for intrusion detection in IoT-based systems, which is based on federated machine learning, combined with distributed ledger technologies for access control, allowing a mechanism to minimize security risks.

1 Introduction

Internet of Things (IoT) consists of machines and devices, internet-connected, cooperatively performing tasks and creating data to support computational systems in solving problems in many application areas [11]. IoT-based systems as a kind of distributed system (interconnected and collaborative autonomous devices) inherit all related problems: How to guarantee confidentiality, integrity, availability, authenticity, and non-repudiation? How to tolerate computation and communication failures?

These questions are becoming even more relevant with the growing demand for IoT-based systems in all areas [11], for example, in healthcare, smart home, smart cities, logistics, and many others, which require a high level of data privacy [16], and robust data access control not to allow unauthorized access to information and available resources and services.

Security requirements for IoT are challenging to satisfy because many of the devices that can be part of an IoT-based system have low processing power and small memory capacity (Problem P.1). Moreover, most of the devices are designed with no concerns about security issues, and so are vulnerable to many

L. Barolli et al. (Eds.): AINA 2022, LNNS 450, pp. 256–268, 2022.
https://doi.org/10.1007/978-3-030-99587-4_22

kinds of security threatens and criminal attacks at all system architecture levels (Problem P.2) [11].

One of the traditional strategies to deal with these problems involves intrusion detection techniques [2], where the traffic network is monitored. Based on some specific rules, inadequate behaviors may indicate a security threat, triggering preventive actions to minimize risks. Unfortunately, it is usual to implement them in a centralized way. In addition to not being scalable for IoT systems with an increasing number of distributed components, it implies an unacceptable single point of failure in the security of IoT systems. Besides, sending all collected data to a centralized server in the cloud poses a great risk to the privacy of information (Problem P.3).

Intrusion detection systems have been used in IoT to cope with security risks, applying Machine Learning (ML) techniques to improve them [11]. Since many of these ML techniques depend on manually selected features (i.e., properties that characterize data to be analyzed), they are becoming obsolete due to the many new IoT applications and the different kinds of network traffic they imply (Problem P.4) [2].

Another very essential strategy to guarantee security and also privacy in IoT is the use of efficient and effective access control mechanisms to prevent unauthorized access to information (confidentiality), to avoid information modification without adequate authorization (integrity), and to guarantee access to information to authenticated users at any needed time (availability) [11]. Since all these issues represent challenging problems, intense research activity is dedicated to them, but many are yet open problems [4].

To cope with the above problems, our work has as main contributions: **i)** a totally decentralized federated learning architecture for IoT-based systems, which can predict attacks based on previous incidents, as well on the behaviors of IoT devices, in a privacy-preserving and fault-tolerant way; and **ii)** an architecture for a set of distributed ledgers providing security mechanisms for authentication, authorization, integrity, confidentiality, and high availability, which IoT-based systems can use.

Next Sects. 2 and 3 presents background and related work, respectively. Section 4 describes the main characteristics of our IoT security architecture and Sect. 5 discusses architecture properties and its strategy to IoT security. Final considerations and future work are presented in the last Sect. 6.

2 Background

The term Internet of Things is applied to a network of smart objects, which can intelligently interact with each other and cooperate with the IoT-based system to provide useful services [18]. Cloud, edge, and fog computing paradigms are combined to fulfill IoT complex requirements in its implementation. Figure 1 presents a typical architecture for IoT [4].

By the cloud, the data storage and processing are performed by a network of data centers; In the edge, the data storage and processing are preferentially

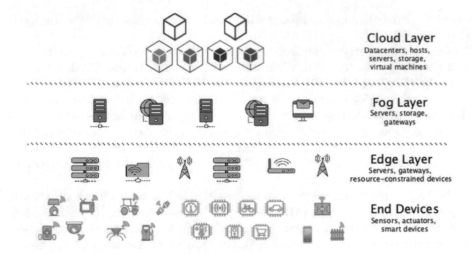

Fig. 1. Typical three-tier architecture for IoT

performed in the data generator (end device) or near from it; and the fog is an intermediate level between edge and cloud, usually collecting data from various edge devices and storing and processing them at more powerful devices than the edge devices themselves.

Some ongoing projects for enhancing IoT security include methods for providing data confidentiality and authentication, access control within the IoT network, privacy and trust among users and things, and the enforcement of security and privacy policies. However, IoT networks are still vulnerable to multiple attacks aimed to disrupt the network, for instance, attacks from malicious traffic and insider attacks [8]. For this reason, another line of defense designed for detecting attackers is needed. Intrusion detection systems can fulfill this purpose [2].

Intrusion Detection Systems (IDS) are systems designed to inspect all inbound and outbound traffic and identify the suspicious traffic and actions of various attackers in a timely and accurate manner, alerting system administrators when they detect a security violation [11]. IDS agents are deployed in nodes with higher computing capacity in traditional networks, and IoT networks are usually composed of nodes with very limited resources. Therefore, finding nodes with the ability to support IDS agents is harder in IoT systems [2].

Traditional IDSs can adopt knowledge-based methods for intrusion detection, using signatures (detect attacks when a system or network behavior matches an attack signature stored in the IDS internal databases) or using specifications (detect intrusions when network behavior deviates from manually specified rules in a set of specification definitions) [11]. An anomaly-based IDS compares the activities of a system at an instant against a normal behavior profile and generates the alert whenever a deviation from normal behavior exceeds a threshold. It is also possible to explore all the above detection methods trying to maximize their advantages and minimize the impact of their drawbacks [8]. It is important to note that signature-based IDS approaches cannot deal with new attacks in which their signatures are unknown, and traffic is encrypted. However, anomaly-based IDS schemes operate according to the expected behavior profiles of the users and can detect the newly unleashed attacks.

Many of the difficulties in developing Intrusion detection algorithms for IoT-based systems are related to their dynamic characteristics: intrusive and normal behavior of users, applications, and networks are always changing over time [11]. These dynamic behaviors are primarily due to the many new IoT devices and applications constantly coming up and even new functionalities offered by current devices and applications. Thus, to be effective and efficient, an IDS must continuously learn and adapt itself to cope with the changes. Machine learning-based intrusion detection is a possible approach to this issue [4].

Machine Learning (ML) is a computational paradigm allowing machines to adjust their behavior based on helpful knowledge inferred from data sets. ML techniques have been used for classification, regression, and estimation tasks, which can be applied to solve problems in fraud detection, computer vision, natural language processing, and in many other areas [11].

An ML-based IDS explores the use of machine learning methods and techniques to detect intrusion based on historical data about previous incidents and behaviors of devices. Many different ML algorithms and techniques have been used to implement IDSs [8].

The classifications produced by the ML-based IDS can be: True Negatives (TN), normal traffic is correctly recognized as normal traffic; True Positives (TP), attacks and malicious behaviors are correctly detected as intrusions; False Positives (FP), normal traffic is falsely detected as attacks; and, False Negatives (FN), attack traffic is falsely detected as normal traffic [11]. Some common evaluation metrics for machine learning based IDSs, in terms of TN, TP, FP, and FN, includes accuracy, Detection Rate (DR), False Positive Rate (FPR), precision, and F1-score, which measure the quality degree of a generated ML model [11].

An efficient ML-based IDS should ideally have 0% FPR and 100% DR. Moreover, to be considered effective, these IDSs should increase the attack detection range, address more IoT technologies, and improve validation strategies [11].

The term *Federated Learning (FL)*, proposed by Google [10], designates an ML technique, where a group of client devices performs model training locally, using their private datasets. A global model is generated by a central server, which only receives updates from local models and never the local datasets. The global model, generated by the central server (so-called aggregator server), is then shared back to all the client devices, which can use it to perform inferences. Figure 2 shows usual architectures for model training.

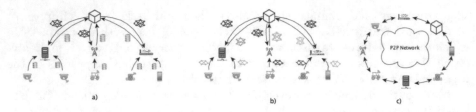

Fig. 2. Model training.

In conventional model training (Fig. 2a), a central server trains a model using data gathered from distributed end devices; The trained model is then shared back with the devices to be used in prediction or inference tasks. In centralized FL (Fig. 2b), local models are trained at devices based on available data on it; A server trains a global model by some type of aggregation applied on collected local models from devices. The trained global model is shared back to devices for updating their local models. In decentralized FL (Fig. 2c), each node can perform locally model training based on local data and models shared by other nodes.

The decentralized FL architecture, shown in Fig. 2(c), corresponds to the ideal concept of federated learning. But, most published approaches adopt a unique central server for aggregation implementation, leading to an undesirable single point of failure in the system [18].

In a decentralized FL architecture, the local model training must converge to a consensus in any global shareable model, depending on how its gradient will be computed. Usually, global shareable models are generated by some aggregation algorithm applied to the locally generated models [10]. One possible aggregation algorithm is based on SGD (Stochastic Gradient Descent), which updates the gradient over tiny subsets (minibatches) of the whole data set, or even by selecting the clients that will contribute with the model updates, calculated independently by each client device [18].

One most common aggregation algorithm, proposed by McMahan et al. [10] and called the FedAvg method, consists of iterative model averaging, i.e., the client devices update the model locally with one-step SGD, and then the server averages the resulting models with weights. This aggregation method allows to treat unbalanced and non-IID (non-Independent Identically Distributed) since the distributed data may come from various sources [10].

Another important issue related to model training in FL is the communication costs. The client devices will send model updates to some aggregator server and eventually receive whole global models from the server. Some approaches to minimize these communication costs are reported in the literature. For instance, Wang et al. [17] propose an adaptive federated learning approach to consider the constrained capacities of the client devices. Based on a given number of available resources, including time and energy, the proposed algorithm determines the best trade-off between the local update and global parameter aggregation.

These approaches to aggregation in federated learning can be applied in an ML-based IDS. Each client device will train a local model based on the network traffic it captures and contribute with model updates for some aggregator server. However, since network traffic patterns are usually specific to given types of devices, it is crucial to train global models per device type.

Since data privacy-preserving is becoming a big issue, due to the recent regulations (in Europe, GDPR [16]), intended to protect users' privacy and provide data security, FL offers a significant advantage: model training is not dependent on access to the device data, which is only available locally at each client device [10]. Moreover, since device data is not transmitted through the network, many security risks are minimized.

However, since an initial global model must be shared with the client devices and further model updates should be transmitted to aggregator servers, some malicious client devices may capture the network traffic (man-in-the-middle attack), threatening data privacy, and even make changes to the models, intending to influence on the final results (poison attack). Thus, some techniques should be adopted to cope with these threats. Differential privacy [1] is one of these techniques, which adds noise to the data or uses generalization methods to obscure model attributes until the third party cannot identify who sent the model updates, thereby making the data impossible to be restored and protecting user privacy. Other approaches to guarantee data privacy-preserving in FL, as well as to avoid poison attacks to the models been transmitted between the nodes in an FL-based system, make use of distributed ledger technologies [3].

In the inference phase, a client device receives an input (a network traffic sequence, in the case of an intrusion detection system). It applies its local model to classify it as normal or attack traffic. Another possibility is the client device to send the input to some aggregator server, which has a trained model, to perform the inference. Then to receive back, the classification [18].

It is also possible to perform part of the inference in a local partial model in the client device that sends the partial result to an edge server, which has another part of the model that finishes the inference, giving back the classification to the client device. Yet another possibility for the inference is that the edge server uses a trained model at a cloud server to perform part of the inference [18]. One of these inference strategies can be chosen according to the available resources for computation, memory, storage, and communication in the client devices and edge servers.

A distributed ledger (DL) maintains transactions or digital interactions transparent, immutable, and auditable. Blockchain, which was first introduced in 2009 by Nakamoto [12], is one of the technologies to implement DLs by storing the set of transactions in a block and chaining every block by using cryptographic techniques. Each transaction can be a transfer of values between different entities that are broadcast to the network and collected into the blocks. The transactions are grouped into a block by the so-called miners or validators, and so they add transactions into the blockchain [3].

The mining or validation process allows the participant nodes to reach a secure, tamper-resistant consensus, even if some of the nodes are unreliable. Many different consensus algorithms have been used by blockchain, including proof of work (PoW), where a miner must complete a given puzzling work to insert a new block in the chain and win processing fees as a reward; and, proof of stake (PoS), where a pseudo-randomly chosen validator can insert a new block [3].

To execute some kind of action associated with registered transactions on a blockchain, it can be used smart contracts [3], which provide the automation mechanism for code execution according to implemented rules that have been agreed upon by the participant nodes.

3   Related Work

Machine learning techniques and, more particularly, deep learning techniques have been widely applied to intrusion detection [7]. However, federated learning application in the area of intrusion detection is recent (last four years), and the combination with distributed ledgers for some aspects of this area is also very recent. One of these recent works is from Liu et al. [9], reporting an intrusion detection system built using a federated learning approach specifically for vehicles, where each vehicle works as an edge server training local models based on its data. The aggregation is performed by RSUs (Road Side Units) to produce global models. All the models are stored in a public blockchain to improve the models' security. Unlike our approach, the adopted aggregation algorithms are mainly oriented to vehicles, not any IoT device. Moreover, using a public blockchain and a PoW consensus algorithm makes the approach resource and time-consuming to be adopted for any kind of IoT system. As discussed previously, the reported results are on the KDDCup99 dataset, which is already obsolete.

DÏoT [13] is a federated learning intrusion detection approach based on representing network packets as symbols in a language. This strategy allows implementing a language analysis technique to detect anomalies, using GRU (Gated Recurrent Neural Network), a kind of Recurrent Neural Network. According to the IoT device type, it adopts a federated learning approach for aggregating anomaly-detection profiles for intrusion detection. Unlike our work, where an entirely decentralized federated learning approach is defined, DÏoT depends on a centralized IoT security service to aggregate all local models generated by security gateways to produce new models later sent back to the security gateways. This situation constitutes a single point of failure, which is not desirable in robust approaches to security.

Diro et al. [6] developed an IDS based on a distributed deep learning approach at the fog layer. A master fog node collects model updates from other fog nodes that perform SGD-based training on their local data to generate local models. So, the master node aggregates the model updates to produce a global model, which is shared back to all fog nodes. The fog nodes directly connected to IoT devices collect network traffic and perform predictions to detect eventual intrusion. This approach assumes a single global model that can demand growing computational resources to be generated proportionally to the increasing number of connected IoT devices. Also, it can not consider the many different network traffic characteristics according to the IoT device types. Moreover, it is assumed a single fog node is a master, representing a single point of failure in the system.

An architecture for access management in IoT based on the use of a blockchain is proposed in [14]. The IoT devices are isolated from the blockchain network and have to register in a manager node, which interacts with the blockchain network. The manager defines access control rules for the IoT devices' resources and can modify and delete policies for the IoT devices. These management operations are performed employing smart contract transactions in the blockchain, which is one of the main limitations of the proposed architecture, since the waiting time for transactions to complete in the blockchain is too much longer, and not appropriated for some management operations, for example, the revocation of some privilege for a given IoT device.

Ding et al. [5] also present blockchain-based access control-oriented to IoT. Authentication and authorization attributes are registered in a blockchain, which validates any posterior request for access to resources. Since any request must go through a global blockchain, an unacceptably high latency may occur for access requests response. In our approach, we have local blockchains, minimizing this risk. Putra et al. [15] also make use of smart contracts in a blockchain to develop an authorization model based on trust, where the evaluation of positive and negative interactions of nodes in the IoT system determines the level of trust of each one. These smart contract-based approaches depend on computational, communication, and storage resources beyond the edge devices capacities, without providing some mechanism to consider these constraints, as our approach proposes.

4   Decentralized Federated Learning-Based IDS for IoT

In our totally decentralized FL approach for intrusion detection in IoT, global models are generated by a set of servers, called aggregators, and not just one central server. Each aggregator handles a cluster of client IoT devices, called detectors, and interacts with the other aggregators to generate their models. Figure 3 shows the components of our proposed decentralized FL-based IDS.

Our architecture consists of a network of Peer-to-Peer (P2P) clusters, each composed of computation, storage, and communication nodes, which can be machines/devices from the cloud, fog, or edge layers. By adopting P2P clusters, we have a fault-tolerant architecture, where each cluster is oriented to specific tasks in the intrusion detection process. Inside each cluster, a lightweight local blockchain is used to register information generated/used by the IDS tasks in a secure and decentralized manner.

Nodes in the clusters are selected from available resources in the IoT system or some provisioned resource specifically for the IDS. Nodes are dynamically

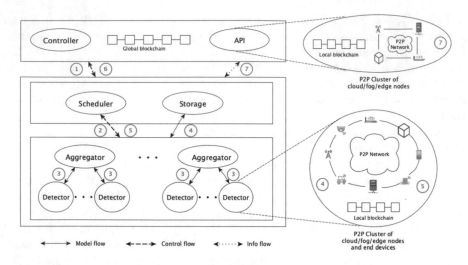

Fig. 3. Our decentralized IDS architecture for IoT.

allocated into the different clusters (controller, scheduler, aggregator, detector, storage, and API) according to the demand from the IDS tasks. Communication between all the nodes, intra- and cross-clusters, makes use of TLS, with a digital identity for each node, generated from the node's public key and managed by the controller cluster.

Each node uses an API service call to register in the controller, providing its characteristic properties (its identity, computational storage, and energy attributes), including its public key from a key pair (public and private keys) generated by the node itself, using a specific distributed client application (DApp). Only then, the new node can participate in the IoT system. Thus, the controller cluster manages the IDS's infrastructure, providing authentication and authorization services for nodes, including the P2P network topology management, by registering the participant nodes in its lightweight local blockchain.

In the IDS workflow, step 1 (indicated by little numbered circles in Fig. 3) consists of the controller activating the scheduler cluster, which performs task scheduling, resource allocation (which specific cluster will perform each task), and operation binding (which device will execute each task operation). The scheduler periodically (step 2 in Fig. 3):

- starts the updates of global and local ML models;
- creates training rounds for global and local models, taking into account computational and storage resources, power consumption, and eventual timing constraints in the model training tasks; and,
- periodically starting inferences in the devices to identify eventual anomalies.

Then, aggregator clusters, which implement the generation of global models, per device type and security threat, oriented to intrusion detection, perform step 3:

- collect local models to be considered in the aggregation, receiving control information from the scheduler and accessing the storage cluster and its eventual available cache nodes;
- perform model aggregation, selecting and executing an aggregation algorithm; and,
- publish generated global models, accessing storage cluster and its eventual available cache nodes and notifying the scheduler cluster.

Detector clusters, which perform basic intrusion detection tasks, in step 4:

- monitor/collect network traffic, periodically, using a sniffer and registering it locally or at storage cluster;
- prepare collected data for training, according to scheduling, by:
  - accessing data locally or in the storage cluster, depending on device resources,
  - applying data transformations to adequate them to the deep neural network algorithm to be used in the model training,
  - registering transformed data, locally or in the storage cluster, depending on the device resources;
- execute training of local models at each end device (sensors, actuators, smart devices) or edge devices (routers, switches, small servers), according to scheduling, by:
  - accessing data locally or in the storage cluster, depending on device resources,
  - executing deep neural network algorithm for training, and
  - registering trained model updates in the storage cluster.

Detector clusters also perform inference, in step 5, to identify eventual intrusions by:

- using current ML model at each device, according to scheduling, by:
  - obtaining transformed data and locally available or from storage cluster,
  - executing deep neural network algorithm for inference,
  - notifying inference result to the scheduler cluster.

As mentioned in the above IDS workflow, the storage cluster implements decentralized storage for global and local models, offering cache resources to minimize communication between nodes of all clusters in the IDS. The storage cluster can (step 6 in Fig. 3) authenticate nodes, which are requesting access, by using information from controller cluster; validate node authorization by using information from controller cluster; execute the operation, as requested by the node, in the case of valid authentication and authorization; and register operation logs in the controller cluster.

Finally, the API cluster handles all requests for provided services in our IDS in a secure and distributed way. It can receive requests, call respective services, and send responses (indicated as step 7 in Fig. 3).

5   Discussion

Our IDS comprises a set of detector and aggregator nodes, as P2P clusters, and provides a specific API for intrusion detection based on the proposed decentralized, federated learning approach. According to available resources in the corresponding host, each edge server may work as a blockchain node to the controller cluster and a detector or aggregator node to the IDS cluster. It would even be possible for an edge server to work as different types of nodes simultaneously if its available resources allow it.

Our decentralized, federated machine learning part allows us to handle problems P.1, P.3, and P.4 (see Sect. 1), since model training and prediction for intrusion are performed locally at edge nodes, selected from the ones that can execute the tasks, according to their available resources, and only models are transmitted to other edge nodes for global models generation.

Our intrusion detection mechanism is based on deep learning techniques, which explore the network traffic patterns to predict possible attacks. Model training and prediction for intrusion are periodically scheduled and performed at each node, generating alerts when some anomalous behavior is detected.

Moreover, distributed ledgers (global and local blockchains) and off-chain storage, included in our proposed approach, guarantee the integrity and high availability of the necessary information for the IoT security platform in a totally decentralized way. They also support authentication, authorization, and auditing mechanisms, which are essential to treat problem P.2 (see Sect. 1), by providing necessary security features in a distributed manner.

One of the main characteristics of our platform is decentralization, so both controller and IDS-specific clusters can be deployed on the different available edge servers. According to the available computation, memory, and energy resources in the corresponding host, instances of the IDS clusters can also be deployed on different edge servers and provide services (e.g., for monitoring network traffic, for intrusion detection, etc.) and requests services (e.g., for obtaining trained models, for performing authentication and authorization, etc.).

Since computation tasks for the IoT security platform will be primarily executed at the edge layer, this decentralized architecture will reduce network traffic between edge servers and fog/cloud servers. Moreover, privacy will be enhanced since the monitored information in the network traffic will not be transmitted to fog and edge servers. By adopting this architecture, it will be possible to cope with problem P.1 (see Sect. 1).

6   Conclusions and Future Work

This work contributes with solutions for critical IoT security issues. It is oriented to developing decentralized security architecture for intrusion detection in IoT-based systems, allowing secure IoT applications. Decentralized, federated machine learning in a data privacy-preserving manner, using deep learning techniques to learn and adapt itself to new kinds of intrusions dynamically, is combined with distributed ledgers to minimize the security risks associated with not guaranteeing integrity, confidentiality, and availability of IoT-based systems.

As future work, the totally decentralized IDS architecture will be implemented in the form of an IoT security platform and validated and evaluated on recent datasets specific to IoT applications to assess its performance and give empirical evidence for its feasibility and efficiency.

References

1. Abadi, M., et al.: Deep learning with differential privacy. In: ACM SIGSAC Conference on Computer and Communications Security, pp. 308–318 (2016)
2. Adeleke, O.: Intrusion detection: issues, problems and solutions. In: International Conference on Information and Computer Technologies, pp. 397–402 (2020)
3. Ali, M.S., et al.: Applications of blockchains in the internet of things: a comprehensive survey. IEEE Commun. Surv. Tutorials **21**(2), 1676–1717 (2019)
4. Chaabouni, N., et al.: Network intrusion detection for IoT security based on learning techniques. IEEE Commun. Surv. Tutorials **21**(3), 2671–2701 (2019)
5. Ding, S., et al.: A novel attribute-based access control scheme using blockchain for IoT. IEEE Access **7**, 38431–38441 (2019)
6. Diro, A.A., Chilamkurti, N.: Distributed attack detection scheme using deep learning approach for Internet of Things. Future Gener. Comput. Sys. **82**, 761–768 (2018)
7. Ferrag, M.A., et al.: Deep learning for cyber security intrusion detection: approaches, datasets, and comparative study. J. Inf. Secur. Appl. **50**, 102419 (2020)
8. Hussain, F., Hussain, R., Hassan, S.A., Hossain, E.: Machine learning in IoT security: current solutions and future challenges. IEEE Commun. Surv. Tutorials **22**(3), 1686–1721 (2020)
9. Liu, H., et al.: Blockchain and federated learning for collaborative intrusion detection in vehicular edge computing. IEEE Trans. Veh. Technol. **70**(6), 6073–6084 (2021)
10. McMahan, B., et al.: Communication-efficient learning of deep networks from decentralized data. In: International Conference on Artificial Intelligence and Statistics, pp. 1273–1282 (2017)
11. Mishra, N., Pandya, S.: Internet of Things applications, security challenges, attacks, intrusion detection, and future visions: a systematic review. IEEE Access **9**, 59353–59377 (2021)
12. Nakamoto, S.: Bitcoin: a peer-to-peer electronic cash system (2008). https://bitcoin.org/bitcoin.pdf
13. Nguyen, T.D., et al.: DÏoT: a federated self-learning anomaly detection system for IoT. In: IEEE International Conference on Distributed Computing Systems, pp. 756–767 (2019)
14. Novo, O.: Blockchain meets IoT: an architecture for scalable access management in IoT. IEEE Internet Things J. **5**(2), 1184–1195 (2018)
15. Putra, G.D., et al.: Trust-based blockchain authorization for IoT. IEEE Trans. Netw. Serv. Manage. **18**(2), 1646–1658 (2021)

16. Voigt, P., von dem Bussche, A.: The EU General Data Protection Regulation (GDPR). Springer, Cham (2017). https://doi.org/10.1007/978-3-319-57959-7
17. Wang, S., et al.: Adaptive federated learning in resource constrained edge computing systems. IEEE J. Sel. Areas Commun. **37**(6), 1205–1221 (2019)
18. Zhou, Z., et al.: Edge intelligence: paving the last mile of artificial intelligence with edge computing. Proc. IEEE **107**(8), 1738–1762 (2019)

Regression Analysis Using Machine Learning Approaches for Predicting Container Shipping Rates

Ibraheem Abdulhafiz Khan$^{(\boxtimes)}$ and Farookh Khadeer Hussain

School of Computer Science, University of Technology Sydney,
Ultimo, NSW 2007, Australia
ibraheemabdulhafizq.khan@student.uts.edu.au,
Farookh.hussain@uts.edu.au

Abstract. Container freight rate forecasts are used by major stakeholders in the maritime industry, such as shipping lines, consumers, shippers, and others, to make operational decisions. Because container shipping lacks a structured forwards market, it must rely on forecasts for hedging reasons. This research is dedicated to investigating and predicting shipping containerised freight rates using machine learning approaches and real-time data to uncover superior forecasting methods. Ensemble models including Random Forest (RF) and Extreme Gradient Boosting (XGBoost), and deep learning, in particular Multi-Layer Perceptions (MLP) have all been used to provide data-driven predictions after initial feature engineering. These three regression-based machine learning (ML) models are used to predict the container shipping rates in the North American TransBorder Freight dataset from 2006 to 2021. It has been found that MLP surpasses ensemble models with a test accuracy rate of 97%. Although our findings are drawn from American shipping data, the proposed approach serves as a general method for other international markets.

Keywords: Deep learning · Random Forest · Extreme Gradient Boosting · Multi-Layer Perceptions · Forecast models · Container freight rate

1 Introduction

In recent decades, the demands on the global shipping industry have increased dramatically. The shipping industry is now responsible for the carriage of 90% of world trade, with around 20 million containers being shipped across the oceans every day [1]. For instance, China's freight trucking market, which is part of the shipping industry, generated about $750 billion in 2017 [2]. Likewise, the trucking industry in the United States (U.S.) is estimated at more than $700 billion [3]. Between 2014 and 2018, 110 ultra-large container ships (ULCS) entered the market, resulting in a significant increase in available shipping capacity [4].

L. Barolli et al. (Eds.): AINA 2022, LNNS 450, pp. 269–280, 2022.
https://doi.org/10.1007/978-3-030-99587-4_23

One of the most essential and most extensive segments in the shipping industry is containerised goods. Container shipping is crucial for finished commodities and merchandise that will be sold to and utilised by individuals instead of large firms and industries.

This means containerised freight serves numerous freight forwarders, logistics companies, and thus end-users of any product. Hence, container shipping rates (i.e., freight rates) have a significant and direct effect on end-users, and it is in the interests of a broader audience, including the container shipping corporations themselves, to make predictions for container shipping freight rates [5]. Accordingly, more work needs to be done on predictive analytics for container shipping rates to optimize the precision of traditional predictive models.

Forecast models have been proposed for various fields, including predicting container freight rates, mobile phone prices, the stock market index, coal price volatilities, the Baltic Dry Index and many more. This paper focuses on predicting container shipping rates using state-of-the-art machine learning models, namely Random Forest (RF), Extreme Gradient Boosting (XGBoost) [6], and Multi-Layer Perceptions (MLP).

2 Related Work

Despite the importance of forecasting freight rates for a shipping company, it has received scant attention from researchers because it is a complicated task with both demand-side and supply-side uncertainties [7]. The freight rate, which is the demand price for shipping cargo by sea from one location to another is determined by multiple factors, such as the cargo's weight and volume and the distance to the destination [8]. Table 1 summarises the available research on container freight rates in the literature.

ML and Artificial Neural Networks (ANNs) are the most popular prediction models in the marine shipping domain which can be found in [7,9–13]. These models were proposed for uncertain spot demand volume [7], for the Australian shipping industry [9,11], for the Chinese shipping market [10], and both Wang and Meng [12] and Viellechner and Spinler [13] predicted the number of containers for incoming journeys.

The above literature on marine shipping forecasting lack the ability to predict container freight rates due to the limited data that can be obtained. In addition, their datasets are relatively small which makes it hard to be trained using DL algorithms such as MLP. Also, MLP is good for prediction models [14].

Different from the existing literature, this study for the first time investigates different ML methods to lead to a much-needed new understanding of the applicability of different ML methods for freight rate prediction. Given the rapid development of ML methods and applications along with increasing data availability, using ML in freight rate prediction has become and will continue to be a fertile research field. This paper contributes to the existing literature by investigating a set of robust and commonly used machine learning models, in particular ensemble learning methods such as XGBoost and RF. Because of the sequential

Table 1. Recent forecasting ML studies in the marine shipping domain

Paper (Author/Year)	Main task	Methods	Main result
Wang and Meng (2021) [7]	Tackles the pricing problem for the spot shipping demand	Novel SB3 algorithm based on NLP and QCP	NLP-based model enhances the profitability and decrease the adverse effect of spot shipping demand uncertainties
Ubaid, Hussain and Charles (2020) [9]	Calculates shipment spot prices based on shipment demand and capacity	Regression and correlation analysis	The proposed model significantly increases the correlation between shipment price and shipment demand from 0.33 to 0.88 and available capacity from -0.12 to 0.35
Munim and Schramm (2020) [10]	Investigates the competing forecasting models of container shipping freight rates for four major trade routes	ARIMA, VAR, and ANN	VAR/VEC models outperform ARIMA and ANN in raining-sample forecasts, but ARIMA outperforms VAR and ANN taking test-samples
Ubaid, Hussain and Charles (2020) [11]	Predicts the price for container shipping based on supply and demand for the Australian container shipping industry	SVR, RFR and GBR	GBR outperforms the prediction accuracy of SVR and RFR with an accuracy of 84%
Wang and Meng (2019) [12]	Forecasts the number of containers to be transported through long-haul legs for the incoming trip of intercontinental shipping service	PLR, AR, and ANN	ANN achieves a better forecasting precision than PLR and AR and it is best to combine the predictions of ANN, PLR and AR
Viellechner and Spinler (2020) [13]	Provides data analytics-based solutions for the container shipping industry	NN, SVM and naive baseline model	NN & SVM outperforms naive baseline model with a prediction accuracy of 77% compared to only 59%

nature of ensemble development, a gradient boosting machine has demonstrated superior performance for prediction over ANN and SVM (e.g., Barua 2021 [15]) but it has not been considered or com- pared with MLP.

3   Data and Methodology

In this section, we demonstrate the workflow exemplar followed in this study. Figure 1 displays the stages of the methodology followed.

Table 2. Dataset parameter description

Feature symbol	Feature description	Feature type
TRDTYPE	Trade Type Code	Input
USASTATE	U.S. State Code	Input
DISAGMOT	Mode of Transportation Code	Input
MEXSTATE	Mexican State Code	Input
CANPROV	Canadian Province Code	Input
COUNTRY	Country Code	Input
VALUE	Value in US Dollars	Input
DEPE	Port/District Code	Input
CONTCODE	Container Code	Input
SHIPWT	Shipping Weight	Input
DF	Domestic/Foreign Code	Input
MONTH	Month	Input
YEAR	Year	Input
FREIGHT_CHARGES	Freight Charges	Output

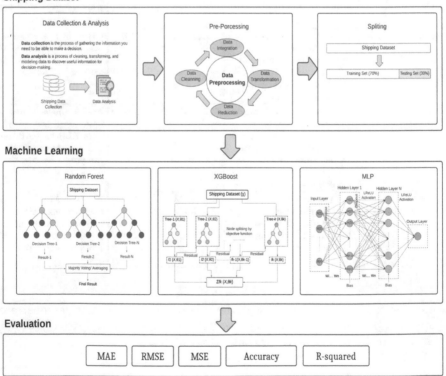

Fig. 1. Proposed methodology

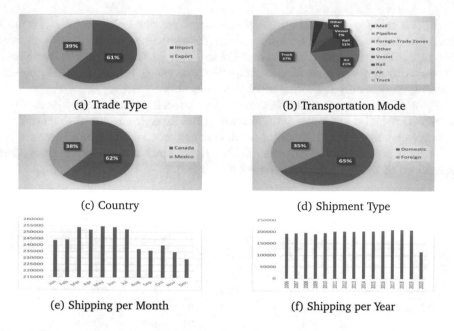

(a) Trade Type

(b) Transportation Mode

(c) Country

(d) Shipment Type

(e) Shipping per Month

(f) Shipping per Year

Fig. 2. Data analysis for the North American shipping industry from 2006 to 2021.

3.1 Dataset

Data Collection and Analysis: This study gathered monthly time series data from June 2006 to June 2021 for freight container rates. This big data has been collected from the North American TransBorder Freight Data with Mexico and Canada [16], with a total of 4714587 shipping transactions. In comparison, these total freight transactions for only container freight contain 180,2847 import transactions and 113,0154 export transactions, as shown in Fig. 2. Table 2 details the dataset parameter description employed along with the feature symbols and its data types.

Data Pre-processing: The process of data pre-processing is considered an introductory step for training the machine learning models. Pre-processing comprises the following steps: examining the dataset for missing values and replacing them with proper values, identifying and excluding any outliers, transforming it into a particularly appropriate framework for implementing the algorithm, the feature scaling process, and feature selection for the model to comprehend and extract useful data information [17]. To do so, first, we performed some data cleaning on our dataset. For example, the dataset uses the number "1" for non-containerised goods, the letter "X" for containerised goods and "NaN" for missing values.

In this case, we dropped all "NaN" rows and replaced "X" with "2" for more readable data. Then, we transformed all categorical variables in our dataset into numerical variables (integers and floats), such as transferring United States names to numbers. The reason for changing these values is because most machine learning models only take numerical variables [17].

After this, we applied the feature scaling process to normalise the range of data features, also known as data normalisation. Since the range of the input values could be disparate, data preparation is considered to be a primary step when using machine learning algorithms. Table 3 details how normalisation normalises the values of the features between −2 and 2 [17].

Table 3. Feature scaling (normalization)

	USASTATE	DISAGMOT	VALUE	SHIPWT	DF	MONTH	YEAR	CONTSTATE
0	−0.1054	0.1934	−0.1035	−0.0698	1.21	1.601	−0.951	−1.4995
1	1.1147	0.1934	−0.1437	−0.0698	−0.81	−0.427	−1.184	1.2638
2	0.4443	0.1934	−0.1548	−0.0698	−0.81	−1.297	−0.016	−1.4995
3	0.6402	0.1934	−0.1488	−0.0698	−0.81	0.731	−0.016	1.2638
4	1.2502	0.1934	−0.1343	−0.0697	1.21	1.601	−0.951	−0.71

Finally, selecting features is based on the attributes contributing the most to the prediction variable (predictive output) and can be executed automatically or manually [17]. However, selecting irrelevant attributes might diminish model precision and cause the model to train based on unrelated features. We used XGBoost as the feature selection procedure. The features are selected by the backward greedy selection procedure: ['Country', 'Mode of Transportation', 'Value in US$s', 'Commodity Code', 'Year']. Only these features are kept for subsequent regression models.

Data Splitting: In this study, we used the common method of splitting datasets to examine our machine learning models. We shuffled the data to induce any imbalance in progressive regression analysis. The dataset was split into a training set and a test set with a ratio of 7:3.

3.2 Machine Learning Models

Machine learning (ML) is becoming widely used in most scientific and industrial fields. It can be defined as a method that automates things by making machines learn from data and make decisions in a particular way based on a specific kind of data as input with minimum human intervention. Commonly, ML is classified into supervised learning and unsupervised learning [18].

Firstly, the supervised learning method utilises a labelled set of training data and outlines the input data to the desired output. This method has two well-known types, namely classification and regression. Classification allocates a data sample into one of the pre-specified separated classes. On the other hand, regression is considered a statistical methodology commonly utilised for a numeric forecast. Secondly, the unsupervised learning method uses a popular technique called clustering. This technique takes an assigned dataset as input and groups it into a limited figure of clusters as per some similarity indication [19].

We adopted three commonly used ML models in this study, which have been applied in many previous studies with a proven high prediction accuracy. These adopted models are RF, XGBoost and MLP. To simplify the concept of these adopted ML approaches, we classify and explain them as follows:

Ensemble Methods: Ensemble learning methods are a branch of ML methods that creates multiple learning algorithms and then merges them to produce a single optimal predictive model. They are based on several tree ensemble methods, but the most broadly applied techniques are bagging and boosting [20]. Ensemble methods are used in two of our adopted ML models: RF and XGBoost. Moreover, the bagging and boosting methods are similar in that they can be applied for classification and regression matters. In contrast, the differences between these two techniques can be defined as follows:

Bagging: (also termed bootstrap aggregating). Bagging is to split the training set into various subsets and feed them to each decision tree. Each subset includes several features along with their observations chosen randomly [21].

Boosting: The principal thought behind the boosting technique is to transform weak learners into powerful learners in sequential iterations. This means that each decision tree attempts to correct its predecessor [21].

– **Random Forest Model:** RF is one of the most prevalent and robust machine learning algorithms. The RF algorithm is resilient, easy to implement, and gives excellent predictive results. This model works with a bagging ensemble technique and can be used for both classification and regression [22]. The predicted shipping rate for a point x is calculated using Eq. (1), where $\hat{f}_{rf}^{B}(x)$ is the intended computation to be reached in x with n data and m variables. This procedure is repeated to grow an RF with $T_b(x)$ tree on bootstrap data.

$$\hat{f}_{rf}^{B}(x) = \frac{1}{B} \sum_{b=1}^{B} T_b(x) \tag{1}$$

– **Extreme Gradient Boosting Model:** XGBoost is another commonly used type of ML model. The important features of XGBoost are its predictive result performance, it is quick to execute, its scalability and it usually outperforms other ML algorithms [23]. Many optimisations have been conducted on the original XGBoost algorithm to ensure model scalability in this research. In comparison, this model works with a boosting ensemble technique and can be applied for classification and regression techniques.
The regularized objective of the XGBoost model at the tth training step is shown in Eq. (2), where $(\hat{y}_i, y_i)$ represents the difference between the prediction of the simulated NULL shipping rate value $\hat{y}$ and the linked ground truth

y_i. Here, $\Omega(f_k) = \gamma T + \frac{1}{2}\lambda\|w\|^2$ is a regularizer denoting the complexity of the kth tree, where T represents the number of leaves and $\|w\|^2$ represents the $L2$ norm of all leaf scores for training instances. When searching the tree, the parameters λ and γ manage the degree of conservatism.

$$\mathcal{L}(\phi) = \sum_i l(\hat{y}_i, y_i) + \sum_k \Omega(f_k) \tag{2}$$

Artificial Neural Network Methods: In contrast to ensemble methods, artificial neural networks (ANNs) make no assumptions about linearity or stationarity and thus, may be used to measure any complex functional relationship. ANNs mimic the behaviour of the human brain, emulating the manner in which biological neurons signal to each other [24]. Hence, it enables machines to recognise patterns and resolve popular issues in the area of shipping forecasting. ANNs are entirely data-driven and are built on the basis of MLP.

- **Multi-Layer Perceptions (MLP):** Equation (3) represents an ANN (x, m, s) model (where x denotes the number of input covariates, m the number of hidden layers, and s the number of output covariates). The container shipping freight rate at time t is represented by z_t; the intercept of the output neuron z_t is represented by y_o; the number of middle layer units is denoted by n; a vector of weights from the middle to output layer units is indicated by y_j; the logistic function is denoted by F, where $F(y) = 1(1 + exp(-y))$; w_{ij} represents a matrix of weights from the input to middle layer at time t; and X_{it} represents the vector of input variables.

$$z_t = n(X_t, y, w) = y_0 + \sum_{j=1}^{n} y_j F\left(\sum_{I=1}^{X} w_{ij}X_{it} + w_{oj}\right) + \varepsilon_t \tag{3}$$

Hyperparameters Fine-Tuning: The hyperparameter optimization aims at finding the optimal combination of values selected for a given dataset before the training starts in a reasonable amount of time (e.g., the number of training cycles and the learning rates) [25]. Through a trial-and-error method, we selected the optimal hyperparameters, as detailed in Table 4, to enhance the performance of the three implemented models. For example, we found that the optimal training iterations are 1000, 500, and 100 for RF, XGBoost, and MLP, respectively. With these optimal values, we were able to reduce the overfitting of the implemented models.

Evaluation of the Models: Evaluating the accuracy of ML models is a fundamental step. In this study, to evaluate the performance of our three models, we employ the following metrics: mean squared error (MSE), mean absolute error (MAE), root mean squared error (RMSE), R-squared, and accuracy. These five

Table 4. Hyperparameter fine-tuning

RF		XGBoost		MLP	
Parameters	Optimal value	Parameters	Optimal value	Parameters	Optimal value
n_estimators	1000	n_estimators	500	epoch	100
max_depth	80	max_depth	10	hidden_layers	3
max_features	3	objective	squarederror	batch_size	3000
min_samples_leaf	4	nthread	4	optimizer	Adam
min_samples_split	10	learning_rate	0.03	learning_rate	0.01

metrics are commonly used to evaluate the performance of ML models. However, the selection of a metric is based on the type model, where a lower MAE, MSE and RMSE value indicates the higher accuracy of a regression model. In contrast, a higher R-squared value indicates the model is outstanding [26].

4   Experiments

It is critical to keep track of the model code, hyperparameters, and outcome metrics for each experiment so we can debate the results, determine where to go next, and replicate the experiments.

In our study, all three models, namely RF, XGBoost and MLP, are used in experiments on the collected dataset. The experiments were performed using the Python programming language [27]. Well-known libraries in Python were used, such as scikit-learn [28] and PyTorch [29]. Also, high-performance computing was used, namely the Google Cloud Platform [30], to perform the experiments where the machine type was n1-highmem-8 (vCPUs, 52 GB memory) and GPUs (4 x NVIDIA Tesla P4).

5   Results and Discussion

Table 5 shows the RMSE, MSE, MAE, R2 and accuracy for the three models on the North American TransBorder Freight dataset. MLP achieved the highest accuracy of 97%, followed by XGBoost with an accuracy of 83% and RF with an accuracy of 80%. However, on the computation side, XGBoost and RF consume fewer resources than MLP and are 3 times faster in terms of training duration.

Most of the world's shipping containers are purchased on the short-term spot market and their economic terms are usually set manually via email and phone calls [31]. Consequently, shipping prices can fluctuate due to a lack of demand and supply. Optimising the freight rate of a spot shipping demand in order to optimise profitability is a difficult operation that is fraught with uncertainty on both the demand and supply sides. To achieve this balance in the spot shipping market, ML models are implemented, which are currently being used in almost every domain. In the shipping domain, if the data is large (i.e., more than 1000K) and the computation resources are available, MLP is the best method to obtain the highest freight rate accuracy.

Table 5. Performance metrics results for RF, XGBoost and MLP models

Model name	RMSE	MSE	MAE	R^2	Accuracy
XGBoost	0.8144	0.6633	0.5691	0.831	0.831
RF	0.8828	0.7793	0.6224	0.8015	0.8015
MLP	2.0144	4.0579	1.6399	0.001	0.9789

6   Conclusion

This paper investigates competing forecasting models of container freight rates using the North American TransBorder Freight dataset. Three promising models have been scrutinized: RF, XGBoost, and MLP.

To check the forecasting robustness of the models, we employed five measures, namely accuracy, $RMSE$, MSE, MAE, and R^2. The empirical results show that MLP outperforms RF and XGBoost in prediction accuracy, but it consumes more computational resources and data. Although our findings are derived from U.S. shipping data, the suggested approach serves as a general method for other international marketplaces.

Researchers should continually strive to create new forecasting methods to enhance the outcomes of previous models. In the future, we will improve the MLP because it has achieved the most accurate prediction for freight rates. In addition, as a container freight pricing volatility indicates shipowners' and investors' behaviour, integrating judgemental and statistical models may be a viable choice.

References

1. Coraddu, A., Oneto, L., Baldi, F., Cipollini, F., Atlar, M., Savio, S.: Data-driven ship digital twin for estimating the speed loss caused by the marine fouling. Ocean Eng. **186**, 106063 (2019)
2. China Transforms the Trucking Business - Bloomberg. https://www.bloomberg.com/opinion/articles/2017-11-30/china-transforms-the-trucking-business
3. The Voice of America's Trucking Industry. https://www.trucking.org/
4. Clarksons Research: Container Intelligence Quarterly (January 2022). https://www.crsl.com/acatalog/container-intelligence-quarterly.html
5. Jeon, J.-W., Duru, O., Munim, Z.H., Saeed, N.: System dynamics in the predictive analytics of container freight rates. Transp. Sci. **55**, 815–967 (2021)
6. Chen, T., Guestrin, C.: XGBoost: a scalable tree boosting system. In: 2016 Proceedings of the 22nd ACM SIGKDD International Conference on Knowledge Discovery and Data Mining, pp. 785–794 (2016)
7. Wang, Y., Meng, Q.: Optimizing freight rate of spot market containers with uncertainties in shipping demand and available ship capacity. Transp. Res. Part B Methodol. **146**, 314–332 (2021)
8. Yan, R., Wang, S., Zhen, L., Laporte, G.: Emerging approaches applied to maritime transport research: past and future. Commun. Transp. Res. **1**, 100011 (2021)

9. Ubaid, A., Hussain, F.K., Charles, J.: Machine learning-based regression models for price prediction in the Australian container shipping industry: case study of Asia-Oceania trade lane. In: Barolli, L., Amato, F., Moscato, F., Enokido, T., Takizawa, M. (eds.) AINA 2020. AISC, vol. 1151, pp. 52–59. Springer, Cham (2020). https://doi.org/10.1007/978-3-030-44041-1_5

10. Munim, Z.H., Schramm, H.-J.: Forecasting container freight rates for major trade routes: a comparison of artificial neural networks and conventional models. Marit. Econ. Logist. **23**, 310–327 (2020)

11. Ubaid, A., Hussain, F., Charles, J.: Modeling shipment spot pricing in the Australian container shipping industry: case of Asia-Oceania trade lane. Knowl. Based Syst. **210**, 106483 (2020)

12. Wang, Y., Meng, Q.: Integrated method for forecasting container slot booking in intercontinental liner shipping service. Flex. Serv. Manuf. J. **31**(3), 653–674 (2019)

13. Viellechner, A., Spinler, S.: Novel data analytics meets conventional container shipping: predicting delays by comparing various machine learning algorithms. In: Proceedings of the 53rd Hawaii International Conference on System Sciences (2020)

14. Le, L.T., Lee, G., Park, K.-S., Kim, H.: Neural network-based fuel consumption estimation for container ships in Korea. Marit. Policy Manage. **47**(5), 615–632 (2020)

15. Barua, L., Zou, B., Noruzoliaee, M., Derrible, S.: A gradient boosting approach to understanding airport runway and taxiway pavement deterioration. Int. J. Pavement Eng. **22**(13), 1673–1687 (2021)

16. TransBorder freight data | bureau of transportation statistics. https://www.bts.gov/transborder

17. Alexandropoulos, S.-A.N., Kotsiantis, S.B., Vrahatis, M.N.: Data preprocessing in predictive data mining. Knowl. Eng. Rev. **34**, E1 (2019)

18. Tsaganos, G., Nikitakos, N., Dalaklis, D., Ölcer, A., Papachristos, D.: Machine learning algorithms in shipping: improving engine fault detection and diagnosis via ensemble methods. WMU J. Marit. Affairs **19**, 51–72 (2020). https://doi.org/10.1007/s13437-019-00192-w

19. Berry, M.W., Mohamed, A., Yap, B.W. (eds.): Supervised and Unsupervised Learning for Data Science. USL, Springer, Cham (2020). https://doi.org/10.1007/978-3-030-22475-2

20. Liu, Y., Browne, W.N., Xue, B.: Adapting bagging and boosting to learning classifier systems. In: Sim, K., Kaufmann, P. (eds.) EvoApplications 2018. LNCS, vol. 10784, pp. 405–420. Springer, Cham (2018). https://doi.org/10.1007/978-3-319-77538-8_28

21. Zhou, Z.-H.: Ensemble Methods: Foundations and Algorithms. Chapman and Hall/CRC (2019)

22. Zhang, W., Wu, C., Zhong, H., Li, Y., Wang, L.: Prediction of undrained shear strength using extreme gradient boosting and random forest based on Bayesian optimization. Geosci. Front. **12**(1), 469–477 (2021)

23. Islam, S., Sholahuddin, A., Abdullah, A.: Extreme gradient boosting (XGBoost) method in making forecasting application and analysis of USD exchange rates against rupiah. J. Phys. Conf. Ser. **1722**(1), 012016 (2021)

24. Xu, A., Chang, H., Xu, Y., Li, R., Li, X., Zhao, Y.: Applying artificial neural networks (ANNs) to solve solid waste-related issues: a critical review. Waste Manage. **124**, 385–402 (2021)

25. Yang, L., Shami, A.: On hyperparameter optimization of machine learning algorithms: theory and practice. Neurocomputing **415**, 295–316 (2020)

26. Chicco, D., Warrens, M.J., Jurman, G.: The coefficient of determination R-squared is more informative than SMAPE, MAE, MAPE, MSE and RMSE in regression analysis evaluation. PeerJ Comput. Sci. **7**, e623 (2021)
27. 3.9.6 documentation. https://docs.python.org/3/
28. scikit-learn: machine learning in Python—scikit-learn 0.24.2 documentation. https://scikit-learn.org/stable/
29. PyTorch. https://www.pytorch.org
30. Cloud computing services. https://cloud.google.com/
31. Sun, H., Lam, J.S.L., Zeng, Q.: The dual-channel sales strategy of liner slots considering shipping e-commerce platforms. Comput. Ind. Eng. **159**, 107516 (2021)

Robust Variational Autoencoders and Normalizing Flows for Unsupervised Network Anomaly Detection

Naji Najari[1,2,3(✉)], Samuel Berlemont[1], Grégoire Lefebvre[1], Stefan Duffner[2,3], and Christophe Garcia[2,3]

[1] Orange Labs, Meylan, France
{naji.najari,samuel.berlemont,gregoire.lefebvre}@orange.com
[2] LIRIS UMR 5205 CNRS, Villeurbanne, France
{naji.najari,stefan.duffner,christophe.garcia}@liris.cnrs.fr
[3] INSA Lyon, Villeurbanne, France

Abstract. In recent years, the integration of connected devices in smart homes has significantly increased, thanks to the advent of the Internet of things (IoT). However, these IoT devices introduce new security challenges, since any anomalous behavior has a serious impact on the whole network. Network anomaly detection has always been of considerable interest for every actor in the network landscape. In this paper, we propose GRAnD, an algorithm for unsupervised anomaly detection. Based on Variational Autoencorders and Normalizing Flows, GRAnD learns from network traffic metadata a normal profile representing the expected nominal behavior of the network. Then, this model is optimized to detect anomalies. Unlike existing anomaly detectors, our method is robust to the hyperparameter selection and outliers contaminating the training data. Extensive experiments and sensitivity analyses on public network traffic benchmark datasets demonstrate the effectiveness of our approach in network anomaly detection.

Keywords: Unsupervised anomaly detection · Robust autoencoders · Dynamic outlier filtering · Network traffic anomaly detection

1   Introduction

Thanks to the recent advances in Internet of Things (IoT) technologies and the steady growth of IT services, IoT devices have become ubiquitous in multiple domains such as Smart Home, Healthcare, Industry 4.0. Although the IoT has played a key role in the enablement of new services and the development of new business value, there is a growing concern about the security of modern networks. IoT devices have numerous technical limitations such as constrained resources, battery failures, connectivity issues, and are vulnerable to diverse cyber threats. Such failures have serious consequences on the Quality of Service (QoS). Therefore, detecting abnormal events is of paramount importance to mitigate risks, prevent system failures.

© The Author(s), under exclusive license to Springer Nature Switzerland AG 2022
L. Barolli et al. (Eds.): AINA 2022, LNNS 450, pp. 281–292, 2022.
https://doi.org/10.1007/978-3-030-99587-4_24

Signature-based Intrusion Detection Systems (IDSs) are commonly used to protect IoT devices from cyber threats. They detect network anomalies by comparing the traffic with known attack signatures. Although they are effective to detect already known attacks, these systems are incapable of mitigating non-malicious anomalies or novel attacks, e.g., zero-day attacks [1]. Unsupervised anomaly detection has been a point of interest to mitigate these limitations and develop reliable and secure networks.

Anomaly Detection (AD) is the task of detecting anomalous data points that significantly deviate from expected normal samples [7]. The most common approaches for AD are based on One-Class Classification (OCC) [7]. OCC consists in learning an accurate representation of the norm, relying only on nominal data points. Once the normal data are well-modeled, the algorithm assigns an abnormality score to each test sample. Finally, a threshold criterion separates inliers and outliers. OCC efficacy depends on the availability of anomaly-free training data, and performance may degrade significantly when this assumption is violated. Unfortunately, this violation is likely to occur in real-world applications. For example, in network traffic monitoring, collected network packets may comprise defective data sent by faulty sensors, damaged fiber connectors, or caused by network congestion [12]. Finally, due to data volumes and potentially unknown anomalies, manual labelling of training samples is not feasible.

In this paper, we propose GRAnD, an algorithm for Generative Robust Anomaly Detection. We introduce a training strategy that alternates between filtering outliers contaminating the training dataset and learning a robust representation of the norm. Our training strategy involves little architectural changes and can be integrated with Variational Autoencoders (VAEs) [11] and Normalizing Flows (NFs) [17]. Unlike recent robust generative methods, our approach makes no assumption about the anomaly distribution, or about the fraction of training outliers. Our method comprises three contributions:

- a robust rejection strategy that filters corrupted training samples, based on Extreme Value Theory (EVT). This strategy separates the training data into three disjoint subsets: an inlier subset containing training data deemed nominal, an outlier subset that comprises the "most anomalous" training samples, and a third subset containing critical undetermined instances;
- a training strategy that leverages filtered anomalies to learn a representation where inliers are well reconstructed and outliers are explicitly corrupted;
- an extensive validation on network traffic datasets, which demonstrates that our approach outperforms some state-of-the-art robust methods and is robust to the hyperparameter selection.

2   Related Work

AD is an active research field that has always been a point of interest in different applications such as network intrusion detection, fraud detection, fault diagnosis, and predictive maintenance. Four families of approaches were proposed: *probabilistic, neighbor, domain and reconstruction-based methods* [7]. *Statistical and probabilistic-based methods* typically model the inlier distribution by learning the parameters of a parametric function. Samples that have low likelihood under this model are considered anomalies. This category includes Gaussian mixture models [22], and kernel density estimators [8]. *Neighbor-based methods*, a.k.a. proximity-based methods, assume that outliers are far from their nearest neighbors, while inliers are close to each other. Well-known proximity-based method include Local Outlier Factor (LOF) [6] and Angle-Based Outlier Detection (ABOD) [13]. *Domain-based methods* estimate a boundary that separates the inlier domain from the rest. Anomalies are samples outside this inlier boundary. One-class SVM (OC-SVM) [19] and Support Vector Data Description (SVDD) [21] are two popular domain-based algorithms. *Reconstruction-based anomaly detection* assumes that, unlike outliers, inliers can be projected into a low-dimensional subspace. The reconstruction error represents a score of data abnormality, as the reconstruction errors of anomalies are higher than inliers. Particularly, AutoEncoders (AEs) have been trained to map nominal input data into a compact latent space, to learn a non-linear representation of the nominal class [7]. Besides, generative models have been profusely proposed for anomaly detection [4]. Furthermore, numerous studies explored Generative Adversarial Networks (GANs) [7] for AD.

The above methods have been applied to detect network traffic anomaly detection [9]. Although they show good results when trained with anomaly-free data, their performance drastically decreases when the training data is contaminated with outliers. In a real-world environment, there is no guarantee that the collected training data are entirely clean. Atypical abnormal traffic may be hidden in the collected data, due to adversarial attacks, or packet collisions. Consequently, it is advocated to develop robust unsupervised anomaly detectors, insensitive to training contaminants [18].

Zhou and Paffenroth [23] proposed Robust Deep Autoencoders (RDAs) to filter sparse corrupted samples from the input data matrix. Robust Subspace Recovery (RSR) [15] is another line of work in robust anomaly detection. RSR assumes that inliers can be projected into a linear low-dimensional subspace, while outliers are not well modeled in this subspace. Lai et al. [14] introduced Robust Subspace Recovery AutoEncoder (RSRAE), where they integrated an RSR-layer in a classical autoencoder. Regarding robust generative autoencoders, Akrami et al. [3] proposed a Robust VAE (RVAE). Their approach uses the robust β-divergence instead of the standard Kullback-Leibler (KL) divergence. Minimizing the β-divergence involves reweighting each sample likelihood gradient with its probability density.

Recently, Kotani et al. [12] used RDA for network flow intrusion detection. Although these approaches proved to reduce the number of false positives on real-world traffic datasets, they involve an explicit regularization, defined by one or many critical hyperparameters. Prior knowledge about the outlier ratio and additional assumptions either on the inlier, the outlier class, or both, are required to select the optimal hyperparameters. Generally, such hyperparameters are empirically tuned with a dedicated validation subset containing manually ground-truth-labeled data. In the context of anomaly detection, labeled outliers are too scarce to form a balanced validation subset. Also, in most situations, the ratio of training outliers is not known. Therefore, hyperparameter selection is prone to misspecification. However, the methods above-mentioned are all sensitive to their hyperparameters, since slightly changing them can drastically degrade their anomaly detection performances. In contrast, our approach does not make any assumptions about outlier distribution. We propose a robust training strategy that jointly performs two tasks. This strategy filters training outliers using EVT. Then, training outliers are leveraged to infer a better representation that can be generalized to unseen anomalies. This strategy can be incorporated with VAEs and NFs, and involves minimal architectural changes.

3   Background

3.1   Generative AEs

We consider the task of unsupervised AD under the standard variational inference setting. Generative models aim to find the optimal parameters θ that maximize the likelihood $p_\theta(x) = \mathbb{E}_{p(z)}[p_\theta(x|z)]$, where z is the model latent variable and $p(z)$ is a predefined prior. However, this likelihood is intractable because of the marginalization over the latent variable z. Variational inference aims to approximate the posterior probability $p(z|x)$ with a parametric distribution $q_\phi(z|x)$, parameterized by ϕ. Regardless of the choice of this distribution, we can reformulate the log-likelihood as follows:

$$\log p_\theta(x) \geq \mathbb{E}_q[\log p_\theta(x|z)] - \mathbb{D}_{KL}[q_\phi(z|x)||p(z)] = -\mathcal{F}(x), \tag{1}$$

where $q_\phi(z|x)$ is the approximate posterior distribution for the latent variables, and $\mathcal{F}$ is the negative free energy, a.k.a., the evidence lower bound (ELBO). This energy comprises two terms. The first term is the reconstruction error, and the second one represents the KL divergence between the approximate distribution and the prior distribution. A common choice of the approximate distribution family is the multivariate Gaussian distribution with a diagonal covariance matrix. Recently, NFs have been used to provide a richer parametric family of approximate posterior to capture complex structures of the latent space. NFs transform an initial simple density function to a more sophisticated one, by applying a sequence of invertible transformations.

3.2   EVT

The objective of EVT is to quantify the probability of occurrence of extreme values in a distribution function. Recently, EVT has been applied to detect anomalies in many applications including network traffic data streams [20]. The Peaks-Over-Threshold (POT) is a typical approach used to model the extreme values of samples that exceed a specific high threshold. This approach is a result of the Picakands-Balkema-de-Han theorem of EVT [5].

Let $(X_1, X_2, \ldots, X_n)$ be n independent and identically distributed (iid) random variables. Let F_u be their conditional excess distribution function, i.e., $F_u(x) = P(X - u > x | X > u)$, where u is a high threshold. The POT method models the extreme values that exceed the threshold u, using the Generalized Pareto Distribution (GPD) parametrized by two parameters, ξ and σ:

$$F_u(x) \to 1 - G_{\xi,\sigma}(x), \text{ as } u \to \infty \quad \text{where} \quad \begin{cases} G_{\xi,\sigma}(x) = 1 - (1 + \frac{\xi x}{\sigma})^{-\frac{1}{\xi}}, \text{ if } \xi \neq 0 \\ G_{\zeta,\sigma}(x) = 1 - e^{-\frac{x}{\sigma}}, \qquad \text{if } \xi = 0. \end{cases} \quad (2)$$

In practice, the two parameters of the GPD are empirically estimated by fitting the GPD to the data. The maximum likelihood estimation is typically used to find these optimal parameters $\tilde{\xi}$ and $\tilde{\sigma}$. Once the extreme values are modeled with the optimal GPD, $G_{\tilde{\xi},\tilde{\sigma}}$, we can identify rare extreme samples that have very low probability [20]. Given a small probability q, we can compute the threshold t_q such that, $P(X > t_q) < q$.

$$P(X - u > t_q | X > u) = \tilde{F}_u(t_q) \sim 1 - G_{\tilde{\xi},\tilde{\sigma}}(t_q). \qquad (3)$$

$$\text{If } \xi \neq 0, \quad t_q \simeq u + \frac{\tilde{\xi}}{\tilde{\sigma}}((\frac{nq}{N})^{\tilde{\xi}} - 1), \qquad (4)$$

where n is the total number of observations, and N is the number of X_i exceeding the threshold u, $X_i > u$. A key question arises as to how to choose the threshold u. Siffer et al. [20] state that "the value of u is not paramount except that it must be high enough". In practice, u is generally selected as a high empirical quantile of the data, e.g., 90% quantile.

4   Contributions

This paper focuses on unsupervised anomaly detection where the unlabeled training data may contain both inliers and outliers, with an imbalanced class distribution. We assume that the majority of the training instances are nominal, along with a small ratio of "contaminants", i.e. outliers. The ratio of these contaminants, which we call γ_p, is not known in advance. In the following, we introduce GRAnD, an algorithm for Generative Robust Anomaly Detection. Our contribution alternates between filtering training outliers and learning a robust distribution of the norm. In the following, we will first explain the rejection strategy that isolates training contaminants. Then, we will detail the objective function to optimize.

4.1   Robust Rejection Strategy

The objective of this rejection strategy is to separate nominal training data points from anomalies. The main idea consists in setting a relevant threshold to segment the reconstruction scores assigned to training samples, in order to reject outliers having extreme scores.

We hypothesize that, early in the training phase, contaminants have larger free energy (cf. Eq. 1), compared to inliers. Consequently, we propose to isolate these extreme values by thresholding the energy with the POT approach, described in Sect. 3.2. The POT approach requires the selection of two parameters: the initial threshold u, and the risk parameter q. In our experiments, we define u as follows:

$$u = Q_3(\mathcal{F}) + \alpha * IQR(\mathcal{F}) \tag{5}$$

where $\mathcal{F}$ is the free energy of the training instances, Q_3 is the third quartile, and IQR is the Inter-Quartile Range, which is defined as the difference between the third and the first quartiles. α controls the scale of the decision rule. In all our experiments, we fixed $\alpha = 1.5$ and $q = 0.001$. In Sect. 5.5, we study the sensitivity of our contribution with respect to α and q.

Using the POT parameters, we propose to split the input data into three subsets $\mathbb{X} = \mathbb{L} \cup \mathbb{S} \cup \mathbb{U}$, as illustrated in Fig. 1. The subset $\mathbb{L}$ contains nominal training samples, having energy lower than the initial threshold u of the POT method. $\mathbb{S}$ contains anomalous data points, with an energy higher than t_q, computed using Eq. 4. $\mathbb{U}$ comprises the remaining critical samples, with an energy higher than u and lower than t_q. These sample energies are neither low enough to be considered nominal, nor high enough to be rejected as anomalies.

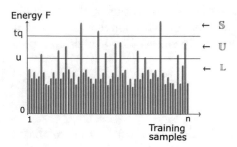

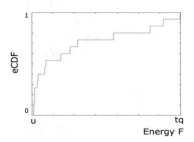

Fig. 1. Illustration of the rejection strategy using the POT approach.

Fig. 2. Empirical cumulative distribution function of $\mathbb{U}$ samples

4.2 Training Loss

The rejection strategy splits the training data into three subsets $\mathbb{L}$, $\mathbb{S}$, and $\mathbb{U}$. We train the autoencoder to jointly perform three tasks: (i) minimize $\mathbb{L}$ sample energy, (ii) badly reconstruct $\mathbb{S}$ samples by maximizing their energy, (iii) maximize a weighted energy function of $\mathbb{U}$ instances, which takes into account the probability of anomalous of these instances. The idea is to associate to each critical instance in $\mathbb{U}$ a weight in $[0, 1]$ that quantifies whether the instance is anomalous or not.

Let $\mathbb{U} = \{X_1, X_2, \ldots, X_n\}$ contain a sequence of n iid instances. We firstly sort these instances in increasing order according to their free-energies

$(\mathcal{F}(X_1), \mathcal{F}(X_2), \ldots, \mathcal{F}(X_n))$. We use the empirical Cumulative Distribution Function (eCDF) to define the anomalousness weight of each $X_i \in \mathbb{U}$.

$$P(X_i \in \mathbb{U} \text{ is anomalous}) = eCDF_n(\mathcal{F}(X_i)) = \frac{1}{n} \sum_{j=1}^{n} \mathbb{1}_{\mathcal{F}(X_j) \leq \mathcal{F}(X_i)} \quad (6)$$

where $\mathbb{1}$ is the indicator function. As illustrated in Fig. 2, $\mathbb{U}$ samples with energies close to the threshold u have a small probability close to 0. Conversely, samples with high scores, i.e., close to t_q, have probabilities close to 1.

GRAnD Objective Function. Given the three subsets of data $\mathbb{L}$, $\mathbb{S}$, and $\mathbb{U}$, respectively generated from the three distributions, D_L, D_S, and D_U, GRAnD optimizes the following objective function:

$$\mathcal{L}(x) = \mathbb{E}_{x \sim D_L}[\mathcal{F}_\mathcal{L}(x)] + |m - \mathbb{E}_{x \sim D_S}[\mathcal{F}_\mathcal{S}(x)]| + eCDF_m(\mathcal{F}_\mathcal{U}(x)) |m - \mathbb{E}_{x \sim D_U}[\mathcal{F}_\mathcal{U}(x)]| \quad (7)$$

The objective function comprises three components:

- $\mathbb{E}_{x \sim D_L}[\mathcal{F}_\mathcal{L}(x)]$ is the expectation of the free energy function of $\mathbb{L}$ samples, defined in Eq. 1. This first component aims to minimize the energy of $\mathbb{L}$ samples.
- $\mathbb{E}_{x \sim D_S}[\mathcal{F}_\mathcal{S}(x)]$ is the expectation of the free energy function of $\mathbb{S}$ samples. $|.|$ is the absolute distance, and $m \in \mathbb{R}^+$ is a margin value. By maximizing this energy, we train the autoencoder to badly reconstruct the potential training contaminants. Since this energy function is positive and unbounded, we propose to fix an upper bound m, to prevent it from diverging in the training. In all our experiments, we fix $m = 100$.
- $\mathbb{E}_{x \sim D_U}[\mathcal{F}_\mathcal{U}(x)]$ is the expectation of the free-energy function of $\mathbb{U}$ samples. We weight the objective function of $\mathbb{U}$ instances according to their anomalousness probability, computed with the eCDF function. These weights account for the uncertainty of the classification of $\mathbb{U}$ instances.

5   Experiments

5.1   Dataset Description

NSL-KDD Dataset. Firstly, we conduct experiments using the NSL-KDD dataset [16], which is one of the most popular datasets used to evaluate network Intrusion Detection Systems (IDSs). Two distinct subsets are provided: the training subset contains 125 973 records and the test subset has 22 544 records. Each data point is represented by 41 features extracted from the network traffic, e.g., the duration of the flow, the TCP flags; and labeled as normal or anomalous. 39 types of network attacks are present in this dataset, ranging from Denial of Services (DoS) to Probe attacks. To investigate algorithm sensitivity with respect to the ratio of anomaly contamination, we vary the anomaly percentage contaminating the training data. We prepare four training subsets, respectively containing 0%, 5%, 10%, and 15% of outliers. These anomalies are selected randomly from all NSL-KDD anomalous training instances. Then, we rescale numerical features to be in the range [0, 1] using the min-max normalization method, and categorical features are one-hot encoded.

MedBIoT Dataset. The MedBIoT dataset [10] is a recent public dataset that contains the network traffic collected from a large network containing 83 real and emulated IoT devices. These devices belong to four categories: switches, light bulbs, locks, and fans. To generate malicious traffic, the authors executed three prominent malware attacks: Mirai, Bashlite, Torii attacks, and labeled the collected training data accordingly. Overall, 17 million network packets were collected: 30% of this traffic is anomalous and the remaining 70% is benign. 61 flow-based features are extracted from the traffic, e.g., flow duration, number and length of packets per flow. A detailed description of each extracted feature is available in [2]. We randomly split the benign data into 60% for training, 20% for validation, and 20% for testing. Similarly to NSL-KDD experiments, we prepare four training datasets with different contamination ratios 0%, 5%, 10%, and 15%. Finally, categorical features are encoded using Count Encoder and numerical features are normalized using the Min-Max normalization method.

5.2   Competing Methods

We compare our approach, GRAnD, against unsupervised AD methods frequently used in the literature: OCSVM with a Gaussian kernel, Isolation Forest (IF), vanilla VAE, vanilla Planar Flow (vanilla PF), Deep Autoencoding Gaussian Mixture Model (DAGMM) [24], and RVAE [3]. In line with prior works, performances are assessed using the Area Under the Curve of the Receiver Operating Characteristics (AUROC).

5.3   Training Parameter Settings

In all experiments, we use the standard Feedforward Neural Network (FNN) architectures for all autoencoders. In NSL-KDD experiments, the autoencoders are composed of a 3-layer FNN with 122-8-122 units. In MedBIoT experiments, the autoencoders are a 5-layer FNN with 61-32-16-32-61 units. All latent layers are followed by ReLU activation function. The last layer of the decoder is followed by a sigmoid function. We use an adaptive learning rate: initially, we use a learning rate of 0.001, which is divided by two if the training loss does not decrease after 20 consecutive epochs. We stop the training when the learning rate is lower than 10^{-6} or the number of epochs becomes higher than 500 epochs. We use a batch size of 256 in all experiments. We initialize model parameters randomly. To limit the impact of random parameter initialization, we repeat each experiment five times and average the results over these five runs.

Our approach comprises three specific hyperparameters: the rejection parameter α that controls the initial threshold u, the risk parameter q, and the margin m. In all experiments, q is fixed to 0.001, α to 1.5, and m to 100. We conduct a sensitivity analysis experiment in Sect. 5.5, to assess our approach robustness regarding the hyperparameters. We use grid search to select competing methods optimal hyperparameters, which maximize their AUROC on the validation subset. The experiments were run on a laptop equipped with a 12-core Intel i7-9850H CPU clocked at 2.6 GHz and with NVIDIA Quadro P2000 GPU.

5.4 Experimental Results and Discussion

NSL-KDD Experimental Results. We show in Fig. 3 the results of the comparison between GRAnD and other competing methods on the NSL-KDD. When the training data are contaminated with anomalies, our approach significantly outperforms competing methods. While the performance of competing methods decreases with higher pollution ratios γ_p, our approach is more stable, with an average AUROC around 94% and very little deviation, for the three contamination ratios 5%, 10%, and 15%. These results mainly highlight the benefit of the robust rejection strategy, where no prior knowledge about the outlier ratio is required in advance.

When the training data are anomaly-free, GRAnD performance slightly degrades, with an AUROC of 92.6% with a standard deviation of 0.8%. This observation can be explained by the fact that GRAnD leverages training outliers to learn a robust projection, where inliers are well reconstructed, while outliers are poorly reconstructed. When training data do not contain anomalies, GRAnD-PF and GRAnD-VAE performances are very similar to vanilla-PF and vanilla-VAE, respectively. Despite this slight increase, GRAnD remains very competitive, with around 6% points better AUROC than IF. Finally, for all contamination ratios, GRAnD-PF slightly outperforms GRAnD-VAE.

MedBIoT Experimental Results. We present the MedBIoT results in Fig. 4. As mentioned previously, we train an anomaly detector for each device type. We obtain similar results for the four device types. Due to space constraints, we report the most representative results in Fig. 4. For the four device types, and for all contamination ratios, GRAnD-PF, GRAnD-VAE, and RVAE outperform other anomaly detectors, with an AUROC of $99.9 \pm 0.1\%$. In particular, we highlight the robustness of our contribution compared to vanilla-VAE and vanilla-PF. While the latter performances are considerably impacted when the contamination ratio is higher than 10%, GRAnD yields stable results. For example, for $\gamma_p = 10\%$, GRAnD-PF and GRAnD-VAE exceed vanilla-VAE and vanilla-PF AUROCs by 19% and 23%, on average. Consequently, the robustness of GRAnD for IoT network traffic anomaly detection is validated on this dataset. Although GRAnD and RVAE yield close results, we will show in the next section that, unlike RVAE, GRAnD is robust to the hyperparameter selection.

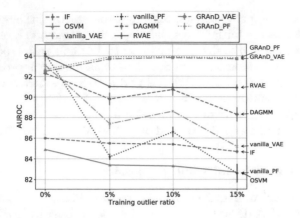

Fig. 3. NSL-KDD experimental results: comparison of AD methods based on average AUROCs and deviations over five runs for multiple contamination ratios.

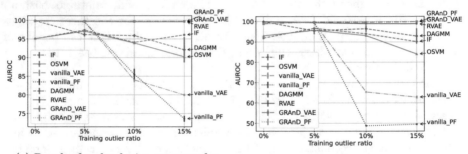

(a) Results for the device category fan   (b) Results for the device category lock

Fig. 4. The MedBIoT experimental results, for the device categories fan and lock. We report the average AUROC with the standard variation over five runs.

5.5   Sensitivity Analysis

As mentioned in numerous works in the anomaly detection community, it is advocated to develop robust anomaly detectors that do not depend on user-defined parameters. The sensitivity to hyperparameters is problematic in unsupervised AD, since outlier labels are scarce, and the selection of the optimal hyperparameters is not guaranteed. We conduct further experiments to assess the sensitivity of our approach regarding its hyperparameters. We train different models with distinct hyperparameters to study the variation of the performance on the same test subset. Due to space constraints, we report in Fig. 5 the results of the sensibility analysis of RVAE and GRAnD-PF on the MedBIoT dataset, with $\gamma_p = 10\%$.

In Fig. 5a, we show RVAE performance for different $\beta \in \{0.0001, 0.001, 0.01, 0.1, 1\}$. Since GRAnD is defined using three hyperparameters, m, q, and α, we run three experiments, where we only vary one hyperparameter and we keep the remaining ones fixed. Figure 5a shows that RVAE is sensitive to the hyperparameter β. For all device types, RVAE AUROC drastically decreases, when β changes. In contrast, GRAnD-PF performance is not impacted by the variation of its hyperparameters, and the AUROC is stable around $99.8 \pm 0.1\%$.

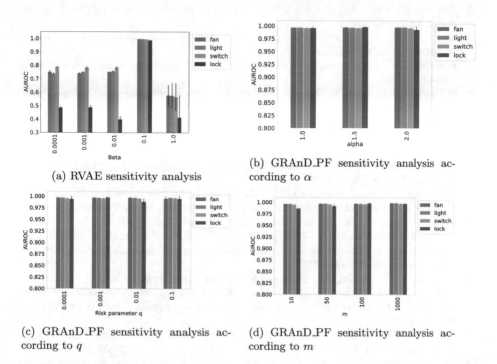

(a) RVAE sensitivity analysis

(b) GRAnD_PF sensitivity analysis according to α

(c) GRAnD_PF sensitivity analysis according to q

(d) GRAnD_PF sensitivity analysis according to m

Fig. 5. The sensitivity analysis of RVAE and GRAnD_PF on MedBIoT dataset. We report the average AUROC with the standard variation over five runs.

6 Conclusion and Future Work

In this paper, we proposed GRAnD, a robust generative method for unsupervised anomaly detection. Our approach uses Extreme Value Theory to filter out outliers contaminating the data and learns a robust representation, where inliers can be accurately reconstructed, while outlier reconstructions are corrupted. Extensive experiments were conducted on benchmark datasets, and showed that our approach outperforms classical anomaly detection methods, all the while showing outstanding robustness to hyperparameter selection. In the future, we will extend GRAnD to detect anomalies in time-series and sequential data. We will adapt our rejection strategy to detect contextual and collective sequential anomalies. Finally, since our contribution involves a minimal change to the underlying model architecture, future studies could fruitfully explore other generative models, such as adversarial autoencoders and GANs.

References

1. Malware and network attacks in 2019. https://www.helpnetsecurity.com/2019/12/13/network-attacks-2019/
2. NFStream - a Network Data Analysis Framework. https://nfstream.org/
3. Akrami, H., Joshi, A.A., Li, J., Aydore, S., Leahy, R.M.: Robust variational autoencoder. CoRR (2020)
4. An, J., Cho, S.: Variational autoencoder based anomaly detection using reconstruction probability (2015)
5. Balkema, A.A., De Haan, L.: Residual life time at great age. Ann. Probab. **2**(5), 792–804 (1974)
6. Breunig, M., Kriegel, H., Ng, R., Sander, J.: LOF: identifying density-based local outliers (2000)
7. Chandola, V., Banerjee, A., Kumar, V.: Anomaly detection: a survey. ACM Comput. Surv. **41**, 1–58 (2009)
8. Desforges, M.J., Jacob, P.J., Cooper, J.E.: Applications of probability density estimation to the detection of abnormal conditions in engineering. Proc. Inst. Mech. Eng. **212**(8), 687–703 (1998)
9. Fernandes, G., Rodrigues, J.J.P.C., Carvalho, L.F., Al-Muhtadi, J.F., Proença, M.L.: A comprehensive survey on network anomaly detection. Telecommun. Syst. **70**(3), 447–489 (2019). https://doi.org/10.1007/s11235-018-0475-8
10. Guerra-Manzanares, A., Medina-Galindo, J., Bahsi, H., Nõmm, S.: MedBIoT: generation of an IoT botnet dataset in a medium-sized IoT network. In: ICISSP, Valletta, Malta, pp. 207–218 (2020)
11. Kingma, D.P., Welling, M.: Auto-encoding variational bayes. CoRR (2014)
12. Kotani, G., Sekiya, Y.: Unsupervised scanning behavior detection based on distribution of network traffic features using robust autoencoders. In: (ICDMW), pp. 35–38 (2018)
13. Kriegel, H.P., Schubert, M., Zimek, A.: Angle-based outlier detection in high-dimensional data. In: ACM SIGKDD, KDD 2008, pp. 444–452 (2008)
14. Lai, C.H., Zou, D., Lerman, G.: Robust subspace recovery layer for unsupervised anomaly detection. In: ICLR (2020)
15. Lerman, G., Maunu, T.: An overview of robust subspace recovery (2018)
16. NSL-KDD: Canadian Institute for Cybersecurity — UNB
17. Rezende, D.J., Mohamed, S.: Variational inference with normalizing flows. In: ICML (2015)
18. Ringberg, H., Soule, A., Rexford, J., Diot, C.: Sensitivity of PCA for traffic anomaly detection. In: ACM SIGMETRICS, pp. 109–120 (2007)
19. Schölkopf, B., Williamson, R.C., Smola, A.J., Shawe-Taylor, J., Platt, J.C.: Support vector method for novelty detection. In: Solla, S., Leen, T., Müller, K. (eds.) Advances in Neural Information Processing Systems, vol. 12. MIT Press (1999). https://proceedings.neurips.cc/paper/1999/file/8725fb777f25776ffa9076e44fcfd776-Paper.pdf
20. Siffer, A., Fouque, P.A., Termier, A., Largouet, C.: Anomaly detection in streams with extreme value theory. In: ACM SIGKDD, pp. 1067–1075 (2017)
21. Tax, D.M., Duin, R.P.: Support vector data description. Mach. Learn. **54**, 45–66 (2004)
22. Yang, X., Latecki, L.J., Pokrajac, D.: Outlier detection with globally optimal exemplar-based GMM. In: SDM, pp. 145–154 (2009)
23. Zhou, C., Paffenroth, R.C.: Anomaly detection with robust deep autoencoders. In: ACM SIGKDD - KDD 2017 (2017)
24. Zong, B., et al.: Deep autoencoding Gaussian mixture model for unsupervised anomaly detection. In: ICLR, p. 19 (2018)

Multiplatform Comparative Analysis of Intelligent Robots for Communication Efficiency in Smart Dialogs

Anna Pogoda[1], Ewa Lyko[1], Michal Kedziora[1(✉)], Ireneusz Jozwiak[1], and Jolanta Pietraszko[2]

[1] Wroclaw University of Science and Technology, Wroclaw, Poland
michal.kedziora@pwr.edu.pl
[2] Military University of Land Forces, Wrocław, Poland
jolanta.pietraszko@awl.edu.pl

Abstract. The objective of the paper is to analyze the algorithms and methods used in dialogue chatbot systems in terms of usability and quality of their functioning. The research used two knowledge corpuses differing in subject matter, number of specific intents, availability of entities, and number of training and test data. The research was performed on three platforms - RASA, Dialogflow, and IBM Watson. The influence of the number of intents and the presence of entities on the quality of the dialogue system using predefined metrics was examined. Additionally, the experiments concerned the analysis of chatbot operation for different ratios of training to test data. Moreover, the influence of the platform and corpus selection on the quality of operation and acceptance level of intent was examined.

1 Introduction

Creating chatbots is associated with many difficulties and challenges [3,10,11, 14]. The basic difficulty is related to natural language understanding (NLU) and natural language processing (NLP) [6,15,16]. Just formulating the same question in several different ways can cause the system not to answer it in the same way or the user will not get any answer [5]. The challenge for modern chatbots is, among others using pronouns in longer statements, which may cause misinterpretation of the thing or person named pronoun [14]. The number and quality of training data is also important. The quality of the data is related to the division into intentions as well as the definition of synonyms and keywords in the user's message. The number of intentions should be skilfully chosen. When there are too many or too few intentions, chatbot will assign bad intentions to the message, which will result in poorer performance and user irritation. Another difficulty is extracting relevant information from the user's message. Depending on the subject of chatbot, the information will be hidden in dates, places, numbers and contact details. Chatbot must be prepared to recognize embedded system objects, but also personalized for the given system [12]. However, the biggest problem with chatbot systems is dialogue management. Recognizing the intentions or system objects of one message is difficult, but even more difficult is managing dialog states or sticking to the context of the conversation. If the system needs specific information to be able to perform an action, it should ask the user for this information until it obtains and stores it [12].

The goal of the paper is to analyze the methods and algorithms of artificial intelligence used in chatbot dialog systems offered by major known software producers. Three platforms will be compared - Dialogflow, IBM Watson and RASA. These tools will be evaluated for their performance. The tested parameters include change in the number of intentions, number of queries per intention, level of acceptance or knowledge base corpus.

2   Related Works

Paper which handle similar aspects of chatbot systems is [9]. As part of this paper, three tools for creating chatbots were tested - Google's Dialogflow, IBM Watson Assistant and Microsoft LUIS. The research was based on two bodies of knowledge differing in subject matter, additionally one of them was tested twice for changed values of intentions and test data. Research shows that all platforms work at a high level, but LUIS and IBM Watson do the best regardless of the corpus. In addition, some improvements in the quality of operation were tested, among others *slot filling* or recognition of new system objects.

Another work worth analyzing is "An Evaluation Dataset for Intent Classification and Out-of-Scope Prediction" [8]. The work concerns testing the chatbot in the event that the user's request is beyond the scope of his intentions. An additional test dataset has been introduced that contains those queries where no intent should be detected by chatbot. Research consisted of comparing chatbot activities before and after adding these queries. In addition, a number of benchmark classifiers were rated, among others SVM, CNN or FastText. Research has shown that some classifiers cope better with queries in the field, but all have a problem with non-thematic messages. In addition, it was noticed that generating non-domain messages is a difficult task and you should do this very carefully.

The work titled "Benchmarking Natural Language Understanding Services for building Conversational Agents" [17] presents a comparison of some of the available NLU natural language understanding tools. The tests were performed on the same body of knowledge, using four platforms - LUIS, Watson, Dialogflow and RASA. The division into learning and testing data has been replaced by the use of 10 times cross-validation for each platform and additionally t-test pairs have been performed to compare the average results for each pair of available platforms. The presented studies show that all platforms have similar functionalities and quality of operation based on the results of F1 values. An interesting conclusion is that IBM Watson is much better at detecting intentions than system objects.

In paper titled "System for Semi-Automated Chatbots Query Classification Training Corpus Generation" [2] solution to the problem of learning chatbot is shown, if there are not many dialog systems on a given topic and the number of learning data is too small for the system to work at a satisfactory level. The approach assumes the possibility of generating sentences using a smaller available corpus by identifying different parts of speech in corpus sentences, and then generating variations for the same sentences using different times or synonyms. The next step is to generate sentences again from already generated sentences, which are subject to human checking and are saved in the training body. Using such a solution saves a lot of effort and allows the system to operate in high quality.

In the field of evaluation and comparison of chatbots, it is worth to mention a paper "Different measurements metrics to evaluate a chatbot system" [1], which presents the metrics used to evaluate chatbot systems, has proved valuable. The first is the effectiveness of the dialog match in terms of atomic match, first word match, most significant match and no match. The metric regarding the quality of the dialogue based on the answer is related to the assessment of whether the obtained answer is senseless, meaningful, strange or incomprehensible. In addition, user satisfaction was assessed on the basis of an open feedback request regarding the received response.

The article "Evaluating and Informing the Design of Chatbots" [4,7] discusses the results of the research, which consisted of 16 people using chatbots on the Facebook Messenger platform. The authors came to the conclusion that these systems should be developed and care especially about chatbot capabilities, guiding users to questions that they can ask to get a satisfactory answer. In addition, you should focus on maintaining the context of the conversation and deal with problems and misunderstanding during the conversation.

The article analyzed is also the article "Evaluating Quality of Chatbots and Intelligent Conversational Agents" [13], which refers to the assessment of the quality and attributes of the system. The quality attributes are divided into three sections - effectiveness, efficiency and satisfaction. In each of them categories were distinguished, among others obtained results, chatbot humanity, availability or effect. In addition, a quality assessment method based on these attributes was proposed and examined, and the AHP "(Analytic Hierarchy Process)" method was presented, which involves multi-criteria analysis of decision problems, enabling problem decomposition and finally creating a final ranking for a set of variants.

The completed literature review shows the level of complexity and interest of other researchers in the topic of chatbot systems. The articles show the complexity of these systems and the difficulty of implementing and testing them to make them work at a satisfactory level. The research has various research problems. Among them was the study of the impact of platform selection (RASA, Microsoft LUIS, Dialogflow, Amazon Lex or IBM Watson) or corpus on chatbot operation [4].

3   Methodology

Research using the same knowledge bodies or parameters will be carried out on different platforms - RASA, Dialogflow and IBM Watson to be able to compare the results of chatbot systems created using them. In addition to NLP methods that are common to platforms, they use hidden algorithms to increase the quality of their chatbots. The study is designed to check whether the results achieved on each platform will be satisfactory. The Ubuntu knowledge corpus with specific intentions and system objects will be used for the study. The test involves testing the body using three environments. The same system parameters were determined on each platform, i.e. what percentage of data will be training and what is test, and the level of intention acceptance was determined to test only the platform's impact on performance. In addition, the level of acceptance of intent has been set to 0.7. The study involves testing two corps of knowledge using two environments. The knowledge bases used in the following study include the Ubuntu and Banking Corps. The same system parameters were determined on each platform, i.e. what percentage of data will be training and what is test, and the level of intention acceptance was determined to test only the influence of the corpus on operation. In addition, the level of intent acceptance was set at 0.75.

The intention recognition model and matching it to the user's message is based on the propabilistic model. This means that the system will very rarely be 100% sure that it has well defined the user's intention. Chatbot assigns the probability of occurrence of each intent to each user's message and sets them in descending order. The intention that comes first in the intention ranking for a given message should be selected. The problem arises when the suggested intention has a very low probability of occurring, which is why chatbot requires a threshold parameter. This parameter determines with which minimum probability the intention can be accepted. The intent acceptance level is between 0 and 1. When the value is 0, chatbot will always assign an intention to the message. However, if the threshold value is maximum, then virtually most messages will have no intention [18]. The research involves testing the level of acceptance of intentions with the help of RASA. To add an intent acceptance threshold, set up the fallback policy.

In the following study, the first threshold value described is relevant, and the Ubuntu knowledge corpus will be used for the analysis. The acceptance value will be tested in the full range from 0 to 1 in 0.1, to get a full picture of the effect of this value on system operation. Each body of knowledge has a specific amount of data, but it is possible to set the division of data into training and test. Training data is used to learn a chatbot system, while test data is designed to assess the quality of this system. The Research aims to verify that the ratio of training and test data is important for chatbot to operate at a satisfactory level. The Banking database will be used for the study. The test adopts a fixed intent acceptance level of 0.7 to test the effect of just splitting data.

4   Results Analysis

Two null hypotheses were put forward before the study. The first concerns the impact of platform selection on the chatbot system operation, while the second assumes that commercial solutions generally work better than open and free solutions. Tables 1, 2 and 3 present the results of testing the impact of platform changes on chatbot system operation. Each of the tables shows the result achieved for a different platform - RASA, Dialogflow or IBM Watson. Based on the results, it can be seen that the platforms use other mechanisms to create and learn chatbot systems, and their results are different for each platform.

Table 1. A table showing the results of testing the impact of the platform on the operation of the chatbot system - for the RASA tool, taking into account individual intentions

Intent	TP	FN	FP	Precision	Recall	F1 score
Make update	28	9	1	0.9655	0.7568	0.8485
Setup printer	9	4	1	0.9000	0.6923	0.7826
Software recommend.	28	11	10	0.7368	0.7179	0.7273
Shutdown computer	11	3	20	0.3548	0.7857	0.4889
None	5	0	0	0.0000	0.0000	0.0000
Average RASA				**0.5914**	**0.5905**	**0.5695**

Table 2. Table presenting the results of testing the impact of the platform on the operation of the chatbot system - for the Dialogflow tool taking into account individual intentions

Intent	TP	FN	FP	Precision	Recall	F1 score
Make update	33	4	1	0.9706	0.8919	0.9296
Setup printer	9	4	0	1.0000	0.6923	0.8182
Software recommend.	3	36	0	1.0000	0.0769	0.1429
Shutdown computer	12	2	0	1.0000	0.8571	0.9231
None	0	5	0	0.0000	0.0000	0.0000
Average DF				**0.6341**	**0.5036**	**0.5628**

It was decided to document the results in the form of graphs showing the difference in chatbot operation for specific intentions and platforms. Figure 1 approximates the impact of platform changes on the number of correctly matched intentions for Ubuntu corpus test data broken down by individual intentions. The conclusions that there are noticeable differences between the intentions of the corps. This is related to the number of training and test data and the occurrence of system objects for some intentions. The best described intention in terms of the number

Table 3. A table showing the results of testing the impact of the platform on the operation of the chatbot system - for the IBM Watson tool taking into account individual intentions

Intent	TP	FN	FP	Precision	Recall	F1 score
Make update	34	3	1	0.9714	0.9189	0.9258
Setup printer	13	0	0	1.0000	1.0000	1.0000
Software recommend.	32	7	2	0.9412	0.8205	0.8767
Shutdown computer	14	0	0	1.0000	1.0000	1.0000
None	1	4	3	0.3334	0.2000	0.1334
Average IBM Watson				**0.8492**	**0.7879**	**0.7872**

of data and system objects is *Make Update*, where the highest results of recognizing this intention come from. In contrast, the worst results for all platforms were obtained for *None*, which contains too little training and test data. Intention *Software Recommendation* has different results depending on the platform. For RASA and IBM Watson platforms, it was very well detected, while the Dialogflow platform did not cope with this intention. Analyzing the operation of Dialogflow for test data of this intention, it was concluded that system objects disrupted the operation of this platform and reduced the number of TP.

The phenomenon of platform change impact on the number of incorrectly classified test data was noted. One of the observations is that the RASA platform classifies definitely more test data for bad intentions. The largest number of chatbot RASA errors were made for the *Shutdown Computer* intention, whose queries did not contain any system objects, and the amount of data was not enough. Other platforms are more cautious in classifying the message for intent. Dialogflow does the best, while the results achieved by IBM Watson are also satisfying. To verify the hypotheses made before the start of the study, statistical analysis was performed. Figure 2 presents basic information about the

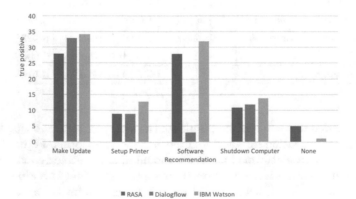

Fig. 1. Graph of the impact of platform changes on the number of correctly matched intentions for Ubuntu corpus test data

three groups, which are the appropriate bot platforms. All the highest values are achieved by the IBM Watson platform. Medium RASA and Dialogflow platforms are very similar, while Dialogflow has a 12% higher median than the other platform. The largest difference of values that are not outliers is visible in the case of the Dialogflow platform, while for other platforms outliers are distinguished.

The p value obtained after the Kruskal-Wallis test is greater than the 0.05 significance level. It is $p = 0.1867$. There are no significant differences between the groups exposed to the test. There are no grounds for rejecting the null hypothesis. It can be assumed that the platform has no significant effect on the operation of the chatbot system. To sum up, commercial solutions do not always have to work better than free and open solutions. All platforms operate at a high level. If in the group of tested platforms there would be a platform that definitely deviates from the quality of operation to those selected above, the test could show the impact of the platform.

4.1   Impact of Corpus Change on Chatbot System Operation

There is one null hypothesis checking whether the body affects the operation of a chatbot system. Two corps of knowledge significantly different in the number of data, intentions or system objects were used for the study. The ratio of training data to testing was the same and was 80:20. The study was conducted on the Dialogflow platform. Based on the results the following conclusions have been reached. There are considerable differences in the results obtained within one corpus for individual intentions. In the Ubuntu corups, the F1 value fluctuates between 0 and 0.9296, and for the second corpus Banking takes the full value. The Ubuntu Corps has far fewer defined intentions than the Banking knowledge base. Each intention has more testing data, hence smaller discrepancies in F1 values. For several intentions of the Banking Corps, the value of F1 depended on only one or no test messages, so it could easily get a score of 0 or 1.

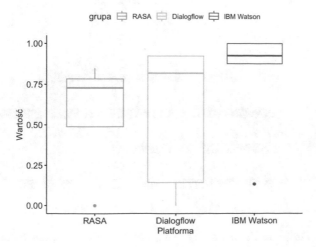

Fig. 2. A graph showing basic statistical data for each platform for the F1 valuea

To assess the validity of null hypothesis, which assumes that the corpus of knowledge affects the operation of a chatbot system, you need to perform statistical analysis. The performance quality is significantly different between the two corpuses. The Ubuntu knowledge base has more than twice the median value. In the case of the Banking corps, it happened that chatbot correctly recognized all messages belonging to one intention. Summing up the results of the survey and statistical analysis, it is stated that the knowledge corpus has an impact on the operation of the chatbot system. No specific influencing elements have been identified, but this will be further analyzed in subsequent studies.

4.2 Impact of System Objects on Chatbot System Operation

The use of personalized system objects affects the detection of intentions in messages from users. Null hypothesis assumes that system objects do not affect operation. Alternative hypothesis recognizes that system objects increase the quality of chatbot operation. Figure 3 aims to show how the value of F1 changed before and after applying system objects. In some cases the value of this coefficient increased, and in some cases it decreased. It was decided to look at the intention of the Setup Printer and the results obtained for the selected test data.

It is worth noting some important observations related to the results of research on the impact of the level of acceptance on the operation of the chatbot system resulting from the above tables. For intentions Software Recommendation, intent detection quality decreased by 16%. Testing data assumed the existence of two system objects for this intention, but the number of their occurrence in the data was far too small to increase the quality of operation. A model example of a system subject is the Make Update intention, for which the existence of only one object is specified - Version. This increased intent detection by 4%. For intentions Setup Printer, two system entities have been specified - Printer and Version. It is important that the subject Printer exists only for this intention,

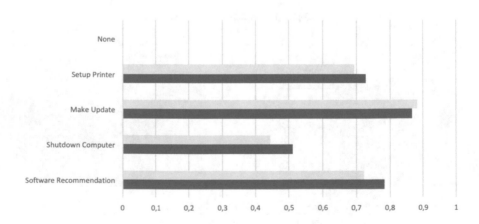

Fig. 3. Graph of the impact of using system objects on the value of F1

so the level of intention recognition increased by 11%. However, the number of cases where this should not be detected has increased.

4.3   Impact of Changing the Level of Intent Acceptance on Chatbot System Operation

There is a relationship between the level of acceptance and the operation of the system which is shown on Fig. 4. As the acceptance level increases, the quality of chatbot operation increases, i.e. the value of the F1 indicator, but after exceeding a certain level the F1 value does not improve or begins to decrease. This indicator begins to increase significantly for the level of acceptance in the range from 0.6 to 0.7, and the highest result is achieved just for the value of 0.7. The optimal value of the acceptance level varies depending on the intentions and ranges from 0.5 to 0.8. The average level of acceptance is 0.66 and the median is 0.7. The level of acceptance depends largely on the quality and amount of data for each individual intentions, hence the discrepancies.

It is worth noting a few important observations related to the results of studies on the impact of the level of acceptance on the operation of the chatbot system. A special case of the intention studied is the intention None, because it only had 4 training queries and several testers. This small amount of training and testing data, additionally very different from each other, was definitely not enough for the system to recognize this intention well. For intentions Setup Printer, the system regardless of the level of acceptance correctly recognized exactly 9 messages from the user, but the differences appeared in the wrongly recognized intention. More errors occurred when the level of acceptance was too low or too high. Very similar training and testing data was used for the Make Update intentions, which is why the system recognized this intention well. Summarizing the results, the best results are obtained for an acceptance level value of 0.7, and the worst for values close to the end of the acceptance level range. In addition, it is found that in many cases the value of the F1 coefficient

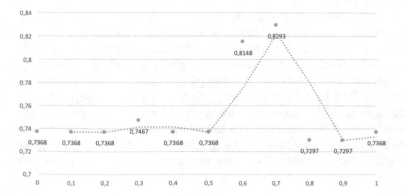

Fig. 4. Graph of dependence of F1 score value on the level of acceptance of Software Recommendation intentions

increases as the level of acceptance increases, and then decreases again when the value of the level of acceptance of intent approaches the value of 1.

4.4 Impact of Training and Testing Data on Chatbot System Operation

The data contained in the knowledge corpus are divided into training and test. The choice of the ratio of these data is very important and can be important in the operation of the system. The established null hypothesis assumes that the ratio of training data to testers has an impact on chatbot operation. Alternative hypothesis recognizes that regardless of the breakdown of data, chatbot works just as well.

Table 4. Table presenting statistical data for testing the effect of the ratio of training data to test data for the Banking corps

Training data	min	1. quantile	median	avg	3. quantile	max
40%	0.0000	0.5427	0.7500	0.6387	0.8286	1.0000
50%	0.0000	0.5714	0.7273	0.6321	0.8000	1.0000
60%	0.0000	0.2000	0.5714	0.5045	0.7333	1.0000
70%	0.0000	0.2000	0.5714	0.5045	0.7333	1.0000
80%	0.0000	0.0000	0.3333	0.3290	0.5834	1.0000
90%	0.0000	0.0000	0.0000	0.2862	0.6667	1.0000

Table 4 presents the results of tests to examine the effect of the ratio of training data to test data on the operation of the chatbot system. The ratio of these data was tested in the proportions from 40:60 to 90:10, but the table shows only the results when the training data is 40% and 80% of all data. All examined groups will be compared in the section of statistical analysis. The analysis was started by establishing basic statistical data. Analyzing the graph of the impact it can be seen that the higher the percentage of this data, the worse the quality of chatbot operation. The highest average and median were achieved when training data accounted for 40–50% of all data. To verify the null hypothesis, it was checked that the sample distribution was close to the normal distribution. The Shapiro-Wilk test performed for each group showed that the distributions are not significantly similar to the normal distribution. The higher the percentage of training data, the worse the results. This is due to the quality of data in the given corpus of knowledge. Increasing training data automatically reduces test data. For 90% of training data, only a few testing queries are left.

5   Conclusion and Future Work

The aim of the paper was to analyze the methods and techniques used in chatbots and to compare the operation of platforms available on the market for creating own bots. The research covered aspects such as - the quality of the corpus of knowledge and selection of optimal parameters during learning and testing of such a system.

In the research, two corps of knowledge were used, differing in the number of intentions, the occurrence of system objects, and the number of queries for each intention. Studies have clearly shown that the body parameters have a significant impact on the operation of the system. Experiments have shown a significant improvement in the quality of the corpus of knowledge with fewer intentions defined and a large amount of training data. The impact of using system objects on the quality of the dialogue system has not been demonstrated. The use of system objects does not apply to small bodies of knowledge with a small number of queries, because for some intentions it causes deterioration of their recognition. System objects work if they are repeatable and identifiable in messages from users. The second part of the research described in the work concerned the chatbot teaching process itself. The tested parameter is, among others level of acceptance of intention. Its range is between 0.0–1.0. The study included a full range with a value jump of 0.1. The test results are consistent with the similar test results available in the literature for other platforms. The quality of the system improved as the level of acceptance increased to a certain level, and then began to decline. The optimal value was set between 0.6 and 0.8. In the case of too high acceptance level, the system in most cases did not assign any intention to messages, while too low level caused that too many different intentions matched the message. Another parameter that was taken into account was the number of training queries for each intention. The influence of the ratio of training data to test on the dialog system was studied. Depending on the specificity of the corpus of knowledge, the optimal division is one in which as much as 80–90 % will be training data. Such a corpus must have well-defined intentions with a large number of repetitive queries. Studies have shown that for a knowledge base with a small amount of data, this breakdown may result in a lack of test data or a small number of them. In this case, the optimal split is 40% of training data and 60% of test data.

Possible directions of development of further research described in the work include increasing the number of parameters such as the number of intentions, the number of queries for each intention or the level of acceptance of intentions. Further research may take into account the parameters related to the full human dialogue with the system and detection of the conversation context. In addition, more knowledge bodies can be used for research, which will show even better diversity and demonstrate the dependence of their operation on their quality. Research can be carried out on more platforms for creating bots, thus detecting differences in the operation of individual tools. Another option for further research development is to study the efficiency of dialog systems depending on the hardware resources used.

References

1. Abu Shawar, B., Atwell, E.: Different measurement metrics to evaluate a chatbot system. In: Proceedings of the Workshop on Bridging the Gap: Academic and Industrial Research in Dialog Technologies, Rochester, NY, pp. 89–96. Association for Computational Linguistics, April 2007
2. Chandna, U., Iyer, M.R.: System for semi-automated chatbots query classification training corpus generation (2018)
3. Hristidis, V.: Chatbot technologies and challenges (2018)
4. Hussain, S., Ameri Sianaki, O., Ababneh, N.: A survey on conversational agents/chatbots classification and design techniques. In: Barolli, L., Takizawa, M., Xhafa, F., Enokido, T. (eds.) WAINA 2019. AISC, vol. 927. Springer, Cham (2019). https://doi.org/10.1007/978-3-030-15035-8_93
5. Cahn, J.: CHATBOT: architecture, design & development, April 2017
6. Weizenbaum, J.: ELIZA—a computer program for the study of natural language communication between man and machine. Commun. ACM **9**, 1, 36–45 (1966)
7. Jain, M., Kumar, P., Kota, R., Patel, S.N.: Evaluating and informing the design of chatbots. In: Proceedings of the 2018 Designing Interactive Systems Conference, New York, NY, USA, DIS 2018, pp. 895–906. Association for Computing Machinery (2018)
8. Larson, S., et al.: An evaluation dataset for intent classification and out-of-scope prediction. In: Proceedings of the 2019 Conference on Empirical Methods in Natural Language Processing and the 9th International Joint Conference on Natural Language Processing (EMNLP-IJCNLP), Hong Kong, China, pp. 1311–1316. Association for Computational Linguistics (2019)
9. Meteer, M., Hickey, M., Rothberg, C., Nahamoo, D., Eide, K.: Are the tools up to the task? An evaluation of commercial dialog tools in developing conversational enterprise-grade dialog systems. In: Proceedings of the 2019 Conference of the North American Chapter of the Association for Computational Linguistics: Human Language Technologies, Volume 2 (Industry Papers), Minneapolis, Minnesota, pp. 106–113. Association for Computational Linguistics (2019)
10. Lee, I., Shin, Y.J.: Machine learning for enterprises: applications, algorithm selection, and challenges. Bus. Horiz. **63**(2), 157–170 (2020)
11. Turing, A.M.: Computing machinery and intelligence. Mind **LIX**(236), 433–460 (1950)
12. Minh, P.Q.N.: Natural language processing problems in chatbot system development (2020). https://techinsight.com.vn/language/en/natural-language-processing-problems-chatbot-system-development/. Accessed 26 Mar 2020
13. Radziwill, N.M., Benton, M.C.: Evaluating quality of chatbots and intelligent conversational agents (2017)
14. Rahman, A.M., Al Mamun, A., Islam, A.: Programming challenges of chatbot: Current and future prospective (2017)
15. Wallace, R.S.: The anatomy of A.L.I.C.E.. In: Epstein, R., Roberts, G., Beber, G. (eds.) Parsing the Turing Test, pp. 181–210. Springer, Dordrecht (2009). https://doi.org/10.1007/978-1-4020-6710-5_13
16. Shum, H.-Y., He, X.-D., Li, D.: From Eliza to Xiaoice: challenges and opportunities with social chatbots. Front. Inf. Technol. Electron. Eng. **19**(1), 10–26 (2018)

17. Liu, X., Eshghi, A., Swietojanski, P., Rieser, V.: Benchmarking natural language understanding services for building conversational agents (2019)
18. WWW. What is a good value for the ML classification threshold in dialogflow? (2020). https://miningbusinessdata.com/good-value-ml-classification-threshold-dialogflow/. Accessed 30 Apr 2020

Using Simplified EEG-Based Brain Computer Interface and Decision Tree Classifier for Emotions Detection

Rafal Chalupnik[1], Katarzyna Bialas[1], Zofia Majewska[2], and Michal Kedziora[1(✉)]

[1] Wroclaw University of Science and Technology, Wrocław, Poland
{rafal.chalupnik,michal.kedziora}@pwr.edu.pl
[2] Clinical Department of Psychiatry and Combat Stress Treatment, 4th Military Clinical Hospital with the Polyclinic in Wroclaw, Wrocław, Poland

Abstract. The aim of the paper was to analyze the possibility to recognize human emotions by using a commercially applicable EEG interface and to check how many distinct emotions it is possible to distinguish. The samples were processed to apply to the classifier training. The AutoML software was used to build the decision tree classifier to check the output accuracy and its reliability. Then, we build the classifier for every possible combination. Every EEG band, without distinguishing the high/low frequency, was included in the training or excluded from it. The output of this research was used to determine which EEG bands are the most important in human emotion recognition from EEG data. The AutoML resulted in an accuracy of recognizing four distinct emotions equal to 99.80%. Later, AutoML experiments on each EEG band have shown that the most important specters are respectively, beta, alpha, and gamma, while delta and theta are the less important ones.

1 Introduction

There are numerous EEG solutions regarding reading human brain waves, some of which can be used to create a brain-computer interface [2]. Various authors have performed experiments which classify the brain waves to detect epilepsy or differentiate emotions, yet they were performed on traditional, full-size EEG interfaces [7, 24, 25].

As there are more and more solutions that are smaller and more convenient than the traditional EEG interface, such as NeuroSky MindWave Mobile 2, it is worth retrying some of these experiments to check the possibilities of smaller interfaces, for example, how well it can classify human emotions [19].

In the paper, we have planned and performed the experiment, in which we have trained the decision tree classifier to find the achievable accuracy of emotion recognition using a convenient and commercially available EEG brain-computer interface, along with forming conclusions from the experiment results.

L. Barolli et al. (Eds.): AINA 2022, LNNS 450, pp. 306–316, 2022.
https://doi.org/10.1007/978-3-030-99587-4_26

The remainder of this paper is structured as follows: first, the EEG is discussed. Then, the related works were analyzed to extract the research methodology and collect the achieved results. After that we described the methodology we used in our research, along with the experiment that was performed. Finally, we described the conclusions drawn from the results.

2   EEG

Electroencephalography is an electrophysiological method to monitor and record brain activity [5]. Commonly, it is a noninvasive examination assuming the presence of electrodes on the subject's head [21]. EEG itself is measuring voltage fluctuations being part of the neurons' work in the brain. EEG output can be divided into bands, which are defined by their frequency, location, and characteristic behavior. The frequencies of particular EEG bands are presented in Table 1. Five common bands can be distinguished:

Table 1. EEG bands and frequencies

Band	Frequency
Delta	Less 4 Hz
Theta	4–7 Hz
Alpha	8–15 Hz
Beta	16–31 Hz
Gamma	More 32 Hz

Delta waves appear especially in the deep sleep state, where are dominant. They can be used to detect e.g. Parkinson's disease, sleep disorders or diabetes [23], and insulin resistance. Theta waves are commonly observed in the front part of the brain. The activity can be noted during the idle states like hypnosis or light dream stage, also with inhibition of elicited responses (trying to suppress the reaction to some stimuli). These waves are reaching higher values in young children. Alpha waves are visible in the best way when there are no visual stimuli (e.g. the subject has closed eyes). They are also associated with the relaxation state and suppressed cognitive activity [15]. Beta waves are correlated with cognitive activity. They are noted to be especially available during the concentration state and can be used to determine whether the subject is being calm, intense, or stressed [16]. Gamma waves are displayed during cross-modal sensory processing (e.g., when the subject is exposed to the stimuli that are affecting at least two different senses, e.g., audiovisual stimulation) [9,11]. They are also displayed while using short-term memory (e.g. memory matching of objects, sounds, or sensations).

The full EEG interface comes with some disadvantages when considering them commercially [13], not in specialistic medical use [8]. Since the full interface requires dozens of the electrodes, which need to be placed all over subject heads and are difficult to wear without help, exploiting the EEG features in e.g. human-computer communication could be a potential impediment.

The MindWave Mobile 2 EEG interface [10] is built upon ASIC (application-specific integrated circuit) module, which are dedicated to the tailored solutions. This particular one, ThinkGear AM, consists of the sensor that is supposed to be placed right above the left eyebrow, and the ear clip acting as a reference allowing to filter out the noises generated by the subject's body or the environment (e.g. electrical devices, computers, etc.). It is especially significant, as EEG devices are sensitive to surroundings interference. The device is applying the set of transforming and rescaling operations on the original voltage measurements, which are volts-squared per Hz. Beside such measures, the chip is also providing ready-to-use values: Attention level, which indicates the focus of the subject; Meditation level, which indicates the subject being relaxed and Blink detection, which can be used in systems that require user interactions. Both Attention and Meditation levels are percentage-like measures, thus they are good candidates for e.g. threshold filtering.

3   Related Works

Detecting emotions using EEG was part of several research papers. In the paper [14] it was decided to use the emotion model which assumed division into two groups: positive and negative [6]. The group of subjects consisted of three women and three men around the age of 22. They have viewed 12 video clips of length around four minutes. The authors decided to use audiovisual stimuli, as they stimulate more than one sense of the subject. After watching each clip, the subject completed the SAM (self-assessment manikin) survey, where they defined emotions by checking the appropriate box. Subjects during watching video clips were wearing the EEG interface consisting of 62 electrodes evenly distributed all over the head. Data from the interface was divided by frequency into alfa, beta, gamma, delta, and theta channels. Recording of brain activity in such a scenario gives 310 characteristics in each sample (62 electrodes * 5 channels). Samples obtained this way samples were pre-analyzed. Results with a dominance value lower than 3 were rejected because it implies an insufficient stimulant effect on the subject. Samples were also visually examined for the presence of anomalies. A spectrogram was then generated using Fast Fourier transform. Later, a transform spectrogram was generated and the energy of individual features was determined. Then samples were smoothed by the Linear Dynamic System. Samples processed in this way were divided into 10 sections and applied to train the classifier with cross-validation in the ratio of training to test 7:3. Separated 5 characteristics group through dividing by channels (alfa, beta, gamma, delta, theta) and established examining each channel 5 classifiers SVM. One output vector consisting of these classifiers went to an additional classifier SVM at the exit. For every characteristic, the correlation coefficient was calculated and ordered from highest to lowest value. Next in sequence examined the effect of each characteristic with testing N characteristics with the highest value of correlation coefficiency. Besides the classification results, the authors have shown the low affection of low-frequency spectres (delta, theta) in emotion recognition. High-frequency channels (alpha, beta, gamma) were crucial in the classification process.

In the second paper, emotion recognition was done while listening to music. Chi et al. [12] decided to use a 2D emotion model consisting of two factors: arousal and valence of the emotion [17,20]. Every one of 26 subjects have listened to the music tracks. The EEG interface had 32 electrodes, distributed evenly through all the head surfaces. The recordings were initially analyzed visually to detect anomalies. Next, the STFT (Short-time Fourier transform) with a Hamming window was used to extract the power spectrum of the gathered signal. Processed samples were used to build the classifier. The authors have tested three approaches to this problem. First was one multi-class SVM classifier directly returning the predicted emotion. The second was the SVM classifier per each emotion and selecting the one with the highest score. The third was the tree of SVM classifiers recognizing valence on the first level, then arousal on the second one. The results of the experiments lead to the conclusion that the best approach to be used is to build the classifier for each emotion separately and then aggregate their outputs. The authors achieved an accuracy of 92.57%.

In the next paper, Danny Oude [18] decided to choose the EEG interface that would be the practical use of BCI (Brain-Computer Interface). Such an approach resulted in using EEG PET (ElectroEncephaloGraphy Personal Efficiency Trainer) device that has only five electrodes available. This device has a two-channel output (from two dipole electrodes and the ground). Danny also chose the 2D emotion model consisting of arousal and valence balances. The stimuli set was created with the usage of two libraries containing emotion-annotated images and sounds, respectively IAPS and IADS. The result was collected from the five subjects varied by gender and age, by using self-assessment manikins. The output accuracy of binary classification exceeded 90% for each tested feature. This result shows that even the EEG interface limited to 5 electrodes can predict the subject's emotions with acceptable reliability.

4   Data Gathering and Processing

As described in the earlier work [3], the data was collected from the study group with age varying between 15 and 45 years. Each subject has been presented with a set of four audiovisual stimuli, each meant to invoke particular emotion (anger, depression, relaxation and happiness), while having the EEG MindWave interface put on together with headphones to reduce the outside noise that could affect the measurements. Each stimuli had a length of about five minutes, having about one minute of break between each other, so the subject was able to clear the mind between the recording.

After the data was gathered, it was processed in order to built the classifier on top of it. The data feed from the MindWave EEG interface consists of eight measures:

- low alpha,
- high alpha,
- low beta,
- high beta,
- low gamma,
- high gamma,
- delta,
- theta.

To reduce the noise present in the data, the Simple Moving Average algorithm [4] was applied. For example, the signal presented in Fig. 1

$$\lambda = [4,9,6,5,2,1,3,10,8,7]$$

Fig. 1. Signal before applying Simple Moving Average algorithm

after applying the Simple Moving Average algorithm with the following parameters, would be smoothed as presented in Figs. 2 and 3

$$width = 4, step = 2$$

$$\lambda\prime = [\frac{\sum_1^4 \lambda_n}{4}, \frac{\sum_3^6 \lambda_n}{4}, \frac{\sum_5^8 \lambda_n}{4}, \frac{\sum_7^{10} \lambda_n}{4}]$$

Fig. 2. Signal after applying Simple Moving Average algorithm - formula

$$\lambda\prime = [6,3.5,4,7]$$

Fig. 3. Signal after applying Simple Moving Average algorithm - values

5　Methodology

The experiment was to try building the decision tree based classifier [22] for recognizing emotions. To build it, FastTreeOva model was used from the ML.NET library. Firstly, the training and test data were prepared. The cross-validation technique was applied, which split the data into 10 batches and evaluated the model for each batch acting as the test set, and the rest acting as the training set. Then, the FastTreeOva model has been built with ML.NET components [1] and trained to fit the training data. Next, the test data were loaded to the trained model and compared the output labels with the real ones. The output data was gathered into a confusion matrix, which was constructed as presented in Table 2. Each row represents the real emotion that was assigned to the samples, and each column represents the predicted emotions for all the samples. The summary confusion matrix, constructed by aggregating confusion matrix for each batch in cross-validation, is presented in Table 2.

Table 2. Summary confusion matrix for tested best classifier

Real/Predicted	Anger	Depression	Happiness	Relax	Recall
Anger	1326	0001	0002	0000	0.997743
Depression	0001	1280	0002	0000	0.997662
Happiness	0000	0000	1267	0001	0.999211
Relax	0002	0002	0000	1024	0.996109
Precision	0.997743	0.997662	0.996853	0.999024	

6   Experiments

After gathering data from the subjects and processing it, the AutoML has been to discover the best classifier it can construct. Feeding the program with collected and processed data and giving it 5 min of the possible runtime resulted in building the classifier reaching the accuracy of 99.80%. What's interesting, the built classifier has reached such a great accuracy without filtering gathered data by Attention. Taking into account that disabling the filter has not affected the experiment outcome noticeably, it is possible to conclude that hiding the felt emotions while being monitored with the EEG interface is a difficult task. However, discarding the Attention filter allowed to use of all gathered data samples in the training process (around 4900 samples).

$$\Omega_s = 2^s \tag{1}$$

$$\Omega - number\,of\,possibilities$$

$$s - number\,of\,spectres$$

$$Q_\Omega = \frac{\Omega_8}{\Omega_5} = \frac{2^8}{2^5} = 2^3 = 8 \tag{2}$$

$$Q_\Omega - reduction\,rate$$

The decision tree training has been additionally run on incomplete data, with each possible variants, to discover the influence of each EEG spectre on human emotion. To reduce the combinatorial explosion effect, the particular spectres high/low division has not been distinguished, so it helped to reduce the number of possibilities 8 times, while not losing any important data.

Running the AutoML on data with only 4 out of 5 spectres available seems to have not tremendously affected the accuracy - it didn't reach the values below 99%. The detailed results are shown in Table 3. It is noticeable that the variants without delta and theta spectres achieved the same accuracy, which is consistent with related works regarding alpha, beta and gamma spectres being the most important ones in emotion recognition (Table 5).

Table 3. Accuracy of the classifier with 4 out of 5 spectres, ordered descending by accuracy

Alpha	Beta	Gamma	Delta	Theta	Accuracy
Yes	Yes	Yes	Yes	No	99.77%
Yes	Yes	Yes	No	Yes	99.77%
Yes	Yes	No	Yes	Yes	99.60%
Yes	No	Yes	Yes	Yes	99.59%
No	Yes	Yes	Yes	Yes	99.38%

Running the AutoML on data with only 3 out of 5 spectres seems to have affected the accuracy more, especially in the variants of alpha, beta and gamma spectres absence - the accuracy felt down to nearly 97%. The detailed results are shown in Table 4.

Table 4. Accuracy of the classifier with 3 out of 5 spectres, ordered descending by accuracy

Alpha	Beta	Gamma	Delta	Theta	Accuracy
Yes	Yes	No	No	Yes	99.58%
Yes	Yes	No	Yes	No	99.58%
Yes	No	Yes	Yes	No	99.42%
Yes	Yes	Yes	No	No	99.41%
No	Yes	Yes	No	Yes	99.39%
No	Yes	Yes	Yes	No	99.02%
Yes	No	Yes	No	Yes	98.94%
No	Yes	No	Yes	Yes	98.63%
Yes	No	No	Yes	Yes	98.26%
No	No	Yes	Yes	Yes	97.83%

Table 5. Accuracy of the classifier with 2 out of 5 spectres, ordered descending by accuracy

Alpha	Beta	Gamma	Delta	Theta	Accuracy
Yes	Yes	No	No	No	99.01%
No	Yes	Yes	No	No	98.79%
Yes	No	Yes	No	No	98.58%
No	Yes	No	Yes	No	97.37%
Yes	No	No	No	Yes	96.65%
No	Yes	No	No	Yes	96.22%
Yes	No	No	Yes	No	96.06%
No	No	Yes	Yes	No	95.49%
No	No	Yes	No	Yes	95.11%
No	No	No	Yes	Yes	81.64%

Table 6. Accuracy of the classifier build per single spectrum, ordered descending by accuracy

Spectrum	Accuracy
Beta	90.04%
Alpha	85.95%
Gamma	84.64%
Delta	56.48%
Theta	48.8%

Training data set containing 2 out of 5 spectres seems to bring more insight into discovering the most important spectres, as the accuracy falls to about 80%. Once again, it seems that the absence of alpha, beta and gamma spectres significantly decreases the accuracy of the built classifier.

The hypothesis about the importance of alpha, beta and gamma spectres seems to be confirmed by running the AutoML per single spectrum, as they achieve higher accuracies than delta and theta spectres. The detailed results are shown in Fig. 4.

To designate the importance of each spectrum, the improvement ratio for each of the spectres has been calculated, which is the ratio of two values: the average accuracy of the classifiers of all variants tested before that contained examined spectre and the variants that did not.

$$Q_\lambda = \frac{\mu_\lambda}{\mu_{\neg\lambda}} = \frac{\frac{\sum^M \lambda_m}{M}}{\frac{\sum^N \neg\lambda_n}{N}} \tag{3}$$

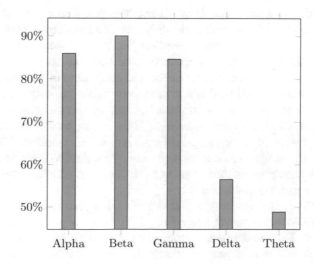

Fig. 4. Accuracy of classifier built per single spectrum

$$Q_\lambda : spectrum\, improvement\, ratio$$
$$\lambda_m : accuracy\, of\, the\, classifier\, containing\, the\, spectre$$
$$\lambda_n : accuracy\, of\, the\, classifier\, \textbf{not}\, containing\, the\, spectre$$

The improvement ratios calculated for each of the spectres according to the explained process can be viewed in Table 7.

Table 7. Average improvement per spectrum

Spectrum	Improvement
Alpha	1.0994
Beta	1.1076
Gamma	1.0917
Delta	1.0231
Theta	1.0109

There is a visible similarity between Table 6 and Table 7. One is visualizing the accuracy of the classifier built per single spectrum, and the other one presents the calculated improvement. While ordering gathered values from highest to lowest, we can conclude that the alpha, beta and gamma spectres are the ones most influenced by the human's emotion, while delta and theta are still influenced, however, significantly less. It's worth mentioning that such conclusions are consistent with the related works described in related work.

7   Conclusion and Future Work

The conducted experiment has shown that it is possible to use the NeuroSky MindWave Mobile 2 device, which has only one EEG electrode placed above the left eyebrow, to predict 4 distinct emotions on a two-dimensional model with an accuracy of 99.80%. Such an accuracy score can lead to the conclusion that the experiment was planned and executed correctly. One of the most important outcomes of this paper is that it is not necessary to use self-assessment manikins (SAM) to gather input from the subjects, as long as the emotion model is clean and universal through all human beings. Since it is possible to declare which stimuli would result in an appropriate quarter of the emotion 2D model, the feedback from the subjects is not required anymore. As the trained classifier did not distinguish the subjects during training, it is safe to conclude that mentioned four emotions are truly universal.

Another interesting outcome is that the filter designed to discard the recorded sample with an Attention level lower than 50% seems to not affect the accuracy of the classifier. That makes sense from the perspective of emotions and Attention level itself, as humans are not always focused on emotion while feeling it. Therefore, it is possible to conclude that it is achievable to correctly detect the emotions without applying any filtering based on focus or relaxation measures.

The last outcome from this research is the consistency of the conclusions drawn out of the collected data. Calculating the improvement ratio for each EEG spectrum and comparing it with the accuracy of the classifier build solely on this spectrum shown that there certainly is a noticeable difference in spectrum influence between two groups: alpha, beta, gamma and delta, theta. The difference between the calculated values can be observed, which is corresponding to the conclusion set by Nie [24], that the high-frequency EEG spectres (alpha, beta, gamma) have a higher influence on emotion recognition than the low-frequency ones (delta, theta).

In the future, it would be promising to explore the output model deeper, to discover the decisions made in the trained decision tree model. It could be possible to define the range of each EEG spectrum that is responsible for recognizing the particular emotion. Also, the resulting accuracy achieved in this research is highly encouraging to retry the experiment with more emotions placed onto the 2D model, and checking how high accuracy it is possible to achieve.

References

1. Ahmed, Z., et al.: Machine learning at microsoft with ml.net. In: Proceedings of the 25th ACM SIGKDD International Conference on Knowledge Discovery & Data Mining, KDD 2019, pp. 2448–2458. Association for Computing Machinery (2019)
2. Bialas, K., Kedziora, M.: Analiza mozliwosci sterowania aplikacja mobilna za pomoca interfejsu mozg-komputer. In: XII Ogolnokrajowa Konferencja Naukowa MLODZI NAUKOWCY W POLSCE_BADANIA I ROZWOJ, Jesien 2020
3. Chalupnik, R., Bialas, K., Kedziora, M., Jozwiak, I.: Acquiring and processing data using simplified EEG-based brain-computer interface for the purpose of detecting emotions. In: ACHI 2021: The Fourteenth International Conference on Advances in Computer-Human Interactions, pp. 97–103 (2021)
4. Clapham, C., Nicholson, J., Nicholson, J.: The Concise Oxford Dictionary of Mathematics. Oxford Paperback Reference, OUP Oxford (2014). https://books.google.pl/books?id=c69GBAAAQBAJ
5. Cohen, M.X.: Where does EEG come from and what does it mean? Trends Neurosci. **40**(4), 208–218 (2017)
6. Ekman, P., Cordaro, D.: What is meant by calling emotions basic. Emot. Rev. **3**(4), 364–370 (2011)
7. He, H., Tan, Y., Ying, J., Zhang, W.: Strengthen EEG-based emotion recognition using firefly integrated optimization algorithm. Appl. Soft Comput. **94**, 106426 (2020)
8. Herrmann, C.S., Strüber, D., Helfrich, R.F., Engel, A.K.: EEG oscillations: from correlation to causality. Int. J. Psychophysiol. **103**, 12–21 (2016)
9. Kanayama, N., Sato, A., Ohira, H.: Crossmodal effect with rubber hand illusion and gamma-band activity. Psychophysiology **44**(3), 392–402 (2007)
10. Katona, J., Farkas, I., Ujbanyi, T., Dukan, P., Kovari, A.: Evaluation of the NeuroSky MindFlex EEG headset brain waves data. In: 2014 IEEE 12th International Symposium on Applied Machine Intelligence and Informatics (SAMI). IEEE (2014). https://doi.org/10.1109/sami.2014.6822382

11. Kisley, M., Cornwell, Z.: Gamma and beta neural activity evoked during a sensory gating paradigm: effects of auditory, somatosensory and cross-modal stimulation. Clin. Neurophysiol. Off. J. Int. Fed. Clin. Neurophysiol. **117**, 2549–2563 (2006)
12. Lin, Y.P., Wang, C.H., Wu, T.L., Jeng, S.K., Chen, J.: EEG-based emotion recognition in music listening: a comparison of schemes for multiclass support vector machine, pp. 489–492 (2009)
13. NeuroSky: Mindwave mobile 2 available now improved comfort. http://neurosky.com/2018/06/mindwave-mobile-2-available-now-improved-comfort/. Accessed 14 Apr 2021
14. Nie, D., Wang, X., Shi, L.C., Lu, B.L.: EEG-based emotion recognition during watching movies, pp. 667–670 (2011). https://doi.org/10.1109/NER.2011.5910636
15. Niedermeyer, E.: Alpha rhythms as physiological and abnormal phenomena. Int. J. Psychophysiol. **26**(1), 31–49 (1997)
16. Niedermeyer, E.: Electroencephalography: Basic Principles, Clinical Applications, and Related Fields. Lippincott Williams & Wilkins, Philadelphia (2005)
17. Osgood, C., Suci, G., Tenenbaum, P.: The Measurement of Meaning. University of Illinois Press, Champaign (1957)
18. Plass-Oude Bos, D.: EEG-based emotion recognition. The Influence of Visual and Auditory Stimuli, January 2006
19. Saganowski, S., et al.: Emotion recognition using wearables: a systematic literature review-work-in-progress. In: 2020 IEEE International Conference on Pervasive Computing and Communications Workshops (PerCom Workshops), pp. 1–6. IEEE (2020)
20. Shiota, M.N.: Ekman's theory of basic emotions. In: The SAGE Encyclopedia of Theory in Psychology (2016)
21. Schachter, S.C., Schomer, D.L.: Atlas of Ambulatory EEG. Academic Press, Cambridge (2005)
22. Sullivan, W.: Machine Learning For Beginners: Algorithms, Decision Tree and Random Forest Introduction. CreateSpace Independent Publishing Platform, 20 August 2017
23. Budzynski, T.H., Evans, J.R., Abarbanel, A: Introduction to Quantitative EEG and Neurofeedback: Advanced Theory and Applications. Academic Press, Cambridge (2008)
24. Wei, C., Chen, L.l., Song, Z.Z., Lou, X.G., Li, D.D.: EEG-based emotion recognition using simple recurrent units network and ensemble learning. Biomed. Signal Process. Control **58**, 101756 (2020)
25. Yin, Y., Zheng, X., Hu, B., Zhang, Y., Cui, X.: EEG emotion recognition using fusion model of graph convolutional neural networks and LSTM. Appl. Soft Comput. **100**, 106954 (2021)

Anomaly Detection from Distributed Data Sources via Federated Learning

Florencia Cavallin[1] and Rudolf Mayer[1,2]([✉]) [iD]

[1] SBA Research, Vienna, Austria
rmayer@sba-research.org
[2] Vienna University of Technology, Vienna, Austria

Abstract. Anomaly detection is an important task to identify rare events such as fraud, intrusions, or medical diseases. However, it often needs to be applied on personal or otherwise sensitive data, e.g. business data. This gives rise to concerns regarding the protection of the sensitive data, especially if it is to be analysed by third parties, e.g. in collaborative settings, where data is collected by different entities, but shall be analysed together to benefit from more effective models.

Besides various approaches for e.g. data anonymisation, one approach for privacy-preserving data mining is Federated Learning – especially in settings where data is collected in several distributed locations. A common, global model is obtained by aggregating models trained locally on each data source, while the training data remains at the source. Therefore, data privacy and machine learning can coexist in a decentralised system. While Federated Learning has been studied for several machine learning settings, such as classification, it is still rather unexplored for anomaly detection tasks. As anomalies are rare, they are not picked up easily by a detection method, and the representation in the model dedicated to recognise them might be lost during model aggregation.

In this paper, we thus study anomaly detection task on two different benchmark datasets, in supervised, semi-supervised, and unsupervised settings. We federate Multi-Layer Perceptrons, Gaussian Mixture Models, and Isolation Forests, and compare them to a centralised approach.

Keywords: Federated Machine Learning · Anomaly detection

1 Introduction

Increasingly, organisations are collecting large volumes of data such as logs, product information, and personal information on clients or customers. The increasing demand for analysing and extracting anomalies, patterns, and possible correlations of these data spurred unprecedented interest in the analysis of this data, propelling some methods to higher effectiveness and efficiency. Alongside, the demand for data sharing and exchange between different parties holding data is increasing, often because different data sets complement each other. Collaborative analysis of data can be beneficial, e.g. to learn from misuse patterns such as fraud that other parties have been exposed to, or in the medical domain.

L. Barolli et al. (Eds.): AINA 2022, LNNS 450, pp. 317–328, 2022.
https://doi.org/10.1007/978-3-030-99587-4_27

However, especially when data contains personal information, regulatory, ethical, and security concerns can restrict the potential to fully leverage data, as distribution and exchange are limited. Thus, means to enable collaborative analysis are required. Federated Learning (FL) [1] is a collaborative learning approach that trains models locally across multiple nodes, which each hold their data. This data never leaves the node, and is thus not exposed to the network and possible attacks. The objective of FL is to obtain models with an effectiveness similar as if trained from centralised data. Anomaly detection is an important task in many domains and applications, and users can benefit from exchanging knowledge on their observed anomalies via collaborative learning. However, FL has not yet received much attention in FL research.

In this paper, we thus investigate whether anomaly detection from distributed data via FL can indeed achieve results comparable to a setting where data is centralised. Differences in how unsupervised, semi-supervised, and supervised learning algorithms are affected by federating are investigated. To this end, we analyse the performance of these approaches on two benchmark datasets, from the medical and fraud detection domain. We consider a setting where data is gathered by multiple organisations, and each has a sizeable number of data records. [2] calls this *cross-silo federated learning*, as each of these organisation operates its own data silo. Data is generated locally, and remains decentralised. Regarding the number of clients in cross-silo FL, [2] e.g. talks about 2–100 clients.

We investigate anomaly detection in tasks in health care for detecting diseases, and in identifying fraudulent behaviour in financial transactions such as credit card payments. These two application areas are prototypes for the importance of privacy and confidentiality of the data analysed, as both medical as financial data contain individual data, and are highly sensitive. Further, these two domains are often characterised by individual data silos collecting parts of the overall available data, and hurdles to exchange or centralise it – either due to regulatory, or also due to reservations for sharing sensitive business data. Thus, they are prime candidates for addressing this task in a federated learning setup.

The remainder of this paper is organised as follows. Section 2 discusses related work, before Sect. 3 describes the federated anomaly detection algorithms we use. Section 4 details the evaluation setup, and Sect. 5 then discusses the results. Finally, we provide conclusions and future work in Sect. 6.

2   Related Work

Anomaly detection is the process by which data points, events, and observations that differ from he normal behaviour within a dataset are identified [3]. Although researchers define an anomaly differently based on the application domain, one widely accepted definition is that of Hawkins [4]: 'An anomaly is an observation which deviates so much from other observations as to arouse suspicions that a different mechanism generated it.' Anomalies can point out significant, but rare, events such as technical malfunctions, accidents, or client behaviour changes.

Anomalies may be caused by variations in machine behaviour, fraudulent behaviour, mechanical defects, human error, instrument error and natural deviations in populations [5]. Anomalies in data lead to important actionable information in many application domains, making it a critical task. An unusual traffic pattern in a computer network, for example, could indicate that a hacked computer is transmitting confidential data to an unauthorised recipient. Anomaly detection is employed e.g. in cybersecurity intrusion detection, defect detection of safety-critical devices, health care, fraud detection, or robot behaviour [3].

Anomalies fall into three main categories [3]. A *point anomaly* is an individual data instance that is anomalous with respect to the rest of data. *Contextual anomalies* are data instances that are anomalous in a specific context (of other data, but not otherwise), while a *collective anomaly* denotes a collection of related data instances that are anomalous with respect to the entire data set. In this paper, we address point anomalies.

Anomaly detection can also be distinguished by the availability of labels in the training data [6]. If we consider two types of instances in the data, namely anomalies and normal (regular) data, then we have the following characteristics of training data. In a *supervised* task, we have labels for both anomalies and regular data; this is most often approached with supervised machine learning (classification). If labels are available only for the regular data, we deal with a *semi-supervised* task, where a model is learned for the normal class, and anomalies are those that deviate from that model. If there are no labels available at all, then we deal with a *unsupervised* tasks. In this work, we consider all cases.

The output of an anomaly detection method can be either directly a label (anomaly or normal data), or a score that measures the degree of anomaly, which is then normally compared to a threshold to arrive at a decision.

The privacy and confidentiality of the training data (resp. the individuals represented by it) has been recognised as an important aspect, and thus, privacy-preserving data mining (PPDM) methods are studied. In [7], PPDM techniques are classified into four main categories: (i) data collection privacy, which refers to data randomisation strategies, before they are sent to a data collector, (ii) Privacy-Preserving Data Publishing (PPDP), (iii) Data Mining Output Privacy (DMOP) and (iv) distributed privacy. PPDP often distorts the data, and includes techniques such as k-anonymity, or ϵ-differential privacy. Data synthetisation, which has been studied for various tasks (e.g. [8]), including anomaly detection [9], can also be seen as a form of PPDP. DMOP, on the other hand, which operates on original, unabridged data, relies on ensuring that the computation does not require the exchange of input data. *Federated Learning* can be considered a distributed DMOP method: FL allows to let data remain distributed at the site where it is created, e.g. on mobile devices, respectively where it is initially gathered. However, it still allows to learn a common model from these data, based on aggregating models learned by local training at each site [10]. The idea of local training is relevant for settings where data sharing brings various regulatory, privacy and technical issues, such as the medical domain, or also when sharing business data, e.g. in a collaborative fraud detection setting.

FL is increasingly used in several domains. In [11], the authors showed that federated learning on medical image data can reach a performance, similar as to when data is centralised before training. For structured data, [12] showed that FL is comparable to centralised learning in several settings. In [13], a framework for applying federated learning to biomedical data was presented. In [14] the authors considered Federated Learning for IoT, and optimised the Federated Averaging algorithm of [10] for Edge Computing.

Several forms of collaboration have been investigated in domains that rely on anomaly detection. Collaborative *intrusion detection systems* (CIDSs) [15] address limitations of conventional systems in terms of scalability and massively parallel attacks, CIDSs comprise several monitoring components in a hierarchical structure that collect and exchange data, to eliminate bottlenecks of a centralised approach. [15] identify *privacy* as one of the requirements of a successful collaborative approach – alerts and data exchanged may contain sensitive information that should not be shared. Exchanging only learned knowledge as e.g. in federated learning would be one approach to mitigate these risks.

Exchanging learned knowledge can also be performed by employing transfer learning and domain adaptation, which knowledge learned in one setting is exploited to improve generalisation in another setting [16]. It can leverage information from labelled examples in one domain to predict labels in another domain. This means that models that are useful for one organisation can be transferred to other, similar cases. Transfer learning shows promising results for several task, but for anomaly detection, an open research question is the degree of transferability [17]. The authors of [18] motivate transfer learning as candidate for detection of unknown attack types, and conclude that semi-supervised methods transfer better than supervised ones, but identified a need for improvement. Opposed to FL, transfer learning generally allows only a one-way transfer from a source to a target, and not collaborative learning.

3   Federated Anomaly Detection Algorithms

We use the following algorithms for federated anomaly detection: Multi-Layer Perceptron (for supervised anomaly detection), Gaussian Mixture Model (for semi-supervised anomaly detection) and Isolation Forest (for unsupervised anomaly detection). We describe these and their federated version below.

Multi-Layer Perceptrons (MLPs), a type of *Artificial Neural Network* (ANN), consist of several *neurons* that are arranged in *layers*. Each neuron computes an activation from its inputs and weights. MLPs are feed-forward, i.e. activations are only passed to the next layer, but not backwards. During training, the weights are updated (learned) iteratively, to minimise the error on the training set, by layer-wise back-propagating the gradient of the error and adapting the weights, e.g. via stochastic gradient descent (SGD). If an MLP contains at least two hidden layers, it is a *Deep Neural Network* (DNN) (though the term DNN can denote any ANN with more than one hidden layer, not just MLPs).

It is relatively straightforward to federate an MLP – the *FedAvg* algorithm [10] e.g. performs averaging of the locally trained models. First, a (global) model is initialised, i.e. the weights are randomly set. They are then sent to each client, where they train the model weights further, each with their local training dataset. Subsequently, the clients send their model parameters updates to the central aggregator. FedAvg combines the updates from the clients by averaging, and replaces the previously randomly initialised model with the new weights. This cycle is normally repeated several times, to allow the model to converge. The local training of the same, randomly initialised model at different clients followed by aggregation and averaging was shown to achieve substantially lower loss compared to independently training models on each subset of the data.

In our evaluation, a hyper-parameter optimisation showed that an MLP with two hidden layers achieves best results on the anomaly detection tasks.

Gaussian Mixture Models (GMM) represent a parametric probability density function as a weighted sum of Gaussian *component* densities [19]. They are commonly used for e.g. clustering purposes. GMM parameters (means, μ, and variances, σ^2, of the Gaussians) are estimated from training data using e.g. the Expectation-Maximisation (EM) algorithm by maximum likelihood estimation techniques that maximise the likelihood of a given data sample with the model parameters. Calculating the solution analytically can be mathematically impossible; expectation maximisation is an iterative algorithm and has the property that the maximum likelihood strictly increases with each subsequent iteration, i.e. it is guaranteed to approach a local maximum or saddle point.

Training mixture models does not require having class labels for the data points. A GMM can thus be used in an anomaly detection task when no anomaly cases are known, i.e. in a semi-supervised algorithm. The model then recognises patterns representative of the normal behaviour. When an anomaly sample is to be predicted, the model will likely not group it in any of the identified clusters, since the clusters were created from the normal samples.

We transfer GMMs to federated Gaussian Mixture Models for anomaly detection as follows. First, the parameters of the global model are randomly initialised for the number of desired Gaussian mixtures (components), and sent to the clients. At each client, the model is trained with the local data, either for a defined number of epochs, or until the certainty of each sample not being part of the assigned cluster is at most a given threshold δ. The averaging to the global model then consists of two steps - finding matching components from each local client, and eventually averaging their parameters.

Isolation Forests (IF) [20] are an unsupervised anomaly detection algorithm. It differs from other approaches, as it is based on *isolating* anomalies, instead of the more common approach of learning a representation of the normal samples. They are based on two assumptions. First, that anomalies are a minority, with very few samples within the dataset. Secondly, that anomalies are different – their values differ notably from normal samples. An Isolation Forest is an ensemble of Isolation Trees, which are binary trees arranging samples by attribute values.

While a Decision Tree, a supervised algorithm, splits data into subsets based on maximising a certain measure (e.g. Information Gain), an Isolation Tree splits based on a random value in the value range of a randomly selected attribute. The number of partitions required to isolate a point is calculated as the length of the path from the root to reach a leaf node. When the Isolation Tree construction is finished, each sample is isolated at a leaf node. Intuitively, anomaly samples are those with a shorter path length in the tree. Based on this, an anomaly score is computed for each instance, and if it is above a predefined threshold (e.g. 0.5), then the sample is labelled as anomaly. The Isolation Forest algorithm has a low linear time complexity, and can be trained with or without anomalies, and in an unsupervised manner [21].

The federated Isolation Forest is implemented as an ensemble of the locally trained Isolation Forests, in a similar manner as in [22].

4   Evaluation Setup

In this section, we describe the setup of our evaluation, including the datasets.

4.1   Datasets

We evaluated our federated anomaly detection algorithms on two benchmark datasets that are frequently used for anomaly detection in centralised settings. *Credit Card Fraud*[1] is a dataset that contains 284,807 credit card transactions made by European cardholders over two days in September 2013. Out of the 284,807 transactions, only 492 transactions (0,17%) are fraudulent, making the dataset heavily skewed. The dataset contains 30 features: the amount and time of the transactions, and 28 features obtained via a PCA on the original input data. There are no missing values. Most fraudulent transactions are very small expenditures, probably unnoticed to the cardholders, while the normal samples exhibit all possible values in the range. For preprocessing, the variable "amount" was scaled via a standard scaler to be in line with the other attributes.

Ann-Thyroid[2] is a medical dataset with 3,772 training and 3,428 testing samples, described by 15 categorical and six numerical attributes. There are three possible target values for each instance, namely *normal* (92.583% of the total samples), *hyperfunction* (5.111%) and *subnormal functioning* (2.306%). The hyper function and subnormal classes are treated as the anomaly classes. For supervised detection (with the MLP), where more than one anomaly class can be identified, they will be treated as two separate anomaly classes. For the Gaussian Mixture and Isolation Forest algorithms, these two anomaly classes are merged, in line with other related work. This could influence the effectiveness of the anomaly detection task, as subnormal and hyperthyroid anomalies may have different behaviours.

---

[1] https://www.kaggle.com/mlg-ulb/creditcardfraud.
[2] https://archive.ics.uci.edu/ml/datasets/thyroid+disease.

4.2   Evaluation Metrics

Anomaly detection can be evaluated by several different metrics. As data is normally heavily imbalanced towards the normal class, measures like accuracy (the number of correct predictions) are not sufficient – as already predicting everything to be of the normal class would score very high. Thus, frequently the so-called F-score, a combination of the precision and recall, is employed. *Precision* denotes the ratio of the number of true positive samples (anomalies identified as such) divided by the number of false positives (normal samples wrongly predicted as anomalies). *Recall* is the ratio of samples classified as positive among the total number of positive samples – i.e. how many of the anomalies have been identified. The *F1 score* then provides a combined score by computing the harmonic mean of precision and recall. On the other hand, the *F2 score* weights recall higher than precision. This makes it suitable for the datasets considered in our evaluation, where identifying most of the anomalies is critical, while a certain amount of false positives can be tolerated. However, the actual preference for F1 or F2 (or other F-scores) depends on the exact application scenario, and how many false positive cases can be tolerated and handled, respectively how critical not identified anomalies are, and needs to be determined by domain experts.

Another measure frequently employed is the *area under curve* (AUC) of the receiver operating characteristics (ROC), which is based on the true positive rate (TPR) and false positive rate (FPR), and indicates if we picked randomly a "normal" and anomaly sample, the anomaly example one will have a higher anomaly score, with a probability that corresponds to the AUC. A perfect model will have its AUC equal to 1, while a poor model will have its AUC score around 0. If the AUC is 0.5, it means that the model has no class separation capacity at all.

4.3   Data Distribution

For federated learning, we test different numbers of clients, namely from two to ten with a step size of one, and then 10, 15, 20, 25, 30. In Sect. 5, due to space limitations we mostly report results for a medium amount of clients, 15, and the largest configuration with 30 clients.

The data is randomly split among the clients to achieve a distributed setting. During our experiments, we use a holdout method to split the data into training and test set in a 90:10 ratio.

5   Results

In this section, we present and discuss our experimental results. We compare the federated results to an idealised, centralised baseline, i.e. where a model can be trained on all data. This represents a glass-ceiling for the federated learning, and is a very difficult baseline to achieve. We thus argue that achieving this glass-ceiling baseline is not necessarily mandated to deem the federated detection as

Table 1. Anomaly detection scores on the credit card fraud dataset

	MLP			GMM			IF		
	Pr	Re	F2	Pr	Re	F2	Pr	Re	F2
Centralized	87.5	85.7	86.1	36.6	39.8	39.1	13.3	4.1	4.7
FL-15 clients	85.4	71.4	73.8	40.7	52.4	49.6	0.9	16.3	3.8
FL-30 clients	78.6	44.9	49.1	78.6	44.9	49.1	0.0	0.0	0.0

successful, as in many real-world settings, gathering all data in a centralised manner will not be possible.

While we present results from supervised, semi-supervised and unsupervised anomaly detection on the same datasets and with the same algorithms, it has to be noted that these approaches are only partially comparable to each other, and each rather constitute a separate task. This is due to the fact that supervised detection has much more information available when learning the model than the other two approaches (labels for both classes), and is thus an easier task. Unsupervised has the least information available, and is thus the hardest task.

5.1 Credit Card Fraud Dataset

Table 1 shows the scores for the anomaly detection algorithms on the credit card fraud dataset, depicting precision, recall and the F2 score. We can observe that anomaly detection on this dataset is difficult already on the original, centralised setting. The easiest task is the supervised setting, which we address with the Multi-Layer Perceptron (MLP), and which consequently achieves the best scores; its precision, recall and F2 values are all in a very similar range, namely 87.5%, 85.7% and 86.1%, respectively. The Gaussian Mixture Model (GMM), which we employ to solve the semi-supervised task, also has all of its scores within a similar range, albeit lower than the MLP, with values of 36.6% for precision, 39.8% for recall, and 39.1% for F2. Isolation Forests (IF) are used to solve the hardest task, i.e. the unsupervised setting. In line with that difficulty, it scores low on the recall, and thus also achieves a low F2 score. The AUC scores for the centralised setting are shown in Fig. 1a, indicating a similar trend.

When comparing the centralised to the federated learning setting, we can notice that the different methods for the tasks are affected to a varying degree. With an increasing number of clients, the MLP loses mostly on recall. Figure 1b shows that also the AUC score drops in the federated setting. This can indicate that the averaging mechanism of FedAvg is not capable of completely preserving the parts of the individual MLPs that learned to represent the anomalies, if the number of clients increases too much, respectively, if there are very few anomaly instances at each client. Strategies to improve this could be e.g. in boosting the weights representing anomalies, similar to the strategy of boosting weights of malicious nodes in the *model replacement* strategy in a federated backdoor attack [23].

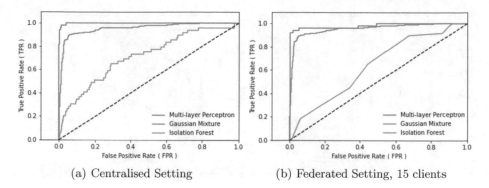

(a) Centralised Setting (b) Federated Setting, 15 clients

Fig. 1. ROC curves for the credit card dataset

Table 2. Anomaly detection scores on thyroid data set

	MLP			GMM			IF		
	Pr	Re	F2	Pr	Re	F2	Pr	Re	F2
Centralized	95.7	83.0	85.3	54.6	37.8	40.3	10.4	18.9	16.2
FL-15 clients	95.6	81.1	83.7	64.9	62.9	63.3	9.8	14.8	13.4
FL-30 clients	95.5	79.2	82.0	12.5	1.9	2.2	12.5	1.9	2.2

For GMMs, we can however notice that the F2 score actually increased in the federated setting, while the AUC score stays roughly the same (cf. Fig. 1b). The cause for this effect is not systematic – sometimes it is due to a higher recall, but other times due to a higher precision.

For Isolation Forests, the initial trend is similar as for the MLP – a larger number of clients in the federation leads to a drop in effectiveness of the detection, for all scores. A too large number of 30 clients, and thus many Isolation Trees used in the Forest, leads to the anomaly detection not properly working anymore, and no anomalies being detected. This is likely due to the fact that if only few clients have data with anomalies, only a few trees representing these anomalies are created, and they are subsequently outvoted by the many trees representing the normal cases.

5.2 Ann-Thyroid Dataset

Table 2 shows the precision, recall and F2 scores for the anomaly detection algorithms on the thyroid dataset. While the overall best scores for the MLP and GMM on this dataset are comparable to the credit card fraud dataset (with a difference in F2 scores of $\pm 1\%$), the Isolation Forests, albeit on a still low score, performs significantly better on the thyroid dataset. We can observe that for MLP and GMM, precision is significantly better than recall, in the range of 10–15%. With Isolation Forests, that observation is inverse. The AUC scores are depicted in Fig. 2a, and indicate a similar trend.

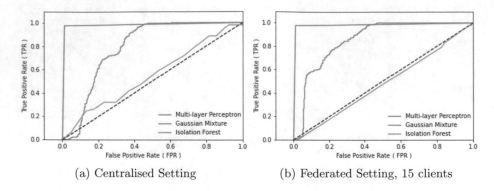

(a) Centralised Setting   (b) Federated Setting, 15 clients

Fig. 2. ROC curve for Thyroid dataset

When comparing the scores from the centralised to the federated setting, Table 2 and Fig. 2b indicate that the trend for the MLP is, albeit dropping, rather stable, i.e. there is a decrease of around 1.5% each when going from centralised to federated learning with 15 clients, and then increasing that number to 30 clients. The drop is mostly due to a drop in recall, as precision stays almost the same for all these settings, just dropping by 0.1% each.

For GMM and Isolation Forests, we can notice that with 30 clients the aggregation into a single model fails – the F2 scores drop to an unusable value, especially due to a low recall. For 15 clients, GMM is however delivering useful results, even better than in the centralised benchmark, mainly due to an increased recall. For Isolation Forests with 15 federated clients, results drop by around 3% from the centralised baseline, but are still significantly better than as for 30 clients. As with the credit card dataset, when too many clients each have only a few anomaly instances, paired with the more difficult semi- and unsupervised tasks, it seem to be a challenge to preserve the knowledge learned on the anomalies during the aggregation process.

6   Conclusions and Future Work

In this paper, we investigated anomaly detection in tasks on medical and financial datasets. We evaluated the performance of three algorithms in central as well as a federated settings, for three settings of availability of labels in the training data – supervised, semi-supervised, and unsupervised. While we observe that especially the supervised method (an MLP) translated very well into the federated setup, the other two methods manged to match their centralise baseline only in some of the settings, especially with a smaller number of federated clients.

We can identify several strands for future work. On the one hand, the adaption of the centralised algorithms into federated versions can be improved, especially for semi- and unsupervised approaches. For Isolation Forest, approaches that e.g. select only the most relevant subset of Isolation Trees might lead to an improvement. Also, further anomaly detection algorithms will be considered.

Further, the centralised baseline where a model is trained on all data is an idealised baseline, and in fact represents a glass-ceiling for the federated learning. It is also an unlikely comparison, as in a real setting, centralising this data is not possible. Another comparison would be to evaluate each locally trained model against the (global) test set; this will simulate how well the anomaly detection works if every client works in isolation, not collaborative, and thus represents a lower bound. To arrive at a realistic judgement on the value of federated anomaly detection, it should be evaluated whether it is well-positioned between these two bounds, and provides a clear advantage over within-silo training.

As studies have shown that unbalanced (in terms of the size of each silo's dataset) and not independent and identically distributed data (non-iid data) can lead to slower convergence, increased communication costs, or lower effectiveness in federated learning [24], we will investigate these for anomaly detection.

Finally, another important aspect to investigate is the vulnerability of federated learning towards inference attacks. Studies such as [25] have shown that federated learning is still vulnerable to e.g. membership inference, and might even open up novel attack vectors. [26] show that outliers might be specifically vulnerable to these attacks, and thus protecting the anomalies is of importance.

Acknowledgments. This work was partially funded from the European Union's Horizon 2020 research and innovation programme under grant agreement No 826078 (Project 'FeatureCloud'). This publication reflects only the authors' view and the European Commission is not responsible for any use that may be made of the information it contains. SBA Research (SBA-K1) is a COMET Centre within the COMET – Competence Centers for Excellent Technologies Programme and funded by BMK, BMDW, and the federal state of Vienna. The COMET Programme is managed by FFG.

References

1. Konečný, J., McMahan, H.B., Yu, F.X., Richtarik, P., Suresh, A.T., Bacon, D.: Federated learning: strategies for improving communication efficiency. In: Workshop on Private Multi-party Machine Learning, Conference on Neural Information Processing Systems (NIPS) (2016)
2. Kairouz, P., McMahan, H.B., et al.: Advances and open problems in federated learning. Found. Trends Mach. Learn. **14**(1–2), 1–210 (2021)
3. Chandola, V., Banerjee, A., Kumar, V.: Anomaly detection: a survey. ACM Comput. Surv. **41**(3), 1–58 (2009)
4. Hawkins, D.M.: Identification of Outliers. Springer, Dordrecht (1980). https://doi.org/10.1007/978-94-015-3994-4
5. Hodge, V., Austin, J.: A survey of outlier detection methodologies. Arti. Intel. Rev. **22**(2), 85–126 (2004). https://doi.org/10.1023/B:AIRE.0000045502.10941.a9
6. Goldstein, M., Uchida, S.: A comparative evaluation of unsupervised anomaly detection algorithms for multivariate data. PLOS ONE **11**(4), e0152 (2016)
7. Mendes, R., Vilela, J.P.: Privacy-preserving data mining: methods, metrics, and applications. IEEE Access **5**, 10562–10582 (2017). https://doi.org/10.1109/ACCESS.2017.2706947
8. Hittmeir, M., Ekelhart, A., Mayer, R.: On the utility of synthetic data: an empirical evaluation on machine learning tasks. In: 2019 International Conference on Availability, Reliability and Security, Canterbury, UK. ACM (2019)

9. Mayer, R., Hittmeir, M., Ekelhart, A.: Privacy-preserving anomaly detection using synthetic data. In: Singhal, A., Vaidya, J. (eds.) DBSec 2020. LNCS, vol. 12122, pp. 195–207. Springer, Cham (2020). https://doi.org/10.1007/978-3-030-49669-2_11

10. McMahan, B., Moore, E., Ramage, D., et al.: Communication-efficient learning of deep networks from decentralized data. In: 2017 International Conference on Artificial Intelligence and Statistics, Fort Lauderdale, FL, USA. PMLR (2017)

11. Sheller, M.J., Reina, G.A., Edwards, B., Martin, J., Bakas, S.: Multi-institutional deep learning modeling without sharing patient data: a feasibility study on brain tumor segmentation. In: Crimi, A., Bakas, S., Kuijf, H., Keyvan, F., Reyes, M., van Walsum, T. (eds.) BrainLes 2018. LNCS, vol. 11383, pp. 92–104. Springer, Cham (2019). https://doi.org/10.1007/978-3-030-11723-8_9

12. Pustozerova, A., Rauber, A., Mayer, R.: Training effective neural networks on structured data with federated learning. In: Barolli, L., Woungang, I., Enokido, T. (eds.) AINA 2021. LNNS, vol. 226, pp. 394–406. Springer, Cham (2021). https://doi.org/10.1007/978-3-030-75075-6_32

13. Silva, S., et al.: Federated learning in distributed medical databases: meta-analysis of large-scale subcortical brain data. Technical report, Inria & Université Cote d'Azur, France (2018)

14. Mills, J., Hu, J., Min, G.: Communication-efficient federated learning for wireless edge intelligence in IoT. IEEE IoT J. **7**(7), 5986–5994 (2020)

15. Vasilomanolakis, E., Karuppayah, S., Mühlhäuser, M., Fischer, M.: Taxonomy and survey of collaborative intrusion detection. ACM Comput. Surv. **47**(4), 1–33 (2015)

16. Weiss, K., Khoshgoftaar, T.M., Wang, D.D.: A survey of transfer learning. J. Big Data **3**(1), 1–40 (2016). https://doi.org/10.1186/s40537-016-0043-6

17. Chalapathy, R., Chawla, S.: Deep learning for anomaly detection: a survey. arXiv arXiv:1901.03407 (2019)

18. Chen, C., Gong, Y., Tian, Y.: Semi-supervised learning methods for network intrusion detection. In: International Conference on Systems, Man and Cybernetics, October 2008. IEEE (2008)

19. Reynolds, D.: Gaussian mixture models. In: Li, S.Z., Jain, A. (eds.) Encyclopedia of Biometrics. Springer, Boston (2009). https://doi.org/10.1007/978-0-387-73003-5_196

20. Liu, F.T., Ting, K.M., Zhou, Z.-H.: Isolation forest. In: 2008 8th IEEE International Conference on Data Mining, Pisa, Italy. IEEE (2008)

21. Liu, F.T., Ting, K.M., Zhou, Z.-H.: Isolation-based anomaly detection. ACM Trans. Knowl. Disc. Data **6**(1), 1–39 (2012)

22. Liu, Y., et al.: Federated forest. IEEE Trans. Big Data (2020). https://doi.org/10.1109/TBDATA.2020.2992755

23. Bagdasaryan, E., Veit, A., Hua, Y., Estrin, D., Shmatikov, V.: How To backdoor federated learning. In: 23rd International Conference on Artificial Intelligence and Statistics (AISTATS), Palermo, Italy. PMLR (2020)

24. Sattler, F., Wiedemann, S., Muller, K.-R., Samek, W.: Robust and communication-efficient federated learning from non-i.i.d. data. IEEE Trans. Neural Netw. Learn. Syst. **31**(9), 3400–3413 (2020). https://doi.org/10.1109/TNNLS.2019.2944481

25. Pustozerova, A., Mayer, R.: Information leaks in federated learning. In: Proceedings 2020 Workshop on Decentralized IoT Systems and Security, San Diego, CA. Internet Society (2020)

26. Choquette-Choo, C.A., Tramer, F., Carlini, N., Papernot, N.: Label-only membership inference attacks. In: International Conference on Machine Learning (PMLR) (2021)

On Predicting COVID-19 Fatality Ratio Based on Regression Using Machine Learning Model

Md. Mafijul Islam Bhuiyan[1]([⊠]), Mondar Maruf Moin Ahmed[2], Anik Alvi[3],
Md. Safiqul Islam[4], Prasenjit Mondal[5], Md Akbar Hossain[6],
and S. N. M. Azizul Hoque[7]

[1] Department of Computational Physics, University of Alberta,
Edmonton, AB, Canada
mbhuiyan@ualberta.ca

[2] Department of Life Sciences, Independent University Bangladesh (IUB),
Dhaka, Bangladesh

[3] Department of Computer Science, New Mexico State University,
Las Cruces, NM, USA
aalvi@nmsu.edu

[4] Department of Medicine, Bikrampur Bhuiyan Medical College and Hospital,
Munshiganj, Bangladesh

[5] Department of Public Health, North South University, Dhaka, Bangladesh

[6] School of Business and Digital Technologies, Manukau Institute of Technology,
Auckland, New Zealand
akbar.hossain@manukau.ac.nz

[7] Memorial University of Newfoundland, Corner Brook, NL A2H 5G4, Canada
ahoque@grenfell.mun.ca

Abstract. The world has been in the grips of the Coronavirus Disease-19 (COVID-19) pandemic for almost two years since December 2019. Since then the virus has infected over a hundred and fifty million and has resulted in over three million deaths. However, fatality rates have been observed to be drastically different in different countries. One reason could be the emergence of variants with differing virulence. Other factors such as demographic, health parameters, nutrition levels, and health care quality and access as well as environmental factors may contribute to the difference in fatality rates. To investigate the level of contributions of these different factors on mortality rates, we proposed a regression model using deep neural network to analyze health, nutrition, demographic, and environmental parameters during the COVID-19 lockdown period. We have used this model as it can address multivariate prediction problems with higher accuracy. The model has proved very useful in making associations and predictions with low Mean Absolute Error (MAE).

© The Author(s), under exclusive license to Springer Nature Switzerland AG 2022
L. Barolli et al. (Eds.): AINA 2022, LNNS 450, pp. 329–338, 2022.
https://doi.org/10.1007/978-3-030-99587-4_28

1   Introduction

COVID-19 pandemic has affected our daily lives, as well as the global economy for almost last two years. In this short period the virus has infected almost a hundred and fifty million and has resulted in over three million deaths [1] and tested the health systems of developed, developing, and under-developed countries and has been extremely disruptive to society and the economy. Though the virus may cause asymptomatic or mild respiratory tract infections resembling influenza, the serious manifestation of disease is characterized by high fever, lung injury, hypoxia, multiorgan failure and death. Some of the other symptoms such as diarrhea, cardiac and kidney issues also manifest suggesting that the disease may be systemic [2]. The pandemic has been one of the biggest challenges faced by mankind in the 21st century and scientists, governments and health systems have been struggling to find respite and have placed their hopes in raising compliance to health guidelines regarding social distancing, mask use and disinfection of surfaces [3]. Global vaccination efforts are also underway, however, there are lingering questions about efficacy and the evolution of new variants may pose further challenges [3].

The causative agent of COVID-19 is a severe acute respiratory syndrome-coronavirus-2 (SARS-CoV-2) or the Novel Coronavirus which is an enveloped positive-sense single-stranded RNA virus. As an RNA virus it must infect a host cell, in this case, human, and infiltrate the cell machinery to make copies of itself [2]. Mutations or changes in its genome during this process give rise to the variants and can even spawn new strains, which are variants, with distinct characteristics [4]. The ancestral Wuhan strain has undergone mutations world-wide and currently, there are many variants. Some of the most virulent include the UK, South African, Indian, and the Brazilian variants and the situations in these UK, South Africa, India, Brazil and the US, are much worse than in other countries like Bangladesh [1]. The different virulence of the variants may be an important reason behind the differing impact of COVID-19 globally [5]. This is apparently the case in Bangladesh. Significant mutations have also occurred in the virus in Bangladesh but 90% have resulted in reduced virulence and may explain the slow spread and low death rate in the country in 2020. The spread of the more virulent UK, and the South African variants may explain the 2021 spike in rates [6].

However, virulence is probably not the only factor since the same variant has showed different spreading and mortality rates in different countries [7]. Health system preparedness, access, nutritional factors, health parameters such as the levels of vitamins (e.g., Vitamins D and B12), blood groups, anemia, Zinc status, age factor, and climatic and environmental conditions such as precipitation and humidity and the level of fine pollutant particulate matter ($PM2.5$) may affect the spread and mortality rates seen worldwide [8]. Several data-driven mathematical modelling with multiple disease transmission parameters have been used to understand the dynamics of COVID-19 outbreak in an heterogeneous region [9]. However, the spread of COVID-19 may depend on the mutated strains [10], healthcare and travel restriction policy with other unspecified parameters [11]. Wang and Wong trained and implemented a Convolutional Neural Network based model to identify the number of COVID-19 patients by using Chest X-Ray images [12]. Pal et al. implemented an Long Short Term Memory (LSTM) model to forecast the spread of COVID-19 from weather data [13]. Tang et al. used the random forest algorithm for understanding the severity of COVID-19 patients by using the Computed Tomography (CT) Scans [14]. Using geographical, travel, health, and demographic data, a fine-tuned Random Forest model enhanced by the Ada-Boost algorithm was used to predict the severity of COVID-19 cases, possible recovery and death [15].

The primary contribution of this research is to develop a regression model using deep neural network to predict the Case Fatality Ratio (CFR) considering all the features related to COVID-19 from diversified locations. We also determined the significance of these features through analyzing the large amount of data collected worldwide during the pandemic which may help to prioritize different measures and to reduce mortality in future.

2   Data Collection and Preprocessing

In this research, we have collected the dataset from diverse sources to train a machine learning model. The dataset consists of 22 input features which have contribution at different ranges on the case spread and fatality ratios. The input features chosen include meteorological (e.g., temperature, humidity, precipitation etc.), health facilities (e.g., health access and quality), pollution (e.g., NO_2, $PM2.5$), demographic (e.g., Gender, percentage of aged population etc.), and geographical (e.g., different locations of the world). The output parameter that we have considered in this paper is the Case Fatality Ratio (CFR). CFR is computed based on the following Eq. (1):

$$CFR = \frac{Cumulative\ number\ of\ deaths}{Cumulative\ number\ of\ cases} \tag{1}$$

Data collection strategy is very important to reduce biasness during the training of the machine learning model. Based on the trend of the CFR until now, we have observed that the most vulnerable are the 65+ aged population [16]. Therefore, we assigned more importance on the percentage of the 65+ aged population while collecting our data. We have chosen 57 locations which include both cities and countries in such a way so that there is an equal number of locations with a similar proportion of 65+ aged population. Figure 1 depicts the distribution of population density of over 65+ aged people. We categorized the numerical aged population data into six different categories to be fed into the machine learning model while training. The categories as follows:

Category 1 (1–5)%: Dubai, Islamabad.
Category 2 (5–10)%: Sasolburg, Delhi, Dhaka, Kuala Lumpur, Ho Chi Minh City, San Agustin Aguascalientes, South Jakarta, Istanbul, Nur-Sultan, San Jose, Sao Paulo, Colombo.

Category 3 (10–15)%: Central Singapore, Beijing, Bangkok, Tel Aviv, Edmonton,
    Los Angeles City, Oklahoma City, Atlanta, New Orleans City, Tallahassee,
    Portland, New York City, Melbourne, Kumanovo, Las Vegas.

Category 4 (15–20)%: Seoul, Vancouver, Tempe Maricopa Phoenix, Auckland,
    Taipei, St. John's, Toronto, Dnipropetrovs'k, Kollarova Trnava, Oslo, Lan-
    caster, Central Hong Kong, London, Madrid, Warsaw, Brussels, Austin, Ams-
    terdam, Vienna, Copenhagen.

Category 5 (20–25)%: Valjevo, Litomerice, Paris, Budapest, Milan Lombardi,
    Helsinki, Thessaloniki.

Category 6 (25–30)%: Tokyo.

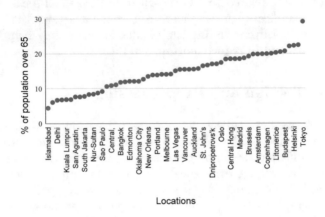

Fig. 1. Distribution of 65 and over aged population at different locations.

We also ensured diversity as much as possible regarding the meteorological,
health facilities, and pollution factors to reduce the degrees of freedom so that the
trained machine learning model can learn the most. For each geographical location, we collected data for different input parameters which include Population
density of 65 or older, vitamin A and vitamin D deficiency, health care quality
and access index, pollutant PM2.5 particle, anemia, proportion of population
of different blood groups (i.e., AB+, AB−, A+, A−, B+, B−, O+, O−), solar
radiation, daily temperature, precipitation, overall population density, average
wind speed, relative humidity, stringency index, and UV index.

There were some missing data points in our training dataset for different
input parameters. To fill in the missing points, in some cases, we have estimated
the missing values based on expert opinion, and in other cases we have computed
the values based on the 4-degree nearest neighbor technique.

3 Methodology

To perform regression analysis, in this paper, we have adopted deep neural network architecture. In the training stage, the goal was to find the optimal weights

($\boldsymbol{w}$) of the neural network to minimize the loss function, $Loss(y,F(w,x))$, where $\boldsymbol{x}$ represents all the input parameters, $\boldsymbol{y}$ represents the vector of ground truth CFR values, and $\hat{F}$ is the trained model. Equation (2) shows the mathematical formulation of the cost function:

$$\hat{F} = \underset{\mathbf{w}}{\text{argmin}} \ \ Loss(y, \hat{F}(w, x)) \tag{2}$$

We have used Root Mean Square Error (RMSE) as the loss function. Equation (3) shows the mathematical expression of the loss function.

$$Loss(y, \hat{F}(w, x)) = \sqrt{\sum_{i=1}^{N} \left| \hat{F}(x_i) - (y_i) \right|^2} \tag{3}$$

To train the regression neural network model, we randomly split that dataset into two parts: training (75%) and testing (25%) datasets. We used the grid search cross-validation technique to determine the optimum values of the different parameters of the neural network. These parameters include activation function, optimizer, sizes of the hidden layers, learning rate, and regularization parameter. For the grid search technique, we used the 5-fold cross validation [17].

Based on the result of grid search, we have adapted a learning rate (0.012575), regularization parameter (0.0001), sizes of the hidden layers (22, 20, 10, 5), activation function (ReLU), and ADAM (Adaptive Moment Estimation) optimizer for the regression model. Various tuning parameters were also used in the regression neural network model.

To train the regression model, we have used the Rectified Linear (ReLU) activation function [18]. This activation function is used to determine the outcome of each single node for the input fed into it. Mathematically, ReLU activation function, f(x), can be expressed as follows (Eq. 4),

$$f(x) = \begin{cases} x & if \ x > 0 \\ 0 & if \ x \leq 0 \end{cases} \tag{4}$$

It maps the resulting values in between 0 to 1 or -1 to 1. We have found that our input training dataset has a non-linear relationship with the corresponding output features. ReLU activation function is widely used to train nonlinear training datasets. Therefore, we adapted an activation function which has an on-linear property. The ReLU activation function provides direct output to the input if it is positive, otherwise, it gives an output of zero [18]. The advantages of using ReLU are manifold such as it does not have the vanishing gradient problem, it permits the training model to converge faster and perform better, and last but not the least, it can compute very fast thus reducing overall training time (Fig. 2).

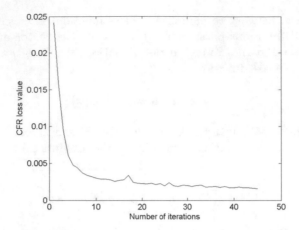

Fig. 2. Convergence of the trained model.

We used ADAM as an optimizer for our regression model. Adaptive moment estimation is an algorithm for optimization technique for Gradient Descent. This method is efficient when working with large problem involving a lot of data and input features. It requires less memory and hence, it is very efficient. Adam is generally used as a replacement of the stochastic gradient descent deep learning algorithm for training deep learning models [18].

It basically combines all the good properties of the Adaptive Gradient Descent (AdaGrad) and Root Mean Square Propagation (RMSP) algorithms to excel the accuracy rate of the training even for sparse gradients on noisy datasets.

We used this solver (i.e., ADAM) as it is very easy to configure though, in most cases, the default configuration parameters perform very well. After choosing all the parameters of the regression model, we have trained the model based on our training dataset. The model converged after 45 iterations and the Mean Absolute Error is 0.0350. Figure 3 depicts the convergence of the training.

4   Feature Analysis

Cross-correlation analysis is performed to measure the strengths of the linear relationship among the independent parameters as shown in Fig. 4. In this paper, we have noticed 'strong positive correlation (i.e., 0.85)' between the blood group A-negative and O-negative, denoted as BG_Aneg and BG_Oneg in the Fig. 5. Therefore, to reduce the degrees of freedom and to improve the training performance of the machine learning model, we have considered only BG_Aneg rather than considering both parameters. Note that strong negative correlation (−0.93) exists between Anemia and HAQ index (denoted as AnemiaData and Haq_Ind in the Fig. 5). This indicates that these two variables are highly inversely correlated.

To perform feature selection, Extra Trees Classifier, an ensemble method (i.e., Random forest) is used. Gini importance indices of each input parameters related to COVID-19 fatality ratio are compared in Fig. 4. In general, environmental parameters, such as, solar radiation, PM 2.5, daily temperature, UV index are less influential parameters, albeit HAQ index, wind and relative humidity may be associated with the CFR. This study indicates that the percentage of population above 65 years, population density, blood groups (except blood group AB negative), and various health parameters (anemia and Vitamin A) are the most influential parameters related to CFR. Based on this outcome, we provided more importance in these parameters with higher Gini index to avoid biasness to any random input features.

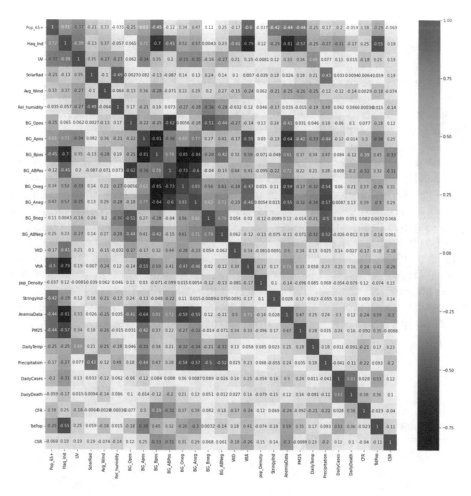

Fig. 3. Cross Correlation matrix of independent feature variables used to train the model.

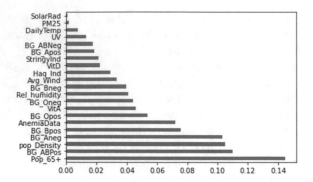

Fig. 4. Feature importance of classification analysis.

5   Results

After training the model, we tested the trained model using the test datasets of 57 different locations. In this paper, we have shown inferred results of a few locations. Figure 5 depicts the training and inferred results for British Columbia, Canada (L) and Japan (R). The black and red '+' signs represent the actual training and testing data respectively whereas the black and red solid circles represent the predicted trained and testing dataset. In both scenarios we can see that the predicted training and tested data are quite aligned with the actual training and testing dataset. It is important to mention here that Japan and British Columbia, Canada datasets have completely different meteorological, demographical, and health access and quality factors. However, it is noticeable that our trained model can predict them very well even for diverse input datasets. Similarly, Fig. 6 shows the training and testing result for Arizona (L) and Kazakhstan (R) respectively. These two locations have similarities in meteorological data, however, all the other input features are quite different. One can see that the prediction results for these two locations are also very accurate. The overall mean absolute error (MAE) of the trained model for the testing dataset is 0.0377.

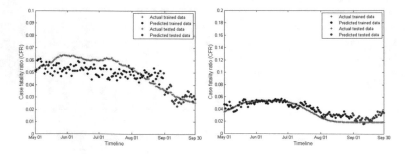

Fig. 5. Actual and predicted results for training and test datasets of BC, Canada (L), and Japan (R). (Color figure online)

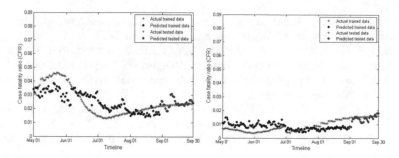

Fig. 6. Actual and predicted results for training and test datasets of Arizona (L), and Kazakhstan (R).

6   Conclusions and Future Work

In this paper, we investigated several parameters in different categories such as health, nutrition, pollution, demographic, and environmental to analyze and predict COVID-19 mortality rate using machine learning model. There are missing data points in different countries and cities due to inadequate resources. We used different data imputation techniques to overcome this impediment and successfully trained our model. We also investigate the importance of different input features using ensemble method (i.e., Random forest) and this helps us to train the model more efficiently. The associations and predictions made by the model have been revealing and may be useful to improve prediction quality and facilitate the policy maker to take proper measurements in advance.

In the future, we will focus on analyzing the impact of the new variants on the CFR due to COVID-19. We will also collect additional data for another 12 months from more diverse places. As CFR has been significantly changed due to a new parameter - rate of partial and full vaccination, we shall include vaccination data to enhance the predictive power of our regression model.

Data Availability. The readers can find the details of the code, training dataset, and inference results at https://github.com/sohelmsc/DataScience.

References

1. Worldometer, 25 April 2021. https://www.worldometers.info/. Accessed 25 Apr 2021
2. Synowiec, A., Szczepański, A., Barreto-Duran, E., Lie, L.K., Pyrc, K.: Severe acute respiratory syndrome Coronavirus 2 (SARS-CoV-2): a systemic infection. Am. Soc. Microbiol. J. **34**, e00133-20 (2021). https://cmr.asm.org/content/34/2/e00133-20
3. World Health Organization, 25 April 2021. https://www.who.int/emergencies/diseases/novel-coronavirus-2019/advice-for-public. Accessed 25 Apr 2021
4. Virology, 25 April 2021. https://www.virology.ws/2021/02/25/understanding-virus-isolates-variants-strains-and-more/. Accessed 25 Apr 2021

5. Council on Foreign Relations, 25 April 2021. https://www.cfr.org/in-brief/how-dangerous-are-new-covid-19-strains. Accessed 25 Apr 2021
6. Tithila, K.K., Antara, N.F.: Experts: South African coronavirus variant more dangerous, unpredictable, Dhaka Tribune, Bangladesh, 8 April 2021. https://www.dhakatribune.com/bangladesh/2021/04/08/experts-south-african-coronavirus-variant-more-dangerous-unpredictable
7. Li, H., Liu, S.-M., Yu, X.-H., Tang, S.-L., Tang, C.-K.: Coronavirus disease 2019 (COVID-19): current status and future perspectives. Int. J. Antimicrob. Agents **55**(5), 105951 (2020). ISSN 0924-8579. https://doi.org/10.1016/j.ijantimicag.2020.105951
8. Alvi, A., Ahmed, M., Hoque, S.N.M.A.: Consequences of lockdown caused by COVID-19 outbreak on the quality of air in Dhaka. In: 2021 International Conference on Automation, Control and Mechatronics for Industry 4.0 (ACMI), pp. 1–6 (2021). https://doi.org/10.1109/ACMI53878.2021.9528097
9. Kermack, W.O., McKendrick, A.G.: Contributions to the mathematical theory of epidemics. II.—The problem of endemicity. Proc. R. Soc. Lond. Ser. A (Math. Phys. Charact.) **138**(834), 55–83 (1932)
10. Wang, Y., Wang, Y., Chen, Y., Qin, Q.: Unique epidemiological and clinical features of the emerging 2019 novel coronavirus pneumonia (COVID-19) implicate special control measures. J. Med. Virol. **92**(6), 568–576 (2020)
11. Chinazzi, M., Davis, J.T., Ajelli, M., Gioannini, C., Litvinova, M., Merler, S., et al.: The effect of travel restrictions on the spread of the 2019 novel coronavirus (COVID-19) outbreak. Science **368**(6489), 395–400 (2020)
12. Wang, L., Wong, A.: COVID-Net: a tailored deep convolutional neural network design for detection of COVID-19 cases from chest radiography images. arXiv arXiv:2003.09871 (2020). Accessed 5 May 2020
13. Pal, R., Sekh, A.A., Kar, S., Prasad, D.K.: Neural network-based country wise risk prediction of COVID-19. arXiv arXiv:2004.00959 (2020). Accessed 7 May 2020
14. Tang, Z., Zhao, W., Xie, X., Zhong, Z., Shi, F., Liu, J., et al.: Severity assessment of coronavirus disease 2019 (COVID-19) using quantitative features from chest CT images. arXiv arXiv:2003.11988 (2020). Accessed 10 May 2020
15. Iwendi, C., et al.: COVID-19 patient health prediction using boosted random forest algorithm. Front. Public Health **8**, 357 (2020). https://doi.org/10.3389/fpubh.2020.00357
16. Kang, S.J., Jung, S.I.: Age-related morbidity and mortality among patients with COVID-19. Infect. Chemother. **52**(2), 154–164 (2020). PMID: 32537961
17. Kingma, D.P., Ba, J: Adam: a method for stochastic optimization. In: 3rd International Conference on Learning Representations (ICLR), 22 December 2014 (2015)
18. Nair, V., Hinton, G.E.: Rectified linear units improve restricted Boltzmann machines. In: Proceedings of the 27th International Conference on International Conference on Machine Learning, pp. 807–814 (2010)

Distributed Training from Multi-sourced Data

Ibrahim Dahaoui[(⊠)], Mohamed Mosbah, and Akka Zemmari

CNRS, LaBRI, UMR 5800, University of Bordeaux, Bordeaux INP,
F-33400 Talence, France
ibrahim.dahaoui@u-bordeaux.fr, {Mosbah,zemamri}@labri.fr

Abstract. A distributed system is a set of logical or physical units capable of performing calculations and communicating with each other. Nowadays, these systems are at the heart of technologies such as the Internet of Things IoT, the Internet of Vehicles IoV, etc. These systems collect data, perform calculations and make decisions. On the other hand, deep learning (DL) has led to enormous progress in the field of artificial intelligence. Since the precision of DL to form a set of reference data on a single machine is known, it becomes more interesting to form several models and distribute the intelligence over the different nodes of the system by different calculation strategies.

In this paper we propose a method using deep neural networks on several machines by distributing the dataset before starting the training, ensuring communication between them in order to improve the calculation time and accuracy.

1 Introduction

Deep Learning (DL) has proven itself in many practical applications. Breakthrough performance has been previously indicated in several domains, ranging from speech recognition [1,2], to visual object recognition [3], or to text processing [4,5].

It has been observed that the scalability of DL in terms of the number of training examples, the number of model parameters or both, can significantly increase the accuracy of the resulting classification. These results have led to considerable success in scaling up the learning and inference algorithms used for these models [7] and in improving the applicable optimisation procedures [8]. A known limitation of the GPUs [3,6,7] approach is that the learning speed is low when the model does not fit in the memory of the GPU. To make effective use of a GPU, researchers often decrease the size of the data or parameters so that transfers between the CPU and GPU are not a major bottleneck. We live in an exciting time where machine learning is a key part of the learning process that has a profound influence on a wide range of applications from text and image understanding and speech recognition to healthcare and genomics. For example, DL techniques are as well known as imaging techniques for identifying

L. Barolli et al. (Eds.): AINA 2022, LNNS 450, pp. 339–347, 2022.
https://doi.org/10.1007/978-3-030-99587-4_29

diseases in diabetics. Many of the recent successes are due to better computational infrastructure and large amounts of training data. Among the many challenges of machine learning, distributing data on various nodes is becoming one of the critical subjects.

There are several forms of computation to improve the accuracy performance of DL, the volume of input data has an impact on the performance of training, the subject of our research is to create strategies using several parameters such as number of machines, distributed data volumes, communication, to have a better performance of DL.

The paper is divided into five sections, the Sect. 1 reviews existing optimization training algorithms and the Sect. 2 focuses on network configuration using docker. The Sect. 3 explores the developed techniques and architecture, data size training and mixed precision training for effective training with multiple Docker containers. The Sects. 4 and 5 are future work and conclusion.

1.1   Motivation

In recent years commercial and academic machine learning data sets have grown at an unprecedented pace. Data is being generated at an unprecedented scale, internet scale companies generate terabytes of data every day which needs to be analyzed effectively to draw meaningful insights. Machine Learning, and in particular DL is rapidly taking over a variety of aspects in our daily lives. As a result, a wide range of authors have investigated the scaling of machine learning algorithms through parallelization and distribution [12–14]. DL has enjoyed considerable success in the recent past, leading to cutting-edge results in various fields such as image recognition and natural language processing. One of the reasons for this success is the increasing size of DL models and the proliferation of vast amounts of training data available. In order to continue improving DL performance, it is important to increase the scalability and communication of DL systems.

In this area, we are conducting a broad and deep investigation into the issues, techniques, and tools of scalable DL on distributed infrastructures and collections, all from different nodes. These include network infrastructures for data distribution, DL collection and training, communication methods between different nodes involved in DL, resource planning and management [18].

1.2   Previous Work

The distribution of neural network training can be approached in two ways, data parallelism and model parallelism. Data parallelism seeks to divide the dataset among the nodes in the system, where each node has a copy of the neural network with its local weights. Each node runs on a unique subset of the dataset and updates its local set of weights, it is this distribution part that we will use to realize our proposed configuration which consists in collecting the distributed data on the different nodes.

Other interesting research on model parallelism then focuses on proper training by separating model engineering into several distinct clusters. Model parallelism is used when the model engineering is too large to be installed on a solitary machine and the model has a few sections that can be paralleled. Model parallelism is used with some models, for example, object detection arrays that have separate bounce and wait heads and no class [17].

Our goal is to distribute data among several servers, establish communication between them via sockets, and then drive these servers to make many decisions to efficiently manage their particular workloads in the chosen environment. The goal is to set up a parameterized framework that allows us to calculate the learning time and accuracy by different algorithms that we will explain later to have a better result.

In the context of DL, most work has focused on training relatively small models on a single machine (e.g., Theano [10]). Suggestions for scaling DL include using an array of GPUs to train a collection of many small models and then averaging their predictions [11], or modifying standard deep networks to make them inherently more parallelizable.

Other work compares deep distributed learning, such as the Pytorch package [16] which is a tensor library optimized for DL using GPUs and CPUs, provides parallel distributed data as a class, where applications provide their model at build time as a submodule. Several techniques are built into the design to provide high-performance training, including compartment gradients, overlapping communication with computation, and synchronization hopping. This solution can be used later for each type of data if we want to have a good training performance. Our method is to distribute the data to a set of nodes.

2 Experiments

In our current proposal, we use docker containers that are configured under a specific network, and then the machines assigned as servers collect data from the neighboring nodes (clients) and run the training.

- **Data Distribution**: In order to distribute the data, we split it into subsets and each node have a part of this data. Before establishing communication between the nodes, we must first make a network configuration of each of them, for this we use tools that exist on docker such as *docker-composes* [20]. The purpose of this part is to ensure data transfers during the collection of these data by specific nodes.
- **Communications**: During this part we are going to work on data transfer (images, text...) so for that we have established a network configuration between a number of containers on dockers. For the communication between the users and the containers, we used a *flask* Api [19].
- **Strategies**: The comparison of the different results with respect to server training time and accuracy is based on a voting strategy between the different server predictions. For example, the prediction value y of a data item x as

input to the three of an aggregation models is the values returned by more than one model. All data is checked with this strategy and compared to the ground truth to determine accuracy.

By using multiple servers when training for distributed data, the training time will be reduced compared to the training time of a single server. And this is something we need to measure on our experiment. We took a basic configuration on docker containers, each has RAM: 1 GB. We have configured the network for this container system on the 172.25.3.0/24 subnet. We need to use the *flask* Api to ensure that the results are transferred to the user's host.

2.1 Distributed Data

After being inspired by some work related to our topic, data distribution based on the number of docker containers has been created to start practicing on its own model. To create them, we used for this version the *docker-compose* which is a tool to define and run multi-container docker applications. With *docker-compose* we can assign to each container an image which is a file, composed of several layers, that is used to execute code in a docker container. An image is essentially built from the instructions of a full executable version of an application, which is based on the kernel of the host operating system. When the user runs a *docker-compose*, each container will execute its code and the data will be distributed to all containers, the distribution will be uniform or non-uniform.

The challenge after this version, is to achieve a good distribution of the data to have a better accuracy of the results and a better time after their formation, because even if we are currently working on docker containers, we will later generalize this distribution to several types of high performance computing nodes.

2.2 Collection and Training

We created ten containers on docker using *docker-compose*, then we distribute the data in two forms (uniform and non-uniform) to the different nodes. Each machine has a training program, so on this program we used a neural network composed of three layers using respectively *sigmoid* activation function, *ReLU* and *softmax* for the last layer. The data used for this version are the handwriting data from *Mnist*. The handwritten digit database consists of a training set of 60,000 examples and a test set of 10,000 examples. This is a subset of a larger set available from Mnist [15]. The figures were normalized in size and centered in a fixed size image. The choice of this data type is made to get hold of a well-processed data type beforehand for a first release. After training, each server will have a model, then all these models will be processed by our majority voting strategy, comparing the *softmax* values, and finally comparing them to the ground truth in order to calculate the accuracy performance of each of them.

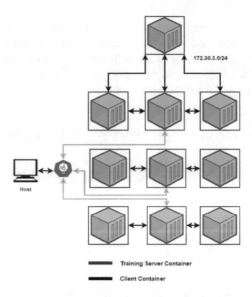

Fig. 1. Distributed training and collection: the three servers boxed in red will collect data from the subset of their neighbors boxed in black, and then begin training. Once complete, our host servers will apply the strategies we developed to compute the accuracy. Communication between the servers and the neighbors is done via sockets, and with the host via the *Api flask* (Color figure online)

2.3 Strategies

This is the main part of the topic. Our goal is to have multi-sourced data on the connected machines, and to work on the created models (three models in our case). This work consists in developing strategies to know the right configuration, necessary to have a good accuracy result like the number of server machines and the form of data distribution (uniform/non-uniform) and with/without replacement. For example, in our case we have a total of ten nodes, three are considered as training and distributed data collection servers, the rest send the sub-data to the neighboring training server. For uniform distribution (with/without replacement) each server will have 6000 data, 60000 training data on 10 servers. For this, we applied strategies for each form of configuration.

2.3.1 Majoritory Vote After having the three models of our learning servers, we apply our developed strategy on each input data and which consists in choosing the majority value between the three results. For example, for an image given as input, each of the models of these three returns a list of ten probability values (because we have ten possible numbers from 0–9), with the *argmax* function of the library in python that returns the index value of the largest value of *softmax* (activation function that returns this probability value).

After that we will have three values *digits* of each data in input, if the first, and the third server return us for example the number **1**, and the second the number **2** thus by the majority vote our program will consider that for this data is the number **1**. In the case where the three values are different we choose the one with the highest probability. The majority vote values of the input data set are stored in a list for comparison with the ground truth values provided by the database in order to calculate the accuracy.

Table 1. Obtain the output of each model for every single input data

Input dataset	Server 01	Server 02	Server 03
Image for number "1"	Number "1"	Number "7"	Number "1"
Image for number "2"	Number "8"	Number "2"	Number "6"
Image for number "3"	Number "9"	Number "3"	Number "3"
Image for number "4"	Number "4"	Number "4"	Number "7"

Table 2. Choosing the majority value among the three results

Input dataset	Server 01
Image for number "1"	Number "1"
Image for number "2"	Number "8" or "2" or "5"
Image for number "3"	Number "9"
Image for number "4"	Number "4"

2.3.2   Comparison Between *softmax* Values For this second strategy, we will compare the probability values returned by the activation function *softmax*. As the first strategy and for each input data the three models return a list of probabilities this program selects the largest value for each list of these three models, then once we have the three vectors of probabilities *P1*, *P2* and *P3*, we compare the greatest value for example *P3* with the sum of the two remaining values, if this value is smaller than the sum of the other two values *(P1+P2)* then we will choose the index of the value of the greatest probability between the two values *agrmax(Max (P1, P2))*. Then we store the results of this comparison between these three values in a list to compare them with the truth in order to calculate the precision value.

Table 3. Obtain the Softmax value of each model for each input data

Input dataset	Server 01	Server 02	Server 03
Image for number "1"	0.22	0.12	0.40
Image for number "2"	0.50	0.27	0.42
Image for number "3"	0.90	0.23	0.55
Image for number "4"	0.012	0.36	0.66

3   Results

We compare the different strategies for which we were able to obtain accuracy results for the two forms of uniform or non-uniform distribution, with or without data replacement during distribution. We can tell from these results that the best configuration and strategies for this data is the uniform distribution without replacement. We also change the volume of shared data in the case of distribution with replacement to test the accuracy. We noticed that if we increase this value, the accuracy improves.

We tested another configuration always in the context of change on parameters, we tested the same strategies with five servers connected to ten machines (containers). What we noticed is that if the number of servers increases, we lose a little bit in accuracy, and this comes back to the fact that the more data we have during the training, the more accuracy we gain, at the moment we gained more since the data are not big and they are shared.

3.1   Time and Limitation

After testing these different strategies, we measured the training time as we went along. The training time for all input data is longer than the training time for multiple servers for distributed data. On the other hand, if we use a powerful computing system, it will be faster because we use a docker and it uses the same machine's kernel, so all operations are done in one processor. We tested the distribution of data with the training on several servers to see the performance of the accuracy and the time taken by this operation, and also to have the best configuration (number of servers and their configuration...).

For this kind of data we noticed that the accuracy performance started to decrease from four servers, and we have a decrease in accuracy until the fifth server so we can say that for a better accuracy with our used strategies we should use only three servers (Docker containers).

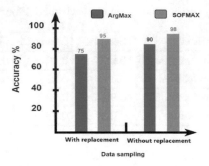

Fig. 2. Uniform dataset

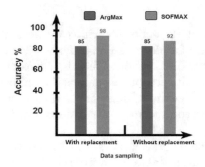

Fig. 3. Non-uniform dataset

4   Next Step

After this release, we plan to work with big data and implement and test as many different communication configurations between servers as possible, and use existing cloud technologies to use new workers, and then we will compare the different results, in order to arrive at an efficient solution and compatible application to use in several projects in the future. Then optimize the data collection so that each node is specialized after training on specific data, and that this optimization takes into account the quantity and quality of the input data. The goal is to have well trained and specialized servers to have a good decision result and to decrease the power and time consumed in a network. After that we are going to build a web application that will take a set of data, then we are going to parameterize it in order to have the different possible results.

5   Conclusion

Our experiments demonstrate the ability of our new data distribution and train-ing methods to use a group of machines to train even modestly sized deep net-works much faster than a single machine, but in terms of accuracy, we could have the same accuracy. To demonstrate the value of being able to train larger models and data, we need to try a model with more than a billion parameters to achieve better performance than the state of the art on the ImageNet object recognition challenge and other.

References

1. Dahl, G., Yu, D., Deng, L., Acero, A.: Context-dependent pre-trained deep neural networks for large vocabulary speech recognition. IEEE Trans. Audio Speech Lang. Process. **20**, 30–42 (2012)
2. Hinton, G., et al.: Deep neural networks for acoustic modeling in speech recognition. IEEE Sig. Process. Mag. **29**, 82–97 (2012)
3. Ciresan, D.C., Meier, U., Gambardella, L.M., Schmidhuber, J.: Deep big simple neural nets excel on handwritten digit recognition. CoRR (2010)
4. Coates, A., Lee, H., Ng, A.Y.: An analysis of single-layer networks in unsupervised feature learning. In: AISTATS 2011 (2011)
5. Bengio, Y., Ducharme, R., Vincent, P., Jauvin, C.: A neural probabilistic language model. J. Mach. Learn. Res. **3**, 1137–1155 (2003)
6. Collobert, R., Weston, J.: A unified architecture for natural language processing: deep neural networks with multitask learning. In: ICML 2008 (2008)
7. Le, Q.V., Ngiam, J., Coates, A., Lahiri, A., Prochnow, B., Ng, A.Y.: On optimization methods for deep learning. In: ICML 2011 (2011)
8. Raina, R., Madhavan, A., Ng, A.Y.: On optimization methods for deep learning. In: ICML 2009 (2009)
9. Martens, J.: Deep learning via hessian-free optimization. In: ICML 2010 (2010)
10. Bergstra, J., et al.: Theano : a CPU and GPU math expression compiler. In: SciPy 2010 (2010)
11. Ciresan, D., Meier, U., Schmidhuber, J.: Multi-column deep neural networks for image classification. Technical report. In: IDSIA 2012 (2012)
12. Shi, Q., et al.: Hash kernels. In: AISTATS 2009 (2009)
13. Langford, J., Smola, A., Zinkevich, M.: Slow learners are fast. In: NIPS 2009 (2009)
14. Mann, G., McDonald, R., Mohri, M., Silberman, N., Walker, D.: Efficient large-scale distributed training of conditional maximum entropy models. In: NIPS 2009 (2009)
15. LeCun, Y., Cortes, C.: The MNIST database of handwritten digits. In: NIPS 1998 (1998). yann.lecun.com/exdb/mnist/
16. Li, S., et al.: PyTorch distributed: experiences on accelerating data parallel trainings. In: VLDB 2020 (2020)
17. He, M.L.Z., Rahayu, W., Xue, Y.: Distributed training of deep learning models: a taxonomic perspective. IEEE Trans. Parallel Distrib. Syst. **31**(12), 2802–2818 (2020)
18. Mayer, R., Jacobsen, H.-A.: Distributed training of deep learning models: scalable deep learning on distributed infrastructures: challenges, techniques and tools. ACM Comput. Surv. **53**, 1–37 (2019)
19. Grinberg, M.: Flask Web Development: Developing Web Applications with Python. O'Reilly Media, Inc. (2018)
20. Merkel, D.: Docker: lightweight Linux containers for consistent development and deployment. Linux J. **2014**(239), 2 (2014)

Viterbi Algorithm and HMM Implementation to Multicriteria Data-Driven Decision Support Model for Optimization of Medical Service Quality Selection

Jolanta Mizera-Pietraszko[1]([X]) [iD] and Jolanta Tancula[2] [iD]

[1] Military University of Land Forces, 51-147 Wroclaw, Poland
jolanta.mizera-pietraszko@awl.edu.pl
[2] Opole University, 45-040 Opole, Poland

Abstract. Medical service quality is one of the major factors upon which Quality of Life Index is calculated. In this paper, we propose a Multicriteria Data-Driven Decision Support (MDDDS) model for a patient who explores medical database to make a decision on selecting the best medical professional specializing in his or her disease. Optimization of the decision relies on the data-driven collection about the particular disease's stage and suitability to the efficiency of the medicament recommended by the doctor. Comparison of the medical service quality is based on popularity of the medicament prescribed by other medical professionals and other patients comments suffering from the disease and being ordered the same medicament. Efficiency of the approach proposed has a pragmatic nature such that it can be applied by medical clinics, hospitals and other healthcare institutions as well. We trust that the data collected by our MDDDS model is found of the greatest importance by the patients while seeking a reliable and high-quality healthcare service. Our findings based on the real data simulation indicate that implementation of HMM and Viterbi's algorithm gives a very promising results in optimization of the patient's decision process as regardless of the data including the disease entities, the number of the medical service providers, we achieve ambiguous outcome, which in each case allows the patient to make a reliable and firm decision of which doctor is the most successful in treating a particular disease entity.

Keywords: Decision support · Data-driven model · Viterbi algorithm · Hidden Markov Model · Stochastic process

1 Introduction

As information is the most crucial factor in our everyday life, since it directs us in the process of decision making at every moment, in particular when we are in need of medical service, the principal motivation of this paper is to propose a model supporting a patient's decision on making the optimal choice based on the criteria such as the disease, the medicaments recommended for the disease by the General Practitioners (GP), the opinions posted publicly in social media by other patients who have taken them just

© The Author(s), under exclusive license to Springer Nature Switzerland AG 2022
L. Barolli et al. (Eds.): AINA 2022, LNNS 450, pp. 348–360, 2022.
https://doi.org/10.1007/978-3-030-99587-4_30

suffering from the same disease and finally how popular the medicament is among the GPs in the sense of prescribing it to the number of other patients as well as it proves that the GP is up-to-date with the top quality and the newest medicaments offered nowadays on the pharmaceutical market. Also, in our opinion, it suggests that the GP studies the medical literature on the research within the scope of his medical specialization.

Information is usually incomplete, such that the decision process relies on the fuzzy knowledge resulting from inference related to the information pieces that has been gathered and some prediction algorithms built on this extended knowledge as a representative one for the observation being the subject of the study.

Viterbi algorithm is used for optimization of the selection process in which a certain number of probable stages are analyzed in their sequential order. Hidden Markov Model (HMM) allows us to build a matrix of transition stages that corresponds to the one which is detected as the optimal. These properties motivated us to build the model.

The reminder of this paper is organized as follow: Sect. 2 provides an literature overview, Sect. 3 presents the fundamental properties of HMM, Sect. 4 describes the Viterbi's algorithm implementation to our MDDDS model. Further we move on to methodology and the details of the decision support model with the aim to show how it works. Thus, in the following section we present some computation to conclude the research project with the features found by us as unique for the MDDDS model.

2 Related Works

Decision support systems have been studied over many years, however recently the area has developed to such an extent that this term seems to be replaced with ML [12, 20, 25], DL [17, 26] or other novel technologies. HMM popularity is integrated with other algorithms for decision optimization. Some examples include; [1] decision tree in real time for alerting the GP about the health condition, [2] auto-negotiation system, or [3] LSTM for creating a language model based on the Viterbi's algorithm. HMM is studied for training telephone speech triphones [4]. But not only, a study on prognosing and monitoring power supply in houses [5] shows that the authors can achieve quite promising results with HMM. Fault detection in supporting diagnosis is studied with application of the Petri net [7]. Financial market is analyzed for prognosing of the trends, which is of a great importance to investors [6]. However, the most widely studied area seems to be speech recognition in which HMM often implemented with Viterbi's algorithm plays an essential role [3–5, 12, 13, 19, 21–23, 26]. It seems the reason behind it is that the concept is relatively simple while it has a wide potential to be applied for prediction or speech analysis. In neurology, a decision about optimum diagnosis can be studied using firefly model which detects Parkinson's disease based on preliminary symptoms [11, 24]. Rough set theory models support decision making in case of fuzzy idea due to the incomplete data [8]. For intrusion detection [15] HMM produces a good results when compared to other studies. NLP in the area of multilingual patients [14] can support decision making in communication between patient and GP in the cases when the patients are foreigners. Quality of food is researched using HMM for safety of the consumers in [17]. Other projects explore cursive script recognition [17] in which the decision on classification correctness relies on HMM and Viterbi's algorithm.

Phonetic classification [13, 20] is proposed for decision making based on HMM and Viterbi's algorithm again. After the overview of the state of the art and on concluding how promising the results can be achieved, we decided to implement HMM and Viterbi's algorithm to our study.

3   Stochastic Process Extended by HMM

A stochastic process X(t) is a random variable, relative to the parameter t. A special class of stochastic processes are Markov processes. The state of the process at the present time does not depend on the path the process has taken to reach a given state.

Let X(t) be a Markov chain with values from the N-element state space E, and the matrix P is the transition matrix.

Consider a complex process (X(t), Y(t)) such that X(t) is a Markov chain and Y(t) is a sequence of random variables. Let y denote the value that Y(t) can take. We call the process (X(t), Y(t)) a hidden Markov process. We observe the process Y(t) and the process X(t) is hidden and affects the process Y(t). We call the values of the set X(t) the state space values of the hidden Markov process, and we call the values of the set Y(t) the observations. Such a model can be represented as a graph shown in Fig. 1

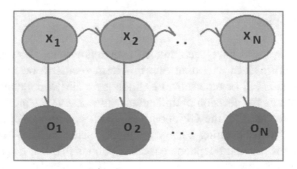

Fig. 1. Hidden random variables X_i and independent variables O_i

Let us assume the following notation:

$O = (o_1, o_2, ..., o_T)$ a sequence of observations

N – the number of the model states

M – the number of the observations

q_f – the state at time t

$A = a_{ij}$ where $a_{ij} = P[q_t = j | q_{t-1} = i]$ is a transition matrix between the states.

$B = b_j$ where $b_j = P[o_t | q_t]$ represents a probability distribution for the state j.

π - vector with the preliminary probability distributions of the model states.

In general, the HMM model can be represented as a triple $\lambda = (A, B, \pi)$ and we will denote the generation of a given sequence of observations by our model λ as $P(O|\lambda)$.

In literature there exist two types of HMM models:

– Ergotic model with arbitrary connections between the states (see Fig. 2)

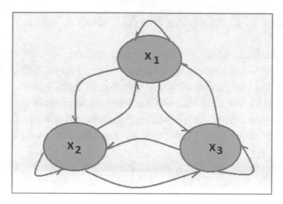

Fig. 2. Ergotic model

– The left to right model known as the Bekis model (see Fig. 3), as the special case of
the ergotic model excluding transition from the current state back to the previous state

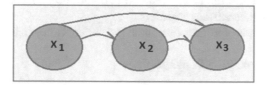

Fig. 3. Bekis model

There are various methods for calculating probability of a sequence of observations for a given model. These can be algorithms of varying degrees of computational
complexity. However, there are no algorithms that give unambiguous solutions.

4 Viterbi Algorithm Implementation to Our MDDDS Model

Creation of the model is the first stage of the research. The next stage is to apply
the model estimation and test it on the developed database. In this paper we will
apply the Viterbi algorithm. Having a given sequence of observations, it is necessary to find the most probable sequence of states. Therefore we define the value of
$\delta_t(i) = \max\limits_{q_1,\dots q_{t-1}} P[q_1,\dots q_t, o_1, \dots, o_t | \lambda]$ where $q_t = i$ while $\delta_t(i)$ is the maximum
probability of a single sequence of states up to t, and ending in i.

Viterbi's algorithm can be described in the following stages:

a) Initialization $\delta_i(i) = \pi_i b_i(o_1)$
b) Induction $\delta_t(j) = \max\limits_{1 \leq i \leq N} [\delta_{t-1}(i)a_{ij}]b_{ij}(o_t)$
c) Termination $Q = \max\limits_{1 \leq i \leq N} [\delta_T]$, where Q denotes the maximum likelihood of a single
state sequence of the model MDDDS

5 Methodology of Developing the MDDDS Model

When making a decision who will be the most suitable doctor or specialist for the particular disease, we often wonder what criteria should be met. Experience and knowledge updating, raising his qualifications by regular participation in the events related to the specialization, studying the details of the new medicaments on the pharmaceutical market, participating in international medical projects, etc., can be found the right criteria. Internet rankings and patients' opinions are equally important for all those seeking healthcare.

Analysis of the efficiency of the medicaments prescribed by the doctor based on the opinions is another right criteria. If the doctor prescribes the medicaments that give spectacular results, i.e. high chance for recovery, we can consider him to be a good specialist. When selecting a medicament, we are guided not only by its effectiveness or price, but also by the lowest possible number of the adverse side effects or the opinion of other doctors recommending the given medicament.

In our model we focus on optimization of such a decision by collecting the information which in our opinion is complete. The data used for this study have been collected from real resources such as DareCash, Healthcare6 and some other US medical websites since just in the US the trend seems to provide the patients with possibly most information about the healthcare providers.

Below is an example of information about the patients under a healthcare of a doctor.

Medicare Patient Condition Demographics

- Percent of Patients with Atrial Fibrillation: **14%**
- Percent of Patients with Alzheimer's Disease or Dementia: **7%**
- Percent of Patients with Asthma: **14%**
- Percent of Patients with Cancer: **10%**
- Percent of Patients with Heart Failure: **14%**
- Percent of Patients with Chronic Kidney Disease: **29%**
- Percent of Patients with Chronic Obstructive Pulmonary Disease: **14%**
- Percent of Patients with Depression: **32%**
- Percent of Patients with Diabetes: **32%**
- Percent of Patients with Hyperlipidemia: **57%**
- Percent of Patients with Hypertension (High Blood Pressure): **74%**
- Percent of Patients with Ischemic Heart Disease: **35%**
- Percent of Patients with Osteoporosis: **11%**
- Percent of Patients with Rheumatoid Arthritis/Osteoarthritis: **63%**
- Percent of Patients with Schizophrenia/Other Psychotic Disorders: **9%**
- Percent of Patients with Stroke: **4%**

In the example given below, we consider two physicians whose patients are grouped by a disease entity. Out of all the health conditions, we decided to analyze the most popular ones such as hypertension, asthma and high blood cholesterol.

In order to present how our model MDDDS works, we assume that the doctors in turn prescribe the best medicaments to their knowledge for each disease entity.

Our criteria include:

- Number of prescriptions for a disease entity
- Number of patients taking the same medicament prescribed by the doctor
- Parametrized and textual opinion of these patients about this medicament
- The medicament's weight determined by its position in the doctor's list to the other doctors who also prescribe this medicament

Tables 1 and 2 present the data collected about the two doctors service quality.

Table 1. The disease entities and the medicaments prescribed by Doctor X

Medicament	Number of prescriptions	Numer of patients	Patients opinion	The medicament's weight
Hypertension				
Lisinopril	110	103	0,62	1/1
Hydrochlorothiazide	155	36	0,48	4/7
Atenolol	108	20	0,56	7/9
Amlodipine Besylate	100	14	0,59	9/5
Furosemide	68	16	0,66	15/8
Metoprolol Succinate	67	11	0,60	16/13
Diovan	59	10	0,66	17/33
Losartan Potassium	54	10	0,56	19/16
Total for hypertension	**721**	**200**		
Cholesterol				
Simvastatin	198	39	0,61	2/2
Atorvastatin Calcium	97	24	0,63	10/6
Total for cholesterol	**295**	**63**		
Asthma				
ProAir HFA	81	17	0,73	13/21
Advair Diskus	59	10	0,65	17/43
Total for asthma	**140**	**27**		
Total	**1156**	**310**		

Table 2. The disease entities and the medicaments prescribed by Doctor Y

Medicament	Number of prescriptions	Numer of patients	Patients' opinion	The medicament's weight
Hypertension				
LISINOPRIL	397	94	0,62	1/1
Warfarin Sodium	262	14	0,61	4/10
Hydrochlorothiazide	162	37	0,48	6/7
Amlodipine Besylate	161	39	0,59	7/5
Furosemide	157	34	0,66	8/8
Metoprolol Succinate	150	38	0,60	12/12
Metoprolol Tatrate	155	35	0,60	11/12
Atenolol	134	28	0,56	13/9
Clopidogrel	55	11	0,58	32/30
Total for hypertension	**1630**	**270**		
Cholesterol				
Simvastatin	302	72	0,61	2/2
Atorvastatin Calcium	123	34	0,63	14/6
Total for cholesterol	**325**	**106**		
Asthma				
Spiriva	66	27	0,51	26/48
Total for asthma	**66**	**27**		
Total	**2021**	**397**		

We implement HMM and Viterbi's algorithm to our MDDDS model. While preparing the model we classify the medicament's weight into three groups (see Fig. 4):

– The patients' opinions (P)
– Position in the doctor's list based on which the medicament is prescribed (R)
– Effectiveness of the given medicament on a disease entity (E)

In Fig. 4 the following notation has been adopted:
H – Hypertension,
C – Cholesterol,
A – Asthma,
X – Doctor X,
Y – Doctor Y.

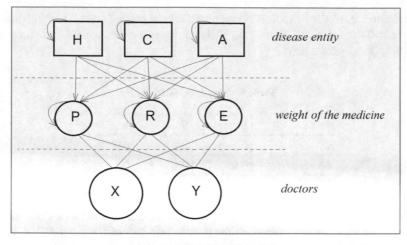

Fig. 4. The MDDDS model

6 Computation of the MDDDS Model Data

At first, we create the transition matrices for each of the stage separately. The computation is presented by the tables below.

Table 3. Transition matrix A

Matrix A	P	R	E
H	0.3	0.6	0.1
C	0.2	0.5	0.3
A	0.1	0.2	0.7

Transition matrix A in Table 3 shows probabilities for the disease entity with reference to the weight of the medicament with the optimal effectiveness based on the patients' opinions.

Table 4. Transition matrix B

Matrix B	P	R	E
P	0.4	0.6	0
R	0.4	0	0.6
E	0	0	1.0

Transition matrix B in Table 4 is computed for the medicaments' weights based on our data in Tables 1 and 2 that is the patients' overall opinion with reference to the position on the doctor's list.

Table 5. Transition matrix C

Matrix C	P	R	E
X	0.2	0.3	0.9
Y	0.8	0.7	0.1

Transition matrix C in Table 5 is computed for Doctor X and Doctor Y. The probabilities in the matrix's cells represent the position on each of the doctor's list, in other words how much they value the particular medicament, here the evidence is referenced on comparison basis to other doctors prescribing the medicament for the same disease entity.

Now, we apply the data to our MDDDS model in order to determine the optimal pathway which suggests the best choice of the doctor. The decision includes the criteria presented the Sect. 5 (Fig. 5).

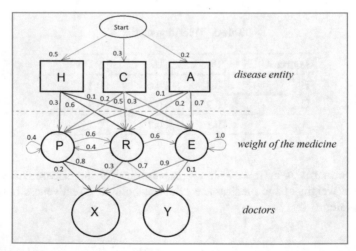

Fig. 5. The MDDDS model with the probabilities of the random variables. Here, the pathways indicate which one is the optimal

The transition matrices between the stages A, B and C are created based on the tables above, in addition we determine also the matrix of the initial states

$$\pi = [0.5\ 0.3\ 0.2]$$

$$A = \begin{bmatrix} 0.3 & 0.6 & 0.1 \\ 0.2 & 0.5 & 0.3 \\ 0.1 & 0.2 & 0.7 \end{bmatrix}$$

$$B = \begin{bmatrix} 0.4 & 0.6 & 0 \\ 0.4 & 0 & 0.6 \\ 0 & 0 & 1 \end{bmatrix}$$

$$C = \begin{bmatrix} 0.2 & 0.3 & 0.9 \\ 0.8 & 0.7 & 0.1 \end{bmatrix}$$

Following Viterbi's algorithm, we estimate the model by moving along the pathways determined by the MDDDS model and calculating the probabilities. Obtaining the maximum value allows us to determine the best solution and to answer the question - which doctor is the best for a given disease entity? A pathway leading to Doctor X looks like the following $P(H,P,R,X) = 0.5 \times 0.3 \times 0.6 \times 0.7 = 0.063$, while the same calculation for Doctor Y gives the outcome of $P(A,R,E,Y) = 0.2 \times 0.6 \times 0.1 \times 0.1 = 0.024$.

Solution for the Optimal Decision as to the Doctor X or Y
While computing each parameter for determining the maximum likelihood our goal is to identify which of the two doctors is better suited for the particular disease which a patient suffers from. In other words, which of these two doctors follows the optimum treatment procedure for the disease entity of our interest. In our example, the pathway denoted as $P(H,P,R,X) = (0.5) \times (0.3) \times (0.6) \times (0.7) = 0.063$ indicates the higher probability value $0.024 < 0.063$ – such a result can be interpreted that Doctor X has the better treatment procedure for curing asthma than Doctor Y. Table 6 presents how the MDDDS model works for other disease entities.

Table 6. Comparison matrix D

Matrix D	Asthma	Cholesterol	Hypertension
Doctor X	P(H,P,R,X) = 0.063	P(C,R,E,X) = 0.27	P(H,R,E,X) = 0.324
Doctor Y	P(A,R,E,Y) = 0.024	P(C,R,Y) = 0.35	P(H,R,Y) = 0.42

The conclusion from our results is that while Doctor X is the best choice for curing asthma, Doctor Y should be selected as a better specialist for curing high blood Cholesterol and Hypertension.

In our simulation, we compare two medical service providers and three disease entities like hypertension, cholesterol and asthma, however by creating a database with the higher number of medical professionals and more disease units, an unsupervised ML method in which clustering can be applied for multicriteria attributes would be recommended.

7 Conclusion

The MDDDS model proposed in this paper for multicriteria classification of the medical professionals, in particular the GPs, has a potential to support the patients in making a difficult decision about the optimal choice in healthcare service quality. We implement Hidden Markov Model and for the MDDDS model estimation we use the Viterbi algorithm. Our research objective was to support the patients with the tool for optimizing the decision based on the data driven model. This way, the patient is able to identify whether or not the doctor we have chosen is the best specialist to treat a given disease entity. The computation of probabilities for each pathway is performed for each parameter separately. After compiling the probabilities, we select the pathway for which we obtain the highest probability value. On this basis, we can determine whether the selected paths with the maximum value gives the answer about which of these two, that is Doctor X or Doctor Y should be chosen.

However, it is essential to stress that the MDDDS model recommends the best medical provider for a particular disease entity, not in a global scale. In our example, we cannot simply infer that due to the fact that Doctor Y turned out to be a better choice for two disease entities, while Doctor X for one only, we can conclude that Doctor Y provides a medical care of a higher quality than Doctor X. Here, it is the patient's individual decision. In our reasoning, each medical provider specializes in a different disease entity, more disease entities does not mean that the GP should be recommended.

In our future work, we plan to extend the MDDDS model by adopting unsupervised ML in order to build the clusters that include the patients preferred criteria, then optimization of the decision indicates the disease entities that are the most crucial in terms of the medical service provider's quality. To our best knowledge, the research community has not considered the problem studied by us so far thus we believe that this niche in science is a challenge due to its practical purpose.

Acknowledgment. This work was supported in part by a grant from COST Action CA18231 Multi3Generation: Multi-task, Multilingual, Multi-modal Language Generation founded by COST (European Cooperation in Science and Technology).

References

1. Keskar, S., Banerjee, R.: Time-recurrent HMM decision tree to generate alerts for heart-guard wearable computer. In: 2011 Computing in Cardiology, pp. 605–608 (2011)
2. Tai-Guang, G., Min, H., Qing, W., Xing-Wei, W., Pei-You, C.: A multi-issue auto-negotiation system based on HMM. In: 2018 Chinese Control and Decision Conference (CCDC), 2018, pp. 3659–3664 (2018)
3. Mizera-Pietraszko, J.: Computer-assisted clinical diagnosis in the official European Union languages. In: Proceedings of the 2016 IEEE 18th International Conference on e-Health Networking, Applications and Services (Healthcom), Munich, Germany, pp. 1–6. IEEE Computer Society (2016)
4. Zhou, W., Schlüter, R., Ney, H.: Full-sum decoding for hybrid Hmm based speech recognition using LSTM language model. In: ICASSP 2020 - 2020 IEEE International Conference on Acoustics, Speech and Signal Processing (ICASSP), pp. 7834–7838 (2020)

5. Kato, T., Kuroiwa, S., Shimizu, T., Higuchl, N.: Efficient mixture Gaussian synthesis for decision tree based state tying. In: 2001 IEEE International Conference on Acoustics, Speech, and Signal Processing, Proceedings (Cat. No.01CH37221), vol. 1, pp. 493–496 (2001)
6. Wang, Y., Phillips, I.T., Haralick, R.M.: A method for document zone content classification. In: International Conference on Pattern Recognition, vol. 3, pp. 196–199 (2002)
7. Makonin, S., Popowich, F., Bajić, I.V., Gill, B., Bartram, L.: Exploiting HMM sparsity to perform online real-time nonintrusive load monitoring. IEEE Trans. Smart Grid 7(6), 2575–2585 (2016)
8. Mizera-Pietraszko, J., Tancula, J.: Rough set theory for supporting decision making on relevance in browsing multilingual digital resources. In: Król, D., Nguyen, N., Shirai, K. (eds.) Advanced Topics in Intelligent Information and Database Systems. ACIIDS 2017. Studies in Computational Intelligence, vol. 710. Springer, Cham (2017). https://doi.org/10.1007/978-3-319-54430-4
9. Palupi, I., Wahyudi, B.A., Putra, A.P.: Implementation of Hidden Markov Model (HMM) to predict financial market regime. In: 2021 9th International Conference on Information and Communication Technology (ICoICT), pp. 639–644 (2021)
10. Aghasaryan, A., Fabre, E., Benveniste, A., Boubour, R., Jard, C.: A Petri net approach to fault detection and diagnosis in distributed systems. II. Extending Viterbi algorithm and HMM techniques to Petri nets. In: Proceedings of the 36th IEEE Conference on Decision and Control, vol. 1, pp. 726–731 (1997)
11. Dash, S., Abraham, A., Kr Luhach, A., Mizera-Pietraszko, J., Rodrigues, J.P.C.: Hybrid chaotic firefly decision making model for Parkinson's disease diagnosis. J. Distrib. Sens. Netw. 16(1), 1550147719895210 (2020)
12. Chien, J.-T., Furui, S.: Predictive Hidden Markov model selection for speech recognition. IEEE Trans. Speech Audio Process. 13(3), 377–387 (2005)
13. Monkowski, M.D., Picheny, M.A., Srinivasa Rao, P.: Context dependent phonetic duration models for decoding conversational speech. In: 1995 International Conference on Acoustics, Speech, and Signal Processing, pp. 528–531 (1995)
14. Mizera-Pietraszko, J., Świątek, P.: Access to eHealth language-based services for multinational patients. In: 2015 17th International Conference on E-health Networking, Application & Services (HealthCom), pp. 232–237 (2015)
15. Gao, F., Sun, J., Wei, Z.: The prediction role of Hidden Markov model in intrusion detection. In: CCECE 2003 - Canadian Conference on Electrical and Computer Engineering. Toward a Caring and Humane Technology (Cat. No.03CH37436), 2003, vol. 2, pp. 893–896 (2003)
16. Bu, K., Li, X., Wang, K., Li, Y.: Data analysis of public food safety cases based on Apriori. In: Chinese Control and Decision Conference (CCDC), pp. 343–348 (2020)
17. Kaltenmeier, A., Caesar, T., Gloger, J.M., Mandler, E.: Sophisticated topology of hidden Markov models for cursive script recognition. In: Proceedings of 2nd International Conference on Document Analysis and Recognition (ICDAR 1993), pp. 139–142 (1993)
18. Junkawitsch, J., Neubauer, L., Hoge, H., Ruske, G.: A new keyword spotting algorithm with pre-calculated optimal thresholds. In: Proceeding of Fourth International Conference on Spoken Language Processing. ICSLP 1996, vol. 4, pp. 2067–2070 (1996)
19. Ozeki, K.: Likelihood normalization using an ergodic HMM for continuous speech recognition. In: Proceeding of Fourth International Conference on Spoken Language Processing, ICSLP 1996, vol. 4, pp. 2301–2304 (1996)
20. Rathinavelu, C., Deng, L.: The trended HMM with discriminative training for phonetic classification. In: Proceeding of Fourth International Conference on Spoken Language Processing, ICSLP 1996, vol. 2, pp. 1049–1052 (1996)
21. Yoma, N.B., Villar, M.: Speaker verification in noise using a stochastic version of the weighted Viterbi algorithm. IEEE Trans. Speech Audio Process. 10(3), 158–166 (2002)

22. Liao, Y.-F., Chen, S.-H.: An MRNN-based method for continuous Mandarin speech recognition. In: Proceedings of the 1998 IEEE International Conference on Acoustics, Speech and Signal Processing, ICASSP 1998 (Cat. No. 98CH36181), vol. 2, pp. 1121–1124 (1998)
23. Hon, H.-W., Wang, K.: Unified frame and segment based models for automatic speech recognition. In: 2000 IEEE International Conference on Acoustics, Speech, and Signal Processing. Proceedings (Cat. No. 00CH37100), pp. II1017–II1020 (2000)
24. Basseville, M., Benveniste, A., Tromp, L.: Diagnosing hybrid dynamical systems: fault graphs, statistical residuals and Viterbi algorithms. In: Proceedings of the 37th IEEE Conference on Decision and Control (Cat. No. 98CH36171), vol. 4, pp. 3757–3762 (1998)
25. Watanabe, Y., Liu, W., Shoji, Y.: Machine-learning-based hazardous spot detection framework by mobile sensing and opportunistic networks. IEEE Trans. Veh. Technol. **69**(11), 13646–13657 (2020)
26. Shah, V., Anstotz, R., Obeid, I., Picone, J.: Adapting an automatic speech recognition system to event classification of electroencephalograms[1]. In: 2018 IEEE Signal Processing in Medicamente and Biology Symposium (SPMB), pp. 1–5 (2018)

Performance Evaluation of a DQN-Based Autonomous Aerial Vehicle Mobility Control Method in Corner Environment

Nobuki Saito[1], Tetsuya Oda[2(✉)], Aoto Hirata[1], Chihiro Yukawa[2], Kyohei Toyoshima[2], Tomoaki Matsui[2], and Leonard Barolli[3]

[1] Graduate School of Engineering, Okayama University of Science (OUS), 1-1 Ridaicho, Kita-ku, Okayama 700-0005, Japan
{t21jm01md,t21jm02zr}@ous.jp
[2] Department of Information and Computer Engineering, Okayama University of Science (OUS), 1-1 Ridaicho, Kita-ku, Okayama 700-0005, Japan
oda@ice.ous.ac.jp, {t18j097cy,t18j056tk,t19j077mt}@ous.jp
[3] Department of Information and Communication Engineering, Fukuoka Institute of Technology, 3-30-1 Wajiro-higashi, Higashi-ku, Fukuoka 811-0295, Japan
barolli@fit.ac.jp

Abstract. In this paper, we present a DQN-based AAV mobility control method and show the performance evaluation results for a corner environment scenario. The performance evaluation results show that the proposed method can decide the destination based on LiDAR for TLS-DQN. Also, the visualization results of AAV movement show that the TLS-DQN can reach the destination in the corner environment.

1 Introduction

The Unmanned Aerial Vehicle (UAV) is expected to be used in different fields such as aerial photography, transportation, search and rescue of humans, inspection, land surveying, observation and agriculture. Autonomous Aerial Vehicle (AAV) [1] has the ability to operate autonomously without human control and is expected to be used in a variety of fields, similar to UAV. So far many AAVs [2–4] are proposed and used practically. However, existing autonomous flight systems are designed for outdoor use and rely on location information by the Global Navigation Satellite System (GNSS) or others. On the other hand, in an environment where it is difficult to obtain position information from GNSS, it is necessary to determine a path without using position information. Therefore, autonomous movement control is essential to achieve operations that are independent of the external environment, including non-GNSS environments such as indoor, tunnel and underground.

In [5–7] the authors consider Wireless Sensor and Actuator Networks (WSANs), which can act autonomously for disaster monitoring. A WSAN consists of wireless network nodes, all of which have the ability to sense events (sensors) and perform actuation (actuators) based on the sensing data collected by the sensors. WSAN nodes in these applications are nodes with integrated sensors and actuators that have the

L. Barolli et al. (Eds.): AINA 2022, LNNS 450, pp. 361–372, 2022.
https://doi.org/10.1007/978-3-030-99587-4_31

high processing power, high communication capability, high battery capacity and may include other functions such as mobility. The application areas of WSAN include AAV [8], Autonomous Surface Vehicle (ASV) [9], Heating, Ventilation, Air Conditioning (HVAC) [10], Internet of Things (IoT) [11], Ambient Intelligence (AmI) [12], ubiquitous robotics [13], and so on.

Deep reinforcement learning [14] is an intelligent algorithm that is effective in controlling autonomous robots such as AAV. Deep reinforcement learning is an approximation method using deep neural network for value function and policy function in reinforcement learning. Deep Q-Network (DQN) is a method of deep reinforcement learning using Convolution Neural Network (CNN) as a function approximation of Q-values in the Q-learning algorithm [14,15]. DQN combines the neural fitting Q-iteration [16,17] and experience replay [18], shares the hidden layer of the action value function for each action pattern and can stabilize learning even with nonlinear functions such as CNN [19,20]. However, there are some points where learning is difficult to progress for problems with complex operations and rewards, or problems where it takes a long time to obtain a reward.

In this paper, we propose a DQN-based AAV mobility control method. Also, we present the simulation results for AAV control using TLS-DQN [21] considering an indoor single-path environment including a corner space.

The structure of the paper is as follows. In Sect. 2, we show the DQN based AAV testbed. In Sect. 3, we describe the proposed method. In Sect. 4, we discuss the performance evaluation. Finally, conclusions and future work are given in Sect. 5.

2   DQN Based AAV Testbed

In this section, we discuss quadrotor for AAV and DQN for AAV mobility.

2.1   Quadrotor for AAV

For the design of AAV, we consider a quadrotor, which is a type of multicopter. Multicopter is high maneuverable and can operate in places that are difficult for people to enter, such as disaster areas and dangerous places. It also has the advantage of not

Fig. 1. Snapshot of AAV.

Table 1. Components of quadrotor.

Component	Model
Propeller	15×5.8
Motor	MN3508 700 kv
Electric speed controller	F45A 32bitV2
Flight controller	Pixhawk 2.4.8
Power distribution board 1	MES-PDB-KIT
Li-Po battery	22.2 v 12000 mAh XT90
Mobile battery	Pilot Pro 2 23000 mAh
ToF ranging sensor	VL53L0X
Raspberry Pi	3 Model B Plus
PVC pipe	VP20
Acrylic plate	5 mm

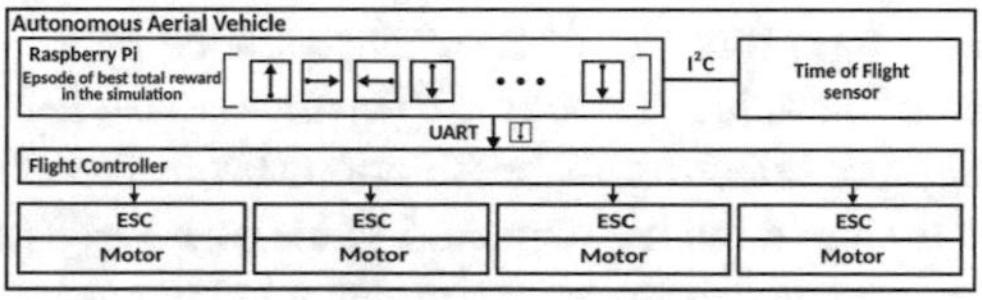

Fig. 2. AAV control system.

requiring space for takeoffs and landings and being able to stop at mid-air during the flight, therefore enabling activities at fixed points.

In Fig. 1 is shown a snapshot of the quadrotor used for designing and implementing an AAV testbed. The quadrotor frame is mainly composed of polyvinyl chloride (PVC) pipe and acrylic plate. The components for connecting the battery, motor, sensor, etc. to the frame are created using an optical 3D printer. Table 1 shows the components in the quadrotor. The size specifications of the quadrotor (including the propeller) are length $87 [cm]$, width $87 [cm]$, height $30 [cm]$ and weight $4259 [g]$.

In Fig. 2 is shown the AAV control system. The raspberry pi reads saved data of the best episode when carrying out the simulations by DQN and uses telemetry communication to send commands such as up, down, forward, back, left, right and stop to the flight controller. Also, multiple Time-of-Flight (ToF) range sensors using Inter-Integrated Circuit (I^2C) communication and General-Purpose Input Output (GPIO) are used to acquire and save flight data. The Flight Controller (FC) is a component that calculates the optimum motor rotation speed for flight based on the information sent from the built-in acceleration sensor and gyro sensor. The Electronic Speed Controller (ESC) is a part that controls the rotation speed of the motor in response to commands from FC. Through these sequences, AAV behaves and reproduces movement in simulation.

2.2 DQN for AAV Mobility

The DQN for moving control of AAV structure is shown in Fig. 3. In this work, we use the Deep Belief Network (DBN), because the computational complexity is smaller than CNN for DNN part in DQN. The environment is set as v_i. At each step, the agent selects an action a_t from the action sets of the mobile actuator nodes and observes a position v_t from the current state. The change of the mobile actuator node score r_t was regarded as the reward for the action. For the reinforcement learning, we can complete all of these mobile actuator nodes sequences m_t as Markov decision process directly, where sequences of observations and actions are $m_t = v_1, a_1, v_2, \ldots, a_{t-1}, v_t$. A method known as experience replay is used to store the experiences of the agent at each timestep, $e_t = (m_t, a_t, r_t, m_{t+1})$ in a dataset $D = e_1, \ldots, e_N$, cached over many episodes into a Experience Memory. By defining the discounted reward for the future by a factor γ, the sum of the future reward until the end would be $R_t = \sum_{t'=t}^{T} \gamma^{t'-t} r_{t'}$. T means the termination time-step of the mobile actuator nodes. After running experience replay, the agent selects and executes an action according to an ε-greedy strategy. Since using histories of arbitrary length as inputs to a neural network can be difficult, Q-function instead works on fixed length format of histories produced by a function ϕ. The target was to maximize the action value function $Q^*(m, a) = \max_\pi E[R_t | m_t = m, a_t = a, \pi]$, where π is the strategy for selecting of best action. From the Bellman equation (see Eq. (1)), it is possible to maximize the expected value of $r + \gamma Q^*(m', a')$, if the optimal value $Q^*(m', a')$ of the sequence at the next time step is known.

$$Q^*(m', a') = E_{m' \sim \xi}[r + \gamma_{a'} \max Q^*(m', a') | m, a]. \tag{1}$$

By not using iterative updating method to optimize the equation, it is common to estimate the equation by using a function approximator. Q-network in DQN is a neural network function approximator with weights θ and $Q(s, a; \theta) \approx Q^*(m, a)$. The loss function to train the Q-network is shown in Eq. (2):

$$L_i(\theta_i) = E_{s, a \sim \rho(.)}[(y_i - Q(s, a; \theta_i))^2]. \tag{2}$$

The y_i is the target, which is calculated by the previous iteration result θ_{i-1}. The $\rho(m, a)$ is the probability distribution of sequences m and a. The gradient of the loss function is shown in Eq. (3):

$$\nabla_{\theta_i} L_i(\theta_i) = E_{m, a \sim \rho(.); s' \sim \xi}[(y_i - Q(m, a; \theta_i)) \nabla_{\theta_i} Q(m, a; \theta_i)]. \tag{3}$$

We consider tasks in which an agent interacts with an environment. In this case, the AAV moves step by step in a sequence of observations, actions and rewards. We took in consideration AAV mobility and consider 7 mobile patterns (up, down, forward, back, left, right, stop). In order to decide the reward function, we considered Distance between AAV and Obstacle (DAO) parameter.

The initial weights values are assigned as Normal Initialization [22]. The input layer is using AAV and the position of destination, total reward values in Experience Memory and AAV movements patterns. The hidden layer is connected with 256 rectifier units in Rectified Linear Units (ReLU) [23]. The output Q-values are the AAV movement patterns.

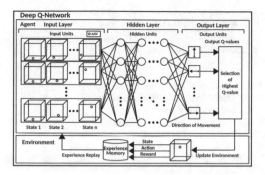

Fig. 3. DQN for AAV mobility control.

3   Proposed Method

In this section, we discuss the AAV mobility control method for DQN-based AAV.

3.1   LiDAR Based Mobile Area Decision Method

The proposed method decides the destination in TLS-DQN within the considered area based on the point cloud obtained by LiDAR and reduces the setting operation for the destination set manually by humans in TLS-DQN. In addition, the proposed method can decide the destination with less computation than using Simultaneous Localization and Mapping (SLAM) or other methods. In Algorithm 1, we consider as inputs the coordinates list of obstacles (*distance*, *angle*) obtained by LiDAR and the coordinates of LiDAR placement. The output is the *Destination* (X, Y), which may be local destination or global destination. The global destination indicates the destination in the considered area, and the local destination indicates the target passage points until the global destination. The Z-coordinate of destination is the median of the movable range in the Z-axis for the destination. In the proposed method, the destination is continuously decided by letting the LiDAR placement be the coordinate of reached destination when the AAV reached the destination.

Algorithm 1 LiDAR Based Mobile Area Decision Method.

Input: *Point Cloud List* ← The coordinates list of obstacles (*distance, angle*) obtained by LiDAR
  (x_{LiDAR}, y_{LiDAR}) ← The coordinates of LiDAR Placement.
Output: *Destination* (X, Y).
1: **for** $i = 0$ to 360 **do**
2:   $x_{Point\ Cloud\ List}[i]$ ← *Point Cloud List*$[i][0] \times cos($*Point Cloud List*$[i][1])$.
3:   $y_{Point\ Cloud\ List}[i]$ ← *Point Cloud List*$[i][0] \times sin($*Point Cloud List*$[i][1])$.
4:   **if** *Point Cloud List*$[i][0] >$ *Any Distance.* **then**
5:     *Distant Point Cloud*$[i]$ ← $(x_{Point\ Cloud\ List}[i], y_{Point\ Cloud\ List}[i])$.
6: (x_{min}, x_{max}) ← Min. and Max. value for X-axis in the *Distant Point Cloud.*
7: (y_{min}, y_{max}) ← Min. and Max. value for Y-axis in the *Distant Point Cloud.*
8: (x_{center}, y_{center}) ← ($\frac{x_{min}+ x_{max}}{2}$, $\frac{y_{min}+ y_{max}}{2}$).
9: $flag$ ← 0.
10: **for** $x = x_{LiDAR}$ to x_{center} **do**
11:   y ← ($\frac{y_{center}- y_{LiDAR}}{x_{center}- x_{LiDAR}}$) $\times (x - x_{LiDAR}) + y_{LiDAR}$.
12:   **for** $i = 0$ to 360 **do**
13:     **if** $\sqrt{(x - x_{Point\ Cloud\ List}[i])^2 + (y - y_{Point\ Cloud\ List}[i])^2} >$ *Any Distance* **then**
14:       *Destination* ← (x, y).
15:     **else**
16:       $flag$ ← 1.
17:       **break**
18: **if** $flag = 0$ **then**
19:   *Destination* is local destination.
20: **else**
21:   *Destination* is global destination.

Algorithm 2 Tabu List for TLS-DQN.

Require: The coordinate with the highest evaluated value in the section is (x, y, z).
1: **if** $(x_{before} \leq x_{current}) \wedge (x_{current} \leq x)$ **then**
2:   $tabu\ list$ ⇐ $((x_{min} \leq x_{before}) \wedge (y_{min} \leq y_{max}) \wedge (z_{min} \leq z_{max}))$
3: **else if** $(x_{before} \geq x_{current}) \wedge (x_{current} \geq x)$ **then**
4:   $tabu\ list$ ⇐ $((x_{before} \leq x_{max}) \wedge (y_{min} \leq y_{max}) \wedge (z_{min} \leq z_{max}))$
5: **else if** $(y_{before} \leq y_{current}) \wedge (y_{current} \leq y)$ **then**
6:   $tabu\ list$ ⇐ $((x_{min} \leq x_{max}) \wedge (y_{min} \leq y_{before}) \wedge (z_{min} \leq z_{max}))$
7: **else if** $(y_{before} \geq y_{current}) \wedge (y_{current} \geq y)$ **then**
8:   $tabu\ list$ ⇐ $((x_{min} \leq x_{max}) \wedge (y_{before} \leq y_{max}) \wedge (z_{min} \leq z_{max}))$
9: **else if** $(z_{before} \leq z_{current}) \wedge (z_{current} \leq z)$ **then**
10:   $tabu\ list$ ⇐ $((x_{min} \leq x_{max}) \wedge (y_{min} \leq y_{max}) \wedge (z_{min} \leq z_{before}))$
11: **else if** $(z_{before} \geq z_{current}) \wedge (z_{current} \geq z)$ **then**
12:   $tabu\ list$ ⇐ $((x_{min} \leq x_{max}) \wedge (y_{min} \leq y_{max}) \wedge (z_{before} \leq z_{max}))$

3.2 TLS-DQN

The idea of the Tabu List Strategy (TLS) is motivated from Tabu Search (TS) proposed by F. Glover [24] to achieve an efficient search for various optimization problems by prohibiting movements to previously visited search area in order to prevent getting stuck in local optima.

$$r = \begin{cases} 3 & (if\ (x_{current} = x_{global\ destinations}) \wedge \\ & (y_{current} = y_{global\ destinations}) \wedge \\ & (z_{current} = z_{global\ destinations})) \vee \\ & (((x_{before} < x_{current}) \wedge (x_{current} \leq x_{local\ destinations})) \vee \\ & ((x_{before} > x_{current}) \wedge (x_{current} \geq x_{local\ destinations})) \vee \\ & ((y_{before} < y_{current}) \wedge (y_{current} \leq y_{local\ destinations})) \vee \\ & ((y_{before} > y_{current}) \wedge (y_{current} \geq y_{local\ destinations})) \vee \\ & ((z_{before} < z_{current}) \wedge (z_{current} \leq z_{local\ destinations})) \vee \\ & ((z_{before} > z_{current}) \wedge (z_{current} \geq z_{local\ destinations}))). \\ -1 & (else). \end{cases} \quad (4)$$

In this paper, the reward value for DQN is decided by Eq. (4), where "x", "y" and "z" means X-axis, Y-axis and Z-axis, respectively. The current means the current coordinates of the AAV in the DQN, and the before means the coordinates before moving the action. The considered area is partitioned based on the local destination or global destination and a destination is set in each area. The tabu list in TLS is used when a DQN selects an action randomly or determined a reward for the action. If the tabu list includes the area in the direction of movement, the DQN will reselect the action. Also, if the reward is "3", the prohibited area is added to the tabu list based on the rule shown in Algorithm 2. The search by TLS-DQN is done in a wider range and is better than the search by random direction of move.

3.3 Movement Adjustment Method

The movement adjustment method is used for reducing movement fluctuations caused by TLS-DQN. The Algorithm 3 inputs the movement of coordinates (X, Y, Z) in the episode of Best derived by TLS-DQN and generates the *Adjustment Point Coordinates List*. In Algorithm 3, the *Number of divided lists* indicates the number of divisions to

---

Algorithm 3 Movement Adjustment Decision.

Input: *Movement Coordinates* ← The movement of coordinates (X, Y, Z) by TLS-DQN
Output: *Adjustment Point Coordinates List.*
  1: *Number of divided list* ← *Any number.*
  2: *Number of coordinates* ← $\frac{Number\ of\ Iterations\ in\ TLS\text{-}DQN}{Number\ of\ divided\ list}$.
  3: $i \leftarrow 0, j \leftarrow 0$
  4: **for** $k = 0$ **to** Number of coordinates in *Movement Coordinates*. **do**
  5:    *Divided List*$[j]$ ← *Movement Coordinates*$[k]$.
  6:    $j \leftarrow j + 1$.
  7:    **if** $j \geq$ *Number of coordinates* **then**
  8:      (x_{min}, x_{max}) ← Min. and Max. values for X-axis in the *Divided List*.
  9:      (y_{min}, y_{max}) ← Min. and Max. values for Y-axis in the *Divided List*.
 10:      (z_{min}, z_{max}) ← Min. and Max. values for Z-axis in the *Divided List*.
 11:      $(x_{center}, y_{center}, z_{center})$ ← $(\frac{x_{min} + x_{max}}{2}, \frac{y_{min} + y_{max}}{2}, \frac{z_{min} + z_{max}}{2})$
 12:      *Adjustment Point Coordinates List*$[i]$ ← $(x_{center}, y_{center}, z_{center})$.
 13:      $i \leftarrow i + 1, j \leftarrow 0$

---

the coordinate movements; the *Number of coordinates* indicates how many coordinates are included in the *Divided List*; and the $(x_{center}, y_{center}, z_{center})$ indicates the center coordinates derived from the maximum and minimum values of coordinates in X-axis, Y-axis and Z-axis included in the *Divided List*.

4   Performance Evaluation

In this section, we discuss the experimental results of LiDAR based mobile area decision method, the simulation results of TLS-DQN and the movement adjustment method.

4.1   Results of LiDAR Based Mobile Area Decision Method

The target environment is an indoor single-path environment including a corner. Figure 4 shows a snapshot of the area used in the simulation scenario and it was taken on the ground floor of Building C4 and C5 at Okayama University of Science, Japan. In Fig. 4, Fig. 5 and Fig. 6, the blue filled area indicates the floor surface and the red filled area indicates the corner space. In Fig. 5, the diagonal line filled area indicates the unmovable area, which is considered as obstacles. In Fig. 5 are shown the visualization results of LiDAR based mobile area decision method, the obstacle obtained by LiDAR and the decided destinations. Figure 6 shows the considered area based on the actual measurements of Fig. 5. The experimental results show that the `global destination` is decided at the end of the path, which is the corner space.

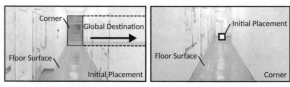

(a) From the initial placement to the corner space.
(b) From the corner space to the global destination.

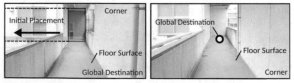

(c) From the global destination to the corner space.
(d) From the global destination to the corner space.

Fig. 4. Snapshot of the considered area.

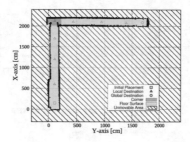

Fig. 5. Visualization results of LiDAR based mobile area decision method.

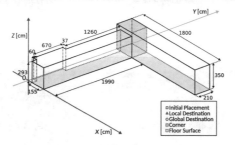

Fig. 6. Considered area for simulation.

4.2   Simulation Results of TLS-DQN

We consider for simulations the operations such as takeoffs, flights and landings between the initial placement and the destination decided by the LiDAR based mobile

Table 2. Simulation parameters of TLS-DQN.

Parameters	Values
Number of episode	30000
Number of iteration	2000
Number of hidden layers	3
Number of hidden units	15
Initial weight value	Normal initialization
Activation function	ReLU
Action selection probability (ε)	$0.999 - (t/\text{Number of episode})$ ($t = 0, 1, 2, \ldots,$ Number of episode)
Learning rate (α)	0.04
Discount rate (γ)	0.9
Experience memory size	300×100
Batch size	32
Number of AAV	1

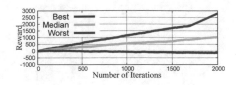

Fig. 7. Simulation results of rewards.

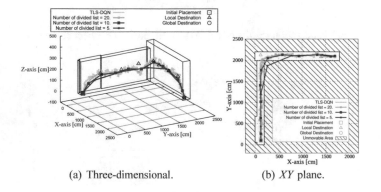

(a) Three-dimensional.   (b) *XY* plane.

Fig. 8. Visualization results of TLS-DQN and movement adjustment method.

Table 3. The distance of movement in the *XY* and *YZ* plane.

Plane	Minimum	TLS-DQN	*Divided list* = 20	*Divided list* = 10	*Divided list* = 5
XY	360.12	936.00	349.37	342.96	327.91
YZ	224.27	985.00	233.02	227.88	220.22

area decision method. The target environment is an indoor single-path environment including a corner. Table 2 shows the parameters used in the simulations. Figure 7 shows the change in reward value of the action in each iteration for Worst, Median, and Best episodes in TLS-DQN. For Best episodes, the reward value is increased much more than Median episodes. While, for Worst episodes, the reward value is decreased.

4.3   Results of Movement Adjustment Method

Figure 8 shows the visualization results on the three-dimensional and *XY* plane for the movement of the Best episodes in TLS-DQN and the results of the movement adjustment method when the *Number of divided list* is 20, 10 and 5, respectively. From the visualization results, the TLS-DQN has reached the destination. On the other hand, the movement adjustment method does not consider the obstacles. Therefore, the movement passes through the unmovable area when the *Number of divided list* is 5 and 10. Table 3 shows the distance of movement derived from the total Euclidean distances between each coordinate. The performance evaluation shows that the movement adjustment method can decrease the distance of movement for both *XY* and *YZ* planes.

5   Conclusions

In this paper, we proposed a DQN-based AAV mobility control method and presented simulation results for AAV control by TLS-DQN considering an indoor single-path environment including a corner space. From performance evaluation results, we conclude as follows.

- The proposed method can decide the mobile area and destination based on LiDAR.
- The visualization results of AAV movement show that the TLS-DQN can reach the destination.
- The proposed method is a good approach for indoor single-path environments including a corner space.

In the future, we would like to improve the TLS-DQN for AAV mobility by considering different scenarios.

Acknowledgement. This work was supported by JSPS KAKENHI Grant Number JP20K19793 and Grant for Promotion of OUS Research Project (OUS-RP-20-3).

References

1. Stöcker, C., et al.: Review of the current state of UAV regulations. Remote Sens. **9**(5), 1–26 (2017)
2. Artemenko, O., et al.: Energy-aware trajectory planning for the localization of mobile devices using an unmanned aerial vehicle. In: Proceedings of The 25-th International Conference on Computer Communication and Networks (ICCCN 2016), pp. 1-9 (2016)
3. Popović, M., et al.: An informative path planning framework for UAV-based terrain monitoring. Auton. Robot. **44**, 889–911 (2020)
4. Nguyen, H., et al.: LAVAPilot: lightweight UAV trajectory planner with situational awareness for embedded autonomy to track and locate radio-tags. arXiv:2007.15860, pp. 1–8 (2020)
5. Oda, T., et al.: Design and implementation of a simulation system based on deep Q-network for mobile actor node control in wireless sensor and actor networks. In: Proceedings of The 31-th IEEE International Conference on Advanced Information Networking and Applications Workshops (IEEE AINA 2017), pp. 195–200 (2017)
6. Oda, T., et al.: Performance evaluation of a deep Q-network based simulation system for actor node mobility control in wireless sensor and actor networks considering three-dimensional environment. In: Proceedings of The 9-th International Conference on Intelligent Networking and Collaborative Systems (INCoS 2017), pp. 41–52 (2017)
7. Oda, T., Kulla, E., Katayama, K., Ikeda, M., Barolli, L.: A deep Q-network based simulation system for actor node mobility control in WSANS considering three-dimensional environment: a comparison study for normal and uniform distributions. In: Barolli, L., Javaid, N., Ikeda, M., Takizawa, M. (eds.) CISIS 2018. AISC, vol. 772, pp. 842–852. Springer, Cham (2019). https://doi.org/10.1007/978-3-319-93659-8_77
8. Sandino, J., et al.: UAV framework for autonomous onboard navigation and people/object detection in cluttered indoor environments. Remote Sens. **12**(20), 1–31 (2020)
9. Moulton, J., et al.: An autonomous surface vehicle for long term operations. In: Proceedings of MTS/IEEE OCEANS, pp. 1–10 (2018)

10. Oda, T., et al.: Design of a deep Q-network based simulation system for actuation decision in ambient intelligence. In: Proceedings of The 33-rd International Conference on Advanced Information Networking and Applications (AINA 2019), pp. 362–370 (2019)

11. Oda, T., et al.: Design and implementation of an IoT-based E-learning testbed. Int. J. Web Grid Serv. **13**(2), 228–241 (2017)

12. Hirota, Y., Oda, T., Saito, N., Hirata, A., Hirota, M., Katatama, K.: Proposal and experimental results of an ambient intelligence for training on soldering iron holding. In: Barolli, L., Takizawa, M., Enokido, T., Chen, H.-C., Matsuo, K. (eds.) BWCCA 2020. LNNS, vol. 159, pp. 444–453. Springer, Cham (2021). https://doi.org/10.1007/978-3-030-61108-8_44

13. Hayosh, D., et al.: Woody: low-cost, open-source humanoid torso robot. In: Proceedings of The 17-th International Conference on Ubiquitous Robots (ICUR 2020), pp. 247–252 (2020)

14. Mnih, V., et al.: Human-level control through deep reinforcement learning. Nature **518**, 529–533 (2015)

15. Mnih, V., et al.: Playing atari with deep reinforcement learning. arXiv:1312.5602, pp. 1–9 (2013)

16. Lei, T., Ming, L.: A robot exploration strategy based on Q-learning network. In: IEEE International Conference on Real-time Computing and Robotics (IEEE RCAR 2016), pp. 57–62 (2016)

17. Riedmiller, M.: Neural fitted Q iteration - first experiences with a data efficient neural reinforcement learning method. In: Proceedings of The 16-th European Conference on Machine Learning (ECML 2005), pp. 317–328 (2005)

18. Lin, L.J.: Reinforcement learning for robots using neural networks. In: Proceedings of Technical Report, DTIC Document (1993)

19. Lange, S., Riedmiller, M.: Deep auto-encoder neural networks in reinforcement learning. In: Proceedings of The International Joint Conference on Neural Networks (IJCNN 2010), pp. 1–8 (2010)

20. Kaelbling, L.P., et al.: Planning and acting in partially observable stochastic domains. Artif. Intell. **101**(1–2), 99–134 (1998)

21. Saito, N., et al.: A Tabu list strategy based DQN for AAV mobility in indoor single-path environment: implementation and performance evaluation. Internet Things **14**, 100394 (2021)

22. Glorot, X., Bengio, Y.: Understanding the difficulty of training deep feedforward neural networks. In: Proceedings of The 13-th International Conference on Artificial Intelligence and Statistics (AISTATS 2010), pp. 249–256 (2010)

23. Glorot, X., et al.: Deep sparse rectifier neural networks. In: Proceedings of The 14-th International Conference on Artificial Intelligence and Statistics (AISTATS 2011), pp. 315–323 (2011)

24. Glover, F.: Tabu search - part I. ORSA J. Comput. **1**(3), 190–206 (1989)

OpenAPI QL: Searching in OpenAPI Service Catalogs

Ioanna-Maria Stergiou, Nikolaos Mainas, and Euripides G. M. Petrakis[✉]

School of Electrical and Computer Engineering, Technical University of Crete (TUC),
Chania, Greece
istergiou@isc.tuc.gr, {nmainas,petrakis}@intelligence.tuc.gr

Abstract. An OpenAPI description details the actions exposed by a
REST API. Existing query languages (e.g. JSONpath or N1QL for
JSON) are not designed for OpenAPI and suffer from several drawbacks the most important of them being that, queries are complicated
expressions and users must be familiar with the underlying OpenAPI
representation. We introduce OpenPI QL, a query language for OpenAPI service descriptions. OpenAPI QL relies on a model that identifies
the features that can be used to query a service description. OpenAPI
QL employees SQL syntax and maintains the necessary simplicity of
expression (i.e. using properties of REST services in simple SQL statements) regardless of service complexity and ignoring the nested structure
of OpenAPI. Although independent from OpenAPI, queries can address
the most important features of a service description in a database. The
run-time performance of SOWL QL has been assessed experimentally in
a database with OpenAPI descriptions of real services. A critical analysis
of its performance is also presented along with several query examples.

1 Introduction

Web services are published in service registries on the Web by various software
vendors and cloud providers so that they can be easily discovered and used by
users. OpenAPI Specification [4] is a powerful framework for the description
of REST APIs which is supported by a large user community and well-known
software vendors like Google, Microsoft, IBM, Oracle, and many others. An
OpenAPI service description is a machine-understandable technical document
with instructions about how to effectively use and integrate an API into an
application. A service is described by a JSON (or YAML) document describing
requests, responses, and security information such as authentication and authorization rules for an action.

Despite its rigorous language format, OpenAPI service descriptions can
become particularly complex. The syntactic restriction of OpenAPI to JSON (or
YAML) complicates the expression of queries in OpenAPI catalogs. OpenAPI-
to-GraphQL [1] generates a GraphQL interface for a given OpenAPI schema
that allows clients to write queries on API collections and retrieve descriptions
of service resources. A way around to the same problem is to store OpenAPI

L. Barolli et al. (Eds.): AINA 2022, LNNS 450, pp. 373–385, 2022.
https://doi.org/10.1007/978-3-030-99587-4_32

service descriptions in a NoSQL database for JSON (e.g. CouchDB, Couchbase MongoDB, ElasticSearch) and use JSONpath[1] syntax to express queries. In particular, JSONPath can be used to select and extract sub-sections from JSON documents matching the query selection criteria.

N1QL[2] is the JSON query language of Couchbase. It is an SQL-like query language for JSON, employs a query engine with a built-in optimizer and indexer and, promises real-time (or close to real-time) performance for millions of concurrent interactions. Despite its advantages (i.e. compliance with SQL and very good performance) querying OpenAPI catalogs with N1QL suffers from several drawbacks the most important of them being that queries become particularly complicated and users must be familiar with the underlying OpenAPI representation.

OpenAPI QL introduces an abstract layer to N1QL that allows users to express queries in OpenAPI catalogs with minimal or no knowledge of OpenAPI. Queries are specified using a combination of HTTP properties of a REST service. The user does not need to use OpenAPI syntax in queries. OpenAPI QL is a high-level query language that removes all syntax complexity from the queries. The key idea is a query model that abstracts OpenAPI properties in the form of key-value pairs ignoring the hierarchical structure of OpenAPI. Queries are issued as SQL statements involving key-value pairs. OpenAPI QL adopts the basic SQL syntax (SELECT - FROM - WHERE) and treats HTTP properties (which are mapped to OpenAPI objects) almost like columns of a relational database. A set of basic operations (e.g. AND, OR, NOT) is introduced for expressing conditional statements in WHERE clauses to retrieve descriptions of REST services that match a combination of selection criteria on service properties. OpenAPI QL is independent of OpenAPI syntax and can work with any other description for REST services (e.g. [5]).

The translation of OpenAPI QL queries to equivalent N1QL ones is also a focus of this work. These N1QL queries are executed afterward to search in OpenAPI catalogs using the N1QL query engine. OpenAPI QL is fully implemented and supported by a Graphical User Interface. The run-time performance of OpenAPI QL has been assessed experimentally in a database with 150 OpenAPI descriptions of real services. A critical analysis of its performance is also presented. More information about OpenAPI QL, its translation to N1QL, and the analysis of its performance can be found in [6].

Related work is discussed in Sect. 2. The proposed query model and how OpenAPI query statements are expressed using key-value pairs of HTTP properties are discussed in Sect. 3. OpenAPI QL syntax is discussed in Sect. 4. Section 5 outlines the translation OpenAPI QL to N1QL, followed by experimental results and issues for future research in Sect. 6 and Sect. 7 respectively.

---

[1] https://restfulapi.net/json-jsonpath/.

[2] https://www.couchbase.com/products/n1ql.

2   Background

An OpenAPI service description comprises many parts (objects). Each object specifies a list of properties that can be objects as well. Objects and properties defined under the Components unit of an OpenAPI document can be reused by other objects, and they can be linked to each other (e.g. using keyword $ref). Figure 1 illustrates the structure of an OpenAPI service description.

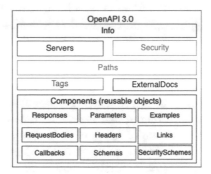

Fig. 1. OpenAPI document structure.

The Servers object provides information about where the API servers are located. Servers can be defined for different operations (locally declared servers override global servers). The service description contains an Info object with some non-functional information for the service, an External Documentation object, and all possible Tag objects (i.e. tags refer to resources described by a Schema object and are used to group operations either by resources or, by any other qualifier). Tags are optional in OpenAPI and are commonly used to group services by endpoints.

The Path object holds all the available service paths (i.e. endpoints) and their operations. It provides information about expressing HTTP requests to the service and about the responses of the service. It describes the supported HTTP methods (e.g. get, put, post, etc.) and defines the relative paths of the service endpoints (which are appended to a server URL in order to construct the full URL of an operation).

The Response object describes the responses of an operation, its message content, and the HTTP headers that a response may contain. The Parameters object describes parameters of operations (i.e. path, query, header, and cookie parameters). OpenAPI responses can include custom headers to provide additional information about the result of an API call. Every operation must have at least one response. A response is described using its HTTP status code and the data returned in the response body or headers.

Service operations accept and return data in different formats (e.g. JSON, XML, text, or images). These formats are defined by Media Type objects within a request or response. A Media Type object is identified using the *content* keyword. Request bodies in operations (e.g. put, post, get) specify the message that will be sent in a request. It is defined using the *requestBody* keyword and its contents are media types or Schema objects (using the *Schema* keyword). The response body can be a Schema object which is defined under the Components object. The Schema object describes the request and response messages based on JSON Schema[3]. A Schema object can be a primitive (string, integer), an array or a model, or an XML data type and may have properties on its own (i.e. externalDocs). Schema properties do not have semantic meaning, and their meaning can be vague [3].

The Components object lists reusable objects and includes (among others) definitions of schemas, responses, headers, parameters, and security schemes. The Security object lists the security schemes of the service (declared using keyword 'security scheme'). The specification supports HTTP authentication, API keys, OAuth2 common flows or grants (i.e. ways of retrieving an access token), and OpenID Connect. If an operation authentication scheme is 'oauth2', the value of a flow is a string with values one of 'authorizationCode', 'implicit', 'password' and 'clientCredentials'. The value of a scope is a string defining an action like 'read' or 'write'. OpenID Connect security declares a sign-in flow that enables a client application to obtain user information via authentication. If the security scheme is of type 'openIdConnect' then the value is a list of scope names required for the execution. An API key is actually a token that a client provides when making API calls. In OpenAPI 3.0, API Keys are described as a combination of name and location, defined as 'name' and 'in' properties, respectively. Finally, if the security scheme is HTTP, OpenAPI 3.0 includes numeric cases of security definitions built in HTTP protocol (e.g. bearer authentication). Each HTTP scheme defines the property 'type' with value 'http' and 'scheme' with values like 'bearer', 'basic', or another arbitrary string.

3   Query Model

Each API document includes information spread among objects and properties in a deeply nested (i.e. tree) structure. Therefore, an expression for accessing a particular property is a JSON path. Expressing queries using JSON paths (e.g. using N1QL) is possible but requires that a user be familiar with the peculiarities of the service and with the OpenAPI document nested structure. The Query model is a simplified representation of the properties described in OpenAPI. It is an attempt to map the semi-structured document format of a service description to a structured representation. The query model suggests the *flattening* of nested JSON objects so that all OpenAPI objects and their values are expressed as key-value pairs at a single level. More specifically, even though the desirable information can be located deep in the JSON data tree, flattening the JSON tree description will allow addressing OpenAPI properties as key-value pairs.

---

[3] https://json-schema.org.

Not all OpenAPI objects and properties are equally useful for expressing queries. The most essential properties of a service description are defined under the Path object. A request is made up of four components such as endpoint, HTTP method, headers, and body. In some cases, a request requires a security definition for authentication and authorization purposes. Objects Info and Servers, represent general information about the service and where the servers are installed. This information is worth displaying in query results, but not to be used as selection criteria in queries.

OpenAPI Query model focuses on properties that are worth to be searched. An assumption made in this work is that useful information about an API is described within the Path object. Paths object contains the information needed to execute a request (in Request object) and information returned in response to that request (in Response object). Listing 1 is an example Path object from Petstore service[4].

Listing 1. Path object with request body response and authentication scheme.

```
 1    {
 2      "paths": {
 3        "/pet": {
 4        "put": {
 5        "tags": [
 6        "pet"
 7        ],
 8        "summary": "Update an existing pet",
 9        "operationId": "updatePet",
10        "requestBody": {
11          "$ref": "#/components/requestBodies/Pet"
12        },
13        "response": {
14          "400": {
15          "description": "Invalid ID supplied"
16          },
17          "404": {
18          "description": "Pet not found"
19          }
20        },
21        "security": [
22        {
23          "petstore_auth": [
24          "write:pets",
25          "read:pets"
26          ]...
27        }...
28        }
```

The flattening of nested JSON descriptions suggests that the Path object (i.e. execution of a request and its corresponding results) be unpacked to a flat structure. These are Operation object, Parameter object, Tag object, Header object, Request Body object, Response object, Media Type object, Schema object, Security Requirement object, and Reference object. The goal is to convert these elements to a flat list of fields and values [6]. Figure 2 illustrates the

---

[4] https://petstore3.swagger.io.

hierarchical structure of objects in Path objects (left) and their key-value structure for expressing queries on these objects. Table 1 summarizes the key-value pairs (i.e. OpenAPI properties and their values) that can be used in queries. For each OpenAPI object, a set of properties is identified along with the set of values it may take. Each element in the table is a reserved word of the model.

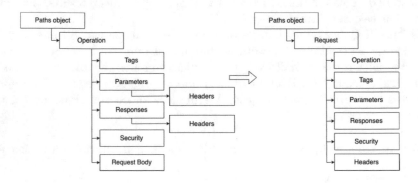

Fig. 2. From OpenAPI 3.0 model (left) to OpenAPI query model (right).

4   OpenAPI-QL

OpenAPI QL supports most of an SQL language syntax and clauses the most important of them being: (a) SELECT: specifies the service identifiers or Service objects to be returned. (b) FROM: declares the OpenAPI objects to query from; it always follows SELECT. (c) WHERE: includes logic operations and comparisons between object property values that restrict the number of answers returned by the query; it always follows FROM. OpenAPI QL supports the following operators: (a) AS: a property alias (i.e. renames the data parts of a query) in FROM or in SELECT: it allows using an instance of an OpenAPI object in a query. (b) AND, OR, NOT: connect logically two expressions in WHERE. (c) BETWEEN: searches for property values in a given range in WHERE (only for searching 'status code' property expressions). (d) EQUALSIGN (or '='): defines the meaning of equality in WHERE. (e) WILDCARD (or '*') in SELECT: all values in data source satisfying the conditions in FROM will be returned. Listing 2 summarizes OpenAPI QL syntax:

Listing 2. OpenAPI QL syntax.

```
1  SELECT <service identifier> AS <service identifier name>
2  FROM AS <OpenAPI object> AS < OpenAPI object name> ... AS
        ...
3  WHERE  OpenAPI object property <condition> OpenAPI object
        property ...   <response status code> BETWEEN <response
        status code>
```

Table 1. OpenAPI query model.

OpenAPI object	Key	Value
Service	Service.id	Service identifier
	Service.*	Service description
Request	Request.method	One of 'get', 'put', 'post', 'delete' etc.
	Request.media_type	One of 'application/json', 'application/xmlapplication/', 'x-www-form-urlencoded' etc.
	Request.schema	Schema name in Request body
	Request.tag	Value of property 'tag'
PathParam	QueryParam.name	Value of property 'name' in Parameters object, where 'in' property has value 'path'
	QueryParam.required	Value of property 'required' in Parameters object is 'true', where 'in' property has value 'path'
	PathParam.schema	Schema object in Parameters object, where 'in' property has value 'path'
CookieParam	CookieParam.name	Value of property 'name' in Parameters object, where 'in' property has value 'cookie'
	CookieParam.required	Value of property 'required' in Parameters object is 'true', where 'in' property has value 'cookie'
Header	Header.name	Value of property 'name' in Header object or, name of Schema object in Parameters object where property 'in' has value 'header'
Tag	Tag.name	Name of Tag in Request object
Response	Response.media_type	One of 'application/json', 'application/xmlapplication/', 'x-www-form-urlencoded' etc.
	Response.status_code	HTTP status code
	Response.schema	Schema object in Response object
Schema	Schema.property	Value of property 'properties' in Schema object
	Schema.type	Value of property 'id' in Schema object (one of 'integer', 'array' etc.)
QueryParam	QueryParam.name	Value of property 'name' in Parameters object, where 'in' property has value 'query'
	QueryParam.schema	schema object in Parameters object, where 'in' Property has value 'query'
OAuth2Auth	OAuth2Auth.name	Value of property 'name' in Security scheme and property 'type' has Value 'oauth2'
	OAuth2Auth.flow	One of 'implicit', 'password', 'clientCredentials', 'authorizationCode'
	OAuth2Auth.scope	One of 'read', 'write'
ApiKeyAuth	ApiKeyAuth.name	Property 'name' is in Security scheme where property 'type' has value 'apiKey'
	ApiKeyAuth.in	Value of property 'in' is in Security Scheme Object and property 'type' has Value 'apiKey'
OpenIdConnectAuth	OpenIdConnectAuth.name	Value of property where security Scheme object has type 'openIdConnect'
	OpenIdConnectAuth.scope	Value of property where security Scheme object has type 'openIdConnect' (i.e. one of 'read', 'write')
	OpenIdConnectAuth.openId-ConnectUrl	Value of property (i.e. a URL string) where security Scheme object has type 'openIdConnect'
HTTPAuth	HTTPAuth.name	Value of property 'name' in Security scheme object and property 'type' has value 'http'
	HTTPAuth.scheme	Value of property 'scheme' is in Security scheme object and property 'type' has value 'http'

The query of Listing 3 addresses the description of the Response object of stored services and requires that the Response.status_code property takes values in the range [200 : 400]. Although this is a simple query with one condition, its translation is more complicated than it appears to be. For example, in Listing 1 the Response object is deeply nested within Path object of the OpenAPI document. The statements in the FROM clause attempt to constraint the search within Response objects of each operation (i.e. a service may specify more than one operation). Otherwise, the query would search for property Response.status_code in the entire OpenAPI description of each stored service (i.e. cannot know which objects to search).

Listing 3. OpenAPI QL query example with one condition.

```
1  SELECT s.id (AS) id
2  FROM Service (AS) s,
3    Request (AS) req,
4    Response (AS) resp
5  WHERE s.request = req
6    AND req.response = resp
7    AND resp.status code BETWEEN 200 AND 400
```

The query of Listing 4 specifies multiple conditions and addresses services whose response body has media type application/json (i.e. Response.media_type = 'application/json') and constraints clients access to data by applying an OAuth2.0 authorization protocol. The last condition of the WHERE clause requires that the OAuth2.0 applies an implicit authorization flow (i.e. the client has to retrieve an access token before issuing the request).

Listing 4. OpenAPI QL example with two conditions.

```
1  SELECT s.id (AS) id
2  FROM Service (AS) s
3    Request (AS) req
4    OAuth2Auth (AS) oa2
5  WHERE s.request = req
6    AND req.media_type = "application/json"
7    AND req.OAuth2Auth = oa2
8    AND oa2.flow = "implicit"
```

5   Query Translation

Figure 3 illustrates the architecture of the OpenAPI QL query translation system. The input to the system can be a query or an OpenAPI service description to be stored. Before storage, the *flattening* service searches the OpenAPI document for objects defined using keyword $ref. These point to reusable objects defined in the Components section of the OpenAPI document. Each such definition is replaced by its actual JSON object using Jsoref library[5]. OpenAPI QL queries are issued using a Graphical User Interface (GUI) and they are translated to equivalent N1QL queries.

---

[5] https://jsonref.readthedocs.io/en/latest/.

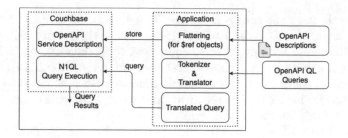

Fig. 3. OpenAPI QL system architecture.

The OpenAPI database is implemented using Cauchbase[6]. Python SDK[7] connects the Python translator and Couchbase. OpenAPI QL uses this module to execute queries expressed in N1QL[8]. The database stores raw OpenAPI service descriptions in JSON (no metadata is stored).

Algorithm 1. Query tokenization.

1: **procedure** TOKENIZE_QUERY(query_string)
2:     SPLIT_QUERY(query_string, select_string, from_string, where_string);
3:     TOKENIZE_FROM_STRING(from_string);
4:     TOKENIZE_SELECT_STRING(select_string);
5:     TOKENIZE_WHERE_STRING(where_string);
6: **return** FROM, SELECT, WHERE token lists;

Query translation takes place in two stages namely, tokenization and translation. Algorithm 1 scans the query (as a string) to extract the elements (tokens) from SELECT, FROM, and WHERE clauses. All extracted tokens are characterized by their type and are inserted in pairs of the form (token, type) in separate lists for each clause. The tokens in the FROM clause can be of any of the following types (i.e. OpenAPI objects): Service, Request, Response, QueryParam, PathParm, CookieParam, Header, ApiKeyAuth, OAuth2Auth, OpenidConnectAuth, HttpAuth. The WHERE clause specifies conditions on pairs of tokens (i.e. OpenAPI object properties for the objects in FROM). The condition types supported by OpenAPI QL can be any of AND, OR, NOT, or a numerical comparison operator. The tokens extracted from the SELECT clause can be OpenAPI objects, object properties, or service identifiers and are those to be displayed.

The second stage is translation to N1QL for each one of the three query clauses. The translation starts with the elements where OpenAPI QL searches into (which are actually those extracted from the FROM clause) followed by the translation of the elements in WHERE and SELECT clauses. First, the

---

[6] https://www.couchbase.com.
[7] https://docs.couchbase.com/python-sdk/current/hello-world/start-using-sdk.html.
[8] https://www.couchbase.com/products/n1ql.

translation checks if the tokens in SELECT and WHERE clauses are in the list of tokens extracted from the FROM clause. Then, the conditions specified in the WHERE clause are translated to equivalent N1QL conditions. In N1QL, the elements in a FROM expression are those in a 'bucket' containing the OpenAPI services. Its translation is a rather simple process as long as the query requests that only service identifiers are returned. The three translated query clauses are combined to form the final N1QL query.

The OpenAPI QL identifier (i.e. 'id') is mapped to the 'id' field of N1QL. It is can be queried using the 'META( )' function of N1QL. In OpenAPI QL, the wildcard ∗ selects all services in a database. In N1QL,'∗' selects all the OpenAPI documents in a bucket. For WHERE, the translator first detects the conditions involved. Each condition is mapped to its equivalent in N1QL. Each translated condition involves an 'ANY' predicate that checks entire OpenAPI document (i.e. over all OpenAPI objects) for properties satisfying the query condition. The final N1QL query is the composition of the SELECT, FROM, WHERE translated expressions.

Table 2. Query to retrieve services whose request body is 'application/json'.

	OpenAPI QL	N1QL
SELECT	s.id AS id	META(m).id service_id
FROM	Service, Request req, Response resp	OASBucket m
WHERE	s.request = req	ANY v WITHIN OBJECT_VALUES(m.'paths')[*] SATISFIES OBJECT_VALUES(m.'paths')[*] IS NOT MISSING
	AND req.response = resp	AND ANY s IN OBJECT_NAMES(v.'responses') AND resp.status_code BETWEEN 200 AND 400 & SATISFIES TO_NUMBER(s) BETWEEN 200 AND 400 END END

Table 2 illustrates a query that retrieves services (i.e. their identifiers) whose response codes are between 200 and 400. The query of Table 3 retrieves services whose request body is 'application/json'. Their equivalent N1QL queries are shown in the right columns of both tables. Notice the complexity of the N1QL queries as opposed to the simplicity of their OpenAPI QL equivalents. More detail on the translation of OpenAPI QL to N1QL along with many examples can be found in [6].

6   Evaluation

The database includes 150 services (400 lines each) collected from SwaggerHub[9] and Amazon. Table 4 shows the average (over 10 queries) execution times for

---

[9] https://swagger.io/tools/swaggerhub/.

Table 3. Query to retrieve services with response codes between 200 and 400.

OpenAPI QL	N1QL
SELECT s.id AS sid, s.*	'META(m).id' AS id, m. *
FROM s.id AS sid, s.*	ANY v WITHIN
WHERE s.request = r	OBJECT VALUES(m.'paths')[*] <optional method declaration> SATISFIES OBJECT_VALUES(m.'paths')[*] IS NOT MISSING AND <nested expression 1> END
AND	<operator does not connect conditions in a N1QL query>
r.media type = "application/json"	nested expression 1: ANY y IN OBJECT NAMES(v.'requestBody'. 'content') SATISFIES y = "application/json" END

six categories of OpenAPI QL queries ordered by response time. The queries are distinguished by the type of OpenAPI object they address (i.e. last column in the table). The 'Filtering time' reports the time to execute the WHERE statements of the translated N1QL query. The translated expressions in N1QL may contain multiple predicates within the WHERE clause even though the original OpenAPI QL query does not specify a condition.

The results reveal that object type impacts retrieval response times significantly. This is related to the level of nesting (in the JSON tree) where each property is located. For example, searching for the method of an operation (i.e. 'get') in Listing 1 will locate this property under Path objects at the top of the JSON tree (there is no need to search the rest of the tree hierarchy). In OpenAPI, the definition of path objects starts with HTTP and encapsulates all other objects deeper in a nested structure. A request for a media type (in a response or a request) will need to search deeper in the tree hierarchy (i.e. a media type is defined under 'content' property in the body of in response to a request).

Queries searching for 'media_type' objects in responses are faster than queries on request objects because not all requests have body objects. The response times for Schema, Parameters or Security objects are about the same (i.e. they are at the same nesting level). Queries on Header objects are slower than all other types of queries: Header objects can be found in more locations in an OpenAPI document (i.e. custom Header objects can be found either in Request or Parameter objects).

Table 5 shows average execution times for queries with one conditional statement (i.e. AND, OR). These queries address the name (i.e. the value of property 'properties') or the type (i.e. the value of property 'id') of a Schema property of a Request object. Queries with AND conditions are slower since their execution involves 'join' operations. Results for conditional queries on different OpenAPI object types (e.g. Schema and Security objects) are not reported. Different object types would require searching at different levels of nesting and is difficult to conclude the performance of such queries.

Table 4. Average query execution times.

OpenAPI QL	Response time (ms)	Filtering time (ms)	Target element
Request.method	48.63	25.25	Operation
Response.media_type	393.73	289.94	Response
Request.media_type	458.28	360.83	Request
Schema.prorpery	464.52	371.32	Schema
\<Parameter\>.name	465.46	393.34	Parameter
\<Security\>.name	476.27	443.15	Security scheme
Header.name	492.42	471.24	Header

Table 5. Average execution times for queries with conditions.

OpenAPI QL	Response time (ms)	Filtering time (ms)	Target element	Operator
Schema.property	464.25	377.15	Request.schema	AND
Schema.type	534.72	444.12	Request.schema	AND
Schema.property	227.17	199.87	Request.schema	OR
Schema.type	274.42	201.08	Request.schema	OR

7 Conclusions and Future Work

OPENAPI QL is a query language for querying OpenAPI service catalogs. Query execution can be seeded-up by indexing (for different types of OpenAPI properties) and can be supported by reasoning for composed or poly-morphed Schema objects in the example of [2]. Extending the syntax with more operators, nested-queries (i.e. queries on results of queries), supporting queries on the full set of OpenAPI properties for Schema objects, and query optimization are in our plans for future work.

References

1. Cha, A., Laredo, J.A., Wittern, E.: OpenAPI-to-GraphQL (2019). https://developer.ibm.com/open/projects/openapi-to-graphql/
2. Anagnostopoulos, E., Petrakis, E., Batsakis, S.: CHRONOS: improving the performance of qualitative temporal reasoning in OWL. In: IEEE International Conference on Tools with Artificial Intelligence, ICTAI 2014, Limassol, Cyprus, pp. 309–315 (2014). https://ieeexplore.ieee.org/document/6984490
3. Karavisileiou, A., Mainas, N., Bouraimis, F., Petrakis, E.G.M.: Automated ontology instantiation of OpenAPI REST service descriptions. In: Arai, K. (ed.) FICC 2021. AISC, vol. 1363, pp. 945–962. Springer, Cham (2021). https://doi.org/10.1007/978-3-030-73100-7_65. http://www.intelligence.tuc.gr/~petrakis/publications/FICC2021.pdf
4. OpenAPI Specification v3.1.0 (2021). https://spec.openapis.org/oas/v3.1.0

5. RAML Version 1.0: RESTful API Modeling Language (2021). https://raml.org
6. Stergiou, I.: Searching in REST service catalogues with OpenAPI descriptions. Technical report, Diploma thesis, School of Electrical and Computer Engineering, Technical University of Crete (TUC), Chania, Crete (2021). https://dias.library.tuc.gr/view/83891

Sensor Virtualization and Provision in Internet of Vehicles

Slim Abbes$^{(\boxtimes)}$ and Slim Rekhis

LR11TIC04, Communication Networks and Security Research Lab, Higher School
of Communication of Tunis (SUP'COM), University of Carthage, Tunis, Tunisia
{slim.abbes,slim.rekhis}@supcom.tn

Abstract. Nowadays vehicles become self-organized mobile computers
thanks to the emerging solutions that are designed to manage interconnection and interaction with the environment such as the IoV. Moreover,
the wide spatial-temporal spread nature of these vehicles attracts cloud
service providers to exploit under-used resources in the aim of growing
their profitability. Virtual sensors represent a promised solution to optimize the exploitation of sensor resources and to offer on-demand services.

We propose a Cloud IoV architecture that integrates functional blocks
of mobile sensor suppliers, Sensor Cloud Service Provider (SCSP), and
service consumers. We design a Markov chain-based solution to predict
the availability of mobile sensors. Furthermore, we propose a sensor virtualization technique to optimize the exploitation of sensor devices. To
allow the SCSPs to maximize their revenue, we model the utility function of SCSP, and deduce the optimal number of allocated sensors. We
also propose a sensor selection algorithm to select the most relevant sensors. A simulation is conducted to assess the efficiency of the proposed
algorithm in terms of allocation blockage rate.

Keywords: Mobile sensor virtualization · Internet of Vehicles (IoV) ·
Sensor Cloud Service Provider · Quality of Information · Optimal
resource allocation · Vehicle mobility prediction

1   Introduction

The Internet of Vehicles (IoV) aims to bring together a huge number of heterogeneous sensors. The mobility of these sensors could lead to inefficiency in resource
exploitation, either because resources are duplicated or are under-used (idle time,
one-time data-processing). Hence, cloudifying the sensor network infrastructure
represents a promising solution to create added-value services and leads to the
virtualization of sensors and the provision of Sensing as a Service (Se-aaS). The
use of sensor virtualization allows to reduce the network overhead and the number of required devices (i.e. if sensors will be configured on-demand). Unfortunately, the high mobility of IoV networks impacts the availability of sensors
leading to the instability of service provision.

© The Author(s), under exclusive license to Springer Nature Switzerland AG 2022
L. Barolli et al. (Eds.): AINA 2022, LNNS 450, pp. 386–397, 2022.
https://doi.org/10.1007/978-3-030-99587-4_33

Several recent studies addressed the allocation of virtual sensors and the provision of cloud-based services. The authors in [2] described a methodology for real-time services provision by virtual sensors (web-based mapping services) and argued the preference of traffic data obtained from virtual sensors compared to those obtained from physical sensors. The authors in [4] designed a cost optimization model for virtual sensors provision in the cloud computing. Their solution addressed cost optimization as a deterministic model without considering the dynamic aspect of vehicle's resource availability. Authors [6] designed a Deep Reinforcement Learning (DRL) solution for resource slicing and provision in vehicular network. Authors in [7] proposed a neural-based solution to estimate the amount of fuel injection in the vehicle's engine using data collected from virtual sensors. The authors in [8] proposed a solution for virtual sensors provision in sensor clouds. Their methods relies on classifying physical sensors with similar properties into separated clusters to optimize energy consumption. Their solutions do not consider the cost optimization. The authors in [3] designed a solution for service discovery and selection with considering mobility characteristics in mobile WSN. The authors in [5] designed a solution for services selection based on the QoS offered by each service. The QoS mainly consists of Energy cost and IoT device availability time. Most of proposed solutions do not treat the virtual sensors allocation considering the mobility of vehicles and thus the availability of sensors to offer the required QoI.

We design in this paper a virtual sensor provision architecture for IoV. We also develop the concept Quality of sensed Information (QoI) and sensor virtualization which supports devices reconfiguration and configuration mapping allowing one sensor to simultaneously serve multiple requests. Due to the fact that the QoI satisfaction depends on the number of allocated sensors in vehicles, we use Markov chain to estimate the availability of their sensors based on the vehicle mobility patterns. Moreover, we define an utility function for the SCSP in order to determine the optimal number of sensors to be allocated. A selection algorithm is then provided to optimally allocate sensors. A simulation is conducted to assess the impact of the proposed sensor selection algorithm on the blockage rate of sensor allocation.

The remaining part of this paper is organized as follows. In Sect. 2, the proposed architecture is presented. Mobile sensor virtualization concept is highlighted in Sect. 3. The Markov model to predict the availability of vehicles is illustrated in Sect. 4. The proposed solution of Se-aaS provision and the utility function for the SCSP are studied in Sect. 5. A sensor selection algorithm is explained in Sect. 6. Before concluding our work in Sect. 8, we conducted in Sect. 7 a simulation to assess the efficiency of our proposed algorithm in reducing the blockage rate.

2   Cloud IoV Architecture and Se-aaS Provision

We propose a Cloud IoV architecture that enables the provision of on-demand, elastic and scalable sensor-based services.

2.1   Cloud IoV Architecture

The proposed architecture includes three layers. The first layer represents the vehicular network composed of a set $\mathcal{O}$ of sensor suppliers representing the vehicles. Each vehicle $o \in \mathcal{O}$ embeds various and heterogeneous reconfigurable (on-demand) physical sensors that can be rented to one or multi-cloud providers. Since vehicles occupy an important geographical area, we propose to split that area into smaller and equal cells in order to facilitate service provision management. We model the cell by a hexagon shape as shown in Fig. 1. The second layer represents the set $\mathcal{C}$ of SCSPs. We consider that each SCSP $c \in \mathcal{C}$ registers a set O_y of sensor suppliers via a SLA contract. The SCSP is responsible for providing on-demand sensing as a service (Se-aaS) with respect to the QoI required by the service consumer. It is assumed that each SCSP has the privilege to request the configuration of the physical sensors to guarantee the elasticity of resources and to track the registered vehicles in order to collect their location history. We suppose that each SCSP c_y installs its infrastructure and offers services within a predefined set of cells. The third layer represents the set $\mathcal{U}$ of service consumers, $u \in \mathcal{U}$, that leverage the services offered by the SCSP with considering their preferences in QoI and service cost.

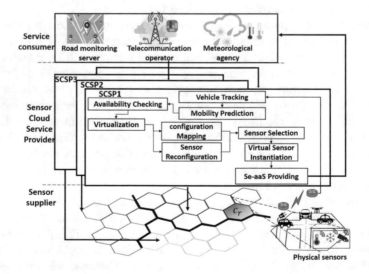

Fig. 1. Proposed Cloud IoV architecture

2.2   Se-aaS Provision

The different actors interact together using the following scheme: 1) The service consumer u_x starts by sending a service request r to the SCSP specifying the required geographical area C_r^x, the service time-period T_r^x and the required Quality of Information Q_r^x; 2) The SCSP must prove its ability to serve r by checking the availability of sensor suppliers registered with it; 3) Thus, the SCSP must firstly divide T_r^x into equal time-slots TS. Based on the history of each registered vehicle $o_z \in O_y$ collected during a sufficient observation period, it predicts the trajectory of each vehicle using the Markov chain; 4) The SCSP assesses its utility for serving r in terms of QoI and allocation cost. In terms of QoI, it has to increase the number of allocated sensors so that data accuracy can be increased. In terms of cost, it has to minimize the total sensor allocation cost and the potential penalties imposed by the SLA (initially concluded between the interactive actors) in case of sensor's unavailability. The SCSP have to compute the required optimal number of allocated sensors to serve r; 5) The SCSP executes a selection algorithm to select the most relevant sensor suppliers with respect to their probability of failure while considering the ability to reuse data from reconfiguarable sensors to serve other requests; 6) Finally, the SCSP instantiates the virtual sensor and associates the selected sensors.

3   Mobile Sensor Virtualization

The sensor virtualization consists in decoupling the physical sensor from its original functionality to serve dynamically one or multi requests. We base our sensor virtualization on two different concepts: a) Sensors reconfiguration: Sensors can run with having different configurations (i.e., threshold, positions, zoom, angle); and b) Configuration mapping: A configuration C_x can be mapped to a configuration C_y (denoted by $C_x \succeq C_y$) if data generated using configuration C_x, can be transformed into the same data that could be generated if C_y is applied. We associate to each sensor a configuration table Tab_p that contains all allowed configurations.

Sensor Reconfiguration: It consists in performing a set of configurations on the sensor by adjusting its parameters. The configurations C_x and C_y can represent two angles of vision performed by the same dashcam. The configuration C_x allows the SCSP to adjust the dashcam with an angle of vision degree α_x to monitor the crossroad congestion in profit of a maps service provider. The configuration C_y can allow an adjustment of the camera lens to focus on vehicle registration plates in profit of road monitoring authorities with another angle of vision α_y, $(\alpha_y \subset \alpha_x)$. In this case, we say that a sensor reconfiguration is applied to serve two requests with different configurations C_y and C_x, $(C_y \preceq C_x)$. Here, $C_y \preceq C_x$ means that the sensor enables a reconfiguration allowing the mapping from C_x to C_y.

Configuration Mapping: It consists in applying a set of software tools and processing operations to better exploit infrastructure resources. We take as an example a request claimed by the road monitoring authority to detect speed violating vehicles. The SCSP can apply a combination of pattern recognition and artificial intelligence algorithms on the video recorded from a dashcam to detect the registration plate of violating vehicles. In configuration mapping, C_x and C_y can represent two resolutions of the same sensor camera. The configuration C_x

allows the camera to generate photos with a higher resolution than C_y. In that case, we say that a mapping from C_x to C_y ($C_x \succeq C_y$) can be done using software tools, allowing the same sensor camera to simultaneously serve two users with two requested configurations C_x and C_y.

4   Predicting Sensors Availability

We propose a Markov chain-based solution to predict sensor devices availability along the service provision.

4.1   Markov Chain for Predicting Vehicle Trajectory

We adopt the Markov chain in the vehicular context to predict the future states of vehicles based solely on their present state (memoryless property). In our model, the present and the future states of a vehicle represent the actual visited cell and the predicted cells to be visited, respectively. The prediction of future vehicle state requires the knowledge of its previous driving history collected during a sufficient observation period. Collecting the history of a vehicle driving data consists in discretizing the time domain into equal TSs and memorizing the frequency of displacement of the vehicle from a cell i to another j ($i, j \in [1\ L]$) based on several observations. The history is used to construct the transition probability matrix P. A state S_t is then obtained in matrix representation as the matrix multiplication of the present state S_0 and the transition probability matrix P to the power t as shown in Eq. 1.

$$S_t = S_0(P)^t = [P_1\ P_2\ ...\ P_L] \begin{pmatrix} p_{11} & p_{12} & \cdots & p_{1L} \\ p_{21} & p_{22} & \cdots & p_{2L} \\ \cdots & \cdots & \cdots & \cdots \\ p_{L1} & p_{L2} & \cdots & p_{LL} \end{pmatrix}^t \quad (1)$$

Where S_t and S_0 are row vectors of L entrees. Each entry represents the probability of apparition of the vehicle in each cell at TS_t. $(P)^t$ represents the transition matrix to the power t. Each element p_{ij} of P represents the displacement probability from a cell i to another cell j as in matrix representation in (1). We obtain $D = [S_0, S_1, ..., S_T]$ as the vehicle trajectory vector that contains its present state (actual cell) and the estimated future states (cells to be visited) during T_r^x. Considering the mobility of the vehicle and the variance in speed we can obtain two successive states that correspond to the same cell if vehicle speed is low or is moving in a closed loop inside the zone, and to non-adjacent cells if the vehicle speed is relatively high.

4.2   Vehicle Availability and Failure Probability

The service can be affected by many factors including bandwidth overhead, transmission error, and data loss [1,6]. In the IoV context, service dissatisfaction can be caused by the faulty estimation of vehicle availability. In our model, we associate to each vehicle a failure probability which represents the risk associated to its unavailability due to mobility and network impairment. After determining the state of the vehicle at each TS_t, the SCSP checks whether that vehicle is available at each TS during the required service period T_r^x within the geographic area C_r^x. Thereby, extracting the probability of availability (denoted by $p_{av,t}^z$ in the sequel) of the vehicle from the state vector S_t in C_r^x at TS_t. Hence, we derive a failure probability $p_{f,t}^z$ from $p_{av,t}^z$ as: $p_{f,t}^z = 1 - p_{av,t}^z$. Accordingly, a vehicle o_z is then considered available in the geographic area C_r^x at TS_t if and only if: $p_{f,t}^z \leq p_{f,th}$. To decide on the involvement of the vehicle in supplying sensors for a given service r, the SCSP must compute a global failure probability p_{fail} related to o_z during the entire service period T_r^x and compares it with a global failure probability P_F. Considering that the travels of the vehicle and its apparition in different cells are independent events, we compute p_{fail} using the following equation: $p_{fail} = 1 - \prod_{t=1}^{t=T} p_{av,t}^z$.

5   Proposed Solution for Providing Se-aaS

We focus on studying the utility of the SCSP to maximize its revenue earned from providing mobile sensors while considering the required QoI.

5.1   Dependency of the Required QoI on the Number of Sensors

In practice, several phenomena could cause the degradation of the QoI, including errors related to data collection, sensor vibration, and sensor uncertainty. Therefore, a single sensor can not solely satisfy the required QoI. In this context, we define $Q_r^x(n)$ that designates the QoI in function of the number of sensors as follows: $Q_r^x(n) = 1 - e^{-\delta n}, 0 \leq Q_r^x \leq 1 \forall n \in [0 \ N]$. Here, $Q_r^x(n)$ represents the QoI required to serve r, where n denotes the number of allocated sensors and δ is a constant that represents the convergence reachability speed that differs from one application to another. For example, a number n_c of dash cams should collect many scenes from different angles and resolutions to give an accurate level of information to assess crossroad congestion. We notice that the more QoI is required the more sensors are to deploy. In fact, the QoI increases decreasingly with the increase of the number of sensors. The smaller value of δ, the slower the convergence reachability and the more number of sensors are required by the application to reach the required QoI as shown in Fig. 2. The ideal case consists in providing only one sensor to capture the required data with high reliability and accuracy (i.e., no data acquisition error).

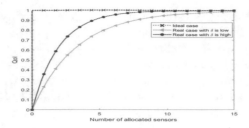

Fig. 2. QoI in function of the number of allocated sensors

5.2   Utility Function for the SCSP

We compute the utility function $U_r^y(Q_r^x, P_r^y)$ to measure the level of satisfaction of the SCSP earned from renting and allocating mobile and re-configurable sensors.

5.2.1   Utility Function Modeling

We suppose that the *SCSP* must consider the penalties caused by vehicle unavailability and prediction errors. Hence, $U_r^y(Q_r^x, P_r^y)$ must satisfy the following properties:

1. Maximizing the revenue: the SCSP seeks to maximize its profit margin G, $(G = P_r^{max} - P_r^y > 0)$. The revenue of the SCSP depends on these parameters: a) The unit allocation price p_r^y payed by the SCSP for each allocated sensor. Thus, the total allocation cost P_r^y is obtained by: $P_r^y = (p_r^y + p_{fail} \cdot C_{pen}) \cdot n + c_m$, where, c_m denotes the fees charged by the SCSP to compensate the sensor supplier c_y in case of sensor damage and b) The unit selling price p_r^c charged by the SCSP for the service consumer for each provided sensor. Based on p_r^c, the SCSP decides on the maximal service cost $P_r^{max} = P_r^c = p_r^c n - c_v$, where c_v denotes the cost of virtual sensor instantiation.
2. Minimizing the probability of sensor failure p_{fail}. Therefore, we consider in the calculation of $U_r^y(Q_r^x, P_r^y)$ a penalty which is imposed by the SLA (initially concluded between o_z and c_y).
3. Maximizing the satisfaction of service consumer by providing the required Q_r^x.

The utility function of the SCSP is defined as follows:

$$U_r^y(Q_r^x, P_r^y) = \alpha \left(\frac{P_r^{max} - P_r^y}{P_r^{max}} \right) + (1 - \alpha) \, Q_r^x \tag{2}$$

where α, $(0 < \alpha < 1)$ is a weight that depend on the preference of the SCSP between paying for sensors allocation and providing a satisfactory QoI to the service consumer. Therefore, reaching the above-mentioned objectives consists in finding an optimal number n_{opt} of allocated sensors subject to:

$arg_{n \in N+} max \ U_r^y(Q_r^x, P_r^y)$. The profit margin G varies according to the strategy of sensors allocation by the SCSP. Therefore, we define two expressions for the total allocation cost P_r^y.

5.2.2 Utility Function Solving for Linear Allocation Cost

We use a first utility function for the SCSP where the expression of the cost P_r^y increases linearly with the number of allocated sensors as in (3):

$$U_r^y(Q_r^x, P_r^y) = \frac{\alpha}{P_r^{max}} \left(P_r^{max} - ((p_r^y + p_{fail} \cdot C_{pen}) \cdot n + c_m)\right) + (1 - \alpha)\left(1 - e^{-\delta n}\right) \tag{3}$$

We search to optimize the utility function $U_r^y(Q_r^x, P_r^y)$ in order to find the optimal number of sensors n_{opt} to serve r while satisfying the service consumer. Firstly, we apply the first derivative test on $U_r^y(Q_r^x, P_r^y)$ with respect to n as in (4):

$$\frac{dU_r^y}{dn} = -\alpha \left(\frac{p_r^y + p_{fail} \cdot C_{pen}}{P_r^{max}}\right) + \delta(1 - \alpha) e^{-\delta n} \tag{4}$$

We obtain the optimal number $(n = n_{opt})$ of allocated sensors, as follows:

$$n_{opt} = -\frac{1}{\delta} \cdot ln\left(\frac{\alpha(p_r^y + p_{fail} \cdot C_{pen})}{\delta \cdot P_r^{max}(1 - \alpha)}\right) \tag{5}$$

Admitting the property of the natural logarithm stating that $ln(x)$ exists if and only if $x \in]0, +\infty[$, and since the value of n_{opt} must be positive $n_{opt} \geq 0$, the following condition needs to be satisfied: $0 < \frac{\alpha(p_r^y + p_{fail} \cdot C_{pen})}{\delta \cdot P_r^{max}(1 - \alpha)} \leq 1$.

5.2.3 Utility Function Solving for Exponential Allocation Cost

We provide another utility function for the SCSP where the expression of the cost P_r^x increase increasingly (exponential cost) with the number of allocated sensors: $P_r^y = e^{(p_r^y + p_{fail} \cdot C_{pen}) \cdot n} + c_m$. The utility function $U_r^y(Q_r^x, P_r^y)$ is then expressed as:

$$U_r^y(Q_r^x, P_r^y) = \frac{\alpha}{P_r^{max}}\left(P_r^{max} - P_r^y\right) + (1 - \alpha) \cdot Q_r^x$$
$$= \frac{\alpha}{P_r^{max}}\left(P_r^{max} - \left(e^{(p_r^y + p_{fail} \cdot C_{pen}) \cdot n} + c_m\right)\right) + (1 - \alpha)\left(1 - e^{-\delta \cdot n}\right)$$

To search for the optimal number n_{opt}, we apply the first derivative on $U_r^y(Q_r^x, P_r^y)$ with respect to n as in (6):

$$\frac{dU_r^y}{dn} = -\alpha \left(\frac{p_r^y + p_{fail} \cdot C_{pen}}{P_r^{max}}\right) \cdot e^{(p_r^y + p_{fail} \cdot C_{pen}) \cdot n} + \delta(1 - \alpha) e^{-\delta \cdot n} \tag{6}$$

Hence, we obtain n_{opt} as given in Eq. (7):

Algorithm 0.1 Sensor Selection Algorithm

Input: Q, tab, C
Output: p_{fail}, n_{opt}, ID, B
Set $p_{fail} \leftarrow \epsilon, ID = Null, B = 0$
While $n_{opt}(p_{fail}) > 0$ and $G(p_{fail}) > 0$
Check if there exist a number N_x of sensors with configuration $C_x = C$
Compute $N_{rem} = n_{opt} - N_x$
Check if $N_{rem} = 0$, then return $p_{fail} = \epsilon, n_{opt}, ID = \{X\}$, else, check if there exist N_y of sensors
with $C_y \succeq C$
Compute $N_{rem} = n_{opt} - (N_x + N_y)$
Check if $N_{rem} = 0$, then return $p_{fail} = \epsilon, n_{opt}, ID = \{X, Y\}$, else, check if there exist N_z of
unexploited sensors
Compute $N_{rem} = n_{opt} - (N_x + N_y + N_z)$
Check if $N_{rem} = 0$, then return $p_{fail} = \epsilon, n_{opt}, ID = \{X, Y, Z\}$, else, there not exist sufficient
sensors with current p_{fail}
Increment p_{fail} with a step and increment B and repeat the selection process until obtaining $N_{rem} = 0$
end while

$$n_{opt} = \frac{ln\left(\frac{(1-\alpha)\cdot\delta\cdot P_r^{max}}{\alpha\cdot(p_r^y+p_{fail}\cdot C_{pen})}\right)}{p_r^y + p_{fail} \cdot C_{pen} + \delta} \tag{7}$$

Based on (7), there exists n_{opt} if and only if: $\frac{(1-\alpha)\cdot\delta\cdot P_r^{max}}{\alpha\cdot(p_r^y+p_{fail}\cdot C_{pen})} > 0$.

6   Sensor Selection Algorithm for SCSP

We propose a selection algorithm for the SCSP based on the failure probability and the configuration allowed by each sensor device as described by Algorithm 0.1. The selection consists in finding the set of sensors (capable of providing data with the required configuration C) associated with the minimal possible failure probability p_{fail} and therefore minimizing the allocation blockage rate.

Algorithm Steps Explanation
The sensor selection algorithm gets as inputs: the required quality of information Q, the sensor configuration table Tab, and the required configuration C. The SCSP computes both n_{opt} and G given the actual p_{fail}. While n_{opt} and G do not reach 0, the SCSP checks whether there exist a sufficient number of sensors to serve the request r. It consults the configuration table Tab_p and selects the sensors that allow the same configuration $C_x = C$ (represents the cheapest alternative and guarantees resources reuse). For each selected sensor, the SCSP checks whether the number N_x of sensors already selected meets n_{opt}. Thereby, computing the remaining number N_{rem} ($N_{rem} = n_{opt} - N_x$) and checks whether it is equal to 0. If the case the goal is reached and the algorithm returns the IDs of selected sensors ID, ($ID = \{X\}$), the corresponding n_{opt}, and p_{fail} with which n_{opt} is obtained. Otherwise, the algorithm continue to check the existence of N_y sensors with configuration $C_y \succeq C$ or N_z of unexploited sensors (reserved by SCSP to serve r). Otherwise, the number of sensors to be allocated is not sufficient, p_{fail} will be incremented by ϵ, the blockage B will be incremented by 1 and the above steps will be repeated until obtaining $N_{rem} = 0$.

7   Simulation and Results

We simulate our model using Simulation of Urban Mobility (SUMO) which is designed to support the mobility of vehicles. We used Traffic Control Interface (or TraCI) to gain mobility information from SUMO and MATLAB to obtain results. We consider a vehicle network with a density of 100 vehicles, network area of $4\,km^2$, a mean speed of $20\,m/s$, a mean service period of $30\,TSs$, and a mean inter-arrival time of requests following the Poisson distribution equal to $35\,TS$. We construct the Markov transition matrix by conducting many simulations to collect the vehicle driving history and predict its future states. We calculate the availability prediction of each vehicle using p_{fail} which varies from 0.05 to 1 with varying step ϵ of 0.05.

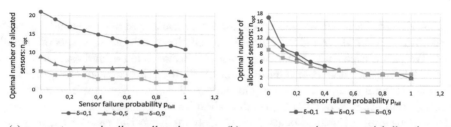

(a) n_{opt} w.r.t p_{fail} using linear allocation cost   (b) n_{opt} w.r.t p_{fail} using exponential allocation cost

Fig. 3. Assessment of n_{opt} w.r.t the sensor failure probability p_{fail} and δ

Firstly, we numerically analyze the impact of δ and p_{fail} on the optimal number of sensors for the cases of linear and exponential allocation costs as depicted in Fig. 3. Based on Fig. 3b, the lower the value of δ, the more the application is needier for sensors to satisfy its required QoI. The curve of n_{opt} (needed to optimally satisfy the SCSP) using a linear allocation cost starts from relatively a high number of sensors (depending on δ) that corresponds to the smallest p_{fail} and keeps decreasing linearly with a slight slope because the impact of adding additional sensors to cope with the increase of p_{fail}, does not have a considerable impact on the allocation cost since it increases linearly as the allocation cost increases. In contrary, the curve of n_{opt} using an exponential allocation cost (as shown in Fig. 3b) starts from a smaller number of sensors (compared to the linear allocation cost) from a low value of p_{fail} and keeps decreasing decreasingly (due to the considerable impact of allocating additional sensors when p_{fail} increases) until reaching a constant n_{opt} from which the SCSP does no longer find it opportunistic to allocate more sensors.

We assess the allocation blockage rate which designates the number of steps (number of times ϵ is incremented) before the number of available vehicles (at a specific p_{fail}) reaches the corresponding n_{opt} compared to the total number of steps. We assign a probability of 0.6 for vehicles equipped with sensors having same configuration as the request and 0.4 for vehicles with incompatible configuration. The obtained Fig. 4 illustrates the blockage rate w.r.t the service area

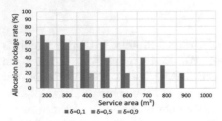
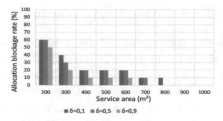

(a) Blockage rate using linear allocation cost   (b) Blockage rate using exponential allocation cost

Fig. 4. Assessment of the blockage rate w.r.t service area surface without using sensor mapping

surface for different δ values. The blockage rate is the highest when the service area is the smallest which corresponds to the most restrictive situation that consists in having the required number of vehicles available in a specific small zone. We also notice that the blockage rate decreases when δ increases since n_{opt} becomes lower and it becomes more probable to find the corresponding number of available vehicles mostly for a higher values of p_{fail}. As shown in Fig. 4a, the blockage rate stills important especially when δ is low because n_{opt} is relatively high when using a linear allocation cost. Contrarily to the case of exponential cost where the blockage rate decreases decreasingly as the surface zone increases and therefore the number of available vehicles increases, as shown in Fig. 4b.

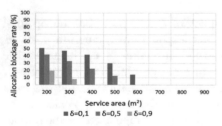
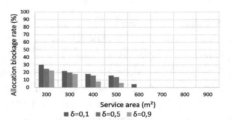

(a) Blockage rate using linear allocation cost   (b) Blockage rate using exponential allocation cost

Fig. 5. Assessment of the blockage rate using the proposed sensor mapping approach

As a next step, we consider the sensor mapping and its impact on the blockage rate. Thereby, setting a probability of 0.6 for vehicles having the same configuration as the request and 0.3 for vehicles with different but enabling the transformation toward the required configuration. The Fig. 5 shows a reduction of blockage that does not exceed 52% using linear allocation cost and 31% using exponential allocation cost even for a service area surface equal to $200\,\mathrm{m}^2$. Moreover, the blockage rate keeps decreasing until it reaches 0 from a service area equal to $700\,\mathrm{m}^2$ for both the linear and exponential allocation costs. As shown in Fig. 5a, the blockage rate keeps decreasing until reaching 0 from a service area equal to $700\,\mathrm{m}^2$ for the lowest value of δ and $400\,\mathrm{m}^2$ for the higher value of δ when using linear allocation cost. Contrarily, when using exponential allocation cost (as shown in Fig. 5b) the blockage rate decreases more significantly because n_{opt} becomes low regardless δ.

8 Conclusions

In this paper, we proposed a Cloud IoV architecture to manage the provision of on-demand, elastic and scalable services. We developed the concept of mobile sensor virtualization to optimize resources exploitation and to reduce system overhead. To cope with mobility issues, we proposed a Markov-based solution to predict the availability of mobile sensors. We defined an utility function for the SCSP that allows its maximal profitability. We proposed a selection-based algorithm to select the most available sensors, reuse already allocated sensors, or call for configuration mapping. Finally, we conducted a simulation that shows the efficiency of our approach to reduce the allocation blockage rate.

References

1. Fiems, D., Vinel, A.: Connectivity times in vehicular networks. IEEE Commun. Lett. **22**(11), 2270–2273 (2018)
2. Morgul, E.F., et al.: Virtual sensors: web-based real-time data collection methodology for transportation operation performance analysis. Transp. Res. Rec. J. Transp. Res. Board **2442**(1), 106–116 (2014). https://doi.org/10.3141/2442-12
3. Osamy, W., Khedr, A.M., Salim, A.: ADSDA: adaptive distributed service discovery algorithm for internet of things based mobile wireless sensor networks. IEEE Sens. J. **19**(22), 10869–10880 (2019)
4. Rachkidi, E.E., Agoulmine, N., Belaâšid, D., Chendeb, N.: Towards an efficient service provisioning in Cloud of Things (CoT). In: 2016 IEEE Global Communications Conference (GLOBECOM), Washington, DC, USA (2016)
5. Singh, M., Baranwal, G., Tripathi, A.K.: QoS-aware selection of IoT-based service. Arab. J. Sci. Eng. **45**(12), 10033–10050 (2020). https://doi.org/10.1007/s13369-020-04601-8
6. Sun, G., Boateng, G.O., Ayepah-Mensah, D., Liu, G., Wei, J.: Autonomous resource slicing for virtualized vehicular networks with D2D communications based on deep reinforcement learning. IEEE Syst. J. **14**(4), 4694–4705 (2020)
7. Wong, E., Schneider, T., Schmitt, J., Schmid, F.R., Kolter, J.Z.: Neural network virtual sensors for fuel injection quantities with provable performance specifications. In: 2020 IEEE Intelligent Vehicles Symposium (IV), Las Vegas, NV, USA, pp. 1753–1758 (2020)
8. Zhang, M.-Z., Wang, L.-M., Xiong, S.-M.: Using machine learning methods to provision virtual sensors in sensor-cloud. Sensors **20**(7), 1836 (2020)

A Secure Data Storage in Multi-cloud Architecture Using Blowfish Encryption Algorithm

Houaida Ghanmi[1(✉)], Nasreddine Hajlaoui[1], Haifa Touati[1],
Mohamed Hadded[2], and Paul Muhlethaler[3]

[1] Hatem Bettahar IResCoMath Research Lab, University of Gabes, Gabes, Tunisia
houaida.ghanmi22@gmail.com, nasreddine.hajlaoui@fsg.rnu.tn,
haifa.touati@cristal.rnu.tn
[2] Institut de Recherche Technologique SystemX, Paris, France
mohamed.elhadad@irt-systemx.fr
[3] INRIA Paris, 2 Rue Simone IFF, 75012 Paris, France
paul.muhlethaler@inria.fr

Abstract. In recent years, cloud computing has played an important role in accessing services around the world. Thanks to its rapid development in the IT industry it has gained huge appeal. It offers users the ability to store, retrieve and share data and resources with other users however and whenever they wish. Availability, confidentiality and integrity are key issues in cloud computing. The open nature of the technology and public access make data and applications in a single cloud vulnerable to multiple security breaches such as single point of failure and malicious insiders. To overcome such issues, the data is stored in a multi-cloud environment. In this paper, we integrate different security approaches in order to provide a new lightweight security framework for authentication and data storage in multi-cloud environment using three main algorithms: certificate authentication for user identification, Blowfish for data encryption, and RSA for the transfer of Blowfish's secret key. The proposed solution is evaluated based on download time, upload time and access time. Analysis of the experimental results shows that the proposed architecture guarantees secure, fast and lightweight access to data while preserving data confidentiality against unauthorized third parties.

1 Introduction

Cloud computing has received a great deal of attention over the past decades due to the extensive use and sharing of data on the Internet. It is a smart way to use or share resources as it provides business to many people remotely and at the same time it allows customers to get applications of any kind in the form of services available electronically on the Internet. The cloud computing paradigm plays a major role in sharing data and resources to other devices through data outsourcing. To secure the information in the cloud storage, data will be encrypted before being stored in the cloud environment. However, these

L. Barolli et al. (Eds.): AINA 2022, LNNS 450, pp. 398–408, 2022.
https://doi.org/10.1007/978-3-030-99587-4_34

increased needs also lead to drawbacks. In general, the users store their data within the cloud, however they need to be more cautious to protect their sensitive information like health care data, bank details, and military data from intruders. Even though the Cloud Service Providers (CSPs) have standard regulations and powerful infrastructures to ensure customers' data privacy, reports of privacy breach and service outage have appeared over the last few years. Hence, the most critical issues to solve in cloud computing are data authentication, integrity, data access control, identity management and confidentiality. These cloud security issues and challenges have triggered a migration or move to multi-cloud. The multi- cloud storage environment makes it possible to divide the user's data block into data elements and by distributing them among the available CSPs, at least a threshold number of CSPs can participate in the successful recovery of the entire block of data, which allows for better data confidentiality and availability. However the necessity to improve security in a multi-cloud environment has become increasingly urgent in recent years.

In this paper, we propose a secure data storage framework that encrypts the data using the Blowfish algorithm [4] before storing them in the multi-cloud. This feature protects the data against internal attacks such as the possibility of a malicious cloud insider. Moreover, our framework proposes to authenticate users using digital certificates. This feature protects sensitive data against external attacks and unauthorized third-party access.

The key contributions of this article are summarized as follows:

- We design a secure and lightweight multi-cloud architecture for authentication and data storage. The main idea of this work is to encrypt the data that will be sent to the multi-cloud platform by the Blowfish algorithm and to encrypt the secret key by the RSA algorithm. Our solution guarantees a low computational cost due to the small symmetric key size.
- We provide a secure authentication process using digital certificates. The advantage of using certificates is to allow mutual authentication, which means that the two parties involved in a communication identify each other before a connection can even be established.
- We provide an analysis of the security and the performance of our system, which shows that our system is secure, fast and lightweight.

The remainder of this paper is organized as follows: Sect. 2 reviews the recent studies related to data storage security in a multi-cloud environment. The proposed framework to secure multi-cloud data storage is explained in detail in Sect. 3. The performance evaluation of the proposed scheme, is presented in Sect. 4. Finally, the conclusion is given in Sect. 5.

2   Literature Review of Recent Cloud Security Solutions

Recently, several security solutions have been proposed to improve the security of data storage in the cloud.

Timothy et al. [8] proposed an approach to securely store data in the cloud using a hybrid crypto-system. The authors designed a three-tiered hybrid approach based on Blowfish, RSA, and SHA-256. Encryption is performed using the Blowfish algorithm, the RSA asymmetric algorithm is used for blowfish key encryption, and data integrity is ensured by the SHA-256 algorithm.

In [9], Viswanath et al. propose a new encryption approach to secure big data storage in the multi-cloud. In this paper, the proposed encryption mechanism combines the functionality of AES and a Feistel network. The Feistel network divides the key into a number of blocks. The cipher key produced by the Feistel network is used in the AES algorithm to achieve a high avalanche effect.

Tayssir et al. suggested in [2] an architecture for realizing fine-grained, flexible, and scalable outsourced data storage and access control in multi-cloud. This architecture uses RSA to encrypt the plaintext, and the Diffie-Hellman algorithm to encrypt the private key of RSA to ensure that the private key has been transferred securely. Moreover, SHA-256 is used to check the authentication and data integrity.

In [5], Mudepalli et al. presented an efficient ciphertext retrieval techniques for large volumes of data. Firstly each file is indexed by Porter stemming. Then the file and the index are encrypted using the Blowfish algorithm and the Elliptic-Curve Cryptography (ECC) is used for secret key sharing.

Pravin et al. [6] presented an efficient framework for secured data sharing in multi-cloud storage. The proposed framework integrates the 3DES and ECC algorithms. 3DES has the advantage of proven reliability and a longer key length that eliminates attacks that can be used to break DES, and ECC can use small number of keys for a high level of security. This approach provides the flexibility of inter-cloud communication with secure transactions. The performance evaluation conducted by the authors shows that, compared to traditional approaches, this solution achieves better performance in terms of resistance to internal attacks, sharing of secret keys and data integrity.

Table 1 analyzes the cloud security solutions that we have presented according to different criteria. More precisely, we summarize the different algorithms used by these approaches and compare them in terms of four security goals, namely authentication, integrity, data access control and confidentiality. An important point that we can note from Table 1, is that none of the existing solutions meets the four security goals. Hence we emphasize the need to design a new solution to secure data storage in multi-cloud environment. Moreover, we compare in the last line of Table 1 our solution to existing approaches. Our qualitative comparison confirms that our method reduces the risk of data leakage by putting a powerful access control mechanism. Moreover, it provides confidentiality, integrity and access control of data stored in the cloud through the use of a multi-cloud concept, digital certificate authentication, symmetric and asymmetric cryptography.

Table 1. Qualitative comparison of cloud security solutions

Solution	Data encryption	Key encryption	Key sharing	Evaluation parameters	Security gols			
					Authentication	Integrity	Data access control	Confidentiality
Timothy et al. [8]	Blowfish	RSA	–			✓		✓
Viswanath et al. [9]	AES	Fistel	Secure channel	Throughput running time				✓
Ismail et al. [2]	RSA	RSA	Diffie-Hellman	–		✓	✓	✓
Mudepalli et al. [5]	Blowfish	ECC	–	Communication cost				✓
Pravin et al. [6]	3DES	ECC	Trusted third party	Latency		✓		✓
Our solution	Blowfish	RSA	RSA	Encryption time Decryption time	✓	✓	✓	✓

3   The Proposed Multi-cloud Security Framework

In this section, we propose a new architecture to secure data stored in a multi-cloud environment. Specifically, we are designing a new hybrid architecture which integrates digital certificates and two popular and widely used asymmetric and symmetric cryptography algorithms which are RSA and Blowfish. The Blowfish algorithm is based on a 16-round Feistel network and it uses large key-dependent S boxes for encryption/decryption purposes. It is a very well-known fast secret key cryptography, with a block size of 64 bits and a variable key length ranging from 32 bits to 448 bits. The main objectives of the proposed framework are:

- **Ensure secure, fast and light data storage:** The experimental results of the comparative analysis [3] of Symmetric Algorithms in Cloud Computing show the superiority of Blowfish encryption over other algorithms in terms of processing time. Moreover, in another study [1], the experimental results show that, compared to other symmetric algorithms, Blowfish reduces both power consumption and execution time. Hence, in our solution we chose to integrate the Blowfish algorithm given the high speed of its encryption/decryption process. Once the data have been encrypted on the client side, a signature is performed by our framework with SHA-256 in order to guarantee the integrity of the data.
- **Ensure secure key exchange:** Even though the asymmetric algorithm is slower, it gives good results for the secure data transfer. That is why we chose to use RSA to encrypt the secret key of the Blowfish algorithm. Once the secret key has been encrypted with the public key, it can only be decrypted with the corresponding private key.
- **Guarantee secure and robust authentication:** Digital certificates are generated for secure, authenticated and accurate access from the cloud.

Figure 1 gives an overview of the proposed framework. As shown in this figure, our solution operates in three main steps. The first step, *the registration and authentication process*, ensures that all users are registered and correctly identified in our framework. The second step, *the storage process* is the result of Shamir's secret sharing approach. Shamir's secret sharing approach [7] is used for the distribution of the data among various CSPs, as well as to ensure the security of the data stored in the cloud (i.e. to anchor data against the curiosity of cloud providers). This approach makes it possible to divide D (Data) into m pieces $D_1...D_m$ so that D is easily reconstructable from any k pieces, but even a complete knowledge of $k-1$ pieces reveals absolutely no information about D. The third step, **the recovery process**, aims to securely transfer data to the right data requester. This is the process of recovering data from the multi-cloud. A detailed description of each step is given in the following subsections.

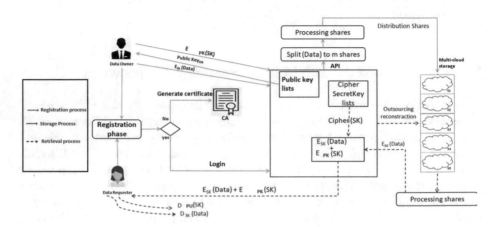

Fig. 1. Flowchart of the proposed framework

3.1   The Registration and Authentication Process

This process ensures the authentication and registration of all users in our API using a login/password as well as with digital certificates. The steps below describe the authentication operation of the proposed methodology. First, the user requests a certificate from a certification authority. Once authenticated, he can register on the platform or he can connect.

- The user initiates an HTTPS connection when connecting to the API.
- The API presents its certificate to the user through the HTTPS protocol.
- The user checks the API certificate. If it is valid, he can continue. Otherwise, the connection will immediately be ended.
- The user sends his authentication certificate to the API. This allows mutual authentication between the user and the API.
- The API checks the validity of the user's certificate. In other words, it contains a digital signature and it has been signed by a trusted certification authority.
- The API checks whether the user has access to the data or not. i.e. it checks if this user is on the list of users with whom the data owner will share his data.
- Finally, if all goes well, the user has access to the secure data.

This process is illustrated in detail in Fig. 2 and the notations used are summarized in Table 2.

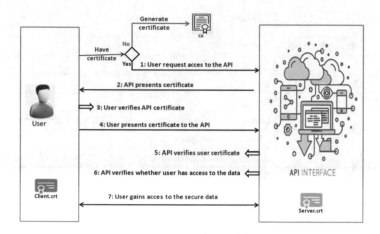

Fig. 2. Authentication process

3.2   The Storage Process

This process helps ensure secure data storage in a Multi-cloud environment. In this process, we will use Shamir's secret sharing approach to split the data into chunks and distribute them to different cloud service providers.

The proposed storage process works as follows:

- The Data Owner (DO) sends a request to the API for data storage.
- When the API receives the request from the DO, it checks that this user has access to the API.
- Once the DO obtains access to the API, the data can be sent along with the list of users (Data Requester) with whom the data can be shared. Therefore, a Secret Key (SK) for the Blowfish algorithm must be chosen to encrypt clear text (D) to produce cipher text (C) using (SK)
  $C = Enc\_Blowfish(SK, D)$
- When the API receives the list of DRs with which the DO will share his data, the API sends to the DO the list of public keys of the DRs.
- The DO encrypts his secret key (SK) with the various public keys of the DRs with whom he will share his data and sends the encrypted list to the API. That is, encrypt (SK) with the RSA algorithm using the list of RSA public keys sent by the API.
  $C\_SK = Enc\_RSA\_PK(SK)$
- When the API receives the encrypted data, it applies the SHA-256 hash function to the data. SHA-256 is intended to ensure the integrity of the message.
  $M = SHA\_256(C)$

After that, the API divides D into m shares $[D_1, ... D_m]$ using the Shamir secret sharing approach, which are the shares to be stored in independent CSPs.

Table 2. Used annotations

Annotation	Definition
E	Encryption
D	Decryption
SK	Secret key
n	Number of cloud service provider to store data
$CSP_1, ... CSP_n$	Cloud service provider
$D_1, ... D_m$	Cloud service provider
m	Number of data chunks
PK	Public key
PU	Private key
H(.)	Hash function

3.3   The Retrieval Process

The proposed recovery process works as follows:

- The data consumer first connects to the API to access the data. If it is authorized, the API reconstitutes the requested data (C') from the various CSPs. (C') must be passed to SHA-256 to produce (M')
  $M' = SHA\_256(C')$
- If $M = M'$ then the data is authenticated.
  Otherwise it will be rejected and the sender will receive a request to resend the message.
- Subsequently, the API assigns the data encrypted with the corresponding secret key encrypted with the public key of the data requester. Afterwards, the latter can decrypt the secret key by his private key. The owner of the corresponding private key will be the only user able to decrypt the encrypted message.
- The original message can be obtained by applying the Blowfish algorithm with its decrypted secret key (SK) on the cipher text (M').
  $D = Dec\_Blowfish(SK, M')$

4   Performance Evaluation

In this section, we implement our solution using Multi-cloud storage with the Cloudera platform to store and retrieve encrypted data. User data is stored using the HDFS service, which is a service offered by the Cloudera. The application used for data encryption and decryption is written in the Python language. We evaluate our solution in terms of three metrics: *download time*, *upload time* and *access time*, as shown in Table 3. We have varied the file sizes, file types and the hash function used with the Blowfish algorithm for the upload/download times:

- Files type: pdf, mp3, txt, mp4
- Files Size: 3.5 KB–243.8 MB
- Hash function: SHA-256 and SHA-512

4.1   Time Evaluation Without Using Certificates for Authentication

In this part, we focus on the upload and download times without taking into account the user authentication. These metrics are defined as follows.

Upload Time: The time it takes to encrypt, upload the data file, and distribute it to different CSPs.

Download Time: The time required to reconstruct data from different CSPs and the time to decrypt that data. A comparison of the upload and download times taken by Blowfish with a key size of 128 and 256 bits for different file sizes are shown in Fig. 3 and Fig. 4. We observe that there is not a big time difference between the two key lengths. Varying the key length of the Blowfish algorithm has no effect on the upload or download time, while changing the file size directly affects data processing time. The analysis of the results shown in Fig. 5 and Fig. 6 shows that the time taken by the SHA-256 hash function is much less than the time taken by SHA-512, and this is especially the case when the file size increases. Since our goal is to develop a secure and fast architecture, we will be using the SHA-256 hash function.

Table 3. Comparative analysis of upload/download time between BLOWFISH-128 and BLOWFISH-256 without certificate integration

File type	Input file size	Upload time (ms)		Download time (ms)	
		BLOWFISH-128	BLOWFISH-256	BLOWFISH-128	BLOWFISH-256
.txt	3.5 KB	0.06	0.08	0.03	0.03
.mp3	700 KB	0.1	0.1	0.09	0.08
.pdf	2.1 B	0.22	0.21	0.19	0.19
.mp3	5 MB	0.88	0.45	0.45	0.43
.mp4	10 MB	0.81	0.92	0.80	0.77
.mp4	17 MB	1.51	1.47	1.41	1.41
.pdf	19 MB	1.67	1.58	1.56	1.56
.mp3	86.4 MB	7.45	6.99	6.72	6.81
.mp3	177.8 MB	15.69	16.40	14.43	14.66
.mp4	243.8 MB	20.14	19.89	19.47	19.48

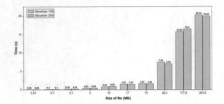

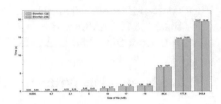

Fig. 3. Upload time taken by (Blowfish-128 vs Blowfish-256)

Fig. 4. Download time taken by (Blowfish-128 vs Blowfish-256)

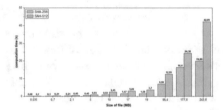

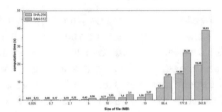

Fig. 5. Upload time taken by (SHA-256 vs SHA-512)

Fig. 6. Download time taken by (SHA-256 vs SHA-512)

4.2 Time Evaluation Using Certificates for Authentication

In this section, we have added the authentication time taken by a user to access our framework. This is the time that our framework takes to verify the validity of the user's certificate.

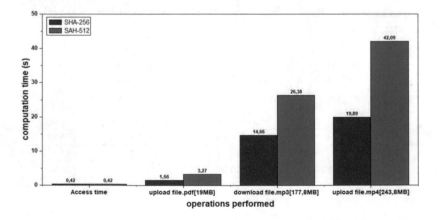

Fig. 7. Access time for upload/download time

As shown in Fig. 7, the average authentication time equals 420 ms for both the SHA-256 and the SHA-512 hash functions.

Figure 8 represents the upload and download time with the access time, the uploading and downloading time without the access time and the total time of 4 files of different types (txt, pdf, mp3 and mp4) using Blowfish with the SHA-256 function.

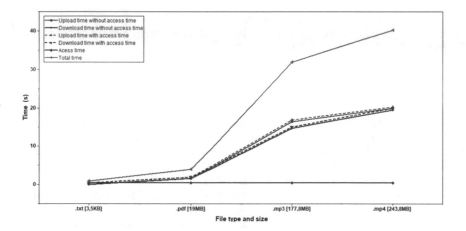

Fig. 8. Comparison of upload, download and total time with SHA-256

According to Fig. 8, we notice that the "upload and download times with the access time" is a little longer than the "upload and download times without the access time", that is to say that we have not lost a lot of time during the authentication phase. The total time is the time required to store and retrieve a file. For a 3.5 KB .txt file, the total time required to store and retrieve this file is 0.95 95 s. It is the same for a .pdf file of 19 MB [total time = 3.98 98 s], .mp3 of 177.8 MB [total time = 31.9 9 s] and .mp4 of 243.8 MB [total time = 40.21 21 s]

5 Conclusion

In a multi-cloud environment, the primary concern for data security is to ensure secure storage and access control of stored-data in the multi-cloud. Therefore, we propose in this paper an efficient and scalable access control approach of hybrid encryption technique based on the Blowfish algorithm, digital certificate for authentication and the RSA algorithm to encrypt Blowfish's secret key to ensure that the secret key has been transferred securely. Additionally, we used SHA-256 to verify authentication and data integrity. We also used Shamir's secret sharing approach for data distribution to various CSP.

Our qualitative comparison confirms that, compared other solutions proposed in the literature, our framework meets the four considered security goals, namely, authentication, confidentiality, data access and integrity. Finally, we evaluated the performance of our security framework by varying data sizes between 3.5 KB and 243.8 MB. The results show that the proposed method is effective in terms of security, speed and efficiency.

References

1. Elgeldawi, E., Mahrous, M., Sayed, A.: A comparative analysis of symmetric algorithms in cloud computing: a survey. Int. J. Comput. Appl. **975**, 8887 (2019)
2. Ismail, T., Touati, H., Hajlaoui, N., Hamdi, H.: Hybrid and secure e-health data sharing architecture in multi-clouds environment. In: Jmaiel, M., Mokhtari, M., Abdulrazak, B., Aloulou, H., Kallel, S. (eds.) The Impact of Digital Technologies on Public Health in Developed and Developing Countries: 18th International Conference, ICOST 2020, Hammamet, Tunisia, June 24–26, 2020, Proceedings, pp. 249–258. Springer, Cham (2020). https://doi.org/10.1007/978-3-030-51517-1_21
3. Mota, A.V., Azam, S., Shanmugam, B., Yeo, K.C., Kannoorpatti, K.: Comparative analysis of different techniques of encryption for secured data transmission. In: 2017 IEEE International Conference on Power, Control, Signals and Instrumentation Engineering (ICPCSI), pp. 231–237. IEEE (2017)
4. Mousa, A.: Data encryption performance based on Blowfish. In: 47th International Symposium ELMAR, pp. 131–134. IEEE (2005)
5. Mudepalli, S., Rao, V.S., Kumar, R.K.: An efficient data retrieval approach using blowfish encryption on cloud ciphertext retrieval in cloud computing. In: 2017 International Conference on Intelligent Computing and Control Systems (ICICCS), pp. 267–271. IEEE (2017)
6. Pravin, A., Prem Jacob, T., Nagarajan, G.: Robust technique for data security in multicloud storage using dynamic slicing with hybrid cryptographic technique. J. Ambient. Intell. Humaniz. Comput., 1–18 (2019). https://doi.org/10.1007/s12652-019-01563-0
7. Shamir, A.: How to share a secret. Commun. ACM **22**(11), 612–613 (1979)
8. Timothy, D.P, Santra, A.K.: A hybrid cryptography algorithm for cloud computing security. In: 2017 International Conference on Microelectronic Devices, Circuits and Systems (ICMDCS), pp. 1–5. IEEE (2017)
9. Viswanath, G., Krishna, P.V.: Hybrid encryption framework for securing big data storage in multi-cloud environment. Evol. Intel. **14**(2), 691–698 (2020)

Micro-Service Placement Policies for Cost Optimization in Kubernetes

Alkiviadis Aznavouridis, Konstantinos Tsakos, and Euripides G.M. Petrakis[✉]

School of Electrical and Computer Engineering, Technical University of Crete (TUC),
Chania, Greece
{aaznavouridis,ktsakos}@isc.tuc.gr, petrakis@intelligence.tuc.gr

Abstract. Kubernetes enables deployment and orchestration of containerized applications across server infrastructures and in the cloud. Services running in separate containers are packed together in pods and are placed in Virtual Machines (VMs) or nodes. Application services must be placed in nodes in a way that minimizes the cost of operating an application. The problem often occurs when deploying applications comprising multiple communicating micro-services which may not fit all together in one node. This work suggests that an application is modeled by means of a graph and handles the problem of service placement as a graph clustering one. Service placement must minimize both, the node resources and the bandwidth consumed by the communicating nodes. Several graph clustering methods are deployed in the Kubernetes infrastructure of the Google Cloud Platform (GCP) to support service placement for two real use cases (an IoT architecture and an e-shop). The experimental results demonstrate that the costs of hosting applications on GCP can be significantly reduced compared to the costs of applications placed using the default Kubernetes method.

1 Introduction

Kubernetes (K8s) has become particularly popular in recent years. Most cloud providers offer Kubernetes as a service to developers to support efficient application deployment, orchestration, and monitoring. Different applications run in clusters of nodes (or VMs). Application services are grouped in pods which may span different nodes within the same or different server infrastructures. The decision of which services should be grouped together and run in the same node has a certain impact on both, the running time and the operation cost of an application [6,9,10]. The need for standardizing technologies for service placement in Kubernetes is also of crucial importance for minimizing the cost of hosting applications in the infrastructure of the provider. It is therefore essential for both, the customer and the cloud provider. Especially for the cloud provider, minimizing infrastructure costs is paramount for user acceptance and market success.

© The Author(s), under exclusive license to Springer Nature Switzerland AG 2022
L. Barolli et al. (Eds.): AINA 2022, LNNS 450, pp. 409–420, 2022.
https://doi.org/10.1007/978-3-030-99587-4_35

Kubernetes orchestrates computing resources per application micro-services at runtime by applying a best-fit (heuristic) policy based on resource requests and user preferences. Service Placement (SP) in Kubernetes is far from optimal in terms of resources consumed and easily results in increased costs for the end-users. Configuring a Kubernetes cluster for multiple applications is not a simple task, if optimization of the node resources and of the network communication between nodes are taken into consideration [16,17].

Most micro-service placement solutions have not been tested using real use-cases or real infrastructures or, are not designed for Kubernetes. In this work, the problem of service placement is formulated as a graph clustering (or graph partitioning) one [3]. Application micro-services form graphs with nodes representing micro-services and edges representing communicating micro-services. Both nodes and edges are labeled by the resources consumed (i.e. mainly CPU and RAM for micro-services) and by the affinities between the micro-services (i.e. network traffic). Graph partitioning suggests a minimum set of weakly connected clusters of nodes each comprising micro-services linked heavily with each other. This guides the placement of service clusters to nodes. The experimental results reveal that this placement results in minimum total usage of resources and reduced cost for the end-user.

Recent approaches (e.g. Hu, Laat and Zhao [8], Sampaio et al. [10]), model the problem of placing micro-services in Kubernetes as a graph partitioning one. As a proof of concept Hu, Laat, and Zhao [8] run several experiments using synthetic workloads on Exogeni [2] cloud simulator; similarly, Sampaio et al. [10] used mock and empirical evaluations and artificial service topologies to assess the performance of their method. Similar to most recent works, service placement methods are tested using simulators rather than real use cases and real infrastructures.

The proposed work forwards these approaches in certain ways: (a) several methods are deployed in the Kubernetes platform of GCP and are evaluated using the Google OnlineBoutique e-Shop[1] application and the iXen IoT architecture [14], (b) cost functions for estimating usage costs per month are proposed inspired by the way GCP charges costs to its users, (c) for reliable monitoring of Kubernetes resources, an architecture based on service mesh[2] is proposed and implemented. The experimental results reveal that the heuristic best-fit service placement method of Kubernetes is far being from optimal (i.e. in most cases, graph partitioning achieved a reduction in cost by at least 25%).

Related work on service placement is discussed in Sect. 2. Section 3 presents the graph partitioning methods in this study. The issue of resource monitoring for Kubernetes using an Istio service mesh is discussed in Sect. 4. The evaluation of four graph partitioning methods and their application to service placement in Kubernetes is presented in Sect. 5, followed by conclusions and issues for future work in Sect. 6.

---

[1] https://github.com/GoogleCloudPlatform/microservices-demo.

[2] https://www.redhat.com/en/topics/microservices/what-is-a-service-mesh.

2   Related Work

Service Placement (SP) is a well-known problem and has been studied extensively in the literature. In recent works by Hedhli and Menzi [6] and Salhat, Desprez and Lebre [15], existing methods are categorized by type of infrastructure (i.e. single or heterogeneous clouds or cloud - fog), and by optimization policy. Farhadi et al. [5] presented a two-scale framework for joint service placement and request scheduling in edge clouds for data-intensive applications. They used simulations for testing. Pallewatta, Kostakos, and Buyya [13] proposed a decentralized micro-service placement policy for heterogeneous and resource-constrained fog environments. Each micro-service is placed to the nearest data center to minimize latency and network usage. Their method improved latency and reduced the delay in network communication. Apat Sahoo and Maiti [1] proposed an SP model for minimizing the energy consumption in a fog environment. They did not test their model on an actual fog environment.

Wang et al. [16] proposed two schedulers for the SP problem in Kubernetes based on historical request data. Zhong and R. Buyya [17] proposed a task allocation strategy for Kubernetes in a heterogeneous environment. Their method relies on a sufficient job configuration policy, cluster size adjustment, and a service re-scheduling mechanism that led to the cost reduction of application hosting by reducing the number of nodes. Sampaio et al. [10] proposed REMaP, a service placement for Kubernetes that proved to increase the response time and reduce the number of nodes of an application. They presented experiment results using simulated application graphs and also Azure cloud. They implemented a monitoring component using Influxdb[3] and Zipkin[4].

Huang and Shen [9] proposed a service deployment method to reduce application response times rather than the cost of application hosting. They modeled an application using graphs that represent the communication costs and the parallelism between services. They handled the problem of service placement in VMs as a minimum k-cut problem. They showed performance results for 4 service deployment methods on Amazon EC2. They did not use Kubernetes and run all experiments on only one VM. Bhahmare et al. [4] dealt with the problem of scheduling micro-services on different types of cloud environments. They showed a reduction in the communication traffic of micro-services and improved response times to requests. Hu and C. Laat and Z. Zhao in [8] modeled SP as a graph partitioning problem and used service affinities to re-arrange services into the nodes. Both methods rely on simulation results and have not been tested in a realistic environment and real use cases.

---

[3] https://docs.influxdata.com/influxdb.
[4] http://zipkin.io.

3   Graph Partitioning for Service Placement

The following methods process the affinity edges of micro-services on the services graph of an application. They partition the micro-services graph into smaller groups of services with high-affinity rates. The final objective is to minimize both, the number of nodes and the traffic rates across nodes. Finally, groups of micro-services are placed in separate Kubernetes nodes. Most solutions work in two stages. A heuristic post-processing stage is applied to the clusters produced by a graph-partitioning method (in the first stage). Its purpose is to determine if placement with even fewer nodes does exist. This will increase the intra-machine affinity and reduce the inter-machine traffic. This is translated to lower Egress traffic for which the end-user is charged. Some nodes are consolidated to one with sufficient capacity by transferring to it their micro-services. The end-user is not charged for high intra-machine traffic.

The Binary Partition (BP) and the K-Partition (KP) algorithms are implemented according to [8]. Both, utilize Karger's (contraction) algorithm [11] to create the application's micro-services partitions. The contraction algorithm randomly chooses an edge with probability proportional to its weight and merges the nodes assigned to this edge into one node. To find a minimum cut, the algorithm iteratively contracts edges until the required number of nodes remains. The Binary-Partition algorithm defines a threshold that represents the upper bound of the resource demands of the output parts. The algorithm attempts to place micro-services into nodes until the resource demands of each part do not exceed the threshold or no part contains more than one service. It divides each part into two sub-parts by applying the contraction algorithm for $K = 2$. The K-Partition algorithm increases K at each iteration.

For both, BP and KP algorithms, a Heuristic Packing (HP) algorithm [12] is run afterward (post-processing) to determine if there is a service placement solution with fewer nodes. Without considering the traffic rate, the problem can be formulated as a multi-dimensional bin packing problem that is NP-hard. The time complexity of the algorithm is reduced to polynomial by applying the (so-called) Traffic Awareness and Most Loaded heuristics [8]. The idea of the Most Loaded heuristic is to improve resource efficiency by packing each part to the most loaded machine. The idea of the Traffic Awareness heuristic is to give priority to machines based on intra-machine traffic. The algorithm terminates when no nodes can host any partition.

Bisecting K-Means (BKM) is a service placement algorithm proposed in this work. Bisecting K-means takes as input the number of clusters to be created and implements a divisive hierarchical clustering process [7]. At each iteration, the algorithm splits the cluster with less traffic into two partitions using k-means clustering (i.e. micro-services inside the selected cluster are assigned to the new clusters). Each service is placed in a cluster if it increases its intra-traffic more than the other cluster. The algorithm terminates when K partitions have been created. The algorithm must run for all values of K in a known range (i.e. available or desired number of machines) and the solution with less cost is selected. Heuristic Packing (HP) is executed afterward to further reduce the number of nodes. This means that the final output of the algorithm can be less than K clusters.

The Heuristic First Fit (HFF) algorithm by Sampaio et al. [10] suggests a placement of application micro-services that minimizes inter-node traffic (i.e. Egress network traffic) and reserves as few nodes as possible. The algorithm re-organizes the micro-services in the host machines of the cluster so that, micro-services with high affinity are co-located. The algorithm produces a complete service placement solution in one step.

4   Monitoring for Service Placement in Kubernetes

Kubernetes is an open-source platform for managing containerized applications and services. Application services run in separate containers. Groups of services are packed in pods that are deployed in nodes (VMs). For deploying applications in Kubernetes, a cluster of nodes is initialized. The Control Plane of Kubernetes is responsible for instantiating and managing nodes and pods.

Scheduling in Kubernetes refers to the process of finding for each micro-service, the most suitable node to place and run its pod (for simplicity of the discussion we assume that each micro-service runs in one pod). Kube-Scheduler[5] is the default scheduler of Kubernetes and runs in the Control Plane. For pods with different resource demands, the scheduler checks existing nodes to decide which nodes have the capacity to host the new pod (feasible nodes). Criteria for placement are resource requirements, affinity relationships with other pods, data locality, user preferences, etc. Placement is determined based on a final score and by applying a heuristic (i.e. best-fit) process.

More elaborate placement decisions can be taken. This requires a pre-processing step on information pertinent to each application and its services. The application's service graph is constructed first. The placement of pods to nodes is determined by applying one of the graph partitioning methods of Sect. 3. The weights on graph nodes (i.e. machine resources) and their affinities (i.e. network traffic) are determined by stressing the application with a large number of requests. The output of a graph partitioning method is a service placement deci-sion. Pods belonging to the same partition are placed in the same node (provided that the node has sufficient resources). Presumably, this placement might not be optimal for different workloads even for the same application, in which case, the whole process (i.e. graph construction and partitioning) has to be repeated when the workload changes.

---

[5] https://kubernetes.io/docs/concepts/scheduling-eviction/kube-scheduler/.

Graph construction requires that the resources and the affinities of each pod (i.e. network traffic between nodes) are carefully monitored. We propose that an additional service layer be deployed in each Kubernetes cluster to add observability and traffic management capabilities to the cluster using a service mesh[6]. Istio[7] is an implementation of this technology. In a service mesh, requests between micro-services are routed using proxies referred to as envoy (sidecar) proxies. An envoy proxy is injected into each pod. Every pod contains application-related containers and an instance of an envoy proxy (i.e. common for all micro-services in the same pod).

Istio components are distinguished into data and control plane. The Data plane consists of every pod's envoy proxy which communicates with all other pods via their envoy proxies. Istio's control plane is responsible for monitoring the status of existing envoys and for injecting envoy proxies into the newly-created pods. Each envoy proxy will send a configuration certificate to Istio's control plane to join the existing service mesh. Istio control plane service is installed in each cluster. It maintains an internal registry containing the set of services and their corresponding service end-points running on the service mesh. Its role is to manage the cluster and push the service mesh configuration to the data plane. Istiod control plane service is installed in each cluster.

Envoy proxies handle inter-node communications, monitoring, and security-related operations via the Istio's data plane. They are responsible for collecting resource metrics using a set of additional services. Kiali[8] is a management console for an Istio-based service mesh. Kiali needs to retrieve Istio data and configurations, which are exposed via Prometheus[9] and the cluster API. Prometheus is an open-source monitoring and alerting service, which collects and stores metrics as time-series data. To pull the desired metrics from the Kubernetes cluster of an application, Prometheus Node Exporter is injected into each node of a cluster. The Node exporter enables measuring various node resources such as memory, disk, and CPU utilization. Prometheus Server extracts these metrics by requesting the data from the Node Exporters. Grafana[10] is used to visualize the collected data in real-time. Grafana can request the data for every node and pod by issuing queries to the Prometheus Server and visualize them in a user interface. Kiali collects these data from the Prometheus Server to create the services graph of an application.

---

[6] https://www.redhat.com/en/topics/microservices/what-is-a-service-mesh.
[7] https://istio.io/latest/about/service-mesh/.
[8] https://kiali.io.
[9] https://prometheus.io.
[10] https://grafana.com.

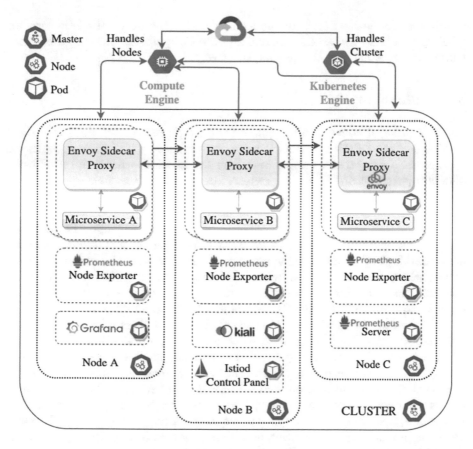

Fig. 1. Kubernetes cluster in GCP with application and Istio services installed.

Each Kubernetes node can host a finite number of pods, including some of the Istio services and Kubernetes components depending on the available size of the node. The installed Istio control plane services, as well as services related to Istio (i.e. Grafana, Prometheus Node Exporter, Prometheus Server, Istiod Control Plane, Kiali), are randomly placed inside the cluster's nodes according to the decision of the Kubernetes Scheduler. These services are not injected with envoy proxies, as they are not part of the application's data plane and their role is only to extract metrics. Istio's service mesh services consume infrastructure resources. It is for the benefit of the cloud provider to support these services free of charge for the end-user (i.e. optimal service placement will result not only in fewer charges for the end-user but also, in more efficient use of resources for the provider).

Istio is responsible for monitoring network traffic (via envoy proxies injected in the pods). Kubernetes cluster is in control of creating and managing nodes and pods. Monitoring data for pods or nodes can be requested from the Kubernetes cluster. Figure 1 illustrates the architecture of a Kubernetes cluster on GCP running an application on three nodes along with the necessary Istio services. GCP provides the Kubernetes Engine and the Compute Engine, to manage the Kubernetes clusters and the nodes respectively.

5   Evaluation

To show proof of concept, all candidate service placement methods are tested on Google Cloud Platform (GCP) using two applications. Each application is deployed in a Kubernetes cluster with Istio service mesh with monitoring capabilities injected according to the architecture of Fig. 1. The first application is iXen [14] an interoperable and expandable server architecture for IoT data management in the cloud. iXen is implemented as a composition of 20 RESTful micro-services in the cloud. The second application is the Online Boutique (e-Shop) of Google[11], a cloud-native micro-services demo application designed for benchmarking and demonstrating purposes. It is a Web-based e-commerce application, where users can browse items, add them to the shopping cart and finally purchase them. Online Boutique consists of 15 micro-services communicating with Remote Procedure Calls (RPC) via gRPC and HTTP.

To construct the services graph, each application is stressed with a synthetic (but realistic) workload using Apache JMeter[12]. Each application is stressed for 15 min. A test plan was designed to simulate the effect of independent users issuing requests simultaneously. Different types of requests stress different services such as login, publication, and subscription to devices and services, search requests for measurements, devices, and users, etc. Similarly, for the Online Boutique application, different types of requests address information such as customer or product indexes, viewing or adding products to chart requests, browsing, and check-out requests. Four thread groups (with 100 threads each) were created for routing different requests. For each application, 10,800 requests in total are issued in 15 min.

All placement algorithms in this study take the services graph information for each application as input and output a group of clusters. The following service placement methods are deployed in Kubernetes:

- Heuristic First Fit (HFF).
- Binary Partition with Heuristic Packing (BP-HP).
- K-Partition with Heuristic Packing (KP-HP).
- Bisecting K-Means - Heuristic Packing (BKM-HP).

5.1   Cost Estimation

The cost of hosting an application on GCP depends on two factors, (a) the network cost for the communication between the nodes of a cluster (i.e. Egress

---

[11] https://github.com/GoogleCloudPlatform/microservices-demo.

[12] https://jmeter.apache.org.

traffic) and, (b) the cost of the resources consumed by the nodes of a cluster (i.e. CPU, RAM and storage). The cost for operating a cluster with 4 VMs is

$$TotalCost = Cost_{CPU} + Cost_{RAM} + Cost_{Traffic} + Cost_{Storage}. \qquad (1)$$

Intra-cluster communication of micro-services within each node is not charged. In the following experiments, each node is allocated 2vCPU and 8 GB RAM. There are no additional volumes attached to the cluster and thus, there is no additional cost for storage. Egress traffic is charged according to the size of data exchanged between the nodes of a cluster. GCP charges each VM per hour of usage. Eg. 2 expresses the cost for hosting an application on GCP Egress cost is computed as a function of the total number and size of the messages exchanged between the micro-services. The Egress cost from node i to node j is expressed $t_e(i \rightarrow j)$. The double summation in the formula takes also into account the cost from node j to node i. The hourly cost of a cluster is computed as the summation of the following terms (i.e. n is the number of nodes):

$$Cost_{cpu} = n \cdot 2vCPU \cdot CPU_{cost} \cdot hours; \quad CPU_{cost} = \$0.028103/vCPU/hour$$
$$Cost_{ram} = n \cdot 8GB(RAM) \cdot RAM_{cost} \cdot hours; \quad RAM_{cost} = \$0.003766/GB/hour \qquad (2)$$
$$Cost_{egress} = Egress_{cost} \cdot \sum_{i=1}^{n} \sum_{j=1, j \neq i}^{n} t_e(i \rightarrow j); \quad Egress_{cost} = \$0.01/GB$$
$$Cost_{storage} \simeq 0.$$

5.2   Experimental Results

The purpose of the following experiments is to determine which is the best service placement method. It is the method that leads to a lower total monetary cost for running an application for a certain period (e.g. one month). According to Eq. (2) this accounts for (a) the number of node resources consumed and (b) Egress traffic (i.e. GCP charges only for communication between services in different host machines). Both applications run on a Kubernetes cluster in the same GCP Region and Zone. A Kubernetes cluster is initialized on GCP for each application. Each cluster comprises 4 nodes (VMs) with 2 virtual CPUs, 8 GB RAM, and 100 GB disk space. In the following, the placement solution of the default Kubernetes scheduler is compared against the placement solution of each method referred to in Sect. 3. The default scheduler will spread the services on all 4 nodes.

Service placement is determined during a pre-processing step by running an algorithm. Therefore, the speed of the algorithm is also a criterion. However, in this study, all methods are relatively very fast and compute a placement solution in less than 14 s. HFF and BKM - HP are the fastest methods (i.e. they run in less than 2 s). This is mostly due to the small size of the service graph in the input. The services graph of the Online Boutique has 15 nodes and that of iXen has 20 nodes.

Figure 2 illustrates that for both applications, most methods suggested a placement with less than 4 nodes. For the Online Boutique, BKM - HP, and BP

- HP utilized only 2 nodes. For iXen (with more micro-services), most methods utilized 3 nodes, and only HFF used only 2 nodes. BKM - HP was run initially for K = 4 but produced 2 clusters after post-processing.

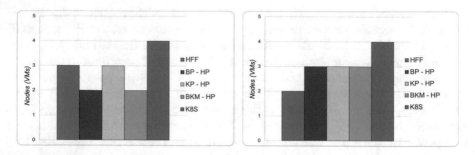

Fig. 2. Number of nodes required by each service placement method for e-Shop (left) and iXen (right).

Figure 3 illustrates that most methods achieved a very significant (up to 80%) reduction of the Egress traffic rate compared to the Kubernetes scheduler. This is an important observation and reveals that the method might be also appropriate for heterogeneous or multi-cloud and fog environments (i.e. where reducing traffic rate between the nodes of different infrastructures is of utmost importance).

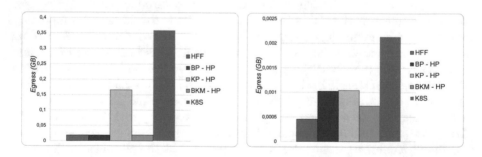

Fig. 3. Egress traffic of each service placement method for e-Shop (left) and iXen (right).

Figure 4 illustrates the overall monetary cost estimation for running each application for different placement decisions (i.e. the output of each method). The cost reduction is at least 25% and up to 50%. HFF reduced the cost by 50% for both applications. BKM - HP performed almost equally well. This is mainly due to the reduction in the number of nodes. The impact of Egress traffic on cost is less important for a homogeneous cloud. GCP charges less than USD $1 per GB when the cost of running more than 2 nodes may rise to USD $120.

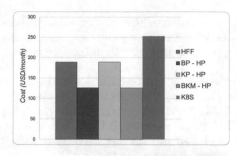 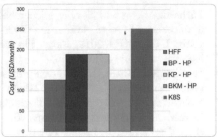

Fig. 4. Monthly monetary cost (right) of each service placement method for e-Shop (left) and iXen (right).

6   Conclusions

This work shows that the cost of hosting micro-service applications in the Kubernetes of the Google Cloud Platform can be reduced by applying elaborate service placement policies. The focus is on methods relying on the partitioning of the services graph of an application that creates clusters of micro-services with high intra affinity. This guides the placement of clusters of micro-services to fewer nodes with less inter-node (i.e. Egress) traffic that reduces the costs for the end-user by at least 25%. This result can have a significant impact on the placement of services in heterogeneous multi-cloud or, fog environments in a way that minimizes application latency. Testing all methods in this study on larger-scale applications (with hundreds of micro-services) is also an important issue for future research. Comparisons with other promising methods [9,10,16] in the same environment and test applications are in our future plans.

References

1. Apat, H.K., Sahoo, B., Maiti, P.: Service placement in fog computing environment. In: International Conference on Information Technology, ICIT 2018, Bhubaneswar, India, pp. 272–277 (2018). https://ieeexplore.ieee.org/document/8724192
2. Baldine, I., Xin, Y., Mandal, A., Ruth, P., Heerman, C., Chase, J.: ExoGENI: a multi-domain infrastructure-as-a-service testbed. In: Korakis, T., Zink, M., Ott, M. (eds.) TridentCom 2012. LNICSSITE, vol. 44, pp. 97–113. Springer, Heidelberg (2012). https://doi.org/10.1007/978-3-642-35576-9_12
3. Banerjee, S., Choudhary, A., Pal, S.: Empirical evaluation of k-means, bisecting k-means, fuzzy c-means and genetic k-means clustering algorithms. In: International WIE Conference on Electrical and Computer Engineering, WIECON-ECE 2015, Dhaka, Bangladesh, pp. 168–172 (2016). https://ieeexplore.ieee.org/abstract/document/7443889
4. Bhamare, D., Samaka, M., Erbad, A., Jain, R., Gupta, L., Chan, H.A.: Multi-objective scheduling of micro-services for optimal service function chains. In: IEEE International Conference on Communications, ICC 2017, Paris, France, pp. 1–6 (2017). https://ieeexplore.ieee.org/document/7996729

5. Farhadi, V., et al.: Service placement and request scheduling for data-intensive applications in edge clouds. In: IEEE Conference on Computer Communications, IEEE INFOCOM 2019, Paris, France, pp. 1279–1287 (2019). https://ieeexplore.ieee.org/document/8737368
6. Hedhli, A., Mezni, H.: A survey of service placement in cloud environments. J. Grid Comput. **19**(3), 1–32 (2021). https://doi.org/10.1007/s10723-021-09565-z
7. Hourdakis, N., Argyriou, M., Petrakis, E.G., Milios, E.E.: Hierarchical clustering in medical document collections: the BIC-means method. J. Digit. Inf. Manage. **8**(2), 71–77 (2010). https://www.researchgate.net/publication/220608815
8. Hu, Y., Laat, C., Zhao, Z.: Optimizing service placement for microservice architecture in clouds. Appl. Sci. **9**(21), 1–18 (2019). https://www.mdpi.com/2076-3417/9/21/4663
9. Huang, K., Shen, B.: Service deployment strategies for efficient execution of composite saas applications on cloud platform. J. Syst. Softw. **107**(C), 127–141 (2015). https://www.sciencedirect.com/science/article/abs/pii/S0164121215001156
10. Jrand, A.R.S., Rubin, J., Beschastnikh, I., Rosa, N.S.: Improving microservice-based applications with runtime placement adaptation. J. Internet Serv. Appl. **10**(4), 1–30 (2019). https://doi.org/10.1186/s13174-019-0104-0
11. Karger, D., Stein, C.: A new approach to the minimum cut problem. J. ACM **43**(4), 601–640 (1996). https://doi.org/10.1145/234533.234534
12. Lewis, R.: A general-purpose hill-climbing method for order independent minimum grouping problems: a case study in graph colouring and bin packing. Comput. Oper. Res. **36**(7), 2295–2310 (2009). https://doi.org/10.1016/j.cor.2008.09.004
13. Pallewatta, S., Kostakos, V., Buyya, R.: Microservices-based IoT application placement within heterogeneous and resource constrained fog computing environments. In: IEEE/ACM 12th International Conference on Utility and Cloud Computing, UCC 2019, Auckland, New Zealand, pp. 71–81 (2019). https://dl.acm.org/doi/10.1145/3344341.3368800
14. Petrakis, E., Koundourakis, X.: iXen: secure service oriented architecture and context information management in the cloud. J. Ubiquit. Syst. Pervasive Netw. **14**(2), 1–10 (2021). https://iasks.org/articles/juspn-v14-i2-pp-01-10.pdf
15. Salaht, F.A., Desprez, F., Lebre, A.: An overview of service placement problem in fog and edge computing. ACM Comput. Surv. **53**(3), 1–35 (2020). https://dl.acm.org/doi/10.1145/3391196
16. Wang, Z., Liu, H., Han, L., Huang, L., Wang, K.: Research and implementation of scheduling strategy in Kubernetes for computer science laboratory in universities. Information **12**(1), 1–10 (2021). https://www.mdpi.com/2078-2489/12/1/16
17. Zhong, Z., Buyya, R.: A cost-efficient container orchestration strategy in Kubernetes-based cloud computing infrastructures with heterogeneous resources. ACM Trans. Internet Technol. **20**(2), 1–24 (2020). https://dl.acm.org/doi/abs/10.1145/3378447

A Differentiated Approach Based on Edge-Fog-Cloud Environment to Support e-Health on Rural Areas

Fernando de Almeida Silva[✉], Walkíria Garcia de Souza Silveira, and Mario Dantas

Departament of Computer Science (DCC), Federal University of Juiz de Fora (UFJF),
Juiz de Fora, Brazil
`fernandoalmeida.silva@ufjf.br`, {`walkiria.garcia`,
`mario.dantas`}`@ice.ufjf.br`

Abstract. The population aging and the concern to maintain primary health care for the elderly population in rural areas requires a differentiated approach to tackle this challenge. In this article, we propose a model for remote control of elderly people with difficult to access internet considering an Edge-Fog-Cloud structure. In other words, the goal is to provide an appropriate environment with an emphasis on prepare and present a study focusing on low cost of data transferring and storage, utilizing a secondary storage device, focusing on the use of Edge-Fog-Cloud environment, and with a focus on data high-performance computing. Our initial experimental results indicate the feasibility of the present proposal and its differential effect, even on computers that initially would not be considered adequate as computational environments.

1 Introduction

The world population is undergoing rapid population aging. Projections carried out by [1] reveal that the number of elderly people will more than double by 2050. Also according to [1], life expectancy has started to increase considerably in recent years and it is estimated that this number will rise more by 2050, Europe and North America being the regions with the greatest increase in life expectancy, as shown in Fig. 1.

This population aging is one of the greatest challenges of contemporary public health, as this population is the one that makes the most use of basic health care [3], in addition to the challenge of ensuring primary health care, we also face the challenge of ensuring quality of life to the elderly [2].

The census carried out by the American government in 2019 showed that for every 5 elderly Americans, 1 resides in rural areas [4], in Brazil we have 15% of elderly people in rural areas [3] and in Europe this number is even higher, being 1 in 3 elderly people today living in rural areas [5].

Elderly people living in rural areas today have some specific difficulties due to their locations, such as difficult access to urban centers and everything they promote, public transport and service provision in general. These difficulties place rural elderly people at a disadvantage on several dimensions, specially health area. In addition, in the many cases the scarcity of health services is resulted from difficulties in recruiting and retaining health professionals for rural areas [5] (Fig. 2).

L. Barolli et al. (Eds.): AINA 2022, LNNS 450, pp. 421–432, 2022.
https://doi.org/10.1007/978-3-030-99587-4_36

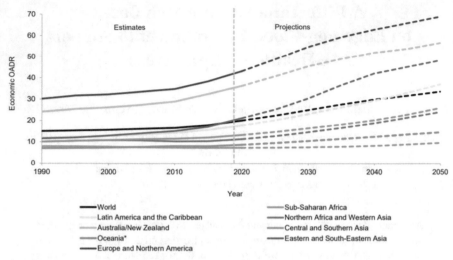

Fig. 1. Estimated and projected economic old-age dependency rations by region,1990–2050 [1].

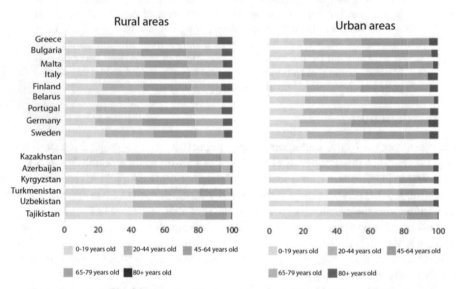

Fig. 2. Age structure in rural and urban areas [5].

On the other hand, there is the rapid technological advancement and digital trans-formation, which have been bringing contributions to health-related problems, such as the assisted home environment (AAL), which aims to contribute to a better quality of life for individuals within their own home environment [6].

The AAL environment has been made possible by the increasingly widespread devel-opment of the Internet of Things (IoT) technology in the most diverse environments and scenarios, for example, from industrial environments to environmental monitoring and the control of human health. This scenario has caused a constant increase in the amount of data generated, consequently a growth in the need for processing and storage, a need that has increased year after year.

According to Balakrishna and Vijender in [7], in the current scenario, around 35 billion IoT devices are connected to the Internet. By 2025, the number is forecast to grow between 80 and 120 billion devices, generating around 180 trillion gigabytes of new data per year through these devices.

However, these contributions and solutions do not arrive at the same speed for the rural population, due to the limited access to the internet, as demonstrated by the last agricultural census, revealing that 28% of rural properties in Brazil have access to the internet [8].

When observing the challenges encountered with the aging of the population and the challenges of rural areas, it was noted the feasibility of a research and development project on techniques and storage devices in secondary memory compared to the performance of conventional main memory, thus analyzing the data transfer rate using current technologies employed in HPC clusters or data centers for use in fog layer devices, with emphasis on primary care for the elderly population residing in rural areas. Thus, providing a differential in the edge fog and cloud layers that allows for an improvement in the quality of life of the elderly in densely populated rural areas.

This work is structured as described below. Section 2 presents the related works. Section 3 presents the materials and methods used in this project. The environment used and the results obtained are presented in Sect. 4. Finally, Sect. 5 concludes and proposes future work.

2   Related Works

The Coutaz's article in [9] addresses one of the challenges of the 21st century: the large and growing volume of data and how to turn it into a useful and customized experience for people. But something that sooner or later would be needed in all areas. Two critical processes are needed for this, the first to recognize the goals of users and their activities, and the second to map these goals and activities adaptively to the population of services and resources available. Thus, through the approach to the context, the interface could be dynamically adapted to offer the user information and services that are necessary for the user without unnecessary confusion or distraction. The author at the time points out that this is an ambitious goal, especially compared to current systems that need extensive manual configuration. In addition, context services are highly complex, as they are made up of a fabric structured in several levels of abstraction, the main layers being detection, perception, situation and the context identification layer. The lack of context-related metrics and, furthermore, due to the capacity of most devices at the time, little progress can be made in a practical sense, due to the large real-time processing power needed and the lack of optimized algorithms for this task. However, the article highlights this urgent need, since, according to him, the context is the key in the development of new services that will have an impact on the social inclusion of the emerging information society.

In his article Manzoor [10] advances in context-related heuristics, from paradigms and general ideals to metrics through the concept of Quality of Context (QoC) and terms such as objective and subjective views of QoC. The objective view of QoC is the quality of context information independent of the requirements of a specific consumer of context, while the subjective view of QoC considers the quality of context information due to its importance for a specific consumer of context. So, dividing the metrics through these two views we have the objective metrics that include reliability, timeliness and integrity of information in context. Subjective metrics, on the other hand, are calculated as the quality of context information compared to the user's requirements for use for a specific purpose. Although the work advances in context quality metrics, little advances in its practical use in Decision Support Systems. This is why we approach this theme, due to the relevance of the quality of the context in health services, highlighted by Madalena [11].

However, due to the large amount of data collected, filtering and cleaning processes are relevant to remove useless information and consequently reduce the collected information. Thus, a cooperative edge, fog and cloud system is necessary to distribute part or all of the tasks that would be focused on just one application, increasing the efficiency of the process [12].

3   Material and Methods

The effort from this study work is to develop an approach that considers Edge-Fog-Cloud environment to provide an appropriate structure for remote monitoring of elderly people with difficult internet access, in an AAL scenario. Asa result, a simulation model was specified for the present research work. The model includes data collections in AAL and Edge-Fog-Cloud technologies, as well as displaying the data to stakeholders.

3.1   Materials

The use of IoT equipment in recent years has allowed the development of wearable devices with sensors that allow monitoring of human health and are increasingly available on the market. Among them we can mention smart band and smartwatches, which are able to capture body temperature, oxygenation level, heart rate, blood pressure, among other vital signs. Some of this information may be important to help detect disease.

Ambient Assisted Living (AAL) is based on the use of IoT devices to ensure that the elderly or anyone in need of assistance can have quality of life and independence in their homes, providing stakeholders with warnings about sudden changes in user patterns or habits, which may be related to the health and safety of the monitored individual. Monitoring is done by wearable devices or additionally with sensors spread across the AAL environment, obtaining information such as temperature, humidity, air quality, luminosity, noise meter, among others. Such information can provide a better context of wearable IoT device information, especially in cases of disease with similar symptoms (Fig. 3).

The Edge is located close to the user, the data is processed locally and then sent to the cloud through a network that connects the edge to the cloud, called Fog, which is a technology that deals with bottlenecks in the network that are generated by the large flow of data collected by IoT devices [14].

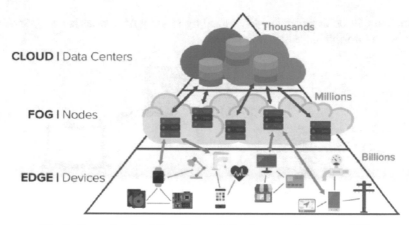

Fig. 3. Example of an Edge, Fog and Cloud cooperative system [13].

A Fog computing architecture aims to improve network management, storage, and application processing. For this purpose, Fog architectures integrate mechanisms that can better distribute resources in a specific infrastructure [15].

Among the main sources of digital data [16] are medicine, geosciences and other areas. This results in the growth of storage, both on-premises, as well as cloud and Fog tier storage between the Edge nodes. For long-term retention, we will need inexpensive mass storage and faster processing and memory technologies to analyze and use data [14].

Among the usually existing storage options, we can mention the Hard Disk Drive (HDD) which can store a large amount of data (high capacity) and the price of an HDD is usually cheaper in relation to the monetary value per storage bytes. However, Solid State Drives (SSD) has gained more space in the market as a storage device. For among the advantages, it is a non-mechanical and non-magnetic storage device in which data is stored using electronic circuitry, so it is less susceptible to physical shocks, quieter and has better overall performance than HDD. For example, it would take more than 15 s to read a 1.8 GB MRI image stored on an HDD compared to just 0.15 s on an SSD storage device. However, by using the SATA III interface, the effective bandwidth peaks at around 555 MB/s, thus limiting the SSD's access and storage speed [17].

3.2 Methods

The AAL must monitor the user's vital signs and environment through wearable devices and sensors that collect, treat and transform these vital signs and environmental data into information, which are accessed by stakeholders through a system, thus monitoring the health of the elderly and its environment, reducing medical appointments and improving the quality of life of users.

The Fig. 4 illustrates the proposed rural AAL model, where we have represented elderly people living in rural areas making frequent use of a wearable device and sensors scattered around the house. The goal is for them to share the collected data with a local device that does advanced processing and analysis of the collected data. In this way, information will be organized and sanitized by sending only what is necessary, so the traffic of the information sent is minimized.

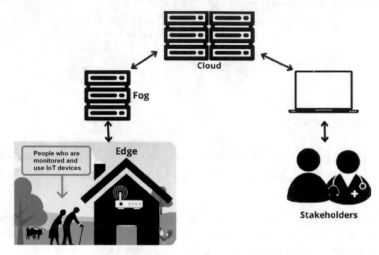

Fig. 4. An ordinary rural AAL scenario.

For this model proposed the Edge-Fog-Cloud Architecture was the best choice. Why In the Edge layer, it is intended to develop an application to collect data from the e-health devices and sensors of residents in the house. The data collected by the applications and sensors will be sent over a local communication network to the Fog layer, which will have a server application collecting data from all Edges in the region. In the Fog layer, the preprocessing and optimizing of these data to a best transmission cost will be carried out and subsequently will be sent to the servers in the Cloud layer. It will receive this previously prepared data, consolidate and do the main processing for use in decision making.

The computer in Fig. 4 represents the web system where all the information received should be displayed, it is through it that the stakeholders will have access to the information.

Due to the Covid-19 pandemic, where the group at greatest risk is made up of the elderly, which are our research focus. Experiments are being carried out through simulations. In addition, based on remote interviews with the target audience, we developed an overview of the Edge, Fog and Cloud architecture in this scenario.

Furthermore, this information allowed for a better understanding of the environment around the house. For even on small properties, raising small animals in family farming is important for feeding the family and generating income, in addition to contributing to soil fertilization. However, this proximity can be a vector of diseases and therefore included in the context by Stakeholders, as illustrated in the Fig. 5.

In this scenario of restrictions, the best solution found was to simulate an Ambient Assisted Living (AAL) with IoT devices and, through this simulation, monitor the generation and sending of data both by the device in the resident and by the devices spread around the housing unit. Among the factors in the decision, it was considered that the

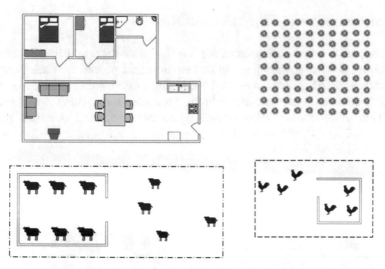

Fig. 5. A typical rural family environment.

use of a simulator allows us to design our proposed IoT paradigm to be adopted and graphically demonstrate it. To achieve these goals, the Siafu simulator was used, similar to Nascimento in [18]. Siafu is an open-source simulator that lets you control location characteristics, agent behavior and the entire context. In addition to having technical resources that allow control of the simulation, the simulator also has a graphical interface and allows the creation and transfer of generated data that can be used in other phases of the project.

Simultaneously, studies and surveys of secondary memory storage device technologies of the last decades were carried out. Evaluating the advantages and disadvantages of each one and how they could contribute to a more efficient proposal. A survey of research related to the evaluation of I/O performance in High Performance Computing between 2015 and 2021 was carried out. Where base articles were identified taking into account the summary and conclusion considering mainly those that address the same type of research, challenges and common references to the proposal, according to the research work indicated by Pioli [19].

Achieving this theoretical proposal, the next step was to implement it, creating a practical example, which could be implemented even in computers that would not have the necessary connection natively, thus allowing to extend the useful life of these equipment, minimizing the environmental impact and cost. Because according to the United Nations (UN) globally each year are generated approximately 50 million tons of electronic waste known as e-waste [20].

Then test, analyze and consider the results obtained, verifying the feasibility of the proposal and its impact on current systems. As well, it can be improved in future research.

4   Environment and Experimental Results

In the proposed solution environment, Siafu is the simulator responsible for simulating a familiar rural AAL environment. In this environment people who need monitoring and use common IoT devices in an Assisted Living Environment. In addition, the environment in question is also monitored at the same time through IoT devices scattered internally in the environment. Since the generated data is captured, stored and processed in an Edge – Fog – Cloud architecture and thus be made available to agents who will use this information (Fig. 6).

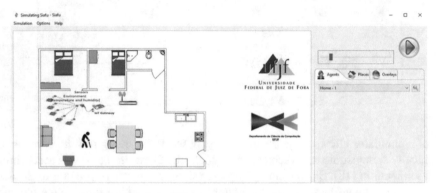

Fig. 6. Siafu simulating IoT devices in a familiar rural environment.

Furthermore, in this project a set of hardware and software was used to implement and evaluate the feasibility of the storage proposal considering different generations of the PCI Express bus, for example, the second generation of the bus that was used in the experiment. Thus, checking the possibility of backwards compatibility, allowing the use of current computers and other models considered obsolete, for example, that do not have a native connector for Non-Volatile Memory express (NVMe) devices.

4.1   Environment

To simulate the viability of the storage proposal, experiments were carried out using a computer with a decade of use having an Intel® Core i5–2500 CPU @ 3.30 GHz, Intel motherboard model DQ67SW and 8 GB of RAM (main memory). Relevantly complementing the hardware, the Reletech branded NVMe P400 device and an NVMe expansion card for PCI Express for x16 slot (Fig. 7).

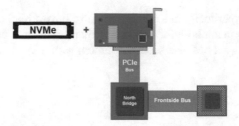

Fig. 7. Example of NVMe Device and NVMe expansion board used in the experiment.

Among the software to implement the proposal, we have the Microsoft Windows 10 Pro (64-bit) Operating System. In addition to this, a similar Clark in [21] was used by the company Dataram the RAMDisk software, which allows using a part of the system's main memory as a virtual disk drive. This being a benchmark for comparison for different secondary storage devices. The version used was 4.4.0.36 (freeware) which allows you to create virtual disks of up to 1 GB in size.

In the battery of reading and writing tests in different scenarios, similar to Clark in [21], the software used was Crystal Disk Mark, the version used being 3.0.4 x64, as it is the latest version compatible with 1 GB partitions. size or smaller, which is the restriction of RAMDisk software in the freeware version. It performs four types of tests: Seq, 512 K, 4 K and 4 K QD32, being: Seq with sequential blocks of 1024 KB; 512 K with 512 KB blocks written at random locations; 4 K tests maintain a write and read queue with a 4 KB block; 4 K QD32 tests maintain a write and read queue with 32 blocks of 4 KB.

4.2 Experimental Results

In the experiment using Crystal Disk Mark software, as shown in Figs. 8 and 9, we have a read and write comparison simulating different storage scenarios between two secondary memory devices, on the left an HD with moving parts and on the right an SSD, both with SATA connection.

	CrystalDiskMark 3.0.4 x64			CrystalDiskMark 3.0.4 x64	
	Read [MB/s]	Write [MB/s]		Read [MB/s]	Write [MB/s]
Seq	105.9	103.3	Seq	510.6	415.1
512K	39.76	53.65	512K	390.6	226.9
4K	0.517	1.025	4K	18.66	46.19
4K QD32	1.120	0.924	4K QD32	216.2	140.7

Fig. 8. Comparison of read and write between HD and SSD.

Continuing the experiment, in a similar way, we have a read and write comparison in different scenarios using both the secondary memory device NVMe on the left and a virtual HD created by RAMDisk software, which would be the theoretical maximum limit.

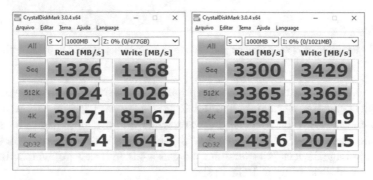

Fig. 9. Comparison of reading and writing between NVMe and virtual HD.

From the results of the experiments, comparisons illustrated in Table 1 were made. Calculating the percentage of efficiency of secondary memory devices, that is, the read and write rates on each real secondary storage device, comparing them with the virtual HD.

Table 1. Comparative percentage between secondary and virtual memories.

Tipo de teste:	HD	SSD	NVMe	Leitura HD virtual (MB/s)	HD	SSD	NVMe	Escrita HD virtual (MB/s)
Seq	3%	15%	40%	3300	3%	13%	35%	3429
512K	1%	12%	31%	3365	2%	7%	31%	3365
4K	0,02%	1%	1%	258	0,03%	1%	3%	211
4K QD32	0,03%	7%	8%	244	0,03%	4%	5%	207

5   Conclusions and Future Works

This research work proposes a contribution to the improvement of problems related to primary health care for rural residents through Edge-Fog-Cloud architecture. The development of the rural AAL model, using the Siafu simulator, allowed the simulation of wearable IoT devices spread across the AAL environment, obtaining information that remotely allows for a deeper and more assertive context for the necessary medical care. However, locally generating more information that are required to be treated and stored before going to the Cloud.

For this, the experiments presented in this work indicate the feasibility of the proposal whose objective was to provide an improvement in the form of storage considering the HDD, SSD and NVMe devices, including in computers that do not have an interface for NVMe devices. This approach has been achieved through adapters that use PCI Express (PCIe) slots available on the motherboard. In view of the results presented, the use of this proposal makes possible a storage with support characteristics for high performance, including in computers that would not be considered as adequate computing environments. Also checking the compatibility in different generations of the PCI Express interface. Obtaining read and write rates higher than those of HDD and SSD, bypassing the limitations of the SATA III interface and using the north bridge, in the same way as the main memory.

As future work, environments other than simulations should be evaluated, considering the diversification of the application. For example, assessing what environmental information needs to be obtained and how often to provide the best context and avoid redundancies. In addition to studying forms of parallelism, allowing simultaneous reading and recording of several NVMe devices, in addition to measuring performance and energy consumption in different scenarios. Evaluate computers with different generations of PCI Express bus and other configurations. In addition, carry out research aimed at developing software solutions that use this secondary memory in their best storage scenarios, for example, in conjunction with main memory. It is intended to investigate in future evaluation the performance loss relationship associated with conventional main memory storage.

Acknowledgments. The authors would like to thanks support from PROQUALI Program from UFJF, CNPq, INESC P&D Brasil and Petrobras.

References

1. United Nations, Department of Economic and Social Affairs, Population Division (2019). World Population Ageing 2019: Highlights (ST/ESA/SER.A/430)
2. Both, C.F., Kunz, A.E., da Costa, L., Pissaia, F.: Percepções sobre a demanda de atendimento médico na população idosa. Res. Soc. Devel. **7**(6), e1176331 (2018). https://doi.org/10.17648/rsd-v7i6.331
3. Penido, A.: Estudo aponta que 75 por cento dos idosos usam apenas o sus. In Fiocruz (2018). https://portal.fiocruz.br/noticia/estudo-aponta-que-75-dos-idosos-usam-apenas-o-sus. Accessed 9 Jan 2022
4. Trevelyan, E., Smith A.S.: In some states, more than half of older residents live in rural areas. United States Census Bureau. https://www.census.gov/library/stories/2019/10/older-population-in-rural-america.html. Accessed 9 Jan 2022
5. Comissão Económica Unece. Resumo de Políticas: Idosos em áreas rurais e remotas. Resumo de políticas da Unece sobre o envelhecimento n° (2017). https://unece.org/DAM/pau/age/Policy_briefs/Portuguese/PB18_V01.pdf. Accessed 9 Jan 2022
6. Nakagawa, E.Y., et al.: Relevance and perspectives of AAL in Brazil. J. Syst. Softw., **86**(4), 985–996 (2013). SI: Software Engineering in Brazil: Retrospective and Prospective Views
7. Balakrishna, S.M. Thirumaran, Vijender Kumar, S.: IoT sensor data integration in healthcare using semantics and machine learning approaches. In: A Handbook of Internet of Things in

Biomedical and Cyber Physical System. Springer, Cham, 2020. pp. 275–300. https://doi.org/10.1007/978-3-030-23983-1_11

8. MILANEZ, Artur, Y., et al.: Conectividade rural: situação atual e alternativas para superação da principal barreira à agricultura 4.0 no Brasil = Rural Connectivity: current situation and alternatives to overcome the main barrier related to agriculture 4.0 in Brazil. BNDES Setorial, Rio de Janeiro, **26**(52), pp. 7–43, set. 2020. https://web.bndes.gov.br/bib/jspui/handle/1408/20180. Accessed 9 Jan 2022

9. Coutaz, J., Crowley, J.L., Dobson, S., Garlan, D.: Context is key. Commun. ACM **48**(3), 49–53 (2005)

10. Manzoor, A., Truong, H.-L., Dustdar, S.: Quality of context: models and applications for context-aware systems in pervasive environments. Knowl. Eng. Rev. **29**(2), 154–170 (2014)

11. Pereira, M., da Silva, A., Gonçalves, L., Dantas, M.A.R.: A conceptual model for quality of experience management to provide context-aware eHealth services. Future Gen. Comput. Syst. **101**, 1041–1061 (2019). https://doi.org/10.1016/j.future.2019.07.033

12. Thomé, T.G., Ströele, V., Pinheiro, H., Dantas, M.A.R.: A fog computing simulation approach adopting the implementation science and iot wearable devices to support predictions in healthcare environments. In: Barolli, L., Woungang, I., Enokido, T. (eds.) AINA 2021. LNNS, vol. 226, pp. 298–309. Springer, Cham (2021). https://doi.org/10.1007/978-3-030-75075-6_24

13. Omni.sci: Fog computing. https://www.omnisci.com/technical-glossary/fog-computing (2021). Accessed November 2021

14. Habibi, P., Farhoudi, M., Kazemian, S., Khorsandi, S., Leon-Garcia, A.: Fog computing: a comprehensive architectural survey. IEEE Access **8**, 69105–69133 (2020). https://doi.org/10.1109/ACCESS.2020.2983253

15. Silva, D.M., da, A., et al.: An analysis of fog computing data placement algorithms. In: Proceedings of the 16th EAI International Conference on Mobile and Ubiquitous Systems: Computing, Networking and Services (2019)

16. Coughlin, T.: A solid-state future [the art of storage]. IEEE Cons. Electron. Magaz. **7**(1), 113–116 (2017)

17. Armoogum, S., Khonje, P.: Healthcare data storage options using cloud. In: Patrick Siarry, M.A., Jabbar, R.A., Abraham, A., Madureira, A. (eds.) The Fusion of Internet of Things, Artificial Intelligence, and Cloud Computing in Health Care, pp. 25–46. Springer International Publishing, Cham (2021). https://doi.org/10.1007/978-3-030-75220-0_2

18. Nascimento, M.G., Braga, R.M.M., José Maria, N., David, M.A., Dantas, R., Colugnati, F.A.B.: Towards an IoT architecture to pervasive environments through design science. In: Barolli, L., Woungang, I., Enokido, T. (eds.) AINA 2021. LNNS, vol. 226, pp. 28–39. Springer, Cham (2021). https://doi.org/10.1007/978-3-030-75075-6_3

19. Pioli, L., et al.: Characterization research on I/O improvements targeting DISC and HPC applications. In: IECON 2020 The 46th Annual Conference of the IEEE Industrial Electronics Society. IEEE (2020)

20. Thakur, P., Kumar, S.: Evaluation of e-waste status, management strategies, and legislations. Int. J. Environ. Sci. Technol. **1**, 10 (2021). https://doi.org/10.1007/s13762-021-03383-2

21. Clark, C.J.: Planning for the Influence of Emerging Disruptive Technologies on IT Systems. Diss. The George Washington University (2017)

Trustworthy Fairness Metric Applied to AI-Based Decisions in Food-Energy-Water

Suleyman Uslu[1], Davinder Kaur[1], Samuel J. Rivera[2], Arjan Durresi[1(✉)],
Mimoza Durresi[3], and Meghna Babbar-Sebens[2]

[1] Indiana University-Purdue University Indianapolis, Indianapolis, IN, USA
{suslu,davikaur}@iu.edu, adurresi@iupui.edu
[2] Oregon State University, Corvallis, OR, USA
{sammy.rivera,meghna}@oregonstate.edu
[3] European University of Tirana, Tirana, Albania
mimoza.durresi@uet.edu.al

Abstract. We propose a *Trustworthy Fairness Metric* and its measurement methodology to evaluate the fairness of different AI-based solutions proposed by the actors utilizing their trust during decision-making in the Food-Energy-Water (FEW) sectors. Since the standardization of the trustworthiness of AI systems is fundamental, the proposed metric is a compelling advance in this process, whereas other approaches stay at the high-level principles. Trust management system is the basis of the measurement methodology as it incorporates human involvement. This metric captures and quantifies the fairness of the solutions evaluated by the actors having different views and is illustrated in decision-making scenarios generated by AI for FEW sectors. Also, the metric and its measurement methodology can be conveniently adapted to various fields of AI. We present that our metric successfully captures the fairness of solutions in multi-stakeholder decision-making.

1 Introduction

Admitting the tremendous and broad prosperity that artificial intelligence (AI) has brought to our lives, still, various applications of AI, such as self-driving cars or hiring algorithms, are being criticized due to causing harm to its users or the society. Therefore, safety, reliability, explainability, and trust have become significant concepts for AI-based systems. These concerns are being tried to be answered by researchers, companies, organizations, and governments. For example, European Union (EU) [50] has published ethical guidelines for trustworthy AI to regulate the development of AI systems. Similarly, Defense Advanced Research Projects Agency (DARPA) [15] started XAI program to motivate the research in explainable and trustworthy AI. Likewise, the National Institute of Standards and Technology (NIST) [41,42] and U.S. Government Accountability Office (GAO) [3] published frameworks for trust and accountability in AI. These

© The Author(s), under exclusive license to Springer Nature Switzerland AG 2022
L. Barolli et al. (Eds.): AINA 2022, LNNS 450, pp. 433–445, 2022.
https://doi.org/10.1007/978-3-030-99587-4_37

publications, reviewed in [24], show the current research and the strong demand in the field of trustworthy, explainable, accountable, and acceptable AI.

Similar to the maturity process of established technologies, AI technologies should also go through the standardization process. It brings various benefits such as enabling consumers to check and compare the product quality, constituting a common language between the users and the manufacturers and limiting the legal liabilities of the providers. International Organization for Standardization (ISO) [18] published approaches to constitute and form trust in AI applications using transparency, accountability, controllability, and fairness. Still, these studies and publications illustrate the problem and present the approaches at a high level, whereas metrics and measurement procedures are required for standardization [26]. Akin to U.S. Food and Drug Administration (FDA) certifications for medications and treatments, we foresee that relevant agencies will use these metrics to accredit AI systems.

As stated in ISO document [18], we adopt the definition that *Trustworthy AI is a framework to ensure that a system is worthy of being trusted concerning its stated requirements based on the evidence. It makes sure that the users' and stakeholders' expectations are met in a verifiable way.* However, present AI approaches lack or have insufficient aspects such as human intelligence, meaning, multi-dimensional data beyond the training set, or ethics. Hence, instead of replacing humans completely, we envision that including human expertise in the loop is an essential requirement of trustworthy AI.

Accountability comes into the picture when computers make the decisions. Hence, they should go through an extensive testing process for acceptance evaluation before the production phase. In [48], we proposed a trustworthy acceptance metric to assess the acceptance of an AI system by the experts considering the trust and presented the results of a simulated environmental decision making. Furthermore, a trustworthy explainability metric is proposed [22] including different criteria for the evaluation. However, having other criteria in evaluating the acceptance raises concerns about the fairness of a solution since a solution can receive varying acceptance rates from different actors. In this paper, we propose a *trustworthy fairness metric* to assess the fairness of AI-based solutions utilizing the trust of actors and present the simulation results of decision makings in Food-Energy-Water (FEW) sectors. Our contributions can be outlined below.

- In Sect. 4, we propose a trustworthy fairness metric to evaluate the fairness of AI-based solutions.
- Measurement procedure for fairness is based on the acceptance of solutions that utilize distance adaptable to a wide range of systems.
- Fairness metric utilizes the trust of actors employing our trust framework summarized in Sect. 3.
- Finally, we illustrate the application of our metric and its measurement methodology using three criteria in environmental decision-making.

2   Related Work

Decision-making is a part of our daily lives and a fundamental aspect of the most consequential fields such as the environment, healthcare, finance, and economics. By illustrating the critical outcomes of faulty or inaccurate decisions such as the world economic crisis, Kambiz [20] analyzed the priority of decision making in such fields. He added that the intricate characteristics of the decisions and multi-stakeholder involvement bringing different backgrounds, expertise, perspectives, or even objectives are behind the hardship of making such critical decisions.

A consensus mechanism is one of the essential requirements of decision making especially when the stakeholders are the experts in the field. In [11], an approach to reduce the adjustments to the solutions at each round is proposed. Moreover, Hegselmann et al. [16] analyzed consensus reaching with mathematical modelings and simulations using agents with varying behaviors and affecting each other. In addition, Babbar-Sebens et al. [4] utilized expert feedback to improve the optimization of the environmental plans. These examples demonstrate the need for advanced algorithmic approaches and frameworks for critical decision-making, especially when human involvement is required.

Trust is considered as one of the fundamental elements of multi-stakeholder decision makings, especially involving humans. Some of the benefits of employing trust are increased integrative behavior and decreased disruptive activities [14,17,25]. Trust management and its frameworks have been discussed in multiple studies [30–32,51] and surveys [10,33,52]. Ruan et al. [38] proposed a measurement theory-based trust management framework for online social networks. Some of the application areas of this framework are social networks [36,53], internet of things [35,37], cloud computing [34,37], healthcare [8,9], emergency communications [12], exposing of fake [21] and damaging users [29], and crime detection [23].

Because of the importance of the decision makings about the sustainability of the environment, a water allocation problem is modeled [1] and a generic model for consensus reaching problem is proposed [2] both of which utilize trust. We proposed a decision support system based on trust for natural resource sharing in FEW sectors [49] and its advanced versions handling precomputed discrete solutions [43], utilizing a game-theoretical approach [44], and having flexible assessment criteria [45]. Moreover, we introduced trust sensitivity and presented its role in environmental decision making [47].

As the algorithmic decision-making mechanisms bring concerns about acceptance and explainability, they also raise questions about fairness. Gajane and Pechenizky [13] claimed that there are various formalizations for fairness without a concrete consensus and surveyed the current formalizations in machine learning. Similarly, Mehrabi et al. [28] analyzed the fairness definitions that machine learning researchers have adopted. Furthermore, Beutel et al. [6] mentioned the recent research trend in algorithmic fairness but also provided a fairness metric that considers distributional differences in the training data. In addition, Srivastaya et al. [40] pointed out that most of the mathematical formulations for fairness deal with the trade-off of different notions of fairness where they decided to ask the participants to find the most compatible notion.

Although the definition of fairness could vary by the domain [54], there are several proposed toolkits to establish fairness in algorithmic approaches. One of them is an open-source one bringing algorithm designers and its users together [5], an audit toolkit that can handle the bias in data before the training phase [39], a technical and socio-political platform bringing scientists, lawmakers, and industry practitioners together [27], and a testing approach to detect statistical and casual discrimination [7]. However, there is still a need and high demand for fairness metrics just as acceptance and explainability to evaluate AI applications involving humans.

3   Trust Management Framework

In this section, we outline our trust management framework [38] since the methodology of the new metric is based on it. There are two main parameters of this framework, namely impression, m, and confidence, c, which show the trust level and its certainty, respectively. As shown in Eq. 1, the arithmetic mean of the measurements is selected as the impression calculation method among the proposed ones in [38], and confidence calculation is based on the standard error of the mean as shown in Eq. 2.

$$m^{A:B} = \frac{\sum_{i=1}^{N} r_i^{A:B}}{N} \tag{1}$$

$$c^{A:B} = 1 - 2\sqrt{\frac{\sum_{i=1}^{N}(m^{A:B} - r_i^{A:B})^2}{N(N-1)}} \tag{2}$$

This framework is also capable of assessing trust between parties with no direct communication using propagation methods, that are transition and aggregation, through other entities in the network. In [38], different calculations for trust propagation methods were proposed for different scenarios. In our work, we have used multiplication for transition and arithmetic mean for aggregation as shown in Eqs. 3 and 4. Their respected error formulations are provided in Eqs. 5 and 6.

$$m^{ST} \otimes m^{TD} = m^{ST} m^{TD} \tag{3}$$

$$m_{T_1}^{SD} \oplus m_{T_2}^{SD} = \frac{m_{T_1}^{SD} + m_{T_2}^{SD}}{2} \tag{4}$$

$$e^{ST} \otimes e^{TD} = \sqrt{(e^{ST})^2(m^{TD})^2 + (e^{TD})^2(m^{ST})^2} \tag{5}$$

$$e_{T_1}^{SD} \oplus e_{T_2}^{SD} = \sqrt{\frac{1}{2^2}((e_{T_1}^{SD})^2 + (e_{T_2}^{SD})^2)} \tag{6}$$

4   Trustworthy Fairness Metric and Its Methodology

This section explains the methodology of our Trustworthy Fairness Metric (TFM). This work assumes that an AI-based system generates a set of solutions to be selected and applied by several parties, possibly after some discussion. Each party has a subset of solutions in this solution set; however, other parties can evaluate and judge a selected solution based on different criteria.

TFM is based on our Trustworthy Acceptance Metric (TAM) [48]. TAM measures the level of acceptance and its confidence based on the distance between a selected solution and a goal solution. A shorter distance results in a higher acceptance and vice versa. As shown in Eq. 7, the distance function is Euclidean and can have multiple dimensions according to the scenario. d_S^T represents the distance from S to T in n dimensions where S_i and T_i are the normalized values of S and T in dimension i. The distance is in $[0, 1]$; therefore, the acceptance, as shown in Eq. 8, is also in $[0, 1]$.

$$d_S^T = \sqrt{\frac{\sum_{i=1}^n (S_i - T_i)^2}{n}} \tag{7}$$

$$A = 1 - d_S^T \tag{8}$$

Trustworthy acceptance, TWA, of the community is calculated as shown in Eq. 9 where A_i and T_i are the acceptance and trust of actor i, and k is the number of actors. Furthermore, we provide the confidence of TWA, c_{TWA}, as shown in Eq. 10. Eventually, TAM consists of trustworthy acceptance, TWA, and its confidence, c_{TWA}.

$$TWA = \frac{\sum_{i=1}^k A_i T_i}{k} \tag{9}$$

$$c_{TWA} = 1 - 2\sqrt{\frac{\sum_{i=1}^k (TWA - A_i)^2}{k(k-1)}} \tag{10}$$

The selected dimensions, which are the parameters used in distance measurement, can change the distance, therefore the acceptance, as much as the chosen solution. Hence, a solution can have high acceptance regarding a subset of dimensions, whereas it can have lower acceptance using another subset. This variation shows different explanations for different solutions. Similar to TAM, Trustworthy Explainability Metric (TEM) provides the acceptance and its confidence for the selected solution but also includes the parameters that are used in distance calculation which explain the reason behind the selection [22]. Each solution can have multiple TEMs based on the different parameter sets used in distance measurement.

In this paper, we propose Trustworthy Fairness Metric (TFM), which measures the fairness of the selected solutions over different explanations, which are parameter sets provided by the actors. First, a fair point for each group of TEM, that is fair TEM or FTEM in short is decided. Then, we normalize the TEMs based on FTEM as shown in Eq. 11. Then, using the fairness index formula [19], we calculate TFM as shown in Eq. 12. This metric shows how fair a proposed solution is considering different evaluation criteria.

$$TEM_i^{norm} = x_i = \frac{TEM_i}{FTEM_i} \tag{11}$$

$$TFM = \frac{(\sum_{i=1}^n x_i)^2}{n \sum_{i=1}^n x_i{}^2} \tag{12}$$

5 Simulated Measurement Results

In this study, the data used for simulated decision-making was generated using an uncalibrated Soil and Water Assessment Tool (SWAT) model specifically for the Willow Creek Watershed in Oregon. The solution set was generated for over 200 water rights exploring different plans regarding various criteria such as irrigation amount, water source, and other crop types and fertilizer usage with outputs such as environmental protection and profit. In our simulations, we selected six fields as a neighborhood, and their owners attended the same decision-making. Although we omitted dynamic trust measurements in this work for simplicity, it is explained in [45–47].

A reference solution is provided together with the solution set for each field. The distance, which is the basis of our acceptance metric, is measured from the selected solution to this reference solution, where the shorter distance means a higher acceptance and vice versa. Figure 1 shows the acceptance of the selected solutions using environmental protection values of solutions. Actors 4 and 6 reached higher acceptance values, whereas the remaining four actors had lower acceptance. Although the selected and the reference solutions are the two main determinants for the acceptance, it is possible to have different acceptance of the same solutions using different parameters. It means that actors can have different concerns on different output parameters of such decisions. The initial results showed that the other four actors could have different objectives than environmental protection.

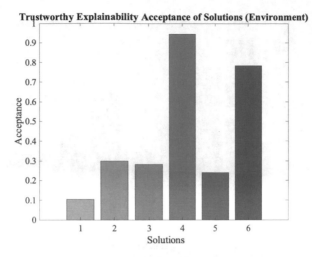

Fig. 1. Actors 4 and 6 reached higher acceptance regarding the environmental values whereas the others had lower acceptance which means that their proposed solutions were not environmental as expected.

After discussing and reviewing the output parameters, the irrigation amount is selected as the new dimension, and the reference solutions are adjusted. Figure 2 shows that actors 1 and 2 have a great concern about the irrigation amount. This could represent a tribe that is sensitive about the rivers and does not want to use excessive water from the rivers to irrigate their land, which would otherwise increase groundwater recharge and result in higher environmental protection.

Similarly, in Fig. 3, we demonstrate the acceptance regarding the profit of the solutions. This shows a more consistent look than the first two parameters, environment and irrigation, because none of the actors sacrificed their profit regardless of the target solution or distance parameter selection.

These three-parameter selections represent three different views for the appropriate solution selection. As explained in Sect. 4, another concern could be the level of fairness of these solutions concerning all three parameters that concern at least one group in the decision making. For simplicity, we selected FTEM values in Eq. 11 as 1 which gives the same flexibility to all parameters to span in [0, 1]. Then, we applied the formula given in Eq. 12. The resulting TFM values are shown in Table 1 together with their acceptance for each parameter.

In Fig. 4, we graphically represent the TFM values where the solution of actor 1 has the lowest fairness. This was expected because it has the lowest environmental acceptance but the highest irrigation acceptance. In contrast, the solution of actor 6 has the highest fairness mainly because of more proportional distribution of acceptance among the parameters.

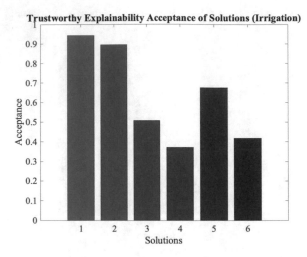

Fig. 2. Actors 1 and 2 reached higher acceptance regarding the irrigation amount since this was their concern while choosing an appropriate solution for their land.

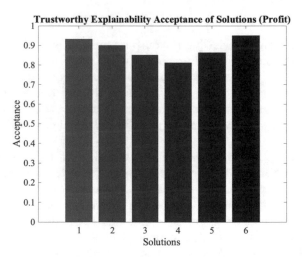

Fig. 3. None of the actors sacrificed their profit which made their solutions to have high acceptance with respect to profit.

Table 1. Trustworthy explainability acceptances of selected parameters and trustworthy fairness of solutions

	Soln. 1	Soln. 2	Soln. 3	Soln. 4	Soln. 5	Soln. 6
Environment	0.1044	0.2996	0.2821	0.9461	0.2408	0.7828
Irrigation	0.9433	0.8955	0.5080	0.3724	0.6751	0.4180
Profit	0.9326	0.8992	0.8500	0.8114	0.8633	0.9500
Fairness	0.7383	0.8599	0.8458	0.8936	0.8381	0.9124

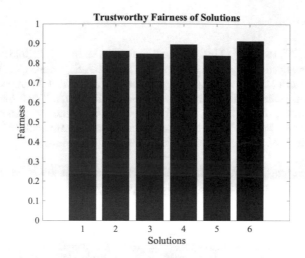

Fig. 4. Trustworthy fairness of the proposed solutions as calculated based on Eq. 12.

6   Conclusions

We presented a *Trustworthy Fairness Metric* and its measurement methodology to evaluate the fairness of proposed solutions generated by an AI system for environmental decision makings as a path toward standardization of Trustworthy AI. Human actors selected the most appropriate solutions among the developed solution set for their region, recognizing their own goals. The distance is measured between the proposed and predetermined optimum solutions, which results in the trustworthy acceptance of the solutions. Due to the varying concerns among the actors other than the optimization criteria of AI, the acceptance using several different parameters is measured and compared, leading to explainable acceptance. The results showed that three groups of actors cared primarily about the environment, irrigation, or profit. However, there is a balance in the acceptance of these criteria for some actors, whereas some solutions are prone to have higher disparity in acceptance levels. Our metric successfully captured the fairness of the proposed solutions respecting their explainable acceptances.

Acknowledgements. This work was partially supported by the National Science Foundation under Grant No.1547411 and by the U.S. Department of Agriculture (USDA) National Institute of Food and Agriculture (NIFA) (Award Number 2017-67003-26057) via an interagency partnership between USDA-NIFA and the National Science Foundation (NSF) on the research program Innovations at the Nexus of Food, Energy and Water Systems.

References

1. Alfantoukh, L., Ruan, Y., Durresi, A.: Trust-based multi-stakeholder decision making in water allocation system. In: Barolli, L., Xhafa, F., Conesa, J. (eds.) Advances on Broad-Band Wireless Computing, Communication and Applications, pp. 314–327. Springer, Cham (2018). https://doi.org/10.1007/978-3-319-69811-3_29

2. Alfantoukh, L., Ruan, Y., Durresi, A.: Multi-stakeholder consensus decision-making framework based on trust: a generic framework. In: 2018 IEEE 4th International Conference on Collaboration and Internet Computing (CIC), pp. 472–479. IEEE (2018)

3. Ariga, T., Sanford, S.: Artificial intelligence: an accountability framework for federal agencies and other entities (2021). https://www.gao.gov/assets/gao-21-519sp.pdf

4. Babbar-Sebens, M., Minsker, B.S.: Interactive genetic algorithm with mixed initiative interaction for multi-criteria ground water monitoring design. Appl. Soft Comput. **12**(1), 182–195 (2012)

5. Bellamy, R.K., et al.: AI fairness 360: an extensible toolkit for detecting and mitigating algorithmic bias. IBM J. Res. Dev. **63**(4/5), 4:1–4:15 (2019)

6. Beutel, A., et al.: Putting fairness principles into practice: challenges, metrics, and improvements. In: Proceedings of the 2019 AAAI/ACM Conference on AI, Ethics, and Society, pp. 453–459 (2019)

7. Black, E., Yeom, S., Fredrikson, M.: FlipTest: fairness testing via optimal transport. In: Proceedings of the 2020 Conference on Fairness, Accountability, and Transparency, pp. 111–121 (2020)

8. Chomphoosang, P., Durresi, A., Durresi, M., Barolli, L.: Trust management of social networks in health care. In: 2012 15th International Conference on Network-Based Information Systems, pp. 392–396. IEEE (2012)

9. Chomphoosang, P., Ruan, Y., Durresi, A., Durresi, M., Barolli, L.: Trust management of health care information in social networks. In: 2013 7th International Conference on Complex, Intelligent, and Software Intensive Systems (CISIS), pp. 228–235. IEEE (2013)

10. Chomphoosang, P., Zhang, P., Durresi, A., Barolli, L.: Survey of trust based communications in social networks. In: 2011 14th International Conference on Network-Based Information Systems, pp. 663–666. IEEE (2011)

11. Dong, Y., Xu, J.: Consensus Building in Group Decision Making. Springer, Singapore (2016). https://doi.org/10.1007/978-981-287-892-2

12. Durresi, A., Durresi, M., Paruchuri, V., Barolli, L.: Trust management in emergency networks. In: 2009 International Conference on Advanced Information Networking and Applications, pp. 167–174. IEEE (2009)

13. Gajane, P., Pechenizkiy, M.: On formalizing fairness in prediction with machine learning. arXiv preprint arXiv:1710.03184 (2017)

14. Gunia, B.C., Brett, J.M., Nandkeolyar, A.K., Kamdar, D.: Paying a price: culture, trust, and negotiation consequences. J. Appl. Psychol. **96**(4), 774 (2011)

15. Gunning, D.: Explainable artificial intelligence (XAI). Defense Advanced Research Projects Agency (DARPA), 2nd Web, vol. 2 (2017)

16. Hegselmann, R., Krause, U., et al.: Opinion dynamics and bounded confidence models, analysis, and simulation. J. Artif. Soc. Soc. Simul. **5**(3), 1–33 (2002)

17. Hüffmeier, J., Freund, P.A., Zerres, A., Backhaus, K., Hertel, G.: Being tough or being nice? A meta-analysis on the impact of hard-and softline strategies in distributive negotiations. J. Manag. **40**(3), 866–892 (2014)

18. Information Technology - Artificial Intelligence - Overview of trustworthiness in artificial intelligence. Standard, International Organization for Standardization, Geneva, CH (2020)
19. Jain, R., Durresi, A., Babic, G.: Throughput fairness index: an explanation. ATM Forum Contribution 99-0045 (1999)
20. Kambiz, M.: Multi-Stakeholder Decision Making for Complex Problems: A Systems Thinking Approach with Cases. World Scientific (2016)
21. Kaur, D., Uslu, S., Durresi, A.: Trust-based security mechanism for detecting clusters of fake users in social networks. In: Barolli, L., Takizawa, M., Xhafa, F., Enokido, T. (eds.) WAINA-2019, pp. 641–650. Springer, Cham (2019). https://doi.org/10.1007/978-3-030-15035-8_62
22. Kaur, D., Uslu, S., Durresi, A., Badve, S., Dundar, M.: Trustworthy explainability acceptance: a new metric to measure the trustworthiness of interpretable AI medical diagnostic systems. In: Barolli, L., Yim, K., Enokido, T. (eds.) CISIS-2021, pp. 35–46. Springer, Cham (2021). https://doi.org/10.1007/978-3-030-79725-6_4
23. Kaur, D., Uslu, S., Durresi, A., Mohler, G., Carter, J.G.: Trust-based human-machine collaboration mechanism for predicting crimes. In: Barolli, L., Amato, F., Moscato, F., Enokido, T., Takizawa, M. (eds.) AINA-2020, pp. 603–616. Springer, Cham (2020). https://doi.org/10.1007/978-3-030-44041-1_54
24. Kaur, D., Uslu, S., Kaley, J.R., Durresi, A.: Trustworthy artificial intelligence: a review. ACM Comput. Surv. **55**(2), 1–38 (2022)
25. Kimmel, M.J., Pruitt, D.G., Magenau, J.M., Konar-Goldband, E., Carnevale, P.J.: Effects of trust, aspiration, and gender on negotiation tactics. J. Pers. Soc. Psychol. **38**(1), 9 (1980)
26. Lakkaraju, S., Adebayo, J.: "NeurIPS 2020 tutorial," in tutorial: (track2) explaining machine learning predictions: State-of-the-art, challenges, and opportunities (2020)
27. Lepri, B., Oliver, N., Letouzé, E., Pentland, A., Vinck, P.: Fair, transparent, and accountable algorithmic decision-making processes. Philos. Technol. **31**(4), 611–627 (2018)
28. Mehrabi, N., Morstatter, F., Saxena, N., Lerman, K., Galstyan, A.: A survey on bias and fairness in machine learning. ACM Comput. Surv. (CSUR) **54**(6), 1–35 (2021)
29. Rittichier, K.J., Kaur, D., Uslu, S., Durresi, A.: A trust-based tool for detecting potentially damaging users in social networks. In: Barolli, L., Chen, H.-C., Enokido, T. (eds.) NBiS-2021, pp. 94–104. Springer, Cham (2022). https://doi.org/10.1007/978-3-030-84913-9_9
30. Ruan, Y., Alfantoukh, L., Durresi, A.: Exploring stock market using Twitter trust network. In: 2015 IEEE 29th International Conference on Advanced Information Networking and Applications (AINA), pp. 428–433. IEEE (2015)
31. Ruan, Y., Alfantoukh, L., Fang, A., Durresi, A.: Exploring trust propagation behaviors in online communities. In: 2014 17th International Conference on Network-Based Information Systems (NBiS), pp. 361–367. IEEE (2014)
32. Ruan, Y., Durresi, A.: Trust management for social networks. In: Proceedings of the 14th Annual Information Security Symposium, p. 24. CERIAS-Purdue University (2013)
33. Ruan, Y., Durresi, A.: A survey of trust management systems for online social communities-trust modeling, trust inference and attacks. Knowl. Based Syst. **106**, 150–163 (2016)

34. Ruan, Y., Durresi, A.: A trust management framework for cloud computing platforms. In: 2017 IEEE 31st International Conference on Advanced Information Networking and Applications (AINA), pp. 1146–1153. IEEE (2017)
35. Ruan, Y., Durresi, A., Alfantoukh, L.: Trust management framework for internet of things. In: 2016 IEEE 30th International Conference on Advanced Information Networking and Applications (AINA), pp. 1013–1019. IEEE (2016)
36. Ruan, Y., Durresi, A., Alfantoukh, L.: Using twitter trust network for stock market analysis. Knowl. Based Syst. **145**, 207–218 (2018)
37. Ruan, Y., Durresi, A., Uslu, S.: Trust assessment for internet of things in multi-access edge computing. In: 2018 IEEE 32nd International Conference on Advanced Information Networking and Applications (AINA), pp. 1155–1161. IEEE (2018)
38. Ruan, Y., Zhang, P., Alfantoukh, L., Durresi, A.: Measurement theory-based trust management framework for online social communities. ACM Trans. Internet Technol. (TOIT) **17**(2), 16 (2017)
39. Saleiro, P., et al.: Aequitas: a bias and fairness audit toolkit. arXiv preprint arXiv:1811.05577 (2018)
40. Srivastava, M., Heidari, H., Krause, A.: Mathematical notions vs. human perception of fairness: a descriptive approach to fairness for machine learning. In: Proceedings of the 25th ACM SIGKDD International Conference on Knowledge Discovery & Data Mining, pp. 2459–2468 (2019)
41. Stanton, B., Jensen, T.: Trust and artificial intelligence (2021). https://nvlpubs.nist.gov/nistpubs/ir/2021/NIST.IR.8332-draft.pdf
42. Team, A.P.: Artificial Intelligence measurement and evaluation at the National Institute of Standards and Technology (2021)
43. Uslu, S., Kaur, D., Rivera, S.J., Durresi, A., Babbar-Sebens, M.: Decision support system using trust planning among food-energy-water actors. In: Barolli, L., Takizawa, M., Xhafa, F., Enokido, T. (eds.) AINA-2019, pp. 1169–1180. Springer, Cham (2020). https://doi.org/10.1007/978-3-030-15032-7_98
44. Uslu, S., Kaur, D., Rivera, S.J., Durresi, A., Babbar-Sebens, M.: Trust-based game-theoretical decision making for food-energy-water management. In: Barolli, L., Hellinckx, P., Enokido, T. (eds.) BWCCA-2019, pp. 125–136. Springer, Cham (2020). https://doi.org/10.1007/978-3-030-33506-9_12
45. Uslu, S., Kaur, D., Rivera, S.J., Durresi, A., Babbar-Sebens, M.: Trust-based decision making for food-energy-water actors. In: Barolli, L., Amato, F., Moscato, F., Enokido, T., Takizawa, M. (eds.) AINA-2020, pp. 591–602. Springer, Cham (2020). https://doi.org/10.1007/978-3-030-44041-1_53
46. Uslu, S., Kaur, D., Rivera, S.J., Durresi, A., Babbar-Sebens, M., Tilt, J.H.: Control theoretical modeling of trust-based decision making in food-energy-water management. In: Barolli, L., Poniszewska-Maranda, A., Enokido, T. (eds.) CISIS-2020, pp. 97–107. Springer, Cham (2021). https://doi.org/10.1007/978-3-030-50454-0_10
47. Uslu, S., Kaur, D., Rivera, S.J., Durresi, A., Babbar-Sebens, M., Tilt, J.H.: A trustworthy human–machine framework for collective decision making in food–energy–water management: the role of trust sensitivity. Knowl. Based Syst. **213**, 106683 (2021)
48. Uslu, S., Kaur, D., Rivera, S.J., Durresi, A., Durresi, M., Babbar-Sebens, M.: Trustworthy acceptance: a new metric for trustworthy artificial intelligence used in decision making in food–energy–water sectors. In: Barolli, L., Woungang, I., Enokido, T. (eds.) AINA-2021, pp. 208–219. Springer, Cham (2021). https://doi.org/10.1007/978-3-030-75100-5_19

49. Uslu, S., Ruan, Y., Durresi, A.: Trust-based decision support system for planning among food-energy-water actors. In: Barolli, L., Javaid, N., Ikeda, M., Takizawa, M. (eds.) Complex, Intelligent, and Software Intensive Systems, pp. 440–451. Springer, Cham (2019). https://doi.org/10.1007/978-3-319-93659-8_39

50. Wachter, S., Mittelstadt, B., Russell, C.: Counterfactual explanations without opening the black box: automated decisions and the GDPR. Harv. J. Law Technol. **31**, 841 (2017)

51. Zhang, P., Durresi, A.: Trust management framework for social networks. In: 2012 IEEE International Conference on Communications (ICC), pp. 1042–1047. IEEE (2012)

52. Zhang, P., Durresi, A., Barolli, L.: Survey of trust management on various networks. In: 2011 International Conference on Complex, Intelligent and Software Intensive Systems (CISIS), pp. 219–226. IEEE (2011)

53. Zhang, P., Durresi, A., Ruan, Y., Durresi, M.: Trust based security mechanisms for social networks. In: 2012 7th International Conference on Broadband, Wireless Computing, Communication and Applications (BWCCA), pp. 264–270. IEEE (2012)

54. Zhang, Y., Bellamy, R., Varshney, K.: Joint optimization of AI fairness and utility: a human-centered approach. In: Proceedings of the AAAI/ACM Conference on AI, Ethics, and Society, pp. 400–406 (2020)

New Security Protocols for Offline Point-of-Sale Machines

Nour El Madhoun[1]([✉]), Emmanuel Bertin[2], Mohamad Badra[3], and Guy Pujolle[4]

[1] Security and System Laboratory, EPITA, 14-16 Rue Voltaire, 94270 Le Kremlin-Bicêtre, France
nour.el-madhoun@epita.fr
[2] Orange Labs, 42 rue des Coutures BP 6243, 14066 Caen, France
emmanuel.bertin@orange.com
[3] College of Technological Innovation, Zayed University, P.O. Box 19282, Dubai, UAE
mohamad.badra@zu.ac.ae
[4] Sorbonne Université, CNRS, LIP6, 4 place Jussieu, 75005 Paris, France
guy.pujolle@lip6.fr

Abstract. EMV (Europay MasterCard Visa) is the protocol implemented to secure the communication between a client's payment device and a Point-of-Sale machine during a contact or an NFC (Near Field Communication) purchase transaction. In several studies, researchers have analyzed the operation of this protocol in order to verify its safety: unfortunately, they have identified two security vulnerabilities that lead to multiple attacks and dangerous risks threatening both clients and merchants. In this paper, we are interested in proposing new security solutions that aim to overcome the two dangerous EMV vulnerabilities. Our solutions address the case of Point-of-Sale machines that do not have access to the banking network and are therefore in the "offline" connectivity mode. We verify the accuracy of our proposals by using the Scyther security verification tool.

Keywords: EMV protocol · EMV vulnerabilities · NFC · Offline · Payment · Security

1 Introduction

EMV is the international protocol implemented to secure contact and contactless-NFC purchase transactions. In order to execute a secure EMV purchase transaction, five EMV actors (see Fig. 1) exchange with each other a sequence of security messages [1]. Indeed, the description of the EMV protocol is non-trivial, due to the high complexity of the EMV specifications in more than 1000 pages in [2–7]. Therefore, in our previous work [8], we introduced a synthetic overview of this protocol: it is essential to refer to this work in order to

© The Author(s), under exclusive license to Springer Nature Switzerland AG 2022
L. Barolli et al. (Eds.): AINA 2022, LNNS 450, pp. 446–467, 2022.
https://doi.org/10.1007/978-3-030-99587-4_38

clarify the roles of the EMV actors and the description of the EMV exchanged messages (EMV phases).

Indeed, in various studies, the operation of EMV protocol has been analyzed in order to verify its reliability: several security vulnerabilities have been identified in this protocol and they represent major risks for our day to day safety [9–14]. In our work [1], we presented a survey of these security vulnerabilities as well as we discussed the possible attacks due to them. Consequently, there are *two EMV security vulnerabilities* that have raised our attention to address them in this paper among the other EMV vulnerabilities [13]:

* **Vulnerability (1)**: the confidentiality of the *Banking-Data* is not ensured: the *PAN* (Primary Account Number) and *ExpDate* (Expiration Date) are sent in clear from *C* (Client's Payment Device) to *P* (Point of Sale). The latter can also store them.
* **Vulnerability (2)**: with the detection of the *Vulnerability (1)*, also the authentication of *P* to *C* is not ensured. *C* can answer any device without authenticating it, by sending the *Banking-Data* (*PAN* and *ExpDate*) in clear. Indeed, since the authentication of *P* to *C* is not ensured, then the non-repudiation for *P* is also not ensured.

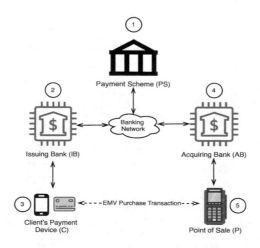

Fig. 1. EMV payment system [8]

Accordingly, in our previous work [8], we proposed a new security solution that solves these *two EMV weaknesses* in the case where *P* can communicate with the banking network and then it is in the "online" connectivity mode. In this paper, we are interested in solving the *two EMV weaknesses* in the case where *P* is in the "offline" connectivity mode and then it cannot communicate with the banking network. More specifically, in this paper, we propose two new security solutions:

– The first solution is designed for the case where P and C are both in the "offline" connectivity mode.
– The second solution is designed for the case where P is in the "offline" connectivity mode and C is in the "online" connectivity mode.

The difference between the two solutions lies in the dependence on the availability of the connection at C. This dependence makes it possible to take advantage of the availability of the Internet connection to communicate with a third party and execute security procedures.

In this paper, we use several acronyms that have been defined by default in existing literature, and in addition, we propose new abbreviations. Therefore, in order to simplify the reading of this paper, we describe in Table 1 the different abbreviations that are hereby used. In Table 2, we present the elements and procedures of security for each EMV actor. It is essential to consult these two tables while reading the paper.

The organization of this paper is as follows. In Sect. 2, we give an essential background of our proposals. In Sect. 3, we describe our proposals that aim to improve the cryptographic layer of the EMV protocol, where P is in the "offline" connectivity mode, by solving *the two EMV vulnerabilities* (discussed above). In Sect. 4, we discuss and evaluate the results of our proposals. In Sect. 5, we analyze our proposals in a formal way using the Scyther tool. The last section provides a brief conclusion.

Table 1. Abbreviations used in the paper

Abb.	Description
AB	Acquiring Bank
$ARQC$	Authorization ReQuest Cryptogram
$AuthorizReq$	Authorization transaction Request
$AuthorizResp$	Authorization transaction Response
C	Client's payment device
CA	Certificate Authority
CAp	Certificate Authority for P
$CertX$	Certificate of X
CK-AC	Symmetric Card Key derived from (IMK-$AC + PAN$). It is only shared between IB and C. It is used to generate cryptograms as $ARQC$
CK-SMC	Symmetric Card Key derived from (IMK-$SMC + PAN$). It is only shared between IB and C. It is used for secure messaging as 'PIN change'
$ConfC$	Confirmation of the authenticity of C, the non-repudiation for C and the integrity of $Banking$-$Data$. In our proposal, $ConfC$ indicates also that IB authorizes the transaction
$ConfP$	Confirmation of the authenticity of P, the non-repudiation for P and the integrity of the signed message
$ConfigInfo$	Configuration Information: results of phases 1 and 2 and other data
EMV	Europay Mastercard Visa
$ExpDate$	Expiration Date
$H(M)$	One way Hashing function of $M = m1, m2...$
IB	Issuing Bank
IdC	Identifier of C in the database of IB. This identifier is not a sensitive information

continued

Table 1. continued

Abb.	Description
IMK-AC	Symmetric Issuer Master Key for Application Cryptogram Computation. It is only shared between IB and C
IMK-SMC	Symmetric Issuer Master Key for Secure Messaging for Confidentiality. It is only shared between IB and C
k(X, Y)	Symmetric key of the current TLS session allows to protect information exchanged between X and Y
MAC	Message Authentication Code
NFC	Near Field Communication
OnX	Specify that X is able to connect to the internet
OffX	Specify that X is not able to connect to the internet
P	Point of Sale
PAN	Primary Account Number
PIN	Personal Identification Number
pk(X)	RSA Public Key of X
PS	Payment Scheme (Visa, MasterCard, etc.)
ReqX	Request for authentication of X and non-repudiation for X
RSA	Rivest-Shamir-Adleman
RX	Unpredictable Random Number generated by X(Unique)
SignX	Digital Signature generated by X using its sk(X) on the hash of a message. It allows to ensure the authentication of X, the non-repudiation for X and the integrity of the message (thanks to the hash)
sk(X)	RSA Secret private Key of X
TD	Transaction Data generated by P (amount, country code, nonce, currency,...). TD are unique for each transaction
TLS	Transport Layer Security
X	It represents: C, P, AB, PS or IB
Y	It represents: C, P, AB, PS or IB

2 Background of Our Proposals

As discussed in the previous section, in this paper, we propose two new security solutions that mainly aim to overcome the *two EMV weaknesses*. Our solutions deal with the case of P machines that do not have access to the banking network and that are therefore in the "offline" connectivity mode. We summarize in Table 3 the characteristics of our security solutions and we list them as follows:

- 1st Security Solution *(OffC/OffP)*: C and P are offline. It allows to secure contact and NFC purchase transactions. C may be either a bank card or an NFC smartphone.
- 2nd Security Solution *(OnC/OffP)*: C is online and P is offline. It allows to secure only NFC purchase transactions because it supports an online C which can be only an NFC smartphone and cannot be a bank card. The latter does not have a Wi-Fi or 4G interface.

Table 2. Elements & procedures of security for EMV actors [8,14]

Actor	Description
PS	• PS acts as a CA to certify IB. • PS has pk(PS)/sk(PS): RSA root keys self-generated. • PS has CertPS: self-signed and contains pk(PS).
IB	• IB generates two symmetric keys IMK-AC and IMK-SMC. • For each PS contracted with IB: — IB stores the CertPS as a trusted anchor CA. — IB securely generates an RSA key pairs pk(IB)/sk(IB). — PS signs pk(IB) with sk(PS) where the CertIB is generated.
C	• C can belong to a single contracted PS at once. • C stores in its secure element the security information: — Client's name. — Banking-Data (PAN, ExpDate): which are generated by IB. They are very sensitive information because they allow to perform purchase transactions. Indeed, the PAN allows to identify C in the database of IB. — Static signature generated by IB using sk(IB): {H(Banking-Data)}sk(IB). — pk(C)/sk(C): they are generated by IB for C. IB also signs pk(C) with its sk(IB) where the CertC is produced. — CK-AC generated by IB and used by C to generate an ARQC. — CK-SMC generated by IB and used by C for secure messaging. — CertPS, CertIB and CertC.
AB	• For each PS contracted with AB: AB stores CertPS as a trusted anchor CA.
P	• For each P, AB stores the CertPS of each contracted PS.

Table 3. Characteristics of the proposed security solutions

Characteristics solution	1st - OffC/OffP	2nd - OnC/OffP
C Connectivity mode	Offline	Online
P Connectivity mode	Offline	Offline
C may be a bank card	✓	–
C may be an NFC smartphone	✓	✓
Secure contact purchases	✓	–
Secure NFC purchases	✓	✓

2.1 Actors

The two main actors in our security solutions are C and P:

– For the 1st solution *(OffC/OffP)*, C and P cannot communicate with a third actor.
– For the 2nd solution *(OnC/OffP)*, C can communicate with the banking network "AB, PS, IB" (see Table 1 and see Fig. 1).

In addition, we suggest to add a new Certification Authority for P: CAp. The latter has a pair of RSA root keys $pk(CAp)$ and $sk(CAp)$ and a certificate $CertCAp$ signed by itself. The role of CAp in this paper is to certify P by generating it an RSA key pair $pk(P)/sk(P)$ and signing $pk(P)$ using $sk(CAp)$ to produce $CertP$. We also assume that CAp is considered by the banking network (AB, PS, IB) as a trusted anchor.

2.2 Authentication Procedure for P

We suggest that the actor P needs to generate an electronic signature $SignP$, using its $sk(P)$, in order to authenticate itself to C. The verification of the authenticity of P, in order to respond to $ReqP$, will be performed by another actor as follows [8]:

- Verification if the issuer of $CertP$ is a trusted certification authority: CAp.
- Verification if $CertP$ is valid (validity period).
- Verification if $pk(CAp)$ validates the signature of $CertP$.
- Verification if $pk(P)$ (obtained from $CertP$) validates $SignP$.

Table 4. Summary of targeted security properties

	Targeted security property	Is it ensured by the EMV protocol?
a)	Integrity of $Banking\text{-}Data$	$\checkmark$
b)	Validity of $Banking\text{-}Data$	$\checkmark$
c.a)	Confidentiality of $Banking\text{-}Data$: They must be sent encrypted from C to IB	Vulnerability (1) (see Sect. 1)
c.b)	Confidentiality of $Banking\text{-}Data$: P must not be able to obtain or store them	Vulnerability (1) (see Sect. 1)
d.a)	Authentication of C to P	$\checkmark$
d.b)	Authentication of P to C	Vulnerability (2) (see Sect. 1)
e.a)	Non-repudiation for C	$\checkmark$
e.b)	Non-repudiation for P	Vulnerability (2) (see Sect. 1)

2.3 Targeted Security Properties

During a purchase transaction, each proposed solution is represented in a set of security messages that are exchanged between the involved actors. Indeed, we aim that these messages guarantee the following security properties:

a) Integrity of $Banking\text{-}Data$: they must not be modified.
b) Validity of $Banking\text{-}Data$: they must not be revoked.
c) Confidentiality of $Banking\text{-}Data$:
    c.a) They must be sent encrypted from C to IB.
    c.b) P must not be able to obtain or store them.
d) Mutual authentication: it is a strong agreement excluding man-in-the-middle and potential replay attacks where an attacker could usurp the identity of one of them.

$\overline{d.a}$) Authentication of C to P.
$\overline{d.b}$) Authentication of P to C.
$\underline{e}$) Non-repudiation of origin: C and P must not be able to deny in the future: strong evidence sent by themselves or their participation in the purchase transaction.
$\underline{e.a}$) Non-repudiation of origin for C.
$\overline{e.b}$) Non-repudiation of origin for P.

In Table 4, we illustrate each targeted security property to be insured by our proposals and we show whether or not it is guaranteed in the EMV protocol.

2.4   Objectives of Our Proposals

Each Security Solution proposed in this paper has several objectives [8]:

- **Objective 1**: to improve the security of the classical EMV protocol by:
  - *Challenge* 1: ensuring the following security properties (see Sects. 1, 2.3 and 4.1):
    * $\{c.a), c.b)\}$ in order to overcome the EMV *Vulnerability (1)*.
    * $\{\overline{d.b), e.b)}\}$ in order to overcome the EMV *Vulnerability (2)*.
  - *Challenge* 2: ensuring the security properties $\{a), b), d.a), e.a)\}$ that are typically well ensured by the EMV standard (see Sects. 1, 2.3 and 4.1).
  - *Challenge* 3: putting P as a trusted element, where all communications are secured/ transparent for P: property $\overline{c.b}$).
- **Objective 2**: to respect the same EMV principle in order to be doable in the real-world in the future by:
  - *Challenge* 4: using the main security keys and certificates of the EMV payment system (see Table 2) and adding a new security layer if judged necessary (new keys, new certificates, etc.).
- **Objective 3**: to effectively use the resources of C, which is poor in the speed of calculation, in order to guarantee a fast run by:
  - *Challenge* 5: avoiding if it is possible that C performs asymmetric cryptographic functions (except for the EMV usual functions). This is because the asymmetric encryption/decryption tends to be 1000 times slower than the symmetric encryption/decryption [15, 16].
  - *Challenge* 6: offloading the execution of the asymmetric functions, that has been avoided to execute in C, to another actor that has more powerful resources than C.

2.5   Assumptions of Our Proposals

- For the two solutions proposed in this paper, we assume that:
  - C is identified by IdC in the database of IB and not by the PAN as in the EMV protocol [8]. IdC is not a sensitive information as the PAN, it does not present a risk if it is sent in clear because it does not allow to make purchases.

- C produces $SignC$ to authenticate itself.
- P produces $SignP$ to authenticate itself.
- $SignP$ allows to ensure the properties $\{\underline{d.b)}, e.b)\}$ (overcoming *Vulnerability (2)*), and to guarantee the integrity of the signed message (see Tables 1 and 4).
- For the 1st Security Solution *(OffC/OffP)*, we assume that:
  - C stores $CertCAp$ as a trusted certificate of a trusted certification authority.
  - $SignC$ does not include the *Banking-Data* and allows to ensure the properties $\{\underline{d.a)}, e.a)\}$ and to guarantee the integrity of the signed message.
  - C proceeds to verify the authenticity of P.
  - P proceeds to verify the authenticity of C.
  - P periodically receives, when it is connected to the Internet, the list of revoked certificates of clients..
  - If $CertC$ is valid, then the $PAN/ExpDate$ are valid.
  - If $CertC$ is revoked or not valid, then the $PAN/ExpDate$ are also revoked or not valid.
  - If $CertP$ is revoked or not valid, then P cannot produce $SignP$.
- For the 2nd Security Solution *(OnC/OffP)*, we assume that:
  - C can securely communicate with IB thanks to the TLS protocol using a TLS session key: $k(C,IB)$.
  - $SignC$ includes the *Banking-Data* and allows to ensure the properties $\{\underline{a)}, d.a), e.a)\}$ and to guarantee the integrity of the signed message (see Tables 1 and 4).
  - AB stores $CertCAp$ as a trusted certificate of a trusted certification authority.
  - AB proceeds to verify the authenticity of P.
  - IB proceeds to verify the authenticity of C.
  - C will not use the $CK\text{-}AC$ to generate an $ARQC$ but it will use the $CK\text{-}SMC$ to generate an enciphered message (see Tables 1 and 2).
  - P will not proceed to verify the authenticity of C but it will wait, from IB, for the confirmation of the authenticity of C (thanks to $ConfC$, see Table 1).

3   Description of Our Proposals

During a purchase transaction, each proposed solution is represented in a set of security messages that are exchanged between the involved actors. For each of our proposals, we suggest dividing these security messages into three phases (*Initialization (Phase 1), Authentication of the Client (Phase 2), Authentication of C and P (Phase 3)*) instead of four phases as in the EMV security protocol: as specified in Sect. 1, it is essential to refer to our work [8] in order to clarify the roles of the EMV actors and the description of the EMV phases.

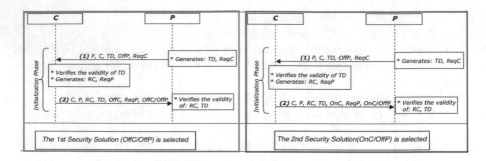

Fig. 2. Initialization (Phase 1)

3.1   Phase 1 and Phase 2

These two phases are more or less the same for the two proposed security solutions.

Initialization (Phase 1). This phase is illustrated in Fig. 2. It allows, firstly, P to inform its connectivity mode to C, and secondly, C to inform its connectivity mode to P. The response from C gives the decision for the **Security Solution** that will be chosen subsequently to execute in phase 3.

(1) P->C:

    1.1 P generates TD that aims to prevent replay attacks and an authentication request for C $ReqC$. Then, it sends to C: TD, P connectivity mode (OnP or $OffP$) and $ReqC$.

    1.2 C receives the message (1), verifies the validity of TD, generates RC to prevent replay attacks and and an authentication request for P $ReqP$.

(2) C->P:

    2.1 C responds to P by sending: RC, TD, C connectivity mode (OnC or $OffC$), $ReqP$ and the decision "C connectivity mode"/"P connectivity mode".

    2.2 P receives (2) and verifies the validity of RC and TD. It will respond to $ReqP$ in phase 3.

    2.3 P also takes into consideration the decision of C, and then, according to this decision, one of the security solutions is selected for the execution in phase 3:

       * OffC/OffP: the 1st Security Solution is selected if C and P are offline.

       * OnC/OffP: the 2nd Security Solution is selected if C is online and P is offline.

Authentication of the Client (Phase 2). This phase is executed in the same manner as in the original EMV protocol thanks to a PIN code or a signature or nothing if the payment is contactless-NFC [8].

3.2   Description of Phase 3 for the 1st Security Solution *(OffC/OffP)*

After the execution of phases 1 and 2, if the 1st Security Solution *(OffC/OffP)* is selected, phase 3 is executed as follows (see Fig. 3):

(1) **P->C**:

   1.1 *P* generates *SignP* with its *sk(P)* principally on the hash of {*TD*, *RC*, *ReqC*, *OffC/OffP*,...} and sends to *C*: *TD*, *RC*, *CertP* and *SignP*.

   1.2 After receiving **(1)**, *C* verifies the validity of nonces and the authenticity of *P* as presented in Sect. 2.2.

   1.3 *ConfP* is a formal message generated by *C* to indicate that *P* is well authenticated (properties {*d.b*), *e.b*)}, overcoming *Vulnerability (2)*).

   1.4 *C* produces *SignC* with its *sk(C)* mainly on the hash of {..., *IdC*, *RC*, *TD*, *ConfP*, *OffC/OffP*, *ConfigInfo*,...} and without including the *Banking-Data*: *PAN*, *ExpDate*.

(2) **C->P**:

   2.1 *C* sends to *P* in clear: *ConfigInfo*, *ConfP*, *CertC*, *CertIB* and *SignC*. *C* does not send the *Banking-Data* in clear.

   2.2 Since the *Banking-Data* are not sent from *C* to *P*, then they cannot be modified or changed (property *a*)). Additionally, in an implicit way, they will remain confidential to *P* and *IB* (properties {*c.a*), *c.b*)}, overcoming *Vulnerability (1)*).

   2.3 *P* receives **(2)**, confirms that it was well authenticated by *C* through *ConfP* and proceeds to authenticate *C* by verifying that:
  - *PS* is the issuer of *CertIB*.
  - *CertIB* and *CertC* are valid today.
  - *pk(PS)* validates the signature contained in *CertIB*.
  - *pk(IB)* validates the signature contained in *CertC*.
  - *pk(P)* validates *SignC*.

   2.4 *P* confirms the authenticity of *C* by generating the formal message *ConfC* (properties {*d.a*), *e.a*)}). *P* also confirms that the *PAN* and *ExpDate* are valid because *CertC* is valid (property *b*)).

(3) **P->C**:

   3.1 *P* sends to *C*: *TD*, *RC*, *ConfC*. Then, *C* verifies the validity of nonces and confirms that it was well authenticated by *P* thanks to *ConfC*.

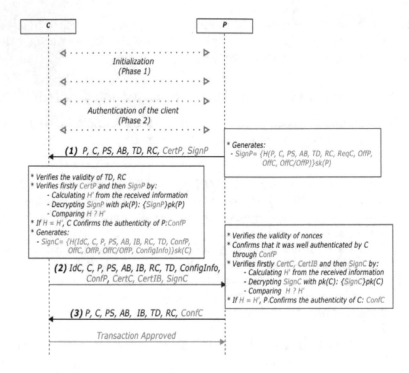

Fig. 3. The 1st Security Solution *(OffC/OffP)* (Phase 3)

3.3   Description of Phase 3 for the 1st Security Solution *(OnC/OffP)*

After the execution of phases 1 and 2, if the 1st Security Solution *(OnC/OffP)* is selected, phase 3 is executed as follows (see Fig. 4):

(1) **P->C:**

   1.1 *P* generates *SignP* with its *sk(P)* mainly on the hash of {*TD, RC, ChallengeTextP, OnC/OffP,...*}. *ChallengeTextP* is a confidential text known only by *P* and *AB*. So, *P* will wait to receive *ChallengeTextP* from *C* to confirm that *C* has well contacted *AB*.

   1.2 *P* also prepares the *MessageP* that mainly contains {*TD, RC, ReqC, OnC/OffP ...*}. Afterwards, it sends to *C*: *CertP, SignP* and the *MessageP*.

   1.3 After receiving **(1)**, *C* will not verify the authenticity of *P*, but it calculates the *ConfidentialityKey* as a symmetric session key by hashing both the *CK-SMC* and *RC*.

   1.4 *C* produces *SignC* with its *sk(C)* mainly on the hash of {*IdC, RC, TD, PAN, ExpDate, OnC/OffP, ConfigInfo,...*}.

   1.5 *C* also creates an authorization transaction request for *IB* '*AuthorizReq*', that mainly contains: {*IdC, RC, TD, PAN, ExpDate, OnC/OffP, ConfigInfo, SignC,...*}. *C* encrypts *AuthorizReq* using the *ConfidentialityKey*.

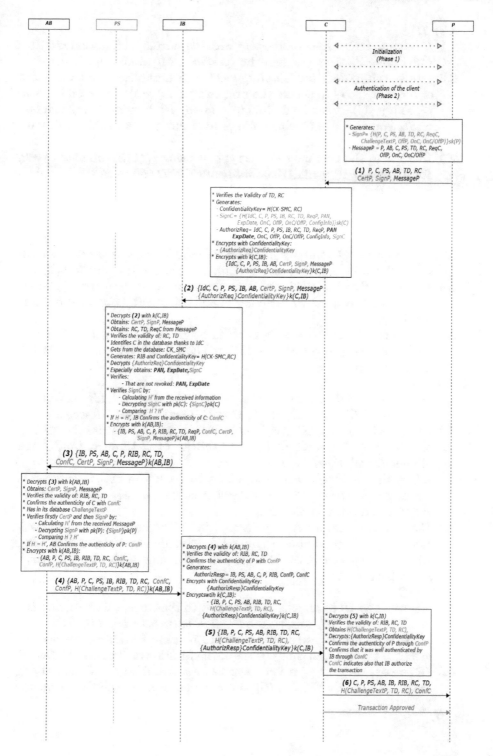

Fig. 4. The 2nd Security Solution *(OffC/OffP)* (Phase 3)

(2) **C->IB**:

2.1 *C* sends to *IB* in an encrypted text with the current TLS session *k(C,IB)*: *CertP, SignP, MessageP* and the {*AuthorizReq*} *ConfidentialityKey*.

2.2 {*AuthorizReq*} *ConfidentialityKey* asks if *IB* authorizes the transaction or not and allows to ensure the properties {c.a), c.b)} (overcoming *Vulnerability (1)*). It can be deciphered only by $\overline{IB}$. In addition, the *AuthorizReq* contains *SignC* which allows to ensure the properties {a), d.a), e.a)}.

2.3 $\overline{IB}$ receives the message (2), decrypts it using *k(C,IB)*, obtains *CertP, SignP, MessageP*, verifies the validity of *RC, TD* and generates a random number *RIB* that serves to prevent replay attacks.

2.4 *IB* then identifies *C* in its database through *IdC* and gets *CK-SMC*. The latter is required to calculate the *ConfidentialityKey* as illustrated in Fig. 4.

2.5 *IB* deciphers {*AuthorizReq*} *ConfidentialityKey* and especially obtains *PAN, ExpDate, SignC* (property c.a), overcoming *Vulnerability (1)*).

2.6 *IB* verifies that the *PAN, ExpDate* are not revoked (property b)) and it will respond to *ReqC* by verifying *SignC* using *pk(C)* obtained from its database.

2.7 *ConfC* is a formal message generated by *IB* to indicate that *C* is well authenticated (properties a), d.a), e.a)) and that the transaction is well authorized by *IB*.

(3) **IB->AB**:

3.1 *IB* sends to *AB* in encrypted text with the current TLS session key *k(AB,IB)*: *RIB, RC, TD, ReqP, ConfC, CertP, SignP, MessageP*.

3.2 After receiving (3), *AB* deciphers it using *k(AB,IB)*, verifies the validity of *RIB, RC, TD*, confirms the authenticity of *C* through *ConfC* and obtains *CertP, SignP, MessageP*.

3.3 *AB* is the bank of *P* and also allows to verify the authenticity of *P*. Then, it will retrieve *ChallengeTextP* from its database and respond to *ReqP* as presented in Sect. 2.2.

3.4 *ConfP* is a formal message generated by *AB* to indicate that *P* is well authenticated (properties {d.b), e.b)}, overcoming *Vulnerability (2)*). Also, *H(ChallengeTextP, $\overline{TD}$, $\overline{RC}$)* is a dynamic hash message which is destined to *C*.

(4) **AB->IB**:

4.1 *AB* sends to *IB* in an encrypted text with the current TLS session key *k(AB,IB)*: *TD, RC, ConfC, ConfP, H(ChallengeTextP, TD, RC)*.

4.2 *IB* receives the message (4), decrypts it using *k(AB,IB)*, verifies the validity of nonces and confirms the authentication of *P* thanks to *ConfP*.

4.3 *IB* creates an authorization response *AuthorizResp* for *C* containing mainly: {*RIB, ConfP, ConfC*} and encrypts it using the *ConfidentialityKey*.

(5) **IB->C**:

5.1 *IB* sends to *C* in an encrypted text with the current TLS session key *k(C,IB)*: *RIB, RC, TD, H(ChallengeTextP, TD, RC)*, {*AuthorizResp*}*Co nfidentialityKey*. *C* receives the message **(4)**, deciphers it, verifies nonces, obtains *H(ChallengeTextP, TD, RC)* and the {*AuthorizResp*}*Confidenti alityKey*.

5.2 *C* decrypts {*AuthorizResp*}*ConfidentialityKey*, confirms the authenticity of *P* thanks to *ConfP*, confirms thanks to *ConfC* that it was well authenticated by *IB* and that the transaction is authorized by *IB*.

(6) ***C->P***:

6.1 *C* sends to *P*: *RIB, RC, TD, H(ChallengeTextP, TD, RC), ConfC*. Then, *P* verifies the validity of nonces, calculates another hash *H'* of *ChallengeTextP, TD, RC*, compares the calculated hash *H'(ChallengeTextP, TD, RC)* with the received hash *H(ChallengeTextP, TD, RC)* and confirms that *C* has well contacted *AB*. *P* also confirms thanks to *ConfC* the authenticity of *C* and that the transaction is well authorized by *IB*.

4   Results of the Proposals

4.1   Discussions

Our two security solutions achieve the *Objectives* presented in Sect. 2.4:

- **Objective 1**: improvement of the security of the EMV classical protocol as follows :
  - In our two proposals, as illustrated in Table 5, the security messages exchanged between the actors, during a payment transaction, meet the targeted security properties {a), b), c.a), c.b) d.a), d.b), e.a), e.b)} (see Table 4 and Sect. 2.3). However, the EMV protocol meets only the security properties {a), b), d.a), e.a)} (see Table 4).
  - In our two proposals, *P* does not receive the *Banking-Data* either in clear nor in an encrypted manner because it is in the offline mode, and then, it is not able also to hack these data: property c.b).
  - In our 2nd solution, *P* does not verify the authenticity of *C* by itself because *SignC* includes the *Banking-Data*. However, in our 1st solution, *P* verifies the authenticity of *C* by itself but *SignC* does not include the *Banking-Data*. Hence, in our two proposals, we place *P* as an element of trust after its authentication by *AB* or by *C*.

Table 5. Achieved security properties in our proposals

Properties	Steps in which they have been achieved (Phase 3)
————The 1st Security Solution *OffC/OffP*————	

Properties	Steps in which they have been achieved (Phase 3)
a)	- In message (2): 2.1, 2.2. The *Banking-Data* are not sent by C to P and then they cannot be modified
c.a), c.b)	- In message (2): 2.2. The *Banking-Data* are not sent by C to P, and then in an implicit way, they remain confidential to P and *IB*
d.a), e.a)	- In message (2): 2.3, 2.4. Thanks to *SignC* that is verified by P
b)	- In message (2): 2.4
d.b), e.b)	- In message (1): 1.1, 1.2, 1.3. Thanks to *SignP* that is verified by C
————The 2nd Security Solution *OnC/OffP*————	
a), d.a), e.a)	- In message (2): 2.5, 2.6, 2.7. Thanks to *SignC* that is verified by *IB*. - In message (6): 6.1. Thanks to *ConfC* that confirms to P the authenticity of C
b)	- In message (2): 2.6
c.a), c.b)	- In message (2): 2.1, 2.2, 2.5. Thanks to the *AuthorizReq* containing the *Banking-Data* and which is sent, from C to *IB*, enciphered by the *ConfidentialityKey*. Only *IB* can obtain the *Banking-Data* by deciphering {*AuthorizReq*}*ConfidentialityKey*. - In message (2): The *Banking-Data* are not sent by C to P, and then in an implicit way, they remain confidential to P
d.b), e.b)	- In message (3): 3.3, 3.4. Thanks to *SignP* that is verified by *AB*. - In message (5): 5.2. Thanks to *ConfP* that confirms to C the authenticity of P

- **Objective 2**: respect of the same EMV principle as follows:
  - As illustrated in Table 6, our proposals globally use the same security elements provided by the EMV payment system (see Table 2). The 2nd Security Solution *OnC/OffP* uses *CK-SMC*, that is classically provided by EMV protocol, instead of *CK-AC* (please refer to [8])). The 1st Security Solution *OffC/OffP* uses neither *CK-SMC* nor *CK-AC* because these keys are specific for the communication with *IB*. Additionally, our proposals add new *CAp*, *pk(P)/sk(P)* and *CertP* that allow to certify P (see Sects. 2.1 and 2.2).

- As illustrated in Table 7, our proposals globally perform the same security procedures as in the EMV security protocol. In the latter, each of the *ARQC* and *SignC* allows to authenticate *C* [8]. For our proposals, we have seen that *SignC* is sufficient to authenticate *C*. For the 2nd Security Solution *OnC/OffP*, the {*AuthorizReq*}*ConfidentialityKey* is needed to encrypt the *Banking-Data*. Then, we replaced the *ARQC* calculation with the {*AuthorizReq*}*ConfidentialityKey* calculation. In addition, the *ARPC*, in the EMV protocol, allows to respond to *C* and to authenticate *IB*. Consequently, for the 2nd Security Solution *OnC/OffP*, since we used the *ConfidentialityKey*, then we replaced the *ARPC* calculation with the {*AuthorizResp*}*ConfidentialityKey* calculation.
- In fact, both the {*AuthorizReq*}*ConfidentialityKey* and {*AuthorizResp*}*Co nfidentialityKey* are unnecessary in the 1st Security Solution *OffC/OffP*, because the *ConfidentialityKey* is derived from *CK-SMC* which is specific for the communication with *IB*.
- **Objective 3**: effective use for *C* resources:
  - As illustrated in Table 8, there is in the EMV protocol only '1' asymmetric function: *calculating SignC*. For the 2nd Security Solution, we successfully succeeded in avoiding performing in *C* other asymmetric functions, with the exception of that which is provided by default by EMV. This success is achieved because we have taken advantage of the availability of the internet connection to offload the asymmetric functions, that are supposed to be executed by *C*, to another actor, such as *AB/IB*.
  - For the 1st Security Solution, since *C* and *P* are both in the offline mode, and *C* needs to authenticate *P*, then, *C* has to additionally execute to the old asymmetric function, a new one in order to verify *SignP*. Therefore, in this solution, it was not possible to avoid the execution of new asymmetric cryptographic functions.

Table 6. Security elements in the EMV protocol and our proposals

	EMV protocol [8]	1st - *OffC/OffP*	2nd - *OnC/OffP*
Elements provided by EMV & used	- *CertIB*	- *CertIB*	- *CertIB*
	- *CertC*	- *CertC*	- *CertC*
	- *pk(C)/sk(C)*	- *pk(C)/sk(C)*	- *pk(C)/sk(C)*
	- *CK-AC*		- *CK-SMC*
Elements provided by EMV & not used	- *CK-SMC*	- *CK-AC*	- *CK-AC*
		- *CK-SMC*	
Elements not provided by EMV: our proposed elements		- New *CAp*	- New *CAp*
		- *CertP*	- *CertP*
		- *pk(P)/sk(P)*	- *pk(P)/sk(P)*

Table 7. Security procedures in the EMV protocol and our proposals

	EMV Protocol [8]	1st - *OffC/OffP*	2nd - *OnC/OffP*
Used EMV procedures	- *SignC*	- *SignC*	- *SignC*
	- *AuthorizReq*		- *AuthorizReq*
	- *AuthorizResp*		- *AuthorizResp*
	- *ARQC*		
	- *ARPC*		
Not used EMV procedures		- *ARQC*	- *ARQC*
		- *ARPC*	- *ARPC*
New procedures: our proposed procedures		- *SignP*	- *SignP*
			- Encipher *AuthorizReq* by *ConfidentialityKey*
			- Encipher *AuthorizResp* by *ConfidentialityKey*

4.2 Performance Evaluation

To the best of our knowledge, our proposals have not been previously proposed with the same ideas and objectives. This make us the first to give better results than the related works as we will discuss in this section. In Table 9, we compare the cost of the cryptographic functions computation, between the existing solutions, EMV protocol and our proposals. In Table 10, we show a robustness comparison. The questions asked in this table are: **Q1**: Is it feasible to actually implement it?, **Q2**: It does not change EMV principle?, **Q3**: Is there an efficient use for C resources?, **Q4**: Does it resist against replays attacks?, **Q5**: Does it resist against impersonation attacks?, **Q6**: Does it resist against man-in-the-middle attacks?, **Q7**: Does it resist if P was stolen ?

The comparison shows the effectiveness of our proposals as follows: from Table 9, the existing solutions [17–19] and [20] are faster than our proposals because the use of the symmetric cryptography takes less time compared to the use of the asymmetric cryptography [16]. From Table 10, our proposals satisfy all properties and are totally robust (answers to questions). In addition, as illustrated in Table 10, the solutions [17–19] and [20] do not satisfy all the security requirements and are not totally robust. The solutions [14,21], compared to the solutions [17–19] and [20], are less rapid (see Table 9), satisfy more security requirements and are more robust (see Table 10). Consequently, we can conclude that our proposals are more interesting than the solutions proposed in literature, and especially when we compare the execution time of our proposals to the EMV execution time: they need, in addition to EMV, two asymmetric operations and two hashing functions to authenticate P (see Sect. 4.1).

Table 8. Cryptographic cost in C

	Symm. encryption	Symm. decryption	Asymm. encryption	Asymm. decryption	Hash functions
EMV [8]	2	–	1	–	3
1st OffC/OffP	–	–	1	1	2
2st OnC/OffP	2	2	1	–	2

Table 9. Cryptographic cost comparison

	Symm. encryption	Symm. decryption	Asymm. encryption	Asymm. decryption	Hash functions	Exchanged messages
[14]	6	6	2	2	6	11
[17]	6	6	–	–	9	8
[18]	7	7	–	–	–	7
[21]	10	10	2	2	4	11
[19]	4	4	1	1	3	6
[20]	7	7	–	–	2	8
EMV [8]	*6*	*6*	*1*	*1*	*4*	*9*
1st *OffC/OffP*	–	–	**2**	**2**	**4**	**3**
2st *OnC/OffP*	**6**	**6**	**2**	**2**	**6**	**6**

Table 10. Robustness comparison

	[14]	[17]	[18]	[21]	[19]	[20]	EMV [8]	1st *OffC/OffP*	2nd *OnC/OffP*
a)	√	√	–	√	√	√	√	√	√
b)	√	–	–	–	–	√	√	√	√
c.a)	√	√	√	√	√	√	–	√	√
c.b)	–	–	–	–	–	–	–	√	√
d.a)	√	√	√	√	–	√	√	√	√
d.b)	√	√	√	√	√	√	–	√	√
e.a)	√	–	–	√	–	–	√	√	√
e.b)	√	–	–	√	–	–	–	√	√
Q1	–	–	–	–	–	–	√	√	√
Q2	√	–	–	–	–	–	√	√	√
Q3	–	–	–	√	√	√	√	√	√
Q4	√	√	√	√	√	√	√	√	√
Q5	√	√	√	√	–	√	–	√	√
Q6	√	√	√	√	–	√	–	√	√
Q7	–	–	–	–	–	–	–	√	√

5   Formal Security Analysis Using Scyther Tool

The verification of the correctness and soundness of a security protocol has proven to this day to be extremely difficult for humans [22]. Hence, we have chosen to verify our solutions by a verification tool called Scyther that allows formal analysis for security protocols by identifying potential attacks and vulnerabilities (man-in-the-middle, data modification, data insertion, etc.) [23]. To the best of our knowledge, Scyther is the most efficient tool in terms of simplicity of implementation and verification of security properties. Furthermore, there are many documents in literature to learn how it works [24]. Additionally, Scyther has been successfully used in both research and teaching fields, and in [25,26], authors prove the performance of Scyther compared to other tools.

			Status		Comments
C	SecondSecurityMechanismOffOff,c1	Nisynch	Ok	Verified	No attacks.
	SecondSecurityMechanismOffOff,c2	Niagree	Ok	Verified	No attacks.
	SecondSecurityMechanismOffOff,c3	Alive	Ok	Verified	No attacks.
	SecondSecurityMechanismOffOff,c4	Weakagree	Ok	Verified	No attacks.
P	SecondSecurityMechanismOffOff,p1	Nisynch	Ok	Verified	No attacks.
	SecondSecurityMechanismOffOff,p2	Niagree	Ok	Verified	No attacks.
	SecondSecurityMechanismOffOff,p3	Alive	Ok	Verified	No attacks.
	SecondSecurityMechanismOffOff,p4	Weakagree	Ok	Verified	No attacks.

Fig. 5. Scyther results for the 1st Security Solution

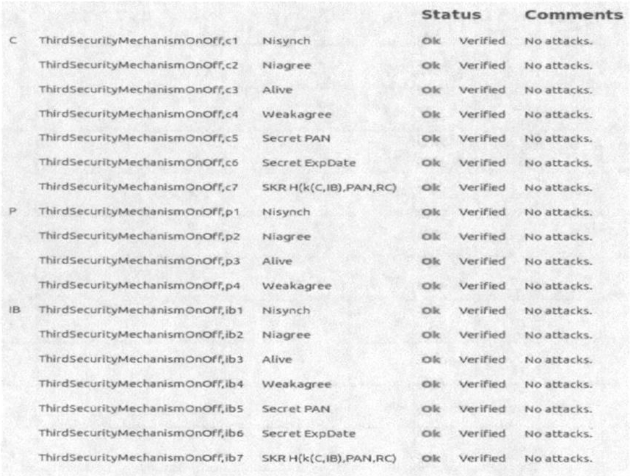

			Status		Comments
C	ThirdSecurityMechanismOnOff,c1	Nisynch	Ok	Verified	No attacks.
	ThirdSecurityMechanismOnOff,c2	Niagree	Ok	Verified	No attacks.
	ThirdSecurityMechanismOnOff,c3	Alive	Ok	Verified	No attacks.
	ThirdSecurityMechanismOnOff,c4	Weakagree	Ok	Verified	No attacks.
	ThirdSecurityMechanismOnOff,c5	Secret PAN	Ok	Verified	No attacks.
	ThirdSecurityMechanismOnOff,c6	Secret ExpDate	Ok	Verified	No attacks.
	ThirdSecurityMechanismOnOff,c7	SKR H(k(C,IB),PAN,RC)	Ok	Verified	No attacks.
P	ThirdSecurityMechanismOnOff,p1	Nisynch	Ok	Verified	No attacks.
	ThirdSecurityMechanismOnOff,p2	Niagree	Ok	Verified	No attacks.
	ThirdSecurityMechanismOnOff,p3	Alive	Ok	Verified	No attacks.
	ThirdSecurityMechanismOnOff,p4	Weakagree	Ok	Verified	No attacks.
IB	ThirdSecurityMechanismOnOff,ib1	Nisynch	Ok	Verified	No attacks.
	ThirdSecurityMechanismOnOff,ib2	Niagree	Ok	Verified	No attacks.
	ThirdSecurityMechanismOnOff,ib3	Alive	Ok	Verified	No attacks.
	ThirdSecurityMechanismOnOff,ib4	Weakagree	Ok	Verified	No attacks.
	ThirdSecurityMechanismOnOff,ib5	Secret PAN	Ok	Verified	No attacks.
	ThirdSecurityMechanismOnOff,ib6	Secret ExpDate	Ok	Verified	No attacks.
	ThirdSecurityMechanismOnOff,ib7	SKR H(k(C,IB),PAN,RC)	Ok	Verified	No attacks.

Fig. 6. Scyther results for the 2nd Security Solution

Indeed, Scyther allows to analyze security protocols thanks to specific Scyther claims (authentication, confidentiality, etc.) with an unbounded number of sessions and guaranteed termination. If it detects an attack corresponding to a mentioned claim, then it produces a graph describing this attack. Consequently, in order to implement our proposals in Scyther, the Security Protocol Description Language (SPDL) is used [26]. In this language, each actor is either written in a Scyther role or is declared as a Scyther Agent. In our work [8], we have given definitions of the Scyther claims/roles/agents. Therefore, we have used Scyther in our solutions as follows:

- For the two solutions: we have used the following Scyther claims to refer to the targeted security properties discussed in Sect. 2.3:
  - *Nisynch, Niagree, Alive, Weakagree*: for the integrity of the *Banking-Data* (*PAN/ExpDate*) and the authentication/non-repudiation of *C, P*: $\{\underline{a)},\underline{d.a)}, \underline{d.b)}, \underline{e.a)}, \underline{e.b)}\}$.
  - *Secret*: for the confidentiality of the *Banking-Data* (*PAN/ExpDate*): $\{\underline{c.a)}, \underline{c.b)}\}$.
- Only for the 2nd Solution: we have used the following Scyther claim:
  - *SKR*: for the confidentiality of the *ConfidentialityKey* that is encoded in scyther by *H(k(C,IB), PAN, RC)* where *k(C,IB)* represents *CK-SMC*.
- For the 1st Solution:
  - Scyther Roles/Agents: we have implemented Scyther roles for $\{C, P\}$, and for $\{AB, IB, PS, CAp\}$, we have used Scyther agents.
  - Scyther Results: as illustrated in Fig. 5, the 1st Security Solution successfully guarantees all the Scyther claims for $\{C, P\}$ and no attacks are found.
- For the 2nd Solution:
  - Scyther Roles/Agents: we have implemented the Scyther roles for $\{C, P, IB\}$, and for $\{AB, PS, CAp\}$, we have used the Scyther agents.
  - Scyther Results: as illustrated in Fig. 6, the 2nd Security Solution successfully guarantees all the Scyther claims for $\{C, P, IB\}$ and no attacks are found.

6   Conclusion

In this paper, we proposed new security solutions aiming to overcome *two security vulnerabilities* that have been detected in the classical EMV payment protocol. According to our previous study in [1], these two EMV vulnerabilities represent major risks for our day to day safety. The idea of our solutions is to improve the security of the classical EMV protocol by solving these two vulnerabilities where P machines are in the "offline" connectivity mode. Consequently, our proposals allow ensuring all the targeted security properties presented in Table 4 and are totally robust compared to the other solutions proposed in literature. They are also verified correctly thanks to the Scyther security tool.

References

1. El Madhoun, N., Bertin, E., Pujolle, G.: The EMV payment system: is it reliable? In: 2019 3rd Cyber Security in Networking Conference (CSNet), pp. 80–85 (2019)
2. EMV - Integrated Circuit Card Specifications for Payment Systems. Book 1: Application Independent ICC to Terminal Interface Requirements, Version 4.3. EMVCo (November 2011)
3. EMV - Integrated Circuit Card Specifications for Payment Systems. Book 2: Security and Key Management, Version 4.3. EMVCo (November 2011)
4. EMV - Integrated Circuit Card Specifications for Payment Systems. Book 3: Application Specification, Version 4.3. EMVCo (November 2011)

5. EMV - Integrated Circuit Card Specifications for Payment Systems. Book 4: Card-holder, Attendant, and Acquirer Interface Requirements, Version 4.3. EMVCo (November 2011)
6. de Ruiter, J., Poll, E.: Formal analysis of the EMV protocol suite. In: Mödersheim, S., Palamidessi, C. (eds.) Theory of Security and Applications, pp. 113–129. Springer, Heidelberg (2012). https://doi.org/10.1007/978-3-642-27375-9_7
7. van den Breekel, J., Ortiz-Yepes, D.A., Poll, E., de Ruiter, J.: EMV in a nutshell. Technical report (2016)
8. El Madhoun, N., Bertin, E., Badra, M., Pujolle, G.: Towards more secure EMV purchase transactions. Ann. Telecommun. **76**(3), 203–222 (2021)
9. Basin, D., Sasse, R., Toro-Pozo, J.: The EMV standard: break, fix, verify. In: 2021 IEEE Symposium on Security and Privacy (SP), pp. 1766–1781 (2021)
10. Emms, M., van Moorsel, A.: Practical attack on contactless payment cards. In: HCI2011 Workshop Heath, Wealth and Identity Theft (2011)
11. Lifchitz, R.: Hacking the NFC credit cards for fun and debit. In: Hackito Ergo Sum conference (April 2012)
12. Cohen, B.: Millions of Barclays card users exposed to fraud (2012). Last connection (24 December 2021)
13. Tubb, G.: Contactless cards: app reveals security risk (2013). https://news.sky.com/story/contactless-cards-app-reveals-security-risk-10443980. Accessed 24 Dec 2021
14. Emms, M.J.: Contactless payments: usability at the cost of security? Newcastle University, Ph.D. Thesis (2016)
15. Elminaam, D.A., Kader, H.A., Hadhoud, M.M.: Performance evaluation of symmetric encryption algorithms. Commun. IBIMA **8**, 58–64 (2009)
16. Badra, M., Badra, R.B.: A lightweight security protocol for NFC-based mobile payments. Procedia Comput. Sci. **83**, 705–711 (2016)
17. Thammarat, C., Kurutach, W., Phoomvuthisarn, S.: A secure lightweight and fair exchange protocol for NFC mobile payment based on limited-use of session keys. In: 17th International Symposium on Communications and Information Technologies (ISCIT), pp. 1–6. IEEE (2017)
18. Ceipidor, U.B., Medaglia, C.M., Marino, A., Sposato, S., Moroni, A.: KerNeeS: a protocol for mutual authentication between NFC phones and POS terminals for secure payment transactions. In: International ISC Conference on Information Security and Cryptology (ISCISC), pp. 115–120. IEEE (2012)
19. Lee, Y.S., Kim, E., Jung, M.S.: A NFC based authentication method for defense of the man in the middle attack. In: Proceedings of the 3rd International Conference on Computer Science and Information Technology, pp. 10–14 (2013)
20. Al-Tamimi, M., Al-Haj, A.: Online security protocol for NFC mobile payment applications. In: 8th International Conference on Information Technology (ICIT), pp. 827–832. IEEE (2017)
21. Pourghomi, P., Ghinea, G., et al.: A proposed NFC payment application. Int. J. Adv. Comput. Sci. Appl. **12**, 173–181 (2013)
22. Basin, D., Cremers, C., Meadows, C.: Model checking security protocols. In: Clarke, E.M., Henzinger, T.A., Veith, H., Bloem, R. (eds.) Handbook of Model Checking, pp. 727–762. Springer, Cham (2018). https://doi.org/10.1007/978-3-319-10575-8_22
23. Cremers, C.J.F.: The Scyther tool: verification, falsification, and analysis of security protocols. In: Gupta, A., Malik, S. (eds.) CAV 2008. LNCS, vol. 5123, pp. 414–418. Springer, Heidelberg (2008). https://doi.org/10.1007/978-3-540-70545-1_38

24. Kahya, N., Ghoualmi, N., Lafourcade, P.: Formal analysis of PKM using scyther tool. In: International Conference on Information Technology and e-Services, pp. 1–6. IEEE (2012)
25. Cremers, C., Lafourcade, P.: Comparing state spaces in automatic protocol verification. In: International Workshop on Automated Verification of Critical Systems (AVoCS) (2007)
26. Cremers, C., Mauw, S.: Operational Semantics and Verification of Security Protocols. Springer, Heidelberg (2012). https://doi.org/10.1007/978-3-540-78636-8

Building a Blockchain-Based Social Network Identification System

Zhanwen Chen[1](✉) and Kazumasa Omote[1,2]

[1] University of Tsukuba, Tennodai 1-1-1, Tsukuba 305-8573, Japan
s2130132@s.tsukuba.ac.jp, omote@risk.tsukuba.ac.jp
[2] National Institute of Information and Communications Technology, 4-2-1 Nukui-Kitamachi, Koganei, Tokyo 184-8795, Japan

Abstract. With the rise of social network service (SNS) in recent years, the security of SNS users' private information has been a concern for the public. However, due to the anonymity of SNS, identity impersonation is hard to be detected and prevented since users are free to create an account with any name they want. Until now, there are few studies about this problem, and none of them can perfectly handle this problem. In this paper, based on an idea from previous work, we combine blockchain technology and security protocol to prevent impersonation in SNS. In our scheme, the defects of complex and duplicated operations in the previous work are improved. And the authentication work of SNS server is also adjusted in order to resist single-point attack. Moreover, the smart contract is introduced to help the whole system runs automatically. Afterward, our proposed scheme is implemented and tested on an Ethereum test network and the result suggests that it is acceptable and suitable for nowadays SNS network.

1 Introduction

Impersonation problem has been a big concern in social network service (SNS). On the one hand, SNS applications like Twitter and Facebook provide a platform for everyone to publish and receive information. People can use any nickname they like rather than communicate by their real name. It is a reasonable strategy since users may not wish to open their identity on Internet. However, the anonymity of SNS also causes incentives for impersonation. For example, a malicious user can easily create an account with the name "Bob, Apple's recruit staff". After some decorating his account with some official-looking pictures and words, the authenticity of this account can be hardly distinguished. Then, this malicious user pretends to be an Apple's recruit staff and requires applicants to send application forms to him, which contains private information of applicants. Furthermore, such information leakage could lead to Internet fraud. Therefore, a certain authentication mechanism is needed in the SNS environment.

A commonly adopted solution to this problem is to authenticate the identity of account holders by SNS companies. This is usually seen when companies and celebrities register their SNS accounts. After being authenticated, a special icon will appear near their name, representing that this user's identity has been authenticated. Nevertheless, such a solution suffers from the efficiency problem: The authentication process is

L. Barolli et al. (Eds.): AINA 2022, LNNS 450, pp. 468–479, 2022.
https://doi.org/10.1007/978-3-030-99587-4_39

manual, meaning it requires human work and time (probably long). Back to the above-mentioned example, authenticating a company's main account can be acceptable. But if every staff of this company requires authentication to prevent impersonation, this mechanism becomes inconvenient for SNS network. Apart from the efficiency problem, the security of SNS database is also worried. If SNS database is hacked, then an attacker can change the state of any account to authenticated.

To solve the impersonation problem in SNS, Chen et al. proposed a Threshold Identity Authentication Signature (TIAS) scheme in 2020 [3]. They use the concept 'group' to represent the trust relationship among a small group of people, where some group members are publicly trusted. A verifier judges an untrusted user by combining and verifying a threshold group signature. But such scheme suffers the problem of heavy work for verifier and security on SNS company side.

In 2008, the appearance of Bitcoin [11] brought about a new technology, blockchain, which expands the traditional financial system to a new type of form. In the following years, the idea of blockchain also inspires the development and research of other cryptocurrency implementations, e.g. Ethereum [20], Hyperledger Fabric [1] and Litecoin. By exploiting blockchain technology, a digital ledger is constructed and maintained among a distributed network, with the following unique characteristics: Decentralized, distributed, immutable, and traceable. In most blockchain applications, a block is used to store the current state of the whole network including account information and transactions. And a new block will be generated at a certain frequency, updating the new state of blockchain network. The blockchain structure utilizes a hash function to link adjacent blocks so that it becomes a 'chain'. This blockchain is maintained by the whole network. Thus, to manipulate data in blockchain, an adversary is required to overcome both hash link and distributed storage. Apart from that, some blockchain supports smart contract, which can be regarded as a code stored and executed on blockchain, for blockchain user interaction. Smart contract is designed to handle a variety of tasks [19] while guaranteeing the code is always executed correctly. We found that the characteristics of blockchain are perfect to make up for the defects of Chen's scheme. If a list of trusted users is stored in a block rather than an SNS server, malicious manipulation can be hardly conducted. And thanks to smart contract, some cryptographic operations can be transferred to blockchain without the worry about whether those operations are fraudulently executed.

Our Contributions. We build a blockchain-based social network identification system for identity authentication in SNS that supports the group endorsement in this paper. It combines blockchain, smart contract, and the idea of Chen's work. By applying our scheme, the SNS company only needs to authenticate a small number of main accounts. Those authenticated users are enabled to set up a group and endorse other group members in need. Regarding data storage, blockchain with smart contract, rather than SNS server, is introduced to store and maintain the list of trusted users, which prevents the attackers from manipulating the database. And the authentication of other users is done by the interaction between users and smart contract. Therefore, SNS only needs to authenticate a very small number of users. And smart contract also takes the role of the verifier, while in [3] the verification has to be done by users repeatedly. From the evaluation of gas and time cost, our scheme is proved to be acceptable for nowadays SNS applications.

2 Related Work

The impersonation problem in SNS networks has been a severe problem in recent years. Tsikerdekis described three types of identity deception [17]: identity concealment, identity theft and identity forgery. It analyses identity deception in two ways: user and developer. In this case, users have limited power in detecting identity deception while developers are given some strategy in designing SNS based on software engineering. Goga studied 16,572 cases of impersonation and uncovers a new kind of attack [5], which clones the profiles of ordinary people. These malicious accounts are hard to detect by software engineering methods. Nuakoh gave a solution to identity impersonation by using artificial immune systems [12], which is inspired by the principles and processes of the vertebrate immune system. But it is still not reliable as a security protocol.

In cryptography, Pretty good privacy [4] proposed an idea called web of trust, where a central authority is not needed and people set a trust level for others' public key. A trusted public key can be signed by other people. Whether a new public key can be trusted is judged by those signatures. But such an idea is not suitable for the discussed problem in this paper, because only partial users require to be trusted. Moreover, Revoking a signature is also a difficult problem. Chen's TIAS scheme [3] focuses on the situation of identity authentication in SNS network. It uses the concept of *group* to grant the power of authentication to trusted users. More specifically, Every authenticated user is eligible to form a group with other users. And the group represents that authenticated members endorse other unauthenticated users in a group. For any SNS user who needs to verify the identity of an unauthenticated user, he can collect the signature shares of group members and run the verification algorithm to check the identity. However, such scheme suffers several problems on both user side and SNS company side, making it not suitable for real world SNS. On the one hand, this scheme heavily relies on verifier's work. It requires a verifier to conduct large number of operations about fetching data and computation on a finite field. Meanwhile, the result of the verification is not reusable for other users. On the other hand, since signature shares are stored in SNS server, a single point of failure attack is easy to be launched to prevent identification. And a more powerful attacker may hack into the database and manipulate stored data to add or remove group members.

In recent years, blockchain with smart contract has become popular in a wide range of research. Basically, there are two main advantages of using it. One is that the execution of smart contract is always reliable, making it appropriate for conducting algorithms with regard to justice. Another is that smart contract provides memory space and a corresponding interface for data storage. For example, Many schemes [6,8,14] utilized smart contract as a trusted database for storing anti-manipulated data. Wang proposed a large scale election based on smart contract [18]. The smart contract is used to accomplish recording, managing, calculating, and checking since the execution of it is totally reliable. Patsonakis analyzed the practicality of building a smart contract Public key infrastructures (PKI) in order to avoid the problem under centralized authorities [13]. It turned out that conducting cryptographic calculations like elliptic-curve signature algorithm on smart contract is feasible. Bünz proposed a fully-decentralized, confidential payment mechanism called *Zether* [2]. It uses the smart contract for verifying proof and signature. And when it comes to solving trust problem by blockchain, there are little research with regard to such a problem. Robinson proposed the idea to establish trust for Ethereum private sidechains [15]. But it is hardly applied to SNS networks since it is designed for trusted sidechains between organizations. Kochovski provided some methods of trust management [9]. But they heavily depend on machine learning instead of security protocol. Some research in IoT area aims at decentralized authentication for IoT devices. Mohanta focuses on designing a decentralized authentication scheme for IoT devices by Ethereum [10]. However, it requires extra work for device holder. And authenticating IoT devices is different from authenticating SNS users. Similar problems also occurs in Hammi's work [7], where IoT devices cannot be simply replaced by SNS users. Therefore, these ideas are not applicable for SNS networks.

3   Preliminaries

3.1   Secret Sharing

Shamir's secret sharing scheme [16] allows a secret to be shared among some group members in a way that a single member only obtains a piece of share but cannot know the secret by himself. In order to regenerate the secret, multiple shareholders have to combine their shares together. In a (t, n) threshold secret sharing scheme $(1 \leq t \leq n)$, a secret is separated into n shares and those shares are distributed to n shareholders. Any t shares out of n can be combined to regenerate the original secret, but any less than t shares could not provide any information about the secret. In such scheme, t is called threshold. The design of secret sharing is based on polynomial interpolation. For secret S, a trusted authority selects $a_1, ..., a_{t-1} \xleftarrow{R} Z_q$ where q is prime. A polynomial of $t-1$ degree is also defined: $f(x) = S + a_1 x + a_2 x^2 + ... + a_{t-1} x^{t-1}$ so that $f(0) = S$. Then choose $x_1, ..., x_n \xleftarrow{R} Z_q$. For $i = 1, ..., n$ calculate $y_i \leftarrow f(x_i)$. Output shares $S_i = (x_i, y_i)$ for shareholder. t shareholders can regenerate the secret S using Lagrange interpolation by $S = g(0) = \sum_{i=1}^{t} S_i L_i(0)$ where $L_i(x) = \prod_{j=1, j \neq i}^{t} \frac{x - x_j}{x_i - x_j}$ is called Lagrange coefficient.

3.2   Blockchain and Smart Contract

The idea of blockchain was first proposed by Nakamoto [11] and inspired other blockchain implementations afterward. Since transaction data stored on blockchain is distributed and maintained among all blockchain clients, blockchain is resistant to malicious modification on any block data. In other words, changing a distributed ledger maintained by the whole network requires enormous power to successfully attack the consensus protocol. And each block on blockchain consists of a cryptographic hash of the previous block, a timestamp, and transaction data, which provides traceability for any transaction on blockchain.

Based on the original design of blockchain, smart contract technology is proposed to enhance the functionality of blockchain and extend its application. One of the most famous blockchain implementations with smart contract is Ethereum [20]. An Ethereum account consists of the following entities: Address (also serve as public key), Ether balance, Contract code, and Storage. There are two types of Ethereum accounts: externally owned accounts, controlled by users' private keys; and contract accounts, controlled by their contract code. The smart contract of Ethereum is a self-executable Turing-complete program that runs in the Ethereum Virtual Machine (EVM). An externally owned account can send a message to a contract account by creating and signing a transaction. On receiving the message, the contract code executes the inside code and conducts corresponding actions like writing information in storage or sending messages. Each smart contract has its own storage space for reading and writing. Users can call the contract by its particular unique address in a global state and pay a "gas" cost for execution.

4   System Model

In our proposed scheme, we introduce the concept of group to represent a trusted relationship between users. If a set of users belong to the same group, it indicates that they trust each other and share a group key pair within the group. And Each user involved in this scheme has a blockchain account, thus possessing a public key (Address) and corresponding private key. Before interacting with smart contract, all group members, including trusted users and untrusted users, run a group key generation algorithm through secure end-to-end communication and generate a group key pair (MPK, MSK). Notice that MPK is the group public key, and each member only holds a share of MSK but no knowledge of MSK. Based on their secret share, every member generates a signature share by signing on a certain data (could be any date as long as every member signs on the same one) using the share. They also negotiate a threshold value to indicate how many signature shares are required at least during verification. Figure 1 shows an overview of our proposed scheme. In general, there are five main entities in our scheme:

SNS Server. Trusted users register their group name in SNS server by submitting a group name and their signature share. After then, the server launches a transaction to smart contract, storing trusted users on blockchain. The stored information includes *user account (address), group name, group public key (MPK), threshold, signature share*. Compared to [3], SNS server no longer maintains a database by itself. The blockchain is adopted as a database and smart contract provides an interface for SNS server. Therefore, SNS companies manage a trust list on blockchain. And any user in the blockchain network is enabled to access this list.

Trusted User. Trusted users obtain their endorsement from the SNS company. They are authenticated in a manual way (SNS company verifies their proof documents). They are the initiator of group registration. Moreover, while interacting with smart contract, trusted users can update their signature share by calling the corresponding smart contract. Updating a group public key can also be easily achieved by voting if there exist multiple trusted users in a group.

Semi-trusted User. Semi-trusted users are merely trusted by certain trusted users. They are the most concerning part in SNS network since they claim to be trusted by certain

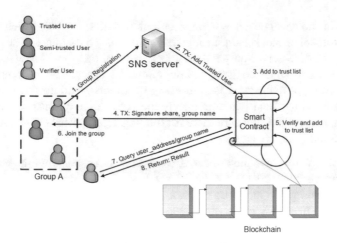

Fig. 1. An overview of our scheme

trusted users without the authentication from SNS company. To become trusted, a semi-trusted user proves that he and some other trusted users belong to the same group by interacting with smart contract. If such proof holds true, the semi-trusted user will be added to the trust list and acknowledged as trusted. This process is executed by smart contract without help from SNS server.

Blockchain. Blockchain with smart contract is applied to handle the data storage. The smart contract achieves the following functions: The SNS company can store and update a trust list on blockchain. And trusted users can update their own signature shares and group public key. To join group, one semi-trusted user launches a transaction containing a group name and a signature share to smart contract for verification. If the verification passes, this applicant is added to the trust list.

Verifier. Anyone in this blockchain network is enabled to check if a target user is trusted by checking whether it is recorded in a trust list on blockchain. This operation is done by querying the smart contract.

5   Threat Model

SNS Server: We assume the SNS server is semi-honest but curious about the group secret key *MSK*. It always authenticates users and adds them to the trust list in the correct way. However, the SNS server may also try to collect as many signature shares from the trust list and recover a valid share, then deciphers it for the group private key. Nevertheless, SNS could be malicious when hacked by attackers. In this case, the server may gain the authority to interact with smart contract and call the server-only functions, thus helping the impersonation attack.

Malicious Users: A malicious user tries to impersonate as a trusted or semi-trusted user. Basically, there are three main ways of achieving the target. The first is to directly attack the blockchain and generate a new admitted block with desired data, which exploits the consensus protocol of blockchain. In another word, it is a 51% attack. The second way is by calling the SNS server-only function in smart contract to add new users. Due to the sender check of smart contract, the malicious attacker needs to compromise the Ethereum account of SNS server for following operations. And the third way is to act like a semi-trusted user and join a group via the verification of smart contract. Under this circumstance, a valid signature share for the target group needs to be forged, which is a signature share forgery attack.

Other Users: Apart from malicious users, we assume other users in Ethereum network are honest but curious about the group secret key of other groups. Their behavior is always correct and group members will not commit fraud since it benefits nothing for their credit. But they may try to obtain the *MSK* of other groups, just like the SNS server.

6 Our Proposed Scheme

We assume that the communication between group members and SNS server is conducted on a secure channel with authenticated end-to-end encryption.

Setup: Given security parameter k, all group members negotiate a threshold value t ($\#trusted < t \leq n$). The SNS server chooses a large prime p, a secure elliptic curve $y^2 = x^3 + ax + b$ with generator P and a one-way hash function $h()$. (n, a, b, p, P, h) are output as public parameters.

Group Key Generation: Denote group key pair of group as (pk_M, sk_M). For every member in a group, it randomly choose a sub-secret s_i with a polynomial $f(x)$ of degree $t - 1$ such that $f(0) = s$. User U_i then calculate a sub-share $ss_{ij} = f_i(j)$ for user U_j ($j = 1, 2...n$) and send it to U_j. After receiving all the sub-share $ss_{ji}(j = 1, 2...n, j \neq i)$ from other members, U_i calculates its share of sk_M in such way: $S_i = \sum_{j=1}^{n} ss_{ji} = \sum_{i=1}^{n} f_i(i)$. S_i will be later used to generate a signature share. The jointly generated group private key $sk_M = \sum_{i=1}^{n} s_i = S$. Then every user broadcast $S_i L_i(0)P$ to other group members where $L_i(0)$ is Lagrange coefficient. pk_M is obtained by choosing t broadcasted messages and recovering $pk_M = \sum_{i=1}^{t} S_i L_i(0)P = SP$. pk_M is then used as group public key for verification.

Group Registration: Trusted users submit a registration query with a group name, threshold value t, and group public key pk_M to SNS server. After confirming that the submitters are authenticated users, SNS server generates a transaction to smart contract containing trusted users' data. More specifically, these data are stored in a trust list where each entity contains *user address, group name, pk_M, signature share*. Threshold t of such group is also stored in blockchain.

Sign: Every user in a group conducts the following calculation: For a time stamp t, $z = h(t)$. Choose a random number $k_i \xleftarrow{R} Z_p^*$ and calculate $K_i = k_i P = (x_{r_i}, y_{r_i})$, $r_i = x_{r_i} K_i$. Let $w_i = k_i x_{r_i} - z S_i L_i(0) \bmod p$. The signature share is $sigs = (r_i, w_i)$. Then it signs on a transaction containing *sigs* to the smart contract with its own private key. The smart contract checks whether a submitter is in the trust list and then stores *sigs* as its signature shares.

Automatic Verify: To prove the attribution of a group, a semi-trusted user generates a transaction, containing information of his group name and signature share, to the smart contract. Smart contract first checks if it is a recorded signature share. Then, by picking $t - 1$ signature shares from trusted accounts in such group and combining them with the applicant's share, the smart contract calculates $R = \sum_{i=1}^{t} r_i$ and $W = \sum_{i}^{t} w_i$. After then, it checks whether $WP + z \cdot pk_M = R$. The signature is accepted if such an equation is satisfied. If true, this semi-trusted user is added to the trust list.

Query: Any user can query the smart contract with certain user id and group name. It returns whether the user belongs to such group. A query on all members of a group is also accepted.

Algorithm 1, 2 explain how functions regarding query and add trusted users are defined and executed in smart contract.

Algorithm 1. Smart contract: Query trusted user

Input: Transaction TX with $userId_q, groupName_q$
  for i = 0 **to** truseList.len() **do**
    if truseList[i].id = $userId_q$ **and** truseList[i].groupName = $groupName_q$ **then**
      return true
  return false

Algorithm 2. Smart contract: Add new user

Input: Transaction TX with $userId_q, sigShare_q, groupName_q$
  if msg.sender = SNSserver **then**
    trustList.push($userId_q, sigShare_q, groupName_q$)
    return msg = "successful"
  else
    sigShares = randPick(trustList, $groupName_q, t$) //Randomly pick t signature shares from this group
    result = Verify(sigShares, $sigShare_q$)
    if result = true **then**
      trustList.push($userId_q, sigShare_q, groupName_q$)
      return msg = "signature share accepted"
    else
      return msg = "signature share rejected"

---

7 Security Evaluation

Theorem 1. *Correctness. If threshold is satisfied and all signature shares involved in the verification are generated correctly, then the signature is always accepted.*

Proof. If all signature shares are generated honestly, $pk_M = \sum_{i=1}^t S_i L_i(0) P$, $R = \sum_{i=1}^t r_i, W = \sum_i^t w_i = \sum_i^t (k_i x_{r_i} - z S_i L_i(0))$.

$$WP = \sum_i^t k_i x_{r_i} P - z \sum_i^t S_i L_i(0) P = \sum_i^t k_i x_{r_i} P - z \cdot pk_M \tag{1}$$

$$WP + z \cdot pk_M = \sum_i^t k_i x_{r_i} P = \sum_i^t K_i x_{r_i} = R \tag{2}$$

Thus the correctness is proved.

Theorem 2. *Secret key hiding. The chance of recovering sk_M from the knowledge of n signature shares is negligible, where n represents the number of members in a group.*

Proof. According to Shamir's scheme, if t secret shares of sk_M is acquired, then it is easy to recover the secret by using Lagrange coefficient. For a signature share $sigs_i$, only $w_i = \sum_i^t (k_i x_{r_i} - z S_i L_i(0))$ contains information of $S_i L_i(0)$. If it can be extracted from such equation, then the secret key is compromised. However, since k_i and x_{r_i} is randomly generated and k_i is discarded after generating a signature share, there is not way to find out the value of it. Thus, the adversary has negligible advantage in obtaining secret shares from the signature shares.

Theorem 3. *Anti-forgery. Even if an adversary gains n signature shares as well as the group public key pk_M from a group, the advantage of generating a valid signature share for an adversary is negligible.*

Proof. We define the following game to demonstrate our proof: The challenger C creates n legal signature shares of a group and give all signature shares with other necessary public parameters to an adversary A. A wins the game if A finally outputs a new signature share $sigs_A$ that can be verified after combined with other $t - 1$ shares.

To solve such problem, A has to recover the $t - 1$ degree polynomial $F(x)$ and calculate a secret share for himself. Now that A knows w_i for $i = 1, 2, ..., n$, any t secret shares could help him recover $F(x)$. However, A needs to correctly guess every randomly chosen k. Thus A has no non-negligible advantage in winning this game.

Theorem 4. *Anti-manipulation. It is quite difficult for an attacker to generate a new admitted block containing malicious data, because it is a 51% attack.*

Theorem 5. *Hacking detection. Even if the SNS server was hacked, a maliciously added user can be detected and removed afterwards.*

Proof. We consider the circumstance that SNS server is hacked and the attacker gains authority to freely add accounts to any group. Because blockchain keeps the state from the genesis to its newest block, after the control of SNS server is retaken, SNS company can easily check its transaction with blockchain during the hacked time. Even if the time is not clear, SNS company can still detect the maliciously added user by verifying the validation of signature shares in trust list since signature share's forgery is hard.

8   Performance Evaluation

Environment. We test our scheme by deploying the corresponding smart contract on Rinkeby, an Ethereum test network. The smart contract is implemented by Solidity 0.8.4. Since the verification of signature requires elliptic curve computation, Elliptic-curve-solidity library[1] is used in our contract. And the adopted elliptic curve is secp256k1 ($y^2 = x^3 + 7$). Our test is based on the following settings: Assume there are 7 test group members on the trust list. In the meantime, there are other 100 irrelevant users in the list, serving as trusted users from other groups. The verification and adding new user operations are only conducted on the test group. The reason for choosing 7 test users and 100 other users is to simulate a real situation that there are plenty of other groups in SNS network apart from the test group.

Table 1. Transaction fee of contract functions

Functions	Transaction fee (ether)	Transaction fee (dollar)
Verify (success)	0.00145465301	$4.92
Verify (failure)	0.00134082201	$4.54
Add one user	0.00013949100	$0.47

Gas Cost. Since calling functions of smart contract requires gas for execution, the gas cost of interaction with smart contract is the most concerning factor for our scheme. Table 1 shows the transaction fee of calling each operation in both ether and dollar. The ether price was obtained on 4th, October 2021. Regarding verification, there are two results: Success and failure. And the cost of success or failure is close. Thus for a new trusted user, it costs less than 5 dollars (0.00145 ethers) to join a group on blockchain. As for SNS server, it only costs 0.47 dollars (0.00014 ethers) to add an authenticated user to the trust list by calling the "Add one user" function. Therefore, the expenses of applying our scheme are completely acceptable.

Time Cost. Rinkeby and Ethereum runs identically: A new block is generated at a fixed speed, which is determined by the network. Thus, we investigate the time cost based on the block time. More concretely, we consider how much time has passed between the generation and the validation of one transaction. The most time-consuming function *Verify* is tested in this experiment. After calling it 10 times, we found out that all transactions are confirmed about 15 s later. The average block time of Rinkeby is 15 s, which is the same as Ethereum. Thus, the result of smart contract is always immediately confirmed at the next block. In addition, regarding the time cost of off-chain parts like generating signature shares, Chen already demonstrated that they cost an acceptable time for an individual without blockchain in his paper [3]. Therefore, it is not a concern in this paper.

---

[1] https://github.com/witnet/elliptic-curve-solidity. Open-source project under the MIT license.

9   Conclusion

To solve the impersonation problem in SNS, we design a blockchain-based social network identification system. Our scheme improves the defects of previous work by applying Ethereum and smart contract technology to handle the storage and verification. Compared to other related solutions, our scheme is based on a protocol scheme, thus security can be guaranteed. Meanwhile, it achieves automatic verification by leveraging smart contract, which largely increases the efficiency compared to previous work. The characteristics of blockchain also make the scheme more resistant to attackers. And detailed security evaluation is provided. We also implement and deploy our system on the Rinkeby test network to evaluate the transaction and time cost of it. The result shows that the cost is totally acceptable.

Acknowledgments. This work was supported by the Grant-in-Aid for Scientific Research (B) (19H04107).

References

1. Androulaki, E., et al.: Hyperledger fabric: a distributed operating system for permissioned blockchains. In: Proceedings of the Thirteenth EuroSys Conference, pp. 1–15 (2018)
2. Bünz, B., Agrawal, S., Zamani, M., Boneh, D.: Zether: towards privacy in a smart contract world. In: Bonneau, J., Heninger, N. (eds.) Financial Cryptography and Data Security. FC 2020. LNCS, vol. 12059, pp. 423–443. Springer, Cham (2020). https://doi.org/10.1007/978-3-030-51280-4_23
3. Chen, Z., Chen, J., Meng, W.: Threshold identity authentication signature: impersonation prevention in social network services. Concurr. Comput. Pract. Exp., e5787 (2020)
4. Garfinkel, S.: PGP: Pretty Good Privacy. O'Reilly Media, Inc., Sebastopol (1995)
5. Goga, O., Venkatadri, G., Gummadi, K.P.: The Doppelgänger bot attack: exploring identity impersonation in online social networks. In: Proceedings of the 2015 Internet Measurement Conference, pp. 141–153. ACM (2015)
6. Gürsoy, G., Brannon, C.M., Gerstein, M.: Using ethereum blockchain to store and query pharmacogenomics data via smart contracts. BMC Med. Genomics **13**, 1–11 (2020)
7. Hammi, M.T., Hammi, B., Bellot, P., Serrhrouchni, A.: Bubbles of trust: a decentralized blockchain-based authentication system for iot. Comput. Secur. **78**, 126–142 (2018)
8. Kirkman, S., Newman, R.: A cloud data movement policy architecture based on smart contracts and the ethereum blockchain. In: 2018 IEEE International Conference on Cloud Engineering (IC2E), pp. 371–377. IEEE (2018)
9. Kochovski, P., Gec, S., Stankovski, V., Bajec, M., Drobintsev, P.D.: Trust management in a blockchain based fog computing platform with trustless smart oracles. Futur. Gener. Comput. Syst. **101**, 747–759 (2019)
10. Mohanta, B.K., Sahoo, A., Patel, S., Panda, S.S., Jena, D., Gountia, D.: DecAuth: decentralized authentication scheme for IoT device using ethereum blockchain. In: TENCON 2019-2019 IEEE Region 10 Conference (TENCON), pp. 558–563. IEEE (2019)
11. Nakamoto, S.: Bitcoin: a peer-to-peer electronic cash system. Decentralized Business Review, p. 21260 (2008)
12. Nuakoh, E.B., Anwar, M.: Detecting impersonation in social network sites (SNS) using artificial immune systems (AIS). In: SoutheastCon 2018, pp. 1–3. IEEE (2018)

13. Patsonakis, C., Samari, K., Kiayiasy, A., Roussopoulos, M.: On the practicality of a smart contract PKI. In: 2019 IEEE International Conference on Decentralized Applications and Infrastructures (DAPPCON), pp. 109–118. IEEE (2019)
14. Ramesh, V.K.C.: Storing IoT data securely in a private ethereum blockchain. Ph.D. thesis, University of Nevada, Las Vegas (2019)
15. Robinson, P.: Using ethereum registration authorities to establish trust for ethereum private sidechains. J. Br. Blockchain Assoc. 1(2), 5055 (2018)
16. Shamir, A.: How to share a secret. Commun. ACM 22(11), 612–613 (1979)
17. Tsikerdekis, M., Zeadally, S.: Detecting and preventing online identity deception in social networking services. IEEE Internet Comput. 19(3), 41–49 (2015)
18. Wang, B., Sun, J., He, Y., Pang, D., Lu, N.: Large-scale election based on blockchain. Procedia Comput. Sci. 129, 234–237 (2018)
19. Wohrer, M., Zdun, U.: Smart contracts: security patterns in the ethereum ecosystem and solidity. In: 2018 International Workshop on Blockchain Oriented Software Engineering (IWBOSE), pp. 2–8. IEEE (2018)
20. Wood, G., et al.: Ethereum: a secure decentralised generalised transaction ledger. Ethereum Project Yellow Paper 151(2014), 1–32 (2014)

Malware Classification by Deep Learning Using Characteristics of Hash Functions

Takahiro Baba[1]($\boxtimes$), Kensuke Baba[2]($\boxtimes$), and Toshihiro Yamauchi[3]($\boxtimes$)

[1] Graduate School of Natural Science and Technology, Okayama University,
Okayama, Japan
[2] Cyber-Physical Engineering Informatics Research Core, Okayama University,
Okayama, Japan
kenbaba983@gmail.com
[3] Graduate School of Natural Science and Technology, Okayama University/JST,
PRESTO, Okayama, Japan
yamauchi@okayama-u.ac.jp

Abstract. As the Internet develops, the number of Internet of Things (IoT) devices increases. Simultaneously, the risk of IoT devices being infected with malware also increases. Thus, malware detection has become an important issue. Dynamic analysis logs are effective at detecting malware, but it takes time to collect a large amount of data because the malware must be executed at least once before the logs can be collected. Moreover, dynamic analysis logs are affected by external factors such as the execution environment. A malware detection method that uses a static property analysis log could solve these problems. In this study, deep learning (DL) was used as a machine learning method because DL is effective for large-scale data and can automatically extract features.

Research has been conducted on malware detection using static properties of portable executable (PE) files, establishing that such detection is possible. However, research on malware detection using hash functions such as Fuzzy hash and peHash is lacking. Therefore, we investigated the characteristics of hash values in malware classification. Moreover, when the surface analysis log is viewed in chronological order, that the data are considered have concept drift characteristics. Therefore, we compared malware detection performance using data with the concept drift property. We found that the hash function could be used to prevent performance degradation even with concept drift data. In an experiment combining PE surface information and hash values, concept drift showed the highest performance for certain data.

Keywords: Malware detection · Deep learning · PE file · Fuzzy hash · peHash

1 Introduction

The number of Internet of Things (IoT) devices installed continues to increase alongside the development of the Internet. Meanwhile the risk of IoT devices

L. Barolli et al. (Eds.): AINA 2022, LNNS 450, pp. 480–491, 2022.
https://doi.org/10.1007/978-3-030-99587-4_40

being infected with malware is increasing as well. Hence, detecting malware is an important issue in the field of cybersecurity.

In recent years, research has been conducted on malware detection using static property analysis logs, and data collection has become an active area of research. Dynamic analysis logs provide useful information for detecting malware because they record the actual operation of malware. Several studies have been conducted to estimate the function of unknown malware from API and categories observed by dynamic analysis [13], as well as to classify malware based on data extracted by dynamic analysis [11].

For dynamic analysis logs, it is necessary that the malware execute at least once before a log of its operation is collected. Moreover, it is necessary to prepare the execution environment, and it takes time to collect a sufficiently large amount of data. Because dynamic analysis logs are affected by the type of execution environment created and collected, the execution environment must be considered in the case of logs collected in multiple environments. The detection of malware using static property analysis logs, can solve the aforementioned problem.

In addition, machine learning (ML) [8] has been successful owing to improvements in computer processing power. Above all, numerous studies have found that the application of deep learning (DL) [16] to large-scale labeled data is effective [10]. In ML, it is often necessary to manually extract features from the data to be analyzed. The selection of specific features has a significant effect on classification performance. By contrast, DL has the advantage that features can be automatically extracted.

Therefore, in this study, we classified malware and cleanware by applying DL to the static property analysis log of malware. Hash values have been used to cluster malware, but few studies have focused on detecting malware. Therefore, in this study, we verified whether a hash value is effective at detecting malware and investigated its characteristics. It was found that it is particularly effective for data with concept drift. An evaluation experiment was conducted using the static property analysis log provided by the FFRI Datasets [2] from 2019, 2020, and 2021. First, we applied a multi-layer perceptron model to learn the PE's surface information using a multi-layer perceptron and verified that DL is effective for malware detection. In addition, a character-level CNN was used to classify hash value, which are character strings, as malware or safe files [23]. Then, we evaluated the learning performance in terms of classification accuracy, precision, and recall. Thus, we verified that some types of hash values are particularly effective at detecting malware compared to others.

Because specific malware packages that are prevalent evolve over time, the FFRI Dataset is said to have the concept drift property. Therefore, by combining the FFRI Datasets 2019 and 2020 as training data and using the FFRI Dataset 2021 as testing data, we verified the effectiveness of PE surface information and hash values for the security analysis of data with concept drift.

Experimental results show that high performance was obtained for both the PE surface information and hash value. The program impfuzzy had the highest performance among hash values. In addition, in experiments with data with the concept drift characteristic, the hash value using the Import API showed high performance. Therefore, combining the hash value of the data and the PE surface layer information with the concept drift property will provide higher performance.

2   Malware Classification Method

2.1   Purpose

The purpose of this study is as follows.

- Verification that the hash value is effective for malware detection
- Clarify the characteristics of hash values

This study's purpose was to verify the effectiveness of the hash value at detecting malware, and to clarify the characteristics of the hash value. So far, no research has been conducted to verify which hash value is useful for malware detection, and it has not been clarified. Moreover, we evaluated the malware detection performance for data with concept drift to clarify the characteristics of hash values.

2.2   Dataset

As described in Sect. 1, the use of static property analysis logs has advantages over using dynamic analysis logs in terms of risk and other factors. In this study, we conducted experiments using the FFRI Datasets 2019, 2020, and 2021 as the datasets for static property analysis logs, as shown in Table 1. PE surface information and hash values for 225,000 malware and 225,000 cleanware were provided. We used the FFRI Datasets 2019, 2020, and 2021 as the datasets for our experiments. Of these, 405,000 (90%) were used as training data, and the remaining 45,000 (10%) were used as testing data for evaluation.

To verify the effectiveness of the proposed approach on data with concept drift properties, FFRI Datasets 2019 and 2020 were used as training data and FFRI Dataset 2021 was used as testing data, as shown in Table 2.

Table 1. Datasets used for the experiment

	Cleanware			Malware		
	All	Train (90%)	Test (10%)	All	Train (90%)	Test (10%)
FFRI Dataset 2019	75,000	67,500	7,500	75,000	67,500	7,500
FFRI Dataset 2020	75,000	67,500	7,500	75,000	67,500	7,500
FFRI Dataset 2021	75,000	67,500	7,500	75,000	67,500	7,500
Total	225,000	202,500	22,500	225,000	202,500	22,500

Table 2. Dataset to verify the validity of data with concept drift

	Cleanware			Malware		
	All	Train	Test	All	Train	Test
FFRI Dataset 2019	75,000	75,000	0	75,000	75,000	0
FFRI Dataset 2020	75,000	75,000	0	75,000	75,000	0
FFRI Dataset 2021	75,000	0	75,000	75,000	0	75,000
Total	225,000	150,000	75,000	225,000	150,000	75,000

2.3 Classification Method Using PE Surface Information

In the first experiment, we classified malware using the PE surface information extracted from the dump file obtained from a program for the purpose called PE file [3]. Table 3 shows the data content of the PE surface information. The learning method used was the multi-layer perceptron [21], which is a type of DL. The advantage of the multi-layer perceptron is its ability to automatically adjust the weights using the training data, and it can train on any type of training data. For this reason, we used a multi-layer perceptron for classifying the PE surface information.

When vectorizing, all the values of each field of the PE surface information were investigated, as shown in Fig. 1, and the overall vector was converted into a binary vector depending on the presence or absence of the value; i.e., one-hot encoding was performed. The reason for making such a vector is that there are many fields that are not binary values of "yes" and "no". For example, DLL values have four patterns of character strings: "yes", "no", "yes(no)", and "no(yes)". Therefore, when the DLL is "yes" the vector is "[1,0,0,0]". These vectors were obtained for each field and combined to form the entire vector.

2.4 Classification Method Using Hash Values

Character-level CNN [23], which is a convolutional neural network applied at the character level, was used as a learning method for classification using hash values. A character-level CNN can perform convolution together with the information of connections before and after each character. Because the hash value used in

Table 3. Data content of PE surface information

Field name	Contents
PE	32bit or 64bit
DLL	DLL or not
Packed	With or without packing
Anti-Debug	With or without Anti-Debug
GUI program	GUI program or not
Console program	Console program or not
Mutex	With or without mutex
Contains base64	With or without
PEiD	Matched PEiD signature name
AntiDebug	Anti-Debug method

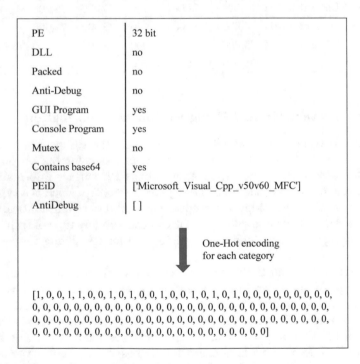

Fig. 1. Example of vectorization of PE surface information

this study was a similar string when the files were similar, we assumed that character-level CNN was an effective method.

In performing ML, the characters of the hash value were made to correspond to integers, as shown in Fig. 2, and vectorization was performed. The hash values used were close to one another in the case of similar files, and the relationship before and after the characters was important; therefore, they were quantified for each character.

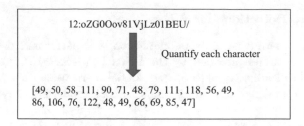

Fig. 2. Hash value vectorization example

2.5 Verifying with Concept Drift

Concept drift is an unexpected change in statistical characteristics over time. To verify this, experiments were conducted using the FFRI Datasets 2019 and 2020 as training data and the FFRI Dataset 2021 as test data. For each data set, the same method as in Sects. 2.3 and 2.4 was used.

2.6 Verifying with a Method that Combines PE Surface Information and Hash Values

From the experimental results in Sect. 2.5, it was found that when using data with concept drift, classification using hash values prevented the deterioration of classification performance. Therefore, we decided to verify whether malware could be classified by combining the hash value and the PE surface information. First, the probability of malware was calculated in the range of 0 to 1 using the hash value by the method described in Sect. 2.4. This value was added to the vector of the PE surface information and calculated with the method described in Sect. 2.3 to determine whether it was malware.

3 Experimental Results

3.1 Malware Detection Using PE Surface Information

Table 4 shows the experimental results using the method described in Sect. 2.3. Classification accuracy, precision, and recall are all high performance.

Table 4. Classification performance of learning using PE surface information

	Accuracy	Precision	Recall
PE surface information	0.9723	0.9723	0.9723

3.2 Malware Detection Using Hash Value

In the second experiment, malware detection was performed using hash values. Of the hash values provided by the FFRI Datasets 2019, 2020, and 2021, the fuzzy hashes impfuzzy, ssdeep, and TLSH were used. In addition to the fuzzy hash, we used peHash, imphash, totalhash, endgame, anymaster, crits, and pehashng.

A comparison was performed for all nine types of hash values. Table 5 shows the results of the experiment. The impfuzzy hash function had the highest performance in terms of classification accuracy, precision, and recall.

Table 5. Classification performance of learning by hash value

fuzzy hash

	Accuracy	Precision	Recall
impfuzzy	0.8871	0.8869	0.8874
ssdeep	0.6993	0.6996	0.6979
TLSH	0.6187	0.6187	0.6189

peHash

	Accuracy	Precision	Recall
imphash	0.7284	0.7284	0.7277
totalhash	0.6705	0.6704	0.6704
endgame	0.6701	0.6698	0.6680
anymaster	0.6959	0.6959	0.6959
crits	0.6238	0.6237	0.6242
pehashng	0.6836	0.6836	0.6833

3.3 Experimental Results with Concept Drift

Experiments using data with concept drift were conducted using the FFRI Datasets 2019 and 2020 as training data and the FFRI Dataset 2021 as test data. Table 6 lists the experimental results in terms of the PE surface information, and Table 7 shows the hash value types.

In this case, the performance was lower than that found in Sects. 3.1 and 3.2. In particular, the hash value belonging to peHash deteriorated in performance. However, when impfuzzy was used, the rate of decrease was small. In addition, impfuzzy and imphash, which calculated the hash value using the Import API of pefile, showed the highest performance in each fuzzy hash and peHash.

Table 6. Classification performance of PE surface information with concept drift

	Accuracy	Precision	Recall
PE surface information	0.9148	0.9148	0.9148

Table 7. Classification performance of hash value with concept drift

fuzzy hash			
	Accuracy	Precision	Recall
impfuzzy	0.8450	0.8454	0.8443
ssdeep	0.6081	0.6078	0.6100
TLSH	0.6060	0.6060	0.6064

peHash			
	Accuracy	Precision	Recall
imphash	0.6423	0.6423	0.6423
totalhash	0.5628	0.5627	0.5627
endgame	0.5769	0.5770	0.5770
anymaster	0.5722	0.5723	0.5720
crits	0.5487	0.5487	0.5482
pehashng	0.5769	0.5769	0.5768

3.4   Experiment Combining Hash Value and PE Surface Information

Table 8 shows the experimental result when the data with the concept drift property were classified by combining the hash value and the PE surface layer's information. The performance was higher than that shown in Table 6, which was the result of using the PE surface information alone.

Table 8. Classification performance of PE surface information and hash value with concept drift

fuzzy hash			
	Accuracy	Precision	Recall
impfuzzy	0.9180	0.9203	0.9180
ssdeep	0.9183	0.9206	0.9183
TLSH	0.9198	0.9199	0.9198

peHash			
	Accuracy	Precision	Recall
imphash	0.9131	0.9151	0.9131
totalhash	0.9179	0.9201	0.9179
endgame	0.9184	0.9186	0.9184
anymaster	0.9191	0.9194	0.9191
crits	0.9176	0.9201	0.9176
pehashng	0.9178	0.9179	0.9178

4   Discussion

In the experiments using PE surface information, high performance was obtained in terms of classification accuracy, precision, and recall. In the experiments using hash values, the best results were obtained using the impfuzzy hashing function. In contrast to ssdeep, impfuzzy did not hash the entire PE file but used only the Import API for hashing. Similarly, imphash, which exhibited the highest performance among peHash, used the Import API for hash value calculation as did impfuzzy. Therefore, in malware detection, performance improvement can be expected by emphasizing on the Import API rather than the entire PE file. Although it was possible to detect malware using the hash value, the performance obtained using the hash value alone did not exceed the performance when the

PE surface information was used. Based on the experimental results, it was possible that high or higher classification accuracy could be obtained by combining impfuzzy with the PE surface information.

In addition, for the data with concept drift property, the decrease in performance was small when impfuzzy was used. Therefore, impfuzzy was effective for unknown data, and it was considered that some performance could be guaranteed without taking measures to account for concept drift. Impfuzzy, which is a type of fuzzy hash, calculates the hash value using fuzzy hashing from the Import API of the PE file. Similarly, imphash, which is a type of peHash, also calculates the hash value from the Import API of PE file. As shown in Table 7, impfuzzy and imphash performed best with fuzzy hash and peHash. Hence, the Import API of the PE files is effective for data subject to concept drift.

By classifying the data with concept drift property via combining the PE surface layer information and the hash value, a higher classification performance was realized compared to the case of the PE surface layer information alone.

5   Related Work

5.1   Research on Malware Detection Using Deep Learning

DL automatically extracts features when large amounts of data are available. In cybersecurity research, the problem of malware detection and classification deals with large datasets and has been the subject of various studies [22].

A study on malware detection by dynamic analysis [12] proposed a neural network enabling malware detection based on detailed dynamic analysis reports of malware behavior.

In a study on Android malware detection [25], features extracted by the static analysis were used for training based on ML. As a result, it may be considered that DL is particularly suitable for malware detection and exhibits high performance.

Therefore, in this study, we conducted classification experiments using PE surface information with DL. In addition, we compared the classification performance of each hash value.

5.2   Research on Malware Detection Using Static Properties

Detecting malware using static properties has several advantages over dynamic analysis. First is the ease of data collection. In dynamic analysis, malware must be executed once, and it takes time to prepare the execution environment and operate it, which makes data collection more difficult than the analysis of static properties. Second, static property analysis is not affected by external factors, while dynamic analysis is subject to external factors such as the execution environment. Third, malware could be detected without executing it if static properties were used for detection. There is a risk of infection with dynamic analysis because dynamic analysis can only detect malware after it has been executed. Therefore, in this study, we conducted a static property analysis.

Research has been conducted on the use of static properties for various malware detections [20]. High performance has been achieved in detecting malware variants [14] by applying ML to static properties. In addition, in [26], a malware detection model using static properties was proposed, and a model was constructed using ML that showed high robustness. In [18], ML was applied to surface information by applying natural language processing, demonstrating high performance against unknown new malware and packed malware. In addition, as a static property, a previous study proposed an efficient signature generation method for IoT malware [7].

5.3 Research on Malware Detection Using Hash Functions

By using methods such as fuzzy hash [15] and peHash [24], malware analysis could be accelerated compared to using PE surface information.

When the data size is large, if malware can be classified using hash values, which are character strings, it can be detected more quickly. In a study on malware detection using kNN [9], high-speed malware detection was performed using a classifier that used similarity hashes. In addition, a comparative study on malware detection based on similarity hashes was conducted [19]. However, to the best of our knowledge, the present work is the first to verify which hash value types are useful for DL in malware detection.

In this study, we examined the usefulness of each hash value type. In a fuzzy hash, the hash values are similar if the files or sentences are similar. Fuzzy hash is a method that combines a method of hashing substrings while shifting them. The hashing functions used in this study included impfuzzy [1], ssdeep [5], and TLSH [6]. We used impfuzzy to calculate the hash values from the Import API. The, ssdeep function is a hashing method that generates strings in order from the beginning of the file, while TLSH is an ELF file format, which is a Linux executable.

Next, peHash is a hash value designed for malware clustering. The hash value is designed to be the same for a given development environment. Studies have been conducted to cluster into groups of similar malware using hash values [17]. The data used in this study include imphash, totalhash, endgame, anymaster, crits, and pehashng, which were collected using the pehash program [4]. These hash values are computed from PE files and Import APIs using hash functions such as md5 and sha1.

6 Conclusion

In this study, to clarify that a hash value is effective for detecting malware, we clarified the characteristics of the hash value and verified how it could be used to contribute to malware detection.

First, DL was applied to each PE surface information and hash value for evaluation. High performance was obtained when the PE surface information was used. In addition, when each hash value was used, impfuzzy achieved the best performance. From these results, it was concluded that DL is an effective learning method for static property analysis logs.

We also conducted experiments using data with concept drift properties. It was found that impfuzzy was effective for data of this property type. Therefore, the experiment was conducted in which the hash value and the PE surface information were combined. As a result, we were able to obtain higher performance than performing malware detection using PE surface information alone. Specifically, when only PE surface information was used as training data, the classification accuracy was 0.9148. However, when the PE surface information and the hash value were used in combination, the highest classification accuracy was 0.9198. These results suggest that it may be useful to use hash values when detecting undiscovered malware.

Acknowledgments. A part of this research is supported by JST, PRESTO Grant Number JPMJPR1938 and JSPS Grants-in-Aid for Scientific Research JP19H05579.

References

1. Classifying Malware using Import API and Fuzzy Hashing - impfuzzy. https://blogs.jpcert.or.jp/en/2016/05/classifying-mal-a988.html. Accessed 3 Aug 2021
2. FFRI Dataset. https://www.iwsec.org/mws/datasets.html. Accessed 3 Aug 2021
3. pefile. https://github.com/erocarrera/pefile. Accessed 3 Aug 2021
4. peHash. http://github.com/knowmalware/pehash. Accessed 3 Aug 2021
5. ssdeep. https://ssdeep-project.github.io/ssdeep/index.html. Accessed 3 Aug 2021
6. Trend micro locality sensitive hash. https://github.com/trendmicro/tlsh. Accessed 3 Aug 2021
7. Alhanahnah, M., Lin, Q., Yan, Q., Zhang, N., Chen, Z.: Efficient signature generation for classifying cross-architecture IoT malware. In: 2018 IEEE Conference on Communications and Network Security (CNS), pp. 1–9 (2018)
8. Bishop, C.M.: Pattern Recognition and Machine Learning (Information Science and Statistics). Springer, Heidelberg (2006)
9. Choi, S.: Combined KNN classification and hierarchical similarity hash for fast malware detection. Appl. Sci. **10**(15), 5173 (2020)
10. Dargan, S., Kumar, M., Ayyagari, M.R., Kumar, G.: A survey of deep learning and its applications: a new paradigm to machine learning. Arch. Comput. Meth. Eng. **27**, 1071–1092 (2019)
11. Huang, W., Stokes, J.W.: MtNet: a multi-task neural network for dynamic malware classification. In: Caballero, J., Zurutuza, U., Rodríguez, R.J. (eds.) DIMVA 2016. LNCS, vol. 9721, pp. 399–418. Springer, Cham (2016). https://doi.org/10.1007/978-3-319-40667-1_20
12. Jindal, C., Salls, C., Aghakhani, H., Long, K., Kruegel, C., Vigna, G.: Neurlux: Dynamic malware analysis without feature engineering. In: Proceedings of the 35th Annual Computer Security Applications Conference, ACSAC 2019, pp. 444–455, New York, NY, USA, 2019. Association for Computing Machinery (2019)
13. Kawaguchi, N., Omote, K.: Malware function classification using APIs in initial behavior. In: 2015 10th Asia Joint Conference on Information Security, pp. 138–144. IEEE (2015)

14. Kita, K., Uda, R.: Malware subspecies detection method by suffix arrays and machine learning. In: 2021 55th Annual Conference on Information Sciences and Systems (CISS), pp. 1–6. IEEE (2021)
15. Kornblum, J.: Identifying almost identical files using context triggered piecewise hashing. Digital Invest. **3**, 91–97 (2006)
16. LeCun, Y., Bengio, Y., Hinton, G.: Deep learning. Nature **521**(7553), 436–444 (2015)
17. Li, Y., et al.: Experimental study of fuzzy hashing in malware clustering analysis. In: 8th Workshop on Cyber Security Experimentation and Test (CSET 2015), Washington, D.C. USENIX Association, August 2015
18. Mimura, M., Ito, R.: Applying NLP techniques to malware detection in a practical environment. Int. J. Inf. Secur. 1–13 (2021)
19. Namanya, A.P., Awan, I.U., Disso, J.P., Younas, M.: Similarity hash based scoring of portable executable files for efficient malware detection in IoT. Future Gener. Comput. Syst. **110**, 824–832 (2020)
20. Ngo, Q.-D., Nguyen, H.-T., Le, V.-H., Nguyen, D.-H.: A survey of IoT malware and detection methods based on static features. ICT Express **6**(4), 280–286 (2020)
21. Noriega, L.: Multilayer perceptron tutorial. School of Computing. Staffordshire University, January 2005
22. Qiu, J., Zhang, J., Luo, W., Pan, L., Nepal, S., Xiang, Y.: A survey of android malware detection with deep neural models. ACM Comput. Surv. **53**(6), 1–36 (2020)
23. Saxe, J., Berlin, K.: eXpose: a character-level convolutional neural network with embeddings for detecting malicious URLs, file paths and registry keys. CoRR, abs/1702.08568 (2017)
24. Wicherski, G.: peHash: a novel approach to fast malware clustering. In: 2nd USENIX Workshop on Large-Scale Exploits and Emergent Threats (LEET 2009), Boston, MA. USENIX Association, April 2009
25. Yuan, Z., Lu, Y., Wang, Z., Xue, Y.: Droid-Sec: deep learning in android malware detection. SIGCOMM Comput. Commun. Rev. **44**(4), 371–372 (2014)
26. Zheng, W., Omote, K.: Robust detection model for portable execution malware. In: ICC 2021-IEEE International Conference on Communications, pp. 1–6. IEEE (2021)

Toward a Blockchain Healthcare Information Exchange

Ryuji Ueno[1(✉)] and Kazumasa Omote[1,2]

[1] University of Tsukuba, Tennodai 1-1-1, Tsukuba 305-8573, Japan
s2020526@s.tsukuba.ac.jp, omote@risk.tsukuba.ac.jp
[2] National Institute of Information and Communications Technology, 4-2-1
Nukui-Kitamachi, Koganei, Tokyo 184-8795, Japan

Abstract. The introduction of healthcare information exchange (HIE) systems has been limited, although its compelling social need has been recognized. This could be due to medical institutions' concerns about information leakage and decreasing profit. The existing research proposals suppose that the medical institutions would accept the system without incentives for medical institutions to be proactive in it. Herein, we propose the HIE system that provides incentives to both medical institutions and patients using blockchain technology. By using the system, we expect that both patients and medical institutions will proactively share medical information. We also evaluate the implementation of the proposed system and measure time required to share medical information to demonstrate that the system could function in a reasonable time. The system will help reducing concerns about information leakage and decreasing profit.

1 Introduction

In recent years, the rapid development of information technology has led to the widespread use of electronic medical records in medical institutions, thus reducing the workload of medical workers. The conventional process of patient information retrieval is time-consuming and laborious. The application of electronic medical records has significantly reduced this burden.

Additionally, blockchain technology, which is applied to cryptographic assets, is expected to also be applied to other fields that require secure information (e.g., records) exchange. Blockchain is known as the technology behind cryptographic assets, such as Bitcoin. Blockchain relies on a distributed ledger that is reproduced across a peer-to-peer (P2P) network without the need for a centralized authority. Using blockchains to decentralize the management of medical information would, therefore, make healthcare information exchange (HIE) possible without the need for a centralized entity. Such research is already underway [4–6,8]. By sharing medical information of patients among different medical institutions, the patients can avoid duplication of medical exams and reduce their medical and other costs. Furthermore, medical institutions will be able to provide more appropriate healthcare by leveraging the HIE, as they can quickly gain access to patients' full medical history from anywhere.

However, medical institutions nowadays are concerned that sharing medical information will not only increase their responsibility for control of patients' medical information but also decrease their profit of medical exam fees. Extant research assumes that medical institutions will accept the HIE proposal, which is a bit presumptuous. There is, in fact, no research on the mechanism by which medical institutions will accept such a proposal. Gan et al. [3], for example, introduced incentives to patients to motivate them to share their medical information. However, they did not consider incentives for the medical institutions.

In this research, we propose the HIE system that provides incentives to both medical institutions and patients using blockchain technology. From this, we expect both patients and institutions to be proactive in medical information sharing. Furthermore, we evaluate the implementation of the proposed system and measure the time required to share medical information to demonstrate the system is capable of sharing medical information within a time whereby no hindrance should be placed on practical use. This should help reduce the aforementioned concerns so that the driving social need can be fulfilled.

2 Background

2.1 HIE

Electronic medical records have revolutionized the medical profession. From related advances, research on HIE has been introduced based on the compelling social need for shareable, secure medical information. The followings are some of the advantages of HIE:

- Improving the quality and the speed of medical care through the ubiquity of detailed medical history.
- Reducing medical costs by avoiding duplication of medical exams and related documentation.
- Increasing the speed of lifesaving emergency services.
- Developing healthcare treatments and diagnoses through the secondary use of medical data.

There has also been considerable research on patient-centered HIE systems in recent years, based on the idea that patients have the authority of ownership of their medical information [3, 10]. Medical information is personal information that must be protected. Further, it is necessary to obtain permission from patients in advance to share their medical information. In this research, we designed the system that allows patients to take the initiative in access control.

2.2 Blockchain

In a paper published by Satoshi Nakamoto [7] in 2009, blockchain was proposed as the underlying technology for the Bitcoin crypto asset. The model now supports many additional crypto assets. Blockchain relies on a distributed ledger spread across a P2P network; it records transactions that occur within the blockchain network in the form of connected blocks. When a transaction is generated, it is digitally signed to validate its issuance.

Since all nodes of the blockchain P2P network hold the same replicated data, there is no single point of failure, even if one or more nodes stop working. Each block in the blockchain contains 1) transactions, 2) a hash value of the previous block, and 3) a nonce (i.e., the result calculations) required to generate blocks. The structured connection of blocks makes the blockchain difficult to tamper with because all subsequent blocks would have to be subsequently tampered with to cover the trail; otherwise, the hash values will become inconsistent. The computation required to tamper with all the blocks in a blockchain is currently beyond possible. However, when large amounts of data (e.g., medical information) are stored in the blockchain, a scalability problem occurs because blocks are small. Therefore, it is more effective to store the uniform resource locator (URL) of the large data instead [8]. In the proposed method, URLs of medical information are shared on the blockchain, and actual access to medical information is provided in an off-chain fashion.

Among the various kinds of crypt assets, Ethereum [1], the second-highest within the market, has the main function called smart contract, which we utilize for the system in this research. In smart contract scheme, contracts for transactions on the blockchain are automatically executed when specific criteria and authentications are met. Its main feature is that it can execute contracts between traders alone if the contract meets the necessary conditions based on a pre-designed program.

3 Existing Research

Zhuang et al. [10]. proposed, implemented, and verified a patient-centric HIE system in which the ownership of medical information changed from the medical institution to the patient. They proposed a system in which patients managed the access rights of healthcare providers as an allowed list using smart contracts and only authorized healthcare providers could access patients' medical information. Here, actual medical information was not exchanged on the blockchain, but its encrypted form was exchanged between the databases of medical institutions, and the decryption key was shared on the blockchain. The decryption key was encrypted with the recipient's public key via smart contract so that only the recipient could decrypt and view the medical information.

Buzachis et al. [2]. implemented, proposed, and verified a prototype that combined blockchain and the InterPlanetary File System (IPFS) for an HIE possessing confidentiality, integrity, interoperability, and privacy protection among medical institutions. IPFS is a hypermedia protocol that works on a P2P network and shares the hash value of URLs at which medical information exists on the IPFS. For verification, they measured the registration, login, authorization, and revocation times of the service.

Toyoda et al. [9]. proposed, implemented, and verified the cost of a product ownership management system in a supply chain using radio-frequency identification (RFID) tags and smart contracts to prevent RFID counterfeiting. This research provided incentives for manufacturers to send correct transactions, aiming to ensure that all participants generated correct transactions without counterfeiting.

Gan et al. [3]. proposed an HIE system that allowed patients to control their medical information and authorize medical institutions to use their medical information. They emphasized the importance of strong patient-centric decision-making power to the extent that it would not interfere with ordinary medical examinations. They applied an access control scheme to protect patients' ownership of their medical information. They also proposed an incentive system to encourage patients to share their information alongside a method to determine the value of incentives.

Refs. [2] and [10] assumed the medical institutions would accept the HIE system. On the other hand, Ref. [3] proposed a system for patient-centric HIE in which incentives were given to patients so that they could be proactive in sharing. The common issue of these proposals is the lack of enough consideration towards medical institutions.

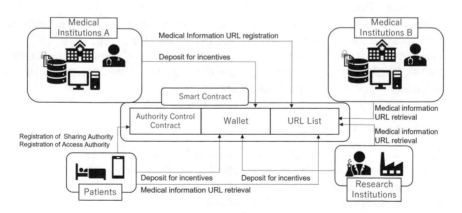

Fig. 1. Overall view of our HIE system

4 Proposed Method

Gan et al. [3]. proposed a system to encourage patients to share their medical information by providing them with incentives. However, they did not consider concerns, such as decreased profit for medical institutions. This research proposes the HIE system that provides incentives to both institutions and patients, and it aims to promote the sharing of medical information by reducing their concerns.

4.1 Entities and Their Role in HIE Systems in This Research

Figure 1 shows the overall view of the HIE system assumed in this research. The proposed method assumes a consortium blockchain network that considers the confidentiality of medical information. The entities in the system include smart contract, medical institutions, patients, and research institutions. In this research, we assume that Medical Institutions A is the primary group of medical institutions used by the patient, and Medical Institutions B is the group of other medical institutions. Each medical institution, patient, and research institution is assigned a global identifier (ID) that uniquely identifies each medical institution, patient, and research institution; herein, we use an Ethereum address as the global ID.

Next, we describe the main roles of each entity.

- The system owner (deployer) deploys smart contract on the blockchain and sets up the initial HIE system configuration.
- Patients provide Medical Institutions A the authority to share their own medical information with Medical Institutions B and research institutions (hereinafter called "Sharing Authority") and provide Medical Institutions B and research institutions the authority to access the patients' medical information stored by Medical Institutions A (hereinafter called "Access Authority"). In addition, the patients record own intentions regarding the secondary use of medical information on the blockchain.
- Medical Institutions A records URLs of patients' medical information on the blockchain after being given the Sharing Authority by the patients.
- Medical Institutions B obtains URLs of the patients' medical information on the blockchain after being given the Access Authority by the patients.
- The research institutions obtain URLs of the patients' medical information on the blockchain for research purposes after the patients consent to the secondary use of the medical information.

4.2 Providing Incentives for HIE

In this section, we describe the model of incentivizing HIE as proposed in this research. We assume that the medical consultation fees are paid in cash and that the incentives are paid automatically by cryptographic assets using the smart contract. We propose two types of incentives.

- Sharing authority incentive
  The sharing authority incentive is paid by Medical Institutions A to patients to obtain the Sharing Authority. The payment is made according to the following steps:
  1. Medical Institutions A deposits cryptographic assets into the smart contract in advance and sets sharing levels, which refer to the category (e.g., internal medicine, surgery, or dermatology) of the patients' medical information to be shared.
  2. The patients register the Sharing Authority in the smart contract.

3. When both Medical Institutions A and the patients register the above, incentives corresponding to the sharing levels are automatically paid to the patients from the cryptographic assets deposited by Medical Institutions A. Medical Institutions A then obtains the Sharing Authority.

This incentive is expected to motivate patients to participate in HIE.

- Medical information retrieval incentive
  The medical information retrieval incentive is paid by patients and research institutions to Medical Institutions A. The patients pay incentives to share their medical information with Medical Institutions B when they visit there. Research institutions pay incentives for the secondary use of patients' medical information. The payment is made according to the following steps:
  1. Patients and research institutions deposit cryptographic assets to the smart contract in advance.
  2. The patients register the Access Authority and own intentions regarding the secondary use of medical information to the smart contract. At this time, the patients set sharing levels for which the medical information is to be shared. This allows only authorized research institutions and Medical Institutions B to obtain the medical information URLs based on the sharing levels set by the patients.
  3. When research institutions or Medical Institutions B are allowed to obtain the medical information URLs, incentives corresponding to the sharing levels are automatically paid to the Medical Institutions A from the cryptographic assets deposited by the patients or research institutions.

We expect that this incentive will resolve medical institutions' concerns of profit and will motivate them to be more proactive in sharing medical information.

4.3   The flow of Healthcare Information Exchange

In this section, we describe the flow of medical information, as it is shared via the proposed method shown in Fig. 2. Note that, in the proposed method, access to medical information is controlled by general authentication (e.g., public key infrastructure).

1-1 A patient registers the Sharing Authority for Medical Institutions A.

1-2 Medical Institutions A sets the sharing levels to obtain the Sharing Authority. At the time of the incentives payment, the patient must register the Sharing Authority for Medical Institutions A in the smart contract, and both Medical Institutions A and the patient must agree to share the medical information. If both parties agree, the incentives are automatically paid to the patient corresponding to the sharing levels from the cryptographic assets deposited by Medical Institutions A in advance.

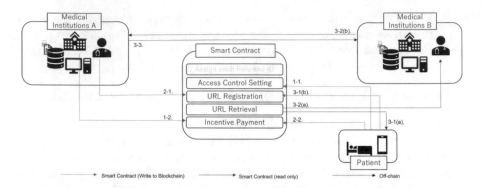

Fig. 2. The flow of our HIE system

2-1 Medical Institutions A, which has been given the Sharing Authority by the patient, registers the patient's medical information URLs through the smart contract in an encrypted form. This encryption is performed using the public key registered by the patient in advance in the smart contract. However, encryption is performed off-chain because the computational complexity of the smart contract is insufficient.

2-2 The patient sets the sharing levels when one visits Medical Institutions B and requires access to medical information which Medical Institutions A manages. In addition, the patient must register the Access Authority for Medical Institutions B in the smart contract. By this, the incentives are automatically paid to the Medical Institutions A corresponding to the sharing levels from the cryptographic assets deposited by the patient in advance.

3-1 When the patient needs to share medical information, (3-1(a)) the patient decrypts the encrypted medical information URLs using the private key and (3-1(b)) encrypts it with the public key of the Medical Institutions B, registering it on the blockchain. This encryption/decryption activity is performed off-chain.

3-2 (3-2(a)) Medical Institutions B, which is given the Access Authority, obtains the encrypted medical information URLs in (3-1) and decrypts it using the private key. This decryption is performed off-chain. (3-2(b)) Then, Medical Institutions B accesses the patient's medical information, which Medical Institutions A manages, off-chain.

3-3 If Medical Institutions B passes the authentication off-chain, it can obtain the patient's medical information which Medical Institutions A manages.

5   Implementation

In this section, we describe the implementation of the smart contract of the proposed HIE method.

- About Modifier of Solidity
  The modifier is the Solidity-specific function that allows only the correct user

to execute functions by attaching modifiers to each. For the modifier to work, the msg.sender, the Solidity-specific function, is important. The msg.sender represents the Ethereum address of the transaction sender, and the sender cannot tamper with the value of msg.sender. In our proposed method, access is controlled by the global ID and the role of each entity using a modifier.

When a patient invokes a function, if the patient's global ID is equal to the caller's msg.sender, the function can be executed. When a medical institution or a research institution invokes a function, if its global ID is equal to the caller's msg.sender and the role associated with the caller's Ethereum address is equal to the valid role of each of the above entities, the function can be executed.

- Registration of Medical Information URLs
  Medical Institutions A, which is given the Sharing Authority, can register patients' medical information URLs on the blockchain. When Medical Institutions A registers medical information URLs on the blockchain, the URLs are encrypted off-chain with the patients' public keys, registered in the blockchain. Thus, the medical information URLs published on the blockchain can be prevented from being read by unauthorized third parties. Since the medical information URLs encrypted with the public key are only a few kilobytes, there is no need to worry about overloading the blockchain network.

- Retrieval of Medical Information
  For Medical Institutions B to obtain medical information URLs, patients must decrypt the encrypted medical information URLs using the private keys. The smart contract checks whether the incentives are mutually paid between Medical Institutions A and patients among the list of Medical Institutions A. The only institutions among the list, confirmed as paid, are extracted as sharable institutions. For each of the sharable institutions, the smart contract then extracts the medical information URLs of patients set to the sharing levels owned by each institution and combines them into a medical information URL list. The list is encrypted with the public key of Medical Institutions B and registered by the patients. Then, if Medical Institutions B is given Access Authority, the institution can obtain the patients' medical information URLs by executing the function to obtain the URLs.

  For research institutions to obtain medical information URLs, patients must first consent to the secondary use of medical information. When the research institutions execute the function to obtain the URLs, the smart contract checks whether the research institutions have paid incentives to Medical Institutions A among the list of Medical Institutions A. The only institutions among the list, confirmed as paid, are extracted as sharable institutions. For each of the sharable institutions, the smart contract extracts the medical information URLs of patients set to the sharing levels owned by each institution and combines them into a medical information URLs list. Thus, research institutions can obtain patients' medical information URLs.

Table 1. Verification parameters

Parameter	Constant values	Parameter values
# of medical institutions	# of URLs: 5, # of patients: 1	5, 10, 15, 20
# of URLs	# of medical institutions:5, # of patients:1	5, 10, 50, 100

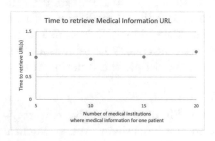

Fig. 3. Number of medical institutions and URLs retrieval time

Fig. 4. Number of URLs and URLs retrieval time

6   Experiments

6.1   Implementation Environment

This section describes the implementation environment of the research. The proposed system is implemented in an Ethereum environment. Using the Truffle, we deployed smart contract to the Rinkeby, the Ethereum test net, via the Infura. We also used Solidity as the language for smart contract development. The computer we used was a Dell Inspiron Intel(R) Core (TM) i7-8565U CPU @ 1.80 GHz 1.99-GHz, Memory: 8.00 GB. The experiment was conducted on the Windows Subsystem for Linux.

6.2   Experimental methods

In our proposed method, the medical information URLs of patients are recorded for each Medical Institutions A associated with the patients' global ID. Then, the recorded URLs were retrieved. Therefore, we measured the time between the execution of the URLs fetching command and the output of the result five times for each parameter shown in Table 1. Then, we calculated the average.

6.3   Results of Experiments and Discussion

Figure 3 shows the result of measuring the URLs retrieval time when the number of medical institutions with medical information for one patient was changed to 5, 10, 15, or 20. In this case, the number of medical information URLs for a patient was set to five. As a result, the URLs retrieval time was approximately 1 s for all parameter values of the number of medical institutions. This shows that the number of medical institutions where the medical information for one patient is stored does not affect the retrieval time of medical information URLs. Therefore, our proposed method is highly effective. Figure 4 shows the result of URLs retrieval time measured when the number of medical information URLs for one patient in one medical institution was changed to 5, 10, 50, or 100. In this case, the number of medical institutions visited by a patient was set to five. The measurement result shows that the URLs retrieval time was approximately 1 s when the number of URLs was five and approximately 2 s when the number of URLs was 100 with a monotonous increase. This shows that, as the number of medical information URLs for one patient in one medical institution increases, the URLs retrieval time is expected to increase linearly. This suggests that the number of medical information URLs significantly impacts the URLs retrieval time. However, it is assumed that this effect can be reduced by the structure of the medical information URLs. That is, by assigning one URL for each department in one medical institution, the number of URLs does not increase enormously; hence, the medical information URLs can be retrieved in as many as 2 s, as shown in Fig. 4. Therefore, our proposed method is still considered highly effective.

Additionally, in our proposed method, it is necessary to ensure that only valid user receives incentives. In the following, we describe security against attacks targeting incentives. There are several types of such attacks: (1) one in which a fraudulent user impersonates a legitimate user and steals incentives; (2) one in which an attacker illegally transfers money, although a legitimate user processes it; and (3) one in which an attacker steals incentives without sharing medical information.

In our proposed method, the global ID of the executor is authenticated when the incentive transaction is executed. At this time, the msg.sender of Solidity represents the address of the executor, and the executor has no control over its value. Therefore, the risk of an attacker impersonating patients or medical institutions to steal incentives is extremely low. An attack in which a fraudulent user transfers money illegally occurs when the fraudulent user embeds a malicious program but cannot pass the authentication with a global ID. Therefore, the risk of this attack is extremely low. Additionally, when users receive only incentives without sharing medical information, the transaction history is stored in the blockchain, and fraud can be identified. Thus, authentication with a global ID as a precautionary measure and the use of records on the blockchain as a post-measure can be used to deal with attacks targeting incentives.

7   Conclusions and Future Directions

In this research, we proposed the HIE system that provides incentives to both medical institutions and patients using blockchain technology. By evaluating the implementation of the proposed system and measuring the time required to share medical information, we found that the scheme was practical.

In the future, we plan to conduct a questionnaire survey of medical institutions to further develop our proposed method.

Acknowledgements. This work was partly supported by the Grant-in-Aid for Scientific Research (B) (19H04107).

References

1. Buterin, V.: A next-generation smart contract and decentralized application platform (2014)
2. Buzachis, A., Celesti, A., Fazio, M., Villari, M.: On the design of a blockchain-as-a-service-based health information exchange (BaaS-HIE) system for patient monitoring. In: ISCC, pp. 1–6 (2019)
3. Gan, C., Saini, A., Zhu, Q., Xiang, Y., Zhang, Z.: Blockchain-based access control scheme with incentive mechanism for ehealth systems: patient as supervisor. Multimed. Tools Appl. **80**, 30605–30621 (2021)
4. Jiang, S., Cao, J., Wu, H., Yang, Y., Ma, M., He, J.: BlocHIE: a blockchain-based platform for healthcare information exchange. In: SMARTCOMP, pp. 49–56 (2018)
5. Kim, J.W., Lee, A.R., Kim, M.G., Kim, I.K., Lee, E.J.: Patient-centric medication history recording system using blockchain. In: BIBM, pp. 1513–1517 (2019)
6. Kumar, R., Tripathi, R.: Secure healthcare framework using blockchain and public key cryptography. Blockchain cybersecurity, trust and privacy. Adv. Inf. Secur. **79**, 185–202 (2020)
7. Nakamoto, S.: Bitcoin: a peer-to-peer electronic cash system (2008)
8. Sun,Y., Zhang, R., Wang, X., Gao, K., Liu, L.: A decentralizing attribute-based signature for healthcare blockchain. In: ICCCN, pp. 1–9 (2018)
9. Toyoda, K., Mathiopoulos, P.T., Sasase, I., Ohtsuki, T.: A novel blockchain-based product ownership management system (POMS) for anti-counterfeits in the post supply chain. IEEE Access **5**, 17465–17477 (2017)
10. Zhuang, Y., Sheets, L.R., Chen, Y.-W., Shae, Z.-Y., Tsai, J.J.P., Shyu, C.-R.: A patient-centric health information exchange framework using blockchain technology. IEEE J. Biomed. Health Inform. **24**(8), 2169–2176 (2020)

A Design Thinking Approach on Information Security

Lukas König[(✉)] and Simon Tjoa

Institute of IT Security Research, St. Pölten University of Applied Sciences,
Saint Pölten, Austria
{Lukas.Koenig,Simon.Tjoa}@fhstp.ac.at

Abstract. More than ever before, the economic success of companies depends on the use of information and communication technologies. Along with this development, cyber security plays a vital role to ensure the continuous and secure operation of critical applications and IT-services. The human factor represents one especially important aspect for ensuring cyber security in organizations, which has taken a turn for the worse in recent time. Security awareness activities, such as security training, newsletters or quizzes, are often performed to try to improve the situation, but the effects are slow to materialize and often do not bring lasting change.

This paper therefore gets to the root of the problem using a different approach, which is centered around the people involved. The introduced framework combines the domains of design thinking and information security and presents a creative and human-centered way towards cyber security. We highlight building blocks, tools and techniques, which support the implementation of the presented framework. In order to demonstrate the applicability of the approach, we present our evaluation results of start-up company, which used our approach.

Keywords: Design thinking · Information security · Problem solving · Human factor

1 Introduction

The importance of human factor in cyber security well-known. A large amount of security breaches begin with specific attacks against the human operator such as phishing and social engineering [6,13].

Especially during the global pandemic, when working from home gained massive popularity, it became clear that the devices used to manage this shift are often personal devices, which in most cases cannot provide the same level of security. Since many organizations were unprepared, their cyber defences are not able to keep up with the shifting IT-landscape. Additionally, according to Deloitte [27] the security levels of organizations decreased by 20% and an overwhelming majority of 80% of organizations have no concept or contingency plan on how to respond to a possible cyber attack.

L. Barolli et al. (Eds.): AINA 2022, LNNS 450, pp. 503–515, 2022.
https://doi.org/10.1007/978-3-030-99587-4_42

This decrease or general lack of security in organizations can become a serious problem. Most non-malicious actors do not have a formal IT (security) education and are therefore unaware of cyber security and not very proficient IT operations. Even worse, the majority of users neither sees information security as one of their concerns, nor would they be able to react in case of emergency [12,13,26].

The human factor is therefore a crucial element, which must be secured to increase cyber defenses, and should be dealt with accordingly. To seize this opportunity however, it requires a focus on the needs and understanding of humans in a human-centered approach on cyber security [32].

The rising numbers of incidents suggest that current approaches on securing the human factor are either insufficient or heading in the wrong direction, as the problem of human vulnerability remains at large. A new, human-centered way of dealing with information security can prove itself useful to an organization.

The major contribution of this paper is the introduction of an iterative human-centered information security framework based on the principles, approaches and best practices of design thinking and information security. Combining the fundamental elements of cyber security and design thinking into one overarching framework (Security Design Thinking Framework - SecDT) enables organizations to address security challenges in a human-centered, innovative and flexible way, without being overwhelmed by the entirety of information security at once.

The remainder of this paper is structured as follows: First there will be a short overview of related work in Sect. 2. Following that, we also provide an overview of the related areas of design thinking and information security. The next section outlines our main contribution - the proposed framework - in Sect. 3 with the finalizing step being an evaluation of the framework in Sect. 4. We conclude our paper in Sect. 5 and present open limitations, and future research directions.

2   Related Work

Many research fields (e.g. software engineering, digital transformation) already demonstrated the positive effects of using design thinking tools and techniques. At the time of writing, no approach could be found, which combined cyber security with design thinking methods. Therefore, in the following we provide a brief overview on the usage in other disciplines before we provide more details on the related work in design thinking and information security, which inspired our approach.

The importance of human understanding and the focus on the human in human-computer-interaction (HCI) as opposed to a purely technical perspective is seen as a driving factor [1,4,9,25,33]. In some instances, design thinking is even portrayed as one of the, if not the most important asset of any software development endeavour, which also leads to the inclusion of design thinking into software development lectures and frameworks [9,19,24].

When it comes to a human-centric approach in information security, several authors have pointed out the importance of properly integrating humans in the cyber defenses, as well as current flaws [2,16,18,23]. The socio-cognitive behaviour of humans and security awareness is seen as what pivots the success of information security. The lack thereof consequently means failure. This situation makes it even more surprising that there is an overwhelming lack of regard for the individual overall. The general purport is that by securing the human factor, the rest will find its way automatically.

2.1 Design Thinking

There are multiple and slightly varying definitions for what design thinking is, with some authors even stating that design knowledge should be a third pillar next to arts and science knowledge.

Authors, such as Meinel et al. [20], define it as a way of thinking and a labour culture with which human and societal needs will be satisfied through innovation. This view is similar to what is stated in [1], where it is pointed out that design thinking is the way designers think and work when solving problems for customers.

Other views on the matter like that of Tim Brown [4] see it as a way to address the needs of the people in regards to technical possibilities and consumer value with a strong focus on market capitalization and the practical use of design thinking as a business enabler.

A third point of view is stated by Dave Kelly [4], where he defined design thinking as a way of finding human needs and creating new solutions using the tools and mindsets of design practitioners.

Although the presented definitions account for only a brief number of available definitions, they illustrate two essential points, which are found in most of the design thinking literature.

First of all, design thinking is a vivid practice and a way of thinking creatively and progressing non-linear. Secondly, depending on the background or field of application it is highly possible that design thinking is understood and practiced differently. For this reason, this paper will aggregate the previously stated definitions and form it into one concluding definition:

> Design thinking is a work mode that allows finding alternative solutions to human-centred needs and problems by using creative ways of thinking, personalized tools and iterative process methods.

2.2 Design Thinking Processes and Models

A clearly defined process is helpful when it comes to teaching and expressing the individual work stages of design thinking. However, it is emphasised that models and processes are merely supportive tools and not meant to be followed strictly, as creativity should not be hampered by strict rules [8,20].

There are however several models that emerged as some of the most common or useful process models. One of them is the Five-Phases-Model [4,8,14,15,20], with different variations of it existing simultaneously. Common phases are Emphasize/Understand, Define, Ideate, Prototype, and Test. At first, a problem or its impact on humans must be understood, then exact needs, partial problems and requirements can be defined. As a solution to these problems, ideas are generated, of which the best or most promising ones will be refined into prototypes and ultimately tested, whether the provided solution solves the problem.

2.3   Design Thinking Tools and Techniques

As mentioned by [4], it is impossible to describe every tool in existence that could be used for design thinking. It is also stated that design thinking practitioners usually draw from a toolbox consisting of more than a hundred different tools for their design thinking work, depending on which would be suited best to deal with a certain situation. There are however certain tools that are rather well-known and that can be seen as a main entry in every design thinking toolbox.

To provide an example, one of such tools is the stakeholder/community map [4,17]. A stakeholder map is used to identify any person or group of people that might by affected by the project in any way or form. It is advantageous to create this map with a broad and liberal understanding of the term stakeholder, as it is better to include more and extreme edge cases than to omit a group that turns out to be immensely important in the end. Especially end users are an important stakeholder.

Since there is not one single set of tools that must be used, as each of the tools caters to a slightly different setting and the correct use of the most fitting tool is what can really influence the outcome. When working with design thinking, it is completely viable to create a tailored toolbox, just as it is a viable strategy to use already existing off-the-shelf design thinking toolboxes as for example [3].

2.4   Information Security

When it comes to information security on a higher level, a lot of effort goes into explaining the importance of implementing a proper information security management system (ISMS). In the light of ever increasing numbers and varieties of cyber crime [5,6,11,27,32], protecting information adequately should become a priority for both private individuals and companies alike. Information security itself may be a fuzzy term for those unfamiliar with it. The foundation of it all are the principles of confidentiality, integrity, and availability, also known as the CIA Triad [22,28–31]

It must be mentioned though that there is a second triad revolving around information security. While the classical triad of confidentiality, integrity, and availability mostly focuses on how information security is to be implemented and operated, the second triad with people, technology, and processes highlights what it is that needs to be secured and which domains are affected [10,31].

Technology is the most obvious part of these three, as it usually are ICT-systems that come to mind when considering cyber security.

The domain People might come as a surprise at first, but as already mentioned in Sect. 1, the human factor is without a doubt responsible for a major part of all information security issues as it is often times the weakest and easiest to crack link. Securing the human factor is an important undertaking for organizations of all sizes.

Processes and Policies on the other hand show that there is a higher maturity, as these are usually elements of an organized information security effort with a systematic approach that also considers roles and responsibilities with a high regard for risk-assessments.

3 The Security Design Thinking Framework

The proposed SecDT-Framework (Security Design Thinking Framework) for combining design thinking with information security, as seen in Fig. 1 is comprised of various elements that should be familiar to practitioners of both worlds. Each of the elements will be described in this section.

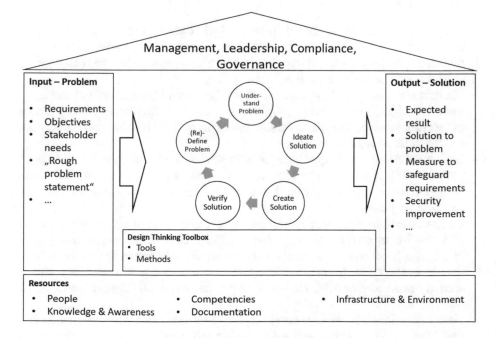

Fig. 1. Proposed SecDT-Framework

Input - Problem: The input is what steers the outcome of the whole framework. It is here where requirements like security objectives, needs of stakeholders, and further propelling forces enter the system. An initial and rough problem statement that describes the starting situation or position is sufficient at this point. There is no need for fully specified procedures, but rather pointing to a general direction to know what should be solved.

Output - Solution: The output, or solution of the framework is a general approach on how to reach a security objective, or an expected result of the process. This output could be the establishment of a whole ISMS, the implementation of a single process, refining of existing components and procedures, or anything that is meant to improve the overall levels of information security and privacy within an organization.

Problem-Solution-Space: The Problem-Solution-Space is the element where the heart of design thinking practices lies. It consists of a Problem-Solution-Process, which takes a problem as input and is meant to deliver a solution to that problem. This cycle becomes even more apparent when looking at the components of this process. It can be generally split into two spaces, the problem space and the solution space. The solution provided in such a process is experimental at first and will only become a complete security measure or another final product once it is verified successfully. The individual phases of this process are the following:

- (Re)-Define Problem: A clear definition of what the specific problem actually is and what it means to the organization is the needed foundation for the following phases and should therefore receive an adequate definition in regards to the overall security objective.
- Understand Problem: It is a phase of analysis and information gathering to understand the extent of the problem and the stakeholders involved. Types of relevant information can be the security impact or risk of a problem, root causes, influencing factors, etc.
- Ideate Solution: Once a problem is fully understood and specified, the next phase is to find solutions to the problem. The important part is that ideas are generated quantitatively so that there is a large pool of ideas available for the coming phase.
- Create Solution: The most promising or realistic ideas of the previous phase will now be selected for their implementation. That means experimenting or building a prototype. The results created in this phase are the experimental output of the whole framework. They can come as a specific security control, security processes, or other elements which are part of an organization's cyber security efforts.
- Verify Solution: The experimental implementations of solutions must be verified however. If a solution solves the problem, it shall be documented accordingly. If a solution does not solve the problem, the whole cycle starts anew and the initial problem statement must be re-defined.

Design Thinking Toolbox: For a seamless operation of the Problem-Solution-Process, a fitting toolbox is crucial. This means a catalogue of fitting design thinking techniques and tools that are needed in each of the phases of the process, which ideally should work well with the tasks and environment of information security. Such a catalogue can come in the form of an off-the-shelf toolbox like the 75 Tools for Creative Thinking [3], or the Mitre Innovation Toolkit [21], but there is no strict requirement on how it can or must be assembled. Continuous use of the framework will illustrate the effectiveness of one tool over the other for specific application on security matters. There are however several considerations to make about the choice of tools, i.e. some of the tools have a limited or minimum requirement of participating persons and materials.

Resources: The availability of the right and needed resources is fundamental for the success of any design thinking or information security effort. This includes people, infrastructure, environments, materials, knowledge, skills and competencies, awareness, communication, and of course documented information.

Management: Without management support, all efforts and measures being implemented are reduced to a mere nice-to-have without clear commitment from the top management of an organization. Especially when it comes to reporting, it is the top management that is being reported to. So their involvement is inevitable. It is also them who are responsible for providing needed resources and supporting a successful campaign.

4   Conducting the Experiment and Evaluation of the SecDT-Framework

The applicability of design thinking methods for information security has been tested in an experiment with a participating start-up organization active in the field of software development for education in Austria with no prior knowledge about information security and consequently no existing implementations of security controls or mitigations. The participants consist of the three co-founders aged between 20 and 25, one of them being female. The collaborative online whiteboard tool miro[1] was used during the workshop to facilitate the simultaneous involvement of all participants.

The initial starting point or input was defined to be the introduction of information security to the organization and to assess the current situation with regard to active measures and precautions. The solution or expected output that has been agreed on is defined as a proposed action plan about how identified issues should be addressed as a first step towards securing the organization. The Problem-Solution-Process with the support of design thinking tools and other needed resources, mostly information at this stage, is used to achieve the expected result.

The Problem-Solution-Process was held in the form of a workshop. The workshop is structured following the principles of design thinking work as outlined in The Workshopper Playbook [7]. It is split in the phases Collect, Choose,

---

[1] https://miro.com.

Create, and Commit, with each of these phases utilising design thinking tools. Before that however, the participants were introduced to information security in the form of a presentation on the principles of information security and the information security management system/ISO 27001.

In the collect phase the focus lies on gathering information, data, challenges and problems. For that, a brainstorming for stakeholders and a fishbone diagram for causes of insufficient information security was used. The result of the stakeholder map can be seen in Fig. 2, to give a visual representation of one of the tools in action.

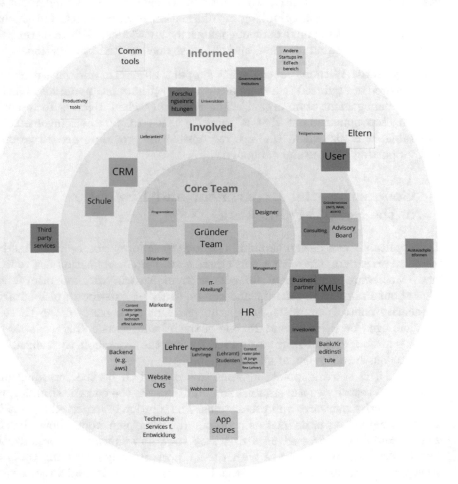

Fig. 2. Overview of the stakeholder map of the workshop

The second phase was used to filter all gathered information and to select the most important parts. In a continuation, the stakeholder brainstorming was structured as a proper stakeholder map, and the identified causes on the fishbone diagram were voted on with a process called dot voting to identify the most pressing issues.

During the create phase, possible solutions and ideas on how to solve the previously identified and prioritized problems were created. This was done by a series of *How-Might-We-Questions* that each first described a problem and what the solution should actually solve, and then a mind map for the ideation of multiple solutions to the specified problems, which were again dot voted to identify the most valuable solution to a problem.

In the last phase, the commit phase, the chosen solutions were placed on an impact-effort-matrix. Such a matrix is used to map a possible solution/implementation or other tasks according to their impact and effort, which allows for a quick identification of which solution is the most viable for a possible implementation. According to their positioning based on the previously mentioned relation of impact and effort, these solutions can be prioritized and collected into one action plan.

The action plan itself contained the "solution experiment", which is the actions taken as part of the solution to solve the problem at hand, a pre-defined success criteria for each action. These success criteria can be used at a later stage to evaluate any implemented solution and to assess whether it was successful or whether another iteration is needed with a new solution. Additionally, each experiment needs to be assigned to a person.

According to the partner organization, the SecDT-Framework worked very well for them. Information security has never been part of their concerns prior to the experiment, thus they were not familiar with it in any way. In the end however the partner organization stated that they now feel confident enough to tackle the first set of issues and that they now have at least the needed basic understanding of the matter.

The SecDT-Framework is likely to behave slightly differently based on the background of the participants. For example, if the participants are all security specialists themselves, there won't be a need for introductions to the topic or further literature, which was a vital part in this work. It is also very likely that the tools used to determine problems and solutions would vary as well. This variation is because of technical or organizational depth of an issue that a certain group of participants is simply not able to reach. So while the framework was tested with and approved by design thinkers, another experiment with security or technical experts could yield additional insights about the effectiveness or applicability of the framework.

The creative design thinking approach worked very well for the introduction of the rather complex topic of information security and the partner organization stated that they strongly believe that such a framework works very well for start-ups or smaller companies that do not have the resources for dedicated security expert personnel, as this approach is promising in making such a complicated topic much more accessible. Nonetheless, the participants also stated that multiple iterations and a continuous usage of the proposed cycle are definitely necessary and desired, so that the organization can gradually improve, solve, and overcome any prevalent issues, and to ensure ongoing security operations.

To add to that, in an interview shortly after the experiment, the COO of the partner organization even stated that cyber security implementations would never have happened in the first place, if not with the help of design thinking. The reason that was given for such an absolute statement is that simply "doing" security is tedious and dry. It also sparked their interest in trying a similar approach for other rigid topics. Additionally, it came to a big surprise how much can be done in a relatively short amount of time.

5   Conclusion

Although humans make up a major part of all information security flaws and breaches and have been repeatedly identified as the weakest link in any cyber security program, there has been little focus on securing the human, except for the usual awareness programs or training.

There are hardly any approaches trying to understand the problems at hand as to why humans act the way they do in a security environment. The approach chosen for this task was to combine design thinking with the world of information security. Design thinking as it is has a strong focus on the individual and tries to fully grasp what the actual problem is, so it can provide a fitting solution that is tailored to affected persons. By doing that, this new school of thought on information security elevates the individual so that it becomes the centerpiece of all security efforts and implemented measures will be created according to the needs of humans.

The key contribution of this work was therefore the creation of a framework, which accounts for the lack of empathy in traditional information security environments. The SecDT-Framework offers a new way of introducing complex topics like information security. By combining "the best of both worlds" it offers a structured and measurable approach for organizations, while it grants the affected individuals working on it the freedoms needed for creative problem-solving. The problem-solving for information security issues was tested in an experimental workshop with a partner organization.

The experiment and subsequent interviews affirmed that the SecDT-Framework is well received and motivates an organization that had previously had no association with information security to implement safeguards and measures, and to continuously improve their levels of security. Nonetheless, the fact that such a serious topic could be approached in a creative way that is only made possible by using design thinking and it still resulted in a proper action plan of measures and safeguards is seen as an extremely important outcome.

According to the current research results it is assumed that the SecDT-Framework works best in an environment where creative and agile work methods are already in use, preferably design thinking in particular, and where there is at least one security professional for the provision of knowledge and guidance during the implementation and for the operation available, or the resources to consult an external expert.

However, the framework has only been tested with a rather small organization where information security had not focused before. Assessing the viability of the SecDT-Framework with a larger organizations where there are multiple security experts already and where security processes have been determined before could yield different results.

References

1. Adikari, S., McDonald, C., Campbell, J.: Reframed contexts: design thinking for agile user experience design. In: Marcus, A. (ed.) DUXU 2013. LNCS, vol. 8012, pp. 3–12. Springer, Heidelberg (2013). https://doi.org/10.1007/978-3-642-39229-0_1
2. Asquith, P.M., Morgan, P.L.: Representing a human-centric cyberspace. In: Corradini, I., Nardelli, E., Ahram, T. (eds.) AHFE 2020. AISC, vol. 1219, pp. 122–128. Springer, Cham (2020). https://doi.org/10.1007/978-3-030-52581-1_16
3. Booreiland: 75 tools for creative thinking (2019). http://75toolsforcreativethinking.com/. Accessed 02 July 2021
4. Brenner, W., Uebernickel, F., Abrell, T.: Design thinking as mindset, process, and toolbox. In: Brenner, W., Uebernickel, F. (eds.) Design Thinking for Innovation, pp. 3–21. Springer, Cham (2016). https://doi.org/10.1007/978-3-319-26100-3_1
5. BSI: Why is information security needed (2021). https://shop.bsigroup.com/Browse-By-Subject/ICT/Information-security-standards-and-publications/Why-is-information-security-needed/. Accessed 17 May 2021
6. Bundeskriminalamt: Lagebericht cybercrime 2018-entwicklungen, phänomene und schwerpunkte. Technical report, Bundeskriminalamt (2019)
7. Courtney, J.: The Workshopper Playbook–How to Become a Problem-Solving and Decision-Making Expert. AJ&Smart (2020)
8. Dam, R.F., Siang, T.Y.: What is design thinking and why is it so popular? (2021). https://www.interaction-design.org/literature/article/what-is-design-thinking-and-why-is-it-so-popular. Accessed 23 Mar 2021
9. Dobrigkeit, F., de Paula, D., et al.: The best of three worlds-the creation of innodev-a software development approach that integrates design thinking, scrum and lean startup. In: DS 87–8 Proceedings of the 21st International Conference on Engineering Design (ICED 17), vol. 8: Human Behaviour in Design, Vancouver, Canada, 21–25 August 2017, pp. 319–328 (2017)
10. Ebrary: Pillars of security, people, organization of information security-the infosec handbook (2021). https://ebrary.net/26644/computer_science/pillars_security. Accessed 24 June 2021
11. EC-Council: What is information security and why it is important (2021). https://blog.eccouncil.org/what-information-security-is-and-why-it-is-important/. Accessed 17 May 2021
12. Hadlington, L.: Human factors in cybersecurity; examining the link between internet addiction, impulsivity, attitudes towards cybersecurity, and risky cybersecurity behaviours. Heliyon 3(7), e00346 (2017)

13. Hadlington, L.: The human factor in cybersecurity: exploring the accidental insider. In: Research Anthology on Artificial Intelligence Applications in Security, pp. 1960–1977. IGI Global (2021)

14. Ingle, B.R.: Design Thinking for Entrepreneurs and Small Businesses: Putting the Power of Design to work. Apress, New York (2013)

15. Jensen, M.B., Lozano, F., Steinert, M.: The origins of design thinking and the relevance in software innovations. In: Abrahamsson, P., Jedlitschka, A., Nguyen Duc, A., Felderer, M., Amasaki, S., Mikkonen, T. (eds.) PROFES 2016. LNCS, vol. 10027, pp. 675–678. Springer, Cham (2016). https://doi.org/10.1007/978-3-319-49094-6_54

16. Kassicieh, S., Lipinski, V., Seazzu, A.F.: Human centric cyber security: what are the new trends in data protection? In: 2015 Portland International Conference on Management of Engineering and Technology (PICMET), pp. 1321–1338. IEEE (2015)

17. Kettunen, J.: The stakeholder map in higher education. Int. Proc. Econ. Dev. Res. **78**, 34 (2014)

18. Klein, J., Hossain, K.: Conceptualising human-centric cyber security in the arctic in light of digitalisation and climate change. Arct. Rev. **11**, 1–18 (2020)

19. Lucena, P., Braz, A., Chicoria, A., Tizzei, L.: IBM design thinking software development framework. In: Silva da Silva, T., Estácio, B., Kroll, J., Mantovani Fontana, R. (eds.) WBMA 2016. CCIS, vol. 680, pp. 98–109. Springer, Cham (2017). https://doi.org/10.1007/978-3-319-55907-0_9

20. Meinel, C., Von Thienen, J.: Design thinking. Informatik-Spektrum **39**(4), 310–314 (2016)

21. MITRE: Mitre innovation toolkit (2021). https://itk.mitre.org/. Accessed 02 July 2021

22. Nieles, M., Dempsey, K., Pillitteri, V.Y.: An introduction to information security. NIST Spec. Publ. **800**, 12 (2017)

23. Nieto, A., Rios, R.: Cybersecurity profiles based on human-centric iot devices. HCIS **9**(1), 39 (2019)

24. Palacin-Silva, M., Khakurel, J., Happonen, A., Hynninen, T., Porras, J.: Infusing design thinking into a software engineering capstone course. In: 2017 IEEE 30th Conference on Software Engineering Education and Training (CSEE&T), pp. 212–221. IEEE (2017)

25. Pereira, J.C., de FSM Russo, R.: Design thinking integrated in agile software development: a systematic literature review. Procedia Comput. Sci. **138**, 775–782 (2018)

26. Proctor, R.W., Chen, J.: The role of human factors/ergonomics in the science of security: decision making and action selection in cyberspace. Hum. Factors **57**(5), 721–727 (2015)

27. Ruzicka, A., Niederbacher, A.: Deloitte cyber security report österreich 2020-eine studie von deloitte österreich in kooperation mit sora. Technical report, Deloitte österreich (2020)

28. Samonas, S., Coss, D.: The cia strikes back: redefining confidentiality, integrity and availability in security. J. Inf. Syst. Secur. **10**(3), 21–45 (2014)

29. Seker, E.: Confidentiality, integrity, availability (cia triad)–the backbone of cybersecurity (2020). https://medium.datadriveninvestor.com/confidentiality-integrity-availability-cia-triad-the-backbone-of-cybersecurity-8df3f0be9b0e. Accessed 24 July 2021

30. Singh, H.: 3 principles of information security — definition (2021). https://thecyphere.com/blog/principles-information-security/. Accessed 17 May 2021

31. Stamp, M.: Information Security: Principles and Practice. John Wiley & Sons, Hoboken (2011)
32. Team, S.R.: Sophos 2020 threat report. Technical report, Sophos (2020)
33. Valentim, N.M.C., Silva, W., Conte, T.: The students' perspectives on applying design thinking for the design of mobile applications. In: 2017 IEEE/ACM 39th International Conference on Software Engineering: Software Engineering Education and Training Track (ICSE-SEET), pp. 77–86. IEEE (2017)

Modeling Network Traffic via Identifying Encrypted Packets to Detect Stepping-Stone Intrusion Under the Framework of Heterogonous Packet Encryption

Noah Neundorfer, Jianhua Yang(⊠), and Lixin Wang

TSYS School of Computer Science, Columbus State University, 4225 University Avenue, Columbus, GA 31907, USA
{neundorfer_noah,yang_jianhua,wang_lixin}@ColumbusState.edu

Abstract. As the capability to detect network intrusion has increased, so has attackers' ability to avoid detection. Commonly, attackers use Secure Shell (SSH) to hide their identity. SSH securely connects two hosts together and encrypts their interactions. In stepping-stone attacks, one connection of SSH leads to another on a different host, and again until the attacker becomes untraceable from a victim host. In attempts to explore a method to detect stepping-stone attacks, we are pinned between two difficult issues. First, we desire to be able to detect a stepping-stone attack from any "stone" in the link from an attacker to a victim, which runs against our other issue; we must do this without being able to access any encrypted data. This means we only have the data "surrounding" the encrypted information. The previous work detecting a stepping-stone by viewing the lengths of encrypted packets assumed that the encryption algorithm used for both encryptions would remain the same. Our work is to determine whether we can use of the same effect, across algorithms, by focusing on the length sequences of incoming and outgoing packets. We expect that these sequences can be found when considering stream ciphers and comparing the length sequences coming in and out of a host. If this is the case, we will be able to develop a detection algorithm that can identify these sequences and determine if a host is being used as a stepping-stone to access another host.

1 Introduction

Most intruders prefer to launch their attacks using stepping-stones [1]. The primary reason of exploiting stepping-stones is that intruders' identity can be well protected from a long connection chain, which spans multiple compromised hosts. In order to prevent such attacks from happening, one direct way is to detect if stepping-stones are used to launch attacks. From a victim's side, it is hard to know if an access to a victim host is through a long connection chain. However, from any host on a connection chain between the attacker and the victim, it is possible to know if the host is used as a stepping-stone. Such a host is called a sensor. The method used in most popular approaches proposed to detect stepping-stone intrusion (SSI) from a sensor was to determine if the sensor is used as a stepping-stone.

© The Author(s), under exclusive license to Springer Nature Switzerland AG 2022
L. Barolli et al. (Eds.): AINA 2022, LNNS 450, pp. 516–527, 2022.
https://doi.org/10.1007/978-3-030-99587-4_43

There have been many algorithms developed to decide if a host is used as a stepping-stone by checking its incoming and outgoing network traffic since 1995. If a host is used as a stepping-stone, its incoming and outgoing traffic must be relayed. Staniford-Chen [2] et al. proposed a "content-thumbprint" method for SSI by comparing the content of the packets from an incoming connection with an outgoing one of a host. If the content matches, then they are a relayed pair of connections. However, this "content-thumbprint" method for SSI only works when the network traffic is not encrypted.

To overcome the shortcoming of the "content-thumbprint" approach proposed in [2], a "time-thumbprint" method was proposed in another seminar work [1] by Zhang et al. This method makes the decision for whether there is a matched pair of connections by analyzing the timestamps of packets. Since packets' timestamps are not encrypted during data communication, this "time-thumbprint" method works well for Stepping-stone Intrusion Detection (SSID) when the network traffic is encrypted. However, this approach of using "time-thumbprint" does not work effectively if the sessions are manipulated by intruders using evasion techniques such as chaff-perturbation or time-jittering.

K. Yoda et al. [3] proposed a deviation-based detection approach by setting up monitors for packets at many nodes on the Internet to store attackers' activities. This method is similar to the one proposed in [1]. Yoda's method in [3] used the deviation between two sessions. If a machine is employed as a stepping-stone to gain access to a remote target system, the packets information at the host we recorded are compared to find the closest match. The deviation for one packet stream on a connection from another is defined and computed. If such a deviation is small, the two connections should belong to the same chain. The less the deviation, the larger the chance for the two sessions to be relayed. This approach attempts to find a set of data streams that might match the one directly sent from the original attacker's machine.

[1] and [3] have some common issues when the network traffic are manipulated by intruders using evasion techniques such as either chaff perturbation or time-jittering, both the time-based method and the deviation-based approach for SSID in [1] and [3], respectively, are significantly affected. Since then several SSID approach were developed to overcome these issues.

D. Donoho et al. [4] proposed a method for SSID by monitoring the outgoing and incoming network traffic of a gateway router. His paper used a different approach by considering a computer network as a stepping-stone. A pair of outgoing and incoming connections of a gateway router is referred to as a stepping-stone pair of connections if those connections are used for a stepping-stone intrusion. The detection method developed in [4] is in the category of a host-based approach. That is, the incoming and outgoing connections of a gateway router are compared to see if there is a relayed pair of connections. An advantage of this method is that the network traffic can be encrypted as well as it is theoretically resistant to intruders' session manipulation to a certain degree.

[5] by A. Blum et al. proposed a Detect-Attacks-Chaff stepping-stone detection algorithm (DAC) for SSID by counting the number of packets of a connection. This paper [5] employed the ideas from Computational Learning Theory and conducted analysis using the concept of random walks. Blum's method for SSID used the idea that two sessions are relayed if and only if the difference between the packet numbers in the two sessions is bounded above. This method was known to be a good SSID approach that

could be resistant to intruders' evasion manipulation using the time-jittering technique, but Blum's method does not work effectively in resisting to intruders' evasion manipulation with chaff-perturbation as the upper bound for the number of monitored packets may be huge.

Another detection method by J. Yang et al. in [13] was developed by using random-walk to resist attackers' session manipulation with chaffed meaningless packets. [13] modeled the differences between the number of responses and the number of requests as a random-walk process. It was verified by the authors in [13] that if the two connections are a matched pair, then the behavior of the above difference follows a random-walk process.

The paper [6] by J. Yang et al. improved the ideas in [13] and proposed an RTT-based random-walk detection method for SSI, in which whether an outgoing connection and an incoming one are a matched pair is determined by applying the number of RTTs in a connection as well as the modelling of using random-walk method. The experimental results obtained in [6] showed that the RTT-based random-walk approach can defeat intruders' session manipulation more effectively than the one proposed in [13] by using the number of monitored packets, with either jittered-timing or chaff perturbation evasion technique.

A software to inject chaff-packets into an interactive TCP connection was developed in [7] by Yang et al. in 2018. This software can be easily used to determine whether or not a proposed SSID method can work effectively to resist session manipulations with chaff-perturbation by attackers. A framework to test whether a detection algorithm for SSI is resistant to time-jittering evasion was developed in [8] by L. Wang et al. Network security researchers can use the tool developed in [7] or the framework proposed in [8] to conduct network experiments by manipulating a TCP connection with either chaff-perturbation or time-jittering evasion technique, and determine whether their proposed approaches for SSID are resistant to intruders' session manipulation.

In [9] by Y. Zhang et al. developed a SSID method using context-based packet matching aiming to resist session manipulation by intruders. The simulation results obtained in this paper verified that if attackers send attacking packets to a network with chaff-rate 100%, this SSID method proposed in [9] is still able to match the outgoing and incoming connections for SSID as well as working effectively in term of resisting intruders' chaff-perturbation evasion.

In [10] by J. Yang et al. developed a new approach for SSID that can effectively resist session manipulation by intruders using the ideas of packet cross-matching as well as the concept of random walks. Prior approaches proposed for SSID can only work effectively in resisting session manipulation with limited capabilities. The technical analysis conducted in this paper showed that SSID approach proposed in [10] may effectively defeat intruders' session manipulation, such as chaff-perturbation evasion technique with unlimited number injected packets.

A recent work [11] by H. Clausen et al. proposed a framework to simulate stepping-stone behaviors in a realistic scenario by using effective evasion tools such as time-jittering and/or chaff-perturbation to release a large dataset. The datasets are analyzed to produce the effective rates for SSID of eight selected known state-of-the-art detection algorithms in the literature. The framework proposed in this paper may help network security experts to create a model to characterize the SSI behaviors.

While several solutions have been suggested, the work in this project is based off J. Yang et al.'s work [12]: detecting being used as a stepping-stone by viewing the lengths of encrypted packets. Their work suggested that by finding pairs of packets, one entering the host, and one leaving the host, both with identical encrypted lengths x, we could determine that a host was being used as a stepping-stone. However, their work assumed that the encryption algorithm used for both encryptions would remain the same. Our work is to determine whether we can use of the same effect, across algorithms, by focusing on the length sequences of incoming and outgoing packets. We expect that these sequences can be found when considering stream ciphers and comparing the length sequences coming in and out of a host. If this is the case, we will be able to develop detection algorithms that can identify these sequences and determine if a host is being used as a stepping-stone to access another host.

The rest of the paper is laid out as the following. Preliminaries will be given in Sect. 2. In Sect. 3, we present the way to identify an encrypted packet. We will discuss the approach to model network traffic in Sect. 4. The detection algorithm will be introduced in Sect. 5. Experimental results and its analysis are presented in Sect. 6. The whole paper is concluded in Sect. 7.

2 Preliminary

Professional Hackers' propensity to use stepping-stone intrusion is a well-documented facet of cybersecurity. In our attempts to detect this method of hiding an attacker's IP, it is useful to have an example model of a connection chain to more accurately refer to the parts of it. The figure below is one such model. For this example, where Host 0 is the attacker and Host N is the victim (Where malicious activity is likely occurring), host i is what we consider our "sensor" node, that is, the node where we attempt stepping-stone detection (Fig. 1).

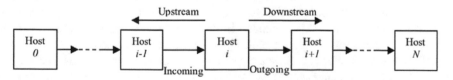

Fig. 1. An example connection chain

2.1 SSH Connection Chain

The SSH protocol allows secure, encrypted connections to a host running an SSH server. Attackers use Sequential SSH protocols to create a "chain" of hosts that transmits commands from an attacker to a victim. Each connection in a chain is separately encrypted, and the unencrypted data cannot be accessed at any stage. The benefits for an attacker of a stepping-stone connection is that the attacker's IP is incredibly difficult to trace back from the victim.

Each "stone" in the connection only knows it has an incoming and outgoing SSH connection, it does not know if the connections match, or the IP addresses of stones not adjacent to itself. An attacker who wishes to use stepping-stone intrusion usually will use 3 or more indusial "stones", often in geographically diverse locations, to create a very difficult to trace path leading back to them.

2.2 Send and Echo Definition

Important topics to discuss for studying SSH connections are TCP headers, IP headers, and SSH packet formation and encryption. Both the TCP and IP header, since they are below the application layer on the OSI model, remain unencrypted, and can be used to match packets.

A TCP header contains the essential information for a reliable connection, like ports, sequence and acknowledgement numbers and other flags. The IP header contains the source and destination IP, as well as other data. Since we know where the TCP header ends, it is trivial to calculate the length of the enclosed data, that is, the length of the SSH packet.

While all SSH packets are essentially created the same, we need to define crucial differences between packets that will be used in this research. The first and most important difference is between Send and Echo packets. When plaintext is enclosed into SSH packets, we then will distinguish them based on where they depart from. A **Send packet is a packet that travels from Host 0 towards Host N, that is in the direction of the victim. A packet traveling from Host N towards Host 0, or from the victim towards the attacker, is an Echo packet.** An echo packet usually contains a repeat, or confirmation, of the message sent by the corresponding Send packet. So, a Send packet leaves Host 0, travels to Host N, and then a corresponding echo packet, with the exact same, or different plaintext, is sent back to Host 0 from Host N.

The other type of Echo packet is a "Data-Echo packet" (or just Data packet). This is a packet that returns from the victim host where a command is executed containing information that is a response for the command execution (ex. The plaintext: "Document Downloads Pictures etc...." is contained in the Data-Echo packets returned when the victim host executes an "ls" command). For the purpose of this paper, the term "Echo packet" will always refer to a packet repeating the command from a send packet. The term Data packet or Data-Echo packet will be used to refer to packets containing the response for an executed command.

2.3 Packet Encryption

In SSH, packet encryption is performed using one of several ciphers. Before performing this encryption, the application creates a basic SSH packet. This packet contains several fields. The first is a 4-byte field that contains the length of the packet in bytes. Second is an one-byte field for padding length, containing the length of the padding in bytes, then two fields, payload and padding, of variable length. It may also contain a MAC field (Fig. 2).

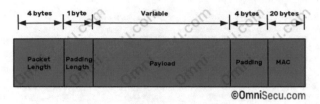

Fig. 2. The structure of an SSH packet

The Payload field is where the data, or the plaintext, of the message is stored, with each letter of plaintext being stored as a standard one-byte char. The padding field is slightly more complex. A certain amount of random padding is added in order to lengthen the packet to a certain total length. **The amount of padding added should increase the total length of the first four fields combined (excluding a MAC) to a multiple of either 8 or the cipher block size, whichever is larger.** So if the cipher block size is 16, the total length must be a multiple of 16 (16, 32, 48, etc.). If the cipher block size was 4 or had no block size (see stream cipher's), the total length must be a multiple of 8 (8, 16, 24, 32, 40, etc.). Once the packet is generated in plaintext, it is encrypted using the session key agreed upon by the client and the server. The first four fields must be encrypted, and as it often is, the MAC varies.

Once this packet is fully generated, it is then passed down the OSI model, adding a TCP header, IP header, etc. Then it is sent to the next host along the connection chain. When an encrypted packet is received at a sensor or an intermediate host, the packet is not fully unencrypted down to its plaintext, but a new SSH packet is made, and the payload is moved to the new packet's payload field. New padding is generated, and the packet is encrypted again **Using the session key from the next connection.** This means that each individual connection ($i - 1$ to i, and i to $i + 1$) is separately encrypted.

For example, say we had an attacker, Host A, set up a connection using chacha20 to our sensor, Host B, and then another connection using aes256-gcm to the victim, Host C. When a packet is sent, it would be encrypted with chacha20, sent to host B, where it is then unencrypted, re-encrypted in aes256-gcm, and then sent to Host C.

This process essentially means that we have no guarantee that a packet entering and leaving a host will have the same length, even if the unencrypted plaintext is the exact same.

2.4   The Length of an Encrypted Packet with Different Encryption Algorithms

What we do know when looking at SSH packets, is that when multiple packets with the same amount of plaintext (the same number of chars) are encrypted, **they will have the same encrypted packet length.** If we know these lengths, we can estimate the number of characters present in any given packet. This occurs because of the combination of two factors: the length of any given character is always one-byte, and the length of a packet will always be padded to a multiple of a certain, knowable value. This means that upon testing, we get results like the ones below (Fig. 3).

Text length	chacha20-poly1305@openssh.	Text length	aes128-gcm@openssh.com	Text length	aes256-gcm@openssh
1-2 Characters	.Length(36)	1-2 Characters	.Length(36)	1-2 Characters	.Length(36)
3-10 Characters	.Length(44)	3-10 Characters	.Length(52)	3-10 Characters	.Length(52)
11-18 Characters	.Length(52)	11-18 Characters	.Length(52)	11-18 Characters	.Length(52)
19-26 Characters	.Length(60)	19-26 Characters	.Length(68)	19-26 Characters	.Length(68)
27-34 Characters	.Length(68)	27-34 Characters	.Length(68)	27-34 Characters	.Length(68)
35-42 Characters	.Length(76)	35-42 Characters	.Length(84)	35-42 Characters	.Length(84)
43-50 Characters	.Length(84)	43-50 Characters	.Length(84)	43-50 Characters	.Length(84)
51-58 Characters	.Length(92)	51-58 Characters	.Length(100)	51-58 Characters	.Length(100)
59-66 Characters	.Length(100)	59-66 Characters	.Length(100)	59-66 Characters	.Length(100)
67-74 Characters	.Length(108)	67-74 Characters	.Length(116)	67-74 Characters	.Length(116)

Fig. 3. An example of packet lengths across encryption algorithms

As we can observe in the above figure, the number of plaintext characters used in any packet encrypted by a given cipher has a predictable, consistent packet length. What's more, when we compare the differing packet lengths of the 6 ciphers that are available for use in the baseline installation of OpenSSH (the most commonly used SSH software), we find that **the total difference in the length of an encrypted packet with the same amount of plaintext will never exceed 8 bytes.** This again is due to the combination of two factors: the padding will always increase the packet length to a multiple of a certain value, and the maximum that this value can be with these baseline ciphers is 16. This second fact is based on all included ciphers having a block size of 16 bytes or less. While this holds true for all useable ciphers, then we know for a fact that the difference in length not exceeding 8 bytes will stay true. While seemingly inconsequential, this deduction is incredibly important in our ability to match packets and will be used often.

2.5   Critical Definitions of Packets

Finally, we will briefly review some critical definitions for the following analysis. The first of these is a Match packet. A match packet refers to a packet that is observed entering a sensor, and then leaving the sensor (After having been decrypted, reformed, and re-encrypted as described Sect. 2.3). Our ability to find match packets is critical in determining if a sensor is being used as a stepping-stone. Knowing that a packet matches another means it must be part of a connection chain, which usually means it is part of a stepping-stone intrusion.

A Minimum Length packet is one that has the minimum number of characters (1 character/one-byte) present as the payload of the packet. These packets are the shortest length possible for their respective cipher. This terminology is important because, **when typing normally in a terminal into a connection chain, each individual character will be sent as a separate minimum length packet.** This behavior is important to packet sequencing and will be discussed further in Sects. 4 and 5.

Finally, it is important to note the distinctions of an exit packet. An exit packet is a packet with the SSH command "exit". This packet is executed slightly different to other packets. If a chain from host 0 to host N is created, when the exit command is issued (the enter key is pressed with "exit" currently in the terminal), it will travel down the connection chain as a normal packet until reaching host N-1. **The exit command will then be executed on host N-1, not host N. This means the exit packet will not be sent to the last host and will not necessarily have a matching echo packet.** This is important to note, as not understanding this can cause confusion when some of the last packets in a sequence can match up differently from a standard send-echo relationship.

3 Identifying Encrypted Packet

In order to successfully modelling a sensor's traffic in a way that can be used to determine if it is being used as a stepping-stone, we must focus on finding Match packets (described in Sect. 2.5). To find these Match packets, we use certain identifying features of packets. We focus on four key characteristics: whether a packet is a send or echo packet, packet length, the number of identical length sequential packets, and the timestamp of the packet. All four of these characteristics are simple to find, and the process of finding them will be expanded upon later.

The Send or Echo state of a packet is the simplest field to use, an incoming send packet can only ever be matched with an outgoing send packet, vice versa. Packet Length is slightly more complex to use, as different algorithms present different length packets for identical plaintext, but as stated above: two packets with identical plaintext, when encrypted with different algorithms, will always have a difference in length of 8 bytes or less[1]. Therefore, any packet with a greater difference is not a matched packet.

The crux of this method is the number of Identical Length Sequential Packets: the number of consecutive packets with the same length. So, 4 minimum length consecutive packets would be recorded as such. By separating these sequences whenever a command is executed, we can create highly accurate sequences of packets that identify two connections.

The timestamp is important because we can match packets up across a sensor because they perform a process of "nestling". Nestling is the process where Host B receives a send packet, sends its own send packet to Host C, received Host C's echo packet, and then sends its own echo packet back to Host A. The important factor we can observe in this nesting behavior is that: an incoming send packet can only be matched to an outgoing send packet with a later timestamp, and an incoming echo packet can only be matched to an outgoing echo packet with a later timestamp. Based on this behavior, we can accurately identify encrypted packets and their associate match packets (Fig. 4).

```
265 70.195240225  168.27.2.107 168.27.2.106  SSH  110 Client: Encrypted packet (len=44)
266 70.195669822  168.27.2.106 168.27.2.103  SSH  118 Client: Encrypted packet (len=52)
267 70.196655384  168.27.2.103 168.27.2.106  SSH  118 Server: Encrypted packet (len=52)
269 70.197168392  168.27.2.106 168.27.2.107  SSH  110 Server: Encrypted packet (len=44)
```

Fig. 4. An example of packet nestling

---

[1] This assumption always holds true given that none of the allowed ciphers have a block size exceeding 16 bytes. All standard ciphers are block size 16 or less. This difference value must be increased if allowed ciphers have a larger block size.

4 Modelling Network Traffic

We modelling network traffic by creating sequences of Identical Length Sequential Packets. An example sequence created with this model will be a list as such:

> *{...*
> *{SEND/ECHO, numPackets, packetLength, startTS, endTS},*
> *{SEND/ECHO, numPackets, packetLength, startTS, endTS},*

A single member of the above list is referred to as an element. An element is composed of the following fields. SEND/ECHO simply lists whether the element is a sequence of Send packets or Echo packets. Current work can only use Send packet sequencing, due to complex and inconsistent behavior exhibited by send packets[2], but the implementation leaves room for future improvement. The 'numPackets' field lists the number of sequential, identical, packets of the element. Four identical minimum length packets would be noted by a "4" in this field. The field 'packetLength' lists the length of the packets. The 'startTS' field notes the timestamp of the first packet that makes up this element, while the 'endTS' field notes the last packet's timestamp. If the 'numPackets' field is 1, then these two fields will have the same value.

When creating the elements, but not when recording them, they also have an additional field with one of 3 values: UNCAPPED, PARTCAPPED, or CAPPED. This value is used to determine whether the next packet looked at is eligible to be included in this element. When we determine that a command has been executed, we change the values to CAPPED, so that the next sequence of packets begins a new element. In this way, we divide the elements up by whenever a command is invoked.

To create the sequence, we look at each packet individually. If the packet is a Send packet, it is added to the sequence. If the packet is an Echo packet, it effects the packets CAPPED values. For Send packets, if the packet matches the length of the current UNCAPPED or PARTCAPPED element, then the 'numPackets' field of the element is increased, and the 'endTS' field is set equal to this packet's timestamp. In any other case, we create a new element using this packet. The filed 'numPackets' is initialized to one, 'packetLength' is the length of the packet, and both 'startTS' and 'endTS' are set equal to this packets timestamp. Either way, we then move to the next packet.

There are two criteria for echo packets causing element capping. The first is the receiving an Echo packet with a different length than the associated Send packet. When this occurs, it means data has been sent that is not a strict echo message, usually a response to a user's command (ex. "Documents Downloads Pictures etc…" as a response to an "ls" command). Receiving this Echo packet immediately switches the current element to CAPPED. The second criteria is if multiple correct length packets are received. When an Echo packet with a correct length is received, we designate the current element as PARTCAPPED. If another Echo packet with a correct length is received, we then change it to CAPPED. The reason for this is that sometimes the data echoed as a response to a

---

[2] Echo packets sometimes will "separate" or "reform" at nodes. This is believed to be based on the time in between a packet being received and sent out, but as of yet occurs into inconsistent a manner for effective use of Echo packet matching.

command has the same packet length as the previous Send packet. This causes an issue where even though response data is received, it is only seen as a strict Echo packet, so the element would not be CAPPED.

One issue to this method lies with exit packets (A packet with an "exit" message). An exit packet will be executed on the Sensor node, instead of the destination node. So, if an attacker at Host A sends an exit packet to a sensor at Host B, the packet is sent as plaintext normally, but the command is then performed at Host B to close the connection, not at Host C. This means that when there is an exit packet, the sequences of an incoming connection and an outgoing connection will look slightly different at the end of the sequences. This issue is difficult to manage, as it is impossible to know what an exit packet is just by viewing the incoming connection. Thankfully, these account for a small fraction of the total packets sent, so the Exit packet lowers the accuracy with a negligible amount for sequences with more than 20 packets.

5 Detecting Stepping-Stone Intrusion

Based on our model of network traffic, the way to determine if an intrusion is being carried out is a slightly less complex process of packet comparison.

Once we have all sequences prepared, we then compare each incoming sequence to each outgoing sequence, line by line. For each set of two element, the 'numPackets' fields must match exactly, the 'packetLength' fields must be within 8 of each other, and both incoming timestamps (startTS and endTS) must be before their respective outgoing timestamps of the other element.

If all the above fields are within their respective parameters, then the elements are considered to be matching, the total number of matching packets for this specific sequence comparison is increased by an amount equal to the 'numPackets' field. Regardless of the matching status, the total number of compared packets is increased by the 'numPackets' field as well.

Once each comparison is processed, the data is outputted, showing each incoming sequences comparison to each outgoing sequence, and the percent of packets that matched between these sequences. A percent of matched packets exceeding 95% is considered to be a likely stepping-stone intrusion.

6 Experimental Results and Analysis

A series of three labs were designed in order to test the effectiveness of this algorithm at differing amounts of traffic. At the moment, the first test has been performed, and results have been gathered. The first series of labs was a simple test of two separate connections, done at separate times. These resulted in an above 95% match of packets for all correct connections.

The second series test is planned to involve five machines, two attacker, two victims, and one sensor. Both attacker will go through the sensor to their respective victim and performed separate commands. Finally, we will repeat the above experiment, but adding an incoming SSH connection that ends at the sensor and adding an outgoing SSH connection directly from the sensor.

For the found results, an average relayed connection had a packet match rate of 97–98%. This slight below average can be attributed to Exit packets, as discussed above. All non-relayed connections ended up having a match rate of 0–40%. Because of SSH's propensity for minimum length packets, non-relayed connections will often have some amount of matching, but not a large enough amount to spill over into a false positive.

7 Conclusion

In this paper, the method proposed and experiments conducted have given a clear picture of a highly effective algorithm that is resistant to cipher change in a way that the previous version of this method was not. We have developed a method that can be used not only to check connection records, but also actively monitor connections. Finding a connection that has a 95% + match rate after several executed commands is a very likely sign of a stepping-stone intrusion. This method also has a significant advantage over other monitoring methods by being totally position-independent: it does not matter where the sensor is located in a connection chain when performing detection.

Finally, this method is incredibly durable against False Positive errors due to its matching process. Our future work will be completing the further experiments outline in Sect. 6 and performing further experiments related to the method's ability to run when certain anti-stepping-stone intrusion detection measures are in place by an attacker.

Acknowledgments. This research has been funded by NSA grant H98230-20-1-0293.

References

1. Zhang, Y., Paxson, V.: Detecting stepping-stones. In: Proceedings of the 9th USENIX Security Symposium, Denver, CO, pp. 67–81 (2000)
2. Staniford-Chen, S., Heberlein, L.T.: Holding intruders accountable on the internet. In: Proceedings of the IEEE Symposium on Security and Privacy, Oakland, CA, pp. 39–49 (1995)
3. Yoda, K., Etoh, H.: Finding a connection chain for tracing intruders. In: Cuppens, F., Deswarte, Y., Gollmann, D., Waidner, M. (eds.) ESORICS 2000. LNCS, vol. 1895, pp. 191–205. Springer, Heidelberg (2000). https://doi.org/10.1007/10722599_12
4. Donoho, D.L., Flesia, A.G., Shankar, U., Paxson, V., Coit, J., Staniford, S.: Multiscale stepping-stone detection: detecting pairs of jittered interactive streams by exploiting maximum tolerable delay. In: Wespi, A., Vigna, G., Deri, L. (eds.) RAID 2002. LNCS, vol. 2516, pp. 17–35. Springer, Heidelberg (2002). https://doi.org/10.1007/3-540-36084-0_2
5. Blum, A., Song, D., Venkataraman, S.: Detection of interactive stepping stones: algorithms and confidence bounds. In: Jonsson, E., Valdes, A., Almgren, M. (eds.) RAID 2004. LNCS, vol. 3224, pp. 258–277. Springer, Heidelberg (2004). https://doi.org/10.1007/978-3-540-30143-1_14
6. Yang, J., Zhang, Y.: RTT-based random walk approach to detect stepping-stone intrusion. In: IEEE 29th International Conference on Advanced Information Networking and Applications, pp. 558–563 (2015)
7. Yang, J., Wang, L., Lesh, A., Lockerbie, B.: Manipulating network traffic to evade stepping-stone intrusion detection. Internet Things 3, 34–45 (2018). https://doi.org/10.1016/j.iot.2018.08.011

8. Wang, L., Yang, J., Workman, M., Wan, P.-J.: A framework to test resistency of detection algorithms for stepping-stone intrusion on time-jittering manipulation. Wirel. Commun. Mob. Comput., 1–8 (2021). https://doi.org/10.1155/2021/1807509
9. Zhang, Y., Yang, J., Bediga, S., Huang, S.-H.: Resist intruders' manipulation via context-based TCP/IP packet matching. In: The Proceedings of 24th IEEE International Conference on Advanced Information Networking and Applications (AINA 2010), Perth, Australia, pp. 1101–1107, April 2010
10. Yang, J.: Resistance to chaff attack through TCP/IP packet cross-matching and RTT-based random walk. In: Proceedings of the 30th IEEE International Conference on Advanced Information Networking and Applications, Crans-Montana, Switzerland, 23–25 March 2016, pp. 784–789 (2016)
11. Clausen, H., Gibson, M.S., Aspinall, D.: Evading stepping-stone detection with enough chaff. In: Kutyłowski, M., Zhang, J., Chen, C. (eds.) NSS 2020. LNCS, vol. 12570, pp. 431–446. Springer, Cham (2020). https://doi.org/10.1007/978-3-030-65745-1_26
12. Yang, J., Wang, L., Shakya, S., Workman, M.: Identify encrypted packets to detect stepping-stone intrusion. In: Barolli, L., Woungang, I., Enokido, T. (eds.) AINA 2021. LNNS, vol. 226, pp. 536–547. Springer, Cham (2021). https://doi.org/10.1007/978-3-030-75075-6_43
13. Yang, J., Lee, B., Huang, S.S.-H.: Monitoring network traffic to detect stepping-stone intrusion. In: The Proceedings of 22nd IEEE International Conference on Advanced Information Networking and Applications (AINA 2008), Okinawa, Japan, pp. 56–61, March 2008

A Study on Enhancing Anomaly Detection Technology with Synthetic-Log Generation

Takumi Yamamoto[✉], Aiko Iwasaki, Hajime Kobayashi, Kiyoto Kawauchi, and Ayako Yoshimura

Mitsubishi Electric Corporation, Kamakura, Japan
Yamamoto.Takumi@ak.MitsubishiElectric.co.jp

Abstract. Anomaly detection techniques based on unsupervised training have been attracting attention because it is difficult to prepare a large amount of actual attack logs. On the other hand, in anomaly detection, if some attack data is available, more appropriate selection of the parameters and functions to use can be made, and if there are more variations of attack data, it is expected that it will be possible to adjust to various types of attacks. In this study, we investigate the possibility of improving the performance of anomaly detection systems by generating synthetic attack logs in the neighborhood of real attack logs and including them in the verification data to adjust the parameters. Using the same approach, we generate logs in the neighborhood of false positive logs and include them in the training data and the verification data. We confirm the effectiveness of our methodology using a simple anomaly detection system and public attack data.

1 Introduction

Targeted attacks against specific companies and organizations are as serious as ever. Even in recent years, government agencies and companies in Japan have been subject to targeted attacks and suffered damage, and there is a demand for countermeasures [1]. In addition, with the networking of control systems, cyber-attacks on critical infrastructure such as power generation plants and gas plants are becoming a threat, and it is a serious concern that could shake national security [2]. It is expected that events like the Olympic and Paralympic Games, which will attract worldwide attention, will be a prime target for attackers. If critical infrastructure were to stop functioning due to cyber-attacks during the Games, the operation of the Games would be severely hampered.

On the other hand, in the field of security monitoring, the situation of there not being enough staff with the required specialist knowledge has become normalized. According to a research report [3], there was a shortage of one million cyber security staff as of 2013, and it was expected to rise to a shortage of 3.5 million in 2021. Therefore, it is necessary to have a technology that can detect cyberattacks with high accuracy and efficiency, even with a small number of staff.

Rule-based detection techniques using rules to detect known attack methods and attacker behaviors are well known as techniques to monitor cyberattacks [4]. However, recently, it has become difficult to define rules in advance due to the sophistication of attacks and increase in unknown attacks, which causes problems for the monitoring staff of the Security Operation Center (SOC). In addition, it is necessary to manually adjust the rules for each target system, and rule-based detection technology is reaching its limits. Therefore, it is desirable to have an advanced detection technology that does not require the definition of rules in advance, or that can automatically determine the boundary between what is normal and abnormal.

Artificial intelligence (AI) and other machine-learning-based attack detection technologies have been attracting much attention in recent years. AI can learn from data of one or more classes prepared in advance and automatically find the boundary between classes. If a large amount of data can be prepared for each class, the AI can find the boundaries appropriately. If AI can be applied to cyberattack monitoring, it is expected that AI will take over the task of defining and updating rules that has been performed by staff with specialized knowledge and skills thus far.

However, in network security, the difficulty of preparing a large amount of data for each class, which is most important for AI, is a challenge. In particular, the occurrence of attacks is rare, making it very difficult to prepare large amounts of attack data for training purposes. Therefore, it is necessary to develop a technology that can effectively detect attacks as anomalies, even in environments with little or no attack data.

Anomaly detection technology is known as a typical example of such a technology. Anomaly detection technology learns normal traffic and behavior of the monitored system as a normal model, and detects the traffic and behavior that deviates from the normal model as being abnormal. Whitelisting, which defines a normal model with rules, is also an anomaly detection technology in a broader sense.

On the other hand, it has been reported that attackers today have a lot of knowledge about the target organizations and their systems. For example, according to the literature [5], the developers of Stuxnet are said to have had extensive knowledge about SCADA and the control system of the attack target. From this, it is expected that the number of attackers who are familiar with their target organization's information will increase in the future, and the number of attacks that are designed to hide behind normal business operations or normal system operations or that bypass security systems will increase. However, if the threshold of the anomaly detection system is simply adjusted to detect these attacks, the possibility of false positives, in which normal events are erroneously detected as abnormal, will increase.

In order to solve the problems mentioned thus far, this study aims to improve the performance of anomaly detection systems by generating synthetic attack logs that should be in the neighborhood of a small amount of real attack logs and including them in the verification data and adjusting the parameters of the anomaly detection system, based on the anomaly detection techniques using machine learning. Using the same approach, we will also aim to improve the performance of the anomaly detection system by generating synthetic normal logs that should be in the neighborhood of the normal logs (false positive logs) that the anomaly detection system erroneously detects, and including them in the training data and verification data and adjusting the parameters of the anomaly detection system. As a result, it can be expected to follow attacks designed to hide in the normal operation of the system and attacks designed to evade the security system while suppressing erroneous detection.

2   Related Studies

We briefly introduce the existing technologies related to the proposed method.

Reference [6]. In this study, the authors assume a strong attacker and evaluate the robustness of the classifier for malware detection. For this purpose, the authors proposed a technique for automatic generation of malware that evades the classifier. In this study, the authors focus on PDF malware and use genetic algorithms to automatically generate malware variants that evade the detection system. The detection systems targeted by this study are based on machine learning, and the malware variants are generated in such a way that they cross the classification boundaries of the classifier obtained by machine learning.

Reference [7]. We have previously published paper [7] as an existing method for generating attack logs that evade detection. In the paper, the we proposed a technique to automatically generate logs of sophisticated attacks that are designed to have characteristics very similar to normal conditions for the evaluation of security products. In the technique, the feature vectors of attacks are modified to straddle the decision boundary of the normal model, which is learned from the behavior of normal data, and the attacks are generated on the simulated environment so that they have features corresponding to the feature vectors that cross the boundary. This technique modifies feature vectors to straddle the decision boundary of the detection system on the feature space handled by the anomaly detection system, and search for samples that avoid detection.

3   Proposed System

3.1   Concept

The proposed system aims to improve the accuracy of the anomaly detection system by generating synthetic attack logs which can avoid detection by the anomaly detection system based on a small number of real attack logs. It also aims to improve the accuracy of the anomaly detection system by synthetically generating false positive logs based on the normal logs (false positive logs) which are erroneous detected by the anomaly detection system.

The more nonlinear and high-dimensional the feature space becomes, the more difficult it is to inverse-transform the feature space representation (feature vectors) into real space information (logs), and even if the inverse transformation is possible, there is a concern that unnatural logs or synthetic attack logs that are not valid as actual attacks may be generated.

Therefore, in this study, instead of searching for attack logs or false positive logs that avoid detection synthetically by modifying feature vectors on the feature space, we modify each item of the log on the real space, convert them into a feature vectors, and check whether the detection result changes on the feature space. In order to avoid this ad hoc approach, we identify true-negative normal logs in the neighborhood of the log to be modified (attack logs or false-positive logs) in the newly created feature space and modify the logs to include the tendency of features that are often found in true-negative normal logs.

Hereafter, we will describe it using mainly proxy logs as an example, but the proposed concept can be applied to logs other than proxy logs. The minimum unit of input data handled by the anomaly detection system is called one sample, and we assume that one event in the log is one sample. A normal sample represents one event (record) in the normal log, and an attack sample represents one event (record) in the attack log.

3.2 Procedure for Generating Synthetic Attack Logs

We describe a procedure for identifying true-negative normal samples in the neighborhood of the attack sample to be modified, and modifying the attack sample to include features that are often found in neighborhood true-negative normal samples.

(1) Using the anomaly detection system, we determine whether the attack sample to be modified is normal or abnormal. If it is normal, we register the attack sample as an attack to be avoided. If it is determined to be abnormal, it is registered as an attack sample to be modified, and the following procedure is performed.

(2) Using the normal samples determined to be normal by the anomaly detection system (true-negative normal samples), we extract true-negative normal samples in the neighborhood of the attack samples to be modified (neighborhood true-negative normal samples) (see **Nearest Neighbor Extraction** for details). The true-negative normal samples can be prepared in advance.

(3) We calculate the trend of the extracted features of the neighborhood true-negative normal samples (see **Calculation of trends** for details).

(4) Based on the obtained trend, we modify each field of the attack sample to include more feature values of the neighborhood true-negative normal samples (see **Correction of features** for details).

(5) We determine whether the modified attack sample is normal or abnormal using the anomaly detection system to be avoided. If the modified attack sample is determined to be normal, we register the attack as a synthetic attack sample to avoid detection. We repeat steps (4) and (5) until the amount of feature modification exceeds the specified value, and create and collect synthetic attack samples that avoid detection.

(6) Steps (1) to (5) are repeated the specified number of times while changing the original attack sample.

The extraction of nearest neighbors, the calculation of trends, and the modification of features are described in more detail in **Nearest Neighbor Extraction**, **Calculation of trends**, and **Correction of features**, respectively.

Nearest Neighbor Extraction. Nearest neighbor extraction extracts specified features from X attack samples and Y true-negative normal samples, and converts the feature information into a format (feature vector) that can be easily processed by machine learning algorithms. Y is a sufficiently large number compared to X. In the proxy log, categorical data such as domain, method, status code, etc., are converted to one-hot encoding or frequency-based numeric representation [8], for example. The numerical data should be normalized and the sizes should be aligned among feature types. Using the feature vector of attack samples and the feature vector of true-negative normal samples, for X attack samples, we identify K true-negative normal samples in the neighborhood (neighborhood true-negative normal samples). The features and feature representations used can be different from those of the target anomaly detection system. In practice, we extract K′ neighborhood true-negative normal logs from each attack sample and combine them into one (K = X ∗ K′).

Calculation of Trends. To calculate the trends, we extract the specified features from the obtained K neighborhood true-negative normal samples and X true-positive attack samples, and convert the feature information into a format (feature vector) that can be easily processed by machine learning algorithms. The features and feature representations used can be different from the target anomaly detection system and nearest neighbor extraction.

Using the obtained feature vectors, we train a classifier (C) that separates K true-negative normal samples from X true-positive attack samples. We calculate the feature importance of the features when the classifier classifies the K true-negative normal samples and the X true-positive attack samples, and extract the top N features with the highest importance (F_1 to F_N). One of the classifying algorithms that can calculate importance is Random Forest.

For the obtained top N features, we obtain the statistics in the normal log of K true negatives. In the case of categorical data, for example, the most frequent value (mode, mod_i, $i = 1$ to N), and in the case of numerical data, for example, the mean value (μ_i, $i = 1$ to N) and standard deviation (σ_i, $i = 1$ to N) are calculated.

Correction of Features. The following procedure is used to calculate the candidate values for modification for each of the features F_1 to F_N.

(1) Obtain the actual value of the feature F_i (d_i) from the field corresponding to the feature F_i of the true positive attack sample.
(2) When feature F_i is categorical data, it is converted into a numerical expression (d_i) based on frequency. The frequency is calculated in advance from the training data of the machine learning model of the anomaly detection system. We also refer to the statistical information (s_i) of K neighborhood true-negative normal samples. If the feature F_i is categorical data, s_i is mode which stands for the value that occurs most often (mod_i). If the feature F_i is numerical data, s_i is the mean value (μ_i).
(3) If d_i is greater than s_i, we update d_i so that d_i approaches (becomes less than) s_i by Δ_i and add the updated value to the list $list_i$. This is repeated for as long as d_i is greater than s_i. If d_i is less than s_i, we update d_i so that d_i approaches (becomes greater than) s_i by Δ_i and add the updated value to the list $list_i$. This is repeated for as long as d_i is less than s_i. If d_i is equal to si no action is taken. Δ_i is a specified value.
(4) Steps (1) to (3) are repeated for all F_1 to F_N ($i = 1$ to N).
(5) Based on the list $list_i$ ($i = 1$ to N) created from the features F_i ($i = 1$ to N) to be modified, we create all combinations of candidate features to be modified and generate samples corresponding to each pair. If the length of the list $list_i$ ($i = 1$ to

N) is len_i ($i = 1$ to N) respectively, the type of samples generated is $N = \Pi len_i$. The fields of the samples that do not correspond to feature F_i ($i = 1$ to N) (the fields that are not to be modified) retain the same values as the original attack samples.

3.3   Generating Synthetic False Positive Logs

Without going into details, we generate synthetic false positive logs by using the same approach as the procedure for generating synthetic attack logs in Sect. 3.2. The difference from Sect. 3.2 is that the synthetic false positive samples are generated based on the normal samples (false positive samples) that were erroneously detected by the anomaly detection system, rather than based on the attack samples. We modify the false positive sample to include features that are often found in true negative normal samples in the neighborhood of the false positive sample to be modified. If the modified false positive sample is determined to be abnormal in step (5) of Sect. 3.2, we register the sample as a synthetic false positive sample.

By increasing the proportion of normal samples in the training or validation data that are prone to erroneous detection, it is expected that the training will reflect the characteristics of normal samples that are prone to erroneous detection.

4   Evaluation

Evaluation tests were conducted to confirm the feasibility and effectiveness of the proposed method.

4.1   Data Used in the Tests

Normal Data. We used proxy logs collected in the past, at a large-scale organization with about 1000 employees, as normal data that does not contain any attacks. We used three consecutive weekdays' worth as training data, one weekday after the training data as validation data, and one weekday after the validation data as test data. The number of events (records) in the proxy log for one weekday is about 5,000,000. Because the number of events is too large, we randomly sample 30,000 events in this test. For convenience, we refer to the training data as train_data.

Attack Data. For the attack data to be included in the verification and test data, we used the malware traffic data [9] published by CTU University. The CTU dataset contains bot executable files and PCAP files of actual traffic. In this test, we used traffic data which is assumed to have been acquired after 2018[1], the data which is converted to log format (with the.weblog extension) and contains more than 10 events[2].

Attack data assumed to have been acquired between February and March 2018 were mixed into the validation data, and attack data assumed to have been acquired between

---

[1] The acquisition date was determined based on the date assigned to the file name.

[2] Data from malware that issued a large number of requests (e.g., more than 100,000 requests per day) were excluded from the data in this evaluation.

April and October 2018 were mixed into the test data. When mixing, we randomly selected an IP addresses for each type of attack log and inserted them into the validation and test data during work hours while maintaining the relative time interval[3]. In order to reduce the processing time, we used the verification and test data between 10:00AM and 11:00AM in this experiment. For convenience, we refer to the validation and test data as val_data and test_data, respectively. Both val_data and test_data are labeled data, and each event (sample) has a label indicating whether it is a normal sample or an attack sample.

4.2 Basic Model Creation

The basic model is described as an anomaly detection system for the avoidance target described in Sect. 3.2. In order to confirm the feasibility and effectiveness of the proposed method, we prepared a model of an anomaly detection system using simple features and conventional machine learning algorithms, instead of complex features and deep learning. The machine learning algorithm we used is Local Outlier Factor (LOF). The extracted features are the information of the fields in the normal data (proxy logs) described in Sect. 4.1, such as domain, extension, status code, response size, and access interval. The features extracted for each event were converted to a format that can be input into a machine-learning algorithm and made into a feature vector.

LOF was used as implemented in scikit-learn, an open source machine learning library for Python [10].

The model created by using train_data and val_data for training and validation data, respectively, and adjusting hyperparameters with a grid search is called the basic model. The hyperparameters adjusted by the grid search are n_neighbors and contamination.

4.3 Generating Synthetic Logs

This section describes the synthetic attack logs and the synthetic false positive logs prepared for this test.

Synthetic Attack Log Generation. We set the threshold θ_0 so that the false positive rate for val_data in the basic model is less than 1%, and extracted true positive attack samples from val_data and test_data. In addition, we modified the true positive attack samples according to the procedure described in Sect. 3.2, extracted the synthetic attack samples using the basic model and the threshold θ_0, and prepared the synthetic attack logs. For convenience, we refer to the synthetic attack logs created from val_data and test_data as fn_val and fn_test, respectively. The labels assigned to the synthetic attack samples in fn_val and fn_test are the same as those of the original attack samples.

In this test, 20 neighborhood true-negative normal samples were extracted for each attack sample ($X = 1, K = K' = 20$), and the attack samples were modified using the

---

[3] We inserted the attack log so that the attack would start at a random time between 8:30AM and 10:30AM during business hours. We chose this time period because it is the time when people start their work and start reading their emails.

information of the 20 neighborhood true-negative normal samples. Note that the attack logs determined to be normal from the original attack logs (prior to modification) in step (1) of Sect. 3.2 are not included in this test. This is because we determined that it would be difficult to obtain undetected attacks in actual operation, because we cannot notice them in the first place.

The machine learning algorithm used in Nearest Neighbor Extraction is the nearest neighbor method, which is implemented with scikit-learn the same way as LOF. The hyperparameters of the default settings were used.

The machine learning algorithm used in Calculation of trends is a random forest, which is implemented using scikit-learn in the same way as LOF. We used the hyperparameters of the default setting. All features with a feature_importance greater than 0 were used to calculate the trend.

For Δ_i in step (3) of Correction of features, when the feature to be modified is numerical data, we defined $\Delta_i = 6 * \sigma_i/10$ based on the standard deviation (σ_i) of the neighborhood true-negative normal samples calculated by the procedure of Calculation of trends. When the feature to be modified was categorical data, we defined it as $\Delta_i = 1$, and when the feature value was made larger (smaller) by Δ_i, we used the next most frequent (lowest) categorical data in the category.

Synthetic False Positive Log Generation. The synthetic false positive logs were generated following the procedure in Sect. 3.3. The detailed settings are the same as the procedure for Synthetic attack log generation.

We set the threshold θ_1 so that the false positive rate for val_data by the basic model was less than 1%, and extracted the false positive normal samples from train_data and val_data. In addition, following the procedure in Sect. 3.3, we modified the false positive normal samples, extracted synthetic false positive samples using the basic model and the threshold θ_1, and prepared a synthetic false positive log. For convenience, we refer to the synthetic false positive logs created from train_data and val_data as fp_train and fp_val, respectively. The labels given to the synthetic false positive samples are the same as the original false positive normal logs.

4.4 Improved Model Creation

We describe an improved model created using new training data including synthetic logs and validation data.

Data Using Synthetic Logs. We will explain the new training data, validation data, and test data.

Training Data Mixed with Synthetic False Positive Logs. The data obtained by mixing the false positive log (fp_train) with the training data (train_data) consisting of only normal data is train_data_fp. As a result, the total number of events was 31444, of which 1444 were synthetic false positive samples. By using this data as the training data, it is expected that more emphasis will be placed on normal data that is prone to erroneous detection.

Verification Data Mixed with Synthetic False Positive Logs. The data obtained by mixing the false positive log with the verification data (val_data) consisting of attack data and normal data is val_data_fp. Random sampling was performed when the data size of fp_val exceeded 10% of the size of val_data so that the proportion of fp_val in val_data_fp did not become too large. As a result, the total number of events was 23122, of which 2112 were false positive samples. By using this data for the validation data, it is expected that the hyperparameters of the anomaly detection system will be adjusted to correctly judge normal data as normal, which is prone to erroneous detection.

Verification Data Mixed with Synthetic Attack Logs. The data obtained by mixing the pseudo-attack log with the verification data (val_data) consisting of attack data and normal data is val_data_fn. As a result, the total number of events is 21148, 138 of which are attack samples and 22 of which are synthetic attack samples. By using this data as validation data, it is expected that the hyperparameters of the anomaly detection system will be adjusted to correctly judge a synthetic attack sample, which is easily judged as normal, as an attack.

Test Data Mixed with Synthetic Attack Logs. The data obtained by mixing the synthetic attack log with the test data (test_data) consisting of attack data and normal data is test_data_fn. As a result, the total number of events was 23111, of which 2131 were attack samples, and 30 of them were synthetic attack samples. By using this data as test data, we can evaluate whether synthetic attack samples, which are easily judged as normal, can be correctly judged as attacks.

Improved Models. Improved models were created by combining the training data and verification data described in Table 1. The basic model is also described for comparison. The basic model (BM) was built using the original training data (train_data) and validation data (val_data) before adding the synthetic logs. The improved model 1 (IM1) was created using the same original verification data (val_data) as the basic model and training data (train_data_fp) with the false positive logs (fp_train) added. The improved model 2 (IM2) was created using the same original training data (train_data) as the basic model and verification data (val_data_fp) with the false positive logs (fp_val) added. The improved model 3 (IM3) was created using the same original training data (train_data) as the basic model and verification data (val_data_fn) with the synthetic attack logs (fn_val) added. In each model, the hyperparameters were adjusted with a grid search to maximize the accuracy of the verification data. Accuracy comparisons were made using the same test data for all models.

Table 1. Training data and validation data used for basic and improved methods

	BM	IM1	IM2	IM3
train_data	train_data	train_data_fp	train_data	train_data
verification data	val_data	val_data	val_data_fp	val_data_fn

4.5 Test Results

The experimental results of the basic model and the improved models 1 to 3 are described in Table 2. It was confirmed that improved model 1 and improved model 3 had higher accuracy than the basic model.

Table 2. Test results

	BM	IM1	IM2	IM3
test data 1	test_data	test_data	test_data	test_data
AUC	0.787	0.816	0.787	0.832
test data 2	test_data_fn	test_data_fn	test_data_fn	test_data_fn
AUC	0.79	0.818	0.790	0.835

4.6 Discussion

In this test, the effect could not be observed even if the false positive log was included in the verification data (IM2); however, in IM1, which included the false positive log in the training data, and IM3, which included the synthetic attack log in the verification data, an accuracy improvement was over BM was observed. As a result, we were able to show that it is possible to improve the accuracy of the anomaly detection system for unsupervised training by mixing false positive logs and attack logs that are generated synthetically. Although it was not tried, further improvement of accuracy can be expected with a combination of training data and the embedding of synthetic logs in the verification data.

In this study, we generated synthetic attack logs and false-positive logs using the characteristics of true-negative normal logs; however, it is also possible to utilize the tendency of the features of false positive normal logs (erroneous detection) and false negative attack logs (detection omissions). For example, false-negative attack logs are thought to contain many features that cause the anomaly detection system to determine the attack log as normal. Therefore, by including such a feature in the attack log to be modified, the efficiency of modification can be expected. In addition, it is considered that the false positive normal log has many features that the anomaly detection system judges the normal log as abnormal. Therefore, by not including such a feature in the attack log to be modified, efficiency of modification can be expected.

Because we used a very simple anomaly detection system in this test, the accuracy of the anomaly detection system (basic model) to be avoided was low, and the conditions made it easy to improve the accuracy with the improved models. In the future, we will evaluate whether the proposed method is effective for anomaly detection systems that use more complex feature information and more advanced machine learning algorithms such as deep learning.

5 Conclusion

In this study, we focused on the anomaly detection system, assuming an environment where it is difficult to prepare a large number of actual attack logs. We verified whether the accuracy of the anomaly detection system could be improved by embedding synthetically-generated attack logs and false positive logs in the training data and verification data. The evaluation tests showed that the performance of the anomaly detection system can be improved by including the synthetically-generated attack log in the verification data and adjusting the parameters. It was also shown that the performance of the anomaly detection system can be improved by training the model by including the false positive logs generated synthetically in the training data. This method can be used not only for anomaly detection systems for unsupervised training, but also for improving the accuracy of attack detection systems that use supervised training.

References

1. NTT-CERT, Cyber Security. Annual Cybersecurity Report (2021). https://www.rd.ntt/e/sil/overview/NTTannual2021_e_web_look.pdf. Accessed Jan 2022
2. NEC, Commercial facilities as targets: New threats to critical infrastructure. https://www.nec.com/en/global/insights/report/2020022506/index.html. Accessed Jan 2022
3. Cybercrime Magazine, Cybersecurity Jobs Report: 3.5 Million Openings in 2025. https://cybersecurityventures.com/jobs/. Accessed Jan 2022
4. Mitsubishi Electric Corporation, Mitsubishi Electric Develops Cyber Attack Detection Technology. https://www.mitsubishielectric.com/news/2016/0217-f.html. Accessed Jan 2022
5. CSMonitor.com: How Stuxnet cyber weapon targeted Iran nuclear plant. https://www.csmonitor.com/USA/2010/1116/How-Stuxnet-cyber-weapon-targeted-Iran-nuclear-plant. Accessed Jan 2022
6. Xu, W., Qi, Y., Evans, D.: Automatically evading classifiers. In: Network and Distributed System Security Symposium (NDSS), 21–24 February 2016, San Diego, CA. https://www.cs.virginia.edu/~evans/pubs/ndss2016/evademl.pdf. Accessed Jan 2022
7. Yamamoto, T., Nishikawa, H., Kito, K., Kawauchi, K.: Proposal of a method for simulating attacks aimed at evading detection techniques. In: Symposium on Cryptography and Information Security (SCIS 2017) (2017)
8. Jan, S.T.K., et al.: Throwing darts in the dark? Detecting bots with limited data using neural data augmentation. In: Security & Privacy 2020. https://people.cs.vt.edu/vbimal/publications/syntheticdata-sp20.pdf. Accessed Jan 2022
9. Stratosphere: Stratosphere Laboratory Datasets, 13 March 2020 (2015). https://www.stratosphereips.org/datasets-overview. Accessed Jan 2022
10. scikit-learn. https://scikit-learn.org/stable/. Accessed Jan 2022

Application of Hybrid Intelligence for Security Purposes

Marek R. Ogiela[✉] and Lidia Ogiela

AGH University of Science and Technology, 30 Mickiewicza Avenue, 30-059 Kraków, Poland
mogiela@agh.edu.pl

Abstract. In this paper we'll present the main idea and areas of application of hybrid human-artificial intelligence methods in security technologies. Such paradigm support creation of user-centered cryptographic procedures, which can be oriented on specific persons, or areas of application thanks to the involvement of AI approaches. Such protocols will use personal preferences and some user's features analyzed by AI.

1 Introduction

Cryptographic protocols can be oriented on particular users or group of persons. It also can be implemented based on biologically inspired solutions like DNA cryptography, or even use selected AI techniques. There are many papers describing security protocols dedicated for particular users i.e. personalized cryptographic procedures [1, 2], which can use personal characteristics or special AI approaches for selection of unique parameters. Among them we can find solutions, in which AI allow to evaluate behavioral features or cognitive skills [3, 4]. Such application of AI procedures allows to define a new branch of IT security called cognitive cryptography [5, 6].

Presently we can also observe development of new hybrid human-AI solutions in technology. This means that also in security areas it will be possible to define hybrid human-AI procedures, which will be oriented to guarantee high security level, and oriented for selected users or participants of protocol. Such solutions will extend traditional user-oriented security procedures towards more extensive and optimized analysis of security patterns (or features), and creation of security procedures strongly oriented for particular persons. Such extensive analysis may be supported by AI solutions, which allow to select the optimal personal parameters or features for created user-oriented security protocols.

Below will be presented areas of application of such hybrid procedures which can be defined for security purposes.

2 Human – AI Approaches in Security Protocols

When thinking about hybrid security protocols the most important areas of application are following:

© The Author(s), under exclusive license to Springer Nature Switzerland AG 2022
L. Barolli et al. (Eds.): AINA 2022, LNNS 450, pp. 539–542, 2022.
https://doi.org/10.1007/978-3-030-99587-4_45

- Content-based information sharing with privileges
- Thematic-based personalized authentication protocols
- Visual cryptography
- Behavioral security procedures

All of such areas allow to create user-oriented security procedures which involve AI classification algorithms. AI procedures introduce optimization stages in selection of any parameters, or evaluation current features implemented in security protocols. For example, when define a personalized cryptographic protocol we often apply personal features or parameters. Considering different biometric patterns as well as others unique personal features we can create for particular user a quite large feature vector with many personal data. Having such personal record, and with application of AI techniques we can easily select in optimal way only few the most distinctive personal features, which can later be involved in personalized or user-oriented protocol.

Considering above mentioned areas of application we can define most important feature of such solution in following manner:

- **Content dependent information sharing** – in this area of application hybrid intelligence can be applied in such manner that user preferences or human factors have influence on selection of the sharing algorithm and starting parameters i.e. number of shares, generation of personalized parts, privileges etc. AI procedures can determine the way of shadow distribution, and optimal dependencies between parts in case of division of secret information over several layers. Connections between different layers can resemble topology of multi-layer neural networks and have regular or quite random topology.
- **Thematic-based authentication protocols** – such secure authentication protocols can be oriented for particular user or group of persons [7, 8]. In such procedure expertise knowledge and experiences may be considered, especially when secure authentication will be related with thematic visual patterns. In such situation human factors may be related with selection of expertise areas or advancement level connected with particular user. Having selected the thematic areas AI approaches allow to efficiently select or find visual patterns, which can be presented to user during security procedures. It is very important especially due to the fact that AI can generate slightly modified patterns e.g. blurred or noisy, which for humans will be sill easily to recognize but for computer systems not.
- **Visual cryptography** – is one of the most important areas of application of hybrid intelligence in security. Visual cryptography allows to split secret information having the form of visual patterns into several parts. Usually personal recognition abilities connected with perception function decide when particular user is able to recognize original secret data. In such protocols is possible to determine an individual perception threshold evaluated independently for different users. AI approaches allow to evaluate such personal perception levels, which also may be dependent on some user's knowledge and expectations. In such procedures knowledge and expectations create human factors, but perception abilities can be defined by artificial intelligence association procedures.

- **Behavioral security procedures** – behavioral parameters allow to define a special type of security protocols, in which different movement feature can be considered [4, 9]. It may contain both simple procedures oriented for personal key generation with included personal features, till more complex protocols oriented for creation behavioral locks. Here we can consider different types of human body movements starting from hand gesture or finger movements, till more complex human motion patterns registered while walking or doing exercises [1].

In all mentioned application areas, the methodology of using hybrid intelligence looks in the same manner. When creating human-AI security protocol, firstly the set of personal features should be defined or personal parameters record should be evaluated. Having well defined personal features the optimal selection of the most important, and most unique features can be done, especially with application of artificial or computational intelligence methods. The selection of optimal features is complex task especially in situation when a very large personal feature vector is available. AI approaches allow in this case to quickly find optimal feature set. Which represent particular user in optimal manner. Additional advantage of application of AI techniques is possibilities to consider changing parameters, which may occur for selected users during the time or in short time period.

3 Conclusions

In this paper we described the application of hybrid intelligence techniques in security protocols. In particular the way of application of personal features in advance cryptographic solution were described. Additionally, selection of the most important personal parameters can be performed with application of AI algorithms. The most important features of such hybrid intelligence-based security methods are efficiency, and personalization towards application by particular users. Such approaches also allow to consider different personal characteristics, and changing parameters associated with users. Such hybrid intelligence methods will enrich the cognitive cryptographic approaches defined to join security methods with semantic features or personal parameters [10].

Acknowledgments. This work has been supported by the AGH University of Science and Technology research Grant No. 16.16.120.773.

References

1. Ogiela, L., Ogiela, M.R.: Cognitive security paradigm for cloud computing applications. Concurr. Comput. Pract. Exp. **32**(8), e5316 (2020). https://doi.org/10.1002/cpe.5316
2. Ancheta, R.A., Reyes Jr., F.C., Caliwag, J.A., Castillo, R.E.: FEDSecurity: implementation of computer vision thru face and eye detection. Int. J. Mach. Learn. Comput. **8**, 619–624 (2018)
3. Ogiela, L.: Transformative computing in advanced data analysis processes in the cloud. Inf. Process. Manag. **57**(5), 102260 (2020)

4. Ogiela, M.R., Ogiela, L., Ogiela, U.: Biometric methods for advanced strategic data sharing protocols. In: 2015 9th International Conference on Innovative Mobile and Internet Services in Ubiquitous Computing IMIS 2015, pp. 179–183 (2015). https://doi.org/10.1109/IMIS.2015.29

5. Ogiela, U., Ogiela, L.: Linguistic techniques for cryptographic data sharing algorithms. Concurr. Comput. Pract. Exp. **30**(3), e4275 (2018). https://doi.org/10.1002/cpe.4275

6. Ogiela, M.R., Ogiela, U.: Secure information splitting using grammar schemes. In: Nguyen, N.T., Katarzyniak, R.P., Janiak, A. (eds.) New Challenges in Computational Collective Intelligence. SCI, vol. 244, pp. 327–336. Springer, Heidelberg (2009). https://doi.org/10.1007/978-3-642-03958-4_28

7. Ogiela, L., Ogiela, M.R., Ogiela, U.: Efficiency of strategic data sharing and management protocols. In: The 10th International Conference on Innovative Mobile and Internet Services in Ubiquitous Computing (IMIS-2016), 6–8 July 2016, Fukuoka, Japan, pp. 198–201 (2016). https://doi.org/10.1109/IMIS.2016.119

8. Guan, C., Mou, J., Jiang, Z.: Artificial intelligence innovation in education: a twenty-year data-driven historical analysis. Int. J. Innov. Stud. **4**(4), 134–147 (2020)

9. Menezes, A., van Oorschot, P., Vanstone, S.: Handbook of Applied Cryptography. CRC Press, Waterloo (2001)

10. Yang, S.J.H., Ogata, H., Matsui, T., Chen, N.-S.: Human-centered artificial intelligence in education: seeing the invisible through the visible. Comput. Educ. Artif. Intell. **2**, 100008 (2021)

Semantic-Based Techniques for Efficient and Secure Data Management

Urszula Ogiela[1], Makoto Takizawa[2], and Lidia Ogiela[1]([✉])

[1] AGH University of Science and Technology, 30 Mickiewicza Ave., 30-059 Kraków, Poland
{ogiela,logiela}@agh.edu.pl
[2] Department of Advanced Sciences, Hosei University, 3-7-2, Kajino-cho, Koganei-shi, Tokyo 184-8584, Japan
makoto.takizawa@computer.org

Abstract. In this paper, new approaches for secure data management will be described. Presented methods will be a semantic-based procedures, which for data handling use a semantic content and meaning. Such methods are designed for efficient data protection in cloud or distributed systems. Application of such procedures allow to significantly increase the security strength of existing solutions.

1 Introduction

In modern security application we can find solutions in which semantic content of secured data can be considered. Such protocols were proposed in cognitive cryptography area proposed by authors in [1–5]. In such procedures the semantic content or sematic meaning of information can be involved into the security protocol, what finally result that encrypted data is dependent on its semantic meaning. This is a very important extension of traditional security procedures, which usually don't make any connection between semantic meaning and final encryption results.

Similar connections between semantic content and procedures we can introduce in data management techniques. Such novel approach will be described in following sections in which we'll define semantic-based secure data management techniques. The main idea of such procedures is to create a new class of strategic data management procedures oriented for information division and distribution over hierarchical management structures. Information division and distribution will be strongly dependent on the content of shared data [6–8]. So, the core stage of such protocols will be connected with semantic content evaluation, which should result with obtaining data feature vector. Feature factors from this vector will be next used for information division task.

2 Semantic-Based Data Management Protocols

To create a semantic-based secure management procedures it is necessary to define two types of protocols. The first one is procedure which allow to describe a semantic content of encrypted information. The second one is efficient data sharing protocol,

L. Barolli et al. (Eds.): AINA 2022, LNNS 450, pp. 543–546, 2022.
https://doi.org/10.1007/978-3-030-99587-4_46

which allow to divide secret data into the particular number of secret parts, which can be distributed among participants of management procedures. Of course, distribution of secret parts should be dependent on the content, and implemented with using semantic factors evaluated at the beginning of procedures.

For evaluation of semantic content, we can consider cognitive information systems proposed in [1, 9]. It was defined several different classes of such systems oriented for evaluation of different data starting from images till financial or classified data. Cognitive information systems allow to extract the semantic information from analyzed datasets, and extracts some important meaning which is present in such data, but requires very extensive analysis with application of AI, Deep Learning or cognitive resonance approaches. As a result of such cognitive analysis it is possible to receive an information record, which contain some semantic description of analyzed data. Such semantic record can have a large number of information describing different i.e. global or local features connected with analyzed data. Depending on the goal of information sharing and management, we can select from this extensive feature record some most important parameters which next cab be used to perform data division and distribution in efficient and secure manner.

When we selected the main semantic features, we can use them in division and management protocol. To perform such task, it is necessary to select one of data division algorithm [1–4, 6–8], and applied it for particular hierarchical management structure. To do it is necessary to determine the number of levels or layers in hierarchical pyramid, as well as the number of persons at each hierarchy level. Having such parameters and semantic descriptors of divided information it is possible to launch division procedures with the following input parameters:

- semantic descriptors of shared information,
- numbers of layers and participants,
- secret information which should be divided,
- starting parameters for secret sharing threshold procedures.

After performing secret data division protocols, we can obtain required number of secret parts which can distributed in particular levels in hierarchical structure, in different manner depending on the number of persons and accessing grants.

It is possible to consider a specially defined distribution topology of the obtained secret parts, which can be placed in irregular manner over different levels in hierarchical structure.

3 Application of Semantic-Based Management Protocols

Presented semantic-based management procedures have many application possibilities. Such techniques enrich traditional management procedures towards considering semantic content, and depending on selected features the same information can be divided and distributed in different manner. The possibility of selection of semantic parameters introduce additional security level because the whole protocol allows to restore original content only in the situation when input parameters are known. So, the knowledge only about the procedure will not be enough to perform unauthorized data restoration from generated parts.

Such security feature allows to apply this type of protocols in many management or security areas. In particular it may be applicable in secure and trusted data management in distributed systems like cloud environment [9–12]. It may be also applicable in distant services management, as well as strategic data storage and distribution. Performing analytics tasks with application of semantic feature on the analyzed data, makes such protocols also applicable in predictive analysis towards forecasting and prognosis of user trends or behaviors [10, 12].

4 Conclusions

In this paper an idea of creation and application of semantic-based protocols was presented. Such techniques can be applied in a broad range of management task, especially connected with secret data division in hierarchical and distributed structures. The main idea of such protocols lays in evaluation of semantic meaning of encrypted data, and application of such information in security procedures. Extraction of semantic meaning can be performed with application of cognitive information systems, and such features can decide about the way of information encoding and distribution. Such techniques can be widely applied in cloud computing and distributed services management, as well as secure information distribution in hierarchical structures. Such techniques enrich traditional management approaches and have influence in creation of new security protocols in cognitive cryptography [1, 2, 4].

Acknowledgments. This work has been supported by the National Science Centre, Poland, under project number DEC-2016/23/B/HS4/00616.

References

1. Ogiela, L., Ogiela, M.R.: Cognitive security paradigm for cloud computing applications. Concurr. Comput. Pract. Exp. **32**(8), e5316 (2020). https://doi.org/10.1002/cpe.5316
2. Ogiela, M.R., Ogiela, L., Ogiela, U.: Biometric methods for advanced strategic data sharing protocols. In: 2015 9th International Conference on Innovative Mobile and Internet Services in Ubiquitous Computing IMIS 2015, pp. 179–183 (2015). https://doi.org/10.1109/IMIS.2015.29
3. Ogiela, M.R., Ogiela, U.: Secure information splitting using grammar schemes. In: Nguyen, N.T., Katarzyniak, R.P., Janiak, A. (eds.) New Challenges in Computational Collective Intelligence. SCI, vol. 244, pp. 327–336. Springer, Heidelberg (2009). https://doi.org/10.1007/978-3-642-03958-4_28
4. Ogiela, U., Ogiela, L.: Linguistic techniques for cryptographic data sharing algorithms. Concurr. Comput. Pract. Exp. **30**(3), e4275 (2018). https://doi.org/10.1002/cpe.4275
5. Nakamura, S., Ogiela, L., Enokido, T., Takizawa, M.: Flexible synchronization protocol to prevent illegal information flow in peer-to-peer publish/subscribe systems. In: Barolli, L., Terzo, O. (eds.) CISIS 2017. AISC, vol. 611, pp. 82–93. Springer, Cham (2018). https://doi.org/10.1007/978-3-319-61566-0_8

6. Menezes, A., van Oorschot, P., Vanstone, S.: Handbook of Applied Cryptography. CRC Press, Waterloo (2001)
7. Ogiela, L., Ogiela, M.R., Ogiela, U.: Efficiency of strategic data sharing and management protocols. In: The 10th International Conference on Innovative Mobile and Internet Services in Ubiquitous Computing (IMIS-2016), 6–8 July 2016, Fukuoka, Japan, pp. 198–201 (2016). https://doi.org/10.1109/IMIS.2016.119
8. Yan, S.Y.: Computational Number Theory and Modern Cryptography. Wiley, Hoboken (2013)
9. Ogiela, L.: Transformative computing in advanced data analysis processes in the cloud. Inf. Process. Manag. **57**(5), 102260 (2020)
10. Ancheta, R.A., Reyes Jr., F.C., Caliwag, J.A., Castillo, R.E.: FEDSecurity: implementation of computer vision thru face and eye detection. Int. J. Mach. Learn. Comput. **8**, 619–624 (2018)
11. Gil, S., et al.: Transformative effects of IoT, blockchain and artificial intelligence on cloud computing: evolution, vision, trends and open challenges. Internet Things **8**, 100118 (2019)
12. Yang, S.J.H., Ogata, H., Matsui, T., Chen, N.-S.: Human-centered artificial intelligence in education: seeing the invisible through the visible. Comput. Educ. Artif. Intell. **2**, 100008 (2021)

CoWrap: An Approach of Feature Selection for Network Anomaly Detection

Anonnya Ghosh[1]([✉]), Hussain Mohammed Ibrahim[2], Wasif Mohammad[2], Farhana Chowdhury Nova[2], Amit Hasan[2], and Raqeebir Rab[2]

[1] Ahsanullah University of Science and Technology (AUST), 141 & 142, Love Road, Tejgaon Industrial Area, Dhaka 1208, Bangladesh
114anonnya@gmail.com
[2] AUST, Dhaka 1208, Bangladesh
raqeebir.cse@aust.edu

Abstract. Feature Selection (FS) is a crucial technique that picks out the most significant features from an augmented and ambiguous feature set to increase the classification accuracy of a model. As per the tremendous growth of the intrusion detection systems (IDS) research field over the past decade, now network anomaly detection can also utilize feature selection techniques to enhance performance. Several solutions for selecting the best features have been proposed, but further investigation is required to strengthen efficiency. We present a hybrid feature selection method combining the filter and the wrapper approaches for anomaly-based intrusion detection systems in this study. The proposed model offers the minimal subset of features for the highest detection accuracy using both feature to feature and feature to class correlations. Evaluated on the DDOS attack of the CICIDS2018 [1] dataset, the experiment reduced the number of features from 79 to 11, which resulted in the classification accuracy of **99.82%**.

1 Introduction

According to the Check Point Research (CPR)'s Cyber Security Annual Report published in 2020 [5], after being housed on the cloud service MEGA, more than 770 million email addresses and 21 million unique passwords were published in a famous hacker forum. 16 compromised websites yielded 620 million account details, which were then offered on the prominent dark web marketplace Dream Market. In addition, Airbus, the world's second-largest commercial airplane maker, had a data breach, revealing the personal information of some of its employees. Unauthorized attackers gained access to the computer systems of Airbus's "Commercial Aircraft Business". All these cyber events took place only in the first two months of the year 2019. If there's one thing that 2019 has shown us, it's that no organization, no matter how big or little, is immune to a devastating cyber assault.

© The Author(s), under exclusive license to Springer Nature Switzerland AG 2022
L. Barolli et al. (Eds.): AINA 2022, LNNS 450, pp. 547–559, 2022.
https://doi.org/10.1007/978-3-030-99587-4_47

In cyber-security, it is essential to identify outliers and determine whether the outlier is a safety threat. An anomalous network flow contains outliers, defined as behavior that is uncommon or does not align with regular traffic patterns for a specific user, business, or entity. Anomalous traffic could be malicious or not but is significant because of possible future business effects. The detection of anomalies plays a key role in detecting fraud, intrusion into the network and other rarities which might be important but difficult to find. Machine learning and data mining techniques are used to collect, analyze, and uncover patterns in network flow, construct models to adapt the data, and categorize each flow in order to detect anomalous network traffic more rapidly and precisely. These anomalous flows, on the other hand, have both a large quantity and a large dimension. This may cause problems such as low accuracy levels or, unrealistically high accuracy levels, high computational cost, long training time and over-fitting.

The features that are most important should be chosen to increase detection performance. Thus, one of the vital factors that impact the quality of anomaly detection is Feature Selection. The goal of feature selection is to remove unnecessary and redundant features from a data set in order to improve a classification algorithm's prediction capacity. Feature selection techniques improve classification accuracy and speed up the learning process by choosing the best subset of features. These approaches are divided into three categories: Filter Method, Wrapper Method, and Embedded Method. The statistical learning data is used as a metric in the filter technique to evaluate the qualities independently of the classification algorithm. It only chooses features that are relevant to the targets. The wrapper method selects the optimal features using classification algorithms. To measure the usefulness of feature subsets, it employs accuracy (or area under curve, i.e. AUC) as feedback. Embedded approaches allow the classifier to train and choose features at the same time, and the classifier may choose which features to utilize. In general, the wrapper method takes a longer time to run and has a high computational cost and is not computationally feasible with the increased number of features. Embedded methods are dependent on classifiers and may be problematic identifying a small set of features. The filter method on the other hand is independent of the classifier and offers better computational complexity than the wrapper method. It is a quick and flexible method and ignores feature dependencies.

Although the filter method is efficient and has a lower computational cost than other methods, it still has a major drawback. This method first selects the subset of features, then on the selected subset classification is applied. Thus, the resulting subset turns out to be less reliable with the class label. Hence, we propose a hybrid approach combining both filter and wrapper methods that ensures the reliability of the selected features.

In this research, we are using CICIDS2018 [1] which is the most recent IDS dataset, and it replicates a real-world scenario. We specifically worked on the sub dataset of the original data containing two types of DDOS Attacks. The two Attacks are the DDOS attack-HOIC and the DDOS attack-LOIC-UDP. Our proposed methodology mainly consists of Data Pre-processing, calculation of Polynomial Based Correlations, calculation of the Symmetrical τ (tau) and Feature Selection with the help of Artificial Neural Network (ANN). The Data Pre-processing resulted in a consistent and compact dataset which is later used for the whole proposed system. The Polynomial Based Correlation and Symmetrical τ (tau) both approaches give out the correlation between two attributes. Here, The Polynomial Based Correlation is applied to get the correlation between one feature with another by creating feature pairs, while symmetrical τ (tau) is used to find a correlation between the generated feature pairs and class labels. The value of Symmetrical τ (tau) for each feature pair determines the relevance of the features with the class label. It ranks the features pairs to reduce the search space. Then, with the help of Artificial Neural Network (ANN), this order is used to find the best feature set.

2 Related Work

Numerous studies have been conducted on feature selection for network anomaly detection. These studies introduced different approaches for the selection of features. Florian Gottwalt et al. [3] utilizes the Mahalanobis distance to generate feature correlations and save them in a matrix. This algorithm results with an accuracy of **98.65%** even on a large number of samples, where the results of the PCA was drastically decreasing. In another study, Aboul Ella Hassanien et al. [12] proposed a method for picking the best feature subset. As a feature goodness metric, they used correlation analysis between two variables. The experiment reveals that the number of features is reduced from 41 to 17, resulting in an increase in classification accuracy to **99.1%**. Jordan Lam et al. [10] presented as a method utilising CNN to gather and analyse normal and abnormal network flows from a simulated environment. The model correctly identified benign traffic at a rate of **100.0%** and aberrant traffic at a rate of **96.4%**. Another method has been proposed by Zhou XJ et al. [4]. This paper uses the symmetrical τ (tau) statistical heuristic criterion and then discusses its consistency using a Bayesian classifier and its built-in statistical test. τ (tau) is defined as the probability of prediction error reduced by knowing the individual's categorization on the second variable compared to the probability of mistake in the absence of such information. V. Kanimozhi et al. [7] have suggested a technique to identify botnet assaults that uses the latest cyber dataset CSE-CICIDS2018. Calibration curve, a classic analytical method which generates reliability diagrams to determine whether the predicted probability of various classifiers are well-calibrated or not. The maximum accuracy was obtained from the Artificial neural network (ANN), which scored **99%**. For the same dataset, Ilhan Firat Kilincer et al. [11] have achieved the best result of **99.92%** accuracy using the DT classifier. Mohamed Amine Ferrag et al. [8] conducted a survey on deep learning algorithms for cyber security intrusion detection. Among other deep learning models they revealed that a deep neural network is the most error free and precise. For seven attack types, the recurrent neural network (RNN) has the best detection rate.

3   Data Preprocessing

The University of New Brunswick originally generated the CICIDS2018 [1] for the purpose of evaluating anomalous network flow. (Table 1) Gives an overview of the dataset. This dataset is accompanied with imbalanced data (Fig. 1). As a solution to this problem SMOTE [13], or Synthetic Minority Oversampling Technique is employed in this study. In this process we synthesize examples from the minority classes which helps to balance the class distribution (Fig. 1).

Figure 2 Shows the preprocessing procedure of the dataset. The procedure firstly contains cleansing the dataset and then handling null, infinite and missing values. Then we encoded categorical data using label encoder. We applied feature scaling to standardize the independent features present in the dataset into a fixed range. In addition, constant and duplicate features were removed. We also divided the dataset into four categories depending on the characteristics of the features. These categories are:

- **Byte-Based**: Features depending on the bytes (Bwd Pkts, PSH FlagCnt etc.).
- **Packet-Based**: Features based on statistics from packets (Bwd PktLen Min, Fwd SegSize Min etc.).
- **Time-Based**: Time-dependent features (Flow byte/s, FLow Pkt/s etc.).
- **Behavior-Based**: Features reflecting distinct flow behavior (Duration, Flow IAT etc.).

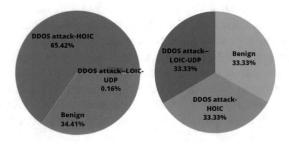

Fig. 1. Attacks found in balanced and imbalanced dataset.

Table 1. Overview of sample features in available Dataset

	Feature	Description
1	Bw Pkt l min	Minimum size of packet in backward direction
2	ACK flag cnt	Number of packets with ACK
3	Tot len packet	Total size of packet
4	Down/up ratio	Download and upload ratio
5	Flow byte/s	Flow byte rate that is number of packets transferred per second

The dataset is categorized to find out the importance of a category in detecting network anomalies. Because using the features of only one or two categories can reduce the training time complexity to a large extent.

4 Proposed Method

The suggested correlation-based feature selection process shown in Fig. 3, involves three phases: generating feature pairs, finding feature association with the class label and selection of best feature subset.

4.1 PHASE 1: Polynomial Based Correlation Between Features

To generate the correlations between any two features, a polynomial-based correlation technique [2] is applied. This correlation is measured using a basic addition function. The addition of the instances of any two features generates a feature pair (fc_p) denoting the feature correlation. For a total set of m features, the process iterates over every instance of a feature F_i,

$$F_i = \{f_i^1, f_i^2, ..., f_i^n\}, \tag{1}$$

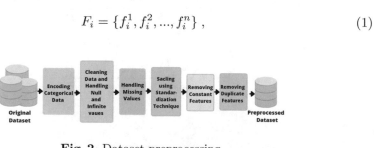

Fig. 2. Dataset preprocessing.

Fig. 3. Proposed methodology for feature selection.

where F_i are the features (m dimensions), $f_i^{\ n}$ are the instances (n dimensions) of each feature and $(1 \leq i \leq m)$. The correlation between any two instances of features F_j and F_k can be denoted as,

$$c_{j,k}^n = (f_j^n, f_k^n) = (f_j^n + f_k^n).$$
(2)

Each feature F_i is individually added to the features from F_{i+1} to F_m to generate every possible combination of feature correlations. Therefore each column of Fig. 3 gives the desired Feature Correlation Pairs (fc_p).

$$\begin{bmatrix} c_{1,2}^1 & c_{1,3}^1 & \cdots & c_{1,m}^1 & c_{2,3}^1 & \cdots & c_{m-1,m}^1 \\ c_{1,2}^2 & c_{1,3}^2 & \cdots & c_{1,m}^1 & c_{2,3}^1 & \cdots & c_{m-1,m}^2 \\ \cdot & \cdot & \cdot & \cdot & \cdot & \cdot & \cdot \\ c_{1,2}^n & c_{1,3}^n & \cdots & c_{1,m}^1 & c_{2,3}^1 & \cdots & c_{m-1,m}^n \end{bmatrix}$$
(3)

Then the feature correlation pairs (fc_p) can be denoted as,

$$fc_p = (fc_1, fc_2, fc_3, ..., fc_p).$$

The feature pairs (fc_p) result with a set of continuous values which then goes through a binning technique to convert these continuous values into discrete categorical values.

4.2   PHASE 2: Correlation Between Feature Pairs and Class Label

The significance of a feature pair in predicting the class label is determined by utilizing the suggested symmetrical τ (tau) [4] correlation measure. The symmetrical τ is defined as,

$$\tau = \frac{a - b}{2 - \sum_{i=1}^{I} P_i^2 - \sum_{l=1}^{L} P_l^2}$$
(4)

$$a = \sum_{l=1}^{L} \sum_{i=1}^{I} \frac{P_{il}^2}{P_l} + \sum_{i=1}^{I} \sum_{l=1}^{L} \frac{P_{il}^2}{P_i}$$
(5)

$$b = \sum_{l=1}^{L} P_i^2 - \sum_{i=1}^{I} P_l^2$$
(6)

where I stands for the bins of a feature correlation pair (fc_p) and L stands for the class labels. P_i represents the marginal probability of (fc_p), whereas P_l represents the marginal probability of l. Moreover, P_{il} is the probability of an instance that falls into bin I and L, where $i = 1, 2, ..., I$ and $l = 1, 2, ..., L$.

The proposed Algorithm (1) describes every step of symmetrical τ (tau) calculation. This calculation is done individually for both the non-categorized and categorized feature pairs. The algorithm begins by taking the feature set

---

Algorithm 1: Finding Feature Pair-Class label Correlation

Input : Feature Set, F

Output: Feature pair-Class label Correlation, τ (tau)

1 **for** For all features F_i **do**
2 | $l \leftarrow i + 1$; //Initialize l with the index of next feature
3 | **for** For all features F_l **do**
4 | |__ $c_{i,l} \leftarrow F_i + F_l$; //Correlations between features F_i and F_l

5 **for** For all $c_{i,l}$ **do**
6 | $I \leftarrow$ number of categories in $c_{i,l}$;
7 | $L \leftarrow$ number of classes of label;
8 | **for** all I **do**
9 | | **for** all L **do**
10 | | |__ Calculate joint probability $P_{i,l}$;
11 | |__ Calculate marginal probability P_i;
12 | **for** all L **do**
13 | |__ Calculate marginal probability P_l;
14 |__ Calculate τ, $T_{i,l}$ of $c_{i,l}$;

(F) as input. All the instances of the feature F_i is added to all the instances of the next feature F_l. These addition values for each feature F_i are preserved in $c_{i,l}$ for the next step. Depending on the bins of $c_{i,l}$ which is I and class label L, for each bin the Joint probability and Marginal probability of $c_{i,l}$ is calculated. Then the marginal probability for the class label L is calculated and all these probability values are stored to calculate the symmetrical τ (tau) in the last step.

Each of the feature pairs generates an individual value of τ (tau) representing their relevance with the class label. These τ (tau) values varied in the range 0.056 to 0.684, where 0.68 is the highest value for the correlation between a feature pair fc_p and the class label la and 0.056 represents the weakest correlation.

4.3 PHASE 3: Selection of Features

The feature selection process of our proposed method is based on the values of the symmetrical τ (tau) for each feature pair and the performance result of a feature subset. Feature subsets giving the highest performance are selected. The symmetrical τ (tau) ranks the feature pairs with the help of its values to reduce the search space using a filter approach. Then this order is used to select the best feature subset with the help of ANN. This part is the wrapper approach. We propose the Algorithm (2) for selecting the best features from all the feature pairs.

The algorithm is initiated by taking the feature pairs and corresponding τ (tau) values from the 1st Algorithm (1) as input. In step 1, the feature pairs are sorted in descending order based on the values of τ (tau). Then, k is initialized

Algorithm 2: Feature selection

Input : Feature pair-Class label Correlation, τ (tau)
Output: Final Feature Set

1 Sort all the values of τ, $T_{i,l}$;
2 $k \leftarrow 1$; //Initialize k with the top most feature pair
3 **while** accuracy not decreasing **do**
4    $F_p \leftarrow$ top k feature pairs;
5    $Topfeatures \leftarrow$ Extract unique features from the feature pairs (Fp);
6    $Finalfeatures \leftarrow$ all features;
7   **for** each features F_i in $Finalfeatures$ **do**
8   **if** F_i not in $Finalfeatures$ **then**
9   **if** F_i not in $Topfeatures$ **then**
10   drop F_i from $Finalfeatures$;

11   Calculate accuracy using ANN using $Finalfeatures$;
12   $k \leftarrow k + 1$; //Increase the number of selected feature pairs

---

as 1, which means the top most feature pair from the sorted pairs is selected as the starting pair. In step 4, top k feature pairs are inserted in Fp then from these pairs the unique features were extracted and preserved in $Topfeatures$ in step 5. In step 6, all the features were assigned in $Finalfeatures$ and in step 7–10 all the features were dropped from the $Finalfeatures$ expect for the features stored in $Topfeatures$. Then in step 11, the $Finalfeatures$ are fed to ANN to generate a new accuracy and then the value of k is increased in step 12. The while loop continues from step 4 if the accuracy keeps increasing.

From the generated feature pairs firstly we took the top most pair and extracted the features and then evaluated using ANN. After that we kept adding more feature pairs to the previously selected pairs and continued evaluation process until the accuracy of the ANN deteriorated. On completion of this procedure we achieved the highest accuracy using the top 18 feature pairs. Inclusion of more than 18 feature pairs resulted with gradual decrease in accuracy (Table 2).

From these 18 feature pairs, removing the common features in multiple pairs only 11 features (Table 3) were extracted and selected as the best feature subset. These 11 features also have the highest values of τ (tau) and are the most correlated features.

Table 2. Selected top 18 feature pairs

Feature pair index	330, 726, 1641, 1683, 1410, 1670, 1656, 625, 1436, 100, 103, 70, 89, 90, 102, 101, 99, 62
τ (tau)	0.684, 0.684, 0.684, 0.684, 0.684, 0.684, 0.684, 0.684, 0.684, 0.684, 0.684, 0.684, 0.684, 0.684, 0.684, 0.684, 0.684, 0.631

Table 3. Selected 11 features using non-categorized τ (Tau)

Feature index	1, 5, 13, 32, 33, 42, 43, 44, 45, 46, 51
Feature names	Protocol, Tot Bwd Pkts, Bwd Pkt Len Min, Fwd PSHFlags, Bwd PSH Flags, RST FlagCnt, PSH FlagCnt, ACK FlagCnt, URG FlagCnt, Down/Up Ratio, Fwd SegSize Min

5   Experimentation and Results

5.1   Evaluation Criteria

Classification models have discrete output, hence a specific metric is needed to compare the discrete classes. The comparison criterion to evaluate the proposed mechanism are the classification: (1) Accuracy, (2) Precision, (3) Recall and (4) F1-Score. These criterion are calculated based on the confusion matrix shown in Table 4 and evaluates a model's performance to decide how good or bad the classification is. True negatives (TN) and True positives (TP) rates are accurate predictions of regular and anomalous behaviors. False positives (FP) rate points to regular behaviors being predicted as anomalous ones, whereas False negatives (FN) are anomalous behaviors that are predicted as regular.

Table 4. Confusion Matrix

	Predicted class	
Actual class	True positives (TP)	False negatives (FN)
	False positives (FP)	True negatives (TN)

$$Accuracy = \frac{TP + TN}{TP + TN + FP + FN} \tag{7}$$

$$Precision = \frac{TP}{TP + FP} \tag{8}$$

$$F1-Score = \frac{2 * Recall * Precision}{Recall + Precision} \tag{9}$$

$$Recall = \frac{TP}{TP + FN} \tag{10}$$

5.2   Result Analysis

In our analysis, our main task was to train the model using the least number of features. We tried selecting a different number of features based on the values of symmetrical τ (tau) and the performance result using Artificial Neural Network (ANN). From All the pairs we got values of τ (tau) ranging from 0 to 0.7.

Using the 11 most important features (Table 3) which were selected we get the best performance result (Table 5) with an accuracy of 99.82% and a minimum loss of 0.0109. Figure 4 Shows that, the accuracy decreases as the number of selected features increases. Performing classification on the feature sets which lies in different tau ranges, we can see that the features retrieved from the feature pairs having higher τ (tau) values result in maximum accuracy and the lowest range results in the lowest accuracy (Fig. 4). So after comparing all the results we choose the top 11 features as the best feature subset. For performance analysis, we again applied the whole feature selection technique in each of the four distinct categories. From each of the category a set of features were retrieved to determine the importance of these categories. Byte Based category yields the highest accuracy of 99.99% (Fig. 5). The second best category is the Packet-based category with an accuracy of 99.86%. If we compare the previously selected top 11 features, we can see that almost all of the features are from these two categories and 9 out of 11 (Table 6) features are Byte-based and Packet-based features. Hence Byte Based and Packet-based features are the most relevant features for predicting the class label and results with the highest accuracy in our proposed system.

Table 5. Result of classification for selected 11 features.

	Precision	Recall	F1-score	Accuracy
ANN	99.82%	99.82%	99.82%	99.822%

Table 6. Common features from the selected 11 features.

Byte based:	Tot Bwd Pkts, Fwd PSH Flags, Bwd PSH Flags, RST Flag Cnt, PSH Flag Cnt, ACK Flag Cnt, URG Flag Cnt
Packet based:	Bwd PktLen Min, Fwd SegSize Min

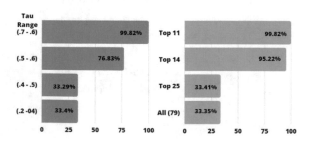

Fig. 4. Accuracy comparison of feature sets on CICIDS 2018 using ANN.

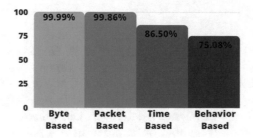

Fig. 5. Category based accuracy comparison.

6 A Comparative Analysis

A lot of works have been done on the dataset CICIDS2018 [1]. This study's findings are compared to those of already available systems. The list of results achieved in various studies is given in Table 7, which is a comparison of existing systems to the suggested technique. The performance of the systems is compared in terms of accuracy, precision, recall and F1-score. In most of the existing approaches, features were not reduced and the model was trained with all the features. We were able to obtain a similar type of result with a reduced number of features. ANN is used for model training and testing in [7] and they obtain an accuracy of 99.97%, while in [6] they first converted the numerical values of the features to images and applied CNN. This study was able to achieve 99.98% accuracy. In [8] 7 Deep Learning models were applied on the dataset among them, RNN performed the best and gave an accuracy of 97.38% and in [9] many ML classifiers were used, but Decision Tree provided the best result giving an accuracy of 99.92%.

Table 7. A comparative analysis of proposed system.

Reference	Model	Accuracy	Precision	Recall	F1-score
[6]	CNN	99.98%	–	–	–
[7]	ANN	99.97	99.96	1	99.98
[8]	RNN	97.38	–	–	–
[9]	DT	99.92	99.94	99.94	99.94
Proposed System	ANN	99.822	99.82	99.82	99.82

It needs to be remembered that we have achieved this accuracy using only 11 features compared to others who used all the 79 features. As a result, it will highly reduce the time complexity of the model training.

7 Conclusion

Cyber-attacks are becoming more common as IT infrastructures and applications become more complicated. To combat this threat, substantial cyber security research has been conducted, particularly in the area of network anomaly detection. Selecting the most significant features from a large dataset is one of the most difficult tasks in anomaly detection. Classification can be done more efficiently with a reduced number of features. In this paper, we showed how to use the Symmetrical τ (tau) measure in a novel approach to help with the feature selection problem. Using the hybrid method we were able to select the 11 most important features from the CSE-CICIDS2018 [1] dataset and achieved an accuracy of 99.82% with a minimum loss of 0.0109. We also discovered that the byte-based and packet-based features are the most important features in detecting anomalies since we were able to get a promising outcome utilizing only these features. Comparable research on this dataset yielded similar results, but with a larger number of features. The higher the number of features, the higher is the time complexity. So, our proposed strategy can aid in the discovery of the best features and in the detection of network anomalies.

References

1. CSE-CIC-IDS2018. A collaborative project between the Communications Security Establishment (CSE) & the Canadian Institute for Cybersecurity (CIC) (2018). https://www.unb.ca/cic/datasets/ids-2018.html?fbclid=IwAR18Ngt-P9p_ndGbJMMCziqGZ1X1WrViQ0VAxQvxtn864pP9pM2HbOmyisA
2. Li, Q., Tan, Z., Jamdagni, A., Nanda, P., He, X., Han, W.: An intrusion detection system based on polynomial feature correlation analysis. In: Proceedings of the 2017 IEEE on Trustcom/BigDataSE/ICESS, pp. 978–983. IEEE (2017)
3. Dillon, T., Gottwalt, F., Chang, E.: CorrCorr: a feature selection method for multivariate correlation network anomaly detection techniques. University of NewSouth Wales, Canberra, Australia. La Trobe University, Melbourne, Australia (2019)
4. Zhou, X.J., Dillion, T.S.: A statistical-heuristic feature selection criterion for decision tree induction. IEEE Trans. Pattern Anal. Mach. Intell. **13**(8), 834–41 (1991)
5. Check Point Research: Cyber Security Report. Check Point Software Technologies Ltd. (2020)
6. Kim, J., Shin, Y., Choi, E.: An intrusion detection model based on a convolutional neural network. J. Multimed. Inf. Syst. **6**, 165–172 (2019). https://doi.org/10.33851/jmis.2019.6.4.165
7. Kanimozhi, V., Jacob, D.T.P.: Calibration of various optimized machine learning classifiers in network intrusion detection system on the realistic cyber dataset CSE-CIC-IDS 2018 using cloud computing, Int. J. Eng. Appl. Sci. Technol. **4**, 209–213 (2019). https://doi.org/10.33564/ijeast.2019.v04i06.036
8. Ferrag, M.A., Maglaras, L., Moschoyiannis, S., Janicke, H.: Deep learning for cyber security intrusion detection: approaches, datasets, and comparative study. J. Inf. Secur. Appl. **50**, 102419 (2020). https://www.sciencedirect.com/science/article/abs/pii/S2214212619305046

9. Dod, J.: Effective substances. In: The Dictionary of Substances and their Effects. Royal Society of Chemistry. Available via DIALOG (1999). http://www.rsc.org/dose/titleofsubordinatedocument. Cited 15 Jan 1999
10. Abbas, R., Lam, J.: Machine Learning based Anomaly Detection for 5G Networks. Macquarie University (2020)
11. Kilincer, I.F., Ertam, F., Sengur, A.: Machine learning methods for cyber security intrusion detection: datasets and comparative study. Comput. Netw. **188**, 107840 (2021)
12. Eid, H.F., Hassanien, A.E., Kim, T., Banerjee, S.: Linear correlation-based feature selection for network intrusion detection model. In: Awad, A.I., Hassanien, A.E., Baba, K. (eds.) SecNet 2013. CCIS, vol. 381, pp. 240–248. Springer, Heidelberg (2013). https://doi.org/10.1007/978-3-642-40597-6_21
13. Karatas, G., (Member, IEEE) Demir, O., Sahingoz, O.K.: Increasing the performance of machine learning-based IDSs on an imbalanced and up-to-date dataset, Marmara University Research Project, FEN-C-DRP-110718-0408, [C-DRP] (2020)

Attack Modeling and Cyber Deception Resources Deployment Using Multi-layer Graph

Amal Sayari[1][✉], Yacine Djemaiel[1], Slim Rekhis[1], Ali Mabrouk[2], and Belhassen Jerbi[2]

[1] Communication Networks and Security Research Lab, SUP'COM, University of Carthage, Tunis, Tunisia
{amal.sayari,yacine.djemaiel,slim.rekhis}@supcom.tn
[2] SAMA PARTNERS, Tunis, Tunisia
{ali.mabrouk,belhassen.jerbi}@samapartners.tn

Abstract. Cyber deception techniques are being used for the proactive defense against attacks. Several techniques were proposed in the literature to address the optimal and intelligent deployment of deception techniques but are unable to consider at the same time the wide sets of threats and defense data (attack tactics, techniques, exploited vulnerabilities, affected machines, generated traces), and the high uncertainty facing the selection of cyber deception resources and their locations. In this work, we provide a multi-layer graph that describes an attack with multi-views: a sequence of vulnerabilities, weaknesses, techniques and tactics. Based on this modeling, we provide algorithms for attack scenarios extraction, metrics for attack scenarios ranking and selection, and analytics for deception techniques assessment and comparison. Finally, a case study is presented to illustrate the efficiency of the proposed model.

1 Introduction

Nowadays, organizations are facing a new breed of cyber attacks which are multi-vectored, multi-staged, polymorphic making detection and mitigation processes becoming complex. Traditional cybersecurity approaches are struggling to defend against modern, dynamic and rapidly evolving cyber-attacks. Faced with these facts, we need active defense techniques capable of confusing attackers and denying them to compromise critical resources. Cyber deception is among these defense techniques that are currently used.

Given the diversity of cyber deception solutions and system components, we have to compare these solutions, choose the right locations and appropriate technique in terms of efficiency, cost, and interoperability within existing security solutions.

Research on defensive deception has focused on diversifying deception techniques, concealing the network assets and deploying intelligent deception resources. Authors in [1] classified deception techniques based on attack phases

© The Author(s), under exclusive license to Springer Nature Switzerland AG 2022
L. Barolli et al. (Eds.): AINA 2022, LNNS 450, pp. 560–572, 2022.
https://doi.org/10.1007/978-3-030-99587-4_48

and layers. In [2], the authors proposed a framework based on concealment techniques to maximize the deception utility and its deployment as a service for cost-effectiveness. In [3], the authors proposed a model for optimal deployment of deception resources using reinforcement learning. However, no attention has been devoted to research on modeling the deception techniques inside graphs of attack scenarios, optimization of deception resources deployment and defense effectiveness maximization.

Several research works dealt with attack scenarios generation and each has benefits and downfalls. In [4], Attack graphs were used to describe the main routes an attacker can take towards a specific network targets, but have scalability and flexibility issues. To overcome these shortcomings, authors in [5] proposed a penetration path generation method based on two-layer threat penetration graph constructed using knowledge graph and penetration information exchange. Nevertheless, this work has focused only on vulnerabilities and CVSS metrics.

In this paper, we propose a powerful graph-based model capable of modeling attack scenarios by considering all aspects related to network vulnerabilities, attack techniques, attack tactics and attacker's preference. Based on that model, we define several correlations between vulnerabilities, weaknesses, techniques, and tactics, we also rank and extract vulnerability exploitation scenarios corresponding to a predefined list of tactic chains. Starting from the extracted scenarios and the already built multi-layer graph, we define a deployment policy of deception resources and assess its efficiency.

Our main contributions can be summarized as follows:

- Design of a multi-layer graph that models an attack scenario using four layers of information showing tactical threat intelligence and vulnerability data (vulnerability, weakness, MITRE ATT&CK techniques, MITRE ATT&CK tactics). It correlates each vulnerability to its weaknesses to attack patterns, and relates techniques of an attack pattern to the potential tactics.
- Design of analytics on multi-layer graph for attack scenarios extraction starting from a set of metrics defined in each layer. Then, potential attack scenarios are selected using ranking metrics that compute the attacker's profitability such as the highest damage, the highest profit, the lowest cost, or the most probable scenario.
- Assessment of deception resources deployment schemes and their effects using a set of proposed metrics and analytics.

This paper is structured as follows. Section 2 describes the attack modeling process. The next section details the approach used to extract attack scenarios and select the potential ones. Section 4 provides a deployment policy for deception techniques and their effects. Section 5 provides a case study to show the effectiveness of our model. Finally, Sect. 6 concludes the paper.

2   Attack Scenarios Modeling

We define the cyber system and the attack model in order to illustrate the effect of the attack on the victim system.

2.1   Requirements for the Proposed Work

Defining a model for efficient deployment of cyber deception resources in a dynamic environment turns out to be tricky. To do so, these requirements should be fulfilled.

Attack description using multi-layer graph: as the data (assets and interconnections, weaknesses and vulnerabilities, techniques, tactics) becomes more heterogeneous, a single graph becomes an oversimplification. Hence, we need a richer representation capable of hosting relations of various scales, a multi-layer graph.

Attack modeling using multi-source data: to model an attack scenario, we need to integrate data from the environment where the attack will be executed (network, security solutions, attack, and defense information) that can be collected using security tools, monitoring systems, MITRE ATT&CK and cyber threat intelligence.

Attack scenarios extraction using multiple datasets: we need to integrate widely used knowledge database of cyber attacks to facilitate the mapping between the above cited data, for example, MITRE ATT&CK. CVE, CWE, CAPEC datasets can be used for mapping vulnerability to attack tactics.

Attack scenarios ranking using cyber threat intelligence: we have to build an adversary profile via CTI. As the most critical vulnerability is not necessarily the most sought after by attackers, CTI give idea on vulnerability in priority treatment.

2.2   Cyber System Modeling

In this work, a cyber system is defined by the following elements:

- A: set of assets (information, software, hardware) to model the cyber system. Each asset is defined by an ID, executed services, vulnerable components.
- NMS: network monitoring system to keep monitoring the status of each normal asset and trigger an alarm when it encounters an attack.
- SS: security solution that protects assets from threats.

As various assets and their interrelations may include inherent weaknesses and vulnerabilities, we are proposing an asset dependency graph via backtracking. To generate it, we can monitor system calls or the default process agent configuration file containing the attributes, values for each process, service and host. Given that a vulnerability scanner provides details about the vulnerability and its related host, by checking the asset dependency graph for hosts connectivity, it can help identify vulnerability connectivity which will be a necessary step to model an attack scenario.

2.3   Attack Scenarios Modeling

An attack scenario is defined by successive actions exploiting vulnerabilities mapped to attack patterns detectable by techniques to perform tactics responding to an attacker's goal. That's why, the multi-layer graph shown in Fig. 1 is used.

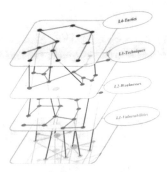

Fig. 1. The multi-layer graph based cyber defense model

Let $G = (V, E, L)$ be a multi-layer graph, where V is a set of vertices, L is the set of considered layers, and $E \subseteq V \times V \times L$ is a set of edges. We consider a multi-layer graph nodes as pairs $V_G \subseteq V \times L$, $E_G \subseteq V_G \times V_G$ then connect pairs $(\langle v, l \rangle), (\langle v', l' \rangle)$ where $\langle v \rangle, \langle v' \rangle \in \langle V_G \rangle$. An edge is often said to be intra- or inter-layer depending on whether $\langle l \rangle = \langle l' \rangle$ or $\langle l \rangle \neq \langle l' \rangle$ [6]. In our model, the vertices for each layer is a security data in {tactic, technique, weakness, vulnerability} that is defined by a set of attributes.

2.3.1   Attack Modeling as a Chain of Vulnerabilities

In the first layer, a vertex v represents a vulnerability defined by a seven-uplet **v=⟨ ID, Host, AV, AC, AU, RAP, OAP⟩** where ID represents the publicly known CVE identifier, Host is the affected hostname, AV is the Access Vector that reflects how the vulnerability is exploited and detailed by CVSS v2 as (Local: 0.395, Adjacent network: 0.646, Network: 1.0), AC is the Attack Complexity required to exploit the vulnerability and detailed by CVSS v2 as (High: 0.35, Medium: 0.61, Low: 0.71), AU is the Authentication Vector detailed by CVSS v2 as (Multiple instances of authentication: 0.45, Single instance of authentication: 0.56, No authentication: 0.704), RAP is the Required Access Privilege to access a victim node (Root: 1, User: 0), OAP is the Obtained Access Privilege after exploitation of the vulnerability.

Two vulnerability nodes are linked together in the graph if they would be exploited sequentially during an attack scenario. In particular two nodes: v_i and v_j are linked together iff: a) No solution is blocking the traffic allowing to exploit v_j; and

b) A physical path exists in the network allowing traffic exchange between $Host_i$ and $Host_j$. Moreover, we have:

If $v_j.AV = \{Network\}$ Then $NR(v_i.Host, v_j.Host) = logical, if v_j.AV = \{Network\_Adjacent\}$ Then $NR(v_i.Host, v_j.Host) = physical$ and if $v_j.AV = \{local\} Then v_i.Host = v_j.Host$ where $\langle NR \rangle$: Neighborhood relation $\in \{logical, physical\}$ (two hosts interconnected to different networks/ same physical network).

2.3.2   Attack Modeling as a Set of Weaknesses

The importance of a weaknesses layer is given by the fact that to exploit a vulnerability, an attacker must have at least one technique that can connect to a weakness. It can be a security policy weakness, technology weakness, configuration weakness.

In the second layer, a vertex v represents a weakness defined by two-uplet **v=⟨CWE-ID, CAPEC-ID⟩** are identifiers of software weakness and attack pattern.

For weaknesses layer, it is not necessary to generate intra-layers edges, the important is that each weakness should be related to vulnerabilities in the bottom layer.

2.3.3   Attack Modeling as a Set of Techniques

In the third layer, a vertex v represents a technique defined by eight-uplet with respect to MITRE ATT&CK framework as **v=⟨ Technique-ID, system requirements, software, platform, permissions, supports remote, defense bypassed, groups⟩** representing respectively the identifier of a technique, the state of the system (patch level), the way to implement a technique, the application or OS, the lowest level of permissions/ level attain by performing the technique, which is required in the privilege escalation, the condition that the technique can be used to execute something on a remote system, which is required for execution, the condition that technique can be used to bypass a defensive tool, which is required for defense evasion and the known groups, and their tactical goals accomplished using a technique.

For techniques layer, it is not necessary to generate intra-layers edges, the important is that each technique is not blocked by a SS and it responds to the attacker's preference because one technique can accomplish more than one tactic in the upper layer and also it can be matched to more than one weakness in the bottom layer.

2.3.4   Attack Modeling as a Set of Tactics

In the fourth layer, a vertex v represents a tactic defined by one-uplet **v=⟨Tactic-ID⟩** representing the identifier of an attack tactic in reference to MITRE ATT&CK.

For tactics layer, it is not necessary to generate intra-layer edges, the important is that each tactic is selected given the attacker's goal: confidentiality/

integrity violation, systems destroying, and tactic dependencies are respected: collection is necessary before exfiltration (attacker needs to read the data before modifying it).

Inter-layer edges connecting 4 layers, generated as shown in Table 1 and the multi-layer graph is created via Algorithm 1 with a bottom-up approach (L1 → L4).

Table 1. Inter-layer edges significance

Layers	Inter-layer edges	Significance
L1 → L2	$\langle e \rangle : \langle v_i \rangle \rightarrow \langle v_j \rangle$ where $\langle v_i \rangle \in L1, \langle v_j \rangle \in L2$	Represents the fact that a weakness makes the vulnerability exploitable by correlating CVE-ID to CWE-ID
L2 → L3	$\langle e \rangle : \langle v_i \rangle \rightarrow \langle v_j \rangle$ where $\langle v_i \rangle \in L2, \langle v_j \rangle \in L3$	Represents the fact that using attack patterns, an attack is connected to how it is executed and to where it is targeted by mapping CWE-ID to CAPEC-ID and CAPEC-ID to Technique-ID
L3 → L4	$\langle e \rangle : \langle v_i \rangle \rightarrow \langle v_j \rangle$ where $\langle v_i \rangle \in L3, \langle v_j \rangle \in L4$	Represents what techniques can be performed to achieve an attacker goal by mapping Technique-ID to Tactic-ID

Algorithm 1: Multi-layer graph generation

Input : $\langle V \rangle$: vulnerabilities, $\langle w \rangle \in \langle W \rangle$: weakness, $\langle te \rangle \in \langle Te \rangle$: technique, $\langle tc \rangle \in \langle Tc \rangle$: tactic, $\langle l \rangle \in \langle L \rangle$: layer multi-layer graph $\langle G \rangle$ (initially empty)

1 **while** $\langle v \rangle \in \langle V \rangle$ **do**
2   **for** *each* $\langle v_G \rangle \in \langle G \rangle$ **do**
3   **if** *Check_dependency* $(\langle v \rangle, \langle v_G \rangle)$ **then**
4   remove_from_V_set($\langle v \rangle$), append_node ($\langle v \rangle, \langle G \rangle, \langle l1 \rangle$)
5   add_link ($\langle v \rangle, \langle v_G \rangle$), $\langle W \rangle \prec$ - fetch_weakness ($\langle w \rangle$), append_node($\langle w \rangle, \langle G \rangle, \langle l2 \rangle$)
6   **if** CVE-id($\langle v \rangle$)is related to CWE-id($\langle w \rangle$) **then** add_link($\langle v \rangle,\langle w \rangle$)
7        $\langle Te \rangle \prec$- fetch_technique($\langle te \rangle$), append_node ($\langle te \rangle, \langle G \rangle, \langle l3 \rangle$)
8   **if** CWE-id($\langle w \rangle$)is related to Technique-id($\langle te \rangle$) **then** add_link($\langle w \rangle, \langle te \rangle$) $\langle Tc \rangle \prec$ - fetch_tactic ($\langle tc \rangle$), append_node($\langle tc \rangle, \langle G \rangle, \langle l4 \rangle$)
9   **if** Technique-id($\langle te \rangle$)is related to Tactic-id($\langle tc \rangle$) **then** add_link($\langle te \rangle, \langle tc \rangle$) break;
10   **end if**
11   **end for**
12 **end while**

---

3 Attack Scenarios Extraction and Ranking

We define in this section an algorithm for attack scenarios extraction and metrics for potential scenarios selection.

3.1 Attack Scenarios Extraction

We can model and extract all the possible attack scenarios that respond to an attacker's goal by achieving a sequence of tactics attainable through the exploitation of a sequence of vulnerabilities. To do so, these steps must be followed:

- We have to set the datasets. CVE details and NVD list the vulnerabilities by CVE-ID with the related weaknesses. NVD contains more than 160,000 vulnerabilities relating to thousands of vendor products. MITRE ATT&CK includes 14 entreprise tactics, 188 entreprise techniques and 379 sub-techniques, gives an idea on the CAPEC-ID related to each technique and it is regularly updated with the latest techniques. Finally, we need a predefined dataset of known attack scenarios mapped to MITRE ATT&CK. For our model, we started with some open source datasets [7].
- We can use a vulnerability scanner to obtain the hosts and vulnerabilities information. Also, we can use a compliance audit to identify the weaknesses.
- We can construct our multi-layer graph using Mully, which is an R package that helps us to create, and visualize graphs with multiple layers, or using GraphX in Spark or others. Then, to facilitate the creation of inter-layer edges, that relate the vertices of L1 to L2 to L3 to L4, we can use the data provided by BRON [8].
- To extract the attack scenarios, we need to anticipate the source node which is the first facing by an attacker and determined by the NMS. We may have several source nodes since the attacker can leverage multiple attack vectors. Target node is anticipated using ATT&CK knowledge enriched by the CTI. In fact, MITRE mapped publicly reported techniques and softwares to groups. So, observations about the used techniques and softwares in the network may help the defender to get an idea about the attacker's goal and the possible attack scenario.
- Based on our known attack scenarios dataset, we can identify the known attack scenarios that can take place in the system in the case of the presence of the required vulnerability and the tools to exploit it. To minimize the false positives and negatives when extracting scenarios, a vulnerabilities scenario must be compatible with a tactics scenario. That's why, we have divided ATT&CK tactics into 3 categories:

- Network access: reconnaissance, resource development and initial access tactics.
- Network propagation: includes persistence, privilege escalation, defense evasion, lateral movement, discovery, credential access and command & control tactics.
- Network impact: includes execution, exfiltration, collection and impact tactics.

Therefore, a significant scenario must include at least one tactic from network impact which reflects at least one of attacker's objectives. One example is the Wannacry attack that considers the following tactics [9]: reconnaissance, resource development, initial access, discovery, lateral movement, command & control and impact.

Given the aforementioned conditions, a scenario is extracted via Algorithm 2.

Algorithm 2: Attack scenarios extraction

Input : $\langle v \rangle \in \langle V \rangle$:vulnerability, $\langle w \rangle \in \langle W \rangle$:weakness, $\langle te \rangle \in \langle Te \rangle$:technique, $\langle tc \rangle \in \langle Tc \rangle$:tactic, $\langle h \rangle \in \langle H \rangle$:host

Output: Attack scenario

1 Check conditions for attack scenarios extraction **Then** Extract un attack scenario $\langle S \rangle$

2 S $= \prec \langle v_1 \rangle, \ldots\ldots, \langle v_n \rangle \succ$ **where**

3 **for** *each* $\prec \langle vi \rangle, \langle vi + 1 \rangle \succ$ **do**

4     $\exists \prec \langle hp \rangle, \langle hp + x \rangle \succ / R(\langle vi \rangle, \langle hp \rangle) = True \, and \, R(\langle vi + 1 \rangle, \langle hp + x \rangle) = True \, and \; R(\langle hp \rangle, \langle hp + x \rangle) = True$

5     $\exists \prec \langle wj \rangle, \langle wj + x \rangle \succ / R(\langle vi \rangle, \langle wj \rangle) = True \, and \, R(\langle vi + 1 \rangle, \langle wj + x \rangle) = True$

6     $\exists \prec \langle tek \rangle, \langle tek + y \rangle \succ / R(\langle wj \rangle, \langle tek \rangle) = True \, and \, R(\langle wj + x \rangle, \langle tek + y \rangle) = True$

7     $\exists \prec \langle tcz \rangle, \langle tcz + d \rangle \succ / R(\langle tcz \rangle, \langle tek \rangle) = True \, and \, R(\langle tcz + d \rangle, \langle tek + y \rangle) = True$

8 **end for**

9 Check the output of this algorithm

3.2 Potential Attack Scenarios Ranking

Based on the correlations between network vulnerabilities and weaknesses, attack techniques and tactics described in our model, several vulnerability exploitation scenarios corresponding to a predefined list of tactic chains can be extracted. In order to select the potential ones, we proposed to calculate the attacker's preference based on a profitability metric (Pr) calculated for each attack scenario (S):

$$Pr(S) = \sum_{i=1}^{n} (\prod_{j=1}^{i} (P_j)(D_i - \alpha C_i)) \tag{1}$$

Where $\langle D_i \rangle$ is the damage impact using a vulnerability $\langle i \rangle$, $\langle C_i \rangle$ is the attack cost using a vulnerability $\langle i \rangle$ and $\langle P_j \rangle$ represents the vulnerability exploitability probability.

$P_t = D_i - \alpha C_i$, where $\langle P_t \rangle$ is the attacker's profit and α is the preference weight of the cost parameter that indicate the attacker's profitability where $0 \le \alpha < 1$

$D_i = 1 - (1 - \mathtt{C}_i) * (1 - \mathtt{I}_i) * (1 - \mathtt{A}_i)$

where $\mathtt{C}_i, \mathtt{I}_i, \mathtt{A}_i$ are confidentiality, integrity, and availability related to i

$C_i = AC_i * AV_i * AU_i \, and \, P_j = AV_i * AC_i * AU_i$ where $\langle AV_i \rangle$ is access vector, $\langle AC_i \rangle$ is access complexity and $\langle AU_i \rangle$ is authentication

So, given the attacker's profitability, we can compute the preference for each attack scenario and hence, select the potential ones.

4 Deployment Policy of Cyber Deception Techniques and Their Effects on the Current Attack Scenario

To detect attackers and make it more time, resources and cost consuming for them to attack, we have to deploy defensive deception solutions.

In this work, the decision to add cyber deception is justified by a need for deflecting the attackers from a potential scenario and leading them through desired paths in our multi-layer graph. Since the attack scenario is a sequence of nodes and edges on the multi-layer graph starting from a source node, leading an attacker through a desired attack path amounts to add a node or an edge to the current attack scenario or remove a node or an edge from the current attack scenario. Based on our multi-layer graph and asset dependency graph, we propose to use these deception techniques: a) Ports and services deception: fake protocols: HTTP, FTP, SMTP, DNS, SSH. b) Host deception: fake directories and configuration, files with fake credentials, fake credentials in browser password manager, fake resources using honeypots, decoy files and documents giving alerts when opened. c) Traffic deception: inject fake URLs and credentials into the network traffic. d) Vulnerability deception: hide targeted vulnerability by blocking access to it or show a new vulnerability. e) Weakness deception: hide a targeted weakness or show a new weakness.

4.1 Cyber Deception Resources Deployment Locations

The deployment of deception can be modeled on our graphs as shown in Fig. 2

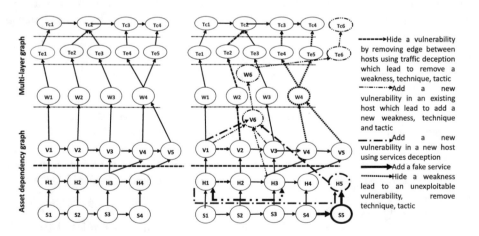

Fig. 2. Cyber deception resources deployment locations

4.2 Cyber Deception Resources Deployment Policy and Effect

Given a scenario $A = \prec (S_1, H_1, V_1, W_1, Te_1, Tc_1), ..., (S_n, H_n, V_n, W_n, Te_n, Tc_n) \succ$, deploying a deception solution on A lead to the bifurcation of the attack scenario.

The most appropriate deception solution is the one that create uncertainty in the attacker's choices causing a deviation from potential scenarios to the desired scenarios (those with lower success probability, difficult exploitability, higher cost). Therefore, we have to check the induced multi-layer graph (with deception) for the new attack scenarios, and select deception solutions that reveal new techniques and tactics making a divergence of the attacker from the preferable potential scenario.

After the cyber deception solutions deployment, two cases are presented:

– The attacker fails to reach the target node in the current attack path, then he is invading in a fake resource where he will lose time and resources which means that the defender traps the attacker and the deception deployment policy is successful.
– The attacker reaches the target node with more effort. The attacker detects the deception resource and try to execute an anti-deception and the game will restart.

5 Case Study

The authors of the research work in [5] proposed a path generation method based on two-layer threat penetration graph. Our case study used the same topology, information of hosts, vulnerabilities and the same service access rules in [5] in order to recalculate the penetration path and compare the results of the two approaches (Fig. 3).

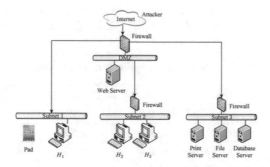

Fig. 3. Experimental network topology

Assuming that an attacker wants to launch an attack against Database Server, the approach presented in the previous work shows that the optimal penetration path is the one with the highest success probability: Attacker → Pad → H2 → Data Server (Table 2).

By exploiting our multi-layer graph, each vulnerability can be associated with the parameters in Table 4. Using the inter-layers relations introduced by

Table 2. Attack paths probability based on the work in [5]

Possible attack paths	Probability
Attacker - Web server - Data server	0.024
Attacker - Web server - File server - Data server	0.00096
Attacker - Pad - Web server - File server - Data server	0.000576
Attacker - Pad - Web server - Data server	0.0144
Attacker - Pad - H3 - File server - Data server	0.004608
Attacker - Pad - H2 - Data server	0.1152
Attacker - Pad - H1 - H3 - Data server	0.055296
Attacker - Pad - H1 - H3 - File server - Data server	0.00221184
Attacker - H1 - H2 - File server - Data server	0.0036864
Attacker - Pad - H1 - web server - Data server	0.006912
Attacker - Pad - H1 - web server - File server - Data server	0.00006912
Attacker - H1 - H2 - Data server	0.09216

our multi-layer graph, we showed that 5 vulnerabilities out of 12 can be part of a scenario: CVE-2017-9798/ CVE-2016-5555/ CVE-2018-8120/ CVE-2015-1769/ CVE-2017-2741, and among the attack paths proposed in [5], 2 are possibles: S1: Attacker → Web server → Data server and S2: Attacker → H1 → H2 → Data server

To find the more potential scenario between S1 and S2, we can calculate the attacker's profitability as defined in Sect. 3 using Table 3.

Assuming that $\alpha = 1/3 * C\_max, \alpha = 1/3 * AV\_max * AC\_max * AU\_max$, we find the attacker's profitability for S1, S2: Pr (S1) = 0.03 and Pr (S2) = 0.2

Table 3. Vulnerabilities information

Host	CVE	Probability	Damage	Cost	Profit
Webserver	cve-2017-9798	0.5	0.275	0.5	−0.06
H1	cve-2018-8120	0.2	0.96	0.2	0.83
H2	cve-2015-1769	0.2	0.96	0.2	0.83
Dataserver	cve-2016-5555	0.28	0.62	0.28	0.43

The defender will aim at altering the attacker's perception of the network and preventing him from accessing the data server. According to Table 4 the attacker can use one of these techniques: T1018, T1046, T1016, T087, T1135, T1120. So, the defender can deploy these deception techniques on the potential scenario [10]: set up fake network services, create breadcrumbs or honeytokens, create fake local or domain accounts, create decoy files or documents that giving alert when opened.

Table 4. Vulnerabilities and ATT&CK metrics relations

Host name	CVE-ID	CWE-ID	CAPEC-ID	Technique-ID	Tactic-ID
Web server	CVE-2014-0226	362	26/29	–	–
	CVE-2017-9798	416	312	T1082	TA0007
Pad	CVE-2016-4729	119	10/100/123/14/42/	–	–
		119	44/45/46/47/8/9/24	–	–
	CVE-2018-8174	787	–	–	–
H1	CVE-2017-0161	362	26/29	–	–
	CVE-2018-8120	404	125/130/131	T1499	TA0040
H2/H3	CVE-2015-1769	264	17	T1574.010	TA0003
					TA0004
	CVE-2018-7573	119	10/100/123/14/42	-	TA0005
			44/45/46/47/8/9/24	–	–
Print server	CVE-2017-2741	284	478/479	T1068/ T1543.003/	TA0003
				T1543.011/ T1553.004	TA0004
			550/551	T1543.003/ T1543.002/	TA0005
			552/556/558	T1543.004/ T1542.003/	
				T1014/ T1546.008/	
			562/564/578	T1080/T1037/T1547	TA0008
File server	CVE-2014-6287	94	242/35/77/22	–	–
	CVE-2012-0002	94	242/35/77/22	–	–
Data server	CVE-2012-0002	94	242/35/77/22	–	–
	CVE-2016-5555	200	13/292/300/309/312/	T1574/T1016/T1082	TA0003
				T1018/ T1046/	TA0004
			573/574/577/	T1574/ T1562.003/	TA0005
				T1018/ T1046/	
				T1016/ T1082	
			60/643/646	T1057/ T1007/ T1087/	TA0007
				T1087.002/ T1120/	TA0008
				T1069/ T1033/T1135	

6 Conclusion

We provided an efficient deployment policy of deception resources by determining the most suitable solution and efficient location. We defined a multi-layer graph capable of modeling attack scenarios by considering network vulnerabilities, attack techniques and tactics, attacker's preference. We proposed algorithm for scenarios extraction, metrics for potential scenarios selection, analytics for comparing deception solutions and evaluating their effectiveness. Finally, we provided a case study to exemplify the proposal. Future work will address the consistency of our multi-layer graph and the intelligent deployment of deception based on reinforcement learning. In fact, the defender will interact with the network and try to select an optimal deployment policy given the deception action, and the network feedback and state.

Acknowledgment. This project is carried out under the MOBIDOC scheme, funded by the Ministry of Higher Education and Scientific Research through the PromEssE project and managed by the ANPR.

References

1. Zhang, L., Thing, V.L.: Three decades of deception techniques in active cyber defense-retrospect and outlook. Comput. Secur. **106**, 102288 (2021)
2. Duan, Q., Al-Shaer, E., Islam, M., Jafarian, H.: Conceal: a strategy composition for resilient cyber deception-framework, metrics and deployment. In: 2018 IEEE Conference on Communications and Network Security (CNS), pp. 1–9. IEEE (2018)
3. Wang, S., Pei, Q., Wang, J., Tang, G., Zhang, Y., Liu, X.: An intelligent deployment policy for deception resources based on reinforcement learning. IEEE Access **8**, 35792–35804 (2020)
4. Barrère, M., Steiner, R. V., Mohsen, R., Lupu, E.C.: Tracking the bad guys: an efficient forensic methodology to trace multi-step attacks using core attack graphs. In: 2017 13th International Conference on Network and Service Management, pp. 1–7. IEEE (2017)
5. Shuo, W., Jianhua, W., Guangming, T., Qingqi, P., Yuchen, Z., Xiaohu, L: intelligent and efficient method for optimal penetration path generation. J. Comput. Res. Dev. **56**(5), 929 (2019)
6. Mcgee, F., Ghoniem, M., Melançon, G., Otjacques, B., Pinaud, B.: The state of the art in multilayer network visualization. In: Computer Graphics Forum, vol. 38, no. 6, pp. 125–149 (2019)
7. https://github.com/scythe-io/community-threats
8. Hemberg, E., et al.: Linking threat tactics, techniques, and patterns with defensive weaknesses, vulnerabilities and affected platform configurations for cyber hunting. arXiv preprint arXiv:2010.00533 (2020)
9. De Blasi, S.: Mapping MITRE ATT&CK to the WannaCry Campaign (2021)
10. https://github.com/0x4D31/deception-as-detection

Quantum-Secure Aggregate One-time Signatures with Detecting Functionality

Shingo Sato[✉] and Junji Shikata

Yokohama National University, Yokohama, Japan
{sato-shingo-zk,shikata-junji-rb}@ynu.ac.jp

Abstract. An aggregate signature (ASIG) scheme allows any user to compress multiple signatures into a short signature called an aggregate signature. While a conventional ASIG scheme cannot detect any invalid messages from an aggregate signature, an ASIG scheme with detecting functionality (D-ASIG) has an additional property which can identify invalid messages from aggregate signatures. Hence, D-ASIG is useful to reduce the total amount of signature-sizes on a channel. On the other hand, development of quantum computers has been advanced recently. However, all existing D-ASIG schemes are insecure against attacks using quantum algorithms, which we call quantum attacks. In this paper, we propose a D-ASIG scheme with quantum-security which means security in a quantum setting. Hence, we first introduce quantum-security notions of ASIGs and D-ASIGs because there is no research on such security notions for (D-)ASIGs. Second, we propose a lattice-based aggregate one-time signature scheme with detecting functionality, and prove that this scheme satisfies our quantum-security in the quantum random oracle model and the certified key model. Hence, this scheme is the first quantum-secure D-ASIG.

1 Introduction

Background. Digital signature is a fundamental primitive in public key cryptography that ensures integrity of data. The range of applications of this primitive is very wide since publicly verification of data is very useful in many situations. However, when checking validity of multiple messages simultaneously, a total amount of signature-sizes transmitted on a channel (e.g., the internet) is too large, since the total size of signatures is proportional to the number of all the messages. In order to reduce such an amount of signature-sizes, we can use an aggregate signature (ASIG) scheme which compresses (or aggregates) multiple signatures into a short signature called an aggregate signature. Boneh et al. introduced the notion of ASIGs and proposed a pairing-based scheme in the random oracle model (ROM) [2]. Under the certified key model in which signers have to prove knowledge of their secret key at key-registration, Rückert and Schröder gave an ASIG scheme using multilinear maps in the standard model [13]. Hohenberger, Sahai, and Waters presented (identity-based) ASIG schemes using multilinear maps in the standard model [7]. Boneh and Kim proposed an aggregate one-time signature (AOTS) and an interactive ASIG based

© The Author(s), under exclusive license to Springer Nature Switzerland AG 2022
L. Barolli et al. (Eds.): AINA 2022, LNNS 450, pp. 573–585, 2022.
https://doi.org/10.1007/978-3-030-99587-4_49

on lattice problems [3]. Hartung et al. [6] proposed a fault-tolerant ASIG that has functionality of both aggregating multiple signatures and identifying invalid messages from an aggregate signature. Sato, Shikata, and Matsumoto introduced the notion of ASIGs with detecting functionality (D-ASIGs) which can detect invalid messages, and proposed D-ASIG schemes by combining ASIGs with group-testing in a comprehensive way [14]. Since the detecting property of D-ASIGs with a total amount of small signature-size is useful, we focus on this topic in public key cryptography.

Furthermore, development of quantum computers has been advanced recently, and many researches have paid much attention to constructing cryptographic protocols secure against attacks using quantum algorithms, which we call quantum attacks. In particular, we focus on the security model where an adversary is allowed to issue a quantum query (i. e., a quantum superposition of queries) to the signing oracle in a security game [4]. This is because this security model expresses a practical situation where sufficiently large quantum computers are realized. For digital signatures, Boneh and Zhandry gave a formalization of security in this model [4], which we call *quantum-security* in this paper, and proposed signature schemes satisfying this formalized quantum-security. As for (D-)ASIGs, however, there is no research on quantum-security. Furthermore, all existing D-ASIG schemes are insecure against quantum attacks by using several quantum algorithms such as Shor's algorithm [15]. Hence, it is important to research quantum-secure D-ASIG schemes.

Contribution. Our goal is to propose a quantum-secure D-ASIG scheme. To this end, we first formalize quantum-security notions of (D-)ASIGs since there is no research on (D-)ASIGs in the security model where an adversary can issue quantum queries to given oracles. Next, we show that a generic construction of quantum-secure D-ASIGs satisfies our formalized security. Then, we present a concrete aggregate one-time signature (AOTS) scheme which can be applied to the generic construction. This implies that the resulting scheme is quantum-secure. Details on our contribution are shown as follows:

- First, we formalize quantum-security notions of ASIGs and D-ASIGs, namely, security notions in the model where an adversary is allowed to issue quantum queries to the signing oracle in a (D-)ASIG's security game. Following a definition of quantum-security for digital signatures [4], we give quantum-security definitions for ASIGs and D-ASIGs.
- Second, we propose a quantum-secure AOTS scheme with detecting functionality (D-AOTS). To this end, we prove that a generic construction starting from a quantum-secure ASIG scheme and a non-adaptive group testing protocol is a quantum-secure D-ASIG. Then, in order to obtain a quantum-secure D-ASIG, we propose a lattice-based AOTS scheme satisfying the formalized quantum-security of ASIGs in the quantum random oracle model [1] and the certified key model [9,13]. Notice that it is widely believed that lattice problems are computationally hard even if it is possible to utilize quantum

Table 1. Comparison of D-ASIG schemes

Scheme	Underlying Primitives	Unforgeability	Identifiability	Certificateless Model ?	Total Aggregate Signature-Size
HKKKR	ASIG [2]	aggEUF Against cCMA	CMP and wSND against cCMA	✓	$O(d^2 \log \ell)\|\sigma\|$
SSM	ASIG [2] and SNARK [11]	aggEUF against cCMA	CMP and SND against cCMA	✓	$O(d^2 \log \ell)(\|\sigma\| + \|\pi\|)$
Our Scheme	AOTS In Sect. 4	aggEUF against qCMA	CMP and wSND against qCMA		$O(d^2 \log \ell)\|\sigma\|$

HKKKR and SSM are concrete D-ASIG schemes which are constructed by applying the schemes described in "Underlying Primitives" to generic constructions proposed in [6] and [14], respectively. SNARK means succinct non-interactive argument of knowledge. The terms "Unforgeability" and "Identifiability" are the security notions of D-ASIGs, which were formalized in [14]. cCMA and qCMA mean "classical chosen message attacks" and "quantum chosen message attacks", respectively. aggEUF means "aggregate existential unforgeability". CMP and SND (resp., wSND) mean completeness and soundness (resp., a weak variant of soundness) of the identifiability of D-ASIGs. Certificateless model means a model where key-registration is unnecessary. ℓ (resp. d) is the total number of messages (resp. the maximum number of invalid messages). $|\sigma|$ (resp. $|\pi|$) is the bit-length of a signature of underlying ASIG (resp. a proof of underlying SNARK).

algorithms. Hence, the resulting D-AOTS scheme is quantum-secure in the quantum random oracle model and the certified key model. We claim that this proposed scheme is the first quantum-secure D-ASIG, and would be useful like [6,14] in a quantum era.

Furthermore, we compare D-ASIG schemes in order to clarify the difference between existing D-ASIG schemes and our scheme. Table 1 shows a comparison of D-ASIG schemes. Notice that ASIGs with interactive tracing functionality [8] are not included in this comparison because the security model of [8] is different from that of [6,14]. Regarding the underlying primitives applied to generic constructions of [6,14], we have selected schemes whose signature/proof-lengths are the shortest of existing schemes. First, the advantage of our scheme is summarized as follows: From the terms "Unforgeability" and "Identifiability" in Table 1, we see that our scheme satisfies the formalized quantum-security while all existing schemes achieve security only in classical security models. As described before, it should be noted that all existing D-ASIGs are insecure against quantum attacks. As another advantage of our scheme, the increment of a total aggregate signature-size of our scheme is not larger than that of any existing one, even though our scheme achieves security in a quantum setting. Second, the disadvantage of our scheme lies in that each signer with a key-pair can generate only one signature, and the security of ours is ensured in the certified key model. However, considering usefulness of traditional one-time signatures, we expect our one-time D-ASIG scheme in the certified key model would be useful in a quantum era.

2   Preliminaries

Notation. In this paper, we use the following notation: For a positive integer n, let $[n] := \{1, \ldots, n\}$. For n values $x_1, \ldots, x_n$ and a subset $I \subseteq [n]$ of indexes, let $(x_i)_{i \in I}$ be a sequence of values whose indexes are in I, and let $\{x_i\}_{i \in I}$ be a set of values whose indexes are in I. For a vector $\boldsymbol{x}$ with dimension n, let x_i be the i-th entry ($i \in [n]$). For a $m \times n$ matrix $\boldsymbol{X}$, let $x_{i,j}$ be the entry at the i-th row and the j-th column ($i \in [m], j \in [n]$). For a function $f : \mathbb{N} \rightarrow \mathbb{R}$, if $f(\lambda) = o(\lambda^{-c})$ for arbitrary positive c, then f is negligible in λ, and we write $\mathsf{negl}(\lambda)$. A probability is overwhelming if it is $1 - \mathsf{negl}(\lambda)$. In addition, we use the following notation for quantum computation as in [4]. We write an n-qubit state $|\psi\rangle$ as a linear combination $|\psi\rangle = \sum_{x \in \{0,1\}^n} \psi_x |x\rangle$ with a basis $\{|x\rangle\}_{x \in \{0,1\}^n}$ and amplitudes $\psi_x \in \mathbb{C}$ such that $\sum_{x \in \{0,1\}^n} |\psi_x|^2 = 1$. When $|\psi\rangle$ is measured, the state x is observed with probability $|\psi_x|^2$. Suppose that we have superposition $|\psi\rangle = \sum_{x \in \mathsf{X}, y \in \mathsf{Y}, z \in \mathsf{Z}} \psi_{x,y,z} |x, y, z\rangle$, where X and Y are finite sets and Z is a work space. For an oracle $\mathsf{O} : \mathsf{X} \rightarrow \mathsf{Y}$, we write quantum access to O as a mapping $|\psi\rangle \mapsto \sum_{x \in \mathsf{X}, y \in \mathsf{Y}, z \in \mathsf{Z}} \psi_{x,y,z} |x, y + \mathsf{O}(x), z\rangle$, where $+ : \mathsf{Y} \times \mathsf{Y} \rightarrow \mathsf{Y}$ is a certain group operation on Y. "Quantum polynomial-time" is abbreviated as QPT.

Group Testing. Group testing (e.g., [5]) is a method to detect positive items among many items with a smaller number of tests than the straightforward individual testing for each item. Applications of group testing include screening blood samples for detecting a disease, and detecting clones which have a particular DNA sequence.

Canonical non-adaptive group testing is designed by a d-disjunct matrix or a d-cover-free family (e.g., see [5]). A non-adaptive group testing protocol with u tests for ℓ items is represented by a $u \times \ell$ binary matrix, and the (i, j)-th element of the matrix is equal to 1 if and only if the i-th test is executed to the j-th item. The d-disjunct property of binary matrices is defined as follows.

Definition 1 (d-disjunct). A matrix $\boldsymbol{G} = [\boldsymbol{g}_1, \ldots, \boldsymbol{g}_\ell] \in \{0, 1\}^{u \times \ell}$ is d-disjunct if for any d columns $\boldsymbol{g}_{s_1}, \ldots, \boldsymbol{g}_{s_d}$ and any $\bar{\boldsymbol{g}} \in \{\boldsymbol{g}_1, \ldots, \boldsymbol{g}_\ell\} \backslash \{\boldsymbol{g}_{s_1}, \ldots, \boldsymbol{g}_{s_d}\}$ ($s_1, \ldots, s_d \in [\ell]$), there exists $z \in [u]$ such that $v_z < \bar{g}_z$, where $\boldsymbol{v} = \bigvee_{i=1}^{d} \boldsymbol{g}_{s_i}$, and $\bigvee$ is the bitwise-OR.

By using a d-disjunct matrix, a non-adaptive group testing protocol can efficiently detect at most d positive items. We simply describe the process of group testing protocol with a d-disjunct matrix $\boldsymbol{G} \in \{0, 1\}^{u \times \ell}$ as follows: Let $S_i(\boldsymbol{G}) = \{j \mid j \in [\ell] \wedge g_{i,j} = 1\}$ for $i \in [u]$ and $\boldsymbol{G} \in \{0, 1\}^{u \times \ell}$. (Step 1) Let $J \leftarrow \{1, 2, \ldots, \ell\}$ be a set of indexes of all items. (Step 2) For each $i \in [u]$, compress items with indexes in $S_i(\boldsymbol{G})$. (Step 3) For each $i \in [u]$, set $J \leftarrow J \backslash S_i(\boldsymbol{G})$ if the test result of the i-th compressed item is non-positive (i.e., negative). Here, note that the test result of a compressed item shows positive if at least one positive item are included, and shows negative otherwise. After the procedure of those steps, the resulting set J is a set of indexes of all positive items due to the d-disjunct property of $\boldsymbol{G}$.

Aggregate Signatures (with Detecting Functionality). We first describe the syntax of aggregate signatures and formalize its quantum-security notion.

Definition 2 (Aggregate Signatures). An aggregate signature (ASIG) scheme ASig consists of five polynomial-time algorithms (KGen, Sign, Vrfy, Agg, AVrfy): For a security parameter λ, let $\mathcal{M} = \mathcal{M}(\lambda)$ be a message space.

- $(\mathsf{pk}, \mathsf{sk}) \leftarrow \mathsf{KGen}(1^\lambda)$: The randomized algorithm KGen takes as input a security parameter 1^λ, and it outputs a public key pk and a secret key sk.
- $\sigma \leftarrow \mathsf{Sign}(\mathsf{sk}, \mathsf{m})$: The randomized or deterministic algorithm Sign takes as input a secret key sk and a message $\mathsf{m} \in \mathcal{M}$, and it outputs a signature σ.
- $1/0 \leftarrow \mathsf{Vrfy}(\mathsf{pk}, \mathsf{m}, \sigma)$: The deterministic algorithm Vrfy takes as input a public key pk, a message $\mathsf{m} \in \mathcal{M}$, and a signature σ, and it outputs 1 (accept) or 0 (reject).
- $\widehat{\sigma} \leftarrow \mathsf{Agg}((\mathsf{pk}_1, \mathsf{m}_1, \sigma_1), \ldots, (\mathsf{pk}_\ell, \mathsf{m}_\ell, \sigma_\ell))$: The randomized or deterministic algorithm Agg takes as input a tuple $(\mathsf{pk}_1, \mathsf{m}_1, \sigma_1)$, $\ldots$, $(\mathsf{pk}_\ell, \mathsf{m}_\ell, \sigma_\ell)$ of public keys, messages and signatures, and it outputs an aggregate signature $\widehat{\sigma}$.
- $1/0 \leftarrow \mathsf{AVrfy}((\mathsf{pk}_1, \mathsf{m}_1), \ldots, (\mathsf{pk}_\ell, \mathsf{m}_\ell), \widehat{\sigma})$: The deterministic algorithm AVrfy takes as input a tuple $(\mathsf{pk}_1, \mathsf{m}_1), \ldots, (\mathsf{pk}_\ell, \mathsf{m}_\ell)$ of public keys and messages, and an aggregate signature $\widehat{\sigma}$, and it outputs 1 (accept) or 0 (reject).

We require that an ASIG scheme meets correctness as follows.

Definition 3 (Correctness). An ASIG scheme ASig $=$ (KGen, Sign, Vrfy, Agg, AVrfy) meets correctness if the following conditions hold:

- For every $(\mathsf{pk}, \mathsf{sk}) \leftarrow \mathsf{KGen}(1^\lambda)$ and every $\mathsf{m} \in \mathcal{M}$, it holds that $\mathsf{Vrfy}(\mathsf{pk}, \mathsf{m}, \sigma) = 1$ with overwhelming probability, where $\sigma \leftarrow \mathsf{Sign}(\mathsf{sk}, \mathsf{m})$.
- For any $\ell = \mathsf{poly}(\lambda)$, every $(\mathsf{pk}_1, \mathsf{sk}_1) \leftarrow \mathsf{KGen}(1^\lambda), \ldots, (\mathsf{pk}_\ell, \mathsf{sk}_\ell) \leftarrow \mathsf{KGen}(1^\lambda)$, and every $\mathsf{m}_1, \ldots, \mathsf{m}_\ell \in \mathcal{M}$, it holds that $\mathsf{AVrfy}((\mathsf{pk}_1, \mathsf{m}_1), \ldots, (\mathsf{pk}_\ell, \mathsf{m}_\ell), \widehat{\sigma}) = 1$ with overwhelming probability, where $\widehat{\sigma} \leftarrow \mathsf{Agg}((\mathsf{pk}_1, \mathsf{m}_1, \sigma_1), \ldots, (\mathsf{pk}_\ell, \mathsf{m}_\ell, \sigma_\ell))$ and $\sigma_i \leftarrow \mathsf{Sign}(\mathsf{sk}_i, \mathsf{m}_i)$ for every $i \in [\ell]$.

Following definitions of quantum-security of digital signatures [4] and classical security of ASIGs [2], we formalize a quantum-security notion of ASIGs, as follows.

Definition 4 (aggEUF-qCMA security). An ASIG scheme ASig $=$ (KGen, Sign, Vrfy, Agg, AVrfy) satisfies aggEUF-qCMA security if for any QPT adversary A against ASig, its advantage $\mathsf{Adv}_{\mathsf{ASig}, \mathsf{A}}^{\mathsf{aggeuf\text{-}qcma}}(\lambda) := \Pr[\mathsf{A} \text{ wins}]$ is negligible in λ. [A wins] is the event that A wins in the following game:

Setup. A challenger generates $(\mathsf{pk}^*, \mathsf{sk}^*) \leftarrow \mathsf{KGen}(1^\lambda)$, and sends pk^* to A.

Queries. Given a quantum signing-query (i.e., a superposition of messages) $|\psi\rangle = \sum_{\mathsf{m} \in \mathcal{M}, t \in \mathcal{S}, z} \psi_{\mathsf{m}, t, z} |\mathsf{m}, t, z\rangle$, the signing oracle SIGN chooses randomness r used in the Sign algorithm, where it does not need to choose randomness r if Sign is deterministic. Then, it returns $\sum_{\mathsf{m} \in \mathcal{M}, t \in \mathcal{S}, z} \psi_{\mathsf{m}, t, z} |\mathsf{m}, t \oplus \mathsf{Sign}(\mathsf{sk}^*, \mathsf{m}; r), z\rangle$. Let q be the number of queries which A submits to the SIGN oracle.

Output. A outputs $(PM^{(1)}, \widehat{\sigma}^{(1)}), \ldots, (PM^{(q+1)}, \widehat{\sigma}^{(q+1)})$, where for $i \in [q + 1]$, $PM^{(i)} = ((\mathsf{pk}_1^{(i)}, \mathsf{m}_1^{(i)}), \ldots, (\mathsf{pk}_{\ell^{(i)}}^{(i)}, \mathsf{m}_{\ell^{(i)}}^{(i)}))$. A wins if it holds that (i) $\mathsf{AVrfy}(PM^{(i)}, \widehat{\sigma}^{(i)}) = 1$ for every $i \in [q + 1]$, (ii) there exists $\mathsf{m}^{*(i)} \in \mathcal{M}$ such that $(\mathsf{pk}^*, \mathsf{m}^{*(i)}) \in PM^{(i)}$ for every $i \in [q + 1]$, and (iii) $(\mathsf{pk}^*, \mathsf{m}^{*(1)}), \ldots, (\mathsf{pk}^*, \mathsf{m}^{*(q+1)})$ are distinct.

Next, following [14], we describe the syntax of D-ASIGs.

Definition 5 (Aggregate Signatures with Detecting Functionality). An aggregate signature scheme with detecting functionality (D-ASIG) consists of five polynomial-time algorithms (KGen, Sign, Vrfy, DAgg, DVrfy) associated with a set $\mathcal{G}$ consisting of d-disjunct matrices: For a security parameter λ, let $\mathcal{M} = \mathcal{M}(\lambda)$ be a message space. The KGen, Sign, and Vrfy algorithms of a D-ASIG scheme are defined in the same way as those of an ASIG scheme. DAgg and DVrfy algorithms are defined as follows:

- $(\widehat{\sigma}_1, \ldots, \widehat{\sigma}_u) \leftarrow \mathsf{DAgg}(\boldsymbol{G}, (\mathsf{pk}_1, \mathsf{m}_1, \sigma_1), \ldots, (\mathsf{pk}_\ell, \mathsf{m}_\ell, \sigma_\ell))$: The randomized or deterministic algorithm DAgg takes as input a d-disjunct matrix $\boldsymbol{G} \in \{0, 1\}^{u \times \ell} \cap \mathcal{G}$, a tuple $((\mathsf{pk}_1, \mathsf{m}_1, \sigma_1), \ldots, (\mathsf{pk}_\ell, \mathsf{m}_\ell, \sigma_\ell))$ of public keys, messages, and signatures, and it outputs a tuple $(\widehat{\sigma}_1, \ldots, \widehat{\sigma}_u)$ of aggregate signatures.
- $J \leftarrow \mathsf{DVrfy}(\boldsymbol{G}, ((\mathsf{pk}_1, \mathsf{m}_1), \ldots, (\mathsf{pk}_\ell, \mathsf{m}_\ell)), (\widehat{\sigma}_1, \ldots, \widehat{\sigma}_u))$: The deterministic algorithm DVrfy takes as input a d-disjunct matrix $\boldsymbol{G} \in \{0, 1\}^{u \times \ell} \cap \mathcal{G}$, a tuple $((\mathsf{pk}_1, \mathsf{m}_1), \ldots, (\mathsf{pk}_\ell, \mathsf{m}_\ell))$ of public keys and messages, and a tuple $(\widehat{\sigma}_1, \ldots, \widehat{\sigma}_u)$ of aggregate signatures, and it outputs a set J of public keys and messages.

We require that D-ASIG scheme satisfies correctness as follows.

Definition 6 (Correctness). A D-ASIG scheme D-ASig = (KGen, Sign, Vrfy, DAgg, DVrfy) satisfies correctness if the following conditions hold:

- For every $(\mathsf{pk}, \mathsf{sk}) \leftarrow \mathsf{KGen}(1^\lambda)$ and every $\mathsf{m} \in \mathcal{M}$, it holds that $\mathsf{Vrfy}(\mathsf{pk}, \mathsf{m}, \sigma) = 1$ with overwhelming probability, where $\sigma \leftarrow \mathsf{Sign}(\mathsf{sk}, \mathsf{m})$.
- For any $\ell = \mathsf{poly}(1^\lambda)$, every d-disjunct matrix $\boldsymbol{G} \in \{0, 1\}^{u \times \ell} \cap \mathcal{G}$, every $(\mathsf{pk}_1, \mathsf{sk}_1) \leftarrow \mathsf{KGen}(1^\lambda), \ldots, (\mathsf{pk}_\ell, \mathsf{sk}_\ell) \leftarrow \mathsf{KGen}(1^\lambda)$, and every $\mathsf{m}_1, \ldots, \mathsf{m}_\ell \in \mathcal{M}$, it holds that $\mathsf{DVrfy}(\boldsymbol{G}, ((\mathsf{pk}_1, \mathsf{m}_1), \ldots, (\mathsf{pk}_\ell, \mathsf{m}_\ell)), (\widehat{\sigma}_1, \ldots, \widehat{\sigma}_\ell)) = \emptyset$ with overwhelming probability, where $(\widehat{\sigma}_1, \ldots, \widehat{\sigma}_\ell) \leftarrow \mathsf{DAgg}(\boldsymbol{G}, (\mathsf{pk}_1, \mathsf{m}_1, \sigma_1), \ldots, (\mathsf{pk}_\ell, \mathsf{m}_\ell, \sigma_\ell))$ and $\sigma_i \leftarrow \mathsf{Sign}(\mathsf{sk}_i, \mathsf{m}_i)$ for every $i \in [\ell]$.

Regarding classical security notions of D-ASIGs, unforgeability and identifiability were formalized in [14]. Following the definitions of [14] and [4], we define these notions in the security model where an adversary is allowed to issue quantum queries to given oracles. First, the signing oracle SIGN which an adversary is given quantum access to is defined in the same way as Definition 4, since the Sign algorithm of a D-ASIG is defined in the same way as that of an ASIG.

Next, we define the two security notions in the quantum security model.

Definition 7 (daggEUF-qCMA security). A D-ASIG scheme D-ASig = (KGen, Sign, Vrfy, DAgg, DVrfy) satisfies daggEUF-qCMA security if for any QPT adversary A against ASig, its advantage $\mathsf{Adv}_{\mathsf{ASig}, \mathsf{A}}^{\mathsf{daggeuf\text{-}qcma}}(\lambda) := \Pr[\mathsf{A} \text{ wins}]$ is negligible in λ. [A wins] is the event that A wins in the following game:

Setup. A challenger generates $(\mathsf{pk}^*, \mathsf{sk}^*) \leftarrow \mathsf{KGen}(1^\lambda)$, and sends pk^* to A.

Queries. A is allowed to issue quantum queries to the SIGN oracle. Let q be the number of queries which A submits to SIGN.

Output. A outputs $(\boldsymbol{G}^{(1)}, PM^{(1)}, \widehat{\boldsymbol{\Sigma}}^{(1)}), \ldots, (\boldsymbol{G}^{(q+1)}, PM^{(q+1)}, \widehat{\boldsymbol{\Sigma}}^{(q+1)})$, where for every $i \in [q+1]$, $\boldsymbol{G}^{(i)} \in \{0,1\}^{u^{(i)} \times \ell^{(i)}} \cap \mathcal{G}$, $PM^{(i)} = ((\mathsf{pk}_1^{(i)}, \mathsf{m}_1^{(i)}), \ldots, (\mathsf{pk}_{\ell^{(i)}}^{(i)}, \mathsf{m}_{\ell^{(i)}}^{(i)}))$ and $\widehat{\boldsymbol{\Sigma}}^{(i)} = (\widehat{\sigma}_1^{(i)}, \ldots, \widehat{\sigma}_{u^{(i)}}^{(i)})$. The challenger computes $J^{(i)} \leftarrow \mathsf{DVrfy}(\boldsymbol{G}^{(i)}, PM^{(i)}, \widehat{\boldsymbol{\Sigma}}^{(i)})$ for every $i \in [q+1]$. A wins in this game if the following conditions hold: (i) $(\mathsf{pk}^*, \mathsf{m}^{*(i)}) \notin J^{(i)}$ for every $i \in [q+1]$, (ii) $(\mathsf{pk}^*, \mathsf{m}^{*(i)}) \in PM^{(i)}$ for every $i \in [q+1]$, and (iii) $(\mathsf{pk}^*, \mathsf{m}^{*(1)}), \ldots, (\mathsf{pk}^*, \mathsf{m}^{*(q+1)})$ are distinct.

Definition 8 (Identifiability against Quantum Chosen Message Attacks). Regarding the identifiability of a D-ASIG scheme D-ASig $=$ (KGen, Sign, Vrfy, DAgg, DVrfy), we define completeness and soundness, which are denoted by cmp-qCMA security and snd-qCMA security, respectively. Let A be a d-dishonest QPT adversary against D-ASig, where a QPT adversary A against D-ASig is d-dishonest if it outputs $(\boldsymbol{G}, (\mathsf{pk}_1, \mathsf{m}_1, \sigma_1), \ldots, (\mathsf{pk}_\ell, \mathsf{m}_\ell, \sigma_\ell))$ such that $|\{(\mathsf{pk}_i, \mathsf{m}_i) \mid i \in [\ell] \wedge \mathsf{Vrfy}(\mathsf{pk}_i, \mathsf{m}_i, \sigma_i) = 0\}| \leq d$, in the following security game:

Setup. A challenger generates $(\mathsf{pk}^*, \mathsf{sk}^*) \leftarrow \mathsf{KGen}(1^\lambda)$ and sends pk^* to A.

Queries. A is allowed to issue quantum queries to the SIGN oracle.

Output. A outputs $(\boldsymbol{G}, (\mathsf{pk}_1, \mathsf{m}_1, \sigma_1), \ldots, (\mathsf{pk}_\ell, \mathsf{m}_\ell, \sigma_\ell))$. The challenger computes $(\widehat{\sigma}_1, \ldots, \widehat{\sigma}_u) \leftarrow \mathsf{DAgg}(\boldsymbol{G}, (\mathsf{pk}_1, \mathsf{m}_1, \sigma_1), \ldots, (\mathsf{pk}_\ell, \mathsf{m}_\ell, \sigma_\ell))$ and $J \leftarrow \mathsf{DVrfy}(\boldsymbol{G}, ((\mathsf{pk}_1, \mathsf{m}_1), \ldots, (\mathsf{pk}_\ell, \mathsf{m}_\ell)), (\widehat{\sigma}_1, \ldots, \widehat{\sigma}_u))$.

The cmp-qCMA security and snd-qCMA security are defined as follows: For a set $\{(\mathsf{pk}_1, \mathsf{m}_1, \sigma_1), \ldots, (\mathsf{pk}_\ell, \mathsf{m}_\ell, \sigma_\ell)\}$, let $D = \{(\mathsf{pk}_i, \mathsf{m}_i) \mid i \in [\ell] \wedge \mathsf{Vrfy}(\mathsf{pk}_i, \mathsf{m}_i, \sigma_i) = 0\}$, and $\bar{D} = \{(\mathsf{pk}_i, \mathsf{m}_i) \mid i \in [\ell] \wedge \mathsf{Vrfy}(\mathsf{pk}_i, \mathsf{m}_i, \sigma_i) = 1\}$.

- **Completeness:** D-ASig satisfies cmp-qCMA security against d-dishonest adversaries, if for any d-dishonest QPT adversary A against D-ASig, its advantage $\mathsf{Adv}_{\mathsf{D\text{-}ASig},\mathsf{A}}^{\mathsf{cmp\text{-}qcma}}(\lambda) := \Pr\left[\exists \mathsf{m}^* \in \{\mathsf{m}_i\}_{i \in [\ell]}, (\mathsf{pk}^*, \mathsf{m}^*) \in \bar{D} \cap J\right]$ is negligible in λ.

- **Soundness:** D-ASig satisfies snd-qCMA security against d-dishonest adversaries, if for any d-dishonest QPT adversary A against D-ASig, its advantage $\mathsf{Adv}_{\mathsf{D\text{-}ASig},\mathsf{A}}^{\mathsf{snd\text{-}qcma}}(\lambda) := \Pr\left[\exists \mathsf{m}^* \in \{\mathsf{m}_i\}_{i \in [\ell]}, (\mathsf{pk}^*, \mathsf{m}^*) \in D \backslash J\right]$ is negligible in λ.

In addition, weak-snd-qCMA security is defined in the same way as snd-qCMA security except that the advantage of a d-dishonest QPT adversary A against D-ASig is defined as $\mathsf{Adv}_{\mathsf{D\text{-}ASig},\mathsf{A}}^{\mathsf{w\text{-}snd\text{-}qcma}}(\lambda) := \Pr[\exists \text{distinct } \mathsf{m}_{k_1}^*, \ldots, \mathsf{m}_{k_t}^* \in \{\mathsf{m}_i\}_{i \in [\ell]}, t \geq q + 1 \wedge (\mathsf{pk}^*, \mathsf{m}_{k_1}^*), \ldots, (\mathsf{pk}^*, \mathsf{m}_{k_t}^*) \in D \backslash J]$, where $k_1, \ldots, k_t \in [\ell]$ are distinct, and q is the number of queries which A submits to the SIGN oracle.

Weak-snd-qCMA security is a weak variant of snd-qCMA security since the winning condition of weak-snd-qCMA security is a special case of that of snd-qCMA security. Furthermore, Proposition 1 shows the relation between daggEUF-qCMA security and weak-snd-qCMA security.

Proposition 1. *If a D-ASIG scheme* D-ASig = (KGen, Sign, Vrfy, DAgg, DVrfy) *fulfills* daggEUF-qCMA *security, then* D-ASig *also satisfies* weak-snd-qCMA *security.*

Proof. Let A be a QPT adversary breaking the weak-snd-qCMA security of D-ASig, and let q be the number of signing-queries issued by A. By using A, we construct a QPT algorithm F breaking the daggEUF-qCMA security of D-ASig, as follows: F takes as input a public key pk^* and sends pk^* to A. When A issues a signing-query, F simulates the signing oracle by using the oracle given in the daggEUF-qCMA game, in the straightforward way. In **Output** phase, A outputs $(\boldsymbol{G}, (\mathsf{pk}_1, \mathsf{m}_1, \sigma_1), \ldots, (\mathsf{pk}_\ell, \mathsf{m}_\ell, \sigma_\ell))$. Then, F computes $(\widehat{\sigma}_1, \ldots, \widehat{\sigma}_u) \leftarrow$ $\mathsf{DAgg}(\boldsymbol{G}, (\mathsf{pk}_1, \mathsf{m}_1, \sigma_1), \ldots, (\mathsf{pk}_\ell, \mathsf{m}_\ell, \sigma_\ell))$. If there exist distinct $\mathsf{m}^*_{k_1}, \ldots, \mathsf{m}^*_{k_t} \in \{\mathsf{m}_i\}_{i \in [\ell]}$ such that $t \geq q+1$ and $(\mathsf{pk}^*, \mathsf{m}^*_{k_1}), \ldots, (\mathsf{pk}^*, \mathsf{m}^*_{k_t}) \in D \backslash J$, then F outputs $(\boldsymbol{G}^{(1)}, PM^{(1)}, \widehat{\Sigma}^{(1)}), \ldots, (\boldsymbol{G}^{(q+1)}, PM^{(q+1)}, \widehat{\Sigma}^{(q+1)})$ by setting $\boldsymbol{G}^{(i)} = \boldsymbol{G}$, $PM^{(i)} = ((\mathsf{pk}_1, \mathsf{m}_1), \ldots, (\mathsf{pk}_\ell, \mathsf{m}_\ell))$, and $\widehat{\Sigma}^{(i)} = (\widehat{\sigma}_1, \ldots, \widehat{\sigma}_u)$ for $i \in [q+1]$. Otherwise, F aborts.

We analyze the output of F. We assume that the A's output fulfills the winning condition $\exists$distinct $\mathsf{m}^*_{k_1}, \ldots, \mathsf{m}^*_{k_t} \in \{\mathsf{m}_i\}_{i \in [\ell]}$, $t \geq q+1$ and $(\mathsf{pk}^*, \mathsf{m}^*_{k_1}), \ldots, (\mathsf{pk}^*, \mathsf{m}^*_{k_t}) \in D \backslash J$. The first winning condition of daggEUF-qCMA security holds since for $i \in [q+1]$, there exists $(\mathsf{pk}^*, \mathsf{m}^*_{k_i}) \notin J^{(i)} = J$. The second condition also holds since for $i \in [q+1]$, $(\mathsf{pk}^*, \mathsf{m}^*_{k_i})$ is included in $PM^{(i)}$. The third condition holds clearly since $\mathsf{m}^*_{k_1}, \ldots, \mathsf{m}^*_{k_t}$ are distinct due to the winning condition of A. Hence, the output of F is a valid forgery in the daggEUF-qCMA security game, and then we obtain the advantage $\mathsf{Adv}^{\text{w-snd-qcma}}_{\text{D-ASig,A}}(\lambda) \leq \mathsf{Adv}^{\text{daggeuf-qcma}}_{\text{D-ASig,F}}(\lambda)$. □

Quantum Random Oracle Model and Certified Key Model. The quantum random oracle model is a model where a hash function is modeled as an ideal random function, and any party is allowed to issue quantum queries to this function as an oracle called a *quantum random oracle*. See [1] for details on this model. In addition, the certified key model is a model where every signer has to provide a key-pair $(\mathsf{pk}, \mathsf{sk})$ in order to certify pk, and $(\mathsf{pk}, \mathsf{sk})$ is added to the list L of registered key-pairs if sk is a valid secret key corresponding to pk. Following [9,13], we use this model in order to construct a quantum-secure AOTS scheme.

3   Quantum-Secure D-ASIG from Quantum-Secure ASIG

We consider a D-ASIG generic construction starting from an ASIG scheme and a non-adaptive group testing protocol. This scheme is the same as a generic construction of [14] except that we assume the underlying ASIG satisfies the formalized quantum-security. Then, we prove that this D-ASIG scheme satisfies our quantum-security. The D-ASIG scheme D-ASig = (KGen, Sign, Vrfy, DAgg, DVrfy) is as follows: Let ASig = (KGen^{asig}, Sign^{asig}, Vrfy^{asig}, Agg^{asig}, AVrfy^{asig}) be an ASIG scheme. For a matrix $\boldsymbol{G} \in \{0,1\}^{u \times \ell}$ and $i \in [u]$, let $S_i(\boldsymbol{G}) = \{j \mid j \in [\ell] \wedge g_{i,j} = 1\}$.

- $(\mathsf{pk}, \mathsf{sk}) \leftarrow \mathsf{KGen}(1^\lambda)$: Output $(\mathsf{pk}, \mathsf{sk}) \leftarrow \mathsf{KGen}^{asig}(1^\lambda)$.
- $\sigma \leftarrow \mathsf{Sign}(\mathsf{sk}, \mathsf{m})$: Output $\sigma \leftarrow \mathsf{Sign}^{asig}(\mathsf{sk}, \mathsf{m})$.
- $1/0 \leftarrow \mathsf{Vrfy}(\mathsf{pk}, \mathsf{m}, \sigma)$: Output $1/0 \leftarrow \mathsf{Vrfy}^{asig}(\mathsf{pk}, \mathsf{m}, \sigma)$.
- $(\widehat{\sigma}_1, \ldots, \widehat{\sigma}_u) \leftarrow \mathsf{DAgg}(\boldsymbol{G}, ((\mathsf{pk}_1, \mathsf{m}_1, \sigma_1), \ldots, (\mathsf{pk}_\ell, \mathsf{m}_\ell, \sigma_\ell)))$: For each $i \in [u]$, generate $\widehat{\sigma}_i \leftarrow \mathsf{Agg}^{asig}((\mathsf{pk}_k, \mathsf{m}_k, \sigma_k)_{k \in S_i(\boldsymbol{G})})$. Output $(\widehat{\sigma}_1, \ldots, \widehat{\sigma}_u)$.
- $J \leftarrow \mathsf{DVrfy}(\boldsymbol{G}, ((\mathsf{pk}_1, \mathsf{m}_1), \ldots, (\mathsf{pk}_\ell, \mathsf{m}_\ell)), (\widehat{\sigma}_1, \ldots, \widehat{\sigma}_u))$: Set $J \leftarrow \{(\mathsf{pk}_1, \mathsf{m}_1), \ldots, (\mathsf{pk}_\ell, \mathsf{m}_\ell)\}$. For each $i \in [u]$, if $\mathsf{AVrfy}^{asig}((\mathsf{pk}_k, \mathsf{m}_k)_{k \in S_i(\boldsymbol{G})}, \widehat{\sigma}_i) = 1$ holds, then set $J \leftarrow J \backslash \{(\mathsf{pk}_k, \mathsf{m}_k)\}_{k \in S_i(\boldsymbol{G})}$. Output J.

Theorems 1 and 2 show the quantum-security of D-ASig.

Theorem 1. *If an ASIG scheme* ASig *fulfills* aggEUF-qCMA *security, then the resulting D-ASIG scheme* D-ASig *satisfies* daggEUF-qCMA *security.*

Proof. Let A be a QPT adversary breaking the daggEUF-qCMA security of D-ASig, and let q be the number of (quantum) signing-queries issued by A. By using A, we construct a QPT algorithm F breaking the aggEUF-qCMA security of ASig, in the following way: F is given a public key pk^* and runs A by sending pk^*. When A issues a (quantum) signing-query, F responds to this query by using the given signing oracle, in the straightforward way. When A outputs $(\boldsymbol{G}^{(1)}, PM^{(1)}, \widehat{\Sigma}^{(1)}), \ldots, (\boldsymbol{G}^{(q+1)}, PM^{(q+1)}, \widehat{\Sigma}^{(q+1)})$ in **Output** phase, F finds distinct $(\mathsf{pk}^*, \mathsf{m}^{*(1)}), \ldots, (\mathsf{pk}^*, \mathsf{m}^{*(q+1)})$ such that $(\mathsf{pk}^*, \mathsf{m}^{*(i)}) \notin J^{(i)}$ and $(\mathsf{pk}^*, \mathsf{m}^{*(i)}) \in PM^{(i)}$ for $i \in [q+1]$ by following the procedure of the challenger of the daggEUF-qCMA game. Then, for every $i \in [q+1]$, it finds an index $j_i \in [u^{(i)}]$ such that $\mathsf{AVrfy}^{asig}((\mathsf{pk}_k, \mathsf{m}_k)_{k \in S_{j_i}(\boldsymbol{G}^{(i)})}, \widehat{\sigma}_{j_i}^{(i)}) = 1$ and $(\mathsf{pk}^*, \mathsf{m}^{*(i)}) \in (\mathsf{pk}_k, \mathsf{m}_k)_{k \in S_{j_i}(\boldsymbol{G}^{(i)})}$, and outputs $((\mathsf{pk}_k, \mathsf{m}_k)_{k \in S_{j_i}(\boldsymbol{G}^{(i)})}, \widehat{\sigma}_{j_i}^{(i)})_{i \in [q+1]}$. If there does not exist distinct $(\mathsf{pk}^*, \mathsf{m}^{*(1)}), \ldots, (\mathsf{pk}^*, \mathsf{m}^{*(q+1)})$ satisfying the above conditions, then F aborts.

F clearly simulates the environment of A. We analyze the output of F. The A's output satisfies the conditions (i) $(\mathsf{pk}^*, \mathsf{m}^{*(i)}) \notin J^{(i)}$ for every $i \in [q+1]$, (ii) $(\mathsf{pk}^*, \mathsf{m}^{*(i)}) \in PM^{(i)}$ for every $i \in [q+1]$, and (iii) $(\mathsf{pk}^*, \mathsf{m}^{*(1)}), \ldots, (\mathsf{pk}^*, \mathsf{m}^{*(q+1)})$ are distinct. Owing to the conditions (ii) and (iii), there exist distinct pairs $(\mathsf{pk}^*, \mathsf{m}^{*(1)}), \ldots, (\mathsf{pk}^*, \mathsf{m}^{*(q+1)})$ such that $(\mathsf{pk}^*, \mathsf{m}^{*(i)}) \in PM^{(i)}$ for $i \in [q+1]$. Furthermore, the condition (i) ensures that for $i \in [q+1]$, there exists an aggregate signature $\widehat{\sigma}_{j_i}^{(i)}$ (where $j_i \in [u^{(i)}]$) such that $(\mathsf{pk}^*, \mathsf{m}^{*(i)}) \in (\mathsf{pk}_k, \mathsf{m}_k)_{k \in S_{j_i}(\boldsymbol{G}^{(i)})}$ and $\mathsf{AVrfy}^{asig}((\mathsf{pk}_k, \mathsf{m}_k)_{k \in S_{j_i}(\boldsymbol{G}^{(i)})}, \widehat{\sigma}_{j_i}^{(i)}) = 1$. Therefore, the tuple of these pairs $\left((\mathsf{pk}_k, \mathsf{m}_k)_{k \in S_{j_i}(\boldsymbol{G}^{(i)})}, \widehat{\sigma}_{j_i}^{(i)}\right)_{i \in [q+1]}$ is a forgery of the aggEUF-qCMA security game. Then, we obtain the advantage $\mathsf{Adv}_{\mathsf{D\text{-}ASig},A}^{daggeuf\text{-}qcma}(\lambda) \leq \mathsf{Adv}_{\mathsf{ASig},F}^{aggeuf\text{-}qcma}(\lambda)$, and the proof is completed. $\square$

Theorem 2. *The resulting D-ASIG scheme* D-ASig *satisfies the following identifiability: Let d be arbitrary positive integer.*

(i) If $\boldsymbol{G}$ is a d-disjunct matrix, and an ASIG scheme ASig *meets* correctness, *then* D-ASig *satisfies* cmp-qCMA *security against d-dishonest adversaries.*

(ii) *If an ASIG scheme* ASig *meets* aggEUF-qCMA *security, then* D-ASig *satisfies* weak-snd-qCMA *security against d-dishonest adversaries.*

Proof. Let A be a QPT adversary against D-ASig. D-ASig fulfills cmp-qCMA security, in the same way as the proof of Theorem 2 in [14], because this proof do not have to use any list of quantum queries. Thus, we have $\mathsf{Adv}^{\mathrm{cmp\text{-}qcma}}_{\mathsf{D\text{-}ASig},\mathsf{A}}(\lambda) \leq \mathsf{negl}(\lambda)$. We can show that D-ASig satisfies weak-snd-qCMA security by combining Proposition 1 and Theorem 1. Thus, we have $\mathsf{Adv}^{\mathrm{w\text{-}snd\text{-}qcma}}_{\mathsf{D\text{-}ASig},\mathsf{A}}(\lambda) \leq \mathsf{Adv}^{\mathrm{aggeuf\text{-}qcma}}_{\mathsf{ASig},\mathsf{F}}(\lambda)$. $\square$

4 Quantum-Secure Aggregate One-Time Signature Scheme from Lattices

In this section, we propose a quantum-secure AOTS scheme which can be applied to the generic construction in Sect. 3. In order to prove the security of this AOTS, we describe the definition of the short integer solution (SIS) problem (or assumption) which is a computationally hard problem related to lattice problems (see [12]).

Definition 9 (SIS assumption). For a security parameter λ, let $k = k(\lambda)$, $q = q(\lambda)$, and $\beta = \beta(\lambda)$ be positive integers, and let $R = R_\lambda$ be a ring with a norm function $\|\cdot\| : R \to \mathbb{N}$. Then, the SIS assumption $\mathsf{SIS}_{R,k,q,\beta}$ is defined as follows: For any polynomial-time algorithm A, its advantage $\mathsf{Adv}_{\mathsf{SIS}_{R,k,q,\beta}}(\lambda, \mathsf{A}) :=$
$$\Pr_{\boldsymbol{a} \xleftarrow{\$} R_q^k} \left[\boldsymbol{a}^\top \cdot \boldsymbol{x} = \boldsymbol{0} \wedge \|\boldsymbol{x}\| \leq \beta \wedge \boldsymbol{x} \neq \boldsymbol{0} \mid \mathsf{A}(\boldsymbol{a}) \to \boldsymbol{x} \in R^k \right] \text{ is negligible in } \lambda.$$

Then, we present an SIS-based AOTS scheme. This scheme is constructed by combining an existing SIS-based one-time signature scheme of [10] and the generic construction of quantum-secure signatures in the quantum random oracle model [4]. Given one-time signatures $\sigma_1, \ldots, \sigma_\ell$, our AOTS generates an aggregate signature by computing $\sum_{i \in [\ell]} \sigma_i$. However, there is a rogue-key attack against this AOTS. Hence, we assume the certified key model in order to prevent this attack.

Our AOTS scheme AOTS = (KGen, Sign, Vrfy, Agg, AVrfy) is constructed as follows: Let k, q, β_s, β_m, and β_{Vrfy} be positive integers, and let R be a ring. For $\beta \in \mathbb{N}$, let $B_\beta = \{r \in R \mid \|r\| \leq \beta\}$. As system parameters of AOTS, choose $\boldsymbol{a} \xleftarrow{\$} R_q^k$ and a cryptographic hash function $H : \{0,1\}^* \to B_{\beta_m}$.

- $(\mathsf{pk}, \mathsf{sk}) \leftarrow \mathsf{KGen}(1^\lambda)$: Let p be a polynomial of λ. Choose $\boldsymbol{s}_0, \boldsymbol{s}_1 \xleftarrow{\$} B_{\beta_s}^k$ and $r \xleftarrow{\$} \{0,1\}^p$, and then compute $v_0 \leftarrow \boldsymbol{a}^\top \boldsymbol{s}_0$ and $v_1 \leftarrow \boldsymbol{a}^\top \boldsymbol{s}_1$. Output $\mathsf{pk} = (v_0, v_1, r)$ and $\mathsf{sk} = (\boldsymbol{s}_0, \boldsymbol{s}_1, r)$.
- $\sigma \leftarrow \mathsf{Sign}(\mathsf{sk}, \mathsf{m})$: Compute $h \leftarrow H(\mathsf{m}, r)$. Output $\sigma \leftarrow \boldsymbol{s}_0 \cdot h + \boldsymbol{s}_1 \in R^k$.
- $1/0 \leftarrow \mathsf{Vrfy}(\mathsf{pk}, \mathsf{m}, \sigma)$: Compute $h \leftarrow H(\mathsf{m}, r)$. Output 1 if $\boldsymbol{a}^\top \sigma = v_0 \cdot h + v_1$ and $\|\sigma\| \leq \beta_{\mathsf{Vrfy}}$, and output 0 otherwise.
- $\hat{\sigma} \leftarrow \mathsf{Agg}((\mathsf{pk}_1, \mathsf{m}_1, \sigma_1), \ldots, (\mathsf{pk}_\ell, \mathsf{m}_\ell, \sigma_\ell))$: Output $\hat{\sigma} \leftarrow \sum_{i \in [\ell]} \sigma_i \in R^k$.

- $1/0 \leftarrow \mathsf{AVrfy}((\mathsf{pk}_1, \mathsf{m}_1), \ldots, (\mathsf{pk}_\ell, \mathsf{m}_\ell), \widehat{\sigma})$: Let $\mathsf{pk}_i = (v_{i,0}, v_{i,1}, r_i)$ and $h_i = H(\mathsf{m}_i, r_i)$ for $i \in [\ell]$. Output 1 if $\boldsymbol{a}^\top \cdot \widehat{\sigma} = \sum_{i \in [\ell]} (v_{i,0} \cdot h_i + v_{i,1})$ and $\|\widehat{\sigma}\| \leq \beta_{\mathsf{Vrfy}}$, and output 0 otherwise.

We assume $R = \mathbb{Z}^{n \times n}$ or $R = \mathbb{Z}[X]/(X^n + 1)$. AOTS satisfies correctness if $\beta_{\mathsf{Vrfy}} \geq L \cdot n \cdot \beta_s (\beta_{\mathsf{m}} + 1)$, where L is the maximum number of signatures which are given to Agg. Furthermore, Theorem 3 shows the quantum-security of AOTS.

Theorem 3. *If the SIS assumption* $\mathsf{SIS}_{R,k,q,2\beta_{\mathsf{Vrfy}}}$ *holds, then the AOTS scheme* AOTS *satisfies* aggEUF-qCMA *security in the quantum random oracle model and the certified key model.*

Proof. Let A be a QPT adversary against AOTS. The tuple $\mathsf{OTS} = (\mathsf{KGen}, \mathsf{Sign}, \mathsf{Vrfy})$ can be seen as a one-time signature (OTS) scheme constructed by applying the OTS scheme of [10] to the generic construction of signatures [4] satisfying the EUF-qCMA security in the quantum random oracle model. Following [4], EUF-qCMA security is defined in the same way as aggEUF-qCMA security except that the winning condition of the adversary is to output $(\mathsf{m}^{(1)}, \sigma^{(1)}), \ldots, (\mathsf{m}^{(q+1)}, \sigma^{(1)})$ such that $\mathsf{Vrfy}(\mathsf{pk}^*, \mathsf{m}^{(i)}, \sigma^{(i)}) = 1$ for all $i \in [q+1]$, and $\mathsf{m}^{(1)}, \ldots, \mathsf{m}^{(q+1)}$ are distinct. Hence, if by using A, we can construct a QPT algorithm S breaking the EUF-qCMA security of OTS, then we can also construct a QPT algorithm breaking $\mathsf{SIS}_{R,k,q,2\beta_{\mathsf{Vrfy}}}$, by combining the proofs of Theorem 3.2 in [10] and Theorem 3.13 in [4]. Hence, we construct this QPT algorithm S against OTS, as follows: Given a public key $\mathsf{pk}^* = (v_0^*, v_1^*, r^*)$ in the EUF-qCMA security game, S sets the list of certified key-pairs $\mathsf{L} \leftarrow \emptyset$ and sends pk^* to A. When A submits a key-pair $(\mathsf{pk}, \mathsf{sk})$ to certify pk, S sets $\mathsf{L} \leftarrow \mathsf{L} \cup \{(\mathsf{pk}, \mathsf{sk})\}$ if sk is the secret key corresponding to pk. When A issues a (quantum) query to the signing oracle or the quantum random oracle H, it responds to this query by using the oracles given in the EUF-qCMA security game. When A outputs $(PM^{(1)}, \widehat{\sigma}^{(1)})$ and $(PM^{(2)}, \widehat{\sigma}^{(2)})$, then S checks whether (i) $\mathsf{AVrfy}(PM^{(i)}, \widehat{\sigma}^{(i)}) = 1$ for $i \in \{1, 2\}$, (ii) $(\mathsf{pk}^*, \mathsf{m}^{*(1)}) \in PM^{(1)}$ and $(\mathsf{pk}^*, \mathsf{m}^{*(2)}) \in PM^{(2)}$ are distinct, (iii) all public keys $\mathsf{pk}_1^{(i)}, \ldots, \mathsf{pk}_{\ell^{(i)}}^{(i)}$ in $PM^{(i)}$ except for pk^* are registered in L for $i \in \{1, 2\}$, and (iv) $\mathsf{Vrfy}(\mathsf{pk}^*, \mathsf{m}^{*(i)}, \sigma^{*(i)}) = 1$ for $i \in \{1, 2\}$, where let $\sigma^{*(i)} = \widehat{\sigma}^{(i)} - \sum_{j \in [\ell^{(i)}] \text{ s.t. } \mathsf{pk}_j^{(i)} \neq \mathsf{pk}^*} \left(\boldsymbol{s}_{j,0}^{(i)} \cdot h_j^{(i)} + \boldsymbol{s}_{j,1}^{(i)} \right)$ for $i \in \{1, 2\}$, $\mathsf{sk}_j^{(i)} = (\boldsymbol{s}_{j,0}^{(i)}, \boldsymbol{s}_{j,1}^{(i)}, r_j^{(i)})$, and $h_j^{(i)} = H(\mathsf{m}_j^{(i)}, r_j^{(i)})$ for $i \in \{1, 2\}$ and $j \in [\ell^{(i)}]$. If so, it outputs the pairs $(\mathsf{m}^{*(1)}, \sigma^{*(1)})$ and $(\mathsf{m}^{*(2)}, \sigma^{*(2)})$. Otherwise, it aborts.

The output of S is a valid forgery of the EUF-qCMA security game because it satisfies the condition that $\mathsf{Vrfy}(\mathsf{pk}^*, \mathsf{m}^{*(i)}, \sigma^{*(i)}) = 1$ for $i \in \{1, 2\}$, and $\mathsf{m}^{*(1)}$ and $\mathsf{m}^{*(2)}$ are distinct. Hence, S breaks the EUF-qCMA security of OTS with at least probability $\mathsf{Adv}_{\mathsf{OTS,A}}^{\mathrm{ageuf\text{-}qcma}}(\lambda)$. Then, there exists a QPT algorithm solving $\mathsf{SIS}_{R,k,q,2\beta_{\mathsf{Vrfy}}}$, by combining Theorem 3.2 in [10] and Theorem 3.13 in [4]. Due to these theorems, the probability of solving $\mathsf{SIS}_{R,k,q,2\beta_{\mathsf{Vrfy}}}$ is at least $O(1/(\eta^2 q_H^6)) \cdot \mathsf{Adv}_{\mathsf{AOTS,A}}^{\mathrm{ageuf\text{-}qcma}}(\lambda) - \mathsf{negl}(\lambda)$, where $\eta = \mathsf{poly}(\lambda)$ is some polynomial of λ, and q_H is the number of (quantum) queries which A issues to the H oracle. $\square$

5 Conclusion Remarks

In this paper, we proposed a quantum-secure D-ASIG scheme. To this end, we did the following: First, we formalized quantum-security notions of ASIGs and D-ASIGs. Second, we showed a D-ASIG generic construction starting from any aggEUF-qCMA secure ASIG scheme and any non-adaptive group testing protocol with d-disjunct matrices, and then we proved that this scheme satisfies our quantum-security notions daggEUF-qCMA security, cmp-qCMA security, and weak-snd-qCMA security. Finally, we proposed a lattice-based AOTS scheme with aggEUF-qCMA security. To obtain a quantum-secure D-AOTS scheme, it is possible to apply this AOTS to the D-ASIG generic construction. Therefore, the resulting D-AOTS scheme is the first quantum-secure D-ASIG.

We should remark that it is possible to construct a D-AOTS with snd-qCMA security, by combining our AOTS with a proof of knowledge system in the security model where an adversary utilizes quantum computations by itself, but issues only classical queries. This construction is the same as the D-ASIG generic construction with the soundness of the identifiability of D-ASIGs [14].

Acknowledgement. This research was conducted under a contract of "Research and development on new generation cryptography for secure wireless communication services" among "Research and Development for Expansion of Radio Wave Resources (JPJ000254)", which was supported by the Ministry of Internal Affairs and Communications, Japan.

References

1. Boneh, D., Dagdelen, Ö., Fischlin, M., Lehmann, A., Schaffner, C., Zhandry, M.: Random oracles in a quantum world. In: Lee, D.H., Wang, X. (eds.) ASIACRYPT 2011. LNCS, vol. 7073, pp. 41–69. Springer, Heidelberg (2011). https://doi.org/10.1007/978-3-642-25385-0_3

2. Boneh, D., Gentry, C., Lynn, B., Shacham, H.: Aggregate and verifiably encrypted signatures from bilinear maps. In: Biham, E. (ed.) EUROCRYPT 2003. LNCS, vol. 2656, pp. 416–432. Springer, Heidelberg (2003). https://doi.org/10.1007/3-540-39200-9_26

3. Boneh, D., Kim, S.: One-time and interactive aggregate signatures from lattices (2020). https://crypto.stanford.edu/~skim13/agg_ots.pdf

4. Boneh, D., Zhandry, M.: Secure signatures and chosen ciphertext security in a quantum computing world. In: Canetti, R., Garay, J.A. (eds.) CRYPTO 2013. LNCS, vol. 8043, pp. 361–379. Springer, Heidelberg (2013). https://doi.org/10.1007/978-3-642-40084-1_21

5. Du, D.Z., Hwang, F.K.: Combinatorial Group Testing and Its Applications, 2nd edn. Series on Applied Mathematics, vol. 12. World Scientific (2000)

6. Hartung, G., Kaidel, B., Koch, A., Koch, J., Rupp, A.: Fault-tolerant aggregate signatures. In: Cheng, C.-M., Chung, K.-M., Persiano, G., Yang, B.-Y. (eds.) PKC 2016. LNCS, vol. 9614, pp. 331–356. Springer, Heidelberg (2016). https://doi.org/10.1007/978-3-662-49384-7_13

7. Hohenberger, S., Sahai, A., Waters, B.: Full domain hash from (leveled) multi-linear maps and identity-based aggregate signatures. In: Canetti, R., Garay, J.A. (eds.) CRYPTO 2013. LNCS, vol. 8042, pp. 494–512. Springer, Heidelberg (2013). https://doi.org/10.1007/978-3-642-40041-4_27

8. Ishii, R., et al.: Aggregate signature with traceability of devices dynamically generating invalid signatures. In: Zhou, J., et al. (eds.) ACNS 2021. LNCS, vol. 12809, pp. 378–396. Springer, Cham (2021). https://doi.org/10.1007/978-3-030-81645-2_22

9. Lu, S., Ostrovsky, R., Sahai, A., Shacham, H., Waters, B.: Sequential aggregate signatures, multisignatures, and verifiably encrypted signatures without random oracles. J. Cryptol. **26**(2), 340–373 (2013)

10. Lyubashevsky, V., Micciancio, D.: Asymptotically efficient lattice-based digital signatures. J. Cryptol. **31**(3), 774–797 (2018)

11. Maller, M., Bowe, S., Kohlweiss, M., Meiklejohn, S.: Sonic: zero-knowledge SNARKs from linear-size universal and updatable structured reference strings. In: ACM Conference on Computer and Communications Security, pp. 2111–2128. ACM (2019)

12. Micciancio, D., Regev, O.: Worst-case to average-case reductions based on Gaussian measures. SIAM J. Comput. **37**(1), 267–302 (2007)

13. Rückert, M., Schröder, D.: Aggregate and verifiably encrypted signatures from multilinear maps without random oracles. In: Park, J.H., Chen, H.-H., Atiquzzaman, M., Lee, C., Kim, T., Yeo, S.-S. (eds.) ISA 2009. LNCS, vol. 5576, pp. 750–759. Springer, Heidelberg (2009). https://doi.org/10.1007/978-3-642-02617-1_76

14. Sato, S., Shikata, J., Matsumoto, T.: Aggregate signature with detecting functionality from group testing. IACR Cryptol. ePrint Arch., 1219 (2020)

15. Shor, P.W.: Polynomial-time algorithms for prime factorization and discrete logarithms on a quantum computer. SIAM J. Comput. **26**(5), 1484–1509 (1997)

Improving Robustness and Visibility of Adversarial CAPTCHA Using Low-Frequency Perturbation

Takamichi Terada[✉], Vo Ngoc Khoi Nguyen, Masakatsu Nishigaki, and Tetsushi Ohki

Graduate School of Integrated Science and Technology, Shizuoka University, 3-5-1 Johoku, Naka-ku, Hamamatsu, Shizuoka 432-8011, Japan
terada@sec.inf.shizuoka.ac.jp

Abstract. CAPTCHA is a type of Turing test used to distinguish between humans and computing machine. However, image-based CAPTCHAs are losing their function as Turing tests owing to the improvement of image recognition using machine learning. This paper proposes an Adversarial CAPTCHA that provides attacking resistance to CAPTCHAs by using Adversarial Example (AE) as well as maintaining visibility by reducing image degradation. The proposed CAPTCHA maintains the difficulty of solving CAPTCHAs using computing machine by adding resistance against the attack using a machine learning classifiers. The proposed CAPTCHA is evaluated using three evaluation experiments, i.e., the attack using a machine learning classifier, the image quality, and the solving workload. The three evaluation experiments show that an Adversarial CAPTCHA is resistant to the attack by machine learning and is as convenient as the existing CAPTCHA.

1 Introduction

CAPTCHA is a fully automated Turing test that can distinguish between humans and computing machines [14]. Previous studies have applied CAPTCHAs to tasks that are easy for humans to identify but difficult for computing machines, such as complex image recognition tasks [2,5]. However, along with recent advances in image recognition algorithms, such as deep neural network-based image recognition algorithms, computing machines have shown a high recognition accuracy for image-based CAPTCHAs [3,13]. The fact that the recognition accuracy of image-based CAPTCHAs no longer differs between humans and computing machines indicates that CAPTCHAs are losing their ability to determine humans from computing machines.

We propose a novel CAPTCHA algorithm for generating CAPTCHA images that are difficult for machine learning classifier to solve but easy for humans uses. Osadchy et al. proposed DeepCAPTCHA as an attempt to create a CAPTCHA that is difficult to attack using a machine learning classifier [10]. With DeepCAPTCHA, an adversarial example (AE) is applied to make it difficult for machine learning classifiers to attack the CAPTCHA [4]. However, DeepCAPTCHA applies perturbations multiple times, and when the perturbations are applied at a level where the image is sufficiently resistant

© The Author(s), under exclusive license to Springer Nature Switzerland AG 2022
L. Barolli et al. (Eds.): AINA 2022, LNNS 450, pp. 586–597, 2022.
https://doi.org/10.1007/978-3-030-99587-4_50

to machine learning classifiers, the image quality tends to degrade, which may make it difficult for humans to solve.

Considering the problems, we limit the number of perturbations based on a basic iterative method. In addition, we propose a low-frequency CAPTCHA image generation algorithm that limits the perturbations to low-frequency components. We evaluated the proposed CAPTCHA from three perspectives, i.e., the attack resistance by machine learning classifiers, visibility, and convenience, and demonstrated its effectiveness and reliability. We then show that the proposed algorithm is not only able to prevent a degradation of the image quality but also generate CAPTCHA images that are robust against known AE removal attacks, such as median filters, with minimal degradation to human vision.

2   Related Works

2.1   Adversarial Example

AE is a method that can make a machine learning classifier misclassify by adding small perturbations to the input. Goodfellow et al. proposed the fast gradient sign method (FGSM) to efficiently generate AE [4]. In Eq. (1), The AE created by FGSM algorithm is defined as X^{adv}, with the inputs X and their labels y, the model parameters θ, and the parameters indicating the number of updates ε.

$$X^{adv} = X + \varepsilon \ \text{sign} \left(\nabla_X J(\theta, X, y) \right) \tag{1}$$

To reduce the degradation of the image quality owing to perturbations, the basic iterative method (BIM), a method that iteratively applies FGSM by limiting the perturbations of the original image for each pixel, was proposed [9]. BIM [9] uses the Clip function defined as Eq. (2) to limit the perturbations added to the input image to within α of the original image.

$$\text{Clip}_{X,\alpha} \left\{ X^{adv} \right\} = \min \left\{ 255, X + \alpha, \max \left\{ 0, X - \alpha, X^{adv} \right\} \right\} \tag{2}$$

The update equation for AE in BIM is defined as Eq. (3):

$$\begin{aligned} X_0^{adv} &= X, \\ X_{N+1}^{adv} &= \text{Clip}_{X,\alpha} \{ X_N^{adv} + \varepsilon \ \text{sign}(\nabla_X J(\theta_N, X_N^{adv}, y_{true})) \} \end{aligned} \tag{3}$$

where N is the number of updates, X_N^{adv} is the input X after the Nth update, and y_{true} is the correct label corresponding to the input X. Using Eq. (3), we can restrict the range of possible values of the input X_{N+1}^{adv} after $N+1$ updates to within the range shown in Eq. (2).

2.2   Adversarial CAPTCHA

Adversarial CAPTCHA is a method that adds resistance against attacks that use machine learning classifiers to attack a CAPTCHA [14] by applying AE techniques to CAPTCHAs that are considered able to distinguish whether the operator is a human or a computing machine. DeepCAPTCHA, one of the Adversarial CAPTCHA methods proposed by Osadchy et al., uses an AE generation method that is resistant to machine learning classifiers as well ass perturbation removal through preprocessing using image processing filters [10]. Osadchy et al. labeled this method immutable adversarial noise (IAN), and we apply this name in the present paper as well. In this paper, we develop an AE based on the IAN algorithm, which is resistant to perturbation removal methods using median filters and can reduce the degradation in image quality caused by multiple perturbations and apply it to a CAPTCHA.

3   Adversarial CAPTCHA Using Low-Frequency Perturbation

Our proposed method consists of the AE generation process using low-frequency perturbation and CAPTCHA challenge generation process. For clarity, we refer to the proposed AE method, the method that generates images for use in CAPTCHAs using AE, and proposed CAPTCHA, the whole system that presents the generated AE images. The AE generation process consists of BIM [9], a method that limits the amount of image modification through perturbations, and a low-frequency perturbation method that limits the perturbations to the low-frequency components of the image. The CAPTCHA challenge generation process presents the user with the challenge of selecting one image from multiple AE images. Here we describe these two processes.

3.1   Basic Iterative Method

To achieve a method with minimal degradation to human vision, we use BIM, shown in Eq. (3), for AE generation. By setting an upper bound on the amount of perturbations, BIM can set a lower bound of image quality. Controlled image quality makes it easier for human to solve CAPTCHAs.

3.2   Low-Frequency Perturbation

Assuming that the attacker knows that the CAPTCHA image is composed of AE, the attacker may remove the perturbations from the CAPTCHA image by using a median filter before inputting the CAPTCHA to the machine learning classifier. Perturbations were added independently of the semantic information of the images. Therefore, the perturbations are distributed within the high-frequency region in the frequency domain and can be easily removed by applying median filters. We add perturbations only in the low-frequency such that the generated perturbations are distributed in the low-frequency region, making it difficult to be removed using a median filter. The low-frequency perturbation $\mathscr{F}(\boldsymbol{X}, c)$ for image X is defined as follows:

$$\mathscr{F}(\boldsymbol{X},c) = \text{IFFT}(\text{LPF}(\text{FFT}(\boldsymbol{X}),c)) \tag{4}$$

where the function FFT is a 2D Fourier transform, IFFT is a 2D inverse Fourier transform, and $\text{LPF}(\boldsymbol{X},c)$ is a Low-pass filter with c as the cutoff frequency within the 2D frequency domain. In the proposed AE method, $\mathscr{F}$ is applied to the image $\text{Clip}_{\boldsymbol{X},\alpha}\{\boldsymbol{X}^{adv}\}$ in Eq. (2). In the following, we refer to perturbations that are limited to low frequencies after applying BIM as low-frequency perturbations.

The algorithm used for creating an AE with high visibility and resistance to machine learning attacks using low-frequency perturbations is shown in Algorithm 1. Here, δ_α is a parameter that indicates the amount of update of α. In this paper, we assumed a median filter as a perturbation removal method. We assumed a median filter as a perturbation removal method as it has been validated as the most efficient perturbation removal method among various image processing filters, as described in the Osadchy's work [10].

Algorithm 1 Generation of low-frequency perturbations

Require: Let Net be a deep learning network; I be the original image; C_i be the true class, C_d be the predicted class; $confidence$ be the confidence value of the network; p be the threshold for the confidence value; M_f be the median filter; $\mathscr{F}$ be the Fourier transform; $\mathscr{C}$ be the cutoff value; L_i be the upper limit of the number of image updates and L_α be the upper limit of the number of perturbation updates.

1:   $adv(I,C_d,p) \leftarrow I$; $\triangleright$ adv denotes an AE generation function
2:   $\eta \leftarrow 0$;
3:   Update the perturbation at most m times until it can change the true label to a different label:
4:   **while** ($Net(M_f(adv(I,C_d,p)))) = C_i \vee (m < L_\alpha)$ **do**
5:      Update the perturbation at most n times using a low-frequency perturbation:
6:      **while** $(Net(adv(I,C_d,p))) \neq C_d \vee (confidence < p) \vee (n < L_i)$ **do**
7:          $\eta \leftarrow$ *run BIM with noise magnitude* α;
8:          $\tilde{\eta} \leftarrow \mathscr{F}^{-1}(\mathscr{C}(\mathscr{F}(\eta)))$;
9:          $adv(I,C_d,p) \leftarrow adv(I,C_d,p) + \tilde{\eta}$;
10:     **end while**
11:     Update perturbation by adding a small value:
12:     $\alpha = \alpha + \delta_\alpha$;
13: **end while**
14: **Output:** η

3.3 CAPTCHA Challenge Generation

Figure 1(a) shows an example of the operation screen of the proposed CAPTCHA. The proposed CAPTCHA is similar to the DeepCAPTCHA format shown in Fig. 1(b). Since the DeepCAPTCHA shows a single AE image as the CAPTCHA challenge, our system allows the user to choose one image from a set of source images with the same label as the challenge. As shown in Fig. 1, the difference between proposed CAPTCHA and

DeepCAPTCHA is the number of A.E. images included in each CAPTCHA challenge. By using multiple images, we aim to improve the machine learning attack resistance of the CAPTCHA system by increasing the number of AEs used in the single CAPTCHA challenge. The images used in the proposed CAPTCHA are Caltech-256 [6], which is a publicly available dataset. Here, images used for options are AEs created using Algorithm 1.

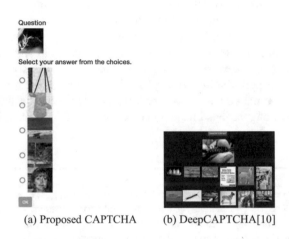

(a) Proposed CAPTCHA   (b) DeepCAPTCHA[10]

Fig. 1. Example of (a) the proposed CAPTCHA and (b) DeepCAPTCHA system.

4   Evaluation

4.1   Preliminary

In our experiment, we trained the CAPTCHA generation model using the MNIST and Caltech-256 datasets [1,6]. For the MNIST dataset, we used a multi-layer perceptron consisting of three fully connected layers with 300, 100 and 10 nodes for the network model. For the Caltech-256 dataset, we used VGG-Net [12]. For the scenario using MNIST with a multi layer perceptron, the number of times the median filter applied t was set to 6. The kernel size of the median filter k was set to 5, and the cutoff size of the low-pass filter c was set to 9. For the scenario using Caltech-256 with VGG-Net, the number of times the median filter applied t was set to 1. The kernel size of the median filter k was set to 7, and the cutoff size of the low-pass filter c was set to 102. We defined cut-off values from preliminary experiment. Note that we applied the low-pass filter only to the proposed AE method for each scenario. We set the threshold p of the confidence value to 0.8. If the confidence value of AE exceeds this threshold, the AE creation process ends. The number of times the median filter applied, the kernel size, and the confidence value were the same as those used by DeepCAPTCHA [10]. In the following part, we will evaluate the proposed CAPTCHA through three evaluations. Among the three evaluations, an attack resistance evaluation using a machine learning classifier is described in Sect. 4.2, an image quality evaluation through a subjective evaluation is detailed in Sect. 4.3.1, and a CAPTCHA workload evaluation is described in Sect. 4.4. The results of the experiments are presented in each section along with the experiment methodology.

4.2 Attack Resistance Evaluation

4.2.1 Procedure

The attack resistance of each CAPTCHA system was evaluated by calculating and comparing the attack resistance retention rate from the results of the classification by the machine learning classifier. We defined the attack resistance retention rate as the ratio of inputs that an attacker cannot solve to the total number of inputs. Here, we assume that the attacker can apply a median filter with parameters k and t to the challenge image before inputting it to the machine learning classifier.

Equation (5) shows the attack resistance retention rate P_d using the total number of inputs N, the number of creation failures n_{mf}, and the number of countermeasure failures n_{sf}. Here, n_{mf} is defined as the number of inputs images for which the CAPTCHA generator failed to create an Adversarial CAPTCHA. n_{sf} is defined as the number of generated Adversarial CAPTCHAs that adversarial perturbation can be removed.

$$P_d = \frac{N - (n_{mf} + n_{sf})}{N} \tag{5}$$

The resistance against solving by machine learning classifiers is evaluated base on the attack resistance retention rate P_d.

4.2.2 Result

Table 1 shows the attack resistance retention rate of IAN and the proposed AE method. From Table 1, we can see that the proposed AE method has a higher resistance retention rate than IAN.

Table 1. Comparison of resistance retention rate varying AE generation methods and attack target models. In each experiment, we used all 10,000 images in the dataset.

Method	Dataset	Model	Resistance retention rate
IAN	MNIST	MLP	79.72%
Proposed AE method	MNIST	MLP	81.29%
IAN	Caltech-256	VGGNet	81.62%
Proposed AE method	Caltech-256	VGGNet	82.10%

4.3   Image Quality Evaluation

4.3.1   Procedure

An image quality evaluation experiment was conducted using a double stimulus continuous quality scale [8]. The dual stimulus method is a subjective quality evaluation method for images and videos defined in ITU-R BT.500-14. In our experiment, 16 university students majoring in computer science between the ages of 21 and 24 were asked to participate in the experiment.

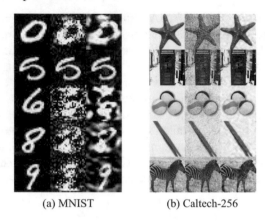

(a) MNIST   (b) Caltech-256

Fig. 2. Example of the output result of AE using (a) MNIST and (b) Caltech-256. The original image, corresponding output of the IAN, and that of the proposed AE method are arranged from left to right. It can be seen that the AE creation the proposed AE method suppresses the degradation of the image quality in some images. In the image quality evaluation, a set of the original image and one of the two types of AE was used.

Prior to the experiment, we provided informed consent to all participants and conducted a practice session to confirm the experimental procedure. Note that we used CAPTCHA images that were completely unrelated to the experiment in a practice. In our experiment, we combined two types of attack methods (IAN and the proposed AE method) and two types of datasets (MNIST and Caltech-256) with four scenarios.

We prepared 10 CAPTCHA challenge pairs for each of the four scenarios and presented them to the participants. Each pair consists of an unperturbed original image and a perturbed AE image correspond to the scenario. The participants scored the image quality of the original and AE images on a 100-point scale for each scenario. The participants are free to switch between the original and AE images as a pair during the evaluation. We show some examples of AE produced by IAN and the proposed AE method in Fig. 2. The original image, corresponding output of the IAN, and that of the proposed AE method are arranged from left to right. We presented the original image and one of the two types of AE to the participants in pairs in the image quality evaluation.

We show the average difference score between the original and AE images with a 95% confidence interval for each scenario. Outliers were detected according to the method applied in ITU-R BT.500-14 Annex.1.

4.3.2 Result

Figure 3 is the result of the image quality evaluation. Each bar represents the mean of difference in image quality evaluation score between the original image and perturbed image in each experimental condition. A small value of the mean of difference indicates that the quality degradation from the original image is small. Note that the error bars show the 95% confidence interval. As shown in Fig. 3, although the difference is insignificant, the mean of difference of MNIST in the proposed AE method is smaller than that of existing method. In addition, the mean of difference in Caltech-256 is smaller in the proposed AE method than in the existing methods. Additionally, we conducted outlier detection following the method specified in the ITU-R document and confirmed that there were no participants whose answers corresponded to the outlier.

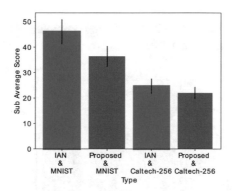

Fig. 3. Results of image quality evaluation. Each error bar is 95% confidence interval. The scores using MNIST are IAN (46.0) and the proposed AE method (36.4). In addition, the scores using Caltech-256 are IAN(24.7) and the proposed AE method (22.0). A small value of the mean of differences indicates that the quality degradation from the original image is small.

4.4 CAPTCHA Workload Evaluation

4.4.1 Procedure

We compared the convenience of the proposed CAPTCHA to existing CAPTCHAs by evaluating the workload using NASA-TLX [7]. An evaluation of NASA-TLX used in this paper was conducted in Japan by evaluating the axes of the Japanese version of NASA-TLX developed by Haga et al. [7,11]. The CAPTCHAs to be compared were GIMPY and reCAPTCHA, which are used in many different websites [5,15]. We recruited 250 participants for this survey using Lancers.jp[1]. The participants were asked to go to the web page of the survey from the URL provided in the work request form on Lancers.jp, and solve the three types of challenges, i.e., the proposed CAPTCHA,

---

[1] URL: https://www.lancers.jp/.

reCAPTCHA, and GIMPY, and then evaluate the workload using NASA-TLX for each of them.

Before conducting the survey, the participants were given an explanation regarding the survey, and were then given a practice session to evaluate the results after they were sufficiently familiar with the operation. Figure 1(a) showed example of challenge images that were used in the practice session. We selected all challenge images from Caltech-256 with 224×224 image size. The survey was conducted by solving the CAPTCHA for a specified number of times. We evaluated the workload using NASA-TLX for each of the question types GIMPY, reCAPTCHA, and the proposed CAPTCHA.

4.4.2 Result

First, the weighted average of NASA-TLX for each CAPTCHA method is shown in Fig. 4. The scores are GIMPY (54.0), the proposed CAPTCHA (40.7), and reCAPTCHA (39.8) in ascending order.

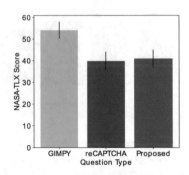

Fig. 4. Results of workload survey. Each error bar is standard deviation. The scores are GIMPY (54.0), the proposed CAPTCHA (40.7), and reCAPTCHA (39.8) in ascending order.

We also tested The NASA-TLX weighted means for each of these three forms using the Shapiro-Wilk test. On running the Shapiro-Wilk test, we set the null hypothesis as the NASA-TLX weighted mean for each participant in each CAPTHCA method follows a normal distribution, with $188°$ of freedom and a significance level of 5%. Table 2 shows the results of the Shapiro-Wilk test. Table 2 shows that the null hypothesis cannot be rejected only for GIMPY, whereas the null hypothesis can be rejected for the other forms. These results indicate that the NASA-TLX weighted mean of GIMPY is not necessarily non-parametric.

Because the NASA-TLX weighted mean values other than GIMPY are non-parametric, we conducted a Friedman test. As the null hypothesis, "There is no difference in the NASA-TLX weighted mean values for each participant in each question format" with $188°$ of freedom and a significance level of 5%. Table 3 shows the results of the Friedman test. Because the null hypothesis is rejected, we can see that there is a difference in the NASA-TLX weighted mean values among the three problem formats.

Table 2. Results of the Shapiro-Wilk test for each question type (note, 188° of freedom and a significance level of 5% were applied). The null hypothesis cannot be rejected only for GIMPY.

Question type	p value
GIMPY	0.480
reCAPTCHA	<**0.001****
Proposed CAPTCHA	**0.030***

*:$p < 0.05$, **:$p < 0.001$

Table 3. The results of the Friedman test(note: 188° of freedom and a significance level of 5% were used). It can be confirmed that the null hypothesis is rejected.

Question type	p value
Proposed CAPTCHA, reCAPTCHA, GIMPY	<**0.001****

**:$p < 0.001$

The NASA-TLX weighted mean values were then tested using Wilcoxon's signed rank test. As the null hypothesis, "There is no difference between the NASA-TLX weighted averages of the targets", and the following three targets were applied: GIMPY and CAPTCHA of the proposed CAPTCHA, reCAPTCHA and CAPTCHA of the proposed CAPTCHA, and reCAPTCHA and GIMPY. In addition, the significance level was 5%. The results of the Wilcoxon's signed rank test are shown in Table 4.

The effect size d showed in Table 4 was calculated using Cohen's effect size d. It can be seen that at a significance level of 1.6%, corrected for the Bonferroni method, the proposed CAPTCHA is significantly smaller than that of GIMPY, but not significantly smaller than that of reCAPTCHA.

Table 4. Results of Wilcoxon's signed rank test (significance level of 5% and 1.6% significance level when corrected using the Bonferroni method). It can be seen that the null hypothesis of the combination of reCAPTCHA and GIMPY and the combination of CAPTCHA used by the proposed CAPTCHA and GIMPY is rejected.

Question type 1	Question type 2	Effect size d	p value (Question type 1 < Question type 2)
reCAPTCHA	GIMPY	0.834	< **0.001****
Proposed method	GIMPY	0.824	< **0.001****
Proposed method	reCAPTCHA	0.064	0.766

**:$p < 0.001$

5 Discussion

In the workload evaluation described in Sect. 4.4, we compared the convenience of the proposed CAPTCHA to other widespread CAPTCHA systems. As we can see from Table 4, Wilcoxon's signed-rank test rejects the null hypothesis in the evaluation using GIMPY and the proposed CAPTCHA but cannot reject the null hypothesis in the evaluation using reCAPTCHA and the proposed CAPTCHA. These results show that the proposed method is more convenient than GIMPY and as convenient as reCAPTCHA while maintaining the resistance to machine learning attacks.

One of the reasons why the participants evaluated reCAPTCHA as so convenient is the inconsistency of the reCAPTCHA procedure. The reCAPTCHA version we used is v2. reCAPTCHA v2 is a method that discriminates between humans and computing machines by selecting objects such as cars and traffic lights scattered in the image only when there is a possibility of access by a bot. In our experiment, we expected that the task of selecting cars or traffic lights would occur at least once in 10 trials, but this did not happen for a small number of participants. Hence, in the open-ended section of the post-survey questionnaire, some participants mentioned that the workload test for the reCAPTCHA was completed only by clicking on the check-boxes. Therefore, it is important to consider that the convenience of reCAPTCHA is highly variable than that of proposed CAPTCHA.

As for the limitations, our proposed CAPTCHA requires users to browse images. In other words, users who cannot see the images, such as the visually impaired users, cannot use the system. As a future challenge, we will extend the CAPTCHA format to include audio and other modalities in addition to images.

6 Conclusion

This paper proposed an Adversarial CAPTCHA with high visibility using low-frequency perturbations. Our proposed method makes it difficult to solve the CAPTCHA by machine learning classifiers and, at the same time, maintains visibility by reducing image degradation.

The proposed CAPTCHA was tested separately from three perspectives: attack resistance against machine learning classifiers, visibility, and usability. We hope that the use and application of the CAPTCHA method proposed in this paper will not only allow the further development of CAPTCHAs as proof of human work, but also contribute to the development of machine learning and various other fields.

References

1. Deng, L.: The MNIST database of handwritten digit images for machine learning research [Best of the Web]. IEEE Signal Process. Mag. **29**, 141–142 (2012)
2. Elson, J., Douceur, J.J., Howell, J., Saul, J.: Asirra: a CAPTCHA that exploits interest-aligned manual image categorization. In: Proceedings of 14th ACM Conference on Computer and Communications Security, pp. 366—374 (2007)
3. Golle, P.: Machine learning attacks against the Asirra CAPTCHA. In: Proceedings of the 15th ACM Conference on Computer and Communications Security, pp. 535–542 (2008)

4. Goodfellow, I.J., Shlens, J., Szegedy, C.: Explaining and harnessing adversarial examples. In: International Conference on Learning Representations (2015)
5. Google: choosing the type of reCAPTCHA—Google developers (2021). https://developers.google.com/recaptcha/docs/versions. Accessed 27 Jan 2022
6. Griffin, G., Holub, A., Perona, P.: Caltech-256 object category dataset. Technical report 7694 (2007)
7. Hart, S.G., Staveland, L.E.: Development of NASA-TLX (task load index): results of empirical and theoretical research. In: Advances in Psychology, vol. 52, pp. 139–183. Elsevier (1988)
8. ITU-R: Recommandation BT.500-14. Methodology for the subjective assessment of the quality of television pictures (2019)
9. Kurakin, A., Goodfellow, I.J., Bengio, S.: Adversarial examples in the physical world. In: International Conference on Learning Representations Workshop (2017)
10. Osadchy, M., Hernandez-Castro, J., Gibson, S., Dunkelman, O., Pérez-Cabo, D.: No bot expects the DeepCAPTCHA! Introducing immutable adversarial examples, with applications to CAPTCHA generation. IEEE Trans. Inf. Forensics Secur. **12**, 2640–2653 (2017)
11. Shigeru, H., Naoki, M.: Japanese version of NASA task load index: sensitivity of its workload score to difficulty of three different laboratory tasks. Jpn. J. Ergonomics **32**, 71–79 (1996)
12. Simonyan, K., Zisserman, A.: Very deep convolutional networks for large-scale image recognition. In: International Conference on Learning Representations (2015)
13. Sivakorn, S., Polakis, I., Keromytis, A.D.: I am robot: (Deep) learning to break semantic image CAPTCHAs. In: IEEE European Symposium on Security and Privacy, pp. 388–403 (2016)
14. Von Ahn, L., Blum, M., Hopper, N.J., Langford, J.: CAPTCHA: using hard AI problems for security. In: International Conference on the Theory and Applications of Cryptographic Techniques, pp. 294–311 (2003)
15. Von Ahn, L., Blum, M., Langford, J.: Telling humans and computers apart automatically. Commun. Assoc. Comput. Mach. **47**, 56–60 (2004)

Comparative Study of Ensemble Learning Techniques for Fuzzy Attack Detection in In-Vehicle Networks

Dorsaf Swessi[✉] [iD] and Hanen Idoudi[iD]

National School of Computer Science (ENSI), University of Manouba,
Manouba, Tunisia
{dorsaf.swessi,hanen.idoudi}@ensi-uma.tn

Abstract. Nowadays, vehicles have become more complex due to the increased number of electronic control units communicating through in-vehicle networks. Controller area network (CAN) is one of the most used protocols for in-vehicle networks. Still, it lacks a conventional security infrastructure, making it highly vulnerable to numerous attacks. The Fuzzy attack is one of the most challenging attacks for in-vehicle networks because of its randomly spoofed injected messages similar to the legitimate traffic and its numerous physical effects on the vehicle. In this paper, we focus on Fuzzy attack detection in the internal vehicle network by investigating the performances of ensemble learning techniques to mitigate this attack. We evaluated their efficiency on realistic datasets and on a new advanced stealthy attack dataset with physical impacts on the vehicle. eXtreme, Light, and Category Gradient Boosting, as well as Bagging ensemble learning techniques, in particular, showed a considerable improvement in detection performance in terms of accuracy, training and testing time reduction, and a decreased false alarm rate.

Keywords: In-vehicle communications · Can-bus · Fuzzy attack · Ensemble learning · Intrusion detection systems

1 Introduction

Intelligent transportation systems (ITS) are critical components of smart cities and play a vital role in constructing future smart highways. Moreover, vehicles have become more complex due to the expensive increase of electronic control units (ECUs) communicating through in-vehicle networks. Controller area network (CAN) is one of the most commonly used networking protocols in vehicles that provide a standard for simultaneous transmission of data between in-vehicle components [1]. Despite its advantages, like including error detection mechanisms and the reduced weight, complexity,and wiring cost, CAN protocol lacks standard security infrastructure and is highly vulnerable to various attacks since all nodes share a single medium and communicate without being able to check the source of the messages [15]. Thus, attackers can easily inject malicious data

© The Author(s), under exclusive license to Springer Nature Switzerland AG 2022
L. Barolli et al. (Eds.): AINA 2022, LNNS 450, pp. 598–610, 2022.
https://doi.org/10.1007/978-3-030-99587-4_51

into the CAN bus to track it or initiate other aggressive attacks, such as DoS, Fuzzy, Masquerade, Replay, and many other attacks [1, 15]. The Fuzzy attack is a common type of complex injection attack launched on the CAN bus because it injects arbitrary messages, creating more complex traffic patterns and requiring more training iterations to create a stable security mechanism [9].

In recent years, researchers are looking for appropriate systems that support vehicular network characteristics and provide robust security mechanisms since most of the security solutions defined for traditional networks are not suitable for vehicular networks [8]. The intrusion detection system (IDS) is a widely used approach that analyses the traffic for indicators of security breaches and creates an alert for any observed security anomaly [8, 15]. Moreover, machine learning is able to realize anomaly-based detection systems capable of detecting unknown and zero-day attacks, learning and trains itself by analyzing network activity and increasing its detection accuracy over time [1, 15].

This work intends to define a complete study about the performances of several ensemble learning techniques on detecting Fuzzy attacks using three well-known realistic datasets, including a new dataset with real advanced stealthy fabrication attacks with physically verified effects on vehicles.

The rest of this paper is structured as follows. In Sect. 2, we expose related works related to ML-based IDS for vehicular networks. In Sect. 3, we explain the properties of the chosen datasets, then, we briefly introduce the ensemble learning techniques used in our work. Section 4 and 5 present the performance metrics, discuss and analyze the results, and provide a performance comparison of the different ensemble learning models. Finally, Sect. 6 gives the conclusion of the study.

2 Related Work

2.1 Fuzzy Attack in In-Vehicle Communications

A Fuzzy, Fuzzing, or Fuzzier attack is a common type of injection attack launched on CAN bus that causes vehicles to display unexpected states or fail. During this attack, an attacker injects spoofed random IDs and DATA fields into the bus at a high frequency to compromise the ECUs [9] then it saves the erroneous data for later analysis. Hence, the medium fills up with mainly injected messages, which obliterate legitimate messages [9]. Figure 1 illustrates the entire process. Besides, a simple Fuzzy attack can cause severe damage with minimal efforts such as the impotent of the accelerator pedal, the activation of the dash, heater, and lights, the movement of the seat positions, the malfunction of the navigation system, and repeated short beeping sound, etc. [7, 12]. A Fuzzy attack can be performed either by injecting CAN IDs generated randomly or by injecting only IDs that appear during regular traffic, which mimics the legitimate traffic [8, 12]. Thus, learning the complex attack patterns is challenging for the ML models since the dataset structure is random, and the attack pattern is almost similar to the legitimate traffic. Furthermore, many works [2, 4, 5, 7, 12, 16] approved that detecting such attack is more difficult compared to other attacks.

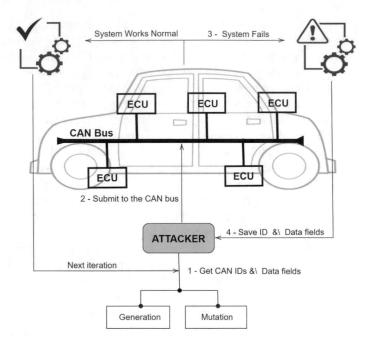

Fig. 1. Fuzzy attack scenarios on vehicle CAN bus network.

2.2 Machine Learning Based IDS for IoV

Many works adopted Machine learning techniques to build efficient intrusion detection systems. Few of them used ensemble learning techniques. A tree-based Intelligent IDS on the internet of vehicles (IoV) that detects DoS and Fuzzy attacks is proposed by Li Yang et al. [16]. Firstly, they tested the performance of Decision tree (DT), Random Forest (RF), Extra trees (ET), and XGradient Boost (XGB) Methods and applied multi-threading to get a lower execution time. Then select the three most models that generate the most inferior execution time as a meta-classifier in the second layer of the stacking ensemble model. Besides that, they used an ensemble feature selection (FS) technique to improve the confidence of the selected features. Finally, the authors tested the model on the Car-Hacking Dataset.

Vuong et al. [14] proposed a decision tree-based method for detecting DoS and command injection attacks on robotic vehicles using cyber and physical features to show up the importance of incorporating the physical features in improving the performance of the model. They tested their model in a collected dataset. In addition to DoS and command injection attacks detection, they also provide in [13] a lightweight intrusion detection mechanism that can detect malware against network and malware against CPU using both cyber and physical input features using decision tree model. In [11], authors developed an intrusion detection system based on Gradient Boosting Decision Tree (GBDT) for CAN-Bus and proposed a new feature based on entropy as the feature construction of

Table 1. Fuzzy attack datasets overview

Dataset	Cars	Injected msg	Normal msg	Injection time	Modified fields	Consequences
Car-Hacking	NA	491,847	3,259,176	every 0.5 ms	CAN ID, DATA	Malfunction of the vehicle
ROAD	A single vehicle	1.061	93,867	every 5 ms	CAN ID	Seat positions movement Dash and lights activation Accelerator pedal impotent
Survival analysis	KIA Soul HYUNDAI Sonata CHEVROLET Spark	33,751	239,246	every 0.3 ms	CAN ID, DATA	Repeated short beeping sound Heater turned on Navigation system error

GBDT and used a dataset from a real domestic car Alsvin CHANA to evaluate the model. Authors in [6] showed that tree-based and ensemble learning models show more performance in detection compared to other models. Random forest, Bagging, and AdaBoosting methods are trained and tested on the Can-hacking dataset, and the DT-based model results in yield performance. Most of the mentioned works focused on the application of some ensemble learning techniques to identify in-vehicle threats including Fuzzy attacks based on certain performance metrics. Authors in [6] utilized three ensemble learning methods that were trained on a single basic dataset and evaluated using four classical performance measures with no execution time. Similarly, [16] examined five ensemble learning models with a single basic dataset and four classical performance metrics and execution time.

To the best of our knowledge, there is no current ensemble learning comparison research for in-vehicle communications and, in particular, fuzzy attack detection. Therefore, the purpose of this research is to examine and compare 11 ensemble learning models in detecting in particular Fuzzy attacks using three distinct datasets.

3   Methodology

3.1   Datasets Description

The effectiveness of a machine learning model depends on a sufficient amount of network traffic trace datasets that can be used to train and validate it. Instead of this, real traced datasets that cover a variety of normal and abnormal samples are needed. To evaluate the ensemble learning models, we considers three different datasets for its implementation in the vehicle's CAN bus. The first two datasets are called "Car Hacking Dataset" and "Survival Analysis Dataset" provided by the Hacking and Countermeasure Research Lab (HCRL) in 2018. The third dataset is called Real ORNL Automotive Dynamometer (ROAD) proposed by Miki et al. in 2020. Datasets characteristics are depicted in Table 1.

Car-Hacking Dataset. Except for the normal run data, this dataset is divided into four car-hacking datasets: "DoS dataset", "Fuzzy dataset", "Gear dataset" and "RPM dataset" [10]. Every dataset has a total of 30 to 40 min of CAN traffic and contains 300 intrusions of message injection, and each intrusion is performed for 3 to 5 s. Moreover, it includes DoS, Fuzzy, spoofing the drive gear, and spoofing the RPM gauge attacks. This dataset is used in numerous CAN IDS studies since it includes long attack captures and a massive number of instances per attack.

Survival Analysis Dataset. HCRL published two different datasets extracted from three different types of vehicles: Hyundai Sonata, Kia Soul, and Chevrolet Spark [4]; one dataset represent normal driving data, and the other includes the abnormal driving data for three attack scenarios: Flooding, Fuzzy, and Malfunction that occurred when attack packets were injected for five seconds every 20 s and each attack capture is 25–100 s long.

ROAD Dataset. It is the first open dataset with real advanced stealthy fabrication attacks that have physically verified effects on the vehicle [12]. All of the data is from a single vehicle, and it contains 12 attack-free captures that last about three hours and 33 attack captures that last about 30 min. The main three attacks are the Fuzzy attack, Targeted ID Attacks, and Accelerator Attacks. They provided two versions for all of the targeted ID attacks: the original fabrication attack and a version slightly modified to simulate a masquerade attack. Since injection attacks are the most frequent CAN bus threats, the CAN data files are logged using the standard can-utils candump format and contain the recorded time (in sec.), the channel, the CAN message identifier (in Hex), and the data field (in Hex).

3.2   Dataset Pre-processing

An effective machine learning model depends not only on a sufficient amount of representative data but also on data that has been appropriately pre-processed. First, As shown in Fig. 2, we must prepare the datasets since they are in various file formats with missing values and different attributes. This can be accomplished by standardizing all of the sample formats, removing redundant fields, and filling in or lowering missing values. The final attributes are: Timestamp which is the recorded time, the identifier of CAN message (CAN ID), the data value decomposed into eight fields (DATA0 to DATA7), and the Flag attribute represents a normal or injected message. For Car-hacking and survival analysis datasets, we dropped missing values in the data fields as well as the number of data bytes "DLC" field. Next, we standardized and normalized each dataset's data because ML training is often more efficient with normalized data, thus, features with numerical values are required. To do this, we converted DATA and CAN ID fields into a numerical values using the "LabelEncoder" function from Sklearn. Then, each feature with numerical values is set to the range of 0.0 to 1.0 using "StandardScaler" and "MinMaxScaler" functions from Sklearn for data standardization and normalization. Finally, we decomposed the dataset into an 80% training set and 20% testing set.

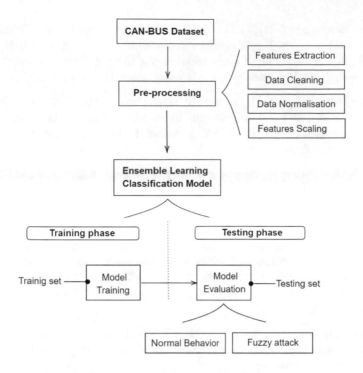

Fig. 2. Ensemble learning-based IDS architecture.

3.3   Ensemble Learning Techniques

After the data preparation phase, 11 ensemble learning models that showed various achievements based on their data structure and the included algorithms were trained and tested. These models are described as below:

Decision Trees (DT). An effective and popular mechanism that reflects a series of decisions that helps to decide afterward. Suitable for larger datasets and composed of three key components: nodes (identified with a feature attribute), arcs (identified with a feature value), and leaves (identified with a class).

Random Forest (RF). It creates, trains, and utilizes a majority of votes of various DTs to classify the data. A forest with 100 trees was generated in this study, and the "Gini" index was used as a learning criterion.

Bagging Tree (BT). Also known as Bootstrap Aggregation, a sampling technique in which the classifiers are weighted and trained by making sampling with a replacement, the plurality vote is used to classify the data. In this study, the decision tree was chosen as a classifier, and 10 trees were built.

Extra Trees (ET). ET is an ensemble model based on a collection of randomized decision trees. It is generated by processing different subsets of the dataset. In this model, 100 trees were built.

Gradient Boosing (GB). GB model is an ensemble learning algorithm suitable for data detection, which has excellent volume and few features. It is designed for speed and performance improvement using the gradient descent method to combine many decision trees. 100 trees were generated in this work.

Adaptive Boosting (AB). An iterative ensemble method that sets the weights of classifiers and trains the data sample in each iteration to ensure accurate predictions of unusual observations. 100 different decision trees are used to classify the intrusion message.

Voting (VO). Voting is one of the easiest ways that employs several classifiers to make predictions based on the most frequent one. We set the voting hyperparameter to "hard" and used GB, BT, DT, and RF estimators because of their high accuracy and low false-positive rate.

Stacking (ST). A common ensemble technique uses the output of various classifiers as input for the meta-classifier to construct a robust classifier. GB was used as a meta-learner, and GB, BT, DT, and RF serve as the base models. We also tested the model (XST) with XGB, BT, LGB, CGB as base models and XGB as meta-learner.

eXtreme Gradient Boosting (XGB). XGBoost is a sophisticated version of the GB technique that is nearly ten times faster, has high predictive power, and aids in reducing overfitting.

Light Gradient Boosting (LGB). LightGB is a fast and distributed version of the GB algorithm that outperforms all other models with its large dataset compatibility, faster training speed, and lower memory usage.

Category Gradient Boosting (CGB). CatBoost is another version of boosting algorithms that can successfully handle categorical variables automatically that reduce the possibility of overfitting.

4   Experimental Results

We used Python for experimental analysis with the free library Scikit-learn for machine learning support. Several metrics are used to evaluate a machine learning model's performance and compare it with other existing Models. The classification produces four outcomes: True Positive (TP) is when the IDS detects an abnormal message as malicious; False Negative (FN) is when the IDS detects an abnormal message as normal; False Positive (FP) is when the IDS detects a normal message as malicious and True Negative (TN) is when the IDS detects a normal message as normal. In the following, we present the performance metrics calculated according to these outcomes and used to calculate the ensemble learning models. The Accuracy (ACC) is the ratio between the number of correct predictions to the total number of predictions made (1). The Balanced accuracy (BA), mainly used in anomaly detection, is based on true positive rate (Recall) and true negative rate (TNR) to calculated the imbalance in classes (2). The False Negative Rate (FNR), also known as false alarm rate, is the number of incorrect positive predictions divided by the total number of negatives (3). F-measure (F-score) is the harmonic mean of both recall and precision (4).

$$ACC = \frac{TP + TN}{TP + TN + FP + FN} \tag{1}$$

$$BA = \frac{\frac{TP}{TP+FN} + \frac{TN}{TN+FP}}{2} \tag{2}$$

$$FPR = \frac{FP}{FP + TN} \tag{3}$$

$$F - score = \frac{2 * TP}{2 * TP + FP + FN} \tag{4}$$

In addition, we measured two advanced performance metrics; the Area Under the Receiver Operating Characteristics (ROC-AUC), which is a single scalar value that measures the overall performance of a classifier and is calculated by adding successive trapezoid areas below the ROC curve [3]. And the area under the precision-recall curve (PR-AUC) that shows the precision (the probability of correct detection of positive values) as a function of recall (the number of accurate positive predictions divided by the total number of positives) [3]. Nevertheless, We compared the models based on their training and testing execution times.

Table 2 lists the detection performance of the different ensemble learning models on ROAD datasets. Overall, ensemble learning models perform well with approximately 99% accuracy. We can see that the XStacking method outperforms all the other models with 99.81% accuracy, 91.96% F-measure, 93.81% AUC, and 92.26% PR-AUC. Moreover, the Gradient Boosting method has shown the lowest false positive rate with 0%. Interestingly, XGB, LGB, and CGB techniques have provided a high predictive power with low false alarms compared to the other models. However, Random Forest and Decision Tree models have shown lower performance. AdaBoost failed to detect Fuzzy attacks with 6.64% PR-AUC, 5.23% F-measure, and 51.65% ROC-AUC and Extra Trees and have the highest false alarm rate 0.14%.

Table 2. ROAD dataset

Model	Classical performance metrics				Advanced metrics	
	ACC	BA	FPR	F-Score	AUC	PR-AUC
DT	99.71	91.66	0.085	87.70	91.66	88.03
BT	99.64	86.16	0.02	83.33	86.16	85.47
ET	99.34	78.87	0.14	68.51	78.87	71.17
RF	99.43	78.07	0.03	70.97	78.07	76.53
GB	99.77	90.85	0	89.93	90.85	90.96
AB	98.28	51.65	0.53	5.23	51.65	6.64
ST	99.75	93.36	0.091	89.47	93.36	89.64
VO	99.68	87.86	0.016	85.58	87.86	87.19
XGB	99.63	86.79	0.042	83.17	86.79	84.76
LGB	99.62	86.57	0.053	82.49	86.57	84.01
CGB	99.69	87.87	0.005	85.99	87.87	87.74
XST	99.81	93.81	0.037	91.96	93.81	92.26

Table 3. Survival analysis dataset

Model	Classical performance metrics				Advanced metrics	
	ACC	BA	FPR	F-Score	AUC	PR-AUC
DT	99.51	98.95	0.305	98.03	98.95	98.14
BT	99.65	98.91	0.11	98.58	98.91	98.71
ET	94.84	81.28	0.709	75.20	81.28	80.24
RF	95.51	83.54	0.56	78.83	83.54	83.06
GB	94.82	80.05	0.33	74.26	80.05	80.81
AB	92.12	69.39	0.41	55.17	69.39	69.91
ST	99.71	98.99	0.061	98.80	98.99	98.93
VO	99.66	99	0.125	98.62	99	98.73
XGB	99.78	99.15	0.02	99.09	99.15	99.20
LGB	99.38	97.68	0.07	97.42	97.68	97.75
CGB	99.72	98.99	0.04	98.87	98.99	99
XST	99.79	99.20	0.023	99.13	99.20	99.23

According to the obtained results in Table 3, we can observe that the detection performance of the models varies, and the XGB, LGB, and CGB methods outperform all the other models taking into account all performance metrics, which make XStacking the most preferment algorithm. In addition, considering the ROAD dataset, the Bagging, Voting, and Decision Tree techniques have a higher performance than the GradientBoost algorithm. Nevertheless, Extra Trees, Random Forest, and AdaBoost still have the worst performances with an imbalance in classes and the highest false-positive rate with 0.709%, 0.56%, and 0.41%, respectively.

As can be seen from Table 4, the detection accuracy is approximately 100% with a low false-positive rate in all models, and the Stacking/XStacking methods achieve the highest performances with a false positive rate equals to 0%. Table 5 shows the training and test time of the models. The results demonstrate that the LightGB technique outperforms the other models on all datasets. The Stacking/XStacking models have the worst execution time even if they have the best performance metrics. Furthermore, we can see that Bagging, Decision Tree, CatBoost, and XGBoost have a sufficient execution time. Still, XGB has the second-lowest training and test times and showed superior performance compared to the other models.

5 Discussion

In the experiments conducted in this study, the ensemble learning models detected the Fuzzy attack very well, with a detection accuracy of approximately 99% with well-balanced classes and an F-measure scale. However, as outlined in Figure 3, the AdaBoost, Random Forest, Gradient Boost, and Extra Tree models performed poorly, especially AdaBoost, which failed to detect the attack. Figures 3,4,5 illustrate the false Positive rates and the execution time bar charts for all the models on the three different datasets. We point out that Bagging, XGB, LGB, and CGB models demonstrated the best detection performance against Fuzzy attacks with reduced training and test times and a low false alarm rate compared to the other models. Thanks to these models, the XStacking model

Table 4. Car-Hacking dataset

Model	Classical performance metrics				Advanced metrics	
	ACC	BA	FPR	F-Score	AUC	PR-AUC
DT	100	99.99	0.0005	99.99	99.99	99.99
BT	100	99.99	0.0005	99.99	99.99	99.99
ET	100	100	0.0002	100	100	100
RF	100	100	0.0002	100	100	100
GB	99.97	99.89	0.0015	99.89	99.89	99.90
AB	99.98	99.92	0.0023	99.91	99.92	99.92
ST	100	100	0	100	100	100
VO	100	100	0.0005	100	100	100
XGB	100	100	0.0002	100	100	100
LGB	100	100	0.0002	100	100	100
CGB	100	99.99	0.0002	99.99	99.99	99.99
XST	100	100	0	100	100	100

Table 5. Training and test time of models on different datasets

Model	ROAD		Survival analysis		Car-Hacking		Avg.
	Train	Test	Train	Test	Train	Test	
DT	0.320	0.005	1.668	0.01	49.21	0.103	8
BT	1.965	0.052	9.914	0.158	367	2.133	63
ET	4.036	0.261	27.31	1.867	560	10.54	100
RF	6.914	0.239	35.96	1.317	1388	11.13	240
GB	12.65	0.046	77.82	0.123	1117	2.391	201
AB	3.627	0.155	18	0.866	381	10.66	69
ST	120.1	0.332	787	2.48	13690	14.02	2435
VO	35.14	0.536	151	2.378	3192	13.92	565
XGB	5.093	0.059	17.51	0.087	239	0.832	43
LGB	1.22	0.109	1.936	0.249	20.56	2.384	4
CGB	34.91	0.109	75.35	0.277	606	3.26	119
XST	207	0.339	582	0.975	6396	8.363	1199

outperforms all performance metrics, but it has the worst execution time, which is inappropriate for Intra-vehicle communications. Consequently, we believe that the Bagging XGB, LGB, and CGB models are the most suitable algorithms for detecting Fuzzy attacks in CAN Bus networks regarding speed, detection rates, and the low false alarm rate.

The datasets used in this study are chosen for different reasons, such as; all the datasets have real CAN data with realistic labeled attacks. Car-Hacking dataset [10] is a big dataset with long attack captures and a massive number of instances per attack. Researchers commonly use it to train and evaluate their works. The survival analysis dataset [4] is the only dataset that contains real attacks that have a real effect on multiple vehicles, and the same set of attacks are repeated multiple times on three different vehicles. The ROAD dataset [12] is

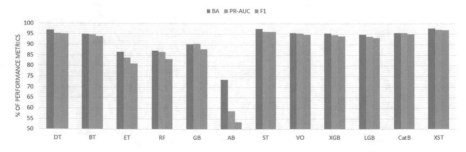

Fig. 3. Ensemble learning models comparison based on BA, PR-AUC and F-Score metrics for all datasets.

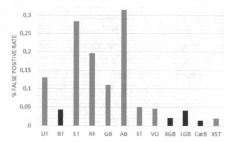
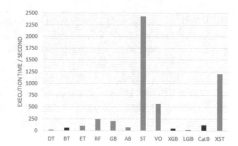

Fig. 4. EL models comparison based on FPR metric for all datasets.

Fig. 5. EL models comparison based on Execution time for all datasets.

a new dataset that describes attacks that have physically verified effects on the vehicle. Nevertheless, these datasets suffer from some limitations; Car-hacking and survival analysis datasets attacks are simple and can be detected with simple frequency-based detectors, and the injected messages contain the same values. That's why the model can easily learn it. For ROAD datasets, the injected messages are not labeled, and all of the data is from a single-vehicle.

6   Conclusion

Securing CAN Bus networks is challenging due to their vulnerability to various attacks such as the Fuzzy attack that arbitrarily injects randomly spoofed messages similar to the legitimate traffic and causes many physical harms to the vehicle. Through this paper, we gave a brief description of the Fuzzy attack, then we examined and compared the feasibility of ensemble learning techniques to detect this attack. Next, we tested these models on three CAN Bus datasets from real vehicles with real Fuzzy attacks. Finally, we discussed and compared these models based on various performance metrics. Our work showed that ensemble learning techniques have very good performances to detect fuzzy attack. Particularly, Bagging, XGB, LGB, and CGB models demonstrated the best detection performances in terms of accuracy, reduced training and test times and a low false alarm. In future work, we intend to build a complete security architecture for a specific vehicular scenario based on investigated techniques that will be able to tackle different attacks in real-time manner.

References

1. Alshammari, A., Zohdy, M.A., Debnath, D., Corser, G.: Classification approach for intrusion detection in vehicle systems. Wireless Eng. Technol. **9**(4), 79–94 (2018)
2. Barletta, V.S., Caivano, D., Nannavecchia, A., Scalera, M.: Intrusion detection for in-vehicle communication networks: an unsupervised kohonen som approach. Future Internet **12**(7), 119 (2020)

3. Boyd, K., Eng, K.H., Page, C.D.: Area under the precision-recall curve: point estimates and confidence intervals. In: Blockeel, H., Kersting, K., Nijssen, S., Zelezny, F. (eds.) Machine Learning and Knowledge Discovery in Databases. LNCS, vol. 8190. Springer, Heidelberg (2013). https://doi.org/10.1007/978-3-642-40994-3_29

4. Han, M.L., Kwak, B.I., Kim, H.K.: Anomaly intrusion detection method for vehicular networks based on survival analysis. Vehic. Commun. **14**, 52–63 (2018)

5. Hossain, M.D., Inoue, H., Ochiai, H., Fall, D., Kadobayashi, Y.: Lstm-based intrusion detection system for in-vehicle can bus communications. IEEE Access **8**, 185489–185502 (2020)

6. Kalkan, S.C., Sahingoz, O.K.: In-vehicle intrusion detection system on controller area network with machine learning models. In: 2020 11th International Conference on Computing, Communication and Networking Technologies (ICCCNT). IEEE, pp. 1–6 (2020)

7. Lee, H., Jeong, S.H., Kim, H.K.: Otids: A novel intrusion detection system for in-vehicle network by using remote frame. In: 2017 15th Annual Conference on Privacy, Security and Trust (PST), pp. 57–5709. IEEE (2017)

8. Lokman, S.F., Othman, A.T., Abu-Bakar, M.H.: Intrusion detection system for automotive controller area network (can) bus system: a review. EURASIP J. Wireless Commun. Netw. **2019**(1), 1–17 (2019)

9. Oehlert, P.: Violating assumptions with fuzzing. IEEE Security Privacy **3**(2), 58–62 (2005)

10. Seo, E., Song, H.M., Kim, H.K.: Gids: gan based intrusion detection system for in-vehicle network. In: 2018 16th Annual Conference on Privacy, Security and Trust (PST), pp. 1–6 (2018). https://doi.org/10.1109/PST.2018.8514157

11. Tian, D., et al.: An intrusion detection system based on machine learning for can-bus. In: International Conference on Industrial Networks and Intelligent Systems, pp. 285–294. Springer (2017) https://doi.org/10.1007/978-3-642-40994-3_29

12. Verma, M.E., Iannacone, M.D., Bridges, R.A., Hollifield, S.C., Kay, B., Combs, F.L.: Road: the real ornl automotive dynamometer controller area network intrusion detection dataset (with a comprehensive can ids dataset survey and guide). arXiv preprint arXiv:2012.14600 (2020)

13. Vuong, T.P., Loukas, G., Gan, D.: Performance evaluation of cyber-physical intrusion detection on a robotic vehicle. In: 2015 IEEE International Conference on Computer and Information Technology; Ubiquitous Computing and Communications; Dependable, Autonomic and Secure Computing; Pervasive Intelligence and Computing. IEEE, pp. 2106–2113 (2015)

14. Vuong, T.P., Loukas, G., Gan, D., Bezemskij, A.: Decision tree-based detection of denial of service and command injection attacks on robotic vehicles. In: 2015 IEEE International Workshop on Information Forensics and Security (WIFS), pp. 1–6. IEEE (2015)

15. Wu, W., et al.: A survey of intrusion detection for in-vehicle networks. IEEE Trans. Intell. Transp. Syst. **21**(3), 919–933 (2019)

16. Yang, L., Moubayed, A., Hamieh, I., Shami, A.: Tree-based intelligent intrusion detection system in internet of vehicles. In: 2019 IEEE global communications conference (GLOBECOM), pp. 1–6. IEEE (2019)

ZeroMT: Multi-transfer Protocol for Enabling Privacy in Off-Chain Payments

Flavio Corradini, Leonardo Mostarda, and Emanuele Scala[✉]

Computer Science, University of Camerino, Camerino, Italy
{flavio.corradini,leonardo.mostarda,emanuele.scala}@unicam.it

Abstract. The privacy problem in public blockchains is a well-know challenge. Despite the robustness and decentralisation properties of the blockchain, transaction information remains visible to everyone in the network. Several proposals aim at solving these issues with advanced cryptographic techniques, such as zero-knowledge proofs, which turned out to be the best candidates. However, previous works are not aimed at off-chain sessions, which often end with a transfer that involves multiple payees, coming from a single payer or from more than one. With this paper, we introduce ZeroMT, a protocol for multiple confidential balance transfers that occur in a single transaction. With this novel approach, parties of an off-chain session gain privacy for their balances and transfer amounts. In addition, all transfers occur within a single transaction that benefits scalability, reducing the number of transactions to be validated in the main-chain. We provide the generic construction of a confidential multi-transfer transaction that can be assembled off-chain and verified by smart contract platforms. As a part of our protocol, we design the multi-transfer proof system by combining the aggregate version of Bulletproofs and an extended Σ-Protocol to n ciphertexts, generalising the Zether-Σ-Bullets protocol.

1 Introduction

With the rise of decentralised cryptocurrencies such as Bitcoin [18], the blockchain technology has gained worldwide interest in both academia and industry [22]. In a decentralised setting, disintermediation is joint with robustness, enabling trust in interactions between untrusted distributed parties. This clearly leads to new class of systems in real-world scenarios, however it introduces several challenges when facing with scalability and privacy. Furthermore, emerging decentralised systems such as Ethereum [9] extend the functionality of a distributed ledger by allowing programmability of transactions with smart contracts. All the transactions contained in a block are executed and validated according to the rules defined within the smart contract. The sequence of instructions in a smart contract produce state updates that are propagated across the network nodes and immutably stored on the blockchain. While consensus mechanisms ensure data integrity and availability, a public announcement of transactions is performed which brings the transaction information publicly visible. Such

L. Barolli et al. (Eds.): AINA 2022, LNNS 450, pp. 611–623, 2022.
https://doi.org/10.1007/978-3-030-99587-4_52

information can reveal the ledger's current state, updates requested by transactions and digital signatures that prove the origin of transactions. This can be a risk factor for those applications where highly confidential data must have privacy protection and anonymity for the parties must not be compromised. Further, well-know attacks that rely on transaction graph analysis can reveal users' real-world identity and transaction information [13,16,20]. In sight of this, the lack of privacy in public blockchain has received huge attention from researchers in the last years. Following a recent major advance in cryptography that makes zero-knowledge proofs applicable, one of the leading privacy applications emerges from the literature: *confidential transactions* (e.g. private financial transfers). In this context, attempts are made to guarantee the anonymity of parties involved in a financial transfer and the confidentiality of the amounts in transfers. However, the techniques used can incur to increased costs in those blockchains where fees are required for executing transactions. This can be mitigated by optimising the transaction size and the number of cryptographic operations to reduce the computational burden on-chain. Another challenge is not to introduce additional trust, as it opposes the pivotal trust model of the blockchain.

UTXO or Account Model. The first thing to consider when designing privacy for blockchains is the choice between the two main balance representations: UTXO (unspent transaction outputs) or account model. Given the simplicity of its model, the majority of privacy proposals follow the UTXO model of Bitcoin. In this, a user that has coins owns a bunch of transaction outputs. To spend coins, current owners issue transactions that take one or more old outputs as inputs and designate new outputs to recipients. Once consumed, old outputs cannot be spent again. The sum of the amounts for each unspent output belonging to a user is considered the balance of that user. In such settings, privacy proposals of Zerocash [21] and Monero [2] move around a simple equation for the validity of a transaction: the total of the input amounts must equal the total of the output amounts. To obfuscate transaction amounts, this equation is then proven in zero-knowledge by using commitment scheme, the *commitment to zero* in Monero, or through zk-SNARK proofs in Zerocash.

On the other side, account model is the balance strategy tailored for smart contracts first adopted by Ethereum. This model features two types of accounts, user and contract account, each with its own balance and address. Account information has its representation in a global state (world state), where balances are expressed as the quantities (or a sub-unit) of coins owned by the address. When a user, say Alice, wants to make a transfer of funds to another user, say Bob, she essentially initiates a transaction that will affect the world state such that Alice's account balance must be debited and the Bob's account balance must be credited. For the transaction to be valid, the payer must have enough money in his balance and the funds must come from his balance. The privacy proposals in account model, such as Zether [7] or the hybrid model of Quisquis [15], have adopted the strategy of keeping the balances encrypted and performing homomorphic updates of that balance. The validity of the transaction is then proved with ZK-proofs on encrypted data. These solutions share the use of more-specific proof systems for confidential transactions, for example by combining Bulletproofs [8] for range proofs and Σ-Protocols to prove algebraic statements related to the discrete logarithm problem. Other proposals rely on proof systems for more general computations inspired by Zerocash, an example is Blockmaze [17] which has designed a specific SNARK circuit for each transaction.

Limitations of Existing Approaches. Despite the benefit of having small size proofs and succinct verification, SNARKs constructions need to generate a large common reference string (CRS) [5] during the setup phase, which must be shared among the participants for producing and verifying proofs. This process can be done by a trusted party or with a costly multiparty computation [3]. However, this makes solutions like Zerocash and Blockmaze not completely trustless and exposes them to security risks. In addition, CRS generation must be done every time the computation behind the SNARK circuit changes. This can be mitigated by using new models that make the CRS updatable, at the cost of a larger size of the proofs.

One limitation of Monero, common to Zerocash, is that addresses are never removed from the UTXO set causing a continuous growth of the set size, when transfers of funds take place [15]. This because an address can be "mixed in" transactions even when the corresponding users are not involved. Instead, in Bitcoin the addresses of the senders are replaced with those of the receivers once the coins have been spent. Anonymity in Monero is also subject of discussions due to the fact that addresses are reused multiple times, leading to de-anonymization attacks based on how often a given address appears in transactions [15]. A solution to these problems is proposed by Quisquis with a new cryptographic primitive called "updatable public keys" (UPKs). With this primitive, all the output addresses are updated with each transaction, generating new addresses that will be consumed in the future. However, this mechanism implies that a transaction could potentially fail due to the "front-running" problem [7]. Because of the greater complexity than the UTXO model, there are few solutions for confidential transactions in the account model, the most notable is Zether. One reason is that in the account model, along with private funds transfer, the accumulated balances for the accounts must be updated. One solution is to keep track of these balances in encrypted form within smart contracts and perform homomorphic updates. However, efficiently countering the linking of addresses is more difficult and may require updating all balances for each transaction [6].

1.1   Contribution of the Paper

Previous works refer to the traditional payment system in which the transfer of funds happen from a single source to a single recipient. The only exception is Quisquis that also considers the case in which a group of entities want to *redistribute wealth* among them. We believe this situation is well suited for off-chain scenarios where there are interactions between multiple entities with different roles. In fact, an off-chain session often ends with a transfer that involves multiple payees, coming from a single payer or from more than one. However, traditional payment systems only allow separate transfers for each recipient, requiring inefficient work from the main chain to execute and validate individual transactions. This can also have an impact on privacy, giving an attacker the possibility to analyze a wider set of transactions coming from the same source. With this paper, we introduce *ZeroMT*, a novel privacy-preserving payment protocol that can be easily integrated with off-chain systems. *ZeroMT* features aggregation mechanisms for constructing a unique transaction that has the effect of multiple transfers. We refer to the account model to take advantage of greater programmability of transactions by means of smart contracts. We develop a protocol that includes the user program (MTU) and the *ZeroMT* smart contract (MTSC), for the creation and execution of a *multi-transfer* transaction (MTX) which guarantees simultaneous confidential transfers. Our contribution can be summarised in two aspects: i) for *privacy*, *ZeroMT* combines trust-free zero-knowledge techniques for confidential transactions in a secure protocol suitable for off-chain scenarios; ii) for *scalability*, *ZeroMT* aims to reduce the number of transactions to be validated in the main-chain by leveraging on aggregation techniques.

2   Motivation Scenario

We motivate our proposal by taking the smart lock case study as in [12] and [11]. In this case study there are untrustworthy small devices and servers performing off-chain interactions, before some data is stored in the blockchain. In particular, a smart lock is installed on different electric bicycles. A Client can start the ebike rental service by requesting the opening of the smart lock and its closing at the end of the service. The Client pays the rental time to the Service Provider which shares the profit and fees with the Insurance. In the following, we generalise this scenario to show how to integrate *ZeroMT* as a standalone protocol. The Client starts a new session making a request to the IoT device, then forwards the received data to the Servers. The IoT device generates session data for the Client. The Servers make their own computations on Client's session data, then assemble a transaction with the data to make the session publicly verifiable. Finally, Servers generate an access key to enable the IoT device. The blockchain is used for the integrity of the Client's session data and also for balance transfers between Insurance and Service Providers. A first attempt for the Client's privacy is not to involve him in payments via the blockchain, nor to store his identity in the blockchain. However, this limit the possibility of having automatic payments, as well as mechanisms to discourage dishonest behaviours through automatic penalties, e.g. when a dispute occurs. There are also privacy concerns for the Service Provider and Insurance: balances and transaction amounts are all in clear in the blockchain side and everyone can access them. In business contexts, it is often preferred not to disclose balances and the amounts exchanged in transactions to competitors. *ZeroMT* aims at solving these privacy issues with the following features and properties: (i) confidential transactions *multi-transfer*, which allow transfers to multiple payees in single transaction without reveal the amounts to the public blockchain network; (ii) tailored for off-chain protocols as the multi-transfer transaction is assembled off-chain; (iii) zero-knowledge balances and homomorphic updates; (iv) no added trust to the off-chain protocol.

Simplistic ZeroMT. The more complex case of a *multi-transfer* is when we want to assemble a single transaction for transfers between a group of senders and a group of recipients. This can be seen as a many-to-many transfer of funds relation. We reduce the problem with the simpler case where one payer creates a transaction for multiple recipients, say a one-to-many transfer relation. Moreover, we claim that this payment transaction must be authorised by the payer, the one to whom the balance will be charged (the Client in our smart lock scenario). Suppose we have an ebike User who has finished the rental service. After a sequence of context-dependent interactions with the Service Provider, the User's personal device has collected the following information, before assembling the *multi-transfer* transaction: (i) a list of the public addresses of the recipients, in our case the public key of the Service Provider (SP) and that of the Insurance (INS), respectively denoted with $\bar{y}_{sp}$ and $\bar{y}_{ins}$; (ii) a list of transfer amounts to be credited to recipients' balances, in our case those of the SP and INS, denoted with a_{sp} and a_{ins}.

To make the *multi-transfer* confidential, we need to hide the amounts in the transaction. We use the homomorphic encryption method of Zether [7], that is an ElGamal encryption for amounts "in the exponent". Moreover, with this encryption scheme we keep the balances of all off-chain users encrypted in the MTSC and perform homomorphic updates of those balances when the transaction is executed. In order to transfer the right amount to the right recipient, the Client encrypts each amount under both its public key and the recipient's public key. Then the Client generates two lists of ciphertexts, $\mathbf{C} = \{C_1, ..., C_n\}$ and $\bar{\mathbf{C}} = \{\bar{C}_1, ..., \bar{C}_n\}$, where $Enc_y(a_i) = C_i$ is the ciphertext under the Client's public key y of each i-th amount in the list of transfer amounts, and $Enc_{\bar{y}_i}(a_i) = \bar{C}_i$ is the ciphertext under the i-th public key in the list of recipient addresses of the i-th amount in the list of transfer amounts. Together with these ciphertexts, the Client also provides with the MTX a ZK-proof, that here we denote $\bigwedge \pi$, to prove the following statements:

1. both the i-th ciphertext in the list $\mathbf{C}$ and $\bar{\mathbf{C}}$ are well-formed and encrypt the same i-th amount in the list of transfer amount;
2. the Client knows the secret key for which the respective public key encrypts the ciphertexts in the $\mathbf{C}$ list and the public key is well-formed;
3. each i-th amount in the list of transfer amounts is positive;
4. the Client's remaining balance, say b', is positive.

These are formally defined in the relation[1] (1) Sect. 3. The general concepts of the *multi-transfer* transaction are collected in Fig. 1. Here we can see the Client

---

[1] The relation specifies a valid pair of instance-witnesses and the relative statements expressed in algebraic form for which the proof is constructed.

generate the ZK-proof $\bigwedge \pi$ and the MTSC verification procedure V validate this proof. If the proof is valid (the procedure outputs 1), all balances encrypted in the MTSC are updated. The MTX and the procedure V also require other information for the creation and verification of the proof that we will see in detail in the Sect. 3.

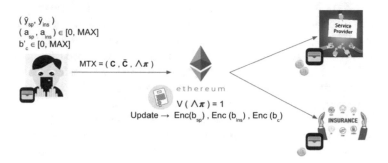

Fig. 1. Multi-transfer transaction

The proof symbol preceded by the logical symbol $\bigwedge$ means that our proof is constructed from the conjunction of statements, satisfied by a combination of two known proof systems: Σ-Protocols [10] and Bulletproofs [8]. The first is used for proving statements on encrypted data and the knowledge of a secret (1. and 2. statements above); the second for range proofs, to prove that transfer amounts and the remaining balance of the sender are non-negative values (3. and 4. statements), e.g. for some value $v \in [0, \text{MAX}]$ where MAX is a fixed maximum value of the form 2^n ($n = $ bit length of v). Our system significantly differs from the Σ-Bullets of Zether [7] in both the two underlying proof systems as follows. Instead of making one range proof for each transfer, we use the *Aggregated Inner Product Range Proof* [8] to construct one range proof of m aggregated values. It turns out that a single range proof for m aggregated values is more efficient than m individual range proofs: considering 32 bytes curve points and scalars and 64 bit ranges, for two values v_1 and v_2, two individual transactions have a total of $2 * 672 = 1344$ bytes range proofs, while aggregating the two values we have 736 bytes range proof in one transaction. Further, we extend the Σ-Protocol of Σ-Bullets to prove statements on the encryptions of multiple transfer amounts together with the sender's balance. Using this combination of proof systems, we obtain a generalisation to the case of n transfers per epoch compared to one transfer per epoch of Zether-Σ-Bullets.

3   ZeroMT System Model

ZeroMT is a protocol for multiple confidential balance transfers that occur in a single transaction. The protocol consists of the user program (MTU) and the

ZeroMT smart contract (MTSC). We do not provide the full protocol description in this paper, which includes a complete list of user functions such as *CreateAddress*, *Fund*, *Burn* and the respective functions for the smart contract. *Front-running* and *replay* protection mechanisms can also be addressed similarly to Zether [7]. Instead, we go directly into the formal detail of the *multi-transfer* transaction creation performed by the MTU.CreateMTX program and the MTSC.Transfer verification procedure.

Notations and Primitives

ELGamal Encryption. We use the same notation of Zether [7] in which a ciphertext for an integer value $a \in \mathbb{Z}_p$ has the form of a tuple $(C = g^a y^r, D = g^r)$ where g is the generator of a group $\mathbb{G}$ of prime order p, $y = g^x$ is the public key, $x \xleftarrow{\$} \mathbb{Z}_p^*$ the private key uniformly sampled from the set of inverses in $\mathbb{Z}_p$, and $r \xleftarrow{\$} \mathbb{Z}_p^*$ is the randomness used in the encryption. We exploit the additive homomorphic property for balance updates and operations between amounts when they are encrypted under the same public key.

Zero-knowledge Proof System is a two-party protocol in which a prover $\mathcal{P}$ can convince a verifier $\mathcal{V}$ that some statement is true, without revealing any information beyond the validity of the statement. This can be done through an interactive protocol between $\mathcal{P}$ and $\mathcal{V}$, until $\mathcal{V}$ accepts that the statement is true. The protocol is *complete* if every valid statement has a convincing proof from the honest prover, is *sound* if invalid statements have no false proofs. The zero-knowledge property holds if the proof reveals essentially nothing, even if $\mathcal{V}$ chooses to trick $\mathcal{P}$ into revealing something. The prover $\mathcal{P}$ can convince about NP statements for which $\mathcal{P}$ and $\mathcal{V}$ know public inputs, and only $\mathcal{P}$ knows the witnesses.

Σ-Protocols are Honest-Verifier Zero-Knowledge (HVZK) interactive proof systems for special purposes, e.g. proof of knowledge of a discrete logarithm and proof of knowledge of the opening of a commitment. Their structure consist of a 3-messages public-coin protocol, where the only messages from the verifier are random coin tosses. The interactive protocol can be made non-interactive (NIZK) and publicly verifiable in the Random Oracle Model by applying the Fiat-Shamir transform.

Zero-knowledge Relation. A relation *Rel* specifies a collection of valid instance-witness of the form $Rel : \{(i_1, ..., i_n \; ; \; w_1, ..., w_m) : f(i_1, ..., i_n, w_1, ..., w_m)\}$, where prover and verifier know the public inputs $i_1, ..., i_n$, and the prover shows the knowledge of witnesses $w_1, ..., w_m$ such that $f(i_1, ..., i_n, w_1, ..., w_m)$ is true. f can be defined in algebraic form to express how the proof is constructed.

Confidential Multi-transfer Relation. In the simplistic version of *ZeroMT* a sender with public key y may transfer funds to n recipients with public keys

$\bar{\mathbf{y}} = \{\bar{y}_1, ..., \bar{y}_n\}$. To this ending, the sender should publish two list of ciphertexts $\mathbf{C} = (C_i = g^{a_i} y^r)_{i=1}^n$ and $\bar{\mathbf{C}} = (\bar{C}_i = g^{a_i} \bar{y}_i^r)_{i=1}^n$ with the same randomness $D = g^r$.

The ciphertexts at the same i-th index of both lists should encrypt the same amount a_i to be credited to the recipient's balance with public key $\bar{y}_i$. We denote with (C_L, C_R) the current balance of the sender y, with b' the value of the remaining balance after the transfer and with $\mathbf{a}$ the list of the transfer amounts $\mathbf{a} = \{a_1, ..., a_n\}$.

The MTU.CreateMTX program acts as a prover, generating a proof π for the following relation:

$$Rel_{ConfMultiTransfer} : \{ (y, \bar{\mathbf{y}}, C_L, C_R, \mathbf{C}, \bar{\mathbf{C}}, D, g; \ x, \mathbf{a}, b', r) :$$
$$(C_i = g^{a_i} y^r \wedge \bar{C}_i = g^{a_i} \bar{y}_i^r \wedge D = g^r)_{i=1}^n \wedge$$
$$C_L / \prod_{i=1}^n C_i = g^{b'} (C_R/D)^x \wedge y = g^x \wedge \qquad (1)$$
$$\{a_1, ..., a_n\} \in [0, MAX] \ \wedge \ b' \in [0, MAX]\}$$

Then the MTSC.Transfer program takes the proof π that attests the prover's knowledge:

- of a secret key x for which $g^x = y$,
- of randomness r for which $g^r = D$,
- that each ciphertext $(C_i, D_i)_{i=1}^n$ and $(\bar{C}_i, D_i)_{i=1}^n$ for which $D_1 = ... = D_n = D$, is well-formed and encrypts the same amount a_i,
- that his balance cannot be overdraft if the equation $C_L / \prod_{i=1}^n C_i = g^{b'} (C_R/D)^x$ holds,
- $\{a_1, ..., a_n\}$ and b' are positive values.

Figure 2 shows the two above programs for the ZeroMT *multi-transfer* protocol. The purpose of the first is to create the MTX transaction, while the second performs the verification of the MTX in zero-knowledge and updates the balances. The programs engage in an interactive proof system for the relation (1) , shown here non-interactive through the **Prove** and **Verify**$_{nikz}$ functions. These both share the public instances $[y, \bar{\mathbf{y}}, C_L, C_R, \mathbf{C}, \bar{\mathbf{C}}, D, g]$ from the relation; the witnesses for our proof system, grouped in $w = (x, \mathbf{a}, b', r)$, are the private key of the sender, the list of transfer amounts, the sender's remaining balance and the randomness used in the encryptions. In order to update the balances, we get the reference to the current balance of the sender with $b[y]$ and to that of the various recipients with $b[\bar{y}_i]$. Then, we perform the homomorphic updates with group-add operations between the balance and the respective amount (for the sender there will be an amount equal to the sum of all amounts).

Signatures, Confidentiality and Anonymity. We have omitted the digital signature to authorise the multi-transfer. Authorisation is requested from the sender, the one to whom the balance will be charged, and therefore only his

MTU.CreateMTX
INPUTS:

- sender public key y
- sender private key x
- receivers public keys
  $\bar{y} = \{\bar{y}_1, \dots, \bar{y}_n\}$
- sender balance b'
- transfer amounts
  $\mathbf{a} = \{a_1, \dots, a_n\}$

OUTPUTS:
MTX$(y, \bar{y}, \mathbf{C}, \bar{\mathbf{C}}, D, \pi)$

1. Let $(C_L, C_R) = b[y]$
2. $r \xleftarrow{\$} \mathbb{Z}_p$
3. Set $C_i = g^{a_i} y^r \quad \forall i \in [n]$
4. Set $\bar{C}_i = g^{a_i} \bar{y}_i^r \quad \forall i \in [n]$
5. Set $D = g^r$
6. Set $w = (x, \mathbf{a}, b', r)$
7. Let $\mathbf{C} = \{C_1, \dots, C_n\}$
8. Let $\bar{\mathbf{C}} = \{\bar{C}_1, \dots, \bar{C}_n\}$
9. $\pi = \mathbf{Prove}(Rel_{ConfMultiTransfer}$
   $[y, \bar{y}, C_L, C_R, \mathbf{C}, \bar{\mathbf{C}}, D, g], w)$

MTSC.Transfer
INPUTS:

- sender public key y
- receivers public keys $\bar{y}$
- ciphertexts $(C_1, D), \dots, (C_n, D)$
- ciphertexts $(\bar{C}_1, D), \dots, (\bar{C}_n, D)$
- proof π

1. Let $(C_L, C_R) = b[y]$
2. Let $\mathbf{C} = \{C_1, \dots, C_n\}$
3. Let $\bar{\mathbf{C}} = \{\bar{C}_1, \dots, \bar{C}_n\}$
4. Set $C_{tot} = \prod_{i=1}^{n} C_i$
5. require:

- $\mathbf{Verify}_{nizk}(Rel_{ConfMultiTransfer}$
  $[y, \bar{y}, C_L, C_R, \mathbf{C}, \bar{\mathbf{C}}, D, g], \pi) = 1$

6. $b[y] = b[y] \circ (C_{tot}^{-1}, D^{-1})$
7. For each $i \in [1, n]$ set
   $b[\bar{y}_i] = b[\bar{y}_i] \circ (\bar{C}_i, D)$

Fig. 2. ZeroMT multi-transfer protocol

signature is required. However, we can easily provide this signature deriving from the transformation process to non-interactive proofs based on the Fiat-Shamir heuristic [1,4]. Indeed, as in Zether, the generated proof can also act as a signature, without an additional signature scheme. This version of ZeroMT does not provide transaction anonymity but only confidentiality. That is, it does not reveal the balances and transferred amounts but senders and recipients are identifiable. In the settings of multi-transfer, hiding the sender and all receivers from the transaction is a challenging task and can entail a more complex ZK-proof. We are conducting research on this, a many-out-of-many proof [14] scheme could be used to build *ring signatures* and strengthen our multi-transfer with anonymity.

4 Related Works

Zerocash [21], is one the first anonymous payment system that strengthens the UTXO model of Bitcoin with strong privacy guarantees. This is achieved with a nested use of commitments in order to hide and mix together the value of a coin and its owner's address. The spend transaction is done without revealing the values of the input and output coins and the owner's address by providing a succinct zero-knowledge proof (zk-SNARK) with the transaction. In this way, the payer proves that: i) he/she owns the input coins, showing that he/she has knowledge of the secret key relating to those coins; ii) each of the input coins was minted in the past or is the output coin of a previous spend transaction; iii) the total value of the output coins equals the total value of the input coins. Despite Zerocash implements constant-size proofs with succinct verification of the proofs, adopting zk-SNARKs in practice require a trusted party to generate parameters for proving and verifying statements. Once these parameters are compromised, it is possible to create valid proofs for false statements.

Monero [2] is a privacy cryptocurrency in UTXO model with hidden amounts, origins and destinations of transactions. It is based on RingCT protocol [19] which uses ring signature on Pedersen Commitment. More in depth, Monero also move around the simple equation for the validity of a transaction: the total of the input amounts must equal the total of the output amounts. To obfuscate transaction amounts, this equation is then formed such that the difference between the sum of the input and output commitments will result in a *commitment to zero*. To hide source and destination of a transaction, the sender mixes his/her unspent outputs with a set of unrelated addresses coming from the blockchain and the commitments of their unspent outputs. Within this ring structure, it will appear that only one term is the right *commitment to zero* expressed above. Being the only one who knows the private keys corresponding to the addresses and the blinding factor of the *commitment to zero*, the sender proves that he/she is the owner of the unspent outputs he/she wants to spend. This mechanism is at the basis of the ring signature scheme MLSAG [19], used in Monero to produce a single signature on a set of public keys so that it is difficult to tell which key belongs to the actual signer. The last ingredient of the RingCT protocol deals with proving that the outputs of a transaction fall within a range of admissible positive values (range proofs).

Quisquis [15] is a design for anonymous cryptocurrencies that addresses the limitations of Monero and Zerocash. In Quisquis, the public keys of the senders are replaced with the public keys of the recipients at each transaction, allowing to have a concise representation of the UTXO set as in Bitcoin. Anonimity is achieved with the cryptographic primitive UPKs, that allows to update public keys without changing the underlying secret. This method is used so that an address appears in the blockchain only when it is generated as an output and when it is consumed as an input of a transaction. Moreover, the balance associated to an address and the transaction amounts are kept hidden through commitments. Finally, vector Pedersen commitments are used in the construction of a zero-knowledge argument: sender proves in zero-knowledge that i) he knew the private key for the public keys replaced with that of the recipients, and he has correctly updated the other keys in the (ii) he does not decrease the balance of the other accounts in the anonymity set. Instead of using zk-SNARKs, the zero-knowledge proof system relies on standard discrete-log assumptions and commitments under public keys are constructed in a group where DDH problem is hard (similar to the scheme proposed in Zether [7]). One problem in Quisquis is that a transaction could potentially fail due to the "front-running" problem [7].

BlockMaze [17] is a privacy-preserving transfer mechanism in account-model. It offers a single account address linked to a dual-balance model composed of a plaintext balance and a zero-knowledge balance. The zero-knowledge balance is a secure commitment on the associated amount, then zk-SNARK is used to prove the validity of a transfer fund to another account without revealing the transaction amounts and account balances. Unlinkability is designed with a two-step fund transfer procedure: first a sender generates a zero-knowledge transaction through a commitment to deposit fund; second the recipient recovers the commitment from sender and generates a zero-knowledge proof to receive the fund, without leaking from which transaction he/she receives the fund. However, the zk-SNARK construction leads the same problem of the trusted setup as in Zerocash.

Zether [7] is privacy-preserving payment scheme for smart contracts. The main contributions of Zether are: i) confidential payment transactions; ii) anonymity in transfers; iii) completely trustless. For confidentiality, Zether keeps account balances encrypted and users provide cryptographic proofs to spend their money. More precisely, a new ZK-proof scheme is proposed, called Σ-Bullets, to perform range proofs for the validity of amounts in transfers and balances, and to create ZK-proofs of correct encryption. This approach is used to efficiently prove statements over ELGamal encryption, without disclosing information. Moreover, the accounts are identified with ElGamal public keys which are stored in the contract's internal state. For anonimity and trustless, Zether hides the sender and receiver of a transaction and does not require any trusted setup. All these properties are implemented in a Zether smart contract (ZSC) that exposes methods to deposit, transfer, withdraw of funds to/from accounts through cryptographic proofs at only small cost. Replay attacks, front-running protection and interoperability of smart contracts are also investigated. However, Zether hides sender and receiver identities among a small anonymity set and the transaction size grows linearly with the anonymity set.

5   Conclusion and Future Work

The ZeroMT protocol allow confidential transfers to preserve the privacy of the parties involved in an off-chain session. Further, all transfers in a session occur within a single transaction, reducing the number of transactions to be validated in the main-chain. However, this version of ZeroMT does not provide transaction anonymity, that is, senders and recipients are identifiable. Future directions are to hide the sender and all receivers from the multi-transfer through a more complex ZK-proof. We are conducting research on this, proposals such as [14] could be used to strengthen our multi-transfer with anonymity.

References

1. Abdalla, M., An, J.H., Bellare, M., Namprempre, C.: From identification to signatures via the Fiat-Shamir transform: minimizing assumptions for security and forward-security. In: Knudsen, L.R. (ed.) EUROCRYPT 2002. LNCS, vol. 2332, pp. 418–433. Springer, Heidelberg (2002). https://doi.org/10.1007/3-540-46035-7_28
2. Alonso, K.M., et al.: Zero to Monero (2020)
3. Ben-Sasson, E., Chiesa, A., Green, M., Tromer, E., Virza, M.: Secure sampling of public parameters for succinct zero knowledge proofs. In: 2015 IEEE Symposium on Security and Privacy, pp. 287–304. IEEE (2015)
4. Ben-Sasson, E., Chiesa, A., Spooner, N.: Interactive oracle proofs. In: Hirt, M., Smith, A. (eds.) TCC 2016. LNCS, vol. 9986, pp. 31–60. Springer, Heidelberg (2016). https://doi.org/10.1007/978-3-662-53644-5_2
5. Blum, M., Feldman, P., Micali, S.: Non-interactive zero-knowledge and its applications. In: Providing Sound Foundations for Cryptography: On the Work of Shafi Goldwasser and Silvio Micali, pp. 329–349 (2019)
6. Bowe, S., Chiesa, A., Green, M., Miers, I., Mishra, P., Wu, H.: Zexe: enabling decentralized private computation. In: 2020 IEEE Symposium on Security and Privacy (SP), pp. 947–964. IEEE (2020)
7. Bünz, B., Agrawal, S., Zamani, M., Boneh, D.: Zether: towards privacy in a smart contract world. In: Bonneau, J., Heninger, N. (eds.) FC 2020. LNCS, vol. 12059, pp. 423–443. Springer, Cham (2020). https://doi.org/10.1007/978-3-030-51280-4_23
8. Bünz, B., Bootle, J., Boneh, D., Poelstra, A., Wuille, P., Maxwell, G.: Bulletproofs: short proofs for confidential transactions and more. In: 2018 IEEE Symposium on Security and Privacy (SP), pp. 315–334. IEEE (2018)
9. Buterin, V., et al.: A next-generation smart contract and decentralized application platform (2014)
10. Butler, D., Aspinall, D., Gascón, A.: On the formalisation of σ-protocols and commitment schemes. In: POST, pp. 175–196 (2019)
11. Cacciagrano, D., Corradini, F., Mazzante, G., Mostarda, L., Sestili, D.: Off-chain execution of IoT smart contracts. In: Barolli, L., Woungang, I., Enokido, T. (eds.) AINA 2021. LNNS, vol. 226, pp. 608–619. Springer, Cham (2021). https://doi.org/10.1007/978-3-030-75075-6_50
12. Cacciagrano, D., Corradini, F., Mostarda, L.: Blockchain and IoT integration for society 5.0. In: Gerber, A., Hinkelmann, K. (eds.) Society 5.0 2021. CCIS, vol. 1477, pp. 1–12. Springer, Cham (2021). https://doi.org/10.1007/978-3-030-86761-4_1
13. Chan, W., Olmsted, A.: Ethereum transaction graph analysis. In: 2017 12th International Conference for Internet Technology and Secured Transactions (ICITST), pp. 498–500. IEEE (2017)
14. Diamond, B.E.: Many-out-of-many proofs and applications to anonymous zether. In: 2021 IEEE Symposium on Security and Privacy (SP), pp. 1800–1817. IEEE (2021)
15. Fauzi, P., Meiklejohn, S., Mercer, R., Orlandi, C.: Quisquis: a new design for anonymous cryptocurrencies. In: Galbraith, S.D., Moriai, S. (eds.) ASIACRYPT 2019. LNCS, vol. 11921, pp. 649–678. Springer, Cham (2019). https://doi.org/10.1007/978-3-030-34578-5_23
16. Fleder, M., Kester, M.S., Pillai, S.: Bitcoin transaction graph analysis. arXiv preprint arXiv:1502.01657 (2015)

17. Guan, Z., Wan, Z., Yang, Y., Zhou, Y., Huang, B.: BlockMaze: an efficient privacy-preserving account-model blockchain based on zk-SNARKs. IEEE Trans. Dependable Secure Comput. (2020)
18. Nakamoto, S.: Bitcoin: a peer-to-peer electronic cash system. Technical report, Manubot (2019)
19. Noether, S., Mackenzie, A., et al.: Ring confidential transactions. Ledger **1**, 1–18 (2016)
20. Ron, D., Shamir, A.: Quantitative analysis of the full bitcoin transaction graph. In: Sadeghi, A.-R. (ed.) FC 2013. LNCS, vol. 7859, pp. 6–24. Springer, Heidelberg (2013). https://doi.org/10.1007/978-3-642-39884-1_2
21. Sasson, E.B., et al.: Zerocash: decentralized anonymous payments from bitcoin. In: 2014 IEEE Symposium on Security and Privacy, pp. 459–474. IEEE (2014)
22. Sekaran, R., Patan, R., Raveendran, A., Al-Turjman, F., Ramachandran, M., Mostarda, L.: Survival study on blockchain based 6G-enabled mobile edge computation for IoT automation. IEEE Access **8**, 143453–143463 (2020)

Prevention of SQL Injection Attacks Using Cryptography and Pattern Matching

R. Madhusudhan[(✉)] and Mohammad Ahsan

Department of Mathematical and Computational Sciences, NITK, Surathkal, Karnataka, India
{madhu,mohammadahsan.202cd016}@nitk.edu.in

Abstract. The internet is rapidly expanding that allow easy access to information, thus attackers develop different methodologies to access it and hence the security related to it becomes priority for all. SQL injection attack (SQLIA) has consistently posed serious threat since its existence. SQLIA is a web security vulnerability through which attackers can give specifically designed input to steal or manipulate sensitive information by interacting with the database. The objective of the research is to provide a defensive mechanism to protect a particular web application against such attacks. The paper acknowledged some existing models and give special attention to models based on encryption and pattern matching techniques. Encryption based models have proven themselves to be very effective against SQLIA by preventing attackers from authentication access. But such model will undermine the integrity of the tables if used in places other than the authentication form. Thus, we employ an additional layer of security based on pattern matching techniques. Our idea differs in a way that it compares a temporary structure generated from the user's query with all defined benign structures created from the benign queries that are usually expected by the web application. The proposed model uses Blowfish algorithm in authentication form which upon simulation is preventing all kind of SQLIA from authentication access and upon the implementation of Knuth-Morris-Pratt pattern matching technique, the model will ensure the prevention of any new and existing kind of SQLIA. The model is under development and is believed to provide a robust environment in preventing all kind of SQLI attacks with overall reduced complexity.

Keywords: SQLI attack · Web application security · Encryption method · Pattern matching technique · KMP algorithm

1 Introduction

In today's digital world, where a lot of devices got connected via internet for ease of communication and for the purpose of driving businesses. The risk involves in the security of such information got high. The abundance of sensitive information such as personal details, transaction details, bank details, etc. which are required by all business firms on the trust of their customers to drive their businesses smoothly attracts cyber criminals. Hence, it is very important to be aware of the ways through which a system can be exploited. In ordered to prevent such attacks, different models have been developed over the period of time which are discussed in the paper.

© The Author(s), under exclusive license to Springer Nature Switzerland AG 2022
L. Barolli et al. (Eds.): AINA 2022, LNNS 450, pp. 624–634, 2022.
https://doi.org/10.1007/978-3-030-99587-4_53

There are different types of web vulnerabilities through which attackers can exploit the system. To categorize the level of risk a threat possess, risk assessments are done by several organizations. Open Web Application Security Project (OWASP) is the well-known non-profit organization, wherein the SQLI attack has always been in the top ten injection vulnerabilities [1]. SQLI attack is a web security vulnerability that allow attackers to inject malicious SQL queries to the vulnerable web applications. The attacker search for input parameters, information about the Database Management System (DBMS) and the database schema to make successful attack. The potential risk involved in it are authentication bypass, authorization access, disruption of confidentiality and integrity of data [2]. The different types of SQLI attacks of which most common are tautology, illegal/logically incorrect queries, union based, piggy-backed and alternate encoding. This paper discusses various types of models with special focus on model based on encryption and pattern matching algorithm. The literature survey done during research reveals that encryption techniques were found to be very effective to prevent injection attacks during authentications but not enough to secure whole system alone. Thus, there is a dire need of providing an additional layer of security for the model using pattern matching technique for the security purposes. The main objective of the paper is to introduce with the idea of developing a hybrid model based on encryption and pattern matching algorithm using Knuth-Morris-Pratt (KMP) algorithm, which we believe would undermine the risk related to all types of SQL injection attack for a particular web application.

Rest of the paper is organized as follows: Section 2 discusses different types of SQLI attacks and how they are done using simple examples. Section 3 explains some of the existing models related to our work along with their advantages and disadvantages. Section 4 discusses the proposed method in detail. Section 5 carries results and future work for the proposed model which in fact is under development. Finally, we conclude with Sect. 6 explaining the implication of the model.

2 Background

The client using a browser, request the server for specific files. Let's consider a case where the client request for some services to the web application which in fact is connected to the database management system that carries sensitive information. Assuming that certain parameters of the web application where user can enter input to places that are vulnerable to SQL injection. The attacker take this opportunity to inject malicious queries that may extract the stored data, modify or delete information, execute remote command such as denial of service, etc. [3,4].

Let's take a simple example, where the attacker intend to bypass authentication page of the web application which has two fields i.e., username and password. At the backend of the authentication page the DBMS process a query to check for the corresponding username and password entered through the input parameter.

This query look like one defined below while execution. Here, 'userinfo' is the name of table and *username, password* are the user inputs and can be specifically designed malicious input injected by attacker to make successful attack.

SELECT * FROM 'userinfo' WHERE username='$username' AND password='$password';

Now, let's see how different types of SQL injection attacks can be performed using mentioned queries [5]:

Tautology
It is a technique of injecting a malicious query to the input parameter which always turns out to be a true condition upon execution. This may bypass authentication page without knowing correct username and password in case the parameter in vulnerable to such attack.
SELECT * FROM 'userinfo' WHERE username=' ' OR 1=1;– AND password='abc';

Illegal/Logically Incorrect Query
Attacker intent to analyze the error message which may reveal some sensitive details related to system such as information about the server, database schema and DBMS. This can be achieved if no error message is displayed upon injecting an incorrect query.
SELECT * FROM 'userinfo' WHERE username=' '; convert('abc' , int) ;– AND password= 'abc';

Union Based Query Attack
The attacker injects malicious queries along with the first query separated using UNION keyword of SQL. This allows the attacker to send multiple queries to process, where any of the query could make the system vulnerable. The malicious query mentioned below will display the information about 'admin' user.
SELECT * FROM 'userinfo' WHERE username=' ' UNION SELECT * FROM 'userinfo' WHERE username= 'admin'; – AND password='abc';

Piggy Backed Query
The attacker writes multiple queries separated by the semicolon that can serve different purposes. If the first query is used to extract information while other query may intent to do potential damage such as deleting records or dropping a table.
SELECT * FROM 'userinfo' WHERE username=' ' ; ALTER TABLE 'userinfo' DROP fullname ;– AND password= 'abc';

Like Based Query Attack
The attackers make use of LIKE keyword of SQL to guess the value of an application parameter. Any correct guess will display the details of related users. The query mentioned below will search for all records where 'a' is present anywhere in the username field of a table.
SELECT * FROM 'userinfo' WHERE username=' ' OR username LIKE '%a%';– AND password='abc';

Inference Attack
The attack is used to identify injectable parameters, database schema, and to extract data. The attacker injects a malicious query and observes the behaviour of web application. Inference attacks are of two types: *Blind Injection*, where the attacker inject SQL statement which results in either true or false values, for true statement there is no change while for false statement application behave differently. *Timing Attack*, where the attacker inject sleep command based on DBMS, if attack is successful then there

will be a time delay and hence help attackers to identify vulnerable parameters of the web application.

Alternate Encoding Based Attack

Attackers may conduct injection attack in hexadecimal, unicode formats of user input which is not obvious but can be interpreted by execution engine of DBMS. The attack combined with other type of attacks as discussed earlier can do some serious damage if not handled properly. Here, exec(char(Ox73687574646j776e)) is the hexadecimal form of SHUTDOWN which is a system call to shut down DBMS.

SELECT * FROM 'userinfo' WHERE username=' '; exec(char(Ox73687574646j 776e)) '; – AND password='abc';

3   Related Works

In order to avoid SQL injection attacks, researchers have proposed different prevention techniques. These techniques can be classified into two broad categories. First is to detect SQLI attack through checking anomalous SQL query structure using pattern matching, query processing, encryption algorithm and instruction set randomization. Other approach uses data dependencies among data items which are less likely to change for identifying malicious database activities. Many researchers proposed different techniques with integrating data mining and intrusion detection systems. This type of approaches reduces the false positive alerts, minimizing human intervention and better detection of attack [6].

The model like AMNESIA (*Analysis And Monitoring For Neutralizing SQL Injection Attacks*) combines static and dynamic approach to detect web application vulnerabilities. It statically builds a query model, the query statement then dynamically compared against already built static query model for detection and prevention purposes [7]. CANDID (*Dynamic Candidate Evaluations For Automatic Prevention Of SQLI Attacks*) is a dynamic candidate evaluation method that compare the user input query structure dynamically with that of benign query structure. This model solves the issue of manually modifying the application to create prepared statements. It partially stops SQL injection attack because of the constraints on the basic approach [4, 8]. There is approach like AutoRand based on Instruction Set Randomization (ISR) method [9] which transforms all SQL keywords in a Java program into a randomized value before query execution. It overcomes the need of user intervention to prevent SQLIA. If no anomaly found the query is derandomized and executed properly. One limitation that it is confined only to Java program [10].

There are existing techniques that are effective in preventing SQLI attack based on data encryption and pattern matching. S. Ali et al. [4, 11] proposed SQLiPA (*SQL Injection Protector for Authentication*) which is an efficient solution for preventing SQLI during authentication. The model generates hash values for the user generated query then compare it against the corresponding parameters of the database, where information is stored using same hashing function during registration. Balasundram et al. [12] proposed a better technique that enhanced the security mechanism of the above method. He uses AES (*Advanced Encryption Standards*) cipher algorithm, where each user is allotted a unique secret key during registration, which is used for encryption of parametric values. Qais Tameiza et al. [13] proposed hashing model using SHA-1, where user's query is hashed after removing query attributes and compared with the stored benign hashed queries without attributes. Thus, it allows only those queries which matches and discard all other queries.

Prabhakar et al. [14] proposed a model based on pattern matching using Aho-Corasick algorithm. The model statically compares the user generated query with the stored list of anomaly patterns and anomalous score is calculated dynamically, the model triggers alarm if it passes certain threshold value set by the admin. This allows the administrator to manually check the user's generated query. The administrator takes action after analyzing the query and update the stored list of anomalous patterns for every new detection of SQLI attack. But such model needed a list of anomalous patterns, which could be in large numbers and hence requires large processing time to detect patterns. Also, it requires manual intervention of the administrator to prevent new kind of SQLI attacks. Georgiana et al. [15] proposed a model that check for the presence of SQLI vulnerabilities inside a web application. It takes input as URL or file location and then perform crawling over application or file. The model identifies the parameters, which later is analysed in testing panel. If the web application is found to be vulnerable, it is sent to exploit panel and report is generated in the report panel. Limitation here is that it requires a lot of time to crawl large files.

Hongcan Gao et al. [16] proposed a ATTAR model that can detect SQLI attacks using grammar pattern recognition and access behavior mining using machine learning. The model extracts and analyzes the features of SQLI attacks in Web Access Logs of web applications. The features are extracted using grammar pattern recognizer and access behavior miner to obtain the grammatical and behavioral features of SQLI attacks, respectively. Finally, based on the two feature vector matrices, machine learning algorithms, e.g., Naive Bayesian, SVM, ID3, Random Forest, and K-means, are used for training and testing of the model to classify SQL queries as malicious or benign. The technology like machine learning has good scope of preventing the injection attacks but due to scarcity of the dataset required to train the model, most of them fails somewhere and report some false negative and positive results.

Benjamin Appiah et al. [17] proposed a signature based SQLI attack detection framework by integrating Rabin fingerprinting method and Aho-Corasick pattern matching algorithm to distinguish genuine SQL queries from malicious queries. Fingerprints are short and compact tags that represent raw data and documents and are generated using some hash functions. The framework monitors SQL queries to the database and compares them against a dataset of signatures from known SQLI attacks. If the fingerprint method does not find a match, then the query is considered as legitimate. In other case, when a match is found, the Aho-Corasick pattern matching algorithm is invoked to ascertain whether attack signatures appear in the queries and thus classify these queries as malicious or benign. Though the model provides high detection rate but still have some false negative and positive results. O. C. Abikoye et al. [18] proposed a pattern matching model using KMP (*Knuth-Morris-Pratt*) algorithm, which is a faster algorithm compared to the existing pattern matching algorithms. The algorithm finds a pattern in linear order from a given string. In this method, the researchers have analyzed the pattern of malicious queries in their extensive research studies and then categorises them based on their patterns. The author uses KMP algorithm to identify such patterns that fall under different categories in the user generated query and then prevent if found by implanting defensive coding for every category of SQLI attacks. The major problem with such kind of model is that it requires deep understanding to categorize different types of SQLI attacks and also it does not guarantees to prevent any new kind of SQLI attacks if ever developed in near future.

The literature survey revealed that the existing models are mostly based on the concept of comparing structure of malicious SQL query with the benign query, encrypting data and randomization of SQL keywords along with others and there also exist certain sophisticated models which may detect attacks using machine learning techniques and other which can make use of database management system to reduce any overhead in the model as implemented in SEPTIC (*SElf-ProtecTIng databases from attaCks*) [19–21]. All of these techniques come with some advantages and disadvantages as mentioned within this section. Hence, we observe that there is a need of one such model that guarantees the detection of SQLI attacks at low processing cost. The next section is dedicated in explaining the idea of achieving such goals.

4   Proposed Method

Any web application that are developed which allow users to interact with the database so as to provide different services for smooth driving of any firm, they all require security to maintain trust with their customers. One of the simple ways it can be done is to limit the action that a normal user can perform to make the application more secure. In that way, we would be having limited number of benign queries to look after. For example, a business model like e-commerce website, which allows the user to search for a specific item, update their profile and allow transactions, etc. have certain constraints upon the user's action. Keeping this in mind, the user generated queries can be compared with only those validated queries required for user's action. It means it would be more relevant for a model to search for patterns from the list of all queries that a user is allowed to perform instead of malicious one, which in fact exists in different forms and there are large possibility of others to come in future. Thus, a kind of security mechanism is required along with some good practices like escaping special characters and making use of prepared statements available within the built-in procedural functions of PHP at all level of security that would make the system more secure.

The proposed security mechanism is a hybrid method of two different preventive techniques based on encryption and pattern matching algorithm, discussed as follows:

1. Encryption based models have proven themselves to be very effective when it comes to prevent SQLI attacks through authentication form. It's effectiveness depends upon the cryptographic algorithm used in it. Here, the user gives input to different parameters of authentication form. The parametric values is then required to be validated with the credentials of the actual user. These parametric values get encrypted before they are sent to the DBMS for execution. It is worth noting that during registration, the records of login credentials for every user are stored in an encrypted format using same cryptographic algorithm. Hence, every user's input is validated by encrypting it first and comparing it against the stored database of user's credentials. If any malicious query is injected through input field of the authentication form, it will first

transform into cipher text and hence it is considered as normal text while execution because DBMS can not interpret it. Also, the encrypted value will not match with any of the stored values of user's login credentials because any difference in the query, results in different cipher text. Hence, all kind of SQL injection attacks can be avoided. This method works well, but it is not sufficient alone. Although encryption can avoid attackers from accessing others account but still make the website vulnerable. Attackers can register themselves as a new user and breaches this security layer which only meant to secure authentication access. Thus, such attacks can even now affect the database from some other vulnerable places within the web application, fields that search for specific items or that allow users to update their profile, etc. In cases, we could not make other parameters encrypted because it may disrupt the integrity of the tables within the database. Also, it would be harder for administrators to understand such information.

2. The limitation with encryption method can be countered by implementing an automated approach, which is modified version of the existing pattern matching model using Knuth-Morris-Pratt (KMP) algorithm [16]. Our work differs in a way that instead of analyzing and categorizing different form of possible SQL injection attacks, which in fact would requires a lot of time. It would be better to focus on the creating a whitelist, which is a list of benign structures created from all benign SQL queries that normally a user is allowed to perform on a particular web application. If any of the benign structure defined in whitelist matches exactly with the temporary structure generated from user's query following the security algorithm explained in the later section of the paper, then the user's query would be considered safe and thus sent for execution to DBMS engine. This method would be more effective as one need not to worry about all type of existing SQLI attacks, instead programmers need to write security codes to check for limited benign structures based on benign queries expected by the web application. The method will not only reduces the complexity but also save processing time at this stage because the pattern needs to be analyzed from limited number benign structures present in the whitelist. Ultimately, it would eliminate the risk of any kind (including new class if developed in future) of SQL injection attacks because every query would be treated as malicious in case the pattern do not match with any of the benign structures defined in the whitelist. Hence, the system would be robust to prevent all kind of SQLI attacks.

Let us consider a simple example where the users after authentication are allowed to update their details. Consider a case where an attacker create a new account by registering himself/herself. Thus, the attacker would not be facing any problem at the authentication page of the web application. This way the attacker would be able to update other records also, simply by giving input to the username field as ' OR fullname LIKE '%s%'.

Thus, the **malicious SQL query** would be generated as follows:

UPDATE 'user_info' SET fullname='Allen', gender='Male', qualification='M.Tech' WHERE username=" OR fullname LIKE '%s%';

To prevent such attack, the idea is to define a benign structure in the whitelist and pattern array containing all possible keywords, identifiers and operators, etc. that are normally used in SQL queries.

The **benign structure** defined for the benign query is as follows:

UPDATE user_info SET fullname= gender= qualification= WHERE username= ;

The temporary structure is generated from the user's query by identifying elements of pattern array using KMP algorithm and concatenating them in ascending order of their indices.

The **temporary structure** thus formed would be as follows:

UPDATE user_info SET fullname= gender= qualification= WHERE username= OR fullname LIKE %%;

In this example, the temporary structure is different from benign structure. Hence, the user generated query would be classified as malicious SQL query. This way malicious query is detected before sending for execution and hence the SQL injection attack is prevented.

5 Result

The model composed of web application with two layers of securities. The web application is developed using technologies such as HTML, CSS, Javascript, PHP, XAMPP Apache Server. The application provides simple functionalities that are common for any applications such as registration of the new user, authentication of the user using login form and if user enters correct values he/she will be redirected to homepage revealing their information. In order to interact with the database, PHP is used that provide procedural functions to interact with the database. These functions are as follows: mysqli_connect(), mysqli_select_db(), mysqli_query(), mysqli_num_rows() etc. PHP also provide library for cryptographic algorithms such as password_hash() and password_verify() functions that can encrypt and verify the user input values using Blowfish algorithm respectively.

The first security layer is based on the encryption method using Blowfish algorithm, which has been implemented successfully preventing all kind of SQL injection attacks during user's authentication. We used Blowfish cryptographic algorithm because it is a more secure and fast encryption algorithm compared to AES, Triple DES, RSA, Twofish as shown in Table 1. Also, there is another advantage of using it, as the decryption speed of Blowfish is relatively slow in comparison to others, it would be very difficult for attackers to decrypt the encrypted parametric values and hence make the system more secure from authentication access [5].

In the Fig. 1, an example of tautology based SQLI attack has been tried through username and password fields of authentication form and the Fig. 2 shows the database of stored login information wherein both the parameter values (i.e. Username and Password) are available in encrypted form using blowfish algorithm for every authentic user. So, it becomes very difficult for attackers to access any detail because any malicious query if injected will gets encrypted differently and hence it will not match with any of the stored credentials. Also, any malicious query after encryption is sent for execution to the execution engine of DBMS, where the encrypted form of malicious query would be considered as plain text and thus save our system from illegal authentication access. Hence, encryption alone is very effective to avoid authentication access using SQLI

Table 1. Comparison of cryptographic algorithms (*Time for Encryption and Decryption are in Micro-Seconds*)

Algorithms	Encryption time	Decryption time
AES	0.000531197	1.09673E−05
Triple DES	0.000544071	3.40939E−05
RSA	0.315055132	0.05095005
Blowfish	0.000361919	9.20296E−05
Twofish	0.003043175	0.000346899

Fig. 1. Login form with malicious SQL query

| Show all | Number of rows: 25 | Filter rows: Search this table |

+ Options

username	password
$2y$10$cr9q9Afa.iNSZlq9TAbhxuJ.e9hG0zuHvZds5WLeUZB...	$2y$10$Hc5Pqd9/AV3/iVwrJ8fJfuZYXI5uCkbnwk0sh
$2y$10$RVj9m.Bvdfi4tAoJcQ34FOHjLu2McAKqCUHMXOA0fEr...	$2y$10$kaqgtTt/oscryUhphr3bM.1KMVeWbqF0mT.S
$2y$10$r1tJRIRrtHBTMvZSfVuSoOEenZ2Syoqc1HdjvVdMlMK...	$2y$10$QoNOoSIExNuiNBdMdlH2xu82QCv3BDcHr

Fig. 2. Table with encrypted values of Username and Password

attacks but this method alone is not sufficient as explained earlier. Thus, there is a need of adding another layer of security based on pattern matching using KMP algorithm based on the idea as discussed earlier and the implementation of it is in our future work.

6 Conclusion

SQL injection attacks has been always remains in the top 10 OWASP injection vulnerability list since it's existence. Also, it is a very commonly seen type of injection attacks through which sensitive information can be revealed, which can sometime do very serious damage to an individual or any business firm. This motivated us to work on security aspect related to SQLI attacks. The research paper meticulously discusses SQL injection attacks and their types. It also discusses how such attacks are done and different models which were developed over the period of time. Based on research studies, a hybrid model is proposed to improve the security using two different methods - Encryption Method, that found to be very effective in preventing all kind of SQL injection attacks during authentication and other is based on pattern matching that compares the temporary structure generated from user's query with the benign structure. If they match the user's query is benign otherwise malicious. The project work for the first security layer has been completed and found to be very effective preventing all kind of SQLI attacks during authentication. The results and the technologies used for first security layer based on cryptographic technique are discussed in details within the paper. The work for the second security layer based pattern matching using KMP is under implementation. The implementation is based on the algorithm as discussed earlier and is a part of our future work. We believe that implementing the second security layer would make the system robust and is capable of preventing all kind of SQLI attacks for a particular web application at low processing cost. It's a major challenge of information security, where the data are stored within storage facilities and SQL queries are widely used to retrieve information. Hence, preventing such injection attacks became inevitable and thats remain a motivating factor throughout our research work.

References

1. Ghafarian, A.: A hybrid method for detection and prevention of SQL injection attacks. In: IEEE London Computing Conference, pp. 833–838 (2017)
2. Tajpour, A., Ibrahim, S., Masrom, M.: SQL injection detection and prevention techniques. In: IEEE 2nd International Conference on Computational Intelligence, Communication Systems and Networks, vol. 3, pp. 216–221 (2011)
3. Voitovych, O.P., Yuvkovetskyi, O.S., Kupershtein, L.M.: SQL injection prevention system. IEEE International Conference of Radio Electronics & InfoCommunications, pp. 1–4 (2016)
4. Alwan, Z.S.: Detection and prevention of SQL injection attack: a survey. Int. J. Comput. Sci. Mob. Comput. **6**(8), 5–17 (2017)
5. Karunanithi, J.S.: SQL injection prevention technique using cryptography. Culminating Projects in Information Assurance (2018)
6. Patel, N., Shekokar, N.: Implementation of pattern matching algorithm to defend SQLIA. Procedia Comput. Sci. **45**, 453–459 (2015)

7. Halfond, W.G.J., Orso, A.: Preventing SQL injection attacks using AMNESIA. In: 28th International Conference on Software Engineering (ICSE), pp. 795–798 (2006)
8. Bisht, P., Madhusudan, P., Venkatakrishnan, V.N.: CANDID: dynamic candidate evaluations for automatic prevention of SQL injection attacks. ACM Trans. Inf. Syst. Secur. **13**(2), 1–39 (2010)
9. Ping, C., Jinshuang, W., Lin, P., Han, Y.: Research and implementation of SQL injection prevention method based on ISR. In: 2nd IEEE International Conference on Computer and Communications (ICCC), pp. 1153–1156 (2016)
10. Perkins, J., Eikenberry, J., Coglio, A., Willenson, D., Sidiroglou-Douskos, S., Rinard, M.: AutoRand: automatic keyword randomization to prevent injection attacks. In: Caballero, J., Zurutuza, U., Rodríguez, R.J. (eds.) DIMVA 2016. LNCS, vol. 9721, pp. 37–57. Springer, Cham (2016). https://doi.org/10.1007/978-3-319-40667-1_3
11. Ali, S., Shahzad, S.K., Javed, H.: SQLIPA: an authentication mechanism against SQL injection. Eur. J. Sci. Res. **38**(4), 604–611 (2009)
12. Temeiza, Q., Temeiza, M., Itmazi, J.: A novel method for preventing SQL injection using SHA-1 algorithm and syntax awareness. In: Joint International Conference on Information and Communication Technologies for Education and Training and International Conference on Computing in Arabic (ICCA-TICET), pp. 1–4 (2017)
13. Balasundaram, I., Ramaraj, E.: An authentication mechanism to prevent SQL injection attacks. Int. J. Comput. Appl. (IJCA) **19**(1), 30–33 (2011)
14. Prabakar, M.A., KarthiKeyan, M., Marimuthu, K.: An efficient technique for preventing SQL injection attack using pattern matching algorithm. In: IEEE International Conference on Emerging Trends in Computing, Communication and Nanotechnology (ICECCN), pp. 503–506 (2013)
15. Buja, G., Abd Jalil, K.B., Ali, F.B.H.M., Rahman, T.F.A.: Detection model For SQL injection attack: an approach for preventing a web application from the SQL. In: IEEE Symposium on Computer Applications & Industrial Electronics, pp. 60–64 (2014)
16. Gao, H., Zhu, J., Liu, L., Xu, J., Wu, Y., Liu, A.: Detecting SQL injection attacks using grammar pattern recognition and access behavior mining. In: International Conference on Energy Internet (ICEI). IEEE (2019)
17. Appiah, B., Opoku-Mensah, E., Qin, Z.: SQL injection attack detection using fingerprints and pattern matching technique. In: IEEE 8th International Conference on Software Engineering and Service Science (ICSESS), pp. 583–587 (2017)
18. Abikoye, O.C., Abubakar, A., Dokoro, A.H., Akande, O.N., Kayode, A.A.: A novel technique to prevent SQL injection and cross-site scripting attacks using Knuth-Morris-Pratt string match algorithm. EURASIP J. Inf. Secur. **2020**(1), 1–14 (2020). https://doi.org/10.1186/s13635-020-00113-y
19. Hasan, M., Balbahaith, Z., Tarique, M.: Detection of SQL injection attacks: a machine learning approach. In: International Conference on Electrical and Computing Technologies and Applications (ICECTA), pp. 1–6 (2019)
20. Ines Jemal, O., Cheikhrouhou, H., Hamam, A.M.: SQL injection attack detection and prevention techniques using machine learning. Int. J. Appl. Eng. Res. (IJAER) **15**(6), 569–580 (2020)
21. Kasim, O.: An ensemble classification - based approach to detect attack level of SQL injections. J. Inf. Secur. Appl. **59**, 102852 (2021)
22. Medeirios, I., Neves, N., Correia, M.: SEPTIC: detecting injection attacks vulnerabilities inside the DBMS. IEEE Trans. Reliab. **68**(3), 1168–1188 (2019)
23. Ping, C.: A second-order SQL injection detection method. In: IEEE 2nd Information Technology, Networking, Electronic and Automation Control Conference (ITNEC), pp. 1792–1796 (2017)

Monitoring Jitter in Software Defined Networks

Jithin Kallukalam Sojan[✉] and K. Haribabu

BITS Pilani, Pilani, Rajasthan, India
{f20170163,khari}@pilani.bits-pilani.ac.in

Abstract. End-to-end jitter of a flow is an important metric that indicates the Quality of Service a user is experiencing, particularly for real-time applications such as video streaming, cloud gaming, and so on. Monitoring the jitter can help controllers make routing decisions for certain flows. This paper discusses methods in which we can estimate/follow important patterns in the end-to-end jitter of a flow. Two main approaches are taken, one using a Software Defined Network controller and the other using P4 programmable data-planes. Results from the simulations of each method are discussed.

1 Introduction

1.1 End-to-End Jitter

End-to-end jitter between hosts in a network has multiple definitions, the most common being the difference in time for consecutive packets to reach from one host to the other over a number of network nodes, i.e., the difference in one-way delays between end hosts for consecutive packets. Jitter is of concern to the Quality of Service realization in a network.

Even when packets are sent from a host at a uniform rate, it is possible that the packets are not received by an end host at a uniform rate (zero jitter). This behaviour is the result of various characteristics of the traffic at different network nodes. According to previous works [3], it can be accounted for by the difference in the in-time and out-time of packets at each node.

1.2 Motivation Behind Monitoring Jitter

Most applications that rely on continuous transmission of packets like video/audio streaming, video chats and cloud gaming [6] are affected by jitter. Jitter at the end host deteriorates the quality of the user experience.

In order to minimize the jitter by taking real time decisions within a network, such as changing the path of a flow, we need to estimate the end-to-end jitter experienced. Being able to closely estimate the jitter (or patterns in the jitter) at multiple hops away from the end-host is thus important.

© The Author(s), under exclusive license to Springer Nature Switzerland AG 2022
L. Barolli et al. (Eds.): AINA 2022, LNNS 450, pp. 635–645, 2022.
https://doi.org/10.1007/978-3-030-99587-4_54

1.3 Approaches Taken to Monitor the Jitter

Our first attempt to monitor the jitter involves using a common controller (Floodlight) that polls for statistics already collected by OVS switches, such as the count of packets that have passed through a switch (or a port of the switch) within a particular interval of time.

Later, we switch our approach to using only local information at the switch level. This involves using programmable data-planes (P4 programmable switches) to estimate the jitter at the data-plane itself. We can then send these estimates to a controller for further use. The switch can also be programmed to make real-time decisions based on the statistics it collects and calculates.

2 Related Work

In [1], RTT estimation at the mid-point of a TCP stream is discussed. Jitter is calculated as the absolute difference between the estimated RTT and the actual RTT calculated at that instant. Mean jitter is defined as the average of all the estimates for that particular stream. The authors give a direct formula to estimate the jitter.

$$Jitter_{t+1} = \frac{7}{8}.Jitter_t + \frac{1}{8}.Jitter_{sample} \tag{1}$$

Compared to the focus of [1], we direct our attention to the jitter experienced in the use of real time protocols such as UDP, as the objective is to monitor jitter that can affect the QoS a user experiences.

In [2], the end-to-end jitter of a tagged stream in a tandem queuing network is discussed. The authors show various results, including the fact that the jitter decreases with increasing the load, and that the total jitter depends on the position of the congested nodes in the path. The authors assume a Poisson distribution for each stream, and mathematically derive the average jitter of an entire flow. Although it discusses estimating the jitter in a multi-node case, traffic with a Poisson distribution cannot be assumed for a modern network. As compared to a model that calculates the average jitter for an entire flow, we aim to monitor the jitter continuously.

In [3], a strong mathematical formulation for the end-to-end jitter is discussed with the assumption of periodic traffic. The end-to-end jitter is given by the expected absolute value of the sum of inter-packet delay variations introduced by each node along the path. It introduces an exponential approximation for the steady state waiting time, which is used in the estimations of the jitter for both single and multi-node cases, but the jitter estimation is limited by the assumption on waiting times between two consecutive packets.

RFC 1889 [4] discusses a Real Time Protocol and some traffic characteristics related to it. The jitter at a packet arrival is calculated as the sum of $1/16$ times the jitter caused by the current and the previous packet and $15/16$ times the previous estimate of the jitter. We use this idea of exponential averaging to estimate the jitter in our P4 (programmable data-plane) implementation.

In [5], the authors analyze the delay and jitter in networks that handle huge traffic volumes. Although the work is not on directly estimating the jitter, the authors give an analytical formulation for jitter with the assumption of Poisson traffic. It discusses the effect on end-to-end jitter as a result of changing network parameters such as increasing the network size, increasing the load, etc.

In [6], the authors discuss latency and jitter in Cloud-Gaming, an example of an application where jitter estimation and actions based on the estimate are important. Although this paper does not directly talk about the estimation of jitter, it proposes a Load-Sharing algorithm that can be employed in an SDN controller to greatly decrease the jitter experienced in certain cases.

3 Mathematical Formulation

3.1 Nodes Adding Variation in the One-Way Delay

Previous works [3] mention the idea that the end-to-end jitter can be modelled as the amount of variation each node in the network adds to the flow. Using SDN controllers we can collect real-time statistics such as the count of packets/bytes that have passed through a port for a particular flow. We collect the count of packets that have been transmitted by a port x of a switch S. Let this count of packets be a. Next, we collect the same statistic after t seconds. Let this count of packets be b. Thus, the rate at which packets were transmitted by port x of Switch S is:

$$r = \frac{(b-a)}{t} \tag{2}$$

Or, over the interval t, the average time of upload of a packet into the link at port x of Switch S was:

$$t_{offload,x,S} = \frac{t}{(b-a)} \tag{3}$$

Consider a network with two switches. Both S1 and S2 have in-ports x1 and x2 and out-ports y1 and y2. Host h1 is connected to S1 and host h2 is connected to S2, a flow from h1 to h2 is considered. Therefore the total time that a packet takes from h1 to h2 (one-way delay) is:

$$t_1 = t_{h1,S1} + t_{load,x1,S1} + t_{offload,y1,S1} + t_{S1,S2} + t_{load,x2,S2} + t_{offload,y2,S2} + t_{S2,h2} \tag{4}$$

Similarly, the one-way delay for the next packet will be:

$$t_2 = t_{h1,S1} + t'_{load,x1,S1} + t'_{offload,y1,S1} + t_{S1,S2} + t'_{load,x2,S2} + t'_{offload,y2,S2} + t_{S2,h2} \tag{5}$$

It is safe to assume that the time it takes for packets to travel across a link is constant. Thus, the only variable times in the above equations would be the load and offload times. The end-to-end jitter as a result of these two packets would be:

$$t_2 - t_1 = (\delta t_{load,x1,S1}) + (\delta t_{offload,y1,S1}) + (\delta t_{load,x2,S2}) + (\delta t_{offload,y2,S2}) \tag{6}$$

This equation is developed with the idea that we can model the jitter as the addition of variations in the one-way delay of a flow by the network nodes in the path. The difference here is that instead of just network nodes, we are considering each port (in-port and out-port) of each network node in the path.

Similar to the equation above, the current estimate of the delay value can be subtracted from the delay value that was calculated during the previous iteration. This would give us a statistic to monitor jitter, as we will see in Sect. 5. This statistic can be easily calculated by a controller using the count of packets that have been transmitted or received at a particular port over an interval t.

3.2 Deviation from the One-Way Delay

As compared to a simple SDN controller setup, it is possible to calculate the actual one-way delay from the end-host to various network nodes in a P4 data-plane. If any node (P4 switch) between the sending host and the receiving host has collected the previous $n-1$ one-way delays that the packets took to reach that switch, the mean one-way delay can be calculated on the current packet arrival as:

$$t_{mean} = \frac{\sum_{i=1}^{n} t_i}{n} \tag{7}$$

As indicated by previous works [1], the jitter can be considered as the standard deviation from the mean of the one-way delay. This standard deviation can be calculated as:

$$\delta t = \sqrt{\frac{\sum_{i=1}^{n} (t_i - t_{mean})^2}{n-1}} \tag{8}$$

Thus, we can calculate an estimate for the jitter using a data-plane programmable node that is able to retain the past $n-1$ one-way delays of packets that it received. At the same time, there are limitations to most P4 programmable switches that we need to account for in order to estimate the jitter.

Firstly, many P4 programmable switches cannot perform normal division. Division by a power of 2 is possible, emulated by bit-shifts to the right. Since the degree of freedom for the variance of a set is n−1 where n is the number of data-points, we consider n in our experiments to be a power of 2 plus 1, i.e., $n = 2^m + 1$. Since the mean of the one-way delays also have to be calculated, we approximate it to the mean of the last n−1 packets.

Secondly, P4 programmable switches cannot be used to calculate square roots. Thus, in our experiments, we compare the squares of the actual value of the end-to-end jitter with the variance of the one-way delay, i.e., $(\delta t)^2$.

Finally, squaring a number is a computationally expensive task that should be avoided in the P4 programmable switch if possible. We thus attempt to use a deviation from the mean that is based on Manhattan distance, rather than Euclidean distance (standard deviation). Manhattan distance is given by the absolute value of the difference in two vectors.

The deviation based on the Manhattan distance will be calculated as:

$$\delta t = \frac{\sum_{i=1}^{n} |t_i - t_{mean}|}{n-1} \tag{9}$$

4   SDN Controller Implementation and Results

4.1   Implementation of the Jitter Estimator

Following the first mathematical formulation, we calculate the average time that it takes for a packet of a flow to offload into a link from a port and to load from a link into a port. We thus consider the average time it takes for a packet to offload from port2 of S1, port3 of S2 and port1 of S3 (refer Fig. 1). We also consider the average time it takes for a packet to load into port1 of S1, port2 of S2 and port3 of S3. We subtract the sum of these average times from the sum of the average times corresponding to the same ports from the previous iteration of statistics collection.

The experiments in this section were run on OVS switches in a Mininet network. A Floodlight SDN controller was used to communicate with the switches. The links between the three OVS switches were configured at a 100 Mbps each.

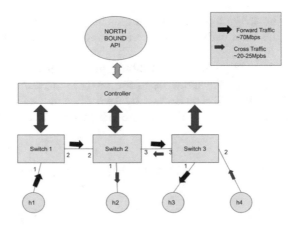

Fig. 1. Topology used in SDN approach (100 Mbps links between switches)

4.2   Single Flow Runs

First, a 90 Mbps flow was run from h1 to h3 for 200 s, with statistics polled every 1 s. On plotting the end-to-end jitter actually experienced along with the results of the estimate against time, we get Fig. 2 (orange represents the estimate while blue represents the actual jitter).

It is impossible to synchronize these two time series using an SDN controller. This is because the estimates of the jitter are completely time-independent of the jitter calculation at the host. Thus, one can only approximately place the two time series together for visual comparison. The measure of similarity between the time series is done using a distance measure called Dynamic Time Warping (DTW). It is a measure of the minimum distance between the two time series, even when they are not synchronized in

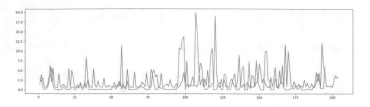

Fig. 2. Single flow 90 Mbps

time. For the above mentioned run, the DTW distance between the estimate and the actual jitter experienced at the host is 327.206 units.

Although the base of the estimate seems to be lower than that of the actual end-to-end jitter, the peaks in the jitter, which is of concern to us, are almost of the same height. Also, the number of peaks are almost the same in the estimate as in the actual jitter, although the peaks have been depicted a bit forward in time, and sometimes at different heights. The ability to capture the surges in jitter is extremely helpful as the controller can use this information to make real time decisions in order to reduce the jitter.

Note that for all plots in this section, the above justifications hold. Thus we will be evaluating the runs based on the similarity of the peaks in the estimate and the actual jitter, and also based on the DTW distance between the two time series.

Next, we run a single flow of 50 Mbps between h1 and h3. The results are shown in Fig. 3.

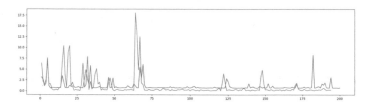

Fig. 3. Single flow 50 Mbps

Although the peaks of the estimate are higher than that of the actual jitter, most of the surges in the estimate seem to have a corresponding surge in the actual jitter. The DTW distance between the estimate and the actual jitter turns out to be 201.516 units. This indicates that when the bandwidth utilization is decreased, the estimates seem to be getting better.

Next, we run a single flow of 100 Mbps between h1 and h3. The results are shown in Fig. 4.

The DTW distance between the estimate and the actual jitter in this run is 345.85 units. Since a 100 Mbps flow is not healthy over a 100Mps link, a number of packets are lost. This again indicates that on increasing the bandwidth utilization of the flow, the estimate tends to become worse.

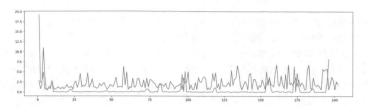

Fig. 4. Single flow 100 Mbps

4.3 Increasing the Period of Statistics Collection

The period of statistics collection is generally increased to counter an inherent problem of statistics collection at smaller time intervals, i.e., the likelihood of spurious values for the traffic statistics collected and the resultant calculations. On increasing the time interval over which the next statistic is collected, we smoothen out the values of the statistics in question.

We run a 90 Mbps flow from h1 to h3 for 200 s, with statistics from the ports of interest collected every 3 s. This is compared with the end-to-end jitter experienced by the flow over periods of 3 s as shown in Fig. 5.

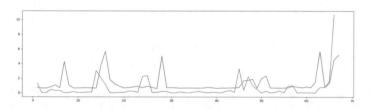

Fig. 5. Single flow 90 Mbps-Interval 3 s

The DTW distance between the two time series is 61.66. Each surge seems to have a corresponding surge in the actual jitter experienced. Since the estimate is calculated over a different period of statistics collection from the other runs, a direct comparison between the DTW distances cannot be made (the DTW distance per point is much lesser in this run than the previous ones).

5   P4 Implementation and Results

5.1   Implementation of the Jitter Estimator

This approach aims at estimating the end-to-end jitter within a programmable data-plane setup in a computationally inexpensive manner. As discussed in Sect. 3, we save one-way delays from the source host at respective switches for $2^n + 1$ packets. These delays are used to estimate the jitter according to the formulae given in Sect. 3. Since the one-way delays are saved until $2^n + 1$ packets arrive, the jitter estimator and thus the extra computation is used only once in these intervals.

The topology used for the experiments below is similar to that in the previous section. Three P4 programmable BmV2 switches S1, S2 and S3 are linearly connected in a Mininet network. We do not need to use P4 run-time as these experiments are run on static flows. A host h1 is connected to S1 and a host h2 is connected to S3. All the runs consists of a flow from h1 to h2. In order to collect the timestamps of when the packets are sent, an extra header dedicated to timestamps is used.

5.2   Euclidean Distance Estimate

As explained in Sect. 3, the variance in mean one-way delay is an estimate of the square of the average jitter experienced over $2^n + 1$ packets. Figure 6 shows a comparison of the estimates from switch S2 and the actual jitter.

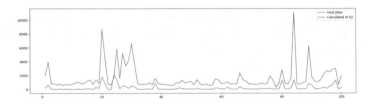

Fig. 6. Euclidean distance variance at S2 (17 packets)

Unlike in the SDN approach, there is no discrepancy regarding where the two series should be placed with respect to each other. This helps with the visual comparison of the two series. Although the base of the estimate is lower than that of the actual jitter, the peaks of the estimate fall right in place with the peaks of the actual end-to-end jitter experienced in the network. The distance between the two time series shown in the plot Fig. 6 is 125805 units. This value is large because we are comparing squares of the value of jitters.

For the same run as the one above, we estimate the jitter at Switch S3 and compare it with the square of the actual end-to-end jitters. Figure 7 shows the results.

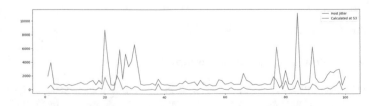

Fig. 7. Euclidean distance variance at S3 (17 packets)

The results are almost the same as in the previous case with the peaks at the same places, except for a few random outliers. The distance between the estimate and square of the actual jitters is better than that of S2, at 123340.37 units. This indicates that as the number of hops to the receiving host decreases, the jitter estimates tend to get closer to the actual jitter.

Next we run a Euclidean Distance estimate for the square of the jitters, where the interval of estimation is at 33 packets. Figure 8 shows the results.

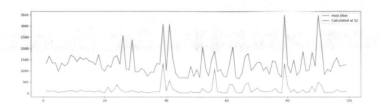

Fig. 8. Euclidean distance variance at S2 (33 packets)

As in the previous case, there seems to be a one is to one correspondence of the peaks. The distance between the two series is 111154.71 units, which is better than the estimates of both S2 and S3 in the 17 packets case. This indicates that as we increase the interval of jitter estimation, the estimate tends to get closer to the actual jitter.

5.3   Manhattan Distance Estimate

We estimate the jitter at switch S2 using a Manhattan distance estimate (ref. Sect. 3) and at an interval of 17 packet arrivals. Figure 9 shows a comparison between the estimate and the actual end-to-end jitter.

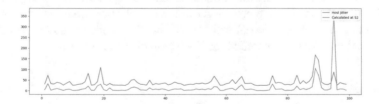

Fig. 9. Manhattan distance estimate at S2 (17 packets)

The base of the actual jitter time series and the estimated jitter is much closer than in the Euclidean distance estimate (since the estimate is not on the squared jitter). There is a one-is-to-one correspondence between the peaks, as in the previous cases. The distance between the two time series is 3230.63 units. For the estimate at switch S3 in the same run, the distance is lower, at 3209.3 units.

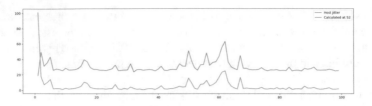

Fig. 10. Manhattan distance estimate at S2 (33 packets)

We also estimate the jitter at switch S2 at an interval of 33 packet arrivals, using a Manhattan distance estimate. Figure 10 shows a comparison between the estimate and the actual end-to-end jitter.

The distance between the two time series is 2558.14 units. This is lower than the distance which results from a 17 packet interval of estimation. Also, as expected, the distance of the estimate at S3 from the actual jitter is slightly lower at 2535.71.

6   Conclusions

Modelling the end-to-end jitter experienced by a flow as the sum of variations in the one-way delay added by each port in the path gives a statistic that lets the SDN controller monitor the jitter. It indicates surges in the jitter, which is the major motivation behind jitter estimation. Also, from the results of the simulations, it can be concluded that with a decrease in the link utilization of the flow, the jitter estimate gets closer to the actual jitter. Increasing the period of statistics collection results in less spurious values in the estimated jitter.

Modelling the end-to-end jitter as the standard deviation of one-way-delays between end-hosts is especially useful in programmable data-planes where the count of packets that have passed through and their timestamps can be stored. On increasing the number of packets over which the standard deviation is calculated, or on decreasing the hops to the receiving host, the estimate gets closer to the actual jitter experienced. Also, using Manhattan distance gives a decent estimate of the jitter with the added benefit of its computational lightness compared to the Euclidean distance estimate.

References

1. But, J., et al.: Passive TCP stream estimation of RTT and jitter parameters. In: The IEEE Conference on Local Computer Networks 30th Anniversary (LCN 2005), 1, 8pp. IEEE (2005)
2. Dahmouni, H., Girard, A., Sansó, B.: An analytical model for jitter in IP networks. Ann. Telecommun. - annales des telecommunications **67**(1–2), 81–90 (2012). https://doi.org/10. 1007/s12243-011-0254-y
3. Brun, O., Bockstal, C., Garcia, J.M.: A simple formula for end-to-end jitter estimation in packet-switching networks. In: International Conference on Networking, International Conference on Systems and International Conference on Mobile Communications and Learning Technologies (ICNICONSMCL 2006), p. 14. IEEE (2006)

4. Network Working Group, et al.: RFC 1889-RTP: a transport protocol for real time applications (1996)
5. Mesbahi, N., Dahmouni, H.: Delay and jitter analysis in LTE networks. In: 2016 International Conference on Wireless Networks and Mobile Communications (WINCOM), pp. 122–126. IEEE (2016)
6. Amiri, M., et al.: An SDN controller for delay and jitter reduction in cloud gaming. In: Proceedings of the 23rd ACM International Conference on Multimedia, pp. 1043–1046 (2015)

Designing and Prototyping of SDN Switch for Application-Driven Approach

Diego Nunes Molinos[1(✉)], Romerson Deiny Oliveira[2], Marcelo Silva Freitas[3],
Natal Vieira de Souza Neto[1], Marcelo Barros de Almeida[4],
Flávio de Oliveira Silva[1], and Pedro Frosi Rosa[1]

[1] Faculty of Computing, Federal University of Uberlandia (UFU),
Joao Naves de Avila Avenue, Santa Monica, Uberlandia 2121, Brazil
{diego.molinos,natalneto,flavio,pfrosi}@ufu.br
[2] High Performance Networks Group, University of Bristol,
Woodland Road, Bristol, UK
romerson.oliveira@bristol.ac.uk
[3] Department of Exact Sciences, Federal University of Jataí (UFJ), Jataí, Brazil
msfreitas@ufg.br
[4] Electrical Engineering Faculty (FEELT), Federal University of Uberlandia (UFU),
Joao Naves de Avila Avenue, Santa Monica, Uberlandia 2121, Brazil
marcelo.barros@ufu.br

Abstract. Currently, the Internet has become a limiting factor for its evolution. Applications are being developed from a new perspective of network utilization, demanding more Quality of Service (QoS) and Quality of User Experience (QoE). Approaches that aim to redesign the architecture, for example, Software Defined Networks (SDN), have become popular in the field of computer networks in an attempt to minimize the problems experienced by TCP/IP. In theory, SDN Networks naturally leave all the flexibility and programmability of the network in charge of the control plane, disregarding the data plane's ability to improve the QoS and QoE for users. This work presents the specification and the development of a prototype of *Switch* with *MAC* driven by the applications for SDN networks. Although it is possible to reconfigure the behavior of the network element, the *MAC* remains the same. Compared to other similar approaches, this proposal can expose, through the fine-grained, the low-level forwarding rules logic to the control plane through an orchestrator module, systematically allowing reprogramming. We carried out a case study using the Entity Title Architecture (ETArch) to show the ability of *Switch* to handle parameters, such as priority and bandwidth, in real-time.

1 Introduction

Internet applications are demanding more and more resources from the network. Although those applications require mobility, security, energy efficiency, bandwidth, QoS, and QoE, they usually leave to the network the best scenario to satisfy their requirements. Therefore, SDN networks have offered better and more

L. Barolli et al. (Eds.): AINA 2022, LNNS 450, pp. 646–658, 2022.
https://doi.org/10.1007/978-3-030-99587-4_55

efficient control over networking devices from a logically centralized controller to improve network management.

Due to the decoupling of the data plane and control plane in SDN, a distance from the functions implemented in hardware and software was created, making the infrastructure level a single granular network fabric. Furthermore, it contributed so that all flexibility and programmability were carried over to the control plane, disregarding the data plane's ability to improve the QoS and QoE for users.

Some proposals aim to explore the flexibility of the network through software solutions, for example, [14] and [9]. However, these solutions typically assume link-level connectivity and bypass data plane resources to provide QoS and QoE. In general, SDN implementations use OpenFlow-based switches or some virtualization technique. Although it is possible to reconfigure the behavior of the network element, these actions do not offer flexibility, making it challenging to manipulate QoS and QoE parameters at the *MAC* level.

This work presents a new switch for SDN networks with *MAC* driven by the application. This prototype is Linux-based, and it enables flexibility and programmability in the network's data plane. With our approach, the *MAC* remains the same, offering backward compatibility, and we go further to provide control to QoS and QoE parameters that impact the *MAC* decisions. In addition, we present a case study using the Entity Title Architecture (ETArch) to demonstrate and validate our switch. ETArch uses the SDN paradigm and is an architecture with features and support to context-oriented algorithms that allow adjustments according to the application requirements [2]. With our work, ETArch can modify the switches parameters changing their operation's context to guarantee the QoS and QoE driven by the application [2].

The remainder of the document is structured as follows: Section 2 presents the overview of flexibility in the SDN Data Plane and the background of the ETArch. Section 3 offers Switch prototype design details. Section 4 provides the results, and finally, Sect. 5 presents the conclusion and future work.

2   Background

This section provides a background about the programmability and flexibility in the SDN field and a brief about ETArch architecture.

2.1   Programmability and Flexibility in SDN Data Plane

SDN-based networks present an innovative way of rethinking computer networks in the Future Internet field. According to [13], this paradigm aims to create a well-defined logical interface for network control, abstracting the network and providing reconfigurability of services at runtime. Its uniqueness lies in the fact that it gives programmability for the network by decoupling the control and data planes in the architectural design of the network.

Since the beginning of SDN, there has been an adhesion to use the OpenFlow technology. According to [4], OpenFlow enables the creation of configurable flows from a central software-based controller. It separates the data plane from the control plane and presents a centralized programming model to manage the network elements. It offers an easy way to develop an L2 solution to support SDN scenarios through the abstraction of the network infrastructure level.

The interface between the centralized control plane and the forwarding elements is established by the OpenFlow protocol [4,14]. It is an out-of-band interface in which the controllers are connected to the network devices through physically dedicated links used exclusively for control traffic [7]. Figure 1 illustrates OpenFlow from the point of view of communication interfaces with details of the data plane.

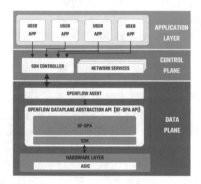

Fig. 1. OpenFlow architecture interface. Figure adapted from [10]

As shown in Fig. 1, there is a dependency on using OpenFlow on legacy hardware platforms, making it difficult to improve any low-level network features [6]. From the point of view of actions implemented by flow, the OpenFlow has a coarse-grained concerning managing flows. There is an entry of flows and one action associated with each entry, and the action is applied to the entire flow. There is no treatment for each primitive but a limited set of actions per stream.

The drawback observed in the OpenFlow-based switches is that it requires, before startup, a manual configuration of your controller's IP address and TCP port [7]. Besides, it is necessary to evolve OpenFlow, expanding the range of equipment, protocols, content headers to support new network architectures, and this requires continuous modification in its specifications [8].

Solutions OpenFlow-based make it challenging to implement new features at the network's low level. Their policies are applied to the entries flows and not per primitives. In addition, the low level of OpenFlow switch the MAC remains the same.

In response to OpenFlow difficulties, the Programming Protocol-Independent Packet Processors (P4) [9] emerged, a high-level programming language for programming network devices in the data plane. The drawback in using P4 is that the solution requires a translation infrastructure to be implemented down to the processor level. In addition, using high-level synthesis tools to generate custom hardware can be an approach that introduces additional complexity into the design, with results that are still not so satisfactory.

2.2   Programmability and Flexibility in ETArch Data Plane

The ETArch has been designed to approximate the Application layer semantically to the lower layers, allowing the requirements defined by the applications to permeate through the architecture layers. ETArch is based on SDN and presents innovations related to addressing network elements by the separation of location and identification, which intrinsically enables multicast transmissions and network mobility [1].

From the ETArch perspective, it's essential to comprehend the concepts of entity, Workspace, and the control agent to understand how the architecture works.

Entities are elements that want to communicate in a distributed environment [2]. It can be applications, devices, computers, network elements, etc. Entities are only identified by a title and a set of capabilities.

The Workspace represents an instance of communication driven by the application and carries a set of capabilities of the communication domain. As an example of the application requirements, can be mentioned the VoIP applications, where the transmission delays can be smaller than 150 ms. However, delays above 400 ms can impair transmission. Therefore, the Workspace created to serve a VoIP communication must operate with a transmission delay between 150 ms and 400 ms. For it to happen, all the architecture layers must be able to understand the application's requirements [2].

2.2.1   Domain Title Service - DTS

A Domain Title Service (DTS) represents a set of ETArch controllers in a distributed system. The goal of DTS agents is to separate the network into manageable subnets within ETArch [2]. A Domain Title Service Agent (DTSA) acts in the control plane, managing the network elements. In addition, DTSA is responsible for creating, managing, and dropping Workspaces on network elements. A typical ETArch environment is illustrated in Fig. 2.

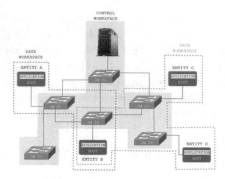

Fig. 2. Typical ETArch Environment with Data Workspace, Control Workspace, DTSA and entities. Figure from [11].

As shown in Fig. 2, DTSAs act on the network control plane and all communication between DTSA's and network elements is performed through a Control Workspace. In turn, the network elements act on the network data plane level, allowing entities to communicate within a communication domain, a Data Workspace.

3 Designing and Prototyping of SDN Switch Application-Driven Approach

This section will be present the design and prototyping of SDN Switch.

3.1 ETSCP Protocol

ETArch Switch Control Protocol (ETSCP) was designed by [11] to manage the control communication between DTSA and Switches in the ETArch network. There is no application entity involved in this control communication, which means, applications do not use ETSCP. The ETSCP format is similar to Ethernet protocol (802.3 [12]) to maintain compatibility with the current network interface cards, allowing the reuse of available technology. The changes in the content of the Ethernet header are in the source *MAC* address and destination *MAC* addresses field to support the identification of the *Title (Workspace)* + *Source Entity* and the control message type added in the payload. Figure 3 illustrates the message format used by ETSCP.

Fig. 3. ETSCP message format

About Services and Vocabularies, the ETSCP messages are *Add Workspace*, *Edit Workspace*, *Update Workspace* and *Remove Workspace*. All the ETSCP services are connectionless with acknowledgments, and ETSCP messages follow the taxonomy of confirmed services, Request/Indication, and Response/Confirmation. A new service in the ETSCP that acts on the scheduling policies to ensure QoS through non-ETArch communication was instituted, the Update Regular Connection service. Table 1 offers a clear association between the protocol services and their messages.

Table 1. ETSCP services and vocabulary

Service	Messages	Functional description
Create	*AddWkpREQ*	Confirmed service
Workspace	*AddWkpRESP*	To create new Workspaces in the switch
Edit	*EditWkpREQ*	Confirmed service
Workspace	*EditWkpRESP*	To modify parameters of the Workspace
Remove	*RemWkpREQ*	Confirmed service
Workspace	*RemWkpRESP*	To remove a Workspace in the Switch
Update	*UpdateWkpREQ*	Confirmed service
QoS Parameters	*UpdateWkpRESP*	To modify the QoS parameters of the switch's schedulers
Regular Connection	*UpdateRegConREQ*	Confirmed service
QoS Parameters	*UpdateRegConRESP*	To modify the QoS parameters of the switch's schedulers

3.2 SDN Switch FSM - Finite State Machine

Modeling a solution means building processes capable of interpreting messages and acting according to the context. The design of the SDN Switch was specified using a finite state machine (FSM). The diagram in Fig. 4 formally describes the behavior of the services defined by the ETSCP protocol.

The state ACTIVE is the initial and final state. All the other states are signalized in different colors according to each service. The state PARSER is responsible for analyzing all the messages and forwarding them to the other states according to each service. In the sequence are presented a brief description of the modeled services.

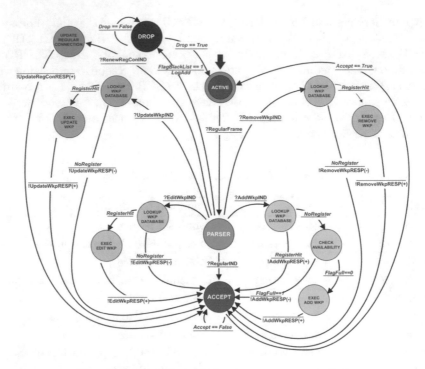

Fig. 4. Switch SDN - Finite State Machine (FSM)

1. *Create Workspace:* add support to the new schedulers' policies when the DTSA needs to create a control logic domain to respond to an entity request to create a data Workspace. When DTSA wants to create a Workspace, it sends an *AddWkpREQ* message to SDN Switch (*AddWkpIND*). Then, it verifies the existence of Workspace on the *WorkspaceDataBase* and checks the Switch availability to support this new Workspace, whether both of those actions before were true the Workspace will be created, and an ACK is sent or a NACK if not.

2. *Edit Workspace:* attach or remove a *Entity* in the existing Workspace, extending the Workspace's domain to reach out to the new *Entity* and update the scheduler's policies. When DTSA needs to edit a Workspace, it sends an *EditWkpREQ* message to SDN Switch (*EditWkpIND*). It verifies the existence of Workspace on the *WorkspaceDataBase*, the Workspace will be edited, and an ACK is sent or a NACK if not.

3. *Update Workspace:* change the QoS parameters of the existing Workspace. The QoS parameters used for schedulers policies are *Priority* that define the frame's prioritization per port, *Rate and Ceil* define the assigned bandwidth to each Workspace, *Burst and CBurst* determine the size, in bytes, to control the duration of the bursts.

4. *Remove Workspace:* remove the existing Workspace from the SDN Switch. When DTSA needs to remove a Workspace, it sends a *RemoveWkpREQ*

message to SDN Switch (*RemoveWkpIND*). Then, it verifies the existence of Workspace on the *WorkspaceDataBase*, whether it was true, the Workspace will be removed, and an ACK is sent or a NACK if not.

5. *Update Regular Connection:* change the QoS parameters of the existing regular traffic. The QoS parameters used for schedulers policies are the same. Any frame that is not a control frame or an ETArch frame will match the general forwarding rule.

3.3   Architecture of SDN Switch

From the point of granularity, the SDN Switch proposed has a fine granularity concerning the legacy switches. It allows forwarding policies oriented to frame and not the physical or logical ports as legacy equipment. Figure 5 present a view of the architecture and organization of SDN Switch.

Fig. 5. SDN switch architecture and organization

The Switch proposed in this work is a Linux-Based, the use Bridge module, support for 802.1 standards including 802.1P/Q and VLANs, and managing multiple network interfaces (NIC) justify the adoption of the Linux operating system. Furthermore, Linux Kernel natively supports several L2/L3 network functions, and the Bridge and Traffic Control modules are essential to promote traffic aggregation and manipulate forwarding traffic.

3.4   NEA-SWITCHD Module

This module is responsible for orchestrating all the elements in the SDN Switch, aiming to offer all the SDN Switch programmability and flexibility. The NEA SWITCHD module comprises NEA-Control, NEA-Parser, and NEA-Scheduler. The entire SDN Switch architecture operates sequentially. All primitives are queued and handled one by one by the NEA-SWITCHD module. Figure 6 present a view of the architecture and organization of the NEA SWITCHD module divided into software (Kernel Space and User Space) and hardware layer.

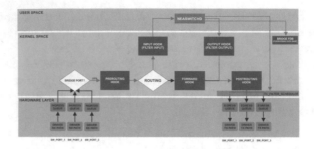

Fig. 6. NEA SWITCHD module architecture and organization

The NEA SWITCHD module interacts with Linux Kernel. In general terms, this module analyzes all frames that pass through the Bridge, schedules them, and orchestrates the policies within the scheduler defined by the network controller.

4   Tests and Prototyping Experimentation

The test aims to validate the Switch's ability to modify its control parameters to ensure the QoS of the applications. As depicted by Fig. 7, sequence diagrams show when messages are sent and received between Entities, Switch, and DTSA. For Workspace creation or edition, an entity sends the first message, and it is a data message addressed to a random Workspace. When Switch receives it, there is no forwarding policy for this message addressed, and the controller is triggered to configure or edit forwarding rules on the Switch. After that, DTSA sends a `AddWkp_Req` or `EditWkp_Req` to configure the Switch.

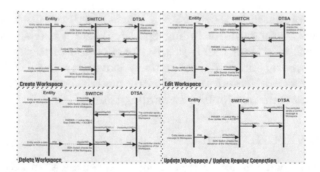

Fig. 7. Message Exchange to Service Executions. All message exchange was performed by a lightweight application that sends frames directly from the L2 layer. This application was written using the Python language and institutes ETSCP type messages.

Regarding removing Workspaces and updating QoS parameters, it begins with the DTSA. DTSA sends a `RemoveWkp_Req`, `RenewReg_Con_Req`, or `UpdateWkp_Req` to reconfigure the Switch, and after this, the Workspace is completely removed from Switch, or its QoS parameters are updated.

Figure 8 illustrates the environment developed for the verification of the SDN Switch's ability to configure and reconfigure the application-oriented QoS parameters. The ETSCP control messages are sent to Switch from the lightweight version of the DTSA. In addition to DTSA, there are two entities attached in a Workspace, and one of them runs a stream application.

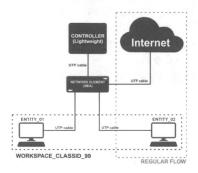

Fig. 8. Scenario to validate the SDN Switch performance concerning the control messages sent by the DTSA. The Entities are applications written in Python that exchange ETArch messages. At the same time, Entity 02 has been running an application streaming.

The metrics observed are (i) the ability to understand control messages from DTSA and (ii) the ability to configure application-oriented QoS parameters. Furthermore, the DTSA will act on the Workspace and regular flow to adjust the throughput through the QoS parameters.

In Workspace communication, the Entity_01 sends frames with a payload size of 58 bytes destined for the Workspace_Classid_99. Parallelly, the Entity_01 sends frames with a payload size of 1253 bytes with the target workspace_Classid _99. Although the frame sizes are different, stricter QoS policies allow less traffic in bytes by Workspace, and less strict QoS policies allow more traffic through Workspace.

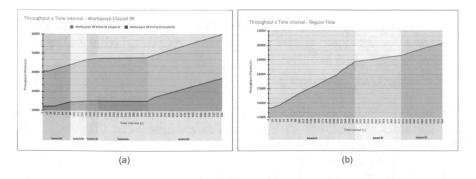

(a)   (b)

Fig. 9. Throughput (f/s) × Time interval (s) - Workspace and Application Streaming.

The SDN Switch throughput related to the Workspace during the time interval of 544 seconds is shown in Fig. 9 (a), the QoS parameters are changed five times (Scenarios 1, 2, 3, 4, and 5), and in Fig. 9 (b), the QoS parameters were modified three times (Scenario 1, 2, and 3). A clear association between the Traffic rate and QoS parameters configuration about Workspace communication and Application streaming is observed in Table 2. It is essential to point out that during the tests, the *Burst* and *Cburst* parameters were not changed, just the priority and available bandwidth do. Both scenarios (a) and (b) presented in Fig. 9 were evaluated concomitantly using the same time interval, and all the results were collected in real time.

Table 2. Traffic rate and QoS parameters configuration

Entity/Workspace communication	Traffic rate (f/s)	QoS parameters
Entity_01	*26,91 frames/s*	Scenario 01
Entity_02	*39,11 frames/s*	Priority: 2, Rate: 100 Mbit, Ceil: 100 Mbit
Entity_01	*10,04 frames/s*	Scenario 02
Entity_02	*53,79 frames/s*	Priority: 7, Rate: 100 Kbit, Ceil: 100 Kbit
Entity_01	*1,0 frames/s*	Scenario 03
Entity_02	*10,4 frames/s*	Priority: 7, Rate: 10 Kbit, Ceil: 10 Kbit
Entity_01	*0,10 frames/s*	Scenario 04
Entity_02	*2,10 frames/s*	Priority: 7, Rate: 1 Kbit, Ceil: 1 Kbit
Entity_01	*53,73 frames/s*	Scenario 05
Entity_02	*56,70 frames/s*	Priority: 2, Rate: 1 Mbit, Ceil: 1 Mbit
Entity/Application streaming	Traffic rate (f/s)	QoS parameters
Entity_02	*119,86 frames/s*	Priority: 6, Rate: 100 Mbit, Ceil: 100 Mbit
Entity_02	*30,84 frames/s*	Priority: 7, Rate: 100 Kbit, Ceil: 100 Kbit
Entity_02	*66,41 frames/s*	Priority: 4, Rate: 300 Kbit, Ceil: 300 Kbit

Both scenarios (a) and (b) presented in Fig. 9 were evaluated concomitantly using the same time interval, and all the results were collected in real-time.

5   Concluding Remarks

This work presented the design and prototyping of SDN switch with an application-driven approach. This proposal is a Linux-based SDN switch that allows the creation and modifies of forwarding rules in the SDN data plane. This solution has a fine granularity concerning other solutions used in SDN networks, allowing adjusting the application-oriented QoS parameters, not by flow or port as most legacy switches.

The flexibility added to the Software layer makes it easy to add new features and new support to new SDN architectures. The ETSCP guides a Switch to Workspace support and changes in operation time on ETArch architecture. The NEA_SWITCHD module acts in the orchestration of the Linux Kernel modules responsible for the programmability.

The results presented in this work show the ability of the SDN Switch to receive control messages from DTSA and act in the configuration of QoS policies (priority and bandwidth) in real-time. To the road ahead, we look forward to using high-performance hardware as the switch fabric implementation co-designed in a Kernel Linux embedded software environment.

References

1. de Oliveira Silva, F., Gonçalves, M.A., de Souza Pereira, J.H., Pasquini, R., Rosa, P.F., Kofuji, S.T.: On the analysis of multicast traffic over the entity title architecture. In: 2012 18th IEEE International Conference on Networks (ICON), pp. 30–35. IEEE (2012)
2. de Oliveira Silva, F., de Souza Pereira, J.H., Rosa, P.F., Kofuji, S.T.: Enabling future internet architecture research and experimentation by using software defined networking. In: 2012 European Workshop on Software Defined Networking, pp. 73–78. IEEE (2012)
3. Kalyaev, A., Melnik, E.: FPGA-based approach for organization of SDN switch. In: 2015 9th International Conference on Application of Information and Communication Technologies (AICT), pp. 363–366. IEEE (2015)
4. McKeown, N., et al.: OpenFlow: enabling innovation in campus networks. ACM SIGCOMM Comput. Commun. Rev. 38(2), 69–74 (2008)
5. Farhad, H., Lee, H., Nakao, A.: Data plane programmability in SDN. In: 2014 IEEE 22nd International Conference on Network Protocols, pp. 583–588. IEEE (2014)
6. Bifulco, R., Rétvári, G.: A survey on the programmable data plane: abstractions, architectures, and open problems. In: 2018 IEEE 19th International Conference on High Performance Switching and Routing (HPSR), pp. 1–7. IEEE (2018)
7. Freitas, M.S., Rosa, P.F., de Oliveira Silva, F.: ConForm: in-band control flows self-establishment with integrated topology discovery to SDN-based networks. In: Barolli, L., Amato, F., Moscato, F., Enokido, T., Takizawa, M. (eds.) WAINA 2020. AISC, vol. 1150, pp. 100–109. Springer, Cham (2020). https://doi.org/10.1007/978-3-030-44038-1_10
8. Oliveira, R.D., Molinos, D.N., Freitas, M.S., Rosa, P.F., de Oliveira Silva, F.: Workspace-based virtual networks: a clean slate approach to slicing cloud networks. In: CLOSER, pp. 464–470 (2019)
9. Bosshart, P., et al.: P4: programming protocol-independent packet processors. ACM SIGCOMM Comput. Commun. Rev. 44(3), 87–95 (2014). https://doi.org/10.1145/2656877.2656890
10. Mehra, M., Maurya, S., Tiwari, N.K.: Network load balancing in software defined network: a survey. Int. J. Appl. Eng. Res. 14(2), 245–253 (2019)
11. Oliveira, R.D., Freitas, M.S., Molinos, D.N., Rosa, P.F., Mesquita, D.G.: ETSCP: flexible SDN data plane configuration based on bootstrapping of in-band control channels. In: 2021 IFIP/IEEE International Symposium on Integrated Network Management (IM), pp. 711–715. IEEE, May2021

12. IEEE: IEEE standard for ethernet, IEEE Std 802.3-2018 (Revision of IEEE Std 802.3-2015), pp. 1–5600, August 2018
13. Kim, H., Feamster, N.: Improving network management with software defined networking. IEEE Commun. Mag. **51**(2), 114–119 (2013)
14. OF-CONFIG 1.2. In: OpenFlow Management and Configuration Protocol. Open Networking Foundation. Available via DIALOG (2014). https://opennetworking.org/wp-content/uploads/2013/02/of-config-1.2.pdf

SD-WAN: Edge Cloud Network Acceleration at Australia Hybrid Data Center

Junjie Wang[1] and Lihong Zheng[2(✉)]

[1] University of Canberra, Canberra, ACT, Australia
u3191448@uni.canberra.edu.au
[2] Charles Sturt University, Wagga Wagga, NSW, Australia
lzheng@csu.edu.au

Abstract. The recent pandemic has accelerated the development of remote connectivity around the world. Fast global access is essential to Australia's cloud data centre. SD-WAN offers an intelligent solution to improve data exchange speeds when connecting multiple remote sites to provide a much more stable and faster network. SD-WAN optimises the connections of your network. It uses the technology of software-defined networking (SDN) to intelligently route network traffic between branch offices and data center sites. So when your network expands and demands increase, the data can be organised well and handled with less latency. Fast and reliable global access is then guaranteed. This paper investigates the two types of SD-WAN architectures in detail, namely On-POP-Overlay and On-Premises-Overlay, and simulates the Edge Cloud Network Acceleration service scenario On-POP-Overlay architecture, constructs measurement messages. Analyse the impact of the Edge Cloud Network Acceleration on the forwarding delay through actual test data comparison in Australia.

1 Introduction

A Datacenter platform based on edge and cloud computing results in producing a large amount of data; however, the data itself has only limited value since data interpretation is essential. An Australian-based hybrid cloud data centre needs to consider the network latency impact in providing global access to the data centre. This is even more evident when considering the globalized scale on which data production, sharing and re-use are occurring.

Overlay network structure (On-Premises-Overlay) is the most basic SD-WAN (Software-Defined Wide Area Network) architecture [15]. Branches need to establish tunnels directly, which is not suitable for large-scale expansion. When multiple branches worldwide need Full-Mesh links with data centre sites, the On-Premises-Overlay architecture cannot be expanded quickly.

Establishing edge cloud network acceleration based on POP (Point of Presence) cloud network architecture (On-POP-Overlay) is the best architecture for large-scale branch deployment of SD-WAN [7]. The On-POP-Overlay architecture uses cloud POP nodes or ISP POP nodes to end the CPE. When designing and deploying, it will choose to deploy multi-line POP nodes in multiple cloud data centres in countries worldwide. Branches need first to detect and select the best POP node to establish a connection. Deploy vPE (virtual provider edge) router or gateway equipment in the POP. Establish a VPN tunnel between the CPE and the POP node. Solve the intercommunication across Internet service providers and public cloud providers at the POP layer to improve interconnection quality. Establish multiple links backbone network between POPs to ensure SD-WAN remote transmission service quality between different countries or continents.

Section 2 of this article introduces the basic architecture and functional model of establishing edge cloud network acceleration based on the POP cloud network architecture (On-POP-Overlay). Section 3 proposes the measurement method of processing delay for establishing edge cloud network acceleration in Australia hybrid cloud data centre. Section 4 measures the impact of the network architecture using edge cloud network acceleration on the network delay between Australia's hybrid cloud data centre and Asian countries.

2   Introduction of Edge Cloud Network Acceleration Based on ON SD-WAN

SD-WAN creates a mature private network and increases the ability to share network bandwidth dynamically. It also achieves central control, zero-touch configuration, integrated analysis, and on-demand configuration, thereby achieving policy-based centralized security and management. In addition, SD-WAN increase bandwidth at a lower cost. This configuration can achieve the best speed and has the ability to restrict low-priority applications. SD-WAN allows centralized management of branch networks, eliminating physical access to WAN routers and manual configuration by on-site IT personnel. It also provides more visibility and a standard network view for IT staff and middle managers. Since the network uses both private and public transmission media to route traffic, this also provides more choices for the type of transmission media and the choice of transmission providers. The advantages of SD-WAN can be summarized into three categories: flexibility, manageability and low cost [13].

Feature of SD-WAN

1. Flexibility: SD-WAN chooses between available transmissions, select the most suitable transmission, and increase the rerouting of applications to different transmissions during peak usage periods Function. It also set policies to choose more expensive or cheaper transmissions as needed. The most critical point is that SD-WAN controls bandwidth allocation rather than the ISP [11].

2. Manageability: SD-WAN enhances the network connection by using multiple transport/network operators, if one of them fails, the other one can be used, and this failover can be completed within a few seconds [11].
3. Low cost: SD-WAN technology reduces or replaces expensive MPLS lines to centrally manage and reduce the need for IT personnel to reduce the overall cost of WAN connections [18].

Standardized Definition of SD-WAN

1. The SD-WAN controller [12] provides centralized management for SD-WAN implementation. The entire network can be viewed through the central console or user interface. The SD-WAN controller can be deployed internally or implemented in the cloud. Since it only pushes network overlays and policies to SD-WAN Edge devices, it does not actually perform packet inspection, and its network usage is minimal. Through the console, IT staff can set policies, and then the orchestrator will execute these policies.
2. The SD-WAN orchestrator [12] is a virtualized network manager that monitors traffic and applies policies and protocols. SD-WAN orchestrator usually also includes SD-WAN controller function, used to set up the centralized strategy, use these strategies to make forwarding decisions for application flows. An application flow is an IP data packet classified to determine its user application or application grouping associated with it. Application flow grouping based on general types (for example, meeting applications) is called AFG (Application Flows Group) in MEF-70 [4]. SD-WAN is a policy-driven structure in which IP data packets are divided into AGF. AGF can be classified based on the 2nd to 7th layers of OSI. In addition, the AGF can block or allow the forwarding of IP packets based on the availability of the route to the target SD-WAN UNI on the remote SD-WAN Edge.

This is the conceptual representation of the SD-WAN Edge function. The policy process and the IP Forwarder are displayed in parallel [1]. In a given implementation, they can run in parallel (as shown in Fig. 1) or run sequentially in either order. The related point is that when an IP data packet arrives, it is either discarded or assigned to the TVC according to the IP data packet's policy and IP forwarding requirements. There are many different architectures for SD-WAN, the following are the two main different architectures used for edge cloud network acceleration based.

2.1   On-Premises-Overlay Architecture

On-Premises-Overlay architecture is the typical architecture of the initial SD-WAN (as shown in Fig. 2). Under this architecture, each location has an SD-WAN Edge (CPE) device. Fast networking with small and medium-sized businesses or small-scale organizations. Deploy SD-WAN CPEs in local branches, and deploy SD-WAN controllers in the cloud to interconnect with CPEs in each

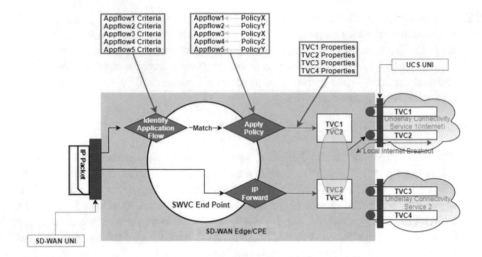

Fig. 1. Application flows and policies

branch. CPEs use the VPN technology to achieve more reliable links. (The underlying technology is GRE+IPSEC, VxLAN, etc.) [9]. It can meet the flexible deployment and networking of small business branches based on the Internet, realize SDN control and unified strategy, ZT-PNP automatic deployment, plug and play. The networking mode includes Hub-Spoke or small-scale Full-Mesh.

On-Premises-Overlay architecture is suitable for the interconnection of small and medium-sized branches. Because branches need to establish tunnels directly, it is not ideal for large-scale expansion. If there is a large-scale Full-Mesh demand, this architecture is a disaster. This architecture mainly solves the problems of automatic deployment and VPN interconnection. In the case of global deployment, it is necessary to consider the bottleneck problem of interconnection across multiple operators. On-Premises-Overlay architecture is completely uncontrollable when interconnecting across multiple Internet operators.

2.2   On-POP-Overlay Architecture

On-POP-Overlay architecture is a reliable architecture for large-scale branch deployment of SD-WAN. This architecture uses the cloud or the operator's POP node to terminate CPE (as shown in Fig. 3). When designing and deploying, it will choose to deploy multi-line POP nodes in multiple computer rooms in various places. Branches need first to detect and select the best POP node and establish a connection. Deploy vPE or gateway equipment in POP [8], establish VPN tunnel between CPE and POP node, solve cross-operator intercommunication at the POP layer to improve interconnection quality, and build a dedicated line backbone network between POPs to ensure the quality of SD-WAN remote transmission services.

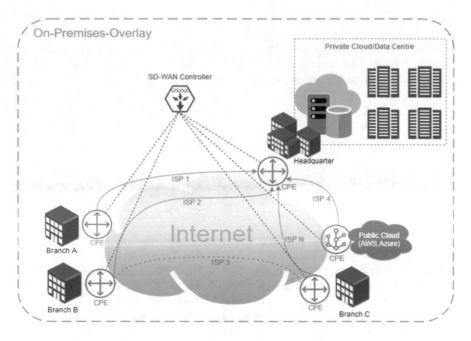

Fig. 2. On-Premises-Overlay architecture

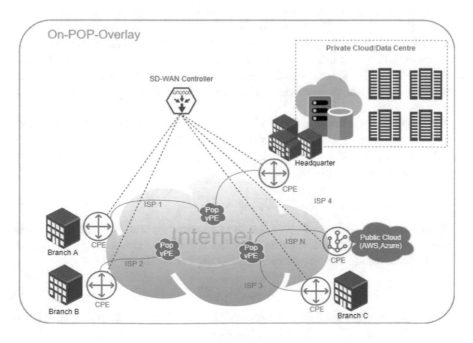

Fig. 3. On-POP-Overlay architecture

On-POP-Overlay architecture is very suitable for large-scale SD-WAN deployment. Due to the POP approach, the number of tunnels is greatly simplified. It unified strategy, minimal operation and maintenance, plug and play. The networking mode can be Full-Mesh or Hub-Spoke. It can solve the Internet bottleneck problem of global deployment across multiple operators and improve the quality of SLA service based on Internet networking. On-POP-Overlay operates by measuring basic network traffic metrics such as delay, packet loss, jitter, and availability. On-POP-Overlay can actively respond to real-time network conditions and select the best path for each data packet through these data.

3   Edge Cloud Network Acceleration Delay Measurement Method

On-POP-Overlay is to sink the business access equipment of the cloud network into the IDC where the edge cloud resource pool is located to form the network edge access area. It is a logical collection of the network edge access area plus the network exit area equipment in the cloud. On-POP-Overlay is deployed along with the edge cloud and is not only the endpoint of entering the cloud but also the endpoint of the inter-cloud network. The design architecture of On-POP-Overlay is shown in Fig. 3. On-POP-Overlay includes the network egress area in the cloud and the network edge access area. The network exit zone in the cloud realizes various cloud-native gateway functions of the overlay on the cloud, and the network edge access zone is connected to the edge routers of the cloud private network provider, packet transport network, slice packet network, optical transport network, metropolitan area network on demand and Basic network.

Latency is the time it takes for a packet to travel through the network to reach its destination [10]. The types of delays **D** encountered in an underlay packet-switched network are propagation delay, transmission delay, queuing delay and processing delay.

Propagation delay is the time it takes for a bit to get from one end of the link to the other end. The delay depends on the distance **d** between the sender and the receiver, and the propagation speed (**s** propagation speed $(2 \times 10^8 \text{ m/s})$) of the wave signal. It [5] is calculated as:

$$D_{prop} = d/s \tag{1}$$

Transmission delay refers to the time required to transmit the data packet to the output link. If a data packet contains L bits and the capacity of the link is R bits per second, the transmission delay [5] is calculated as:

$$D_{trans} = L/R \tag{2}$$

Queuing delay refers to the time a data packet waits for processing in the router's buffer. The delay depends on the incoming packets rate and the outgoing link's transmission capacity.

Processing delay is the time it takes for the router to process the packet. The delay depends on the processing speed of the router.

Then the delay of the packet delay **D** [5] is as in:

$$D = D_{proc} + D_{queue} + D_{trans} + D_{prop} \tag{3}$$

The experimental is based on Edge Cloud Network Acceleration of On-POP-Overlay networking. An experimental cross-continent PoP node environment was built based on the Australian headquarters and a Chinese branch (as shown in Fig. 4).

Fig. 4. Edge cloud network acceleration of on-POP-overlay networking

The single controller solution in the current OpenFlow specification is not scalable for large multi-domain networks [17] due to the limitation of the processing capacity of a single controller, the delay caused by remote network equipment, and the messages between the controller and the switch. It is caused by a large amount of overhead caused by delivery. Routing end-to-end QoS requires collecting the latest global network state information, such as each link's delay, bandwidth, and packet loss rate. Service Level Agreements, SLA is used to establish QoS parameters to manage application traffic [14].

In the global On-POP-Overlay-based SD-WAN design and deployment, it is necessary to consider the Internet bottleneck of each operator. The SDN controller uses the POP line SLA detection technology PolicyCop and realises the collaborative cloud and network integration deployment [2]. PolicyCop-an open, flexible and vendor-agnostic QoS policy management framework for SDN based on OpenFlow. PolicyCop provides interfaces for specifying QoS-based SLAs and implements them using OpenFlow API. It monitors the network and automatically re-adjusts network parameters to meet the customer's SLA.

To improve resilience, DCs may be connected via many links to a single or several ISPs [3]. POP deployment design chooses to deploy multi-line POP nodes in multiple computer rooms, each POP node deploys vPE equipment, and the vPE of each POP node establishes a backbone network through a dedicated line Guarantee the SLA of the traffic converged by SD-WAN. In the Edge and Core of SD-WAN, it is necessary to consider the SLA guarantee of the line, and then use the SDN controller to deploy the unified routing, security and QoS strategy deployment and control of the whole network to solve the problem based on Internet access latency issues and network-wide unified policy deployment issues.

Overlay POPs based on cloud data centres are built in Japan, China, Hong Kong, Singapore, Malaysia, Australia. By presetting different levels of business levels between PEs, central nodes, and edge nodes in the cloud PoP VPN, which uniformly carries all VxLAN tunnels between clouds. Establish Edge Cloud Network Acceleration [6] based on On-POP-Overlay networking. In this experimental environment, the test traffic will simulate the Internet packet forwarding delay test and Edge Cloud Network Acceleration delay test based on On-POP-Overlay networking to test the overall delay of the entire network tunnel. At the same time, the third-party testing instrument Pingplotter was selected to verify the measurement results. The first test implementation is based on the On-Premises-Overlay architecture's Internet packet forwarding delay test (as shown in Fig. 2). The data centre in Australia uses ICMP to test the network status [16] of the branch in China. The return includes delay, jitter, TTL value and other information. The second implementation method is the Edge Cloud Network Acceleration delay test based on On-POP-Overlay networking(as shown in Fig. 3). The data centre in Australia uses ICMP to test the network status of the branch in China and returns information including delay, jitter, TTL value.

After implementation of the network delay test, the data will be collected for further quantitative analysis for knowledge discovery. The experiment will compare the network delay that the same edge node accessed the Australian hybrid cloud computing centre in an On-Premises-Overlay architecture network and Edge Cloud Network Acceleration environment based on On-POP-Overlay networking. The positive or negative impact of the SD-Wan based On-POP-Overlay architecture network on the Australian hybrid cloud data centre network latency will be analysed in the end.

4 Edge Cloud Network Acceleration Delay Measurement Results

We conducted experiments to evaluate the performance of the two proposed SD-WAN-based architecture solutions. Compare the network delay that the same edge node accessed the Australian hybrid cloud computing centre in an On-Premises-Overlay architecture network and Edge Cloud Network Acceleration environment based on On-POP-Overlay networking.

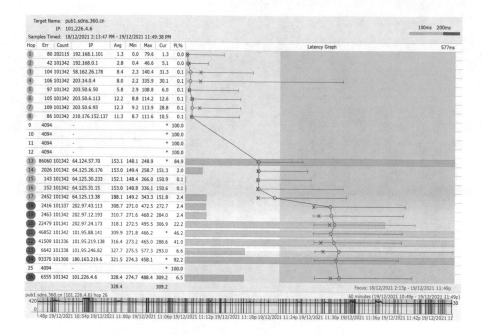

Fig. 5. On-Premises-Overlay instrument measurement results

4.1   On-Premises-Overlay Delay

The physical location of the starting point of the measurement was the Australia X cloud data centre, and the physical location of the endpoint of the measurement was the Beijing Y data centre in China. In measuring network delay, the data centre test host records the timestamps of sent and received messages and does not perform any security processing on traffic. The detection traffic constructed based on the ICMP protocol generates a UDP datagram with a TTL of 1 and was directly forwarded by the data centre CPE to the remote CPE. The number of measurements was about 100,000 data packets, and the interval of each test data packet was 2.5 s.

The constructed packet measurement test results and instrument measurement results (as shown in Fig. 5). Based on the On-Premises-Overlay architecture, there are 26 hops from the Australian data centre to the Chinese data centre. The lowest latency of the packet measurement results was 274.7 ms, and the highest latency was 488.4 ms, with an average delay of 328.4 ms. In 101,342 data packet tests, 6,555 errors occurred, and the packet loss rate was 6.5%.

4.2   On-POP-Overlay Delay

The physical location of the starting point of the measurement was the Australia X cloud data centre, and the physical location of the endpoint of the measurement was the Beijing Y data centre in China. In measuring network delay, the

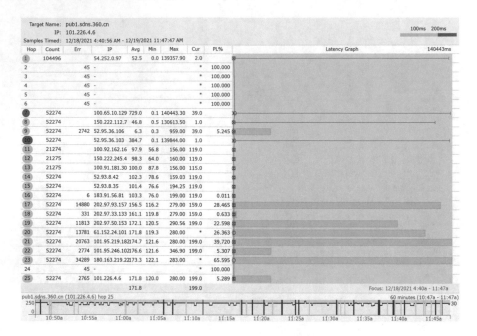

Fig. 6. On-POP-Overlay instrument measurement results

data centre test host records the timestamps of sent and received messages and does not perform any security processing on business traffic. The detection traffic constructed based on the ICMP protocol generates a UDP datagram with a TTL of 1, which was then forwarded by the data centre CPE to the nearest POP node, and the POP node vPE completes the WAN routing and data packet forwarding. The number of measurements was about 50,000 data packets, and the interval of each test data packet was 2.5 s.

The constructed packet measurement test results and instrument measurement results (as shown in Fig. 6). Based on the On-POP-Overlay architecture, there are 25 hops from the Australian data centre to the Chinese data centre. The packet measurement result has a minimum delay of 120,0 ms, and a maximum delay was 280.0 ms, and the average delay was 171.8 ms. In 52,274 data packet tests, 2765 errors occurred, and the packet loss rate was 5.29%.

5   Conclusion

By constructing SD-WAN-based On-Premises-Overlay and On-POP-Overlay architectures, this paper verifies the impact of the two architectures on the forwarding delay of transcontinental and transnational transmission links through analysing the traffic between Australia and China. Comparing the 328.4 ms latency of the On-Premises-Overlay architecture and the 171.8 ms latency of the On-POP-Overlay architecture, the delay of On-POP-Overlay is only about half of that of On-Premises-Overlay. The actual measurement results show that Australia's Edge Cloud Network Acceleration has apparent advantages in link selection and optimization based On-POP-Overlay SD-WAN. Therefore, the future research direction is to study selecting and establishing SD-WAN-based POP nodes worldwide. Ensure routing optimization between multiple continents and Australia data centre, reducing delay in critical services. Other SD-WAN research fields should focus on SD-WAN's ability to understand applications better and serve applications, such as AI-based intelligent traffic analysis, providing accurate, intelligent scheduling capabilities (especially in POP detection selection algorithms). AI-based Intelligent operation and maintenance tools provide fine-grained operation and maintenance for Underlay and Overlay. In addition, the openness, common standards and interoperability of SD-WAN are also future research directions.

References

1. Aweya, J.: IP router architectures: an overview. Int. J. Commun. Syst. **14**(5), 447–475 (2001). https://doi.org/10.1002/dac.505
2. Bari, M.F., Chowdhury, S.R., Ahmed, R., Boutaba, R.: PolicyCop: an autonomic QoS policy enforcement framework for software defined networks. In: SDN4FNS 2013 - 2013 Workshop on Software Defined Networks for Future Networks and Services (2013). https://doi.org/10.1109/SDN4FNS.2013.6702548
3. Dulinski, Z., Stankiewicz, R., Rzym, G., Wydrych, P.: Dynamic traffic management for SD-WAN inter-cloud communication. IEEE J. Sel. Areas Commun. **38**(7), 1335–1351 (2020). https://doi.org/10.1109/JSAC.2020.2986957
4. Forum, M.: MEF 70 - SD-WAN Service Attributes and Services. Technical Report. MEF, July 2019
5. Johari, L., Mishra, R.K.: An improved QoS technique to minimize delay for multimedia application in MANET. Int. J. Adv. Res. Comput. Eng. Technol. (IJARCET) **6**, 1249–1252 (2017)
6. Karras, K., Zotos, N., Bogdos, G.: A cloud acceleration platform for edge and cloud. In: ENeSCE 2017, Stockholm (January 2017). https://doi.org/10.475/123_4. https://www.researchgate.net/publication/313236609_A_Cloud_Accelation_Platform_fro_Edge_and_Cloud
7. Mine, G., Hai, J., Jin, L., Huiying, Z.: A design of SD-WAN-oriented wide area network access. In: Proceedings - 2020 International Conference on Computer Communication and Network Security, CCNS 2020, pp. 174–177 (2020). https://doi.org/10.1109/CCNS50731.2020.00046
8. Minoves, P., Frendved, O., Peng, B., MacKarel, A., Wilson, D.: Virtual CPE: enhancing CPE's deployment and operations through virtualization. In: CloudCom 2012 - Proceedings: 2012 4th IEEE International Conference on Cloud Computing Technology and Science, pp. 687–692 (2012). https://doi.org/10.1109/CloudCom.2012.6427560
9. Naranjo, E.F., Salazar Ch, G.D.: Underlay and overlay networks: the approach to solve addressing and segmentation problems in the new networking era: VXLAN encapsulation with Cisco and open source networks. In: 2017 IEEE 2nd Ecuador Technical Chapters Meeting, ETCM 2017, January 2017, pp. 1–6 (2018). https://doi.org/10.1109/ETCM.2017.8247505

10. Papagiannaki, K., Moon, S., Fraleigh, C., Thiran, P., Diot, C.: Measurement and analysis of single-hop delay on an IP backbone network. IEEE J. Sel. Areas Commun. **21**(6), 908–921 (2003). https://doi.org/10.1109/JSAC.2003.814410
11. Pratiwi, W., Gunawan, D.: Design and strategy deployment of SD-WAN technology: in Indonesia (Case Study: PT. XYZ). In: 2021 International Conference on Green Energy, Computing and Sustainable Technology, GECOST 2021 (2021). https://doi.org/10.1109/GECOST52368.2021.9538796
12. Scarpitta, C., Ventre, P.L., Lombardo, F., Salsano, S., Blefari-melazzi, N.: Every-WAN - an open source SD-WAN solution. In: 2021 International Conference on Electrical, Computer, Communications and Mechatronics Engineering (ICEC-CME), October, pp. 7–8. IEEE (2021)
13. Segec, P., Moravcik, M., Uratmova, J., Papan, J., Yeremenko, O.: SD-WAN- architecture, functions and benefits. In: ICETA 2020 - 18th IEEE International Conference on Emerging eLearning Technologies and Applications, Proceedings, pp. 593–599 (2020). https://doi.org/10.1109/ICETA51985.2020.9379257
14. Tong, V., Souihi, S., Tran, H.A., Mellouk, A.: SDN-based application-aware segment routing for large-scale network. IEEE Syst. J., 1–10 (2021). https://doi.org/10.1109/jsyst.2021.3123809
15. Troia, S., Zorello, L.M., Maralit, A.J., Maier, G.: SD-WAN: an open-source implementation for enterprise networking services. In: International Conference on Transparent Optical Networks, July 2020, pp. 2019–2022 (2020). https://doi.org/10.1109/ICTON51198.2020.9203058
16. Wenwei, L., Dafang, Z., Jinmin, Y., Gaogang, X.: On evaluating the differences of TCP and ICMP in network measurement. Comput. Commun. **30**(2), 428–439 (2007). https://doi.org/10.1016/j.comcom.2006.09.015
17. Wibowo, F.X., Gregory, M.A., Ahmed, K., Gomez, K.M.: Multi-domain software defined networking: research status and challenges. J. Netw. Comput. Appl. **87**, 32–45 (2017). https://doi.org/10.1016/j.jnca.2017.03.004
18. Wood, M.: How to make SD-WAN secure. Netw. Secur. **2017**(1), 12–14 (2017). https://doi.org/10.1016/S1353-4858(17)30006-5

Decision Tree Based IoT Attack Detection in Programmable Data Plane Using P4 Language

Rahul Poddar[✉] and Hari Babu

Department of Computer Science and Information Systems,
Birla Institute of Technology and Science Pilani, Pilani, India
{f20170746,khari}@pilani.bits-pilani.ac.in

Abstract. The Internet of Things (IoT) is a massively growing domain. With this the threats are also growing. Software Defined Networking (SDNs) is an emerging architecture which separates the control plane and the data plane of a network. It is being put to practice in networks around the world to mitigate issues. With growing heterogeneity in IoT protocols, it is cumbersome and costly to use SDNs. The Programming Protocol-independent Packet Processors (P4) is an open source, domain-specific programming language for network devices, specifying how data plane devices (switches, routers, NICs, filters, etc.) process packets. To overcome the challenges of IoT, P4 language is ideal as it provides flexibility for programming the data plane. We propose a light and fast approach to use decision tree to detect attacks from network traces and form small header fields to implement high accuracy attack detection in the programmable data plane using the P4 language.

1 Introduction

The Internet of Things (IoT) is the description of a network of devices that are embedded with sensors, software, and other technologies for connecting and exchanging data with other devices and systems over the internet. There are more than 7 billion connected IoT devices today and the number is growing exponentially with time. With such a network of interconnected devices, is a major concern. There is no standard protocol that all the devices use and hence, catering to individual nodes becomes a challenge.

The traditional methods of strengthening the physical and application layer of a network involves firmware and hardware changes. These changes take a huge amount of time and are extremely costly to put into effect. Therefore, there is a need for network layer to mitigate these attacks.

One of the ways to implement network level is to use Software Defined Networking (SDN). SDN is an architecture that makes the network more flexible and easier to manage. There are three main components of SDNs: SDN Applications, SDN Controller and SDN Networking Devices. The flexibility of separating the control and data planes provides opportunity to develop dynamic systems.

L. Barolli et al. (Eds.): AINA 2022, LNNS 450, pp. 671–683, 2022.
https://doi.org/10.1007/978-3-030-99587-4_57

Openflow protocol is one of the most popular protocols used in SDNs. It is an open standard network protocol. It is used to manage network traffic. It has been used to develop centralised securitised systems. The limitation that Openflow poses is that it contains predefined headers. In the IoT domain there are various network protocols for which the Openflow doesn't have predefined headers. Programming Protocol-independent Packet Processors (P4) language solves this by making the data plane programmable and defining custom headers. P4 a programming language for network devices, specifying how data plane devices process packets.

P4 is an open source programming language. With the help of P4, we can program the data plane and execute functions on the data plane itself. Unlike OpenFlow, it is not restricted by certain number of protocols. User can define custom headers and extract custom fields from the packets to meet their requirements. This flexibility allows us to implement much robust network level systems.

Machine learning methods have been applied for many dynamic problems and given outstanding results. Machine learning has been applied to various problems as well. The problem with most machine learning algorithms is that specific features need to be identified and fed to the algorithms for required results. But decision tree is one such algorithm that makes classifications based on importance of features and provides easy mechanisms for pruning the tree.

By combining the power of machine learning and the flexibility of the P4 language, we intend to provide a robust network application to be deployed in the data plane. The contributions of this paper are:

- Machine learning algorithm that learns from network traffic bytes and determines whether the particular traffic is an attack.
- Extracting particular bytes as header fields for making a custom header and implementing flow rules in the P4 language to detect real time network attacks in the data plane.

2   Related Works

IoT security has attracted wide attention. Comprehensive surveys of IoT attacks and classifications are provided in [2] and [7]. New attack types including attack methods in IoT protocols is discussed in [8–10]. [11] and [12] target the physical attacks on the IoT devices. These researches indic- ate that authentication methods are required to prevent attacks. But network level securities are important along with authentication mechanisms that will act as a better shield towards attacks.

Machine learning has been applied to on specific packet headers [14]. [15] applies machine learning on 6LoWPAN headers, [16] applies machine learning to detect IoT attacks, [17] and [18] apply machine learning to identify various IoT device types. Although these approaches are effective, these require knowledge of header fields and defining different methods for different protocols.

Another approach is to classify packets based on raw packet bytes [4, 19, 20]. They have great merit that they don't rely on any assumptions of protocols etc.

and have high accuracy but these are mainly deployed on remote servers or hosts and not on switches. Therefore they are unable to process the packets at line speed.

Our proposition is to develop a firewall at the switch such that it is protocol independent and processes the packets at line speed. Developments in SDN has inspired researchers to implement switch functions for example [21] and [22] use OpenFlow in this direction. However P4 provides greater customisability for handling heterogeneous IoT protocols. Aggregation of sensor data from multiple packets has been implemented by [23] using P4 header operations. [6] provides switching between IoT services in multiple protocols using P4 enabled switches.

3   Discussion

We propose a system that has two components - the control plane that is the SDN controller and the data plane that is an IoT gateway. We consider that the IoT gateway is P4 programmable.

The aim is to use the IoT gateway to detect malicious traffic before it is routed. To achieve this, we program the IoT gateway to execute a firewall function first. tables are implemented which record the values of certain packet header fields. These fields are checked against the fields of incoming packets and marked as normal or malicious. The packets which are marked as normal are passed to the routing function and the malicious packets are blocked, dropped or forwarded to a honeypot. The SDN controller generates the flow rules by the help of a classifier which determines whether a packet is normal or malicious. The controller can be used to convert classification results into header field definitions and flow rules to install them in the firewall at the IoT gateway.

The classifier is made with the help of Decision Trees. Decision tree is a supervised learning algorithm. The nodes in the tree correspond to a class label. Any boolean function can be represented with the help of a decision tree. It can be used in classification and regression problems. The tree is built on a series of comparisons. Each node selects the most appropriate and important attribute to compare against a record data and further branch out.

The decision tree branches nodes based on all the available attributes and selects the branch which results in the most homogeneous sub-nodes. The criteria to evaluate the attribute that is most suitable may be one of the following: entropy, information gain, gini index, gain ratio, reduction in variance and chi square. In our study, we have used gini index as the criteria for splitting the nodes.

Raw packets from publically available datasets are used to train the decision tree. The decision tree results in a series of comparisons and classifying whether the packet is malicious or normal. We prune the decision tree to a suitable number of nodes. From the pruned decision tree, we extract the specific bytes the decision tree used to make classifications.

The P4 language is a flexible language which allows the formation of custom headers for extraction of packets. With help of the bytes we identify in the decision tree, we form a custom header. We extract that header and make flow rules based on the comparisons of the decision tree. If the comparisons lead to a conclusion that the packet is malicious, we drop the packet. If the comparisons lead to the conclusion that the packet is normal, we forward it ahead.

Thus with the help of the P4 language, we can make flow decisions in the data plane at line speed. The flow rules are installable from the control plane and hence, if there are any changes required, the flow rules can be altered and installed on the programmable switch thus reducing costs.

4   Experiment

To perform the research, we use the datasets created and provided by [43] available in their public github repository.

The **Cooja Network Simulator Dataset (Cooja)** contains different types of attacks in the 6LoWPAN networks. It contains the packets captured from simulations in the Contiki OS using the RPL routing protocols. The attacks that the dataset packets comprise of are Increased Rank Attack and Version Number Attacks.

The **Wapstone IoT Sensor Dataset (Wapstone)** contains packets sent form real IoT devices. Temperature, humidity and luminosity sensors are installed on the Waspmote Smart Cities Pro board. These sensors periodically send 6LoWPAN frames to the gateway. Physical attack like impeding electrical connection on the board is performed so that the sensors send wrong data.

The datasets contain the raw packet bytes in rows. These bytes have been converted to the integer equivalents for the purpose of experimentation. Data pre-processing is done on the datasets by normalising each byte of data and padding or limiting the packet size to 128 bytes. Further the data is divided into 80% for training and 20% for testing.

The **ISCX Botnet 2014 dataset** is also used. It contains the botnet traffic from various different and well known datasets. The dataset mainly contains HTTP, P2P and IRC traffics. This dataset is very large and hence only 10% of the dataset is used. The dataset contains malicious traffic originating from specific mapped IPs, hence we randomise all the IPs. The bytes of each packet are converted to integer equivalents in a text file and the entire packet is truncated to 128 bytes. Since the dataset already has training and testing datasets divided, we don't divide the datasets further.

There are 2 stages to this experiment.

Stage 1: The Decision Tree algorithm is used to classify the traffic as malicious or normal. Decision Tree is an algorithm which takes features as inputs and classifies based on the importance of the features. Thus, the algorithm determines which feature is more important in the training phase. Therefore, knowledge about the protocols and packets is not necessary to classify the packets as malicious or normal.

Each byte of a packet acts as a separate feature for the decision tree and hence, the decision tree classifies the packets based on the importance of the bytes. Therefore, we train the decision tree algorithm with the training data which we collected and pre-processed from the datasets. All the training is done on Google Colab.

This method has two advantages:

- The packets are classified as malicious or normal by the algorithm.
- We get the sequence of comparisons necessary for classification as malicious and normal.

This decision tree is deployed in the control plane of the SDN for classifying the packets. This is the end of stage 1.

Stage 2: We prune the resultant decision tree and select the tree which gives the best trade-off between the accuracy and the number of comparisons. Post pruning the tree, the most important bytes are used to form a p4 header. This P4 header is used to extract the specific bytes from the incoming packets in a switch. Match and action tables are formed based on this header and flow control rules are formed. Based on the sequence of matches, if the decision is that the packet is malicious, we drop the packet, else we forward the packet normally. This experimentation is done on mininet along with BmV4 switches.

Thus, by using a machine learning technique i.e. the decision tree, we can classify packets at line speed in the data plane of an SDN enabled, P4 programmable switch.

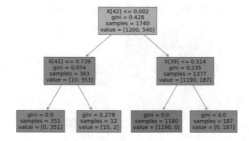

Fig. 1. Decision tree for wapstone database

5   Results

In Table 1, we have shown the comparison between the time taken train and test the classification model by our algorithm that is the decision tree and the state of the art neural network model [43].

In Table 2, we have shown the number of nodes in the decision tree and the accuracy derived from it before pruning the tree and after pruning the tree.

In Table 3, we have shown the accuracy derived from the state of the art neural network [43] and the number of parameters that need to be trained for it.

The result of stage 1 is showed in Table 1 and the pre-pruning column of Table 2. Table 1 shows the time taken in seconds. As is evident from the table, the decision tree algorithm 10^5 times faster in training and 10^3 times faster in testing the datasets than the neural network algorithm.

The accuracy derived from the decision tree in pre-pruned stage shown in Table 2 is better in all the cases than the deep neural network accuracy shown in Table 3. The number of nodes in decision tree is 10^6 less than the number of trainable parameters in the neural network.

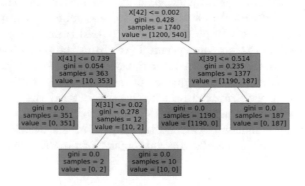

Fig. 2. Decision Tree for Cooja database

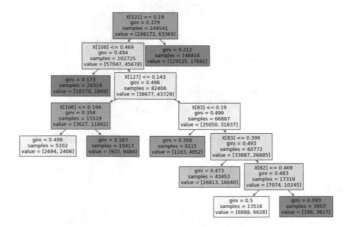

Fig. 3. Decision tree for ISCX database

Post pruning we achieve much lighter decision trees with a loss of just 4-5% in accuracy while still achieving the accuracy of 95%.

We use the Adaboost Classifier on top of the decision tree algorithm to enhance the performance of the algorithm on the ISCX dataset. With this, we are able to achieve a detection accuracy of 88.8

Figure 1 shows the post pruned decision tree for the Wapstone database, Fig. 2 shows the post pruned decision tree for the Cooja dataset and Fig. 3 shows the post pruned decision tree for the ISCX dataset. Each of the figures depict the series of comparisons on specific bytes of the packets to determine the nature of the packet i.e. normal or malicious. The left node indicated the comparison in the parent node resulted true. The right branch indicated the comparison in the parent node resulted in false. The final classification is made on the majority class of the leaf node. If a series of comparisons lead to a leaf node which has a majority of normal class, the packet is classified as normal. If the series of comparisons leads to a leaf node that has a majority of malicious class, the packet is classified as malicious.

In stage 2, we use the pruned decision tree to extract the specific bytes from the packet to make a decision in the data plane. We wrote a custom P4 program designed to extract the specific bytes from the packet. The flow rules are based on the output of the decision tree. If the based on the flow rules the decision is that the packet is malicious, we drop the packet else we forward the packet normally.

We send both the attack and normal packets of the ISCX test database and achieve an accuracy of 88.29% which is consistent with the accuracy we achieved with the decision tree. Table 4 depicts the results of this stage. The accuracy is calculated as follows: Let N be total normal packets sent, NR be normal packets received, A be total attack packets sent, AR be attack packets received

$$Accuracy = \frac{NR + (A - AR)}{N + A} \tag{1}$$

We have NR as 114864, A as 72128, AR as 13634 and N as 124221. Substituting these values in Eq. 1 we get the accuracy as 0.8829.

Table 1. Training and testing times

Datasets	Decision tree		Neural network	
	Train time	Test time	Train time	Test time
Wapstone	0.0036	0.00034	479.242	0.3239
Cooja	0.0565	0.00324	6579.1788	3.8835
ISCX	37.1393	0.03817	35713.6477	17.1624

The comparison between the training time of the algorithm that we have used i.e. the Decision Tree vs the training time for the state of the art neural network [43] is shown in Fig. 4. It is clear from the chart that compared to

Table 2. Decision tree performance metrics

Datasets	Pre-pruning		Post-pruning	
	No. of Nodes	Accuracy	No. of Nodes	Accuracy
Wapstone	9	1.0	7	0.998
Cooja	43	0.998	7	0.959
ISCX	53	0.9021	15	0.8880

Table 3. Deep neural network performance metrics

Dataset	Number of trainable parameters	Accuracy
Wapstone	659217	0.9954
Cooja	659217	0.9891
ISCX	659217	0.911

Table 4. Performance of stage 2

Type of packet	Packets sent	Packets received
Attack	72128	13634
Normal	124221	114864

deep neural network, our algorithm takes negligible time to train and still gives better attack detection accuracy in the case of Wapstone and Cooja dataset and comparable accuracy in the case of the ISCX dataset.

The comparison between testing time for decision tree and deep neural network is shown in Fig. 5. It follows the same trend as in training the two algorithms.

We compare the accuracy of attack detection between our decision model and the neural network model. We also compare our decision tree approach in

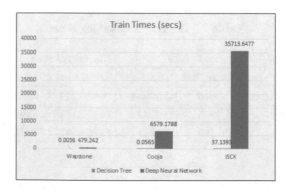

Fig. 4. Comparing training time

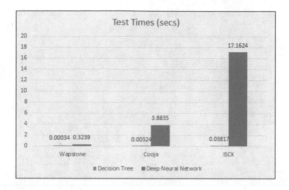

Fig. 5. Comparing testing time

the programmable data plane with the OpenFlow Protocol which represents an SDN without programmable dataplane. Here we again use a decision tree (DT) and SVM which only take the values of certain header fields as input. Since OpenFLow does not support header for 6LoWPAN, we cannot test the Cooja dataset on DT and SVM. Also the abnormalities in the Wapstone can only be detected in the payload and not in the header. Hence, even this dataset cannot be used with DT and SVM. Therefore, we show the comparison of detection accuracy for ISCX dataset for all the methods in Fig. 7 and only neural network and our proposed method for the Cooja and Wapstone in Fig. 6. It is evident from Fig. 6 that the accuracy derived from our decision tree based security model is better than the neural network based model in the case of Wapstone and Cooja dataset. It is only slightly less accurate than the neural network based model in the case of the ISCX dataset.

As is evident from Fig. 8, it is much cheaper and lighter to use the decision tree algorithm as we get comparable if not better accuracy with the decision tree using 10^6 times lesser parameters space to deploy in the SDN controller.

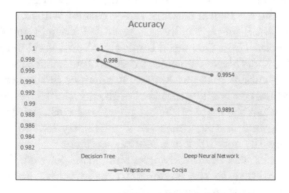

Fig. 6. Comparing accuracy

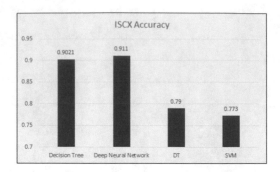

Fig. 7. Accuracy comparison between all the methods

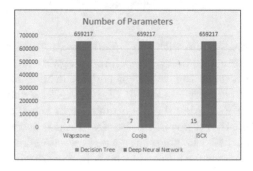

Fig. 8. Comparing number of parameters

6   Conclusion

In this study we approached network security in the data plane to prevent attacks at the data plane. We took a two stage approach we first used the decision algorithm to classify malicious and normal traffic from publically available databases. With the help of the decision tree we could achieve the achieve higher accuracy than the state of the art neural network in the case of Wapstone Datasets which contains packets sent from physically tampered IoT devices and Cooja database which contains packets from Increased Rank attack and Version Number attacks. In the case of ISCX database we achieve comparable accuracy.

The major advantage that our approach provides is the compact and light weight and hence less costly algorithm that is the decision. The number of parameters is 10^5 lesser than the state of the art neural network. As already mentioned above the training and testing times are significantly less as compared to the decision tree being 10^5 and 10^3 times less respectively.

The model works well practically in a simulation as well. The P4 language provides great flexibility to the engineer to make robust security systems. It provides the option to extract specific bytes and make flow rules based on these bytes.

For future studies, the models can be improved with the improvement in the research of machine learning algorithms. The models can be deployed in real life to see the actual performance and load handling.

References

1. Kolias, C., Kambourakis, G., Stavrou, A., Voas, J.: DDoS in the IoT: mirai and other botnets. Computer **50**(7), 80–84 (2017)
2. Andrea, I., Chrysostomou, C., Hadjichristofi, G.: Internet of things: security vulnerabilities and challenges. In: 2015 IEEE Symposium on Computers and Communication (ISCC), pp. 180-187. IEEE (July 2015)
3. McKeown, N., et al.: OpenFlow: enabling innovation in campus networks. ACM SIGCOMM Comput. Commun. Rev. **38**(2), 69–74 (2008)
4. Lotfollahi, M., Siavoshani, M.J., Zade, R.S.H., Saberian, M.: Deep packet: a novel approach for encrypted traffic classification using deep learning. Soft. Comput. **24**(3), 1999–2012 (2020)
5. Bosshart, P., et al.: P4: programming protocol-independent packet processors. ACM SIGCOMM Comput. Commun. Rev. **44**(3), 87–95 (2014)
6. Uddin, M., Mukherjee, S., Chang, H., Lakshman, T.V.: SDN-based multi-protocol edge switching for IoT service automation. IEEE J. Sel. Areas Commun. **36**(12), 2775–2786 (2018)
7. Alaba, F.A., Othman, M., Hashem, I.A.T., Alotaibi, F.: Internet of things security: a survey. J. Netw. Comput. Appl. **88**, 10–28 (2017)
8. Cao, X., Shila, D.M., Cheng, Y., Yang, Z., Zhou, Y., Chen, J.: Ghost-in-zigbee: energy depletion attack on zigbee-based wireless networks. IEEE Internet Things J. **3**(5), 816–829 (2016)
9. Pongle, P., Chavan, G.: A survey: attacks on RPL and 6LoWPAN in IoT. In: 2015 International Conference on Pervasive Computing (ICPC), pp. 1–6. IEEE (January 2015)
10. Mayzaud, A., Badonnel, R., Chrisment, I.: A taxonomy of attacks in RPL-based internet of things. Int. J. Netw. Secur. **18**(3), 459–473 (2016)
11. Fu, K., Xu, W.: Risks of trusting the physics of sensors. Commun. ACM **61**(2), 20–23 (2018)
12. Shoukry, Y., Martin, P., Yona, Y., Diggavi, S., Srivastava, M.: Pycra: physical challenge-response authentication for active sensors under spoofing attacks. In: Proceedings of the 22nd ACM SIGSAC Conference on Computer and Communications Security, pp. 1004–1015 (October 2015)
13. Antonakakis, M., et al.: Understanding the mirai botnet. In: 26th USENIX Security Symposium (USENIX Security 17), pp. 1093–1110 (2017)
14. Li, C., et al.: Detection and defense of DDoS attack-based on deep learning in OpenFlow–based SDN. Int. J. Commun. Syst. **31**(5), e3497 (2018)
15. Napiah, M.N., Idris, M.Y.I.B., Ramli, R., Ahmedy, I.: Compression header analyzer intrusion detection system (CHA-IDS) for 6LoWPAN communication protocol. IEEE Access **6**, 16623–16638 (2018)
16. Midi, D., Rullo, A., Mudgerikar, A., Bertino, E.: Kalis-A system for knowledge-driven adaptable intrusion detection for the Internet of Things. In: 2017 IEEE 37th International Conference on Distributed Computing Systems (ICDCS), pp. 656–666. IEEE (June 2017)

17. Nguyen, T.D., Marchal, S., Miettinen, M., Fereidooni, H., Asokan, N., Sadeghi, A.R.: DÏoT: a federated self-learning anomaly detection system for IoT. In: 2019 IEEE 39th International Conference on Distributed Computing Systems (ICDCS), pp. 756–767. IEEE (July 2019)

18. Miettinen, M., Marchal, S., Hafeez, I., Asokan, N., Sadeghi, A.R., Tarkoma, S.: IoT sentinel: automated device-type identification for security enforcement in iot. In: 2017 IEEE 37th International Conference on Distributed Computing Systems (ICDCS), pp. 2177–2184. IEEE (June 2017)

19. Wang, W., Zhu, M., Wang, J., Zeng, X., Yang, Z.: End-to-end encrypted traffic classification with one-dimensional convolution neural networks. In: 2017 IEEE International Conference on Intelligence and Security Informatics (ISI), pp. 43–48. IEEE (July 2017)

20. Wang, Z.: The applications of deep learning on traffic identification. BlackHat USA **24**(11), 1–10 (2015)

21. Luo, T., Tan, H.P., Quek, T.Q.: Sensor OpenFlow: enabling software-defined wireless sensor networks. IEEE Commun. Lett. **16**(11), 1896–1899 (2012)

22. Galluccio, L., Milardo, S., Morabito, G., Palazzo, S.: SDN-WISE: design, prototyping and experimentation of a stateful SDN solution for WIreless SEnsor networks. In: 2015 IEEE Conference on Computer Communications (INFOCOM), pp. 513–521. IEEE (April 2015)

23. Lin, Y.B., Wang, S.Y., Huang, C.C., Wu, C.M.: The SDN approach for the aggregation/disaggregation of sensor data. Sensors **18**(7), 2025 (2018)

24. Dang, H.T., et al.: Whippersnapper: a p4 language benchmark suite. In: Proceedings of the Symposium on SDN Research, pp. 95–101 (April 2017)

25. Haykin, S.: Neural Networks: A Comprehensive Foundation, Prentice Hall PTR. Upper Saddle River, NJ, USA (1994)

26. Oord, A.V.D., et al.: Wavenet: A generative model for raw audio. arXiv preprint arXiv:1609.03499 (2016)

27. Yu, R., et al.: Nisp: Pruning networks using neuron importance score propagation. In: Proceedings of the IEEE Conference on Computer Vision and Pattern Recognition, pp. 9194–9203 (2018)

28. Verhelst, M., Moons, B.: Embedded deep neural network processing: algorithmic and processor techniques bring deep learning to iot and edge devices. IEEE Solid-State Circuits Mag. **9**(4), 55–65 (2017)

29. Roffo, G., Melzi, S., Cristani, M.: Infinite feature selection. In: Proceedings of the IEEE International Conference on Computer Vision, pp. 4202–4210 (2015)

30. Bertsekas, D.P.: Dynamic Programming and Optimal Control, I and II, Athena Scientific, Belmont. Massachusetts, New York-San Francisco-London (1995)

31. Beigi, E.B., Jazi, H.H., Stakhanova, N., Ghorbani, A.A.: Towards effective feature selection in machine learning-based botnet detection approaches. In: 2014 IEEE Conference on Communications and Network Security, pp. 247–255. IEEE (October 2014)

32. Lashkari, A.H., Kadir, A.F.A., Gonzalez, H., Mbah, K.F., Ghorbani, A.A.: Towards a network-based framework for android malware detection and characterization. In: 2017 15th Annual conference on privacy, security and trust (PST), pp. 233–23309. IEEE (August 2017)

33. Le, A., Loo, J., Luo, Y., Lasebae, A.: The impacts of internal threats towards routing protocol for low power and lossy network performance. In: 2013 IEEE Symposium on Computers and Communications (ISCC), pp. 000789–000794. IEEE (July 2013)

34. Libelium. (n.d.) Waspmote. http://www.libelium.com/products/waspmote/
35. Nair, V., Hinton, G.E.: Rectified linear units improve restricted Boltzmann machines. In: Icml (January 2010)
36. Nanda, S., Zafari, F., DeCusatis, C., Wedaa, E., Yang, B.: Predicting network attack patterns in SDN using machine learning approach. In: 2016 IEEE Conference on Network Function Virtualization and Software Defined Networks (NFV-SDN), pp. 167–172. IEEE (November 2016)
37. Pedregosa, F., et al.: Scikit-learn: machine learning in python. J. Mach. Learn. Res. **12**, 2825–2830 (2011)
38. Abadi, M., et al.: Tensorflow: a system for large-scale machine learning. In: 12th USENIX Symposium on Operating Systems Design and Implementation (OSDI 16), pp. 265–283 (2016)
39. Lantz, B., Heller, B., McKeown, N.: A network in a laptop: rapid prototyping for software-defined networks. In: Proceedings of the 9th ACM SIGCOMM Workshop on Hot Topics in Networks, pp. 1–6 (October 2010)
40. Consortium, P.L., et al.: Behavioral model (bmv2) (2018)
41. Neural Network Source codes and datasets (2020). https://github.com/vxxx03/ICDCS2020
42. Biondi, P., et al.: Scapy (2011). https://scapy.net/
43. Qin, Q., Poularakis, K., Tassiulas, L.: A learning approach with programmable data plane towards IoT security. In: 2020 IEEE 40th International Conference on Distributed Computing Systems (ICDCS), pp. 410–420. IEEE (November 2020)

Prevention of DrDoS Amplification Attacks by Penalizing the Attackers in SDN Environment

Shail Saharan[✉] and Vishal Gupta

Birla Institute of Technology and Science (BITS) Pilani, Pilani, India
{p20170404,vishalgupta}@pilani.bits-pilani.ac.in

Abstract. Distributed Denial of Service (DDoS) attacks is one of the most prevalent and dangerous cyber-attacks that can bring down the targeted part of Internet
infrastructure in a short amount of time, thus resulting in significant economic
losses. As a defense strategy against these attacks, attack detection followed by
attack mitigation is not enough as there will always be a time lag between detection and mitigation. Instead, attack prevention is a more promising strategy. This
paper focuses on preventing such attacks that save the targeted (or victim's) network from any harm and penalize the attacker's network for making the attack. We
propose two DDoS prevention techniques named Port-Mapping, and PortMergeIP,
considering DNS amplification attack as a specific and one of the most dangerous
types of DDoS attack. All the methods are proven to prevent the victim from the
attack altogether. The packet loss is up to 98% at the attacker's side when the
proposed algorithm is implemented during a DDoS attack. The delay introduced
due to the proposed algorithms is approximately 30% lesser than an existing work
based on authentication.

1 Introduction

With the proliferation of Internet activities in one's personal and professional life, network security has become one of the most important topics for discussion. It protects
the integrity, confidentiality, and accessibility of computer networks and data. Countless
security breaches attempting to steal user data and hack into private spaces have made
this topic even more critical. One such category of a breach is the Distributed Denial of
Service (DDoS) attack. It aims at disrupting the flow of genuine traffic by overwhelming
the resources of the targeted network by sending a huge amount of network traffic in a
distributed fashion to (mainly) a single victim.

Distributed Reflection-based Amplification Denial of Service Attacks (DrDoS) is
a specific type of DDoS attack wherein a network client sends a request packet to the
server and gets (generally amplified) response packet in return. The attacker spoofs the
source IP address to that of victim's in the request packet. Since the underlying Internet
architecture does not validate the source IP address while forwarding the packets; as a
result, all amplified responses are directed towards the victim. DNS amplification attack
is an example of such an attack. Such attacks are hard to detect and mitigate as it is tough
to differentiate between attack traffic and legitimate traffic. Therefore, attack prevention
is a more promising strategy, and this paper proposes the strategies for the same. The

L. Barolli et al. (Eds.): AINA 2022, LNNS 450, pp. 684–696, 2022.
https://doi.org/10.1007/978-3-030-99587-4_58

effectiveness of the proposed techniques is shown using DNS amplification attacks, although it can be modified for other types of attacks too.

The word prevention in the context of DDoS attacks is used with multiple definitions in the literature. For the techniques presented in this paper, we conform to the following definition of prevention (called True-Prevention) as provided in [1]:

Let $\mathbf{B}$ be the network bandwidth in which an attacker (or a bot in control of an attacker) resides, $\mathbf{V}$ be the victim, $\mathbf{I_V}$ be the IP address of victim $\mathbf{V}$ and, $\mathbf{I_A}$ is an attacker's IP address. True prevention is defined as a set of techniques embedded into the network routers which prevent the attack traffic from reaching $\mathbf{V}$, even though the destination IP address of network packets belonging to DDoS attack is $\mathbf{I_V}$, and automatically mitigates the attack for some constant time $\mathbf{T}$ where $\mathbf{T}$ is directly proportional to $\mathbf{B}$.

Because of the flexibility and programmability aspects provided by Software Defined Network (SDN) in terms of controllers, it is used to show and validate the required network intelligence to be induced in the underlying network infrastructure to prevent DrDoS attacks. Also, DNS based DrDoS attack is used to show the effectiveness of the proposed techniques. Of course, the proposed prevention algorithms can be applied to other types of DrDoS attacks and traditional network infrastructure. The remaining part of the paper is structured as follows—Sect. 2 highlights related work in DDoS prevention. Section 3 describes the prevention techniques in detail. Section 4 contains results and experimental setup. Finally, Sect. 5 concludes the paper.

2 Related Work

Kalkan et al. [2] have proposed that DDoS defense can be categorized into detection and prevention. Prevention can be further classified into capability-based prevention and filtering. According to Bhatia et al. [3], DDoS defense consists of Traffic monitoring and analysis, Prevention, Detection, Traceback, Characterization, and Mitigation modules. The remainder of this section primarily shows the DDoS defense mechanisms close to the definition of True prevention.

Duan et al. [4] proposed a defense technique against infrastructure DDoS attacks on specific flows. This technique works for critical links to a server by a proactive routing mutation. The problem with this approach is that the victim can still be affected when the attack traffic is switched to non-critical links. Keromytis et al. [5] proposed secure overlay services as a prevention technique. Here, authentication of the packets is done at secure overlay points known as SOAPs, which forward the packets through overlay nodes to beacon nodes. The primary issue with this approach is additional infrastructure requirements, thus difficult to scale. Luo et al. [6] proposed prevention against DDoS attacks with dynamic Path Identifiers (PIDs). PIDs identify the path between network entities as inter-domain routing objects. This approach is difficult to implement, as it needs much change in the existing underlying network architecture and, if a link breaks in between the communication, the response packet is lost. Basheer et al. [7] have also used the concept of DPID with Get message logging in Bloom Filters. Get message logging helps in prevention, as while normal users respond to get messages, attackers do not. Wu et al. [8] proposed Source Address Validation Improvement (SAVI), and it complements ingress filtering by adding IP address validity to an individual source. SAVI is defined as network-based so that there is no dependency on a host. SAVI instances may face reliability issues due to loss of bindings in SAVI devices through a restart of SAVI devices or binding information for a new link not reaching the SAVI device. Several SAVI documents have been standardized based on the different address assignment techniques (e.g., FCFS-SAVI [9], DHCP-SAVI [10], SEND-SAVI [11], and MAA-SAVI [12]. Guangwu et al. [13] proposed to store the sender's information in IPV6 extension headers. For inter-domain accountability, SAVI is used, and before the packet leaves the network, the gateway router/switch embeds users' credibility information in the packet. The user's private keys will be stored in the extension header of the IPV6 protocol. The limitation of this technique is the use of SAVI devices, and it is also susceptible to DOS attacks. A new IPV6 address generation algorithm was proposed by Ying et al. [14]. Firstly, SAVA (Source Address Validation Architecture) is used to authenticate source addresses. After this, an address is generated using NID (network identity) and time, and this address is assigned to the host. This newly generated IP address will be used for communication. Yang et al. [15] have proposed to use Message Authentication Codes (MACs), which are used by ASs (autonomous systems) to verify the origin of a packet. These MACs are stored in a new proposed PASSPORT field in the IP header. A secret key is shared between the source AS and each AS between source and destination. The limitation of this technique is that Diffie-Hellman is used for exchanging the keys, which is not secure. The accountable Internet Protocol proposed by Andersen et al. [16] provides self-certify addresses without depending on a third party. This technique proposes Accountability Domains (AD), and each host in that AD is given a unique EID for authentication so that each host will have a combination of AD and EID. The problem with this technique is the deployability and refurbishment of Internet Protocol. Christos et al. [17] proposed Forwarding Accountability in which the receiver decides the policies that the sender must adhere to. The inter-between ASs are also responsible for sending the traffic as they mark the packets that pass through them. Kim et al. [18] have proposed a framework using SDN. A DNS response is only accepted at the client-side when there is a request; otherwise, the packet is dropped. The DNS request information is stored in the switch or the memory of the SDN controller. The problem with this approach is that the attack traffic still reaches the victim's network. The packets allowed on a particular link are controlled in the proposed approach proposed by Park et al. [19] and Li et al. [20]. These techniques can still be susceptible to intelligently spoofed IP addresses and do not support the topologies' dynamic changes. Jessica et al. [21] have proposed a Wireless Intrusion Prevention System known as WIDIP, and its focus is to protect the internal network from attacks. In this work, the attackers are identified and blacklisted using IP addresses and MAC addresses. The problem with this technique is that innocent machines controlled by bots are also blacklisted for a long time and cannot avail any service as they are blacklisted. Sahri et al. [22] have proposed an authentication approach to prevent DNS amplification attacks in SDN as an underlying architecture. In their authentication approach, the DNS server, before sending the response, sends a query back to the client, asking whether the query was sent or not, and if the client responds, only then is the response provided to the client. Because of this, a delay of 1 extra RTT is introduced in the response packet.

3   Techniques for True Prevention

This section explains the two techniques (Port-mapping and PortMergeIP) for DrDoS prevention.

3.1   Port-Mapping

Using Port-Mapping, the forward path of the request packet is stored in the packet itself to enable the corresponding response packet to use the same path to return to the source. According to CAIDA's Skitter Map, the average-path length of any packet lies between 10 to 30 hops [23]. This means it is required to store a path of about 30 hops within the packet. Also, the fields that can be used to store path information in an IP packet are [24]: Identification Field (16 bits), TOS field (8 bits), Flags Field (3 bits) and, Option and padding field. We propose the options and padding field (hereafter called options field) in the IP packet as it is typically not used in a DNS query and can be up to 40 bytes long. The Port-mapping technique assumes that (a) an intermediate router has 8 to 48 interfaces or physical ports; thus, a router would require 3 to 6 bits to uniquely identify such port (b) The underlying network is SDN-enabled.

   When a DNS Request packet passes through a router, it inscribes the packet with the input interface (the in-port of the router) in the options field, and the packet is forwarded through out-port(outgoing interface) as per IP forwarding rules. When a DNS Response packet arrives, the router removes the corresponding in-port number engraved earlier and routes the packet to that port. Rather, if the packet is not the DNS packet (i.e. packet.dest-port ! = 53 or packet.source-port ! = 53), it is treated as per normal IP forwarding rules. The options field is considered as stack memory. So for all UDP packets with a destination port of 53, the incoming interface of the router will be pushed on the options field. Rather, if the source port is 53, the top element will pop out, and the packet will be forwarded on that interface. The steps to be taken by an SDN switch to implement Port-Mapping are explained in the Port-Mapping algorithm.

3.2   PortMergeIP

This technique combines the strength of Port-Mapping and an IP-Switching technique. IP-Switching is switching the source IP address of a request packet with the downstream network's IP address known by ISP. The first-hop router of an ISP does this switching. The information about the path from an end-host to the organization's gateway router is stored in the options field of the IP packet in terms of physical port numbers, as proposed in Port-Mapping. Formally, the technique is explained in the PortMergeIP algorithm. By doing this, the packet does not require to go to the controller every time it hops; thus, the delay is reduced. The packet also reaches the attacker's network without getting lost.

   Let n = number of routers in-between source and destination
   r^i = i^{th} router $(0 < i \leq n)$
   m = total number of organizations or access networks
   p = switches or routers between the host machine and ISP's first hop router
   t = number of interfaces a router has
   $r_j{}^q$ = j^{th} organization's q^{th} switch or router between host and ISP's first-hop

router ($0 < j \leq m, 0 < q \leq p$)

$r_j{}^h$ = ISP's j^{th} organization's h^{th} first hop router ($0 < h \leq m$)

$r_j{}^{hk}$ = ISP's j^{th} organization's h^{th} first hop router's k^{th} incoming interface ($0 < k \leq t$)

$r_j{}^{qk}$ = j^{th} organization's q^{th} router's k^{th} incoming interface

$r_j{}^{ql}$ = j^{th} organization's q^{th} router's l^{th} outgoing interface ($0 < l \leq t$)

ipv4-src, ipv4-dst, src-port, dst-port and *option* is source IP address, destination IP address, source port number, destination port number and options field of the packet respectively

rtr-id = ID of router

in-port- The interface of router where the packet arrives

out-port- The outgoing interface

SRC-IP-ADDR = Source IP address of the downstream network known by ISP

Port-Mapping Algorithm (Switch-Side Processing)
   INPUT- Packet coming at r^i
  for i= 1 to n **do**
          if packet is udp and dst-port==53 then
               Send the packet to controller
          else if packet is udp and src-port==53 then
               Send the packet to controller
          else Forward the packet to out-port
          end if
  end for
Port-Mapping Algorithm (Controller-Side Processing)
  for each packet-in from switch **do**
     if packet is udp **and** dst-port==53 **then**
         if option==none then
              new-option=in-port
        else
              new-option=new-option + in-port
        end if
       // Create a packet with new-option field
        packet=create-new-packet (new-option)
        Forward the packet to out-port
    else if packet is udp and src-port==53 **then**
       // remove last field from option and put it in outgoing port
        out-port= option [-1]
       // create new option by removing last field
        new-option = option -1
        packet =create-new-packet (new-option)
        Forward the packet to new out-port
      end if
  end for

```
PortMergeIP Algorithm (Switch-Side Processing)
INPUT- Packet coming at rⁱ
for i= 1 to n do
    while rtr.id = rⱼq and in-port = rⱼqk or rtr.id = rⱼq and in-port= rⱼql do
        if packet is UDP and src-port==53 then
            Send the packet to Controller
        else if packet is UDP and dst-port==53 then
            Send the packet to Controller
        else
            Forward the packet to out-port
        end if
    end while
    while rtr.id = rⱼh  and in-port = rⱼhk do
                ipv4-src = SRC-IP-ADDR
                Forward the packet to out-port
    end while
    Forward the packet to out-port
end for
Port-MergeIP Algorithm (Controller-Side Processing)
    Same as Port-Mapping
```

The Port-Mapping technique has the advantage of correctly implementing True-prevention, and no attack traffic reaches the victim network. On the other hand, it also has a few limitations. These are (a) the existing free space in the IP packet is a bottleneck, (b) since the processing time at each intermediate router increases, it is comparatively a bit slower than its counterpart, (c) it requires a change in the functionality of each router along the path, and (d) when the response packet follows the corresponding request packet path and some link along the reverse path gets broken, the packet is lost. On the other hand, the PortMergeIP technique also correctly implements true-prevention, no attack traffic reaches the victim, reduces delay, correctly routes the reverse-path traffic to the attacker's machine or legitimate user, overcomes the bottleneck in terms of free memory requirement for storing physical port numbers, and requires minimal change in ISP network.

4   Results and Discussion

4.1   Experimental Setup

To implement and validate the proposed algorithms, a topology, as shown in Fig. 1, was created in an SDN environment using mininet. It shows four access networks (Access N/W-1 to Access N/W-4). Access N/W-1 belongs to the victim's organization, and Access N/W's 2, 3, and 4 are in control of an attacker. Any number of hosts in these access networks can launch the DDoS attack; we have used A1, A2, and A3 for launching the attack by spoofing the source IP address field of the packets to that of a host in the victim's network. Each access network is connected to the ISP's first-hop router through its gateway router. These first-hop routers are responsible for switching the IP addresses in the PortMergeIP algorithm. Besides this, DNS servers to generate a DrDoS attack and HTTP servers to generate legitimate traffic are also connected to the network. We used Ryu as the SDN controller and OpenFlow as a protocol for communication between the SDN controller and switch. We used two machines to emulate the same topology and perform the experiments. a) Dell™ PowerEdge™ T710 High-Performance Intel 2S Tower Server Intel (R) Xeon (R) Processor 2.40 GHz, 64 GiB memory, b) Intel® Core™ i5–4590 CPU @ 3.30 GHz × 4, 31.3 GB memory. The CICDDoS2019 dataset was used to generate the attack traffic. It contains benign and the most up-to-date common DDoS attacks, resembling actual real-world data (PCAPs) [25, 27]. Legitimate traffic is also generated based on the features extracted from the dataset. These features are the total number of flows of DDoS attack packets, the total number of flows of benign packets, average number of packets in attack flow, the average number of packets in normal traffic flow, the average size of attack packets, the average size of normal packets, and the duration of the attack. DrDoS attack traffic of 1 Gbps and 500 Mbps, 1 Mbps, and 500 Kbps, are generated using Scapy, Iperf, and Nping. The attackers combinedly generate this attack. After attack generation, both the algorithms are tested on various parameters described in more detail later in the paper.

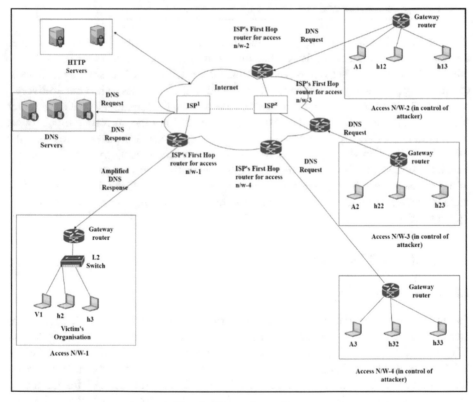

Fig. 1. Network topology

4.2 Results and Analysis

The two proposed prevention algorithms, Port-Mapping and PortMergeIP, were tested on different parameters to check their validity against DDoS attacks generated using the features from the CICDDoS attack dataset. Comparative analysis was done with the authentication approach [22] because it was also proposed to prevent DNS amplification attacks, and their underlying architecture is an SDN environment.

A. Loss of Packets Due to Attack

This section shows the loss of packets due to an attack for all the cases, i.e. when the prevention technique is in place when it is not. This percent loss of packets is calculated per Eq. (1) and Eq. (2) as in [28].

If LP is loss of packets, N^{TX} is the total number of packets sent, and N^{RX} is the total number of packets received, then

$$L.P. = N^{TX} - N^{RX} \tag{1}$$

$$\% \text{ of lost packets} = \left[L.P./N^{TX} \right] \times 100 \tag{2}$$

DNS request packets are generated simultaneously by the three attackers A1, A2, and A3 with spoofed source IP of victim V_1, as shown in Fig. 1. Corresponding to these, amplified DNS response packets are generated, thus leading to a DrDoS attack. Link bandwidth between the gateway router and ISP's first-hop router for all four access networks is varied and shown on X-axis. Y-axis shows the percentage of lost packets.

Figure 2 shows the percentage of lost packets with 500 Mbps and 1 Gbps attacks when no prevention technique exists. The complete attack traffic reaches the victim and chokes its bandwidth. For 500 Mbps, the packets loss percent at victim varies from 97% to 53% as the link bandwidth of victim is varied from 10 Mbps to 500 Mbps, respectively. Similarly, for 1-Gbps, it varies from 99% to 62%.

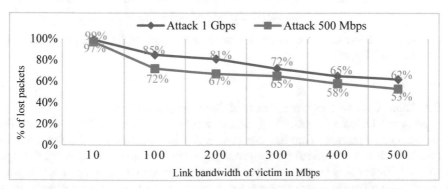

Fig. 2. Percentage of lost packets at the victim during an attack of 500-Mbps and 1-Gbps without any prevention algorithm.

Figure 3 and Fig. 4 show the percent of lost packets with port-mapping and PortMergeIP when DDoS prevention algorithms are in place. A 500-Mbps and 1-Gbps DDoS attack was generated for PortMergeIP (Fig. 4). Instead, only 500 Kbps and 1 Mbps attack could be generated for port-mapping (Fig. 3) because with this technique, each packet is to be forwarded to the controller, which became the bottleneck for generating a large attack in the simulated environment. With Port-Mapping, as shown in Fig. 3, the loss percentage of packets for 1Mbps attack varies from 73% to 23% for all three attackers as link bandwidth is varied from 100 kbps to 500 kbps. For a 500 kbps attack, this variation is from 49% to 6.9% for all three attackers. The victim's network is completely safe from attack traffic, and the attackers are penalized. When the proposed prevention algorithm PortMergeIP is implemented in the underlying network, as shown in Fig. 4, the percentage of lost packets for 1Gbps attack varies from 99% to 46% for all three attackers as link bandwidth is varied from 10 Mbps to 500 Mbps. For a 500 Mbps attack, this variation is from 99% to 32% for all three attackers. The percentage loss of packets at the victim's side is due to the legitimate traffic, not attack traffic (between 6.8% to 7.3%). It proves that the PortMergeIP prevention algorithm not only saves the victim but also penalizes the attackers.

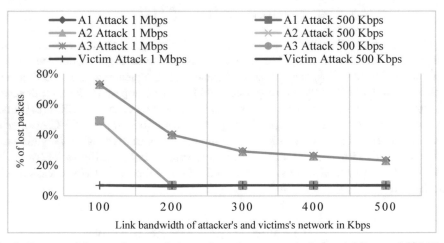

Fig. 3. Percent of lost packets at victim and attackers network during 1-Mbps and 500-Kbps attack with Port-Mapping.

B. The Delay in DNS Response Due to Prevention Algorithms

The proposed DDoS prevention algorithms would introduce a delay between DNS request-response packets. Another aspect that requires attention is the measurement of this delay. It was calculated for the proposed algorithms and the authentication approach [22]. Figure 5(a) shows the delay between DNS request-response packets when the prevention algorithms are implemented, but there is no attack in the network. Baseline delay indicates the average delay without any prevention algorithm, which is almost negligible, nearly 0.13 s. Then PortMergeIP has slightly more delay, as the packet goes to the controller till IP is switched. The delay in the PortMapping technique is slightly more as all the packets go to the controller at each router before they are forwarded. The highest delay is in the authentication algorithm [22], as it needs one additional

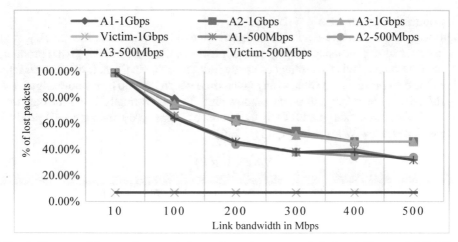

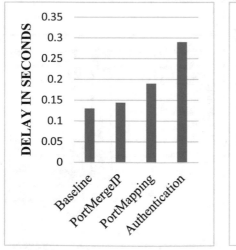
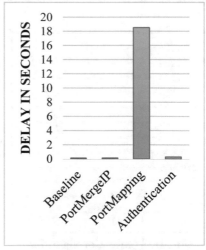

Fig. 4. Percent of lost packets at victim and attackers network during 1-Gbps and 500-Mbps attack with PortMergeIP.

RTT before the client receives the DNS response. Figure 5(b) shows the delay introduced due to algorithms when the underlying network is under attack. The delay in response from DNS servers during the attack is almost the same as when there is no attack for PortMergeIP and authentication techniques [22]. For PortMapping, the delay in DNS response increases during the attack as we are using only one controller, and the controller gets busy due to the attack, hence the increase in DNS response.

(a) Network is not under attack   (b) Network is under attack

Fig. 5. DNS response delay due to prevention algorithms.

C. Congestion in Attacker's Network

Here we show the congestion in the attacker's network to prove that the attack would mitigate for some time using the proposed algorithms. Bandwidth-delay (B-D) product, as the name suggests, is the product of bandwidth and delay in the network [26]. It tells about the data present in the link, at any given time which is yet to be acknowledged, the in-flight data often referred to as the window size. To put it formally, if c is the data rate of a link (i.e., bandwidth) and RTT is the round-trip time delay, then the B-D product defines the window W given by Eq. (3) in [26].

$$W = c \times RTT \tag{3}$$

Through this, we can find the congestion in the network. With no prevention algorithm in place, the B-D product increases gradually until the allowed TCP window size of all the links in between is reached, as shown in Fig. 6(a). Because the attacker is constantly sending request packets to the DNS server, which is continuously sending responses to the victim, the victim's network is in continuous congestion. A change in the behavior of the B-D product at the attacker's side can be seen in Fig. 6(b). It shows that the B-D product in the network does not increase gradually till it becomes constant; instead, it oscillates. When the attacker's network is congested due to all the DNS responses coming to it, the attacker cannot send any more request packets; as a result, there is no response from the DNS server, and for this duration, the attacker's network becomes congestion free. Now the attacker can send packets again, leading to another congestion, and the same process repeats; thus, oscillatory behavior is observed. This process repeats till the attacker stops sending the packets. The attack is mitigated for the amount of time the attacker cannot send request packets. It conforms to the definition of True-Prevention. Based on such behavior, the attacker can be blocked.

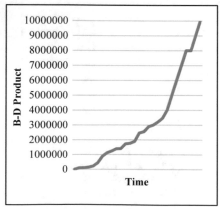

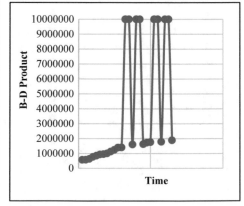

(a) No DDoS prevention algorithm implemented

(b) Network implements DDoS prevention algorithm

Fig. 6. Congestion in victim's network

5   Conclusion

Prevention of DDoS amplification attacks is a more promising idea than its detection and mitigation. For this, the underlying network infrastructure should be made intelligent enough to prevent such attacks in some constant amount of time. Such network infrastructure would take away the burden from end-users to deal with such attacks. This paper proposed two techniques for preventing DDoS amplification attacks, with implementation and results specific to DNS amplification attacks. All the proposed techniques use the common philosophy to equip underlying network infrastructure with enough rules/intelligence to enforce reverse path forwarding the same as that of corresponding forward-path forwarding of the network packet. With such rules, an attacker would attack itself. The implementation results show that the targeted victim is always safe as no spoofed traffic reaches the victim. Any such attack automatically gets mitigated away (thus prevention) for some constant time. This time depends majorly upon the bandwidth of an attacker. Moreover, the attacker will always attack itself; hence, the techniques will penalize the attacker. We want to propose prevention techniques for DDoS flooding attacks for future work.

Acknowledgement. We thank Anmol Naugaria, Nishchay Agrawal, and Nilay Arora for their constructive feedback and for implementing a part of some of the modules of the presented work. With their help, we were able to accomplish the results within the justifiable time frame.

References

1. Saharan, S., Gupta, V.: DDoS prevention: review and issues. In: Patnaik, S., Yang, X.-S., Sethi, I.K. (eds.) Advances in Machine Learning and Computational Intelligence: Proceedings of ICMLCI 2019, pp. 579–586. Springer Singapore, Singapore (2021). https://doi.org/10.1007/978-981-15-5243-4_53
2. Kalkan, K., Gur, G., Alagoz, F.: Filtering-based defense mechanisms against DDoS attacks: a survey. IEEE Syst. J. **11**(4), 2761–2773 (2017). https://doi.org/10.1109/JSYST.2016.2602848
3. Bhatia, S., Behal, S., Ahmed, I.: Distributed denial of service attacks and defense mechanisms: current landscape and future directions. In: Conti, M., Somani, G., Poovendran, R. (eds.) Versatile Cybersecurity, pp. 55–97. Springer International Publishing, Cham (2018). https://doi.org/10.1007/978-3-319-97643-3_3
4. Duan, Q., Al-Shaer, E., Chatterjee, S., Halappanavar, M., Oehmen, C.: Proactive routing mutation against stealthy distributed denial of service attacks: metrics, modeling, and analysis. J. Def. Model. Simul. **15**, 219–230 (2018)
5. Keromytis, A.D., Misra, V., Rubenstein, D.: SOS: an architecture for mitigating DDoS attacks. J. Sel. Areas Commun. **22**, 176–188 (2004)
6. Luo, H., Chen, Z., Li, J., Vasilakos, A.V.: Preventing distributed denial-of-service flooding attacks with dynamic path identifiers. IEEE Trans. Inf. Forensics Secur. **12**, 1801–1815 (2017)
7. Al-Duwairi, B., Özkasap, Ö., Uysal, A., Kocaoğullar, C., Yildirim, K.: LogDoS: a Novel logging-based DDoS prevention mechanism in path identifier-based information centric networks. Comput. Secur. **99**, 102071 (2020). https://doi.org/10.1016/j.cose.2020.102071
8. Wu, J., Bi, J., Bagnulo, M., Baker, F., Vogt, C.: Source address validation improvement (SAVI) framework (2013)

9. Nordmark, E., Bagnulo, M., Levy-Abegnoli, E.: FCFS SAVI: FirstCome, First-Served Source Address Validation Improvement for Locally Assigned IPv6 Addresses (2012)
10. Bi J, Wu J, Yao G, Baker F. Source Address Validation Improvement (SAVI) Solution for DHCP (2015)
11. Bagnulo, M., Garcia-Martinez, A.: SEcure Neighbor Discovery (SEND) Source Address Validation Improvement (SAVI) (2014)
12. Bi, J., Yao, G., Halpern, J.M., Levy-Abegnoli, E.: Source Address Validation Improvement (SAVI) for Mixed Address Assignment Methods Scenario (2017)
13. Hu, G., Chen, W., Li, Q., Jiang, Y., Xu, K.: TrueID: a practical solution to enhance Internet accountability by assigning packets with creditable user identity code. Futur. Gener. Comput. Syst. **72**, 219–226 (2017)
14. Liu, Y., Ren, G., Wu, J., Zhang, S., He, L., Jia, Y.: SCIENCE CHINA Inf. Sci. **58**(12), 1–14 (2015). https://doi.org/10.1007/s11432-015-5461-0
15. Liu, X., Li, A., Yang, X., Wetherall, D.: Passport: Secure and adoptable source authentication. NSDI **8**, 365–378 (2008)
16. Andersen, D.G., Balakrishnan, H., Feamster, N., Koponen, T., Moon, D., Shenker, S.: Accountable internet protocol (AIP). In: ACM SIGCOMM 2008 Conf. Data Commun, p. 339–50 (2008). https://doi.org/10.1145/1402958.1402997
17. Pappas, C., Reischuk, R.M., Perrig, A.: FAIR: forwarding accountability for Internet reputability. In: EEE 23rd Int. Conf. Netw. Protoc, pp. 189–200 (2016). https://doi.org/10.1109/ICNP.2015.22
18. Kim, S., Lee, S., Cho, G., Ahmed, M.E., Jeong, J.P., Kim, H.: Preventing DNS amplification attacks using the history of DNS queries with SDN. In: Foley, S.N., Gollmann, D., Snekkenes, E. (eds.) ESORICS 2017. LNCS, vol. 10493, pp. 135–152. Springer, Cham (2017). https://doi.org/10.1007/978-3-319-66399-9_8
19. Park, K., Lee, H.: On the effectiveness of route-based packet filtering for distributed DoS attack prevention in power-law internets. ACM SIGCOMM Comput Commun Rev **31**, 15–26 (2001)
20. Li, J., Mirkovic, J., Wang, M., Reiher, P., Zhang, L.: SAVE: source address validity enforcement protocol. In: Twenty-First Annu. Jt. Conf. IEEE Comput. Commun. Soc., pp. 1557–66 (2002)
21. Goncalves, J.A., Faria, V.S., Vieira, G.B., Silva, C.A., Mascarenhas, D.M.: WIDIP: wireless distributed IPS for DDoS attacks. In: 2017 1st Cyber Secur. Netw. Conf. CSNet 2017, vol. 2017- Janua, 2017, pp. 1–3. https://doi.org/10.1109/CSNET.2017.8241996
22. Sahri, N.M., Okamura, K.: Protecting DNS services from IP spoofing-SDN collaborative authentication approach. In: ACM Int Conf Proceeding Ser 2016, 15–17-June, pp. 83–9. https://doi.org/10.1145/2935663.2935666
23. CAIDA's Skitter MAP n.d. https://www.caida.org/~bhuffake/papers/skitviz/. Accessed 1 Oct 2021
24. Ehrenkranz, T., Li, J.: On the state of IP spoofing defense. ACM Trans. Internet Technol. **9**(2), 1–29 (2009). https://doi.org/10.1145/1516539.1516541
25. DDoS Evaluation Dataset (CIC-DDoS2019). Univ Brunswick n.d. https://www.unb.ca/cic/datasets/ddos-2019.html. Accessed 11 June 2021
26. Medhi, D., Ramasamy, K.: IP Traffic Engineering. In: Network Routing, pp. 214–258. Elsevier (2018). https://doi.org/10.1016/B978-0-12-800737-2.00009-0
27. Sharafaldin, I., Lashkari, A.H., Hakak, S., Ghorbani, A.A.: Developing realistic distributed denial of service (DDoS) attack dataset and taxonomy. In: IEEE 53rd International Carnahan Conference on Security Technology, Chennai, India (2019)
28. Lee, Y.L., Loo, J., Chuah, T.C.: Modeling and performance evaluation of resource allocation for LTE femtocell networks. In: Modeling and Simulation of Computer Networks and Systems, pp. 683–716. Morgan Kaufmann (2015)

Author Index

A

Abbes, Slim, 386
Adachi, Hiroyuki, 231
Ahmed, Mondar Maruf Moin, 329
Ahsan, Mohammad, 624
Akash, Bathini Sai, 243
Alvi, Anik, 329
Ampririt, Phudit, 1
Arifin, Farhadur, 19
Aznavouridis, Alkiviadis, 409

B

Baba, Kensuke, 480
Baba, Takahiro, 480
Babbar-Sebens, Meghna, 433
Babu, Hari, 671
Badra, Mohamad, 446
Baldwa, Siddarth, 167
Barolli, Leonard, 1, 10, 361
Battula, Ramesh B., 43
Berlemont, Samuel, 281
Bertin, Emmanuel, 446
Beydoun, Kamal, 219
Bhuiyan, Md. Mafijul Islam, 329
Bialas, Katarzyna, 306
Biradar, Kuldeep M., 43
Bourgeois, Julien, 118, 130
Bousselmi, Khadija, 204
Bylykbashi, Kevin, 1

C

Cabusas, Renz M., 154
Cavallin, Florencia, 317

Chalupnik, Rafal, 306
Chen, Zhanwen, 468
Corradini, Flavio, 611
Cuzzocrea, Alfredo, 141

D

Dahaoui, Ibrahim, 339
Dantas, Mario, 421
Dantas, Mario A. R., 81
de Almeida Silva, Fernando, 421
de Almeida, Marcelo Barros, 646
de Oliveira Silva, Flávio, 646
de Souza Neto, Natal Vieira, 646
de Souza Silveira, Walkíria Garcia, 421
Dedu, Eugen, 219
Dhaouadi, Asma, 204
Dhoutaut, Dominique, 219
Di iorio Silva, Gabriel, 81
Djemaiel, Yacine, 560
do Nascimento, Francisco Assis Moreira, 256
Duffner, Stefan, 281
Durresi, Arjan, 433
Durresi, Mimoza, 433

E

El Houd, Anass, 130
El Madhoun, Nour, 446
Elme, Khan Md., 19
Epp, Brenna N., 154

F

Fariya, Khadija Yeasmin, 19
Freitas, Marcelo Silva, 646

L. Barolli et al. (Eds.): AINA 2022, LNNS 450, pp. 697–699, 2022.
https://doi.org/10.1007/978-3-030-99587-4

G

Gammoudi, Mohamed Mohsen, 204
Garcia, Christophe, 281
Gautam, Ishan, 130
Ghafoor, Abdul, 180
Ghanmi, Houaida, 398
Ghosh, Anonnya, 547
Giacomin, João Carlos, 68
Gouge, Justin M., 154
Gupta, Vishal, 684

H

Habib, Bachir, 118
Hadded, Mohamed, 398
Hajlaoui, Nasreddine, 398
Hammoudi, Slimane, 204
Haribabu, K., 635
Hasan, Amit, 547
Heimfarth, Tales, 68
Hessel, Fabiano, 256
Hirata, Aoto, 361
Hoque, S. N. M. Azizul, 329
Hossain, Md Akbar, 329
Hourany, Edy, 118
Huang, Chung-Ming, 55
Hussain, Farookh Khadeer, 269

I

Ibrahim, Hussain Mohammed, 547
Idoudi, Hanen, 598
Ikeda, Makoto, 1
Isichei, Bamibo C., 141
Islam, Md. Safiqul, 329
Iwasaki, Aiko, 528

J

Jambavalikar, Shreya Manish, 167
Jerbi, Belhassen, 560
Jozwiak, Ireneusz, 293

K

Kaufmann, Tyson N., 154
Kaur, Davinder, 433
Kawauchi, Kiyoto, 528
Kedziora, Michal, 293, 306
Khan, Ibraheem Abdulhafiz, 269
Khelghatdoust, Mansour, 192
Kobayashi, Hajime, 528
König, Lukas, 503
Koyama, Tomoyuki, 105
Krishna, Aneesh, 167, 243
Kumar, Lov, 167, 243
Kushida, Takayuki, 105

L

Lefebvre, Grégoire, 281
Leung, Carson K., 141, 154
Liu, Yi, 10
Lyko, Ewa, 293

M

Mabrouk, Ali, 560
Madhusudhan, R., 624
Mahdavi, Mehregan, 192
Mainas, Nikolaos, 373
Majewska, Zofia, 306
Makhoul, Abdallah, 118
Matsui, Tomoaki, 361
Matsuo, Keita, 1
Mayer, Rudolf, 317
Medlej, Ali, 219
Meyer, Jerome, 130
Mizera-Pietraszko, Jolanta, 348
Mohammad, Wasif, 547
Molinos, Diego Nunes, 646
Mondal, Prasenjit, 329
Monnet, Sébastien, 204
Morrow, Luke B., 141
Mosbah, Mohamed, 339
Mostarda, Leonardo, 31, 611
Muhlethaler, Paul, 398
Murthy, Lalita Bhanu, 167, 243

N

Najari, Naji, 281
Navarra, Alfredo, 31
Neundorfer, Noah, 516
Ngo, Anh Tuan, 141
Nguyen, Lam Thu, 141
Nguyen, Vo Ngoc Khoi, 586
Nishigaki, Masakatsu, 586
Nova, Farhana Chowdhury, 547

O

Ochii, Hiroji, 231
Oda, Tetsuya, 361
Ogiela, Lidia, 539, 543
Ogiela, Marek R., 539
Ogiela, Urszula, 543
Ohki, Tetsushi, 586
Okazaki, Mikio, 231
Oliveira, Romerson Deiny, 646
Omote, Kazumasa, 468, 492

P

Pagnotta, Fabio, 31
Petrakis, Euripides G. M., 373, 409
Pham, Trang Doan, 141
Pietraszko, Jolanta, 293

Piranda, Benoit, 118, 130
Poddar, Rahul, 671
Pogoda, Anna, 293
Pujolle, Guy, 446

Q
Qafzezi, Ermioni, 1

R
Rab, Raqeebir, 547
Rekhis, Slim, 386, 560
Riaz, M. Mohsin, 180
Ridoy, Prince Mahmud, 19
Rivera, Samuel J., 433
Rosa, Pedro Frosi, 646
Ruthvik, Bokkasam Venkata Sai, 243

S
Saha, Arajit, 19
Saha, Pranta, 19
Saharan, Shail, 684
Saito, Nobuki, 361
Sakamoto, Shinji, 10
Sato, Shingo, 573
Sayari, Amal, 560
Scala, Emanuele, 611
Sergio, Wagno Leão, 81
Sharma, Deepti, 43
Shikata, Junji, 573
Sojan, Jithin Kallukalam, 635
Stergiou, Ioanna-Maria, 373
Ströele, Victor, 81
Swessi, Dorsaf, 598

T
Takano, Yasunao, 231
Takeda, Sena, 231
Takizawa, Makoto, 543
Tancula, Jolanta, 348
Terada, Takamichi, 586
Tjoa, Simon, 503
Touati, Haifa, 398
Toyoshima, Kyohei, 361
Tsakos, Konstantinos, 409
Tully, James R. A., 154

U
Uehara, Minoru, 95
Ueno, Ryuji, 492
Uslu, Suleyman, 433

V
Vipparthi, Santosh K., 43

W
Wang, Junjie, 659
Wang, Lixin, 516
Wang, Meng, 95

Y
Yamamoto, Takumi, 528
Yamauchi, Toshihiro, 480
Yang, Jianhua, 516
Yang, Kai-Jiun, 55
Yannam, Pavan Kumar Reddy, 243
Yoshimura, Ayako, 528
Yukawa, Chihiro, 361

Z
Zemmari, Akka, 339
Zheng, Lihong, 659
Zia, Usman, 180

Printed in the United States
by Baker & Taylor Publisher Services